S0-BTZ-084

OSHA

History, Law, and Policy

OSHA
History, Law, and Policy

Benjamin W. Mintz

The Bureau of National Affairs, Inc.

Library of Congress Cataloging in Publication Data

Mintz, Benjamin W.
OSHA: history, law, and policy.
Includes bibliographical references and index.
1. Industrial safety—Law and legislation—United States. 2. Industrial hygiene—Law and legislation—United States. 3. United States. Occupational Safety and Health Administration. I. Title. II. Title: O.S.H.A.
KF3570.M56 1984 344.73′0465 84-11333
ISBN 0-87179-435-7 347.304465
0-87179-455-1 (pbk)

Printed in the United States of America
International Standard Book Number: 0-87179-435-7 (regular ed.)
0-87179-455-1 (student ed.)

To
Rachel, Jared, Adam, and Harriet*
and
Max J. and Malcha Mintz**

---

*Ecclesiastes, 9, 9.
**Proverbs, 17, 6.

Preface

At least since the time that Moses directed the Children of Israel to construct a parapet for their roofs, "that thou bring not blood upon thine house, if any man fall from thence,"[1] the matter of workplace safety and health has been on the agenda, sometimes high on the agenda, of civilized societies. Early efforts in this country to confront the problem of occupational injuries were sporadic and mostly ineffective and until fairly recently little seemed to be known by regulatory agencies about occupational illnesses. It was not until 1970, when the Federal Occupational Safety and Health Act was passed, that any comprehensive and serious attempt was made to protect the health and safety of American workers.

This book is the story of the implementation of that law. It comprises "readings" from a variety of sources, with introductory material and commentary. The title defines the scope of the work: the "history," "law," and "policy" of OSHA.[2] The book is not a systematic recital of events as they took place, such as might be written by a professional historian (which, of course, I am not). Rather, it is historical in its attempt to trace the development, over time, of the views of the Agency, the courts, and the public on critical issues of law and policy under the occupational safety and health program. The distinction between "law" and "policy" has not always been easy to discern, if, indeed, the distinction has any meaning in the context of a government regulatory program. In any event, this book deals with issues traditionally handled by lawyers as well as the many issues that primarily are confronted by the Assistant Secretary of Labor responsible for administering the OSHA program. Legal issues have been emphasized largely because most critical OSHA issues have been resolved in a legal framework (for example, the use of cost-benefit analysis). Also, I was personally involved as the head of the OSHA legal division in the unfolding of most of these legal events. But, I have gone beyond the traditional legal case book ap-

[1]Deuteronomy, 22, 8. But see *Diamond Roofing Co., Inc. v. OSHRC*, 528 F.2d 645 (5th Cir. 1976) (holding that the OSHA perimeter guarding standard is only applicable to floors and not to roofs).

[2]The term "OSHA" has been used variously to describe the Act, the Agency (Occupational Safety and Health Administration), and the entire occupational safety and health program. I will continue this usage, believing that the context will clarify the meaning intended.

proach in offering, wherever possible, the underlying, or should I say, overriding, policy questions that inform and often direct the legal positions involved. Thus, for example, I deal at length with the question of OSHA and the Fourth Amendment, which is a "legal" question, and relate that discussion to the issue of OSHA "policy" in selecting workplaces for inspection.

For whom, then, is the book intended, and what will be its use? In major part, the book is designed as an introduction to OSHA for students. I use the term "student" expansively to include not only individuals in school but anyone who knows little about OSHA and wants to know more. And, importantly, I have not directed my attention exclusively to law students but also to students in public health and administration, in business schools and colleges. Also, the book should be of use to individuals—both lawyers and nonlawyers—who are familiar with OSHA, and who will find it convenient to have source materials—cases, standards, testimony—available in one place. Finally, I hope the book will be of interest to readers who wish to know more about OSHA because the topic is important and interesting.

In format, the book has been divided into parts, chapters, and sections. The parts represent what I consider the major topic areas in the OSHA program (Overview of Standards; OSHA Standards and the Courts; OSHA Enforcement; Employee Rights and Responsibilities; OSHA and the States). I have added two chapters, one at the beginning and one at the end, describing the interplay between the OSHA program and Congress (legislative background and OSHA and the Congress). In selecting this subject matter, I have given much fuller attention to topics not ordinarily covered in OSHA texts—notably, standards issues, state plans, and the Agency's relationship with Congress. At the same time, I have not attempted to deal exhaustively with the numerous issues that have arisen in OSHA enforcement cases.[3] Enforcement issues have been included mostly where they implicate major policy components; as examples, OSHA's enforcement of the general duty clause, and interpretations of §4(b)(1) of the Act which delineates the jurisdiction of various agencies in regulating occupational safety and health.

This book is what is commonly called a text of "cases and materials." As noted, a major portion of the text is excerpted material. I have added introductions and continuity, extensive footnotes, citations and references to the material used, as well as brief summaries of a wide variety of books and articles directly touching upon OSHA or dealing with related issues. There are, of course, substantial excerpts from cases; this is essential, for what is a book about OSHA

[3]These have been fully discussed in several treatises on OSHA law which are devoted largely to enforcement issues. E.g., B.A. Fellner and D.W. Savelson, OCCUPATIONAL SAFETY AND HEALTH—LAW AND PRACTICE (1976); G.Z. Nothstein, THE LAW OF OCCUPATIONAL SAFETY AND HEALTH (1981); M.A. Rothstein, OCCUPATIONAL SAFETY AND HEALTH LAW, 2d ed. (1983).

without the decisions in the *Benzene, Cotton Dust, Barlow's, National Realty,* and other "classic" OSHA cases. In addition, many other types of excerpts have been included, for example, from OSHA standards,[4] testimony before congressional committees, advisory committee meetings, press releases, comments submitted in rulemaking proceedings, briefs, OSHA's field operations manual and enforcement directives, and a range of other materials.

Two technical notes: footnotes from the excerpts have been deleted for the most part; only those containing critical material have been retained, and they are numbered as in the original. Also, all deletions in the excerpts have been indicated except for deletions of brief cross references to transcripts or to the record in a case.

I have used this broad range of materials for several reasons. First, because they are not generally available at present, and some will be even less available in time, and I wish to preserve in published form at least a sampling of these "historic" texts. Further, the most effective presentation of material is to allow those making the point to speak for themselves. I have found legislative material particularly useful in this connection. In my experience, the sharpest statement of a point of view is often made under the pressure of a congressional hearing; and, similarly, remarks of members of Congress on OSHA are frequently the clearest and most uninhibited reflections of the state of public opinion. I hope it will be apparent that I have sought to present fairly all points of view on OSHA issues. I also wish to emphasize that all materials excerpted and cited are public documents; no internal memoranda or other confidential material has been used.

Finally, I hope that this material will present a verbal canvas of the fascinating and often tumultuous history of the OSHA program and of a government agency at work—Assistant Secretary Bingham telling a congressional oversight committee to watch the *Federal Register* for more health standards; Assistant Secretary Stender explaining his impatience with bureaucratic delays; Senator Schweiker calling OSHA "probably the most despised Federal Agency in existence"; the Court of Appeals for the Fifth Circuit incredulously inquiring why it took OSHA seven years to correct a simple error in the *Federal Register*; Commissioner Moran caustically suggesting that the word "Review" be dropped from the name of the Review Commission; and Congressman Skubitz saying on the floor of the House that OSHA should be renamed as the Agency that mandated "the privy on the plains." These and many more items are an integral part of OSHA history. Some of this material may at times seem humorous, but the issue at stake is deadly serious; for, so long as uncounted numbers of workers are dying of occupationally related cancer and

[4]Substantial excerpts from the vinyl chloride standard and from its preamble have been included in order to acquaint the reader with the substance of a "typical" OSHA health standard.

other diseases, and are killed or badly hurt in workplace accidents, the book on the history of OSHA obviously cannot be closed.

This book reflects, of course, my experience as an attorney in the Office of the Solicitor of the Department of Labor, and originated in part from materials which I prepared in teaching courses on OSHA at the Washington College of Law, American University, the Harvard School of Public Health, and (with co-Adjunct Professor George H. Cohen) at the Georgetown Law Center. I have learned more from my students than from anyone, and they have my gratitude. I am particularly thankful to William Lesser and Thomas Feldman, former student, and student, of the Washington College of Law, American University, for persistent devotion and commitment in working with the manuscript and proofs. The Department of Labor awarded me the Secretary of Labor's Career Service Award, entitling me to a sabbatical year which enabled me to embark on this project. A word of thanks to officials of the Department of Labor—Undersecretaries, Solicitors, and Assistant Secretaries—who, during my sometimes turbulent career in OSHA, provided me with supporting words; and to Dean Thomas E. Buergenthal of the Washington College of Law, American University, who provided me with a worksite during the period that a good part of the book was being written. My colleagues in the Occupational Safety and Health Division in the Office of the Solicitor, with whom I worked for 11 years, stimulated my perceptions on OSHA in countless ways. As friends they have my affection, as professionals, my respect. I particularly thank Daniel M. Jacoby, Daniel Mick, Charles Gordon, Nathaniel Spiller, Delores Gilliam, and John Hynan for many helpful comments on the manuscript. I also appreciate the assistance of a number of former colleagues from OSHA, particularly Barbara Bryant, Ruth Knight, and James Foster. I also thank Marthe B. Kent, Dr. Leslie Boden, Thomas Brown, and Daniel M. Krauskopf for comments on the manuscript; Shyrlee V. Goodman for encouragement and assistance from the earliest stages of the project; and Wilai Mahasirimongkol for her support activities as the book neared completion.

As part of this Preface, I have included two additional sections: a Historical Note, giving a broad overview of the history of OSHA which may be lost in light of the subject-matter organization of the book, and a Bibliographical Note. Appendix 1 contains a Table of Health Standards Proceedings, tracing the history of OSHA's principal health standards rulemakings and resulting litigation. This table includes the most recent information on the status of the various health standards projects.

Historical Note

The Occupational Safety and Health Act was approved on December 29, 1970 and became effective on April 28, 1971. As stated in the legislation, its goal was "to assure so far as possible every working man and woman in the Nation safe and healthful working conditions." Thus, while earlier laws were principally designed to provide compensation for workplace injuries and illnesses, the purpose of the OSHA act was *prevention* of occupational injuries and illnesses. Growing concern over the increasing incidence of occupational illnesses was one of the primary motivations of the new Act.[5] The Act set up an extensive institutional structure to implement its various provisions for enforcement and educational activity. The Occupational Safety and Health Administration, headed by an Assistant Secretary, was established in the Department of Labor with major responsibility for the development and enforcement of occupational safety and health standards. (The Secretary of Labor has since delegated to the Bureau of Labor Statistics, also in the Department of Labor, his responsibilities under the Act for the implementation of a program of collection, compilation, and analysis of occupational safety and health statistics.) The National Institute for Occupational Safety and Health (NIOSH) was established in the Department of Health and Human Services with responsibility for research and training activity in the area of occupational safety and health.[6] The Occupational Safety and Health Review Commission (OSHRC), with three members appointed by the President, was established to adjudicate the Department of Labor's enforcement actions under the new Act. OSHRC is entirely separate from the Department of Labor. The coverage of the Act is pervasive; the only exclusions are government employees. However, federal employees are covered by a separate program under §19 of the Act,[7] and state and local employees are covered under state plans approved under §18(c)(6).[8]

The first Assistant Secretary of OSHA was George C. Guenther, previously head of the Bureau of Labor Standards in the Department of Labor, which had responsibility for administering most federal occupational safety and health programs that preceded the 1970 Act. Mr. Guenther served under Secretary of Labor James Hodgson during the term of office of President Richard Nixon. During the early part of the administration of Assistant Secretary Guenther, OSHA issued a large body of start-up standards under §6(a) of the Act, an emergency temporary standard (ETS) and a "permanent" standard lowering the permissible exposure limit for asbestos fibers in the workplace. The "permanent" standard was later upheld for the most

[5]For a discussion of the magnitude of the problem of occupational illnesses, see, in addition to the materials cited in the text, P.S. Barth and H.A. Hunt, WORKERS' COMPENSATION AND WORK-RELATED ILLNESSES AND DISEASES, particularly at 1–60 (1980).

[6]For a discussion of the activities of NIOSH, see among other works, Lehman, *The National Institute for Occupational Safety and Health: Expanding the Frontiers of Knowledge* in PROTECTING PEOPLE AT WORK 163 (U.S. Dept. of Labor 1980).

[7]29 U.S.C. §668, as implemented by Exec. Order No. 12,196, 46 FED. REG. 12,769 (1980), amended as to effective date, Exec. Order No. 12,223, 45 FED. REG. 45,235 (1980), and by 29 C.F.R. pt. 1960 (1983). Sec. 19 was amended in 1982, deleting the requirement that the President report to the Congress on the federal agency programs. Pub. L. 97-375.

[8]29 U.S.C. §667(c)(6), implemented in part by 29 C.F.R. pt. 1956 (1983). On state plans generally, see chapters 19 and 20.

part by the Court of Appeals for the District of Columbia Circuit.[9] During Mr. Guenther's administration, state participation in the OSHA program was strongly encouraged; this policy often drew the opposition of labor unions. Mr. Guenther left office in January 1973, and considerable controversy later arose concerning the so-called "Guenther Memorandum," which, it was asserted, demonstrated subordination of OSHA regulatory goals to political considerations.[10]

The next Assistant Secretary was John H. Stender, a former union official and a Republican legislator from the State of Washington. He served from April 1973 until mid-1975, when he left office during the administration of Secretary of Labor Dunlop. Mr. Stender issued emergency temporary standards for 14 carcinogenic substances and vinyl chloride. The carcinogen ETS was partially vacated by the court of appeals on procedural grounds. The ETS was later replaced by "permanent" standards, all but one of which was upheld by the court. The vinyl chloride permanent standard was ultimately upheld by the Court of Appeals for the Second Circuit in a major decision.[11] Mr. Stender's term of office occasioned considerable controversy, particularly in the area of state programs.

Mr. Stender was replaced by Dr. Morton Corn, who had been a Professor of Industrial Hygiene and Engineering at the University of Pittsburgh. Dr. Corn was confronted at an early date with criticism of the Agency for its handling, before Dr. Corn came to office, of the kepone matter in Hopewell, Virginia. After making significant changes in OSHA's complaint-handling procedures, Dr. Corn concentrated on increasing OSHA's emphasis on health protection, in both the standards and the enforcement areas. A major accomplishment in Dr. Corn's administration was the issuance of a final coke-oven emission standard which was upheld in substantial part by the Court of Appeals for the Third Circuit.[12] At the same time, several other standards proceedings initiated by Dr. Corn were not completed until after the end of his administration—one of the most important of these was the regulation of lead hazards. Dr. Corn also emphasized professionalization of the OSHA staff through better training and more stringent hiring standards.

Dr. Eula Bingham, previously a Professor of Toxicology at the University of Cincinnati, took office early in 1977. She served as Assistant Secretary under Secretary of Labor Ray Marshall in the administration of President Jimmy Carter. She continued the OSHA emphasis on health, and during her administration, OSHA issued health standards regulating benzene, cotton dust, lead, arsenic, as

[9] *Industrial Union Dep't v. Hodgson,* 499 F.2d 467 (D.C. Cir. 1974).

[10] The text of the memorandum appears in N. Ashford, Crisis in the Workplace 543 (1976).

[11] *Society of the Plastics Industry, Inc. v. OSHA,* 509 F.2d 1301 (2d Cir. 1975), *cert. denied,* 421 U.S. 992 (1975).

[12] *American Iron & Steel Inst. v. OSHA,* 577 F.2d 825 (3d Cir. 1978), *cert. dismissed,* 448 U.S. 917 (1980).

well as the rule on employee access to medical and exposure records.[13] She also promulgated OSHA's Carcinogens Policy, and a proposed standard on hazard identification. Many of Dr. Bingham's actions were applauded by major labor unions, particularly for her concern about employee rights. Among these actions were the medical removal protection requirements in the lead standard and the walkaround pay regulation. The administration of Dr. Bingham, however, was strongly criticized by some, for example, representatives of the states with approved plans who claimed that her administration rejected them as "equal partners" in the OSHA program. Some employers argued that the engineering control requirements of health standards by Dr. Bingham were not feasible and would undermine the economic well-being of those industries. During the term of Dr. Bingham, the Supreme Court issued major decisions in *Barlow's, Inc. v. Marshall,*[14] imposing a warrant requirement on OSHA, and *Industrial Union Department v. American Petroleum Institute,*[15] vacating the benzene standard and requiring OSHA to find "significant risk" before promulgating standards. Also in her administration, a major effort by Senator Schweiker to amend OSHA was successfully fought. Dr. Bingham was the first Assistant Secretary to serve a full four-year term in office.

Mr. Thorne G. Auchter, previously a construction executive, became Assistant Secretary in early 1981 under Secretary of Labor Ray Donovan in the administration of President Ronald Reagan. Mr. Auchter was committed to bringing about significant changes in emphasis in OSHA operations, among other things, improving the management of Agency operations. Explaining that OSHA in the past had unduly emphasized the "policeman's role," Mr. Auchter indicated that he would utilize not only enforcement tools but other available OSHA authority to achieve statutory goals. Thus, he continued OSHA's emphasis on on-site consultation and initiated voluntary protection programs to supplement enforcement. After the Supreme Court had issued its *Cotton Dust*[16] decision, rejecting cost-benefit analysis in OSHA health standards development, Mr. Auchter announced an intention to apply cost-effectiveness principles, which had been approved by the Supreme Court for standards proceedings. To this end, the Agency undertook partial reconsideration of several standards, including access to medical records, cotton dust, and the Carcinogens Policy. Mr. Auchter also issued an emergency temporary standard, significantly lowering the permissible exposure limit for asbestos, which was ultimately vacated by a court of appeals,[17] and issued proposed standards for carcinogens ethylene

[13]The lead standard was affirmed in large part in *United Steelworkers of America v. Marshall,* 647 F.2d 1189 (D.C. Cir. 1980), *cert. denied,* 453 U.S. 913 (1981).
[14]436 U.S. 307 (1978).
[15]448 U.S. 607 (1980).
[16]*American Textile Mfrs. Inst. v. Donovan,* 452 U.S. 490 (1981).
[17]*Asbestos Information Ass'n v. OSHA,* 727 F.2d 415 (5th Cir. 1984).

dibromide and ethylene oxide, the latter after a court order.[18] After withdrawing the proposed standard issued by Dr. Bingham, Mr. Auchter repromulgated a modified version of the hazard communications standard. Mr. Auchter also stated his intention to make the states full partners in OSHA programs and stated that he wished to move towards final approval of some state plans. While Mr. Auchter was applauded by some groups for "bringing order and stability" to OSHA, the AFL-CIO and some international unions were strongly critical of his policies. Mr. Auchter resigned in March 1984.

The history of OSHA has been marked by an expansion of the size, budget, and operations of the Agency. According to OSHA statistics, in fiscal year 1973 the OSHA budget (in current dollars) was $69,274,000, with 1,699 personnel positions. The fiscal year 1981 budget was $208,854,000, with 2,922 positions, and the fiscal 1982 budget was somewhat less. The number of OSHA workplace inspections has been variable, with the number in fiscal 1981 somewhat more than the number in fiscal 1973 (56,998 compared with 48,409). The number of workplace inspections had reached a high point in fiscal 1976 (90,482) during the Administration of Assistant Secretary Stender, but the thoroughness of OSHA inspections at that time has been questioned. A highly significant change occurred in the number of OSHA health inspections which had tripled between 1973 and 1982 (3,184 in fiscal 1973, 9,209 in fiscal 1982). The number of OSHA's compliance officers (inspectors) has also grown substantially, from 524 in 1973 to 1,075 ten years later. There has also been a sharp increase in the absolute and relative number of health compliance officers, also referred to as industrial hygienists.

The Act authorizes federal OSHA to approve state plans for occupational safety and health enforcement, with the Federal Government paying 50 percent of the cost of the program. The first state plan was approved by OSHA on November 30, 1972; it had been submitted by South Carolina. At present, 24 states and jurisdictions have approved plans. Eight states[19] have withdrawn plans after they had been approved. In fiscal year 1982, states with approved plans conducted 92,942 establishment inspections; 8,372 or 9 percent of these were health inspections.

The role of the states in the national OSHA program, the agency budget, and the size of the inspectorate have been continuing controversial issues since the earliest days of the OSHA program.

Bibliographical Note

The Occupational Safety and Health Act appears in Title 29 of the *United States Code,* §650, and following sections. (The text of the

[18]*Public Citizen Health Research Group v. Auchter,* 702 F.2d 1150 (D.C. Cir. 1983).

[19]North Dakota, Montana, New Jersey, New York, Illinois, Wisconsin, Colorado, and Connecticut. (Connecticut has an approved plan for public employees.)

Act is reprinted in the back of this volume.) OSHA's regulations issued under the Act appear in several volumes in the *Code of Federal Regulations*; they constitute Title 29, parts 1900 through 1990 of these regulations.[20] These regulations are basically of two kinds: procedural regulations and occupational safety and health standards. Important examples of procedural regulations are in Part 1903 dealing with inspections, citations, and proposed penalties. The bulk, but not all, of OSHA general industry standards appear in Part 1910; OSHA health standards appear in §§1910.1000–1910.1025. Standards for particular industry groups appear elsewhere in the *Code of Federal Regulations* (e.g., construction industry, Part 1926).

OSHA enforcement activity is governed by the Act, by OSHA regulations, by the OSHA-issued Field Operations Manual and Industrial Hygiene Field Operations Manual, and by OSHA's enforcement directives and standards interpretations. These contain instructions to field staff, and notification to the public, of the manner by which OSHA intends to carry out its statutory responsibilities.[21] Cases decided by the Occupational Safety and Health Review Commissions and by federal courts on OSHA appear in the official reports as well as in volumes published by private services.[22]

Under the Act, the President publishes an annual report on OSHA which is a useful source of information on Agency activity. The Secretary of Labor's Annual Report also contains sections on OSHA and the legal activities of the Department, including OSHA. The many documents published by the National Institute of Occupational Safety and Health, including criteria for health standards ("criteria documents"), bulletins, and hazard alerts are valuable. A number of books and articles have been published on the OSHA program and are cited throughout this volume. In addition, several works viewing occupational safety and health from the standpoint of the worker at risk have been published; among these are: F.W. Wallick, *The American Worker: An Endangered Species* (1972), D.M. Berman, *Death on the Job, Occupational Health and Safety Struggles in the United States* (1978), D. Nelkin and M.S. Brown, *Workers at Risk, Voices From the Workplace* (1984). The hearings of congressional committees and reports on OSHA issues have in most cases been published. Many of these hearings contain extremely useful documentary material on the OSHA program, and are referred to extensively in this text. An Interagency Task Force was convened by President Carter, and its comprehensive report, *Making Prevention Pay,* dealing with numerous OSHA policy issues, was issued in De-

[20]Regulations of the Occupational Safety and Health Review Commission appear in 29 C.F.R. pt. 2200.

[21]The documents, of course, are public and are distributed by the Agency as well as by private services. See, e.g., *Occupational Safety and Health Reporter* (OSHR), published by The Bureau of National Affairs, Inc.

[22]*See, e.g., Occupational Safety and Health Reporter Cases* (OSHC), published by The Bureau of National Affairs, Inc.

cember 1978. Another report on OSHA, entitled *Controls for Health and Safety in the Workplace,* is scheduled to be issued in mid-1984 by the Office of Technology Assessment, United States Congress. A draft of the report was consulted but it was not published in time for citation. Reports of the Comptroller General of the United States on OSHA and by the Administrative Conference of the United States on OSHA-related administrative law issues have also been found useful.

B.W.M.

May 1984

Summary Table of Contents

I. OSHA Standards: An Overview

II. OSHA Standards and the Courts

III. OSHA Enforcement

IV. Employee Rights and Responsibilities

V. OSHA and the States

VI. OSHA and Congress

Detailed Table of Contents

I. OSHA Standards: An Overview

II. OSHA Standards and the Courts

III. OSHA Enforcement

IV. Employee Rights and Responsibilities

V. OSHA and the States

VI. OSHA and Congress

1

Evolution of the Occupational Safety and Health Act of 1970

The Occupational Safety and Health Act was signed into law by President Richard Nixon on December 29, 1970.[1] This law, which universally has been considered a landmark in the history of labor and public health legislation, has as its purpose "to assure so far as possible every working man and woman in the Nation safe and healthful working conditions * * *."[2] This chapter discusses state and federal regulatory activity in the field of occupational safety and health that preceded the Act; the major factors that led to the passage of the Act in 1970; and the significant issues debated during the legislative evolution of the Act.

State regulatory activity of industrial safety began in the 1870s in Massachusetts, and by 1921, most states had workers' compensation systems in effect. The federal government became involved in the regulation of occupational safety and health in 1936 with the passage of the Walsh-Healey Act, which imposed limited occupational safety and health requirements on certain contractors with the federal government.[3] These early regulatory activities in occupational safety and health, culminating in 1968 with the active consideration by Congress of a comprehensive federal statute, are described in an article written by Judson MacLaury, staff historian at the U.S. Department of Labor.[4]

---

[1]The Occupational Safety and Health Act will hereafter be referred to as the Act.
[2]Occupational Safety and Health Act of 1970, §2(b), 29 U.S.C. §651(b)(1976).
[3]Walsh-Healey Act, ch. 881, 49 Stat. 2036 (1936) (current version at 41 U.S.C. §§35–45 (1976 & Supp. V 1981).
[4]*The Job Safety Law of 1970: Its Passage Was Perilous,* MONTHLY LAB. REV., Mar. 1981, at 18, 18–21.

MacLaury, *The Job Safety Law of 1970: Its Passage Was Perilous*

***

In the factories that sprang up after the Civil War chemicals, dusts, dangerous machines, and a confusing jumble of belts, pulleys, and gears confronted inexperienced, often very young workers. The reports of State labor bureaus in the 1870's and 1880's were full of tragedies that too often struck the unwary or the unlucky. The Massachusetts report of 1872 described some particularly grisly accidents. These tragedies and the industrial accident statistics that State labor bureaus collected, spurred social reformers and the budding labor movement to call for State factory safety and health laws. In 1870, the Massachusetts Bureau of Statistics of Labor urged legislation to deal with "the peril to health from lack of ventilation." In 1877, Massachusetts passed the Nation's first factory inspection law. It required guarding of belts, shafts, and gears, protection on elevators, and adequate fire exits. Its passage prompted a flurry of State factory acts. By 1890, nine States provided for factory inspectors, 13 required machine guarding, and 21 made limited provision for health hazards.

The labyrinth of State job safety and health legislation covered a wide range of workplace hazards but was badly flawed. There were too many holes in the piecemeal system and numerous hazards were left uncontrolled. The laws had to be amended often to cover new hazards. Many legislatures failed to provide adequate funds for enforcement. Inspectors, who were often political appointees, were not always given the legal right to enter workplaces. States with strong safety and health laws tended to lose industry to those with less stringent ones, which made States competitive and limited their legislative efforts.

The Progressive Era and the growth of mass circulation newspapers and national magazines helped forge a national movement for workers' safety and health. In 1907, 362 coal miners were killed at Monongah, W. Va., in the worst U.S. mine disaster. This widely publicized tragedy shocked the Nation and led to the creation in 1910 of the U.S. Bureau of Mines to promote mine safety.

***

The "Pittsburgh Survey," a detailed study of living and working conditions in Allegheny County, Pa., done in 1907–08, had a special impact on job safety and health. One of the major topics of the investigation, which was sponsored by the Russell Sage Foundation, was industrial accidents. The survey found that the injured workers and the survivors of those killed on the job bore the economic brunt of accidents, even though most were the employers' fault. The authors of the survey agreed that, for reasons of social equity, employers should bear a substantial share of the economic burden, giving them more incentive to eliminate the causes.

WORKERS' COMPENSATION STARTED

Years before the Pittsburgh Survey, the idea of compensating injured workers from an insurance fund to which employers would contribute had gained a foothold in this country, though it was not at first promoted as a preventive measure. Prince Otto von Bismarck had initiated the first workers' compensation program in Germany in 1884, and the idea soon spread throughout Europe. In the United States, a few States tried to establish early compensation systems. Organized labor successfully opposed the concept, precisely because it was intended as a palliative, not a preventive measure. In 1908, Congress passed, with President Theodore Roosevelt's support, a limited workers' compensation law for Federal employees. Encouraged by this example, several States appointed study commissions. However, until the Pittsburgh Survey, compensation was treated mainly as a humanitarian measure.

The survey's call for an economic incentive to encourage accident prevention struck a responsive chord. It quickly became a key part of the rationale for workers' compensation. This seemed to tip the scales. Both labor and business rallied in support. In 1911, Wisconsin became the first State to successfully establish a workers' compensation program. Within 1 year it was joined by nine other States and by 1921 most States had followed suit.

Ironically, it was as a preventive measure that workers' compensation accomplished the least. The general level of this type of insurance premium was already so low that there was no real incentive for a company to invest heavily in safety improvements to be eligible for the slightly lower rates offered firms with good safety records. Very few States included compensation for disease, although much was already known about occupational illness. Still, insurance company safety experts helped improve their clients' safety programs and the establishment of compensation gave the safety movement a moral boost.

An idea that developed alongside of workers' compensation probably produced more significant long-run results. If the States would create industrial commissions with authority to establish specific safety and health regulations, it would not be necessary to go back to the legislatures and amend the factory laws in order to cover new hazards or change requirements. A workers' compensation advocate, John R. Commons of the University of Wisconsin, found this system in use in Europe and urged its adoption in the United States. Wisconsin, in another pioneering move, created the first permanent State industrial commission which developed and enforced safety and health regulations, after hearing comments from labor, management, and others. This idea was widely accepted and became a guide for future State and Federal regulation of occupational safety and health.

***

LABOR STANDARDS

Frances Perkins appointed. In 1933, President Franklin D. Roosevelt selected Frances Perkins as Secretary of Labor and first woman Cabinet member. She brought to the Labor Department long experience in occupational safety and health with the State of New York. To help assure that workplaces would be "as safe as science and law can make them," Perkins created a Bureau of Labor Standards in 1934 as a rallying point for those interested in job safety and health. This was the first permanent Federal agency established primarily to promote safety and health for the entire work force. The Bureau helped State governments improve their administration of job safety and health laws and raise the level of their protective legislation.

Congress enacted three laws as part of Roosevelt's New Deal which augmented the Federal Government's role in protecting people on the job. The Social Security Act of 1935 allowed the U.S. Public Health Service to fund industrial health programs run by State health departments. This made the Public Health Service, which had begun doing industrial health studies in 1914, the national leader in this field. The Fair Labor Standards Act of 1938, which set a minimum wage and banned exploitative child labor, gave the Labor Department the power to bar workers under age 18 from dangerous occupations. The Walsh-Healey Public Contracts Act of 1936 allowed the department to ban contract work done under hazardous conditions.

Maritime rules. By the late 1950's, the Federal-State partnership which Frances Perkins had cultivated was no longer adequate to deal with growing threats to workers' safety and health, so gradually the Federal Government took a more prominent role. In 1958, Congress passed a seemingly minor amendment to the Longshoremen's and Harbor Workers' Compensation Act.* It gave the Labor Department authority to set safety and health standards for the very small work force covered under this law. In addition to protecting workers in one of the Nation's most hazardous industries, the amendment closed "the last remaining 'no man's land'" in safety enforcement. The Secretary of Labor was authorized to seek penalties against willful violators, but not against those who only carelessly broke the rules. After holding public hearings, the department began enforcing standards in 1960. Compliance was good, and the high accident rates declined sharply.

In December 1960, shortly after the congressionally ordered maritime rules became effective, the department issued on its own a set of mandatory safety and health standards under the Walsh-Healey Act. The department had previously issued most of these standards in a "Green Book" of informal guidelines to aid Federal and State inspectors. States had been encouraged to inspect Federal contractors and enforce their own rules. Now they were barred from applying their standards and had to enforce the Federal rules instead. For the first time, the Federal occupational safety and health requirements were applied to the whole range of industry.

The new rules were not popular. Because there had been no hearings or prior announcement, labor and industry were caught by surprise and miffed that they had not been consulted. Business protested strongly to the Labor Department against making the rules mandatory. The National Safety Council deplored this "monumental set of rigid regulations." The department took the criticisms to heart, and in October 1963 it announced proposed revisions, with hearings held in March 1964.

Business opposition had been building up for 3 years and reached a peak at the hearings. They ran for 2 weeks, and the transcript filled 1,347 typed pages. More than 100 witnesses appeared, mostly from industry. Business felt that the new rules were not only illegal, but also technically deficient and would inhibit innovation. By substituting Federal for State regulations, the Labor Department generally undermined State safety programs, it was argued. Business also felt that the new policy weakened its own long-established pattern of voluntary safety efforts.

Coordination of programs. The powerful wave of criticism that climaxed at the 1964 hearings prodded the Department of Labor into a serious examination of all its safety programs in order to develop a more coordinated safety and health policy. A study by an outside consultant found in the department a fragmented collection of safety programs and laws. It recommended consolidation of all these safety programs under a single agency, which was done somewhat in 1966.

A movement to protect the natural environment from the ravages of mankind and technology began growing while the Labor Department was seeking to improve and expand its protection of workers' safety and health. Large-scale Federal air and water pollution control programs were developed, helping to increase awareness and concern about the occupational environment.

Spurred by this movement, in 1965 the Public Health Service produced a report, "Protecting the Health of Eighty Million Americans," which outlined some of the recently found technological dangers. It noted that a new

*[Author's note—Act of Aug. 23, 1958, Pub. L. No. 85-742, §41, 72 Stat. 835 [amending Longshoremen's and Harbor Workers' Compensation Act, ch. 509, §41, 44 Stat. 1424, 1444 (1927)]. The current version of the Act is at 33 U.S.C. §§901–950 (1976 & Supp. V 1981).]

chemical entered the workplace every 20 minutes, that evidence now showed a strong link between cancer and the workplace, and that old problems were far from being eliminated. The report called for a major national occupational health effort centered in the Public Health Service.

The AFL-CIO urged President Lyndon Johnson to support the report's recommendations. On May 23, 1966, Johnson told a meeting of labor reporters that "the time has . . . come to do something about the effects of a workingman's job on his health." The Departments of Labor and Health, Education and Welfare promptly set about to develop legislation for such a program. A joint task force was then to combine both departments' ideas and submit a proposal to the President. However, Labor and HEW could not agree on which department would control a national program and by late 1966 the task force was deadlocked.

Mining tragedy breaks deadlock. In 1967, it was revealed that almost a hundred uranium miners, an abnormally high number, had died of lung cancer since the 1940's. Up to a thousand more such deaths were expected. In 1947, when large-scale uranium mining was getting underway, the Atomic Energy Commission discovered that radiation levels in these mines were dangerously high. The Commission, in cooperation with the Public Health Service, began a long-term health study of the miners. A number of Federal agencies had limited jurisdiction over uranium mines, but none had clear responsibility for them, and there was very little enforcement.

The lack of action took on tragic overtones with the revelations of 1967, and public attention focused on the Federal Radiation Council. Created in 1959 to advise the President on protective measures to take against all types of radiation hazards, the council was composed of representatives from concerned agencies. In 1967, it had just completed a study of the uranium mines and was expected to recommend a standard shortly. However, when the council met on May 4, 1967, it became deadlocked between a standard that the Atomic Energy Commission recommended and a tougher one preferred by the Labor Department.

The next day, Secretary of Labor Willard Wirtz, impatient with inaction, announced a bold step. Previously, Wirtz had been reluctant to act because he felt that uranium mining was not properly a Department of Labor area. However, without holding public hearings, Wirtz adopted under the Walsh-Healey Act the standard he had unsuccessfully advocated before the Federal Radiation Council.

This move had a decisive impact on the shaping of a national job safety and health program in 1967, as the Departments of Labor and HEW promoted their competing proposals. The Bureau of the Budget accepted the Department of Labor's recommendations.

***

On January 23, 1968, President Lyndon B. Johnson proposed "[t]he Nation's first comprehensive Occupational Health and Safety Program to protect the worker on the job."[5] Bills embodying the administration position were introduced by Congressman O'Hara of Michigan[6] and Senator Yarborough of Texas.[7] Secretary of Labor Willard Wirtz testified dramatically on behalf of these bills. He re-

---

[5]President's Message to Congress on Manpower and Occupational Safety and Health Programs, 4 WEEKLY COMP. PRES. DOC. 104 (1968).

[6]H.R. 14816, 90th Cong., 2d Sess. (1968).

[7]S. 2864, 90th Cong., 2d Sess. (1968).

ferred to the yearly statistics, "14,000 to 15,000 dead, over 2 million disabled, over 7 million hurt" as a result of industrial accidents, and emphasized that although some improvement had taken place prior to 1958, since then the accident rate had "moved up steadily," reaching a "new high" in 1966. Secretary Wirtz then addressed himself to the effectiveness of "present" safety programs. His view was that "private" programs, while inadequate to deal with the problem, demonstrate that "accidents *can* be significantly reduced."[8]

Statement of Secretary of Labor Willard Wirtz

Mr. Chairman and Members of the Committee, while we sit here talking, from now until noon, seventeen American men and women will be killed on their jobs.

Every minute we talk, 18 to 20 people will be hurt severely enough to have to leave their jobs—some of them never to work again.

In the time these two sentences have taken, another 20 people—one every second—have been injured on the job—less seriously, but in most cases needlessly.

Today's industrial casualty list—like yesterday's—and tomorrow's—and every working day's, week after month after year—will be 55 dead, 8,500 disabled, 27,200 hurt.

The figures for the year will be 14,000 to 15,000 dead, over 2 million disabled, over 7 million hurt.

These are all average figures, but there is a disheartening consistency about what happens day after day. Death takes no holidays in industry and commerce.

But to rely on aggregate statistics in this area demeans our humanity. If this kind of human tragedy touched our own families just once it would make us committed crusaders. It cheapens us as individuals to let ourselves—especially if we carry public responsibility—find refuge in our personal good fortune.

Several photographs have been placed here in the hearing room. Each of the pictures shows safe working conditions, the way it is for most of us. But there is imposed on each of these pictures another—smaller—which it takes a strong person to look at and not turn away from quickly.

I hesitated—and then felt smaller for hesitating—about displaying these pictures: of what a hand looks like lying alone by a power saw; of what half a man looks like when a falling girder has cut him in two.

But this is what S. 2864 is about, and it presents really only one central issue. That issue is as much about democracy as it is about occupational safety and health.

The issue is not "States rights"—although S. 2864 will be opposed by those who confound that principle with their interest. It will be a fair question of them whether it is their belief that a comparable degree of responsibility may be expected from the States, and whether they will support the exercise of such responsibility—or hurry from opposing it here to some State capitol to oppose it there. S. 2864 is carefully drawn to maximize State and Federal cooperation.

The issue in S. 2864 is not cost—the expense of protecting employee safety and health. It will be opposed on that basis, although guardedly. A fair question of those who so oppose it will be what price they put on a life, or

[8]*Hearings on S. 2864 Before the Subcomm. on Labor of the Senate Comm. on Labor and Public Welfare,* 90th Cong., 2d Sess. 69, 69, 71–73 (1968).

a limb, or an eye—and whether that price is the same for every member of the family of America as it would be for a member of their own family. S. 2864 is carefully drawn to recognize that *absolute* safety is impossible, but that *no avoidable risk* is to be taken with human welfare in the conduct of commercial enterprise.

The clear, central issue in S. 2864 is simply whether the Congress will act to stop a carnage which continues *only because people don't realize its magnitude,* and can't see the blood on the things they buy, on the food they eat, and the services they get.

***

PRESENT SAFETY PROGRAMS AND THEIR EFFECTIVENESS

It would be totally misleading to suggest or imply that American commerce and industry have been developed without regard to the problem of occupational safety and health, or that American government—state and federal—has been oblivious to it.

To the contrary.

S. 2864 is carefully designed to build on, and to complement, what enlightened employers and responsible State governments—and in narrow areas the Federal Government—have done and are doing.

These efforts are reviewed here with particular emphasis on (i) the evidence of effectiveness of affirmative occupational safety action, and (ii) the telling contrasts between the consequences of attention and inattention to this problem.

A. PRIVATE PROGRAMS

Responsible private employers have demonstrated that where safety is given top priority, positive and significant results can be achieved.

The National Safety Council consists of about 5,000 member employers, from all industries, who are particularly safety conscious. Their record shows it. In 1966, the injury frequency rate in manufacturing for the NSC employers was 5.1 injuries per million man hours worked. This compared with a rate of 17.5 for non-members. For the past ten years the rate for NSC member employers has averaged 4.6, compared with 15.6 for non-members.

***

B. STATE PROGRAMS

The record of State programs is also one of a significant variability in results—reflecting how much can be done.

The work-related accidental death rate in this country is 7.4 deaths per 100,000 industrial workers; but in one State it is as low as 1.6, in another as high as 10.0.

The injury frequency rates vary from 9.3 injuries per million man hours worked in one State, to 18.7 in another.

The kinds of occupations which are principally involved in various states obviously affect the comparative safety records.

So, also, and very clearly, does the type and scope of safety program.

Expenditures for safety standards and enforcement vary considerably. The highest expenditure in a State is $2.11 per nonagricultural worker; the lowest only 2 cents. The rest of the States fall in between, with the overall average being only 40 cents per nonagricultural worker.

In the States that have mounted the most extensive programs, at costs averaging $1.1 million per State, the death rate from work accidents is 19 deaths per 100,000 workers. In the 10 States with the smallest programs, with average costs of $240,000 per State, the death rate averages 110 per 100,000 workers.

The State programs vary greatly. Some agencies lack rule-making authority. Many safety and health standards are outdated and inadequate. Safety manpower is frequently insufficient. Penalties are often too lenient and fail as deterrents. Effective programs are often limited to specific areas, such as boiler and elevator codes.

C. FEDERAL PROGRAMS

Federal efforts in the field of occupational safety have been piecemeal, fragmented, and generally incomplete.

The major law—Walsh-Healey—was passed in 1936, over 30 years ago. It applied to workers performing Government contracts and then only while they are actually working on these contracts.

The Federal programs, too, bear witness, at the same time, what *can* be done.

The Atomic Energy Commission has placed its full support behind an active safety program. About 300 engineers and technical employees, including 26 who work on construction projects, are employed in the safety program. The results are impressive. In construction projects under the AEC, the injury rate in 1965 was 4.68 injuries per million man hours. This compares with an overall construction industry rate, as reported by BLS, of 28.3.

This past year the serious problem of lung cancer resulting from employment in uranium mines was brought to the public's attention. A concerted safety drive, with industry and State cooperation, has resulted in more than a 10 percent overall reduction in the cancer-causing conditions, with specific mines showing a much larger improvement.

In 1960, the Department of Labor promulgated comprehensive safety and health regulations under the 1958 amendments to the Longshoremen's and Harbor Workers' Compensation Act. At that time the injury frequency rate for shipyards was 39 disabling injuries per million man hours worked. In longshoring it was 132. An effective program was initiated, with positive results. Through the first 9 months of 1967, the rates in the shipyard and longshore industries were 21.7 and 82, respectively. This represents a 44 percent reduction in injuries in shipyards and a 38 percent reduction in longshoring over the seven-year period.

It is conservatively estimated that from 1960 through 1966 over 22,000 injuries were prevented in the shipyards. Since the average direct cost has been estimated at $1,500 per injury, this means that a total savings of over $33 million was realized in 6 years. In addition, individual employers have reported millions of dollars in insurance rebates from reduced injury experience.

Safety pays. The Federal maritime industry safety program has yielded substantial returns for many employers:

A stevedore on the Gulf Coast saved $300,000 in insurance premiums during 1965;

A Florida shipyard received a premium rebate of over $600,000;

A shipyard subsidiary of a large general contractor received a $250,800 insurance refund.

The sum of our experience clearly demonstrates that accidents *can* be significantly reduced.

***

Secretary Wirtz, while pointing out the improvements in occupational safety achieved by "responsible" private employers, made it clear that voluntary action by the business community had been inadequate to meet the serious workplace injury problem. There has been considerable discussion on the question of why the free market has not provided sufficient incentives to businessmen to avoid accidents by improving working conditions. The explanation most often given is that most of the costs of workplace injuries and illnesses are "spillover" costs, or "externalities," which are not paid by the employer but rather by the victim, his/her family, and society as a whole. These costs, therefore, would not normally be reflected in the price of the goods produced.[9]

Nor had the threat of negligence suits by employees against their employers, when these suits were allowed, provided any credible incentive for workplace protection. In pressing their claims, workers faced "almost insurmountable obstacles": a low duty of care owed by the employer as established by common law; the reluctance of coworkers to testify against the employer; and the high cost of litigation.[10] In any event, these suits were barred by the newly enacted workers' compensation laws, which are now in effect in all states and jurisdictions.[11]

The workers' compensation system itself did not provide a significant economic incentive for protective improvements in the workplace. Since workers' compensation benefits do not reflect the full cost of accidents, most firms pay premiums that are less than substantial, making the savings from successful safety programs minimal. Few of the insured firms, usually only the larger ones, pay merit-related premiums, with premiums relating to the dollar amount of claims for which the firm is liable. In any event, workers' compensation costs overall are small, amounting in 1970 to one percent of payroll. With respect to occupational illnesses, the Secretary of Labor has concluded that only five percent of those severely disabled from work-related disease receive workers' compensation benefits.[12] As a result an employer, acting on the basis of strict economic motives, has little incentive to eliminate workplace injuries and illnesses.[13]

---

[9]For a discussion of "spillover" costs as a rationale for regulation, see S. Breyer, REGULATION AND ITS REFORM 23–26 (1982); N. Ashford, CRISIS IN THE WORKPLACE 346–53 (1976). See also Stewart, *Regulation, Innovation and Administrative Law: A Conceptual Framework,* 69 CALIF. L. REV. 1259, 1263 (1981).

[10]NAT'L COMM'N ON STATE WORKMEN'S COMPENSATION LAWS, COMPENDIUM ON WORKMEN'S COMPENSATION LAWS 12 (1973).

[11]The historical development of workers' compensation is discussed *id.* at 11–19.

[12]U.S. DEP'T OF LABOR, INTERIM REPORT TO CONGRESS ON OCCUPATIONAL DISEASES 3 (1980). See also Note, *Compensating Victims of Occupational Disease,* 93 HARV. L. REV. 916, 925 (1980) (relatively few disease claims enter state workers' compensation system and even fewer succeed); Chelius, *The Influence of Workers' Compensation on Safety Incentives,* 35 INDUS. & LAB. REL. REV. 235 (1982).

[13]E.g., NAT'L COMM'N ON STATE WORKMEN'S COMPENSATION LAWS, COMPENDIUM ON WORKMEN'S COMPENSATION LAWS 289–90 (1973). See generally N. Ashford, CRISIS IN THE WORKPLACE 388–423 (1976); M.S. Baram, ALTERNATIVES TO REGULATION 77–83 (1981). The debate over an "injury tax," which, it is argued, would provide an economic incentive for safety and health compliance, is discussed in Chapter 11, Section A. Important arguments have also

The provisions of the 1968 administration bills[14] differed in important respects from the law as it was eventually enacted. Employers under these bills were required to comply with the standards issued by the Secretary and, in addition, to "furnish employment and a place of employment which are safe and healthful";[15] this provision is in marked contrast to the narrower "general duty" clause, §5(a)(1), currently in effect.[16] The bills provided that standards would be promulgated by the Secretary of Labor in accordance with the "informal" rulemaking requirements of the Administrative Procedure Act;[17] there was no provision for public hearings prior to the issuance of standards. The bills assigned enforcement responsibility to the Secretary of Labor,[18] and unlike the present Act, no independent commission was established to adjudicate the validity of the Secretary of Labor's enforcement actions.[19] Under these bills, the Secretary would have been authorized to "immediately issue an order providing for the immediate cessation of a . . . violation" which "may result in imminent harm to the safety and health of workers."[20] Criminal penalties were provided for "willful" violations of employer obligations.[21] In addition, government contractors would have been debarred for failing to comply with safety and health standards.[22]

The administration bills were strongly supported by labor unions, public interest groups, and health professionals. Industry representatives, however, for the most part opposed the legislation. Typical of the arguments presented by employer representatives were those in the testimony of Leo Teplow, Vice-President, Industrial Relations, of the American Iron and Steel Institute.[23]

Statement of Leo Teplow

***

Basic progress in occupational safety and health has been made, primarily on the basis of voluntary action on the part of employers and employees, supplemented by State regulations and administration of State safety programs.

been made to justify Government involvement in regulation of occupational safety and health on the basis of moral concerns. See, for example, McCarthy, *A Review of Some Normative and Conceptual Issues in Occupational Safety and Health* 9 Envtl. Aff. 773, 781–92 (1981) (discussed workers' rights, distributive justice, and public values as justification for OSHA).

[14]See *Occupational Safety and Health Act of 1968: Hearings on S. 2864 Before the Subcomm. on Labor of the Senate Comm. on Labor and Public Welfare*, 90th Cong., 2d Sess. 2–20 (1968); *Occupational Safety and Health: Hearings on H.R. 14816 Before the Select Subcomm. on Labor of the House Comm. on Education and Labor*, 90th Cong., 2d Sess. 1–7 (1968).

[15]S. 2864 & H.R. 14816, 90th Cong., 2d Sess. §3(a)(1968).

[16]29 U.S.C. §654(a)(1).

[17]S. 2864 & H.R. 14816, §3(b).

[18]*Id.* at §6.

[19]29 U.S.C. §659(c).

[20]S. 2864 & H.R. 14816, §6(a)(2); cf. 29 U.S.C. §662.

[21]S. 2864 & H.R. 14816, §9(c).

[22]*Id.* at §10(d).

[23]*Hearings on S. 2864 Before the Subcomm. on Labor of the Senate Comm. on Labor and Public Welfare*, 90th Cong., 2d Sess. 347, 349–52 (1968).

This voluntary progress has received international recognition. Less than five years ago, an Assistant Secretary of Labor, speaking at the Annual Meeting of Industrial Hygiene Foundation in Pittsburgh (October 23, 1963) stated:

> "The one question these people (foreign visitors) often ask us is, 'Can you put a safety law violator in jail?' Our answer, of course, is that under many of our laws we can, but don't. At first, this puzzles them. But when they see for themselves how we rely mainly on information and persuasion—instead of compulsion—they begin to understand us better. A safety director from a Near Eastern country wrote, 'I am impressed by how tact and personality may be exercised in encouraging employers and employees to comply with safety requirements.' I think this summarizes very well the approach we follow. It's an export that should constitute a big credit item in our international balance of payments."

Unfortunately, the present Bill departs from the encouragement of voluntary progress and relies primarily on compulsion.

Up to this point the statutory Federal role has been relatively minor except in maritime employment; more recently with respect to mining; and to the extent that occupational safety and health have been receiving some attention in connection with the Walsh-Healey Act and most recently under the Service Contracts Act.

We have read the testimony presented before this Committee as well as that before the House Select Subcommittee on Labor with care, but find no evidence of any sudden or national emergency requiring overwhelming Federal intervention into every aspect of State and non-governmental operations contemplated under this law, especially in light of the relatively limited proportion of all accidents that might be affected. On the contrary, comparison with other countries in which the central government has primary responsibility for occupational safety and health indicates that the record of this country, based on decentralized administration, merits respect.

Enforcement of Federal standards through Federal inspectors would result in the most intimate involvement of the Secretary of Labor in all operations affecting interstate commerce. There are few production and service activities which are not in some way related to occupational safety and health. The knowledge that the Secretary of Labor can be called upon can easily result in blowing up the most minor grievances to very substantial proportions. A minor complaint can very well become a "federal case".

Practical operating managers know that in connection with a wide variety of grievances stemming from a change in job, change in temperature, or change in crew size, safety or health issues are frequently alleged merely to build up a case. If the Federal government has responsibility in the area, it may very well be called in in thousands of grievances which are now satisfactorily resolved either between employee representatives and management or by resort to arbitration. Provision of this kind of authority in the Federal government would tempt many an employee representative to boost his stock by calling on the Federal government, since the very presence of a Federal inspector could be used to demonstrate his importance and influence.

There are many industries in this country in which, in case of an authorized or unauthorized strike, management personnel are called upon to continue the plant in operation, at least until materials in the pipe line have been processed, or until the plant can be closed down safely. Such emergency operations might very well be prevented on the plea that a reduced crew makes the operation unsafe. This would be especially true, if as these bills provide, there is granted to the Secretary of Labor the power to close down an operation because of "imminent hazard." It is not a question of the number of times that the Secretary might use such authority. The problem would be raised by the threat of invoking the Secretary's authority unless concessions are granted—even though such concessions may have nothing to do with occupational safety and health.

***

It is only good judgment to construct the administrative organization to fit the mission to be accomplished. In this case the mission of providing expert assistance and review of safety and health provisions in hundreds of thousands of small undertakings can best be achieved by a decentralized administrative organization. Consequently, good management requires that, if possible, the administration of [a] national program to insure occupational safety and health be done through the State governments, if they are encouraged, and, if necessary, assisted to do the job.

It may be argued that the State governments are neither able nor willing to undertake such a responsibility. Such a generalization is an injustice to a number of States of industrial importance [which] have effective legislation, standards, administrative organizations and financing to carry out State occupational safety and health programs. The status of State programs is rather difficult to evaluate, but some information is available, and it indicates that the perfunctory dismissal of State governments as disinterested, unwilling or unable to carry out this assignment is a superficial and unfair judgment.

My own State of New York has a fine program of occupational safety and health. It devotes constructive assistance to highly hazardous and smaller operations through a consultative program that seeks to encourage the promotion of safety rather than limit itself to penalizing those who fail to meet safety and standards.

In any event, much can be done to assist and encourage the States to measure up to their responsibilities in occupational safety and health. Because we believe it would make a constructive contribution, we endorse those sections of these bills which provide for research, education and training. We believe it would be especially valuable to have the Department of Health, Education and Welfare conduct research into human motivation that would help all concerned in training and motivating employees to work safely.

It has been noted over and over again that in the vast majority of occupational accidents, human failure is wholly or partly responsible. If we could learn from scientific research how we can more effectively get people to work in a way they know to be safe rather than take short cuts or yield to impulse, the whole cause of occupational safety would be given a great push forward.

Similarly, there is a great shortage of adequately trained professional people to fill the need in occupational medicine, occupational safety and occupational hygiene. The training of such people by the Department of Health, Education and Welfare should have a high order of priority. The availability of such trained people would do more for occupational safety and health than hundreds of regulations and thousands of Federal inspectors.

It is our considered conclusion that if State governments and private business are given the benefit of federally sponsored research, a federally sponsored education and training program for physicians, safety engineers, hygienists, nurses and technicians, and if the State governments are given access to additional funds to support and experiment with a variety of State occupational safety and health programs, there will be substantial and prompt progress on the part of the States that would accomplish far more than would a program of Federal penalties and other attributes of overwhelming Federal authority reaching into hundreds of thousands of large and small business operations.

***

Teplow's testimony, which was typical of the employer view in 1968, made several key points in opposing the administration bills. First, the workplace injury and illness problem was not sufficiently serious (no "sudden or national emergency requiring overwhelming federal intervention"); second, adequate improvement in occupational safety and health can be achieved through "voluntary action on the part of employers and employees" and by state regulation (assistance and review of "thousands of small undertakings can best be achieved by a decentralized administrative organization"); third, federal regulatory intervention would be "intrusive," making a "federal case" out of many matters that could be satisfactorily resolved at the plant level; and fourth, the federal presence at the workplace would "tempt" "many an employee representative" to utilize OSHA for promoting ulterior objectives. These arguments, which ultimately did not prevail, have reappeared on occasion since 1970 in statements critical of the implementation of the Act.

S. 2864 and H.R. 14816 failed to pass the Congress in 1968. As explained by MacLaury, President Johnson's decision not to run for reelection, domestic violence in the inner cities, demonstrations against the Vietnam War, and many other events diverted congressional and national attention from dealing with workers' safety and health.[24] In November 1968, Richard Nixon was elected President. He asked for the enactment of a federal occupational safety and health law in a message to Congress in August 1969. Administration bills were introduced, differing from the earlier Democratic administration bills, primarily and significantly, in that they limited the scope of the authority of the Secretary of Labor in the standards and enforcement areas. Industry switched its position and supported the administration bills. It soon became clear that a federal law would be passed during the 91st Congress (1969–1970). During that Congress, job safety and health had become a significant issue. The Construction Safety and Health Act was passed, as was the Coal Mine Health and Safety Act. The only remaining and overriding question was which bill would pass, the administration bill or the substitutes introduced by Democratic members of Congress and supported by the labor unions. The development of occupational safety and health legislation in Congress during 1969–1970 was described in the following analysis, published by The Bureau of National Affairs, Inc. soon after the Act was passed.[25]

---

[24]MacLaury, *The Job Safety Law of 1970: Its Passage Was Perilous,* MONTHLY LAB. REV., Mar. 1981, at 18, 22.

[25]THE BUREAU OF NATIONAL AFFAIRS, INC., THE JOB SAFETY AND HEALTH ACT OF 1970: TEXT, ANALYSIS, AND LEGISLATIVE HISTORY 16–21 (1971).

The Job Safety and Health Act of 1970: Text, Analysis, and Legislative History

HOW THE LAW EVOLVED

The 91st Congress was more receptive to occupational safety and health legislation, and early in the session passed coal mine, construction, and railroad safety acts.

A primitive version (H.R. 3809) of the bill eventually reported by the House committee was introduced by Congressman James O'Hara January 6, 1969, with 17 cosponsors.

On May 16, Senator Harrison Williams (D-NJ), with Senators Edward Kennedy (D-Mass), Walter Mondale (D-Minn), and Ralph Yarborough (D-Texas) dropped a job safety bill (S. 2193) in the Senate hopper.

Although President Nixon called for "proposals to help guarantee the health and safety of workers" in his Domestic Programs and Policies Message to Congress on April 14, 1969, it was not until August 6 that the Administration proposals were spelled out. Following the President's August 6 message, Administration bills (H.R. 13373 and S. 2788) were introduced.

The early bills, although considerably less detailed than subsequent versions, contained many of the seeds of conflicts to come.

PRINCIPAL DISPUTES

President Nixon recommended that a new National Occupational Safety and Health Board set and enforce standards. Democrats wanted the same authority vested in the Secretary of Labor. This division posed formidable problems during congressional debate.

Almost all of the conflicts which plagued the legislators divided along labor-management lines, with management favoring the Administration approach and labor endorsing the Democratic version.

The important and troublesome issues included provisions relating to the general duty requirement for employers, "walk-around" rights of employers and employees during federal inspection tours, plant closings in cases of "imminent danger," and citation and posting requirements.

Fears that employers' rights under due process might be abrogated on the one hand, or that employees might be denied their right to protection or access to relevant information on the other, also hampered agreement.

THE HEARINGS

Intensive hearings were conducted by both the House and Senate Labor Subcommittees.

In contrast to testimony at the 1968 hearings, management abandoned its opposition to federal involvement in favor of the Nixon Administration's approach to separation of standard-setting, investigatory, and enforcement powers. Organized labor went straight down the line for vesting full responsibility in the Labor Department and lent its full weight to Mr. O'Hara's bill.

The House Subcommittee held 13 days of hearings in Washington, D.C., in September, October, and November 1969, with an additional two days in November in San Francisco.

Senate hearings ran intermittently between September 1969 and May 1970 for a total of 12 days. Field hearings were conducted in New Jersey, Pennsylvania, and South Carolina in April 1970.

EVENTS IN THE HOUSE

Early in 1970 the long process of compromise began. With the fall 1969 hearings completed and H.R. 3809 as a basis, Congressman Dominick V. Daniels (D-NJ), House Labor Subcommittee chairman, began preparation of a new bill.

Mr. Daniels' proposal went through seven versions before final presentation to the House Subcommittee.

The bill was approved by the Democrats, and the Subcommittee met March 19. Republican amendments were rejected and the Subcommittee concluded its business on March 25. In retaliation for alleged Democrat unwillingness to compromise, Republican Committee members boycotted the final session.

H.R. 16785, the clean bill reported out of the Subcommittee in the face of the Republican hold-out, was taken up by the full Committee on April 14. Republican hopes for compromise appeared dim, but a revised Administration bill, worked out by Congressman William A. Steiger (R-Wis), was presented to the full Committee at its first working session.

Mr. Steiger's compromise was defeated by a straight party vote, and Committee sessions were suspended as the two sides sought to work out their differences.

The Democrats scheduled an open meeting for June 3, but the Republicans failed to attend. Subsequent quorum calls by the Committee also fell short.

In a last-ditch effort to bring the adversaries together, the Republicans devised another draft, in cooperation with Labor Department representatives, and Congressmen William A. Hathaway (D-Maine) and Phillip Burton (D-Calif). This version, which failed, would have set up an Occupational Safety and Health Court, similar to the United States Tax Court, to adjudicate violations.

On June 13, H.R. 16785 was approved by the House Labor Committee, along straight party lines, save for one Republican defector, Congressman Alphonzo Bell of California, who said he voted just to get the bill out.

In an [appendix] to the Committee report, 12 of the 15 Republican members declared the bill unacceptable. Six also castigated separately the Committee bill as a "tragedy without equal." Another three, presumably acquiescing in the report, signed neither statement. Congressman Phillip Burton (D-Calif) offered a concurrent view, with the reservation that the Committee bill might prove too weak. (The text of the report appears in Appendix C.)

Congressman William H. Ayres (R-Ohio) later hoped to induce the Rules Committee to consider a substitute to be introduced by Mr. Steiger in place of the Committee bill. This tactic failed, however.

The Rules Committee refused the Committee bill a hearing prior to the House summer recess, running from August 14 to September 10. It was not until September 22 that H.R. 16785 was granted a rule, a week after H.R. 19200, a modification of the Administration bill, was introduced by Mr. Steiger, with Congressman Robert Sikes (D-Fla) as co-sponsor. Mr. Steiger made clear his intention to bring up his bill on the floor and have it replace the Committee measure.

On November 23, almost a week after the Senate version of the bill had been passed, H.R. 16785 reached the House floor.

Mr. Daniels presented a series of amendments designed to accommodate most Republican objections—except for divesting the Secretary of Labor of his standard-setting authority.

Resentment to the tardy compromise ran high. Furthermore, as Mr. Sikes pointed out, the proposed amendments left many controversial issues untouched.

The fundamental dispute of who was to set standards was one. Criteria for standards to be developed was another. In addition, walk-around provisions, i.e., the right of employees and employers to accompany the federal inspector, had not been resolved.

Mr. Steiger read a letter of support for H.R. 19200 from the Secretary of Labor, after which further action was postponed until the following day, when the Steiger-Sikes substitute was accepted first by a teller vote of 185-114, and then by a 220-172 roll call vote, as an amendment to the Committee bill.

EVENTS IN THE SENATE

The Senate hearings were under way during March and April, while the House Subcommittee was struggling to come to agreement. The Senate Labor Subcommittee had no intention of waiting for the House to complete action on the measure, but no executive meetings to consider the bill were expected until July 1970.

Like his counterpart in the House, Senate Labor Subcommittee Chairman Harrison J. Williams (D-NJ) had quorum troubles, and for a while it appeared he would have to bypass the Subcommittee if he hoped to report out a bill at all. At the end of August he managed to attract a quorum and by September 9, some problems had been hammered out at two executive sessions. The bill was sent to the full Committee, where consideration began September 16.

Through diligence and hard work, and perhaps with the unfortunate example of the House proceedings before them, the Senate Committee members effected substantial changes in the bill, many significant ones with Republican sponsorship.

Important as these amendments were, however, they still left pending the questions of basic administrative arrangements and aspects of imminent danger closings.

The Committee did encounter some difficulties. On several occasions, a rarely used parliamentary device prevented executive sessions from being held while the Senate was in session. The tactic only delayed the inevitable; there was sufficient agreement within the Committee to report the bill on September 25, by a 17 to 2 vote.

On September 29, Senator Peter H. Dominick (R-Colo) introduced S. 4404, the Senate equivalent of H.R. 19200, with the intention of offering it on the floor as a substitute for the Committee bill.

Originally scheduled for floor debate during the week of October 5, action on S. 2193 was postponed for a week on request of the minority. On October 13, Senator Mike Mansfield (D-Mont) requested unanimous consent to set aside other pending business, stating that additional requests for delay by the Republicans violated a tacit understanding on procedures made at the time of their initial request.

A Republican filibuster against the unanimous consent motion followed.

On October 14, the Senate agreed to make the bill the pending business of the Senate following the election recess.

On November 16, Mr. Williams moved almost immediately to table the Dominick substitute, and his motion carried 41 to 39. It was then agreed to vote on amendments to the Committee bill.

The spirit of compromise was in the air, although the first amendment, offered by Senators William B. Saxbe (R-Ohio) and Richard S. Schweiker (R-Pa), to require a court order for closing a plant due to imminent danger was rejected 42 to 40.

Mr. Javits introduced his proposition for a three-man appeals commission, joined by Mr. Dominick as cosponsor. Mr. Javits accepted the descriptive title of "Commission" from the Dominick bill to replace his proposed terminology of a "Panel." The amendment passed, 43 to 38.

The Senate difficulty with the imminent danger provision finally was resolved by a Javits amendment. This, coupled with an earlier amendment, enabled the Williams bill to come to a final vote and win Senate approval.

THE FINAL AGREEMENT

The conference to iron out differences between the Senate and House bills began December 8. It took five sessions, some of which lasted for more than seven hours to conclude what Congressman Carl D. Perkins (D-Ky) termed a "difficult conference."

At a long meeting on December 11, tentative agreement on the general duty clause was reached, with the compromise eventually accepted. The conferees also came together on provisions for inspection, investigations, and record-keeping.

About midway through the conference, the Senate conceded to the House on the imminent danger clause.

Mr. Steiger had proposed retention of the House provision on standards setting, but with the Secretary of Labor as chairman, and the Secretary of HEW as a member. When that tactic failed, he suggested that the Secretary of Labor set the standards, with review by a separate appeals board.

The Democrats appeared adamant on giving standards setting to the Secretary of Labor, and the House conferees caved in. Later accusations by Mr. Sikes that a longer holdout by the House could have saved the House provision partially were confirmed by admission of some Senators that a potential Senate counterproposal never had a chance to be heard.

A second issue which hamstrung the conferees during the final meeting on December 14 concerned formal *vs* informal rulemaking procedures. The Senate view of informal procedures prevailed over the objections and attempts at compromise by Mr. Steiger.

At the end of the day on December 14 there was some question whether the Republican House conferees would sign the conference report. A key figure was Mr. Steiger. Although still dismayed by some aspects of the report, Mr. Steiger announced his support of the conference report on December 15. He denied, however, that a letter from Secretary of Labor James Hodgson endorsing the compromise or pressure from the White House had influenced his decision.

The compromise went first to the Senate, which accepted it after only 10 minutes of discussion.

It was not quite so easy in the House, where debate lasted an hour. Congressman Sikes and William J. Scherle (R-Ind) argued for voting against the measure, amidst a series of laudatory speeches. At the end of debate, the conference report was adopted, 308 to 60 in the early afternoon of December 17.

The Occupational Safety and Health Act of 1970 was ready for President Nixon's signature.

In a signing ceremony held at the Labor Department, Mr. Nixon noted the compromise that had been made at each step. "This bill," he said, "represents in its culmination the American system at its best."

In reporting favorably the bill introduced by its chairman, Senator Harrison Williams of New Jersey, the Senate Committee on

Labor and Public Welfare in 1970 set forth in considerable detail the reasons why it believed a comprehensive federal occupational safety and health bill should be enacted. The committee emphasized the toll from industrial accidents, the growing problem of occupational diseases, and the inadequacy of voluntary and state programs in combating the "grim current scene" of injuries and illnesses in the workplace.

Senate Report No. 1282

91st Cong., 2d Sess. 2–4 (1970)

***

BACKGROUND

The problem of assuring safe and healthful workplaces for our working men and women ranks in importance with any that engages the national attention today.

As former Secretary of Labor Shultz pointed out during the hearings on this bill, 14,500 persons are killed annually as a result of industrial accidents; accordingly, during the past four years more Americans have been killed where they work than in the Vietnam war. By the lowest count, 2.2 million persons are disabled on the job each year, resulting in the loss of 250 million man days of work—many times more than are lost through strikes.

In addition to the individual human tragedies involved, the economic impact of industrial deaths and disability is staggering. Over $1.5 billion is wasted in lost wages, and the annual loss to the Gross National Product is estimated to be over $8 billion. Vast resources that could be available for productive use are siphoned off to pay workmen's compensation benefits and medical expenses.

This "grim current scene", Secretary Shultz further pointed out, represents a worsening trend, for the fact is that the number of disabling injuries per million man hours worked is today 20% higher than in 1958. The knowledge that the industrial accident situation is deteriorating, rather than improving, underscores the need for action now.

In the field of occupational health the view is particularly bleak, and, due to the lack of information and records, may well be considerably worse than we currently know.

Occupational diseases which first commanded attention at the beginning of the Industrial Revolution are still undermining the health of workers. Substantial numbers, even today, fall victim to ancient industrial poisons such as lead and mercury. Workers in the dusty trades still contract various respiratory diseases. Other materials long in industrial use are only now being discovered to have toxic effects. In addition, technological advances and new processes in American industry have brought numerous new hazards to the workplace. Carcinogenic chemicals, lasers, ultrasonic energy, beryllium metal, epoxy resins, pesticides, among others, all present incipient threats to the health of workers. Indeed, new materials and processes are being introduced into industry at a much faster rate than the present meager resources of occupational health can keep up with. It is estimated that every 20 minutes a new and potentially toxic chemical is introduced into industry. New processes and new sources of energy present occupational health problems of unprecedented complexity.

Recent scientific knowledge points to hitherto unsuspected cause-and-effect relationships between occupational exposures and many of the so-called chronic diseases—cancer, respiratory ailments, allergies, heart disease, and others. In some instances, the relationship appears to be direct: asbestos, ionizing radiation, chromates, and certain dye intermediaries, among others, are directly involved in the genesis of cancer. In other cases, occupational exposures are implicated as contributory factors. The distinction between occupational and nonoccupational illnesses is growing increasingly difficult to define.

In 1966–67, the Surgeon General of the United States studied six metropolitan areas, examining 1,700 industrial plants which employed 142,000 workers. The study found that 65 percent of the people were potentially exposed to harmful physical agents, such as severe noise or vibration, or to toxic materials. The Surgeon General further examined controls that were in effect to protect workers from such hazards, and found that only 25 percent of the workers were adequately covered.

California, a state with more rigorous occupational safety and health reporting procedures than most, showed 27,000 occupational diseases in 1964, a rate of 4.8 per 1,000 workers. Projected nationally, there were an estimated 336,000 cases of occupational diseases that year, a figure which by all indications continues to grow. Based on limited reporting experience, the Public Health Service now indicates that there are 390,000 new occurrences of occupational disease each year.

Studies of particular industries provide specific emphasis regarding the magnitude of the problem. For example, despite repeated warnings over the years from other countries that their cotton workers suffered from lung disease, it is only within the past decade that we have recognized byssinosis as a distinct occupational disease among workers in American cotton mills. Recent studies now show that this illness, caused by the dust generated in the processing of cotton, and resulting in continuous shortness of breath, chronic cough and total disablement, affects substantial percentages of cotton textile workers. In some states as many as 30% of those in the carding or spinning rooms have been affected, and it has been estimated that as many as 100,000 active or retired workers currently suffer from this disease.

Asbestos is another material which continues to destroy the lives of workers. For 40 years it has been known that exposure to asbestos caused the severe lung scarring called asbestosis. Nevertheless, as an eminent physician and researcher, Dr. Irving J. Selikoff, testified during the hearings on this bill:

> It is depressing to report, in 1970 that the disease that we knew well 40 years ago is still with us just as if nothing was ever known.

It has also since been found that manufacturing and construction workers exposed to asbestos suffer disproportionately from pulmonary cancer and mesothelioma. Because nothing has been done about the hazards of asbestos, even after the association of asbestos and lung cancer was first reported in 1935, 20,000 out of the 50,000 workers who have since entered one asbestos trade alone—insulation work—are likely to die of asbestosis, lung cancer or mesothelioma. Nor is the potential hazard confined to these workers, since it is estimated that as many as 3.5 million workers are exposed to some extent to asbestos fibers, as are many more in the general population.

Pesticides, herbicides and fungicides used in the agricultural industry have increasingly become recognized as a particular source of hazard to large numbers of farmers and farmworkers. One of the major classifications of agricultural chemicals—the organophosphates—has a chemical similarity to commonly used agents of chemical and biological warfare, and exposure, depending on degree, causes headache, fever, nausea, convulsions, long-term psychological effects, or death. Another group—the chlorinated hydrocarbons—are stored in fatty tissues of the body, and have been identified as causing mutations, sterilization, and death.

While the full extent of the effect that such chemicals have had upon those working in agriculture is totally unknown, an official of the Department of Health, Education, and Welfare stated, during hearings of the Migratory Labor Subcommittee, that an estimated 800 persons are killed each year as a result of improper use of such pesticides, and another 80,000 injured. Despite the unmistakable danger that these substances present, no effective controls presently exist over their safe use and no effective protections against toxic exposure of farmworkers or others in the rural populace.

Although many employers in all industries have demonstrated an exemplary degree of concern for health and safety in the workplace, their efforts are too often undercut by those who are not so concerned. Moreover, the fact is that many employers—particularly smaller ones—simply cannot make the necessary investment in health and safety, and survive competitively, unless all are compelled to do so. The competitive disadvantage of the more conscientious employer is especially evident where there is a long period between exposure to a hazard and manifestation of an illness. In such instances a particular employer has no economic incentive to invest in current precautions, not even in the reduction of workmen's compensation costs, because he will seldom have to pay for the consequences of his own neglect.

Nor has state regulation proven sufficient to the need. No one has seriously disputed that only a relatively few states have modern laws relating to occupational health and safety and have devoted adequate resources to their administration and enforcement. Moreover, in a state-by-state approach, the efforts of the more vigorous states are inevitably undermined by the shortsightedness of others. The inadequacy of anything less than a comprehensive, nationwide approach has been exemplified by experience with the chemical betanaphthylamine—a chemical so toxic that any exposure at all is likely to cause the development of bladder cancer over a period of years. The Commonwealth of Pennsylvania discovered this extreme effect of betanaphthylamine and banned its use, manufacture, storage or handling in that State, but production of this lethal chemical has begun in another State where legislation is inadequate. The exposure of workers to betanaphthylamine continues today.

In sum, the chemical and physical hazards which characterize modern industry are not the problem of a single employer, a single industry, nor a single state jurisdiction. The spread of industry and the mobility of the workforce combine to make the health and safety of the worker truly a national concern.

***

The committee report mentions three specific toxic substances as constituting examples of the magnitude of the health problem: cotton dust, asbestos, and pesticides. After the Act was passed, OSHA addressed regulations to each of these substances. A cotton dust standard was issued in 1978 and was ultimately affirmed by the Supreme Court, as it applied to the textile industry.[26] A health standard regulating asbestos was issued by OSHA in 1972 and was affirmed in substantial part by the Court of Appeals for the District of Columbia

[26] *American Textile Mfrs. Inst. v. Donovan*, 452 U.S. 490 (1981). For a discussion of OSHA regulation of cotton dust, see Chapter 10.

Circuit.[27] OSHA's attempt to regulate pesticides, however, was not successful. Its issuance of temporary standards establishing time limits for workers reentering pesticide-treated fields was set aside by the Court of Appeals for the Fifth Circuit,[28] and subsequently the Environmental Protection Agency preempted OSHA in regulating reentry times.[29]

The most controversial question during the legislative debates was the scope of the Secretary of Labor's authority. Under the bills reported by the Senate and House labor committees, the Secretary would have had authority to issue occupational safety and health standards (in most cases, after rulemaking proceedings); to conduct inspections and impose sanctions; and, if an employer contested the enforcement action, to adjudicate the appeal. Strong opposition was voiced to what was viewed as a concentration of authority in the hands of a single government official. As an alternative, proposals were made to establish two independent boards (or commissions), to be appointed by the President, one with authority to issue the standards, the other with authority to decide enforcement appeals. During the Senate debates, Senator Peter Dominick of Colorado presented a substitute bill (S. 4044), which would have significantly limited the Secretary of Labor's authority in these respects.[30]

Statement of Senator Dominick

***

Mr. President, I think it is important in making the record here that we point out that the substitute is not in any way designed to downgrade the need and the desirability of affording health standards and occupational safety standards for the workers of this country. I start out by saying that immediately after the substitute starts, section 2(a) reads:

> The Congress finds that personal injuries and illnesses arising out of work situations impose a substantial burden upon, and are a hindrance to, interstate commerce in terms of lost production, wage loss, medical expenses, and disability compensation payments.

Subsection (b) reads:

> The Congress declares it to be its purpose and policy, through the exercise of its powers, to regulate commerce among the several States and with foreign nations and to provide for the general welfare, to assure so far as possible every working man and woman in the Nation safe and healthful working conditions and to preserve our human resources.

[27] *Industrial Union Dep't v. Hodgson,* 499 F.2d 467 (D.C. Cir. 1974). In 1983, OSHA issued an emergency temporary standard on asbestos. See discussion in Chapter 4.

[28] *Florida Peach Growers Ass'n v. Department of Labor,* 489 F.2d 120 (5th Cir. 1974). In issuing the emergency temporary standard, OSHA relied in large part on the labor committee's statements regarding the pesticide hazard to employees. The court, however, found the OSHA "findings" insufficient to establish "grave danger" to employees, as is required by §6(c) of the Act, 29 U.S.C. §655(c).

[29] 39 FED. REG. 16,888 (1974). See discussion on the pesticide ethylene oxide in Chapter 6, Section C.

[30] 116 CONG. REC. 37,336 (1970).

So it ought to be made crystal clear that we are not trying to do anything to downgrade health standards. We are trying to provide legislation which will, in a more equitable way, permit development of safe and prudent health and safety standards by one body, to have the administration of those standards by another body, and to have the enforcement by a third group, so that we do not have so-called star chamber proceedings, which would be permitted under the bill as it was reported out of committee, and to which we object.

Mr. President, the law which we pass must be workable and effective. This means the bill must establish realistic mechanisms which call not only for the development and enforcement of strong standards but will also be fair and accord all parties, including employers subject to regulation.

In my opinion, Mr. President, the committee bill fails to do that. I think it is only fair to ask why. The reason is that it does not balance the need for regulation with the requirements of fairness and due process. The concentration of all authority for the promulgation of standards, the inspection and investigation of complaints, the prosecution of cases, and the adjudication of cases, totally in the hands of the Secretary of Labor is not a balanced approach. It is this structure, this procedural mechanism, which is objectionable to me, and I believe objectionable to many people around the country. It is objectionable because concentration of power gives rise to a great potential for abuse. A single man is easier to harass than an independent board or commission. Political pressure can be concentrated to achieve a particular point of view or course of action. The tradition of this Nation has been to place safeguards on power whenever it is granted. One of the greatest safeguards has been the separation of powers. By separating the legislative, executive, and judicial functions, a balance has been achieved which permits progress without abuse of authority. The separation of power proposed in S. 4404 provides a structure which will achieve the goal of safe and healthful working conditions without raising the sceptre of abuse. It is because of this structural deficiency and because the concentration of power does not permit the fullest utilization and coordination of expert opinion that S. 2193 is opposed by the administration, by employers, by State occupational health and safety agencies, and by various groups which are active in producing standards in the occupational health and safety field.

***

While the administration supported the Dominick substitute, labor union representatives, such as Walter J. Burke, Secretary-Treasurer of the United Steelworkers of America, argued strongly against separate standards and adjudicatory boards. In the unions' view, the result would not be the separation of powers, but rather the "pulverization of powers."[31]

Statement of Walter J. Burke

***

(3) THE BOARD

Gentlemen, our union is unequivocally opposed to those provisions in H.R. 13373 in which a National Occupational Safety and Health Board is

[31] *Hearings on H.R. 13373 Before the Select Subcomm. on Labor of the House Comm. on Education and Labor,* 91st Cong., 1st Sess. 927, 935–38 (1969).

established as the mechanism for developing standards and promulgating them. We support rather the provisions in the O'Hara, Hathaway, and Perkins bills in which the Secretary of Labor is the promulgating authority. Among the arguments elicited to support the concept of a safety board are two upon which I should like to comment: (1) The Board would represent expertise in the field, and (2) the Board would represent a separation of power with respect to standards setting and enforcement.

(A) EXPERTISE: WITH OR WITHOUT THE BOARD?

The impression given by the Board's exponents is that expert knowledge can be developed only through a permanently established board or commission. We are not in disagreement with the idea that professional and technical information must precede the decision to establish a standard. But experts can be located and the benefit of their opinions transmitted to the Secretary. Experts should be in an advisory capacity with the decisionmaking function lodged in the Secretary. In this way, the focal point of responsibility is more easily identified. Under a commission system the responsibility is more dispersed. For this reason we feel there will be more delay and, as indicated to this committee by Jerome Gordon, "it may be a device to avoid making decisions."

***

(B) SEPARATION OF POWERS: A VALID DIVISION?

I would now like to discuss separation of powers, and whether or not this separation would constitute a valid division.

I believe that we really hit upon the main reason for the Safety Board. It is claimed that the enforcing authority—namely, the Secretary of Labor—should be separated from both the developing and promulgating function. Actually, the proponents of the Board want to divide up the administrative responsibility of the act into such small parts that real administrative function no longer exists. The separation of powers concept in H.R. 13373 isn't so much whether the Secretary should be separated from the power to set standards but whether the Secretary should be separated from the power to act.

***

Gentlemen, this is not a separation of powers. It is a pulverization of powers. This is why we feel that any proposed safety legislation, which contains a safety board as defined in H.R. 13373, is no legislation at all.

***

(D) THE BOARD—WHAT IS IT?

Last year a number of employer-oriented organizations and trade associations testified in opposition to the need for Federal occupational health and safety legislation. This year the field has been reversed. The one substantively new thing, which has been added, is the board.

If the creation of the board in the Ayres bill could turn around such basic opposition to Federal legislation, it is not strange that the labor movement views the board with extreme apprehension because it may well be the one instrument which effectively destroys the impact of a Federal law. Actually, we look upon the board as a conduit, a buffer, and an obstacle. It is a conduit for consensus organizations in that it cloaks the private consensus standards with an aura of public sector respectability. It is a buffer in that it is an additional functional layer between the safety expertise in the private sector and effective public responsibility to make definite decisions on safety standards. It is an obstacle in that the evolution of standards must pass through the anonymity of a board vote and the Secretary of Labor must gear his enforcement authority to its discretion.

Far from considering the board to be nonpolitical in nature, it is precisely because it is appointed by the main political leader in the Nation that its character will be influenced by his selection. It is far better then to lodge the authority in the one appointee of the President whose Department is charged with the responsibility to foster the interest of the Nation's work force as it relates to the public policy; namely, the Secretary of Labor.

***

The Dominick substitute was defeated on the Senate floor by a vote of 41–39. However, Senator Jacob Javits of New York introduced an amendment to the committee bill which would have established an independent commission (Senator Javits referred to it as a "panel") to hear and determine enforcement cases under the law. Although Senator Javits had opposed the separate standards board, he believed that an independent panel for adjudication "more closely accords with traditional notions of due process."[32]

Statement of Senator Jacob Javits

MR. JAVITS. Mr. President, this amendment deals only with enforcement, not with standards, which are left in the hands of the Secretary of Labor. Under the amendment an independent panel of three members, appointed for staggered 5-year terms, with the advice and consent of the Senate, would be established to hear and determine enforcement cases under the act. The members of the panel would be appointed solely on the basis of their professional qualifications.

There are several reasons why enforcement through such a panel is to be preferred to enforcement by the Secretary, as provided in the original bill:

First, under the procedures established by the amendment, speed of enforcement would be greatly increased. In most contested cases, between 6 months and 2 years would be saved under the provisions which provide for true self-enforcing orders and discretionary review of trial examiner decisions.

Under the committee bill, no enforceable order to correct a violation would issue until the completion of all administrative and judicial review proceedings. This would involve, at a minimum in a contested case: First, hearings by a trial examiner; second, mandatory review of the decision by the Secretary or his designee; and third, review by a court of appeals. It is doubtful that this process could be completed in less than 18 months—2 years would be a more realistic estimate—in a seriously contested case.

Under my amendment, an enforceable order would issue at the end of the administrative review stage, rather than after judicial review—unless the court of appeals issued a stay. Furthermore, the * * * administrative review stage itself would be shortened by 3 to 6 months in many cases by making review by the panel of trial examiners' decisions discretionary. If

[32] 116 Cong. Rec. 36,532–33 (1970).

review were denied, the trial examiners' decision would automatically become the final order of the panel and enforceable as such.

Second, hearing and determination of enforcement cases by an independent panel more closely accords with traditional notions of due process than would hearing and determination by the Secretary. In the latter case the Secretary is essentially acting as prosecutor and judge. Any finding by the Secretary in favor of a respondent would be essentially a repudiation by the Secretary of his own Department's employees. While this type of enforcement has been used in connection with other statutes, is contemplated by the Administrative Procedures Act, and is not jurisdictionally defective on due process grounds, the awkward mechanics it imposes upon heads of departments who wish to exercise their adjudicatory power personally in order to preserve due process has not generally been appreciated. What happens is that one official of the Department—such as the Deputy Solicitor—will take the position of prosecutor and another official—such as the Solicitor—will take the position of a neutral in order to advise the Secretary.

More important, because of the awkwardness of this procedure and the heavy burden of personally reviewing hundreds of enforcement cases, it is highly likely that the Secretary of Labor will not even exercise his power under the committee bill personally, but will delegate it to a panel of officials within the Department. That is precisely what the Secretary of Interior has done under the Coal Mine Health and Safety Act of 1969. The net result will be enforcement by a panel anyway, but not one which is independent, and without the benefit of the shortened procedures which my amendment would provide.

These considerations, it seems to me, outweigh any possible benefits which might be gained from the better coordination which would allegedly occur if the adjudicatory power, as well as the prosecutorial and standards setting powers were given to the Secretary. Such coordination as is necessary would seem just as readily attainable with a panel as with the Secretary. It is the prosecutors upon whom this burden will primarily fall and under either approach they will be under the Secretary's control.

In short, the adjudicatory scheme of the committee bill can be made to work, and due process can be preserved under it, but the independent panel approach would do the same job faster, preserve due process more easily, and thereby instill much more confidence in the whole program in workers and businessmen alike.

The Javits amendment passed in the Senate by a vote of 43–38, and the provision creating a separate adjudicatory commission became part of the law as enacted. The commission is called the Occupational Safety and Health Review Commission, comprised of three members appointed by the President, with authority to determine contests of citations and proposed penalties issued by the Secretary of Labor.[33]

Although the separate standards board proposed by Senator Dominick did not pass the Senate, it was part of the Steiger substitute which was passed by the House of Representatives. The question of the separate standards board was the most controversial issue con-

[33]Secs. 10 and 12 of the Act; 29 U.S.C. §§659, 661 (1976).

fronting the Conference Committee; ultimately, the House of Representatives receded and the Senate approach, giving the Secretary of Labor authority to issue occupational safety and health standards, subject to review in the U.S. courts of appeals, was adopted.[34]

Another important issue in the evolution of the Act was the role of employees in the enforcement program. The bill that was passed by the Senate gave employees the right to accompany the inspector during the physical inspection of the workplace ("walkaround" right), the right to obtain a workplace inspection by the Department of Labor upon filing a complaint meeting certain requirements, and the right to participate in Review Commission proceedings as a party. The House bill contained only a limited walkaround right and no complaint or party rights. The House walkaround provision provided:

> If the employer, or his representative, accompanies the Secretary or his designated representatives during the conduct of all or any part of an inspection, a representative authorized by the employees shall also be given an opportunity to do so.[35]

The report by the Senate Committee on Labor and Public Welfare emphasized the importance in the statutory structure of the right of employees to participate in inspection activity.

Senate Report No. 1282

91st Cong., 2d Sess. 11–12 (1970)

***

INSPECTIONS AND INVESTIGATIONS

In order to carry out an effective national occupational safety and health program, it is necessary for government personnel to have the right of entry in order to ascertain the safety and health conditions and status of compliance of any covered employing establishment. Section 8(a) therefore authorizes the Secretary or his representative, upon presenting appropriate credentials, to enter at reasonable times the premises of any place of employment covered by this act, to inspect and investigate within reasonable limits all pertinent conditions, and also to privately question owners, operators, agents or employees.

During the field hearings held by the Subcommittee on Labor, the complaint was repeatedly voiced that under existing safety and health legislation, employees are generally not advised of the content and results of a Federal or State inspection. Indeed, they are often not even aware of the inspector's presence and are thereby deprived of an opportunity to inform him of alleged hazards. Much potential benefit of an inspection is therefore never realized, and workers tend to be cynical regarding the thoroughness and efficacy of such inspections. Consequently, in order to aid in the inspection and provide an appropriate degree of involvement of employees themselves in the physical inspections of their own places of employment, the committee has concluded that an authorized representative of employees

[34]Procedures for the issuance of standards are contained in §6 of the Act, 29 U.S.C. §655.
[35]H.R. 19200, 91st Cong., 2d Sess. §9(b), 116 Cong. Rec. 31,873, at 31,876 (1970).

should be given an opportunity to accompany the person who is making the physical inspection of a place of employment under section 9(a). Correspondingly, an employer should be entitled to accompany an inspector on his physical inspection, although the inspector should have an opportunity to question employees in private so that they will not be hesitant to point out hazardous conditions which they might otherwise be reluctant to discuss.

Although questions may arise as to who shall be considered a duly authorized representative of employees, the bill provides the Secretary of Labor with authority to promulgate regulations for resolving this question. Where the Secretary is not able to determine the existence of any authorized representative of employees, section 8(e) provides that the inspector shall consult with a reasonable number of employees concerning matters of health and safety in the workplace. It is expected that such consultation shall be undertaken with a view both to apprising the inspector of all possible hazards to be found in the workplace, as well as to insure that employees generally will be informed of the inspector's presence and the purpose and manner of his inspection.

In order that employees will be informed of any violation found by the inspector, section 10 specifies that citations shall be prominently posted near the place where the violation occurred. In addition, section 8(f)(2) provides that employees or a representative of employees may, before or during an inspection, give written notification to the Secretary or an inspector of any violation which they believe exists, and such employees or representative of employees shall be provided with a written explanation when no citation is issued respecting such alleged violation. The Secretary must also establish informal review procedures for use of employees or employee representatives who wish to question further the refusal to issue a citation.

A further provision, section 8(f)(1), entitles employees or a representative of employees who believe that a health or safety violation exists which threatens physical harm or that an imminent danger exists, to request a special inspection by giving notification to the Secretary, setting forth the basis of the request. If the Secretary determines upon receipt of the notification that there are reasonable grounds to believe that a violation or imminent danger exists, he shall make a special inspection as soon as practicable. If the Secretary determines there are no reasonable grounds to believe that a violation or imminent danger exists he shall so notify in writing those making the request.

By requiring that the special inspection be made "as soon as practicable," the committee contemplates that the Secretary, in scheduling the special inspection, will take into account such factors as the degree of harmful potential involved in the condition described in the request and the urgency of competing demands for inspectors arising from other requests or regularly scheduled inspections.

While the bill provides that a request for a special inspection shall be reduced to writing, the committee intends that notification may first be made by telephone, and that where an immediate harm is threatened, such as in an imminent danger situation, the Secretary should not await receipt through the mail of the written notification before beginning his inspection.

***

The Conference Committee adopted, substantially, the Senate version of all provisions on employee rights.[36]

[36]Employee walkaround rights are set forth in §8(e) of the Act, 29 U.S.C. §657(e); the right to file complaints is in §8(f), 29 U.S.C. §657(f); and the right to participate as parties in Commission hearings is in §10(c), 29 U.S.C. §659(c).

The other major issues that arose during the legislative development of OSHA are the following:

(a) *General Duty Clause.* The Senate bill required employers to furnish employees a place of employment "free from recognized hazards so as to provide safe and healthful working conditions"; no penalties would be imposed for violations of the general duty clause. The House bill, on the other hand, set forth the obligation for employers to furnish employment "free from any hazards which are readily apparent" and are likely to cause "death or serious physical harm"; penalties could be imposed for violation of the clause. In arguing against the Senate version, Senator Dominick stated that the offensive feature of the Senate provision was that "it is essentially unfair to employers to require compliance with a vague mandate applied to highly complex industrial circumstances."[37] The Conference Committee compromised, articulating the employer obligation as applying to "recognized hazards" that are likely to cause death or serious physical harm. Under the conference bill, penalties could be imposed for violations of the general duty clause. The conference compromise version is now §5(a)(1) of the Act.[38] The legislative history of the general duty clause was a basis for the decision of the Court of Appeals for the Eighth Circuit in *American Smelting and Refining Co. v. OSHRC.*[39] In that case, the court of appeals upheld an OSHA citation based on the general duty clause for impermissible levels of lead in the workplace. The court rejected the employer's contention that hazards whose recognition required airborne monitoring were not "recognized" for purposes of §5(a)(1).

(b) *Imminent Danger Situations.* The Senate bill authorized the Secretary of Labor to close down work operations in imminent danger situations if he determines, in light of the "nature and imminence of the danger," that there is insufficient time to seek a court order. Fears were expressed by members of Congress that this authority given to the individual inspector, even if he or she were required to first check with a supervisor or presidential appointee, would be subject to abuse. The House-passed bill required OSHA to seek a court order in all imminent danger situations, and the Conference Committee adopted the substance of the House provision, which is now §13 of the Act.[40] The legislative history of the imminent danger provision was an important element of the Supreme Court decision in *Whirlpool Corp. v. Marshall.*[41] There, the Supreme Court upheld the right of an employee to refuse to work, free from employer

[37] 116 Cong. Rec. 36,531 (1970).
[38] 29 U.S.C. §654(a).
[39] 501 F.2d 504 (8th Cir. 1974). Excerpts from this decision are reprinted in Chapter 14, Section B. Cf. Morgan & Duvall, *OSHA's General Duty Clause: An Analysis of Its Use and Abuse,* 5 Indus. Rel. L.J. 283, 287–98 (1983) (arguing on the basis of legislative history for "limited" application of "general duty"); Drapkin, *OSHA's General Duty Clause: Its Use Is Not Abuse—A Response to Morgan and Duvall,* 5 Indus. Rel. L.J. 322 (1983); Morgan & Duvall, *Reply to Drapkin,* 5 Indus. Rel. L.J. 334 (1983).
[40] 29 U.S.C. §662.
[41] 445 U.S. 1 (1980). Excerpts from this decision are reprinted in Chapter 18, Section B.

reprisal, in a situation reasonably threatening serious physical harm.

(c) *Feasibility of Standards.* The legislative development of the feasibility requirement for standards under §6(b)(5)[42] is discussed fully in the Supreme Court decisions in the *Benzene* and *Cotton Dust* cases.[43]

(d) *Rulemaking Proceedings.* The Senate bill provided that OSHA standards rulemaking would be "informal"; the House of Representatives, on the other hand, provided for "formal" rulemaking, subject, on court review, to the more stringent substantial evidence test. The Conference Committee again compromised, adopting informal rulemaking with public hearings if requested, subject, however, to a substantial evidence test. The first explication of this "anomaly," as it has been called, was made by the District of Columbia Circuit Court of Appeals in the *Asbestos* case.[44] The court, in its landmark opinion, said that OSHA rulemaking was "hybrid" and that standards were subject to substantial evidence review, but not "traditional" substantial evidence review.

(e) *Section 11(c)—Antidiscrimination Clause.* This section prohibits employers from taking reprisals against employees for the exercise of protected rights under the Act. Under this section, OSHA has the power to bring suits in federal district courts to enforce this right of employees. In *Taylor v. Brighton Corp.*,[45] the Court of Appeals for the Sixth Circuit relied in large measure on the legislative history of §11(c) in deciding that it would not imply a private right of action for employees to bring suit to enforce their rights under §11(c).

(f) *State Plan "Product" Proviso.* A proviso to §18(c)(2)[46] limits the authority of states with approved plans to issue standards regulating products which are distributed or used in interstate commerce. In 1983, OSHA approved the State of California's ethylene dibromide standard, which was more stringent than the counterpart federal standard, under the product proviso.[47] Prior to the approval, a proceeding was brought in a federal court to enjoin California from enforcing its ethylene dibromide standard. The federal court, relying extensively on the legislative history of the product proviso, denied the injunction.[48]

(g) *Judicial Review.* An additional item that was not debated in Congress turned out to be significant in retrospect. The House bill provided that challenges to standards promulgated by the "Board"

[42]29 U.S.C. §655(b)(5).

[43]*Industrial Union Dep't v. American Petroleum Inst.*, 448 U.S. 607 (1980), and *American Textile Mfrs. Inst. v. Donovan*, 452 U.S. 490 (1981). Excerpts from these decisions are reprinted in Chapters 9 and 10.

[44]*Industrial Union Dep't v. Hodgson*, 499 F.2d 467 (D.C. Cir. 1974). Excerpts from this decision are reprinted in Chapter 6.

[45]616 F.2d 256 (6th Cir. 1980). Excerpts from this decision are reprinted in Chapter 18.

[46]29 U.S.C. §667(c)(2).

[47]48 FED. REG. 8610 (1983), reprinted in part in Chapter 20, Section C.

[48]*Florida Citrus Packers v. California*, 10 OSHC 1137 (N.D. Calif. 1981).

must be brought in the Court of Appeals for the District of Columbia Circuit and that this review within 30 days of promulgation was the "exclusive" remedy for reviewing standards. The Senate bill allowed challenges to be brought in all courts of appeals and did not contain the exclusivity language. The Conference Committee adopted the Senate version without explanation.[49]

It is apparent that, except on the issue of imminent danger shutdown, the Conference Committee adopted the more stringent Senate provisions on most major issues. In a book published in 1973, the Ralph Nader Study Group discussed the "political" pressures on the Conference Committee and its view of the reasons why the Senate bill prevailed for the most part.

J. Page & M.-W. O'Brien, *The Nader Report: Bitter Wages*[50]

***

When the Senate and House pass conflicting versions of a bill, a final version is hammered out by a joint conference committee, composed of members of both parties from both chambers, and is then sent back to the House and Senate for final passage (or rejection). After nearly three years of testimony, lobbying, and meetings, a conference committee would now decide the fate of the job safety and health bill in the course of several weeks. These conference sessions became crucial.

During the conference itself, which took place in December of 1970, there were several levels of operation and a series of outside influences affecting the participants.

The Democratic House members on the committee were in the awkward position of having to support a bill they had strenuously opposed. They could not completely abandon the Steiger substitute, as this would mean certain defeat for the conference bill in the House. Representative Perkins, the chairman of the House Labor Committee, and Representative Daniels led the Democratic House contingent on the conference committee. They had to evaluate with great care each area of disagreement between the two versions, and decide when they should accept provisions from the stronger Senate bill, and when they had to vote with their Republican colleagues for a weaker provision in order to secure passage in the House. Perkins, who was selected chairman of the conference committee, was in a particularly difficult position. He was at times subjected to strenuous pressure from Congressmen Burton and Hathaway to hold firm for the Senate bill, since the Democrats had the majority in the conference committee if they stood together. But Perkins handled himself well throughout the proceedings. For example, on the issue of formal-vs.-informal rulemaking, the Democrats accepted the Senate provision for informal procedures, but only after extensive debate and soul-searching between Perkins and the other Democrats. The latter maintained that the only way to administer the law was by informal rulemaking, and that it would be too long and arduous a process to require the Labor Department and the unions to go through formal procedures.

The Republican minority on the committee deserved an Oscar. They had put up such resistance all along that most Democratic staff personnel

[49]Sec. 11, 29 U.S.C. §660. See Chapter 7 on the "race to the courthouse," and Chapter 15, Section C, on challenges to standards in enforcement proceedings.

[50]A selection from BITTER WAGES, Ralph Nader's Study Group Report by Joseph A. Page and Mary-Win O'Brien, 176–79. Copyright © 1972, 1973 by The Center for Study of Responsive Law. Reprinted by permission of Viking Penguin Inc.

were certain they would never approve a conference report. Without some Republican support, most observers believed that the conference bill could never receive a majority in the House and that even if it did pass, the President would veto it. The Republicans fostered this impression and complained as vigorously as they could each time a majority on the committee accepted one of the Senate bill's provisions. What they succeeded in concealing was a decision on the part of the administration not to torpedo the job safety and health bill. This amounted to a tacit Presidential recognition of the volatile political potential of the occupational environment issue. The administration did not want to be put in the position of being responsible for blocking the bill. Thus, to the Republicans, the substance of the bill as enacted became relatively unimportant; what was essential was for the Republicans to be able to say, as part of their blue-collar strategy, that the Nixon administration was doing something for the workers. Most of the Democrats wanted a strong bill even more than they wanted to avoid a Congressional defeat. But the Republicans, by referring constantly to the possibility of their refusal to sign the conference report if the bill were too strong, were able to win concessions in the conference.

***

All the pressure from the unions bore down on the issue of who should set standards. Before the conference had begun, the union representatives, led by Jack Sheehan, Anthony Mazzocchi, Arnold Mayer of the Meat Cutters Union, and George Taylor of the AFL-CIO, had decided that what they wanted above all else was that the Department of Labor, and not a board, set the standards. In order to reach this goal, the unions were willing to make certain concessions, which they conveyed to Senator Williams and Representative Perkins. For example, the unions agreed that an inspector, upon discovering an imminent hazard, would have to obtain a court restraining order and could not temporarily shut down a plant on the spot, as authorized by the Williams bill. This compromise was incorporated into the final bill.

What was most apparent throughout these sessions was the role of the Department of Labor officials as spokesmen for business. The unions recognized this and accepted it, hoping for better luck in future administrations. The Department's representative, Solicitor Peter Nash, supplied amendments on behalf of the administration to the Republican conferees, all of which were directed toward weakening the bill in line with proposals by business interests.

***

The issue of the independent board gradually resolved itself as a result of union concessions. The Senate proposal of a review panel (now called a review commission) was adopted, with standard-setting authority itself given to the Secretary of Labor. * * *

***

When Carl D. Perkins, Chairman of the House Committee on Education and Labor and head of the House delegation to the Conference Committee, reported to the full House on the results of the conference, he was faced with the formidable task of explaining why the Conference Committee failed to adopt substantial portions of the House bill. His persuasiveness was evidenced by the fact that the full House of Representatives adopted the conference report by a vote of 308–60 on December 17, 1970.[51]

[51] Chairman Perkins' statement to the House appears at 116 CONG. REC. 42,199 (1970).

The Senate adopted the Conference Committee bill without debate on December 17, and it was signed into law by President Nixon on December 29, 1970, to become effective on April 28, 1971. Because of its genesis as a compromise of two bills, the Occupational Safety and Health Act of 1970 was sometimes referred to as the Williams-Steiger Act; the name of Chairman Dominick Daniels of the House Select Subcommittee on Labor, a major architect of the House Committee bill, was not associated with the law.

Attempts have been made to assess the factors that led to the passage of this landmark legislation. One such assessment was made by Professor Steven Kelman of the Kennedy School of Government, Harvard University.[52]

Kelman, *Occupational Safety and Health Administration*

***

Three broad questions about the legislative history of the law creating OSHA should be raised. First, to what extent did the businesses to be regulated themselves seek the creation of the agency? The answer is rather clear: to no extent. The legislation originated in the Labor Department and the White House, and its major advocates were trade unions representing the workers to be protected.

Second, to what extent were "Naderism" and "environmentalism" important factors in the passage of the legislation? This answer is somewhat more difficult. At neither the 1968 or 1969 hearings did Ralph Nader testify,* and environmentalism did not emerge as a politically potent phenomenon until the spring of 1970. Nader did, however, testify at length and with gusto at the last set of hearings, conducted by the Senate Labor and Public Welfare Committee in late 1969 and early 1970. ("I am grateful to testify regarding the domestic form of violence known as occupational casualties and diseases," Nader began his testimony.) Furthermore, traditional "public interest" health groups, such as the American Public Health Association, testified for the bill from 1968. By the time the final battle over the bill took place in 1970, Nader operatives were involved. And the activist mood of the time, which saw the passage of other pieces of safety legislation, certainly aided the unions in gaining support for new federal legislation. After passage of the tough highway safety legislation in 1966, spearheaded by Nader, general-purpose "public interest advocates" were encouraged to choose safety as an area where chances of success looked good. Furthermore, the auto safety legislation could be cited as a precedent by proponents of legislation in other safety-related areas.

Finally, to what extent were participants in the process aware that "health" rather than "safety" would gradually come to be a major focus of agency mission and agency conflicts? The impression here is that concern over industrial accidents, rather than exposure to chemicals, dominated—though not overwhelmingly. Of the union representatives testifying at the 1968 hearings, two stressed safety concerns, two stressed health concerns, and four stressed both about equally. However, it should be noted that

*[Author's note—Ralph Nader in fact testified on June 28, 1968, in favor of the administration bill. *Hearings on S. 2864 Before the Subcomm. on Labor of the Senate Comm. on Labor and Public Welfare,* 90th Cong., 2d Sess. 509 (1968).]

[52]Kelman, *Occupational Safety and Health Administration,* 236, 242–43, in The Politics of Regulation, James Q. Wilson, ed. © 1980 by Basic Books, Inc., Publishers.

George Meany's testimony concentrated on health problems almost exclusively. "Every year," stated Meany, "thousands of workers die slow, often agonizing deaths from the effects of coal dust, asbestos, beryllium, lead, cotton dust, carbon monoxide, cancer-causing chemicals, dyes, radiation, pesticides, and exotic fuels. Others suffer long illnesses. Thousands suffer from employment in artificially created harmful environments."

***

Another analysis of the genesis of the Act appears in the chapter entitled "Political and Economic Perspectives on the Design of Occupational Safety and Health Policy" in John Mendeloff's book, *Regulating Safety*. Mendeloff suggests that while industry's concern about procedural issues under the bill was not absent, it "was to some degree a screen for more substantive objection to regulatory policy" as well as a concern "about the [Labor] Department's partiality towards unions."[53]

Most of the pertinent documents of OSHA legislative history are included in *Legislative History of the Occupational Safety and Health Act of 1970,* prepared by the Subcommittee on Labor of the Senate Committee on Labor and Public Welfare.[54] The hearings before the Subcommittee on Labor of the Senate Committee on Labor and Public Welfare, and before the Select Subcommittee on Labor of the House Committee on Education and Labor, on the Occupational Safety and Health Acts of 1968, 1969, and 1970 have also been published.

---

[53]J. Mendeloff, REGULATING SAFETY 20 (1979).

[54]STAFF OF SUBCOMM. ON LABOR OF THE SENATE COMM. ON LABOR AND PUBLIC WELFARE, 92D CONG., 1ST SESS., LEGISLATIVE HISTORY OF THE OCCUPATIONAL SAFETY AND HEALTH ACT OF 1970 (Comm. Print 1971).

PART ONE

OSHA Standards: An Overview

2

Start-Up Standards

The duty of employers to provide safe and healthful workplaces is defined primarily by the occupational safety and health standards promulgated by OSHA under §6.[1] Standards contain requirements designed to protect employees against workplace hazards; under §3(8), a standard is defined as requiring "conditions, or the adoption or use of one or more practices, means, methods, operations, or processes, reasonably necessary or appropriate to provide safe or healthful employment and places of employment."[2] Two main arguments have been advanced in favor of the government stating compliance obligations in legally enforceable rules that contain specific requirements with prospective effect: this approach clarifies the law in advance and avoids the imposition of new and often unexpected liability on a case-by-case basis; and it provides the Agency with an opportunity to weigh public views carefully before prescribing the obligations that must be met.[3] Some regulatory statutes, such as the Coal Mine Safety Act of 1969,[4] contained specific standards for compliance. In the Occupational Safety and Health Act, however, Congress did not itself enact occupational safety and health standards. It gave this responsibility to OSHA, authorizing it to promulgate legally enforceable standards, in accordance with the procedures set forth in §6, and consistent with the substantive criteria appearing in various portions of the Act. By transferring this authority to OSHA, Congress provided a flexible framework under which the Agency, on the basis of a broad range of information obtained from a variety of sources, would issue standards which would reflect both the technical

[1]29 U.S.C. §655 (1976). Employers are also required to comply with the general duty clause, §5(a)(1), 29 U.S.C. §654(a)(1). This provision is discussed in Chapter 14, Section B.

[2]29 U.S.C. §652(8). The phrase, "reasonably necessary or appropriate," was a major issue in the litigation involving OSHA's benzene and cotton dust standards. See Chapters 9 and 10.

[3]See *National Petroleum Refiners Ass'n v. F.T.C.*, 482 F.2d 672 (D.C. Cir. 1973), *cert. denied*, 415 U.S. 951 (1974) (detailed discussion of the reasons for the preference for rulemaking over case-by-case adjudication in the establishment of regulatory obligations).

[4]Federal Coal Mine Health and Safety Act of 1969, Pub. L. No. 91-173, 83 Stat. 742 (codified as amended at 30 U.S.C. §§801–962 (1976 & Supp. V 1981)).

and policy expertise of the Agency. Changes in the standards could be made by the Agency, following the prescribed procedures, as warranted by changed circumstances, either brought about by new technical information or because of the reexamination of the policy predicates of the original standards. The major constraint on the Agency's broad discretion is court of appeals review authorized by §6(f).[5]

OSHA standards have been categorized in several different ways. One of the most common distinctions is between "safety" standards and "health" standards. In general, safety standards are intended to protect against traumatic injury; health standards deal with toxic substances and harmful physical agents and protect against illness which often does not manifest itself until many years after initial exposure. The distinction between safety and health was recognized in the regulatory scheme of the Act by §6(b)(5) which prescribes certain specific criteria for standards "dealing with toxic materials and harmful physical agents."[6] In 1979, Senator Schweiker introduced an extensive amendment to the Act, predicated in large measure on the distinction between the enforcement of safety and health standards.[7] Dr. Eula Bingham, Assistant Secretary for OSHA, in arguing against the amendment, rejected the usefulness of this distinction, saying:

> It is difficult, if not impossible, to characterize any workplace problem as being exclusively one of health or safety. Clearly, hazards such as excessive noise or the lifting of heavy objects affect both the safety and health of workers. Then too there are the problems that result from repeated application to body structures of mechanical forces which exceed the allowable stress-level of tissues. Examples of such disabling injuries, which are very prevalent in the workplace, are tendonitis, tenosynovitis, carpal tunnel syndrome, and lower back syndrome. Even in the area of exposure to toxic chemicals, many employees would make the artificial distinction of characterizing acute exposure as a "safety" problem while chronic exposure would be called a health problem. In fact, when you really begin to examine the medical effect of "injuries" and "illnesses," they blur into a single continuum of bodily harm.
>
> ***
>
> One of the major objectives of the persons who worked to pass the act was the integration of safety and health concerns into a balanced program for the workplace. Prior to the passage of the act, there was a very distinct separation in professional education and practice between safety and health. This separation produced a dichotomy in the workplace where many problems were viewed by either safety or health professionals as falling into the other's area of responsibility. Consequently, many problems were not addressed adequately, or at all.[8]

Another distinction has been based on the scope of the standard.

---

[5] 29 U.S.C. §655(f).

[6] The meaning of the first sentence of §6(b)(5), 29 U.S.C. §655(b)(5), was also a major issue in the litigation involving the validity of OSHA's benzene and cotton dust standards.

[7] See *infra* Chapter 13, Section E on the Schweiker amendment, and Chapter 21 on the Schweiker appropriation rider, both of which involve the distinction between enforcement of safety and health standards.

[8] *Occupational Safety and Health Improvements Act of 1980: Hearings on S. 2153 Before the Senate Comm. on Labor and Human Resources,* 96th Cong., 2d Sess. 31–32 (1980).

Standards that apply across-the-board to all, or virtually all, industrial operations are referred to as "horizontal" standards. Most OSHA safety standards are horizontal standards, covering defined hazards that appear in all industrial operations. An example is in 29 C.F.R. Part 1910, Subpart D, covering "walking-working surfaces."[9] The major exceptions are the "vertical" standards applicable to the construction, maritime, agriculture, diving, and telecommunications industries.[10] Health standards in most cases are applicable to specific toxic substances or harmful physical agents; for example, health standards regulate asbestos, vinyl chloride, lead, and noise hazards.[11] So-called generic health standards are not limited in their applicability to particular substances. Examples are the respirator standard[12] adopted in 1971; the rule on access to medical and exposure records[13] issued in 1980; and the hazard communications standard.[14] OSHA's Carcinogens Policy, issued in 1980, does not directly regulate any substance but rather establishes policies and procedures for the identification, classification, and regulation of potential occupational carcinogens.[15]

OSHA standards have also been divided in terms of the procedures by which they are issued. So categorized, there are three kinds of OSHA standards: (1) start-up standards, issued without rulemaking by OSHA during the first two years after the Act became effective; (2) standards issued after rulemaking at any time by OSHA; and (3) emergency temporary standards issued without rulemaking and effective for six months, where a "grave danger" to employees exists and the standard is necessary to protect against that danger. Start-up standards are sometimes referred to as "interim" standards and standards issued after rulemaking as "permanent" standards. This terminology is misleading, however; while Congress expected start-up standards to be replaced eventually with more up-to-date standards, the start-up standards are considered permanent until superseded. The remainder of this chapter deals with OSHA start-up standards.

Section 6(a) of the Act[16] authorizes the Secretary of Labor to adopt, *without rulemaking,* "national consensus standards" and established federal standards. "National consensus standards" are de-

[9]29 C.F.R. §§1910.21–.32 (1983). Sections of regulations are commonly referred to by the Part and Subpart in which they are located, e.g., Part 1910, Subpart D.

[10]Construction Standards, 29 C.F.R. pt. 1926 (1982); Maritime Standards, 29 C.F.R. pts. 1915–1919 (1983); Marine Terminal Standard, 48 FED. REG. 30,886 (1983) (to be codified as 29 C.F.R. pt. 1917); Agriculture Standards, 29 C.F.R. pt. 1928 (1983); Diving Standards, 29 C.F.R. §§1910.401–.441 (1983); Telecommunications Standards, 29 C.F.R. §1910.268 (1983). OSHA's published rule of interpretation is that a horizontal standard will apply to an industry covered by vertical standards (e.g., construction) only "to the extent" that the "particular" vertical standards do not apply to that industry, 29 C.F.R. §1910.5(c) (1983). As to the construction industry, OSHA has identified the horizontal standards that are applicable. See Field Operations Manual, OSHR [Reference File 77:4301].

[11]OSHA health standards appear at 29 C.F.R. §§1910.94–.100, 1910.1000–.1500 (1983).

[12]29 C.F.R. §1910.134 (1983).

[13]29 C.F.R. §1910.20 (1983).

[14]48 FED. REG. 53,280 (1983) (to be codified at 29 C.F.R. §1910.1200).

[15]29 C.F.R. §§1990.101–.152 (1983).

[16]29 U.S.C. §655(a).

fined in §3(9) as those adopted by "nationally recognized standards producing organizations" under procedures allowing "diverse views to be considered" and whereby interested persons have reached "substantial agreement" on their adoption.[17] Established federal standards are those standards previously adopted by a federal agency (notably the Department of Labor) and in effect on April 27, 1971.[18] In both instances, further rulemaking was deemed unnecessary, since the standards were adopted under procedures giving individuals with diverse views an opportunity to be heard. On May 29, 1971, OSHA adopted a large body of national consensus and established federal standards.[19] The national consensus standards had previously been issued by the American National Standards Institute (ANSI) and the National Fire Protection Association (NFPA), two organizations expressly mentioned in the legislative history as "national consensus organizations."[20] The standards are primarily in the safety area and are published in Part 1910 of the OSHA regulations. The established federal standards were adopted from the Walsh-Healey Public Contracts Act[21] (published in Part 1910); the Longshoremen's and Harbor Workers' Compensation Act[22] (published in Parts 1915–1918); and the Contract Work Hours and Safety Standards Act[23] (published in Part 1926).[24]

The provision for swift adoption of start-up standards was in response to several policy imperatives recognized by Congress during the legislative development of the Act, the greatest of which was the strong feeling that OSHA should have a comprehensive basis of standards for a prompt beginning of its enforcement program.[25] At

[17]29 U.S.C. §652(9).

[18]29 U.S.C. §653(b)(2).

[19]36 Fed. Reg. 10,466 (1971).

[20]S. Rep. No. 1282, 91st Cong., 2d Sess. 6, reprinted in 1970 U.S. Cong. & Ad. News 5177, 5182.

[21]Ch. 881, 49 Stat. 2036 (1936) (current version at 41 U.S.C. §§35–45 (1976 & Supp. V 1981)).

[22]Ch. 509, 44 Stat. 1424 (1927) (current version at 33 U.S.C. §§901–950 (1976 & Supp. V 1981)).

[23]Pub. L. No. 87-581, 76 Stat. 357 (1962) (codified as amended at 40 U.S.C. §§327–333 (1976)). These sections are also referred to as the Construction Safety Act of 1969.

[24]In addition to the safety standards adopted under §6(a), OSHA also adopted permissible exposure limits for approximately 400 toxic substances. These health standards, now appearing in 29 C.F.R. §1910.1000 (1983), were derived from both national consensus and established federal standards. The national consensus standards had been issued by ANSI, while the established federal standards had been adopted under the Walsh-Healey Act from the TLVs (threshold limit values) recommended by the American Conference of Governmental Industrial Hygienists (ACGIH). Employers are required to achieve the permissible limits by feasible engineering controls, or if not feasible, by personal protective equipment. These §6(a) standards do not contain the detailed requirements contained in health standards issued after rulemaking. For a comparison of the concept of TLV, which was developed by the American Conference of Governmental Industrial Hygienists, and permissible exposure limit, OSHA's regulatory requirement, see Corn, *Regulations, Standards and Occupational Hygiene Within the U.S.A. in the 1980s,* 27 Ann. Occup. Hyg. 91, 100–01 (1983). For a discussion of the reasons for OSHA's not having adopted the ACGIH's recommended zero tolerances for carcinogens, see D.P. McCaffery, OSHA and the Politics of Health Regulation 73 (1982); N. Ashford, Crisis in the Workplace 154 (1976).

[25]S. Rep. No. 1282, 91st Cong., 2d Sess. 6, reprinted in 1970 U.S. Code Cong. & Ad. News 5177, 5182. The Senate Committee on Labor and Public Welfare acknowledged that the §6(a) standards "may not be as effective and as up-to-date as desirable" and that they would provide only a "minimum level of health and safety." *Id.*

the same time, employers felt that the OSHA enforcement program would be more reasonable if it was based on national consensus standards, in whose development they played a strong role.[26] The start-up standards, as intended, have continued to be an important basis for OSHA enforcement. It was not anticipated, however, that the enforcement, particularly of the national consensus standards, would, for a number of years be a major source of criticism of OSHA. The genesis of the problem was discussed by John Mendeloff in a book written in 1979 analyzing safety and health policy.[27]

Mendeloff, *Regulating Safety*

***

ADOPTION OF THE INITIAL STANDARDS

The thousands of standards OSHA adopted in early 1971 included the consensus standards developed by groups like the American National Standards Institute (ANSI) as well as the federal standards already enforced by the Labor Department under the terms of the Walsh-Healey Contracts Act and other laws. The "consensus" method essentially meant that committees composed primarily of corporate safety engineers, with a light sprinkling of labor and government representatives, overwhelmingly agreed that a certain technique or protective device had desirable safety features. In many states these consensus standards, as well as exposure limits for toxic materials, had been incorporated into state safety codes in the 1950s and 1960s. Although many standards were out of date, they generally represented a model for industry to strive toward.

Every OSH bill had explicitly mandated the adoption of consensus standards. The 1969 Nixon Administration bill also gave ANSI a central role in the ongoing process of establishing new standards, since the Labor department would have had to rely on ANSI to approve new standards. Business spokesmen generally favored this proposal because they wanted to curb the department's discretion. Some had criticized Labor Secretary Wirtz's adoption of standards under the Walsh-Healey Act; he had picked them from "out of the blue" according to one business lobbyist I interviewed. The bills considered in 1970 no longer gave ANSI a special role in new standards, but they still required OSHA to adopt the consensus standards.

***

On May 29, 1971, OSHA promulgated *all* of the consensus and federal standards. The agency still had twenty-three months to go before its deadline expired. The standards occupied three hundred densely packed pages of the *Federal Register,* which still omitted many that were incorporated by reference. The effective dates of most of the standards were delayed for three months; for some, the delay was nine months. During the three months, OSHA inspectors could only cite violations of the act's "general duty" clause, a much clumsier enforcement tool.

[26] *Occupational Safety and Health Act of 1970: Hearings on S. 2193 Before the Subcomm. on Labor of the Senate Comm. on Labor and Public Welfare,* 91st Cong., 1st Sess. 328 (1969) (statement of J. Sharpe Queener, Safety Director, Du Pont Co., representing the Chamber of Commerce of the United States).

[27] J. Mendeloff, REGULATING SAFETY: AN ECONOMIC AND POLITICAL ANALYSIS OF OCCUPATIONAL SAFETY AND HEALTH POLICY 36-41. © 1979 by The MIT Press. Reprinted with permission.

As many observers, including the Occupational Safety and Health Review Commission, have noted, this hasty and wholesale adoption generated many errors. A few standards were actually not relevant to safety or health. Some were so vague as to be unenforceable. In some cases, OSHA had printed the consensus standards improperly, with significantly different wording and requirements. Most importantly, the standards had been adopted in a "wooden and inflexible" manner. No care was given to limiting their application only to particular situations. No attention was paid to how effective they would be, nor to how much compliance with them would cost. As Frank Barnako explained later, no one thought that OSHA would adopt all of the consensus standards. He wondered whether the business community bore some of the responsibility; perhaps it should have established a crash program to review the standards and recommend which should be weeded out.

Why did OSHA promulgate these standards so quickly and indiscriminately? If it had waited eighteen months, it would have had time to do some of the weeding itself. The rush to promulgate standards may have been part of a strategy aimed at achieving a delay in the date when they became effective. The language of the act appeared to provide little support for such delays, but OSHA leaders were eager to give employers time to become acquainted with the standards. Early promulgation may have been a way to appease union critics and to forestall lawsuits against the delayed effective dates. The Nader study of OSHA suggests that this action did hold off legal attacks.

Yet even if it had used the full two years, OSHA would have lacked both the time and the resources for an intelligent assessment of more than a small minority of the consensus standards. As OSHA's standard setters quickly learned, the information needed to make such assessments is highly precise and rarely available. The act had given the consensus standards a presumption of effectiveness; the burden of proof was placed on OSHA to show that they were not effective. This presumption ensured that any widespread pruning of the standards would encounter a wave of lawsuits that OSHA would find very difficult to win.

***

Criticism of OSHA's enforcement of national consensus standards was extensive. Some of this criticism came from members of Congress and was based in large part on the difficulty for employers, particularly small employers, in understanding the mass of material included in the regulations.[28] One of the more outspoken critics of OSHA enforcement of national consensus standards was Robert D.

---

[28]See, e.g., *Occupational Safety and Health Act Review, 1974: Hearings Before the Subcomm. on Labor of the Senate Comm. on Labor and Public Welfare,* 93d Cong., 2d Sess. 52–56 (1974) (statement of Sen. Dominick). Criticism of OSHA's adoption of national consensus standards was by no means limited to employers. Thus, in 1980, Lane Kirkland, President of the AFL-CIO, told an oversight committee that "in retrospect [OSHA] made a major mistake in 1971 when it hastily issued, *en masse,* a ramshackled collection of so-called 'national consensus' standards," developed by NFPA and ANSI, "two business dominated standards making organizations * * * . This hodgepodge collection of standards and OSHA's early efforts to enforce them, probably did more to damage the initial acceptance of the entire program than any other single action * * * . Because of the hue and cry over these 'nit-picking' aspects of the program, the more serious and important goals of the OSHA program became lost in a morass of largely unintelligible debate and political animosity." *Oversight on the Administration of the Occupational Safety and Health Act, 1980: Hearing Before the Senate Comm. on Labor and Human Resources,* 96th Cong., 2d Sess. 730–31 (1980).

Moran, then Chairman of the Occupational Safety and Health Review Commission.[29]

Moran, *Cite OSHA for Violations*

> ***
>
> An occupational safety and health standard, the Act says:
>
> > means a standard which requires conditions, or the adoption or use of one or more practices, means, methods, operations, or processes, reasonably necessary or appropriate to provide safe or healthful employment and places of employment.
>
> Do the OSHA standards fully live up to this requirement of the Act? Or should they be cited for failure to comply?
>
> In my judgment, the answer to the latter question is "yes"—at least for some of them. In fact I think quite a few OSHA standards do "violate" the spirit and purpose of the Act, as well as the letter of the preceding definition. Although I feel that some of the actions OSHA has taken in promulgating some things as "standards" could constitute serious "violations" of fairly high gravity, I'm not going to propose any penalties—but I would like to see some abatement.
>
> HURRY UP. Before turning to a few specific standards that I believe fail to comply with the Act, I'd briefly like to discuss how we came to get standards that don't meet the law's requirements. There are two principal reasons: 1) Congress gave little thought and had little concern for what might be hauled up in the nets which it labelled "any national consensus standard, and any established Federal standard" and 2) the hurry-up atmosphere surrounding OSHA's use of the net when the Act took effect as law in 1971.
>
> The first of these reasons was hardly a major sin. Congress faced many seemingly insurmountable obstacles in hammering out a bill that would be effective—and acceptable—to the broad spectrum interested in this legislation. It hardly occurred to anyone that someday the initial standards might prove to be *the* problem area in the occupational safety and health program. (There may be a few people who still don't see it that way, but few of them have to meet the standards' requirements.)
>
> Even Congressman Steiger, the Act's co-author, felt there wouldn't be much to worry about in such wholesale standards adoption (at least, with regard to ANSI and Federal standards) "because such standards have already been scrutinized. * * *"
>
> RIGOR. But the plain fact is that not even the ANSI standards had been put through the rigorous grilling they would have received had anyone anticipated their being handed down as the Ten Thousand Commandments. Said one ANSI official:
>
> > In the days before OSHA, when standards were developed as advisories, not laws, the committees sometimes tended to incorporate some lofty goals, knowing they would never be held accountable if they didn't achieve them.
>
> The "established Federal standards," (those promulgated under the Walsh-Healey Act are an example) are the source of many of the more troublesome regulations in the OSHA Bible. They were scrutinized even less

[29] Moran, *Cite OSHA for Violations*, Occupational Safety and Health, Mar.–Apr. 1976 at 19–20. Reprinted with permission.

than the ANSI standards when they were first written. Most of them affected few firms. For example, in 1969 there were only about 75,000 firms with any reason to even know there were such things as Walsh-Healey safety and health standards, since that Act applied only to manufacturers doing in excess of $10,000 worth of business with the Federal government.

Compare that to the estimated 5 million workplaces covered by the Job Safety Act! We can safely state that only about 1½ percent of the firms who must now comply with these established Federal standards were aware that they might be regulated by them when they were open for public scrutiny.

LOW PROFILE. Even among those 75,000 firms there was probably little concern over what the standards required, since they were "enforced" using what can only be called a low-profile technique. Only 34 formal complaints were issued against noncomplying companies in the entire 12 months of 1969. That figures out to about 20 percent of the number of citations issued in just one day during the current year.

But, as I said, few people even in the job-safety field anticipated trouble with the initial standards, national consensus or Federal, so Congress really couldn't have been expected to do so either. There were voluminous hearings and many, many witnesses testified. But nobody—not a single soul—mentioned any possible difficulty in abiding by the then existing ANSI or Walsh-Healey standards.

OSHA's haste in casting the net Congress provided could hardly be called a sin either. In fact, the speed with which the standards package hit the streets was commendable—after all, the reason behind this haste was to get into the business of protecting working people. One would be hard pressed to defend delay in an area where lives and their quality are at stake.

DEFICIENCIES. However, because of the rush in which the initial standards were adopted, we got a lot of would-be regulations that didn't fit the Act's definition of what they should be and what they should do. The initial package (and virtually all of it is still around) contained in profusion standards which were:

1. Not binding, not enforced, and not written in terms amenable to enforcement.
2. Not exclusively concerned with worker safety (that is, the safety of equipment, buildings, consumers, the general public, and workers was intermingled).
3. Not applicable to industry as a whole, or in some cases, even to all parts of a single segment of an industry.
4. Not without conflict and inconsistencies.
5. Not specific enough so that an ordinary businessman or employee could understand them.

All of these problems could have been alleviated before OSHA standards were first promulgated in 1971, but speed was the order of the day at that time. It seems to me that OSHA has subjected employers and employees of America to that same frustrating experience endured by millions of military personnel whose routine regularly included what has come to be known as the "hurry-up and wait" syndrome.

Nevertheless, as I said, the "hurry-up" part of OSHA's standard promulgation had its virtuous aspects. The "and wait" of the past four years, however, leaves a lot be desired.

***

One of Mr. Moran's major criticisms of the national consensus standards was that they were "not specific enough so that an ordinary businessman or employee could understand them." This view was translated into early decisions of the Review Commission; for example, in *Santa Fe Trail Transportation Co.,*[30] the Commission held that the OSHA standard[31] requiring that an employer have a person adequately trained in first-aid in the absence of an infirmary, hospital, or clinic "in near proximity" to the workplace was void for vagueness. As stated by Commissioner Van Namee, the standard "does not convey a sufficiently definite warning" and was, therefore, unenforceable.[32] The Court of Appeals for the Tenth Circuit, however, reversed this decision, applying a "reasonable man" standard in upholding the validity of the OSHA regulation.[33]

At the same time that many OSHA national consensus standards were being criticized for being unduly vague, many were also objected to on the ground that they were unnecessarily specific and therefore limited the employer's options in choosing a method of compliance in improving employee safety. This argument introduces another distinction that has been used in describing OSHA standards: "specification" standards and "performance" standards. Specification standards, or "design" standards as they are sometimes called, impose detailed requirements on employers; performance standards, on the other hand, describe the result to be reached, but permit the employer flexibility in determining the means for achieving that result. OSHA's machine-guarding standard, adopted as a national consensus standard in 1971, is an often-cited example of a specification standard:

> (f) *Self-feed circular saws.* (1) Feed rolls and saws shall be protected by a hood or guard to prevent the hands of the operator from coming in contact with the in-running rolls at any point. The guard shall be constructed of heavy material, preferably metal, and the bottom of the guard shall come down to within three-eighths inch of the plane formed by the bottom or working surfaces of the feed rolls. This distance (three-eighths inch) may be increased to three-fourths inch, provided the lead edge of the hood is extended to be not less than 5½ inches in front of the nip point between the front roll and the work.
>
> (2) Each self-feed circular ripsaw shall be provided with sectional nonkickback fingers for the full width of the feed rolls. They shall be located in front of the saw and so arranged as to be in continual contact with the wood being fed.[34]

In 1976, President Ford established a Presidential Task Force whose primary purpose was to "develop a model approach to safety standards which can be used by OSHA to increase the level of safety in the workplace while alleviating the criticisms of current standards." Its cochairmen were Philip J. Harter, Senior Staff Attorney of the Administrative Conference of the United States, and Joseph L. Kirk, Director of Policy and Implementation for OSHA. The Task

[30] 1 OSHC 1457 (1973) (Moran, Chairman, concurring; Cleary, Comm'r, dissenting).
[31] 29 C.F.R. §1910.151(b) (1983).
[32] 1 OSHC at 1458 (citing *Connally v. General Constr. Co.,* 269 U.S. 385, 391 (1926)).
[33] *Brennan v. OSHRC,* 505 F.2d 869 (10th Cir. 1974).
[34] 29 C.F.R. §1910.213(f) (1983).

Force Report recommended a "performance" approach to safety standards and particularly addressed itself to OSHA's machine guarding standards, Part 1910, Subpart 0.[35]

Report of the Presidential Task Force, OSHA Safety Regulation

***

CONSIDERATIONS ENTERING INTO THE RECOMMENDED APPROACH TO SAFETY STANDARDS

The Task Force reached what it believes is the best way to regulate safety in the workplace through mandatory standards. The background and basis for this approach are summarized below.

The Objectives of the New Approach. The Task Force sought an approach which would:

- be specifically directed to safety, without imposing manufacturing requirements not intimately related to worker safety.
- raise the level of safety over that required by the existing standards, especially in those machine groups for which there are no specific standards.
- provide a ready means by which worker[s] can determine whether a standard is being followed, and whether the employer is meeting his obligations.
- clearly define what is required of the employer so that he will know what is expected of him and so that the standard may be fairly enforced.
- ensure that a standard which has not been updated does not hamper the adoption of advanced safety systems and production machinery.
- provide the opportunity for technological innovation while maintaining worker safety.

Analysis of Standards which Specify the Design of Safety Equipment. Design standards specify precisely how a machine must be built or guarded. For example, the current regulations specify the distance between rivets used to attach guards to pulleys.[2] One approach to revising the safety standards would be to attempt to develop a comprehensive code covering all—or nearly all—machines. In this process, the difficulties of the current design standards would be weeded out, and the effort would be directed to writing design standards in a clear, straightforward manner. Thus the Task Force analyzed the benefits and drawbacks of using design specifications as mandatory standards.

The benefits of design standards are:

- They set forth precisely what employers must do to meet the OSHA requirements, so that an employer will know what is expected of him.
- They limit the discretion available to the employer and to the compliance officer, so that employees can be certain the level of safety imposed by the standards is being followed if the standards are fully enforced.

[2] 29 C.F.R. 1910.219(m).

[35] 29 C.F.R. §§1910.211–.222 (1982); REPORT OF THE PRESIDENTIAL TASK FORCE, OSHA SAFETY REGULATION 17–21 (P. MacAvoy ed.) © 1977 by American Enterprise Institute for Public Policy Research. Reprinted with permission.

- They provide the means for incorporating new technological developments in the standards and for requiring their implementation in the workplace.

The disadvantages of design standards are:

- They eliminate flexibility. An employer may not use an alternative approach which provides more protection or provides equivalent protection less expensively without going through a burdensome variance procedure.
- They often contain requirements which are not directly related to worker safety.
- They must be revised every time a technological change takes place. If they are revised, "grandfathering" problems result. If they are not, they retard technological progress and lead to inconsistencies between requirements and industrial practices.
- They may be difficult for compliance officers to enforce, since they require mastering a large amount of technical material and may require sophisticated testing.
- They are frequently so technical that neither employer nor employee can understand and follow them.
- Because of the wide variety of machines, the various models of a particular type of machine, and the extraordinarily wide variety of environments in which machines are used, it is impossible to devise specification standards for each machine in each workplace, let alone keep such standards up to date.

Analysis of Performance Standards. Unlike a design standard which dictates how a machine must be constructed and operated, a performance standard states obligations in terms of ultimate goals which must be achieved. The employer is then free to achieve those goals through any appropriate means. As with design standards, there are benefits and drawbacks of performance standards.

Advantages of Performance Standards:

- They are specifically and directly addressed to the problems to be solved.
- They permit flexibility in devising the solution, thereby reducing costs and inducing innovation.
- They apply equally to all machine groups.
- If properly phrased, employees can readily determine whether employer is complying without mastering difficult technical material.

Disadvantages of Performance Standards:[4]

- The employer must translate the performance criteria into an engineering design suitable for implementation, and he may lack the resources and expertise to do so.
- The employer may desire assurance as to what will be deemed an acceptable undertaking on his part.
- They may require a compliance officer to make subjective judgments.

The key to using a performance standard for regulating safety in the workplace is to design it so that compliance with it can be objectively meas-

[4]Standards do not, of course, fall within dichotomous groups entitled either design or performance. Rather, there is a continuum ranging from explicit detail to abstract performance. An example of an abstract, or pure performance standard would be one that requires an employer to have fewer than one serious injury per million man-hours of operation. With a standard of that level of abstraction, compliance cannot be determined until after the standard has been violated.

ured. Only in that way can employers and employees know what the obligations are before an accident occurs. One such standard would codify into a requirement the fact that a safe workplace can be achieved only by ensuring that employees are not exposed to the hazards associated with the use of machines. Under this standard, the employer would be free to determine the most appropriate manner in which to guard against any hazard which is presented, but his compliance with the requirement is objectively measurable by determining whether or not an employee is exposed to the hazard.

For the reasons developed below, the Task Force believes that such an approach, which it has termed the performance/hazard approach, is the best one for OSHA to use in regulating machine safety. As developed below, the Task Force believes it has the benefits of a performance standard without its drawbacks.

***

The Task Force Recommendation: An Overview. The approach recommended to OSHA by the Task Force has three parts: general performance/hazard obligations, an illustrated guide on machine guarding, and a regulatory guide containing design-oriented standards.

The section on general obligations enumerates the hazards which are associated with the use of machinery and requires each employer to use one of several specific methods of safeguarding to protect employees from injury caused by any of the hazards. Within those constraints the employer is free to meet his obligations in any manner.

Appendix A of the Task Force model is an illustrated guide to machine guarding and contains ideas on general approaches that an employer could consult for guidance. It also lists consensus standards which might be applicable but which have not been adopted by OSHA. Both the illustrations and the list are provided solely for information and neither would have any legal effect.

Appendix B contains more detailed, design-oriented standards applicable to particular machines and machine groups. If an employer follows all the guidelines listed in Appendix B that are applicable to a particular machine, he will be deemed to be in compliance with the general requirements unless there are unusual circumstances in his plant that create hazards for employees. Thus, an employer may rely on the guidelines set out in Appendix B if he so desires.

Because certain situations and machines are extraordinarily dangerous, the proposed regulation contains a section which imposes mandatory design standards. In these few instances, the employer is not free to design his machine guard, but rather must adhere to the requirements set forth in the regulation.

***

In January 1977, OSHA reacted to the Task Force recommendations by publishing a lengthy *Federal Register* notice on machinery and machine guarding. The notice described the Task Force approach, requested information on the issues raised, and announced the holding of public meetings.[36] Although the meetings were held, OSHA did not publish a proposal for revision of the national consensus standards on machine guarding. Several years later, however, in its broad revision of the fire protection standards in 1980, OSHA substantially adopted the Task Force recommendation on perform-

[36] 42 FED. REG. 1742 (1977).

ance standards.[37] Thus, the standard embodies a "format which contains performance oriented standards supplemented by non-mandatory appendices for guidance in compliance." This format, OSHA hoped, would provide employers "with the necessary flexibility to meet the standard in different workplace situations" and yet provide "other employers with specific guidelines" for compliance.[38]

Assistant Secretary Thorne G. Auchter stated on a number of occasions that in issuing new standards, OSHA would emphasize the performance approach. In testimony before an oversight hearing of the Senate Labor Committee, Mr. Auchter described the interim hearing conservation amendment, put into effect by his Administration in August 1981, as "shorter and less complex than the [standard issued by Dr. Bingham in January 1981] and [which] employs a performance approach setting strict regulatory goals, but allowing flexibility in achieving them."[39] On March 8, 1983, OSHA issued its final rule on the hearing conservation amendment.[40] The preamble stated that the new standard was adopting the performance approach "insofar as possible." It said:

> The revised amendment being issued today has adopted a performance approach insofar as possible. This is in marked contrast with the detailed specifications of the January amendment, which did not fully consider the mandate of section 6(b)(5) of the Act that standards be expressed in terms of performance criteria where practicable. The revised amendment generally allows the employer to choose his own method of complying with the obligations imposed by the amendment. This approach is particularly appropriate where the standard applies to many different types of industrial settings and work environments. The flexibility inherent in the performance approach allows the employer, who is familiar with the unique circumstances and problems of his workplace, to use this knowledge to develop the most effective and efficient mechanism to protect his employees. The employer is given suffi-

[37] 45 FED. REG. 60,656 (1980) (codified mainly at 29 C.F.R. §§1910.155–.165 (1983)).

[38] 45 FED. REG. at 60,659. On the issue of performance standards, see also the preambles to the revision of the national consensus on electrical hazards, 46 FED. REG. 4034, 4036 (1981) (codified at 29 C.F.R. §§1910.301–.308 (1983)), and to the proposed amendments to the electrical safety standards for construction, 48 FED. REG. 45,411 (1983).

[39] *Oversight on the Administration of OSHA, 1981: Hearings Before the Subcomm. on Investigations and Gen. Oversight and Subcomm. on Labor of the Senate Comm. on Labor and Human Resources,* 97th Cong., 1st Sess. 27 (1981). The history of OSHA regulation of noise hazards is as follows: OSHA adopted in 1971, as an established federal standard, the noise standard under the Walsh-Healey Act, establishing a permissible level of 90 decibels. 29 C.F.R. §1910.95 (1983). In 1974, OSHA proposed under §6(b) a comprehensive noise standard, specifically raising the question of the appropriate permissible exposure limit and the appropriate method of compliance. 39 FED REG. 37,773 (1974). No final action was taken until January 1981 when OSHA, under Dr. Bingham, promulgated the so-called Hearing Conservation Amendment; this standard, amending the 1971 noise standard, required employers to establish a hearing conservation program, including exposure monitoring, audiometric testing and training, for all employees whose exposure exceeded 85 decibels. 46 FED. REG. 4078 (1981). The effective date of the amendment was stayed by the newly appointed Assistant Secretary Auchter, who issued an interim revised Hearing Conservation Amendment in August 1981, 46 FED. REG. 42,622 (1981); the final Hearing Conservation Amendment, issued in March 1983, 48 FED. REG. 9738 (1983), is described *infra* note 40.

[40] 48 FED. REG. 9738 (1983). In this final standard, OSHA revoked a number of the provisions in the amendment issued by Dr. Bingham, which had been stayed; lifted the administrative stay on other portions; and made other technical corrections. The 1983 standard has been challenged in the court of appeals. *Forging Indus. Ass'n v. Secretary of Labor,* No. 83-1420 (4th Cir. filed Aug. 30, 1983). For a description of this case, see 13 OSHC 355 (1983).

cient leeway to adopt a hearing conservation program which will be compatible with all of the peculiarities of the work environment and the needs of his business, rather than having to implement a number of requirements that would be inadequate, inappropriate or unnecessary in his working environment, merely because they are required by the standard. The performance approach allows and even encourages employers to develop creative and innovative methods of meeting the obligations imposed by the amendment.

It is expected that by providing some flexibility, the amendment will encourage compliance because compliance can be achieved in the manner that is the easiest under the circumstances present in the particular working environment. This is also consistent with one of the purposes of the Act, as stated in section 2(b)(1), of stimulating employers and employees to institute new programs and to perfect existing programs to provide healthful working conditions. In addition, the performance approach adopted herein is consistent with the Supreme Court ruling in *American Textile Manufacturers Institute, Inc. v. Donovan,* 452 U.S. 490 (1981) suggesting that standards must be cost-effective. The revised hearing conservation amendment allows employers to adopt the most efficient method of compliance which will give the protection mandated by the amendment and achieve the goal of the standard, that is, conserving employee hearing.[41]

The performance approach was adopted, for example, in revising the hearing conservation amendment's monitoring requirements.[42] Thus, the final standard permitted area rather than personal monitoring[43] in certain circumstances and deleted the requirements that employees be given an explanation of monitoring procedures, be permitted to observe all the steps involved, and to record the results.[44] However, while in 1983 OSHA adopted a more performance-oriented approach than was utilized by the prior administration, the Agency refused to adopt a considerably shorter version of the hearing conservation amendment, called the "three-paragraph alternative," on the ground that this alternative would not permit OSHA to "enforce the standard vigorously against any recalcitrant employer who fails to provide employees with meaningful protection."[45]

[41]48 FED. REG. at 9741.

[42]*Id.* at 9753.

[43]*Id.* at 9743–46. Unlike area monitoring, personal monitoring consists of measurements taken in proximity to the employee's ear (in the case of noise monitoring) or breathing zone (in the case of toxic substance monitoring). Thus, personal monitoring will more accurately reflect the exposure of employees who work in different plant locations where exposures vary.

[44]*Id.* at 9752–53. Sec. 8(c)(3), 29 U.S.C. §657(c)(3) provides that employees shall have "an opportunity to observe" workplace monitoring and have access to monitoring records.

[45]48 FED. REG. at 9772–75. The first paragraph of the alternative would have required annual audiograms of employees exposed to an 8-hour average noise level in excess of 85 decibels, and that the testing be done in accordance with ANSI standards. The second paragraph would have required review of the results of the audiometric tests by qualified persons to determine which employees' "hearing acuity has diminished more than normal." These employees, the third paragraph required, would be provided with protective devices and training in hearing protection when they were exposed in excess of 85 decibels. See *id.* at 9772–73.

While proposing certain changes to the cotton dust standard in 1983, OSHA retained the requirement that engineering controls be used by the textile industry to meet the permissible exposure limit. OSHA insisted, however, that it was not reversing its "principle" favoring performance standards. This principle, OSHA said, would ordinarily be served by permitting the use of personal protective equipment with medical surveillance as an alternative to engineering controls, if equivalent protection would be afforded; this alternative was not adopted, however, because of the "narrow record" in this proceeding. Proposed Cotton Dust Standard, 48 FED. REG. 26,962, 26,964 (1983). See discussion in Chapter 10, Section D.

Earlier, in 1976, OSHA had announced its intention to undertake a broad review of its national consensus standards, and, after rulemaking, to make all necessary revisions. The *Federal Register* notice described the history of the adoption of the §6(a) standards; set forth six specific criticisms that had been raised against national consensus standards; and asked for public comment to assist in the revision process. As the notice indicates, this review was occasioned in part by the fact that the OSHA appropriations bill[46] in 1976 directed the Agency to undertake a "review and simplification of existing OSHA standards" and to eliminate "nuisance" standards.[47]

Proposed Procedures for Revision of Safety Standards

41 Fed. Reg. 17,100–01 (1976)

***

With regard to the general issues raised concerning OSHA's safety standards, wide public participation is also valuable because the issues are important to many and their resolution should be based on broad experience and wisdom. Among the general issues which have been raised and on which comments are being invited are the following:

1. Simplification or clarification. Some critics have complained about the alleged complexity of OSHA's safety standards. OSHA is aware that particular provisions of its existing standards pose special problems to some employers, especially those who have small numbers of employees, operate with non-fixed places of employment, or use workforces which are highly transient in nature. This awareness was expressed by the Department of Labor in a statement submitted to the House Subcommittee on Environmental Problems Affecting Small Business on 26 June 1975, as follows:

> It has become increasingly evident that the combined body of Federal regulations imposes a substantial, and to some extent, necessary burden upon employers, particularly those who run small businesses. While most of these requirements serve a necessary and useful purpose, a definite potential exists for duplication, conflicting standards, and inappropriate recordkeeping requirements. In an effort to eliminate problems where any exist in the Department of Labor, I have requested my agency heads to assess the small business impact of the laws they administer and determine what can be done to ease the burden on the small employer, while still assuring compliance with the law.

In addition, Congress has in the past recommended that OSHA revise its standards to make them more understandable to small businessmen, and in the Department's appropriation bill for fiscal year 1976, directed OSHA to undertake a "review and simplification of existing OSHA standards * * *".

2. Marginal value in terms of employee safety. Many, including some members of Congress have commented that some of OSHA's standards are unnecessary, have little, if any, relevance to today's workplace, and are directed towards either property protection or safety of the general public. Indeed, Congress in the Department's latest appropriations bill also directed OSHA to undertake:

---

[46]Department of Labor and Health, Education and Welfare Appropriation Act of 1976, Pub. L. No. 94-206, 90 Stat. 3 (1976).

[47]H.R. Conf. Rep. No. 689, 94th Cong., 1st Sess., 121 Cong. Rec. 38,207 (1975).

> * * * elimination of so-called "nuisance standards" or standards which do not deal with workplace conditions that are clearly hazardous to the health or safety of workers or are more properly under the jurisdiction of State Departments of Public Health.
>
> OSHA recognizes that some of the national consensus standards that existed prior to 1970 were intended not just for employee protection but also for the protection of the general public and property. In promulgating these standards under section 6(a), OSHA attempted to avoid inclusion of aspects other than those directed at employee safety. Nevertheless, the contention persists that "nuisance" aspects of the standards must still be eliminated.
>
> 3. Gaps in the existing OSHA standards. Some have complained that there are gaps in the coverage of OSHA's standards, both as to hazards and kinds of workplaces. A recent example of such an alleged gap is the area of commercial diving.
>
> 4. Specification or design versus level of performance provisions. Comment has been generated concerning the performance versus the design or specification type of standard. Many have alleged that OSHA's present safety standards are too design-oriented, and that the design requirements are not always necessary for employee protection. Additionally, it is said that design standards tend to become obsolete quickly and thus are a potential road-block in the way of the growth of new technology. On the other hand, some have stated that performance standards are too general or vague and do not afford sufficient guidance to employers as to what they must do in order to achieve compliance with the standard.
>
> 5. The "Code" problem. Another aspect of the above issue is the dual problem posed by incorporating by reference certain ANSI standards and building, fire, and other specification type codes in the standards, thus making compliance with such codes mandatory. Some commentators feel that OSHA should publish the texts of the codes themselves in the body of its regulations so as to give better notice of the requirements of the standard. Others have suggested that OSHA should immediately update its standards when the referenced codes are revised—sometimes more than once a year.
>
> 6. New Technology. OSHA recognizes that there are many areas where current technology has surpassed existing requirements with the result that section 6(a) standards are no longer effective. Standards must be constantly studied and revised in order to continue to ensure an adequate level of employee safety despite technological advances. As new methods and techniques are developed, new hazards may be created. OSHA, therefore, is seeking ways of developing new standards which continue to protect employees in the face of rapid advances in technology and the introduction of new industrial processes.
>
> ***

The process of broad revision of all national consensus standards was proving to be complex and time-consuming; indeed, to the present time, OSHA has only issued final revisions of the national consensus standards for fire protection[48] and electrical hazards,[49] which were not issued until 1980 and 1981, respectively. In the meantime, the pressure was increasing for OSHA to take positive action and to

---

[48] 45 FED. REG. 60,656 (1980) (codified at 29 C.F.R. §§1910.155–.165 (1983)).
[49] 46 FED. REG. 4034 (1981) (codified at 29 C.F.R. §§1910.301–.308 (1983)).

delete entirely its inconsequential standards. This became a major policy initiative of the new administration of Dr. Eula Bingham, who took office in 1977, and it became known as the "Standards Deletion Project." After receiving public comment, OSHA, in October 1978, deleted approximately 600 safety standards from the Code of Federal Regulations. To be sure, many of these standards had not been actively enforced by the Agency in the past. But, in any event, this action contributed greatly to a reduction of criticism of OSHA. The preamble to the notice of revocation sets forth the seven criteria that were utilized as a basis for determining which standards should be revoked.

Revocation of Selected General Industry Safety and Health Standards

43 Fed. Reg. 49,726–27 (1978)

***

Criterion No. 1—Obsolete or inconsequential. Some provisions have clearly become obsolete due to technological change or the expiration of time set forth within the standard itself. In the Agency's judgment, revocation of these provisions would not jeopardize employee safety in the workplace. Other provisions have been shown to be inconsequential to worker safety because they do not impose substantive requirements appropriate for mandatory enforcement by the Agency.

Criterion No. 2—Concerned with comfort or convenience. Provisions proposed for revocation under this criterion were determined to be directed primarily to the comfort and convenience of employees rather than to safety or health hazards. Although the objectives of these standards may have beneficial consequences, they are not appropriate for enforcement as mandatory requirements under the Occupational Safety and Health Act.

Criterion No. 3—Directed toward public safety or property protection. Some of the national consensus standards promulgated as OSHA standards are explicitly directed toward property protection or the safety of the general public, rather than employee safety and health. However, the Act defines occupational safety and health standards as standards which provide for "safe and healthful employment and places of employment." Section 3(8) of the Act. Therefore, the Agency proposed for revocation under this criterion provisions which were determined to explicitly and primarily affect the general public or property. However, standards which OSHA has ascertained do directly benefit employee safety or health, notwithstanding any public safety or property aspects, have not been revoked.

Criterion No. 4—Subject to enforcement by other regulatory agencies. Standards proposed for revocation under this criterion are those which cover situations over which another Federal agency has exercised its statutory authority consistent with section 4(b)(1) of the Act. This section provides that OSHA's jurisdiction over specific working conditions is preempted whenever other Federal agencies "* * * exercise statutory authority to prescribe or enforce standards or regulations affecting occupational safety or health." 29 U.S.C. §653(b)(1). In addition, section 4(b)(3) of the Act, which reflects the Act's intent to avoid unnecessary duplication of governmental effort, also supports the revocation of these provisions.

Criterion No. 5—Contingent on manufacturer's approvals or recommendations. Another group of OSHA standards proposed for revocation specifically requires employers to obtain the written approval of manufacturers for modifications of equipment. Such provisions may also require the employer to follow the manufacturer's instruction or recommendations for the equipment. Standards of this type were revoked based on the following: (1) the standard contained few or no guidelines for the manufacturers' approvals or recommendations; (2) the manufacturers' approval or recommendations appeared to be motivated by concerns only remotely related to employee safety; and/or (3) the standard gave full discretion to the manufacturer, placing OSHA in a difficult position to challenge effectively the appropriateness of the manufacturer's recommendation in an enforcement proceeding.

Criteria Nos. 6—Encumbered by unnecessary detail and 7—Adequately covered by other general standards. Numerous provisions and standards proposed for revocation contain very detailed design or construction requirements. These details should not be mandated for all situations. However, OSHA recognizes these highly detailed provisions may still be useful guidelines to be followed whenever appropriate. The revocation of these provisions will permit employers greater flexibility in selecting the specific methods to abate these workplace hazards, including the development of new technology. Moreover, other generally applicable provisions in 29 CFR Part 1910 will provide adequate protection without jeopardizing employee safety or health. In selecting the standards under these criteria, consideration was given to the nature of the operation, the type of hazard involved, and the adequacy of coverage under general standards.

***

OSHA's speedy adoption of national consensus standards without full review has given rise to a variety of legal challenges. One of the most troublesome of these challenges resulted from the fact that in the adoption process, OSHA modified in some instances the word "should" appearing in the national consensus standards (an employer "should" take certain action) to read "shall" in the OSHA standards. In some cases, however, OSHA adopted the word "should" verbatim in the OSHA standard. The Review Commission and the courts of appeals consistently held that ANSI, the national consensus organization, had made clear that "should" standards were "advisory" and, therefore, could not validly be modified by OSHA without its first following the rulemaking procedures contained in the Act. The decision of the Court of Appeals for the Tenth Circuit in the *Kennecott Copper* case is typical of this line of cases.[50]

[50]See also *Marshall v. Pittsburgh-Des Moines Steel Co.*, 584 F.2d 638, 643–44 (3d Cir. 1978). Challenges have been made in enforcement proceedings also attacking: (1) the validity of the procedures used by OSHA in promulgating under §6(a), 29 U.S.C. §655(a), a modified form of a national consensus standard or in modifying the standard soon after adoption; and (2) the procedures used by the standards-producing organization in adopting the original standard underlying the OSHA standard. See, e.g., *Noblecraft Indus., Inc. v. Secretary of Labor*, 614 F.2d 199 (9th Cir. 1980) (holding that a safety regulation was a valid "national consensus standard" under the Act although some industries were not represented in the organization formulating the standard, but that the regulation was so modified by OSHA that the "consensus" was destroyed as to some industries and, therefore, did not apply to them).

The validity of §6(a) standards that were preexisting federal regulations have also been challenged in enforcement proceedings as having been improperly adopted, either by the federal agency, see, e.g., *National Indus. Constructors, Inc. v. OSHRC*, 583 F.2d 1048 (8th Cir. 1978), or later by OSHA, see, e.g., *Deering Milliken, Inc. v. OSHRC*, 630 F.2d 1094 (5th Cir. 1980). See Chapter 15, Section C for a discussion of these cases and the scope of challenges made in enforcement proceedings to the validity of standards.

Usery v. Kennecott Copper Corp.

577 F.2d 1113 (10th Cir. 1977)

Barrett, Circuit Judge:
***

I.

The first citation charged Kennecott with failing to comply with a standard mandating the use of guardrails and toeboards in scaffolds:

> Guardrails and toeboards *shall* be installed on all open sides and ends of platforms more than 10 feet above the ground or floor. (Emphasis supplied.)
> 29 CFR 1910.28(a)(3).

This regulation was promulgated shortly after passage of the Act. Accordingly, the Secretary was not required to follow the Act's rulemaking procedures.

The Commission found that there had been no violation of this regulation by Kennecott because it had not been promulgated in accordance with the provisions of the Act and was, therefore, unenforceable. As noted above, the Secretary was granted broad powers to adopt necessary standards during the first two years of the Act's life so that safer working conditions would be provided American employees in a short time. In preparing the interim safety standards for scaffolds, the Secretary turned to standards which had been formulated by the American National Standards Institute (ANSI), which read:

> Guardrails and toeboards *should* be installed on all open sides and ends of platforms more than 10 feet above the ground or floor. (Emphasis supplied.)
> (*American National Standard Safety Requirements for Scaffolding*, American National Standards Institute, 1969, p. 9.)

In promulgating this standard the Secretary changed the language concerning guardrails on scaffolds so that it assumed a mandatory, rather than advisory character:

> Mandatory rules of this standard are characterized by the word *shall*. If a rule is of an advisory nature it is indicated by the word *should* or is stated as a recommendation. (Ibid., p. 7.)

It is the Secretary's adoption of the regulation by use of the word *shall* rather than the word *should* which poses the problem presented here. We must determine whether this usage constitutes such a substantial change that the regulation is not to be considered as a national consensus standard.

The Secretary contends that the change from *should* to *shall* is not significant, in that it is simply a pro forma change, having no substantive effect on the regulation. The Secretary points to this statute which requires an employer to follow health and safety standards promulgated by the Secretary:

> Each employer shall comply with occupational safety and health standards promulgated under this chapter.
> 29 U.S.C. §654(a)(2).

Because of the mandatory nature of the standards under the Act, the Secretary asserts that the substitution of mandatory for advisory language in the adoption of the guardrails and toeboards interim standards was valid. The Secretary argues that if the standards are to be complied with, it makes no difference whether the actual language is advisory or mandatory.

Employers are, of course, required to comply with standards properly established by the Secretary. The Act accords its special interim treatment exclusively to "any national consensus standard or any established Federal standard." The usual procedural due process safeguards accorded to persons who might be adversely affected by government regulations were relaxed only to the extent that standards, which had already been scrutinized and recognized by those to be affected and upon which there existed substantial agreement, would be considered acceptable for adoption as "national consensus standards." If, however, standards which were to be adopted during the two year interim involved modifications of established standards, then a formalized procedure had to be followed. This procedure included: recommendations to the Secretary from an advisory committee, publication of a proposed rule in the Federal Register, allowance of time for comments from interested persons, and public hearing if objections are raised. These procedures are designed to provide those who might be affected the opportunity to acquaint themselves with the proposed rule and to voice any possible opposition thereto. These procedural due process requisites have long been recognized as part of our system of jurisprudence.

We hold that the Secretary did not comply with the statute by reason of his failure to adopt the ANSI standard verbatim or by failure to follow the appropriate due process procedure. The promulgation of the standard with the use of *shall* rather than *should* did not constitute the adoption of a national consensus standard. It is, therefore, unenforceable. In order for the Secretary to have rendered the standard enforceable with the change in language, he was obliged to observe the rulemaking procedures contained in the Act. Administrative regulations are not absolute rules of law and should not be followed when they conflict with the design of the statute or exceed the administrative authority granted. *National Labor Relations Board v. Boeing Co.,* 412 U.S. 67 (1973); *Commissioner of Internal Revenue v. Acker,* 361 U.S. 87 (1959); *Reardon v. United States,* 491 F.2d 822 (10th Cir. 1974).

The Commission, in a decision which posed the same issue as that before us here, articulated the view that only standards which are national consensus standards are valid. *Secretary v. Oberhelman-Ritter Foundry, Inc.,* [1 OSHC 3087 (Rev. Comm'n 1973)].

Further, we observe that the Secretary has recognized that only mandatory standards should be seen as national consensus standards:

> The national consensus standards contain only mandatory provisions of the standards promulgated by those two organizations. [ANSI and National Fire Protection Association.] The standards of ANSI and NFPA may also contain advisory provisions and recommendations the adoption of which by employers is encouraged, but they are not adopted in Part 1910.
>
> Fed. Register 36 No. 165, p. 10466.

We hold that the Secretary improperly promulgated the standard dealing with mandatory guardrails. We affirm the decision of the Commission that Kennecott could not be held to be in violation of an unenforceable standard.

In response to these decisions, OSHA in some cases employed the advisory "should" provisions to demonstrate "recognized hazards" as a basis for violations of the general duty clause.[51] The Review Commission, however, held that so long as the "should" standard was in effect, even though it may not be enforceable, the general duty clause would not apply.[52] Faced with this dilemma, OSHA, in 1982, proposed to delete all §6(a) standards in which the underlying language "should" was changed to "shall," standards in which the "should" terminology was incorporated into the OSHA standard, and certain others. The preamble to the proposal describes as background the difficulties which faced OSHA in enforcing the "should" standards, and its strategy for replacing those standards.

Proposed Revocation of Advisory and Repetitive Standards

47 Fed. Reg. 23,477, 23,478 (1982)

***

TYPES OF PROVISIONS PROPOSED FOR REVOCATION AND REASONS FOR REVOCATION

There are three categories of provisions being proposed for revocation in this notice.

1. Provisions adopted under section 6(a), either verbatim or by incorporation by reference, which use the word "should", or are otherwise advisory in nature.

2. One provision ostensibly creating an obligation through the use of the word "shall", which was changed from "should" in the ANSI standard adopted by OSHA under section 6(a).

3. Other standards which repeat requirements contained elsewhere in Part 1910.

Some of the hazards covered by the "should" standards and other advisory provisions may be serious or potentially serious under certain conditions. The fact that OSHA cannot enforce these provisions either directly or indirectly leaves gaps in coverage, resulting in a decrease in safety and health protection for the nation's employees. Where these provisions cover hazards which may cause death or serious physical harm to employees, the revocation of such provisions will enable OSHA to issue citations for these hazards under the general duty clause.

Ultimately, OSHA intends to promulgate appropriate specific standards under section 6(b) of the Act to deal with these hazards. For example, §1910.97, nonionizing radiation, covers a hazard which OSHA considers to be worthy of consideration for future regulatory action.

Although some of the "should" provisions deal with hazards which may under some conditions be serious, many others have little direct or immediate relationship to employee safety and health. The removal of these other advisory provisions will help to streamline and simplify the existing Part 1910 standards.

The standards contained in the third category, §§1910.166–.168, are not "shoulds" or other advisory provisions, but are merely repetitions of requirements contained elsewhere in Part 1910. These three sections which cover compressed gas and compressed air equipment, contain requirements which are also found in §1910.101 of Subpart H. Future rulemaking under

---

[51]Sec. 5(a)(1), 29 U.S.C. §654(a)(1).

[52]*A. Prokosch & Sons Sheet Metal,* 8 OSHC 2077 (1980). The general rule is that the general duty clause does not apply when a standard is in effect covering the hazard. See Chapter 14, Section B.

section 6(b) to replace these standards, therefore, is not considered necessary.

At its meeting on December 18, 1981, the National Advisory Committee on Occupational Safety and Health (NACOSH) recommended that OSHA not proceed with its deletion of advisory provisions at this time. Instead, the Committee recommended that OSHA wait until the Agency can simultaneously propose mandatory rules to take the place of the advisory provisions wherever necessary, under section 6(b) of the Act. OSHA agrees with NACOSH that rulemaking action may be warranted to provide specific coverage for certain hazards which are currently addressed only by advisory provisions. However, for the reasons set forth above, the Agency has decided to proceed expeditiously with its proposed revocation. This option will strengthen OSHA's current enforcement powers, by permitting the Agency to issue general duty clause citations for serious recognized hazards which are presently covered only by advisory provisions. This will be done in accordance with the new general duty clause policy found in OSHA Instruction CPL 2.50 issued March 17, 1982. OSHA believes that continuation of its ongoing policy of revising its existing standards on a subpart-by-subpart basis will be the most effective way to update these standards while maintaining the greatest degree of protection in the interim.

***

On February 10, 1984, OSHA formally revoked 153 of the 194 provisions in standards which had been proposed for revocation, most of them because they contained advisory language. 49 FED. REG. 5318 (1984).

Since Congress expected that §6(a) standards would be superseded by more up-to-date standards, it provided in §6(b)(8) that when OSHA issues a standard under §6(b) which "differs substantially" from an existing national consensus standard, it must publish a statement of the reasons why the new standard "will better effectuate the purposes of this Act than the national consensus standard."[53] In *AFL-CIO v. Brennan*,[54] the union challenged OSHA's modification, after rulemaking, of the "no-hands-in-dies" machine-guarding standard,[55] which had been previously adopted as a national consensus standard. The Court of Appeals for the Third Circuit found substantial evidence supporting OSHA's modified standard, but remanded the standard to OSHA for a "more complete statement of reasons" under §6(b)(8). The court found that OSHA's explanation of the departure from the national consensus standard was inadequate, noting the statement by Senator Javits that the §6(b)(8) requirement was added "so the people will have an explanation of why the Secretary is doing what he is doing."[56] OSHA subsequently published a supplemental statement of reasons,[57] and the standard was affirmed by the court without opinion.

OSHA's authority to issue national consensus standards without

[53]29 U.S.C. §655(b)(8).
[54]530 F.2d 109 (3d Cir. 1975).
[55]29 C.F.R. §1910.217(d)(1)–(2) (1974).
[56]116 CONG. REC. 37,623 (1970).
[57]41 FED. REG. 40,103 (1976).

rulemaking expired in April 1973. The issue of OSHA's continued interaction with national consensus organizations, particularly ANSI, remained, and in November 1976, OSHA and ANSI entered into an agreement, providing for limited cooperation between the Agency and ANSI in areas of mutual concern, consistent with OSHA's statutory mandate. This agreement, described by an authority as "modest,"[58] has not been a major factor in OSHA standards-setting activity.

[58]Hamilton, *The Role of Non-governmental Standards in the Development of Mandatory Standards Affecting Safety and Health,* 56 TEX. L. REV. 1329, 1388–99 (1978). This article was the basis for the recommendation of the Administrative Conference of the United States on federal agency interaction with private standard-setting organizations in the area of health and safety regulation (Recommendation No. 78-4), 1 C.F.R. §305.78-4 (1983). See also Hamilton, *Prospects for the Nongovernmental Development of Regulatory Standards,* 32 AM. U.L. REV. 455 (1983) (discussing the process for the development of consensus standards and the reasons for bureaucratic hostility toward them).

3

Standards Issued After Rulemaking

A. Rulemaking Procedures

In order to issue new occupational safety and health standards, or to modify or revoke existing standards, OSHA is required by the Act to go through rulemaking proceedings. These proceedings provide an opportunity for the public to participate and thus express views on the development of the standards. They are also intended to make available to the Agency a wide range of information which it would consider in determining the content of the standards, and to give courts of appeals a basis upon which to review the standard issued under §6(f).[1]

Section 6(b)[2] sets forth the rulemaking procedures for the issuance, modification, or revocation of standards. It also includes certain criteria for the contents of these standards (§§6(b)(5) and (7)).[3] The major steps under the Act in OSHA rulemaking are as follows:

1. OSHA initiates the proceeding on the basis of its own information, petitions from interested parties, or recommendations from another government agency. An important basis for OSHA health standards are recommendations from the National Institute for Occupational Safety and Health (NIOSH).[4]

2. In its discretion, OSHA may establish an advisory committee to make recommendations for the development of a standard;[5] if OSHA does so, there are requirements for the composition of the advisory committee (§7(a))[6] and for the time periods within which it must act.

---

[1]29 U.S.C. §655(f)(1976).

[2]29 U.S.C. §655(b).

[3]29 U.S.C. §655(b)(5), (7).

[4]NIOSH was established in the Department of Health, Education and Welfare (now Health and Human Services) to serve as a research arm of OSHA. NIOSH is also responsible for long-term training in the field of occupational safety and health and for the conduct of health hazard evaluations. See §§20–22 of the Act, 29 U.S.C. §§669–671.

[5]By regulation, 29 C.F.R. §1911.10(a)(1983), OSHA has committed itself to consultation with the Construction Safety and Health Advisory Committee on standards development in the construction industry.

[6]29 U.S.C. §656(a).

3. If OSHA decides that a standard should be issued, it must publish a proposed standard and give the public at least 30 days for written comment. If objections are filed to the proposal and a public hearing is requested, a hearing must be held.

4. On the basis of the entire record, OSHA must either promulgate the standard or determine that no standard is needed. Section 6(e)[7] requires that OSHA publish a statement of reasons for its final action.

5. Section 6(b)[8] prescribes time frames for most stages of this process. For example, the final standard must be issued within 60 days of the close of the comment period if no hearing is held, or 60 days from the date of completion of the hearing. The Court of Appeals for the District of Columbia Circuit, however, has held that these time frames are not mandatory so long as delays are a result of OSHA's good faith reordering of priorities.[9]

In 1971, OSHA issued regulations governing rulemaking in standards proceedings.[10] One of the major questions faced by OSHA in these regulations was the manner in which the standards rulemaking proceeding would be characterized. Under traditional principles of administrative law, rulemaking is generally described as either "formal" or "informal"; formal rulemaking is similar to adjudication and involves trial-like procedures, including cross-examination and the imposition of a burden of proof on the agency. By its nature, formal rulemaking is more rigid and time-consuming. In addition, the rule issued as a result of formal rulemaking normally is tested in the courts under the "substantial evidence" rule, which involves relatively strict court review of agency action.[11] Informal rulemaking, on the other hand, was developed to enable regulatory agencies to issue rules, with prospective application, affecting large groups of persons based largely on policy considerations. For informal rulemaking under the Administrative Procedure Act, there are minimal procedural requirements: publication of a proposal, an opportunity for written comments by the public, and a final rule with a brief explanation.[12] Informal rules are normally reviewed in the courts on the basis of the more lenient "arbitrary and capricious" test. The courts and commentators have generally viewed informal rulemaking as being flexible and efficient, fair to interested persons, and effective in providing the Agency with needed information.[13]

On the basis of the statutory language and its legislative history,

[7]29 U.S.C. §655(e).

[8]29 U.S.C. §655(b).

[9]*National Congress of Hispanic Am. Citizens v. Marshall,* 626 F.2d 882, 890 (D.C. Cir. 1979). See discussion in Chapter 6, Section C.

[10]29 C.F.R. pt. 1911 (1983).

[11]Court review of OSHA standards is discussed in Chapter 6.

[12]Administrative Procedure Act, §4, 5 U.S.C. §553 (1982).

[13]For a discussion of informal rulemaking, see, among many other articles, Friendly, *Some Kind of Hearing,* 123 U. PA. L. REV. 1267 (1975); Kestenbaum, *Rulemaking Beyond APA: Criteria for Trial-Type Procedures and the FTC Improvement Act,* 44 GEO. WASH. L. REV. 679, 682–685 (1976). Kestenbaum refers to the view of Professor Kenneth Davis that informal rulemaking is "one of the greatest inventions of modern government." *Id.* at 679.

OSHA in its regulations determined that OSHA rulemaking was "informal" but that "more than the bare essentials of informal rulemaking" should be provided. Specifically, the regulations required that a hearing examiner (now an administrative law judge) preside at the hearing; that there be "an opportunity for cross-examination on crucial issues"; and that a verbatim transcript be kept of the hearing.[14] In *Industrial Union Department v. Hodgson,*[15] which involved review of OSHA's final asbestos standard, the Court of Appeals for the District of Columbia Circuit commended OSHA for its regulations' going beyond the minimum requirements of the statute "by providing an evidentiary hearing in which cross-examination was available."[16] The court further held that the "substantial evidence" test stated in §6(f) of the Act should be construed flexibly in light of the "legislative-type" determinations made by OSHA in issuing standards.[17]

OSHA standards rulemaking, containing elements of both formal and informal rulemaking, has come to be known as "hybrid" rulemaking or "notice and comment plus." Developments in the field of administrative law at least since the early 1960s have involved a movement away from purely "informal" rulemaking and towards the "hybrid" format.[18] The trend towards trial-type procedures in rulemaking, sometimes required by Congress and sometimes by the courts, has been attributed to a number of factors, many of which are present in the OSHA program: broad agency authority to promulgate rules with significant impact on large sectors of the economy; complexity and controversial nature of the issues in the rulemaking; public distrust of the government agency involved; and pressure for fuller public participation in the rulemaking. Other federal social regulatory agencies, such as the Environmental Protection Agency, the Federal Trade Commission, and the Consumer Product Safety Commission, have been affected by these developments.

Over a period of years, there have been numerous developments in the procedures for the conduct of standards proceedings. The most significant has been the increased length and complexity of the process. OSHA's first hearing on a health standard, in asbestos, consumed four days and the record comprised 1,100 pages, of which 400 were the hearing transcript; the Carcinogens Policy hearing held in 1978 was completed in about two months and the transcript consisted of over 8,500 pages and the total record over 250,000 pages. One commentator has referred to an OSHA standards proceeding as being "Byzantine in its complexity."[19]

[14]The pertinent portions of the regulation are quoted in *Industrial Union Dep't v. Hodgson,* 499 F.2d 467, 472 n.12 (D.C. Cir. 1974). Excerpts from this decision appear in Chapter 6, Section A.

[15]499 F.2d 467 (D.C. Cir. 1974).

[16]*Id.* at 476.

[17]*Id.* at 475.

[18]See, e.g., DeLong, *Informal Rulemaking and the Integration of Law and Policy,* 65 VA. L. REV. 257, 262–76 (1979); see also articles cited *id.,* at 260 n.22.

[19]Kelman, *Occupational Safety and Health Administration,* 236, 244, in THE POLITICS OF REGULATION, James Q. Wilson, ed. (1980).

Other significant developments in OSHA rulemaking are as follows:

1. OSHA has encouraged public participation and has requested comment at a number of stages in the rulemaking in addition to the required comment period following the proposal. Thus, for example, OSHA usually issues an advance notice of proposed rulemaking, asking for public views and data to assist it in developing a proposal.[20] Also, OSHA informally consults with affected parties—particularly employer groups and unions—for additional assistance at this preproposal stage.

2. The comment period on a proposal is required by the Act to be no less than 30 days. An executive order issued by President Carter provided that, for major proposals, the comment period should be no less than 60 days,[21] and OSHA normally allows 90 days or more for initial comments on important standards.

3. The Act does not require that OSHA set a public hearing on a proposal unless it was requested; however, OSHA often schedules a hearing when the proposal is issued, anticipating that it will be requested and thus affording the public additional time to prepare for the hearing. In announcing the hearing, OSHA generally requires persons who wish to appear for more than a minimum period of time (10 to 15 minutes, usually) to indicate the amount of time requested and the specific provisions to be addressed, and to provide a copy of the testimony with any documentary evidence to be submitted. OSHA examines the submissions to determine if the person's proposed testimony entitles him or her to the time requested. OSHA also introduces, prior to the hearing, copies of the testimony of the witnesses it intends to present. The purpose of these procedures is to make certain that the time at the hearing will be used productively, and, in addition, to give all participants an opportunity to prepare adequately to question witnesses at the hearing. As a result, standards hearings have focused less on presentation of prepared statements, which are already in the record, and more on the questioning of witnesses. In addition, in several recent safety standards hearings, OSHA has scheduled an "informal" meeting at the time of publication of the proposal; its purpose is to allow the Agency to explain the proposal and to allow an informal exchange of views. These meetings, which are more informal than the statutory informal hearings and which are not conducted by administrative law judges, have in some cases eliminated the necessity for the statutory hearings. Also, in scheduling hearings, OSHA has, whenever possible, limited the issues to be covered at the hearing, particularly where the proposal involves an entire subpart, raising numerous issues.

In proposing a broad revision of the electrical standards, for ex-

---

[20]See, e.g., Advance Notice of Proposed Rulemaking for Ethylene Oxide, 47 FED. REG. 3566 (1982).

[21]Exec. Order No. 12,044, 3 C.F.R. 152 (1978).

ample, OSHA announced an informal meeting to discuss the proposal in its *Federal Register* notice.

Electrical Standards; Proposed Rulemaking

44 Fed. Reg. 55,274, 55,279 (1979)

***

(b) To assist interested persons in submitting their written comments and data, OSHA is scheduling a public meeting during the comment period. The meeting will be held on November 8, 1979 in Room C-5521, Seminar Room #4, Department of Labor, 200 Constitution Avenue, N.W., in Washington, D.C. It will begin at 9:00 a.m., will recess from 12 noon to 1 p.m. and will continue until 5 p.m.

The public meeting is intended as an informal forum for interested persons to present their concerns orally and to seek clarification of the proposal from representatives of OSHA who will conduct the public meeting.

OSHA requests that any person wishing to make an oral presentation at the meeting notify OSHA in advance. Please identify the person and/or organization intending to make a presentation, and the subject matter and a brief summary of the intended presentation, if possible. This written notice should be sent to Docket S108-1, Docket Office, Rm. S-6212, U.S. Department of Labor, 3rd Street and Constitution Avenue, N.W., Washington, D.C. 20210 no later than November 1, 1979. All persons giving advance notice will have time reserved for their oral presentations. Persons wishing to speak who have not filed advance notices are requested to register from 8:30 a.m. to 9:00 a.m. on the morning of the public meeting.

As long as time permits, all persons who wish to be heard will be allowed to speak. However, in the interest of time, persons who have provided advance notice will be given priority.

Detailed minutes or a transcript of the meeting will be prepared and will be made a part of the record of this rulemaking. Copies of the minutes will be available for inspection at the OSHA Docket Office, Room S-6212, U.S. Department of Labor, 3rd Street and Constitution Avenue, N.W., Washington, D.C. 20210.

---

4. At hearings that took place early in OSHA history, such as the asbestos hearing, OSHA permitted public comment but neither presented any significant testimony on the proposal nor asked or answered questions. More recently, however, OSHA has arranged for the testimony of experts to explain the basis for the proposal and, more generally, to answer questions and to help develop a full technical record on the issues in the proceeding. For example, in the Carcinogens Policy hearing, 46 well-known experts in the field of cancer research were presented by OSHA and testified at the hearing.[22] In addition, OSHA now routinely questions the witnesses of other parties to the hearings. These developments have resulted in large measure from the increasing court scrutiny of OSHA standards and the consequent need for "substantial evidence" to support a final

[22]See Preamble to Final Rule on Carcinogens Policy, 45 FED. REG. 5002, 5008 (1980). This large number of experts is unusual. More typically, OSHA will offer fewer than six witnesses.

standard. OSHA's invited experts are commonly viewed as "OSHA witnesses" and this in some measure increases the adversary nature of the hearings.

5. At hearings, OSHA, in consultation with the administrative law judge, establishes an order of witnesses for testimony and, in addition, arranges for an orderly questioning of witnesses; thus, the judge will give a party a specific amount of time for questioning. Also at some recent hearings, voluntary agreements have been reached for the joining of parties for purposes of questioning. For example, at the hearing on the Carcinogens Policy, an agreement was reached whereby industry parties and union parties would generally be represented by a single individual in questioning witnesses.

6. At the close of the hearing, the administrative law judge normally keeps the record open for post-hearing comments and briefs. At recent hearings, the comment period has been bifurcated: parties are allowed approximately 30 days to submit additional data and another period of up to 30 days to file briefs. The purpose of the additional period is to give all participants an opportunity to respond to new evidence presented by other parties.

7. Since 1977, OSHA has made decreasing use of advisory committees in standards proceedings. Between 1971 and 1976, most of the major health standard proposals were based on advisory committee recommendations (asbestos, carcinogens, coke oven emissions), but more recently, no advisory committees have been convened to make recommendations on the major OSHA standards such as those on cotton dust, benzene, and lead. This was due in part to the burdens resulting from the detailed requirements for federal advisory committees issued by the Office of Management and Budget[23] under the Federal Advisory Committee Act[24] and the parallel effort in the administration of President Carter to reduce the number of federal advisory committees. Instead, OSHA has more recently relied on informal discussions with the parties and on expert consultants who have assisted OSHA not only by preparing economic impact statements but also in analyzing the record and preparing standards and preambles. The active participation of consultants in the standards process resulted in a legal challenge in the lead proceeding.[25]

These evolving procedures have avoided redundancy in OSHA hearings and in many respects have improved the quality of the records that come before the Assistant Secretary for decision. Also, the opportunity for public participation in the proceeding has been significantly increased. Rulemaking proceedings are often adversarial; indeed, as long as the issues in rulemaking are controversial, adversary proceedings may be anticipated. And, while adversary proceedings are usually criticized, they are an integral part of American

[23]O.M.B. Circular No. A-63, 39 FED. REG. 12,389 (1974).
[24]5 U.S.C. app. §7 (1982).
[25]*United Steelworkers v. Marshall*, 647 F.2d 1189, 1216–20 (D.C. Cir. 1980), *cert. denied sub nom. Lead Indus. Ass'n v. Donovan*, 453 U.S. 913 (1981); *Lead Indus. Ass'n v. OSHA*, 610 F.2d 70, 80–84 (2d Cir. 1979).

administrative practice and serve the salutary purpose of providing a forum by which evidence is rigorously tested for reliability.[26]

OSHA's announcement that a public hearing on a proposed standard will take place is published in the *Federal Register,* either with the proposal, or after requests for a hearing are received. These notices of hearings set forth the procedures which must be followed by members of the public who wish to participate in the hearing. They also describe the rules for the conduct of the hearing to be followed by the administrative law judge.[27] The following are "Pre-Hearing Guidelines" prepared by OSHA that are currently being circulated by administrative law judges to "parties" participating in OSHA standards hearings.

Pre-Hearing Guidelines

TO: PARTICIPANTS OF THE OSHA HEARING

Your participation in the hearing on a standard for ____________________ is both welcome and appreciated. For your information, the following guidelines for the hearing, Docket No. (Federal Register Notice __________) are issued for the location of the hearing, ____________________.

The hearing will begin at 9:30 a.m. on ____________________ in ____________________ on ____________________, and continue until all presentations have been made. The hearing will also begin at 9:30 a.m. on subsequent days unless special circumstances require a change. Any such change will be announced during the course of the hearing. The hearing day will end when the scheduled testimony and questions for that day are finished.

In the interest of due process, fairness and the development of a complete record, OSHA has requested that any participant who requests more than 15 minutes for his presentation, or who will submit documentary evidence for the hearing, submit the written testimony and documentary evidence he intends to present at the hearing by ____________________. These materials have been placed in the OSHA Docket Office, Room S6212, U.S. Department of Labor, Third Street and Constitution Avenue, N.W., Washington, D.C. 20210, where they are available for inspection and duplication. This will afford all participants full opportunity to review the evidence and to formulate any questions they wish to ask at the hearing. Submission of testimony and other evidence in written form will assist in expediting the hearing process. In the absence of special circumstances, participants who have filed notices of special circumstances, participants who have not substantially complied with the requirements for the submission of written testimony and documentary evidence, will be allowed a maximum time of 15 minutes for their presentations at the hearing.

Persons who have not filed such notice will be permitted to present brief oral statements, to the extent that time permits, after the scheduled testimony has been presented.

---

[26]For a discussion of adversary proceedings and American administrative law, see S. Kelman, REGULATING AMERICA, REGULATING SWEDEN: A COMPARATIVE STUDY OF OCCUPATIONAL SAFETY AND HEALTH POLICY 133–48 (1982). Professor Kelman states: "The history of administrative procedure in the United States is one of the imposition of adversary proceedings on government agencies, conceived as a way to deal with objections to the very legitimacy of administrative decisionmaking." *Id.* at 139.

[27]See, e.g., Proposed Carcinogens Policy, 42 FED. REG. 54,148, 54,183 (1977).

Since all prehearing submission[s] of testimony are received as part of the record in this proceeding and are entitled to equal weight with oral testimony, the presentation of a witness at the hearing should summarize or clarify the written submission. Alternatively, a participant may elect not to make an oral presentation, but may simply offer his written statement as submitted and make himself available for questions. This is recommended to reduce the time and expense required of participants, many of whom attend from distant locations.

Participants may desire to ask questions following a presentation. In order to afford everyone an equal opportunity for questioning, each participant will be expected to limit his questioning to a maximum period of 15 minutes. It is therefore suggested that participants ask their most important questions first. Participants having similar interests are encouraged to designate one representative who can conduct the questioning on their behalf. When an organization is represented by more than 1 questioner of the same discipline only one person should question the witness. If hearings fall significantly behind schedule, some consolidation of questions will be ordered by the Administrative Law Judge.

Questions should be as brief as possible and must be designed to clarify a presentation and/or elicit information that is within the competence or expertise of the witness. To prevent unnecessary duplication, questions that have been asked and responded to by a witness will ordinarily not be permitted a second time. Questions that are argumentative will not be permitted.

A period after the close of the hearing will be designed for the receipt of any additional information or evidence requested from participants during the proceeding. Participants will also be afforded the opportunity to submit written comments, summations or briefs during an additional interval which will be designated for this purpose. It is anticipated that the designated periods will be 30 days after the close of the hearing for additional information and 20 days after that for comments, summations or briefs. At the close of the post-hearing comment period, the entire record of this rulemaking proceeding will be certified to the Assistant Secretary of Labor for Occupational Safety and Health by the Administrative Law Judge.

The foregoing rules are deemed consistent with orderly and fair procedures. Exceptions will be allowed upon a showing of good cause, as determined by the Administrative Law Judge.

B. Rulemaking in Action: Vinyl Chloride and Recent Developments

OSHA's issuance of a final standard regulating the carcinogen vinyl chloride is generally accepted to be an instance of successful OSHA rulemaking. The major steps in this proceeding were as follows:

1. In January 1974, OSHA was advised by NIOSH that employees working for the B.F. Goodrich Chemical Company were suffering from a rare form of liver cancer which may have been related to the production of vinyl chloride.

2. On the basis of this and other information and after an inspection of the Goodrich plant, OSHA held a fact-finding hearing in

Washington, D.C., on possible hazards in the manufacture and use of vinyl chloride and polyvinyl chloride.

3. Petitions for an emergency temporary standard were filed by the Industrial Union Department, AFL-CIO, United Rubber Workers International Union, and the Oil and Chemical Workers International Union. On April 5, 1974, OSHA issued an emergency temporary standard on vinyl chloride and polyvinyl chloride, reducing the permissible exposure limit from 500 parts per million (ppm) to 50 ppm.[28] The emergency standard was not challenged in court.

4. On May 10, 1974, OSHA published a proposed "permanent" standard on vinyl chloride and polyvinyl chloride.[29] Written comments on the proposal were invited, and a public hearing was held from June 25 to 28, 1974, and again from July 8 to July 11. The record remained open for post-hearing comments until August 23, 1974.

5. A final standard on vinyl chloride and polyvinyl chloride was published on October 4, 1974, setting a permissible exposure limit (PEL) of 1 ppm, with a ceiling limit of 5 ppm.[30]

6. The Society of the Plastics Industry, Inc. and individual companies challenged the standard in the Court of Appeals for the Second Circuit. The industry parties sought a stay of the standard in November 1974. After issuing a brief stay, the Court of Appeals for the Second Circuit, in an opinion by former Justice Clark, affirmed the standard in all respects on January 31, 1975.[31] In that decision, the court dissolved the stay and determined that the standard would become effective in 60 days. The Supreme Court refused to hear the case.[32]

A fuller statement of the proceedings leading to the final standard appears in the preamble to the vinyl chloride standard.[33] A description of the vinyl chloride rulemaking was presented by Professor Steven Kelman as one of his rulemaking "case studies" comparing occupational safety and health policy in the United States and Sweden.[34]

Kelman, *Regulating America, Regulating Sweden: A Comparative Study of Occupational Safety and Health Policy*

VINYL CHLORIDE

* * * In January 1974 B.F. Goodrich Chemicals revealed that three workers exposed to the gas vinyl chloride in a plant where it was polymerized into the plastic polyvinyl chloride had died of a rare liver cancer, which

[28] 39 FED. REG. 12,342 (1974).

[29] 39 FED. REG. 16,896 (1974).

[30] 39 FED. REG. 35,890 (1974).

[31] *Society of the Plastics Indus., Inc. v. OSHA*, 509 F.2d 1301 (2d Cir. 1975). Excerpts of the decision appear in Chapter 10, Section A.

[32] *Firestone Plastics Co. v. Department of Labor,* 421 U.S. 992 (1975) *(cert. denied).*

[33] 39 FED. REG. 35,890 (1974) appears in part *infra* in this section.

[34] S. Kelman, REGULATING AMERICA, REGULATING SWEDEN: A COMPARATIVE STUDY OF OCCUPATIONAL SAFETY AND HEALTH POLICY 70–75. © 1981 by The MIT Press. Reprinted with permission.

normally strikes fifteen people a year in the entire United States. Vinyl chloride gas had previously been thought to be harmless except at high concentrations. Polyvinyl chloride plastic was America's second most used plastic, appearing in products from phonograph records to plastic wrap.

There was a broader message in the vinyl chloride revelation. Vinyl chloride went into large-scale production around 1940, along with many other new chemicals. Cancer latency periods are fifteen to twenty-five years. Journalists could thus ask whether the vinyl chloride cases signaled the beginning of a new cancer epidemic.

***

In the late 1960s European producers had sponsored research into the causes of a softening of finger bones observed among some workers exposed to the gas. During the animal experiments, the European researchers discovered that some animals exposed to high concentrations of vinyl chloride developed cancer. On the basis of this finding, a consortium of European producers commissioned further tests, and in 1973 the first results appeared, showing tumors in some test animals at concentrations as low as 250 ppm. The data were kept confidential. Meanwhile the plant doctor at a B.F. Goodrich plant in Louisville learned that a worker had died of rare liver cancer. The doctor remembered that another worker had died of the same cause two years previously and notified his superiors. When a third worker died a month later, B.F. Goodrich made a public announcement.

An even grislier picture began to emerge. Searches of records at other plants confirmed sixteen similar deaths, epidemiological studies showed excess mortality from brain and lung tumors as well, and final results of animal tests commissioned by the Manufacturing Chemists Association showed tumors in a small number of the rats exposed to only 50 ppm of vinyl chloride. Furthermore two of the sixteen liver angiosarcoma deaths had been among workers who had almost certainly been exposed to only low concentrations of vinyl chloride.

Less than a month after the Goodrich announcement, OSHA held an informal one-day hearing so that available information could be exchanged. Most of those present were scientists, and, according to Eugene Regad, the vinyl chloride project manager, there was little dispute from industry people present that vinyl chloride was the culprit. Shortly after, however, attorneys for Union Carbide sent OSHA a letter warning them against adopting an emergency temporary standard: "There is literally no basis in the past history of vinyl chloride monomer exposure for imposing any hastily drawn regulation on such a vital part of the chemical industry. The carcinogenicity of the monomer has not been shown conclusively." And in March two unions petitioned OSHA for an emergency temporary standard.

In April OSHA issued an emergency standard, lowering the threshold limit value from 500 to 50 ppm. It was stated that 50 ppm was designed as an interim measure and that a final decision on a level would be made during rule making. About a week later, test results showing tumors at 50 ppm became available, confirming that 50 was too high. The National Institute for Occupational Safety and Health had meanwhile recommended that OSHA set a no-detectable-level limit. In May OSHA published a proposed final regulation, calling for reduction of permissible exposure from 50 ppm to no-detectable level as determined by a method capable of detecting vinyl chloride at 1 ppm.

In June and July OSHA held eight days of hearings on the proposal. Industry was represented both by individual firms and the Society of the Plastics Industry. The Manufacturing Chemists Association, which had earlier represented the industry, failed to appear. Apparently its officials felt that fighting OSHA on vinyl chloride was a losing proposition in view of the cancer scare; this segment of the plastics industry was small enough in the total membership so they could be snubbed. At this point the firms, organization-less, appeared at the door of the less-inclusive Society of the Plastics Industry and asked the organization to provide a forum for them to discuss a common position, which the Society would then merely repeat to OSHA. (At one point during the hearings, the President of the Society was asked whether his organization had undertaken any studies on the feasibility of reducing vinyl chloride exposures to various levels. "This is not the role of the Society," was the reply. "The role of the Society in an activity like this is to provide these members with a vehicle through which to act, to develop their own activities.")

Industry testimony took two tacks. First, in contrast to the situation where industry people met informally at the fact-finding session, witnesses argued that knowledge about vinyl chloride carcinogenicity was uncertain. Second, witnesses stated that the industry would shut down if OSHA held to its proposal. The Society of the Plastics Industry lawyer specifically noted that the industry was not arguing that meeting the proposal was simply too expensive or unjustified in terms of benefits gained for costs expended; the only economic argument the industry was making, he emphasized, was that compliance was impossible and would thus force an industry shutdown.

The industry was in fact willing to go quite a long way. The majority of the Society of the Plastics Industry committee endorsed a value of 10 ppm for polymerization facilities, a significant decrease from the 50 ppm emergency level. In fact at an early meeting of the industry committee, the company president at whose plants the first deaths had been confirmed threatened to quit unless a decision to push for a 25 ppm value was revised downward, which it was. Many industry people were frightened about the vinyl chloride revelations. They had grown up with the industry and been exposed themselves. Industry adopted its hard stance at the hearings at the urging of the society lawyer. Afterward many industry figures felt the hearings had been a disaster.

Labor testimony came from the AFL-CIO, from Peter Bommarito, president of the United Rubber Workers, and from officials of Oil, Chemical and Atomic Workers Union locals where vinyl chloride was used. Bommarito's testimony stunned the hearings by taking up industry's gauntlet on shutdown: "This country survived for nearly 200 years without polyvinyl chloride and we can survive in the future without it. If polyvinyl chloride cannot be made and used safely, then the proposed standard must be replaced by an orderly procedure to phase out vinyl chloride production and find substitutes for its products, or to phase out the products themselves. This is not an easy position to take. But there is no alternative." The issue, he continued, "is whether life has a higher priority than dollars, whether we will permit corporate structures to run this country without control. * * * The vital organs of our bodies are not for sale."

A union local president spoke in simple, direct terms:

> All I want is a safe place to work, that I might provide for my family without having to die at an early age. * * * The rest of you people probably work in safe offices. Most of you will continue to work for your companies, and be healthy. But we, the blue collar workers, are the ones who are actually going to be affected. It boils down to this: if vinyl chloride isn't controlled, we probably are going to die; if it is controlled, then we will live healthy lives like you people behind the desks.

***

[Assistant Secretary of Labor for OSHA, John] Stender wanted to know what was safe, and he wanted to set the level there. Stender was influenced by comments sent to OSHA by two university-based environmental health researchers in response to requests to the AFL-CIO and the Society of the

Plastics Industry that each name an outside expert to comment on the issues. The scientist named by labor endorsed the strictest possible regulations, but the industry-appointed scientist also submitted a letter that came very close to endorsing a 1 ppm figure and that failed to endorse industry claims that a higher level was safe.

* * * It was decided to require 1 ppm. The lengthy statement of reasons for the standard confirmed the philosophy enunciated for fourteen carcinogens that whether a no-effect exposure level for carcinogens existed could not currently be answered. The statement continued, "We cannot wait until indisputable answers to these questions are available, because lives of employees are at stake. Therefore, we have had to exercise our best judgment on the basis of the best available evidence. These judgments have required a balancing process, in which the overriding consideration has been the protection of employees."

***

The Society of the Plastics Industry challenged OSHA's decision in court, and the court of appeals affirmed OSHA's decision. As it turned out, the industry did not close down. A year after the regulation was promulgated, most polymerization plants were down to 1 ppm, and the cost proved far less than industry had contended.

The final standard on vinyl chloride was published on October 4, 1974.[35] The typical format for an OSHA health standard issued under §6(b)[36] is as follows:

1. An introductory discussion of the substance being regulated, its uses, and toxic properties.

2. A description of the background to the rulemaking proceeding and the history of the rulemaking proceeding itself.

3. A summary of the record and a discussion of the major issues raised by the proceeding. In the case of OSHA health standards, these issues typically are the extent of the risk upon exposure to the substance, the permissible exposure limit, and economic and technological feasibility. The Agency's conclusions on these issues are set forth in this section of the preamble.

4. A discussion of the specific provisions of the standard, section-by-section. This normally includes an explanation why the particular provision was adopted and others were rejected.

5. A statement, as appropriate, on the agency compliance with the requirements of presidential executive orders on regulatory analysis,[37] the National Environmental Policy Act,[38] and the Regulatory Flexibility Act.[39]

6. The text of the standard itself.[40]

[35]39 FED. REG. 35,890 (1974).

[36]29 U.S.C. §655(b).

[37]These executive orders are discussed in Chapter 10.

[38]42 U.S.C. §§4321–70 (1976 & Supp. V 1981).

[39]5 U.S.C. §§601–12 (1982).

[40]For a discussion of the substantive elements of a typical OSHA health standard see Section C, this chapter. The vinyl chloride standard now appears at 29 C.F.R. §1910.1017 (1983), as amended relating to access to records in 45 FED. REG. 35,212, 35,282 (1980).

Vinyl Chloride Final Standard

39 Fed. Reg. 35,890–98 (1974)

I. BACKGROUND—(1) *Vinyl Chloride.*

***

Vinyl chloride (VC) is used primarily in the production of polyvinyl chloride (PVC), a resin which is produced through batch processing. The conversion of the VC monomer into a polymer or copolymer is an incomplete process, i.e., not all of the monomer is reacted.

PVC is fabricated by a variety of techniques, including extrusion, injection molding and calendering, to form a finished product that needs no further chemical handling. The vast majority of employees involved in the VC industry are employed by fabrication firms. Such firms range in size from those with few employees and simple equipment to large plants involving many employees and considerable capital.

Vinyl chloride (VC), a gas at ambient temperature and pressure, is a chlorinated hydrocarbon, which heretofore has been regarded as having moderate liver toxicity. The initial standard, contained in Table G-1 of 1910.93, established a ceiling value of 500 parts of VC per million parts of air.

(2) *The emergency temporary standard.* On January 22, 1974, the Occupational Safety and Health Administration (OSHA) was informed by the National Institute for Occupational Safety and Health (NIOSH) that the B.F. Goodrich Chemical Company had reported that deaths of several of its employees from a rare liver cancer (angiosarcoma) may have been occupationally related. As a result of this notification and after consultation with NIOSH, and a joint inspection of the B.F. Goodrich plant by OSHA, NIOSH and the Kentucky Department of Labor, a fact-finding hearing was announced on January 30, 1974 (30 FR 3874) and held on February 15, 1974.

Information obtained from this hearing, particularly the preliminary reports of experiments conducted by Professor Cesare Maltoni of the Instituto di Oncologia, Bologna, Italy, demonstrated that vinyl chloride induced angiosarcoma in rats at levels as low as 250 ppm, and in other species at higher levels. Experiments performed at lower levels of exposure were not completed at that time. Other testimony from medical witnesses and NIOSH, and the results of autopsies, led to the conclusion that the Goodrich workers had angiosarcoma of the liver and that VC probably was the causal agent in the angiosarcomas observed.

In post hearing comments, additional angiosarcoma deaths were reported among workers who had been exposed to VC in plants operated by Union Carbide Corporation, Firestone Plastics Corporation and Goodyear Tire & Rubber Company.

On the basis of all information available at that time, and the fact that employees were being exposed at levels around the experimentally observed effect level of 250 ppm, an emergency temporary standard (ETS) was promulgated on April 5, 1974 (39 FR 12341) pursuant to section 6(c) of the Act, as 29 CFR 1910.93q.

This standard reduced the permissible exposure level from a ceiling of 500 ppm to a 50 ppm ceiling, and established other requirements, including, for example, monitoring and respiratory protection. It was expressly recognized that this standard limiting exposures to a 50 ppm ceiling was a tentative, interim standard, and that the whole question of exposure to VC would be considered more fully in the light of additional information, especially the results of experiments which were known to be underway at that time.

On April 15, 1974, information and data were presented to representatives of OSHA, NIOSH, and the Environmental Protection Agency by the Industrial Bio-Test Laboratories, Northbrook, Illinois, concerning results of animal exposure studies with VC. These studies were sponsored by the Manufacturing Chemists Association. Although only preliminary in nature at that time, these results revealed that 2 out of 200 mice exposed to VC concentrations of 50 ppm for 7 hours a day, five days a week, for approximately 7 months, had developed angiosarcoma of the liver.

***

(4) *Hearing on the proposal.* The proposal, as published on May 10, 1974, allowed 30 days for interested parties to submit written comments and to request an informal rulemaking hearing. Informal contacts with OSHA staff and early responses indicated that the subject was of great interest and importance to many persons. Because of the limited time available before expiration of the six month period provided in section 6(c)(3) of the Act for promulgation of a final standard, it was decided to hold a hearing as soon as possible. Accordingly, on May 24, 1974, a notice of a hearing was published (39 FR 18303), setting a hearing date of June 25, 1974. The hearing was conducted from June 25 through June 28, and again from July 8, through July 11, before Administrative Law Judge Gordon J. Myatt. All participants were given the opportunity to present testimony and to cross-examine other witnesses. Persons participating in the hearing were given until August 23, 1974, to file additional posthearing comments, including various items of information which were requested during the examination of witnesses.

(5) *Economic and technical impact study.* During the hearing, OSHA determined that additional facts would be needed to determine the practicality of certain aspects of the proposed standard. Accordingly, OSHA contacted an independent consultant, Foster D. Snell Corporation, to conduct studies of the feasibility of compliance at various exposure levels, including those proposed by OSHA and others advanced by industry spokesmen. Snell was also commissioned to collect information regarding the economic costs of compliance. This action was announced at the close of the hearing, and Judge Myatt further announced that the record would be kept open for a period of time beyond August 23, to allow interested persons to comment in writing on the study. On August 26, 1974, OSHA announced that the preliminary study was available and that comments were to be submitted no later than September 6, 1974 (39 FR 30844). On September 13, 1974, OSHA invited comments on both the preliminary and the final study, which was to be received on or before September 25, 1974 (39 FR 33009).

(6) *Environmental impact statements.* A notice of intent to file an environmental impact statement assessing the impact of a proposed standard on occupational exposure to VC was published in the *Federal Register* on April 24, 1974 (39 FR 14522). The notice invited any person having information or data on the environmental impact to submit it to OSHA by May 17, 1974. On June 12, 1974, a draft environmental impact statement was prepared and circulated to all interested persons. Ten copies were forwarded to the Council of Environmental Quality (CEQ), which published a notice of its filing and availability in the *Federal Register* on June 25, 1974 (39 FR 22975). A 45 day period was allowed for the submission of comments on the draft statement. On September 5, 1974, the final environmental impact statement was prepared and a copy of it and all substantive comments were sent to appropriate governmental agencies, private organizations, and other interested persons. CEQ published a notice of availability for the final statement on September 6, 1974 (39 FR 32350). The submission of comment was

invited until September 25, 1974. The final statement and all significant comments have been carefully considered in arriving at the final standard on occupational exposure to VC.

(7) *The record.* The record in this proceeding is one of the most exhaustive ever relied upon by OSHA. It consists of pre and post-hearing comments and testimony received at both factfinding and rulemaking hearings, the studies and inspections conducted by OSHA personnel, the environmental impact statements, the economic and technical impact studies, and all other relevant information. In all, over 600 written comments have been received, with more than 200 separate oral and written submissions made with regard to the two hearings. The record itself exceeds 4,000 pages. Employers, employees, labor unions, public health groups, independent experts, physicians, research scientists, and specialists in many fields have been invited to submit information and have made their views, knowledge and experience available to OSHA. The entire record encompassing these submissions was thoroughly reviewed and evaluated in reaching the determinations set forth below.

II. FINDINGS REGARDING CARCINOGENICITY, EXPOSURE LEVELS AND FEASIBILITY

***

(4) *Conclusions.* The conclusions below are based on a thorough review and evaluation of all the evidence submitted. Where decisions can be based on record evidence, this has been done. Where, however, factual certainties are lacking or where the facts alone do not provide an answer, policy judgments have been made.

There is little dispute that VC is carcinogenic to man and we so conclude. However, the precise level of exposure which poses a hazard and the question of whether a "safe" exposure level exists, cannot be definitively answered on the record. Nor is it clear to what extent exposures can be feasibly reduced. We cannot wait until indisputable answers to these questions are available, because lives of employees are at stake. Therefore, we have had to exercise our best judgment on the basis of the best available evidence. These judgments have required a balancing process, in which the overriding consideration has been the protection of employees, even those who may have regular exposures to VC throughout their working lives.

Based on the available evidence and in view of the above considerations, including feasibility, we believe that employee exposures to VC must be reduced to a 1 ppm time-weighted average (TWA). We also believe that PVC and VC establishments will, in time, be able to attain that level through engineering controls, and that fabricators can do so in the immediate future.

In addition to the TWA requirement, we have established a 5 ppm ceiling (averaged over a 15-minute period) in order to prevent exposure of employees to unacceptable high excursions. From an operation standpoint, this ceiling level is realistic because minor excursions up to the ceiling level are likely to occur on a regular basis.

***

Accordingly, upon consideration of the whole record of this proceeding, Part 1910 of Title 29, Code of Federal Regulations is amended, effective January 1, 1975, by revision of §1910.93q to read as follows:

§1910.93q VINYL CHLORIDE.

(a) *Scope and application.* (1) This section includes requirements for the control of employee exposure to vinyl chloride (chloroethene), Chemical Abstracts Service Registry No. 75015.

(2) This section applies to the manufacture, reaction, packaging, repackaging, storage, handling or use of vinyl chloride or polyvinyl chloride, but does not apply to the handling or use of fabricated products made of polyvinyl chloride.

(3) This section applies to the transportation of vinyl chloride or polyvinyl chloride except to the extent that the Department of Transportation may regulate the hazards covered by this section.

(b) *Definitions.* (1) "Action level" means a concentration of vinyl chloride of 0.5 ppm averaged over an 8-hour work day.

(2) "Assistant Secretary" means the Assistant Secretary of Labor for Occupational Safety and Health, U.S. Department of Labor, or his designee.

(3) "Authorized person" means any person specifically authorized by the employer whose duties require him to enter a regulated area or any person entering such an area as a designated representative of employees for the purpose of exercising an opportunity to observe monitoring and measuring procedures.

(4) "Director" means the Director, National Institute for Occupational Safety and Health, U.S. Department of Health, Education, and Welfare, or his designee.

(5) "Emergency" means any occurrence such as, but not limited to, equipment failure, or operation of a relief device which is likely to, or does, result in massive release of vinyl chloride.

(6) "Fabricated product" means a product made wholly or partly from polyvinyl chloride, and which does not require further processing at temperatures, and for times, sufficient to cause mass melting of the polyvinyl chloride resulting in the release of vinyl chloride.

(7) "Hazardous operation" means any operation, procedure, or activity where a release of either vinyl chloride liquid or gas might be expected as a consequence of the operation or because of an accident in the operation, which would result in an employee exposure in excess of the permissible exposure limit.

(8) "OSHA Area Director" means the Director for the Occupational Safety and Health Administration Area Office having jurisdiction over the geographic area in which the employer's establishment is located.

(9) "Polyvinyl chloride" means polyvinyl chloride homopolymer or copolymer before such is converted to a fabricated product.

(10) "Vinyl chloride" means vinyl chloride monomer.

(c) *Permissible exposure limit.* (1) No employee may be exposed to vinyl chloride at concentrations greater than 1 ppm averaged over any 8-hour period, and

(2) No employee may be exposed to vinyl chloride at concentrations greater than 5 ppm averaged over any period not exceeding 15 minutes.

(3) No employee may be exposed to vinyl chloride by direct contact with liquid vinyl chloride.

(d) *Monitoring.* (1) A program of initial monitoring and measurement shall be undertaken in each establishment to determine if there is any employee exposed, without regard to the use of respirators, in excess of the action level.

(2) Where a determination conducted under paragraph (d)(1) of this section shows any employee exposures, without regard to the use of respirators, in excess of the action level, a program for determining exposures for each such employee shall be established. Such a program:

(i) Shall be repeated at least monthly where any employee is exposed, without regard to the use of respirators, in excess of the permissible exposure limit.

(ii) Shall be repeated not less than quarterly where any employee is exposed, without regard to the use of respirators, in excess of the action level.

(iii) May be discontinued for any employee only when at least two consecutive monitoring determinations, made not less than 5 working days apart, show exposures for that employee at or below the action level.

(3) Whenever there has been a production, process or control change which may result in an increase in the release of vinyl chloride, or the employer has any other reason to suspect that any employee may be exposed in excess of the action level, a determination of employee exposure under paragraph (d)(1) of this section shall be performed.

(4) The method of monitoring and measurement shall have an accuracy (with a confidence level of 95 percent) of not less than plus or minus 50 percent from 0.25 through 0.5 ppm, plus or minus 35 percent from over 0.5 ppm through 1.0 ppm, and plus or minus 25 percent over 1.0 ppm. (Methods meeting these accuracy requirements are available in the "NIOSH Manual of Analytical Methods").

(5) Employees or their designated representatives shall be afforded reasonable opportunity to observe the monitoring and measuring required by this paragraph.

(e) *Regulated area.* (1) A regulated area shall be established where:

(i) Vinyl chloride or polyvinyl chloride is manufactured, reacted, repackaged, stored, handled or used; and

(ii) Vinyl chloride concentrations are in excess of the permissible exposure limit.

(2) Access to regulated areas shall be limited to authorized persons. A daily roster shall be made of authorized persons who enter.

(f) *Methods of compliance.* Employee exposures to vinyl chloride shall be controlled to at or below the permissible exposure limit provided in paragraph (c) of this section by engineering, work practice, and personal protective controls as follows:

(1) Feasible engineering and work practice controls shall immediately be used to reduce exposures to at or below the permissible exposure limit.

(2) Wherever feasible engineering and work practice controls which can be instituted immediately are not sufficient to reduce exposures to at or below the permissible exposure limit, they shall nonetheless be used to reduce exposures to the lowest practicable level, and shall be supplemented by respiratory protection in accordance with paragraph (g) of this section. A program shall be established and implemented to reduce exposures to at or below the permissible exposure limit, or to the greatest extent feasible, solely by means of engineering and work practice controls, as soon as feasible.

(3) Written plans for such a program shall be developed and furnished upon request for examination and copying to authorized representatives of the Assistant Secretary and the Director. Such plans shall be updated at least every six months.

(g) *Respiratory protection.* Where respiratory protection is required under this section:

(1) The employer shall provide a respirator which meets the requirements of this paragraph and shall assure that the employee uses such respirator, except that until December 31, 1975, wearing of respirators shall be at the discretion of each employee for exposures not in excess of 25 ppm, measured over any 15-minute period. Until December 31, 1975, each employee who chooses not to wear an appropriate respirator shall be informed at least quarterly of the hazards of vinyl chloride and the purpose, proper use, and limitations of respiratory devices.

(2) Respirators shall be selected from among those jointly approved by the Mining Enforcement and Safety Administration, Department of the Interior, and the National Institute for Occupational Safety and Health under the provisions of 30 CFR Part 11.

(3) A respiratory protection program meeting the requirements of §1910.134 shall be established and maintained.

(4) Selection of respirators for vinyl chloride shall be as follows: [chart omitted]

(5)(i) Entry into unknown concentrations or concentrations greater than 36,000 ppm (lower explosive limit) may be made only for purposes of life rescue; and

(ii) Entry into concentrations of less than 36,000 ppm, but greater than 3,600 ppm may be made only for purposes of life rescue, firefighting, or securing equipment so as to prevent a greater hazard from release of vinyl chloride.

(6) Where air-purifying respirators are used:

(i) Air-purifying cannisters or cartridges shall be replaced prior to the expiration of their service life or the end of the shift in which they are first used, whichever occurs first, and

(ii) A continuous monitoring and alarm system shall be provided where concentrations of vinyl chloride could reasonably exceed the allowable concentrations for the devices in use. Such system shall be used to alert employees when vinyl chloride concentrations exceed the allowable concentrations for the devices in use.

(7) Apparatus prescribed for higher concentrations may be used for any lower concentration.

(h) *Hazardous operations.* (1) Employees engaged in hazardous operations, including entry of vessels to clean polyvinyl chloride residue from vessel walls, shall be provided and required to wear and use;

(i) Respiratory protection in accordance with paragraphs (c) and (g) of this section; and

(ii) Protective garments to prevent skin contact with liquid vinyl chloride or with polyvinyl chloride residue from vessel walls. The protective garments shall be selected for the operation and its possible exposure conditions.

(2) Protective garments shall be provided clean and dry for each use.

(i) *Emergency situations.* A written operational plan for emergency situations shall be developed for each facility storing, handling, or otherwise using vinyl chloride as a liquid or compressed gas. Appropriate portions of the plan shall be implemented in the event of an emergency. The plan shall specifically provide that:

(1) Employees engaged in hazardous operations or correcting situations of existing hazardous releases shall be equipped as required in paragraph (h) of this section;

(2) Other employees not so equipped shall evacuate the area and not return until conditions are controlled by the methods required in paragraph (f) of this section and the emergency is abated.

(j) *Training.* Each employee engaged in vinyl chloride or polyvinyl chloride operations shall be provided training in a program relating to the hazards of vinyl chloride and precautions for its safe use.

(1) The program shall include:

(i) The nature of the health hazard from chronic exposure to vinyl chloride including specifically the carcinogenic hazard;

(ii) The specific nature of operations which could result in exposure to vinyl chloride in excess of the permissible limit and necessary protective steps;

(iii) The purpose for, proper use, and limitations of respiratory protective devices;

(iv) The fire hazard and acute toxicity of vinyl chloride, and the necessary protective steps;

(v) The purpose for and a description of the monitoring program;

(vi) The purpose for, and a description of, the medical surveillance program;

(vii) Emergency procedures;

(viii) Specific information to aid the employee in recognition of conditions which may result in the release of vinyl chloride; and

(ix) A review of this standard at the employee's first training and indoctrination program, and annually thereafter.

(2) All materials relating to the program shall be provided upon request to the Assistant Secretary and the Director.

(k) *Medical surveillance.* A program of medical surveillance shall be instituted for each employee exposed, without regard to the use of respirators, to vinyl chloride in excess of the action level. The program shall provide each such employee with an opportunity for examinations and tests in accordance with this paragraph. All medical examinations and procedures shall be performed by or under the supervision of a licensed physician, and shall be provided without cost to the employee.

(1) At the time of initial assignment, or upon institution of medical surveillance;

(i) A general physical examination shall be performed, with specific attention to detecting enlargement of liver, spleen or kidneys, or dysfunction in these organs, and for abnormalities in skin, connective tissues and the pulmonary system.

(ii) A medical history shall be taken, including the following topics:

(A) Alcohol intake;

(B) Past history of hepatitis;

(C) Work history and past exposure to potential hepatotoxic agents, including drugs and chemicals;

(D) Past history of blood transfusions; and

(E) Past history of hospitalizations.

(iii) A serum specimen shall be obtained and determinations made of:

(A) Total bilirubin;

(B) Alkaline phosphatase;

(C) Serum glutamic oxalecetic transaminase (SGOT);

(D) Serum glutamic pyruvic transaminase (SGPT); and

(E) Gamma glustamyl transpeptidase.

(2) Examinations provided in accordance with this paragraph shall be performed at least:

(i) Every 6 months for each employee who has been employed in vinyl chloride or polyvinyl chloride manufacturing for 10 years or longer; and

(ii) Annually for all other employees.

(3) Each employee exposed to an emergency shall be afforded appropriate medical surveillance.

(4) A statement of each employee's suitability for continued exposure to vinyl chloride including use of protective equipment and respirators, shall be obtained from the examining physician promptly after any examination. A copy of the physician's statement shall be provided each employee.

(5) If any employee's health would be materially impaired by continued exposure, such employee shall be withdrawn from possible contact with vinyl chloride.

(6) Laboratory analyses for all biological specimens included in medical examinations shall be performed in laboratories licensed under 42 CFR Part 74.

(7) If the examining physician determines that alternative medical

examinations to those required by paragraph (k)(1) of this section will provide at least equal assurance of detecting medical conditions pertinent to the exposure to vinyl chloride, the employer may accept such alternative examinations as meeting the requirements of paragraph (k)(1) of this section, if the employer obtains a statement from the examining physician setting forth the alternative examinations and the rationale for substitution. This statement shall be available upon request for examination and copying to authorized representatives of the Assistant Secretary and the Director.

(l) *Signs and labels.* (1) Entrances to regulated areas shall be posted with legible signs bearing the legend:

CANCER-SUSPECT AGENT AREA AUTHORIZED PERSONNEL ONLY

(2) Areas containing hazardous operations or where an emergency currently exists shall be posted with legible signs bearing the legend:

CANCER-SUSPECT AGENT IN THIS AREA PROTECTIVE EQUIPMENT
REQUIRED AUTHORIZED PERSONNEL ONLY

(3) Containers of polyvinyl chloride resin waste from reactors or other waste contaminated with vinyl chloride shall be legibly labeled:

CONTAMINATED WITH VINYL CHLORIDE CANCER-SUSPECT AGENT

(4) Containers of polyvinyl chloride shall be legibly labeled:

POLYVINYL CHLORIDE (OR TRADE NAME) CONTAINS VINYL CHLORIDE
VINYL CHLORIDE IS A CANCER-SUSPECT AGENT

(5) Containers of vinyl chloride shall be legibly labeled either:
(i)

VINYL CHLORIDE EXTREMELY FLAMMABLE GAS UNDER PRESSURE
CANCER-SUSPECT AGENT

or (ii) In accordance with 49 CFR Part 173, Subpart H, with the additional legends:

CANCER-SUSPECT AGENT

applied near the [label] or placard.

(6) No statement shall appear on or near any required sign, label or instruction which contradicts or detracts from the effect of, any required warning, information or instruction.

(m) *Records.* (1) All records maintained in accordance with this section shall include the name and social security number of each employee where relevant.

(2) Records of required monitoring and measuring, medical records, and authorized personnel rosters, shall be made and shall be available upon request for examination and copying to authorized representatives of the Assistant Secretary and the Director.

(i) Monitoring and measuring records shall:

(A) State the date of such monitoring and measuring and the concentrations determined and identify the instruments and methods used;

(B) Include any additional information necessary to determine individual employee exposures where such exposures are determined by means other than individual monitoring of employees; and

(C) Be maintained for not less than 30 years.

(ii) Authorized personnel rosters shall be maintained for not less than 30 years.

(iii) Medical records shall be maintained for the duration of the employment of each employee plus 20 years, or 30 years, whichever is longer.

(3) In the event that the employer ceases to do business and there is no successor to receive and retain his records for the prescribed period, these records shall be transmitted by registered mail to the Director, and each employee individually notified in writing of this transfer.

(4) Employees or their designated representatives shall be provided access to examine and copy records of required monitoring and measuring.

(5) Former employees shall be provided access to examine and copy required monitoring and measuring records reflecting their own exposures.

(6) Upon written request of any employee, a copy of the medical record of that employee shall be furnished to any physician designated by the employee.

(n) *Reports.* (1) Not later than 1 month after the establishment of a regulated area, the following information shall be reported to the OSHA Area Director. Any changes to such information shall be reported within 15 days.

(i) The address and location of each establishment which has one or more regulated areas; and

(ii) The number of employees in each regulated area during normal operations, including maintenance.

(2) Emergencies, and the facts obtainable at that time, shall be reported within 24 hours to the OSHA Area Director. Upon request of the Area Director, the employer shall submit additional information in writing relevant to the nature and extent of employee exposures and measures taken to prevent future emergencies of similar nature.

(3) Within 10 working days following any monitoring and measuring which discloses that any employee has been exposed, without regard to the use of respirators in excess of the permissible exposure limit, each such employee shall be notified in writing of the results of the exposure measurement and the steps being taken to reduce the exposure to within the permissible exposure limit.

(o) *Effective dates.* (1) Until January 1, 1975, the provisions currently set forth in §1910.93q of this Part shall apply.

(2) Effective January 1, 1975, the provisions set forth in §1910.93q of this Part shall apply.

***

Several observations could be made regarding this relatively early OSHA rulemaking.

1. The proceeding moved rapidly and the statutory time frames for action were met by OSHA. A little more than a year elapsed between the time NIOSH first communicated with OSHA and the court of appeals' decision upholding the standard.

2. OSHA had also moved promptly to issue an emergency temporary standard (January to April 1975). Although there is no statutory requirement to do so, OSHA held a fact-finding hearing before

issuing the emergency standard. At the hearing, Professor Maltoni from Italy testified on his careful animal experiments on vinyl chloride hazards. This hearing, and the exchange of public views, made the emergency standard less controversial; it was not challenged in court.[41]

3. The issues considered at the rulemaking hearing were extremely controversial. On the one hand, industry claimed that the standard was infeasible, both economically and technologically. Unions and public interest groups argued that costs were irrelevant and, in any event, the standard was feasible. OSHA concluded that the standard was feasible for most employers and that others would "in time" be able to reach the permissible exposure limit, using "some new technology and work practices." The court of appeals agreed with OSHA, introducing the concept of "technology-forcing" into the OSHA regulatory vocabulary. The issue of feasibility has, of course, continued to be one of the most difficult in OSHA standards proceedings. In the vinyl chloride proceeding, OSHA for the first time contracted with an independent consultant, Foster D. Snell Corporation, to conduct studies of the feasibility of compliance at various exposure levels and the economic costs of compliance.[42] In addition, OSHA prepared an environmental impact statement, as required by the National Environmental Policy Act.[43]

4. In this proceeding, OSHA was also confronted with the issue of whether there was a safe level of human exposure to carcinogens. Although animal experiments show cancer developing only at the 50 ppm level, OSHA concluded that it would be "imprudent" to assume that either animals or humans would not develop cancer at levels lower than 50 ppm. The standard set a permissible exposure limit of 1 ppm and the court of appeals agreed.[44]

5. While the standard was challenged in court, the litigation was completed quickly and successfully from OSHA's point of view, and the standard remained in effect for most of the period following the issuance of the emergency standard.

6. Industry's pessimistic predictions about the effects of the vinyl chloride standard turned out to be unjustified.[45]

---

[41]See Chapter 4 on OSHA's experience in promulgating emergency temporary standards.

[42]See Chapter 10 on the issue of feasibility of standards.

[43]42 U.S.C. §4332(c)–(d) (1976).

[44]The issue of OSHA's establishing permissible exposure limits in health standards is discussed at length in Chapter 9.

[45]E.g., Doniger, *Federal Regulation of Vinyl Chloride: A Short Course in the Law and Policy of Toxic Substances Control,* 7 ECOLOGY L.Q. 497 (1978); Perry, *Vinyl Chloride Protection: Less Costly Than Predicted,* MONTHLY LAB. REV., Aug. 1980 at 22. For an extensive discussion of the development of OSHA's vinyl chloride standard and of its impact, see H. Northrup, Rowan, Perry, Cassidy, Saviskas & Outlaw, THE IMPACT OF OSHA: A STUDY OF THE EFFECTS OF THE OCCUPATIONAL SAFETY AND HEALTH ACT ON THREE KEY INDUSTRIES—AEROSPACE, CHEMICALS AND TEXTILES 291–418 (Wharton School of Business, Labor Relations, and Public Policy Series No. 17, 1978). A different understanding of the history was set forth by F. Goldsmith & L.E. Kerr in OCCUPATIONAL SAFETY AND HEALTH—THE PREVENTION AND CONTROL OF WORK RELATED HAZARDS 105–16 (1982). Drs. Goldsmith and Kerr refer to the vinyl chloride "coverup," *id.* at 105, and say that "circumstantial evidence suggests that many firms deliberately suppressed evidence of PVC [polyvinyl chloride] dangers in order to avoid costly plant modifications," *id.* at 107.

While the rulemaking on vinyl chloride may properly be considered to be successful both in terms of time consumed and ultimate result in protecting employees, this success was generally not achieved by OSHA in other rulemakings. In large measure, there has been a serious problem in the length of the rulemaking proceedings and the resulting litigation. For example, the lead standard was proposed on October 3, 1975; after extensive hearings, the final standard was not issued until November 14, 1978. The standard was challenged in court and partially stayed. The court decision on the merits was issued on August 15, 1980.[46] While the standard was upheld in substantial part, portions of the standard were stayed and remanded to OSHA for further action, and OSHA issued a supplemental statement of reasons in January 1981 and a revised statement in December 1981.[47] Similarly, the cotton dust proposal was published on December 28, 1976; the final standard was issued on June 23, 1978; and the Supreme Court decision upholding the standard in substantial part was handed down on June 17, 1981.[48] Many of these delays, particularly those at the litigation stage, are understandable in light of the controversial nature of the regulatory action taken. Congress and the public, however, have expressed considerable disappointment regarding the time consumed by OSHA's rulemaking and on the limited number of rulemakings completed, particularly those concerning health hazards.

One of those expressing great disappointment was Dr. Morton Corn, Assistant Secretary for OSHA in 1975–76. In a "Status Report on OSHA," submitted by Dr. Corn to the Secretary of Labor on January 12, 1976,[49] Dr. Corn noted that standards promulgation has required "at best" 18 to 22 months; that it would be a "noble ambition" for OSHA to have a productivity rate of 15 to 20 health standards in 1978 or 1979; and that this rate of production will not adequately address the problem of the thousands of chemicals in the work environment. He therefore concluded that continued development of standards on an individual chemical basis was a "self-limiting" process.[50] In May 1977, the Comptroller General of the United States issued a report, sharply criticizing both OSHA and NIOSH for the slow standards development process, and observing that it "will take more than a century to establish needed standards for substances already identified as hazards."[51]

---

[46]*United Steelworkers v. Marshall*, 647 F.2d 1189 (D.C. Cir. 1980) (remanded in part), *cert. denied sub nom. Lead Indus. Ass'n v. Donovan*, 453 U.S. 913 (1981).

[47]46 FED. REG. 6134 (1981); 46 FED. REG. 60,758 (1981). The court of appeals has not completed its review of these statements.

[48]*American Textile Mfrs. Inst. v. Donovan*, 452 U.S. 490 (1980), *aff'g in part AFL-CIO v. Marshall*, 617 F.2d 636 (D.C. Cir. 1979).

[49]Report from Assistant Secretary of OSHA Morton Corn to the Secretary of Labor (Jan. 12, 1976) (status report on OSHA), reprinted in 6 OSHR 1094–1102 (1977).

[50]*Id.* at 24–25.

[51]COMPTROLLER GENERAL OF THE UNITED STATES, REPORT TO CONGRESS, DELAYS IN SETTING WORKPLACE STANDARDS FOR CANCER CAUSING AND OTHER DANGEROUS SUBSTANCES, HRD-77-71 at 9 (May 10, 1977). See also HOUSE COMM. ON GOVERNMENT OPERATIONS, CHEMICAL DANGERS IN THE WORKPLACE, H.R. REP. No. 1688, 94th Cong., 2d Sess. 19–20 (1976) (criticizing the "slowness" of OSHA standards process). In October 1983, Assistant Secretary Auchter testified

In Senate oversight hearings in 1980, Lane Kirkland, President of the AFL-CIO, testified, among many other issues, on OSHA health standards. While noting a "marked improvement" in OSHA's performance during the administration of Dr. Bingham, he stated that there were a number of problems threatening "timely progress" in health standards development. He referred to "lack of competency among [OSHA] staff," delays in the Office of the Solicitor of Labor, delays occasioned by the regulatory analysis requirements of presidential executive orders, and the "continued unrelenting industry opposition to the promulgation and enforcement of strong standards."[52] Also, in 1980, OSHA in the preamble to its Carcinogens Policy, noted that in its nine-year history, the Agency had concluded only seven rulemaking proceedings regulating carcinogens.[53] The preamble traced the industry rulemaking proceedings, observing that the rulemaking process has taken as long as 67 months for some carcinogenic substances.[54]

Particularly because OSHA has been able to issue relatively few health standards, the process of its setting standards priorities becomes especially important. Section 6(g) of the Act[55] sets forth two criteria for priorities for standards development: first, "the urgency of the need for mandatory safety and health standards for particular industries, trades, crafts, occupations, businesses, workplaces or work environments" (usually referred to as "worst first") and second, recommendations from NIOSH. In *National Congress of Hispanic American Citizens v. Marshall*,[56] where suit was brought to compel OSHA to issue a field sanitation standard, an issue in the case was whether OSHA's ordering of priorities was rational and in good faith. Assistant Secretary Bingham filed statements with the court positing the Agency's bases for establishing priorities. These were ap-

before a congressional committee on OSHA's so-called dormant standards. He said that his staff had attempted to identify "all of the partially completed activities" in OSHA's regulatory program. One hundred sixteen rulemaking projects were found, he said; of these, 23 involved regulatory proposals and 6 involved proposals to change existing permissible exposure limits. The remainder involved either advance notices of proposed rulemaking or requests for information published with the *Federal Register*. Mr. Auchter indicated that "ultimately" he will decide whether to "reactivate" any of these regulatory projects, most of which were initiated between 1973 and 1976. Statement of Thorne G. Auchter, Ass't. Secretary of Labor for OSHA at *Hearings Before the Subcomm. on Health and Safety of House Education and Labor Comm.*, Oct. 4, 1983, pp. 1–7 (mimeo).

[52]*Oversight on the Administration of the Occupational Safety and Health Act, 1980: Hearings Before the Senate Comm. on Labor and Human Resources*, 96th Cong., 2d Sess. 1227–32 (1980). The relationship between OSHA and the Office of the Solicitor has been critiqued by several authorities; for example, the Presidential Task Force on Safety Regulation recommended in 1977 that there be a "closer working relationship" between OSHA and the legal staff. REPORT OF THE PRESIDENTIAL TASK FORCE, OSHA SAFETY REGULATION 33–34 (P. MacAvoy ed. 1977).

[53]45 FED. REG. 5002, 5011 (1980). It noted also that two other health standards not regulating carcinogens had been issued: lead and cotton dust. *Id.* at 5011 n. col. 3.

In the preamble to the 1983 emergency asbestos standard, OSHA noted the extensive delays that had taken place in issuing health standards after rulemaking, suggesting that it is only the "urgency" generated by its issuance of an emergency standard under §6(c) which provides the Agency with the "capability" to produce a "permanent" standard within the statutory deadline. 48 FED REG. 51,086, 51,098 (1983). See discussion in Chapter 4.

[54]45 FED. REG. at 5011–12.

[55]29 U.S.C. §655(g).

[56]626 F.2d 882 (D.C. Cir. 1979). See discussion in Chapter 6, Section C.

proved in principle by the court of appeals as "adequately reflect[ing] the purposes and provisions of the statute, and * * * rational within that context."[57] The court of appeals described OSHA standards' priorities, as follows:

> In response to the interrogatories, the Secretary [of Labor] elaborated upon his earlier statement of the factors entering into the priority-setting process. He distinguished health standards from safety standards. For the former, he said that he considered: the number of workers exposed to particular unregulated hazards; the severity of such hazards; the existence of research relevant to hazard identification and methods of hazard control; the recommendations of the National Institute for Occupational Safety and Health (NIOSH); petitions for standards on the hazards; and court decisions and other factors affecting the enforceability of the standard. Severity was noted as the most important consideration. In the safety area, the factors identified as important by the Secretary were slightly different. They included: the number of workers exposed to particular unregulated hazards; the illness or injury incidence of the affected worker group; the nature and severity of the hazard exposure; the need for classification, simplification and/or revision to improve compliance with the enforceability of an existing standard; changes in the state of the art in work practices; court or Occupational Safety and Health Review Commission decisions affecting the enforceability of existing standards; availability of research and other information relevant to hazard identification and methods of hazard control; the recommendations of NIOSH; Advisory Committee deliberations; and petitions for standards on the hazard. Even after setting forth these lists, however, the Secretary stressed his view that the priority setting process was not susceptible to a "mathematical formula or [a] formal weighing scheme. * * *"[58]

These criteria for health standards' priorities were reaffirmed by Assistant Secretary Auchter in the same proceeding.[59]

Under these criteria, OSHA has given highest priority to carcinogenic substances. All of OSHA's health standards, except for those regulating cotton dust and lead, have related to carcinogens. The priority treatment for cotton dust and lead was based on the severe hazards involved, the large number of employees at risk, and the excellent studies available on the hazards of cotton dust and lead.[60] In 1983 OSHA stated that its highest health priorities were asbestos, ethylene dibromide, both carcinogens, and hazard communication, affecting numerous chemicals.[61]

In addition to OSHA's articulated criteria for standards priori-

---

[57] 626 F.2d at 889.

[58] *Id.* at 886.

[59] *National Congress of Hispanic Am. Citizens v. Donovan,* No. 2143-73 (D.D.C. July 16, 1982).

[60] Between 126,000 and 200,000 workers are exposed to cotton dust in yarn manufacturing alone, 43 FED. REG. 27,350, 27,379 (1978). At least 800,000 workers representing 120 occupations in over 40 industries are exposed to airborne lead. *United Steelworkers v. Marshall,* 647 F.2d 1189, 1204 (D.C. Cir. 1980).

[61] See also *Public Citizen Health Research Group v. Auchter,* 702 F.2d 1150, 1158 (D.C. Cir. 1983) (the court added ethylene oxide, also a carcinogen, to that list of priorities). There has been considerable debate on the contribution of occupational exposures to cancer in the United States. See OFFICE OF TECHNOLOGY ASSESSMENT, CONGRESS OF UNITED STATES, ASSESSMENT OF TECHNOLOGIES FOR DETERMINING CANCER RISKS FROM THE ENVIRONMENT 84–88 (1981). A well-publicized study of the National Cancer Institute, the National Institute of Environment Health Sciences, and NIOSH, published in 1978, estimated that occupationally related cancers

ties, OSHA, quite understandably, has given "due regard"[62] to several other factors in deciding on its priorities. Among them are court orders requiring OSHA to issue standards, for example, the order of the Court of Appeals for the District of Columbia Circuit in *Public Citizen Health Research Group v. Auchter,*[63] involving ethylene oxide; priorities set by the White House, more particularly standards "targeted for review" by the President's Task Force on Regulatory Relief, established under Executive Order 12,291;[64] and standards activity resulting at least in part from congressional oversight and appropriations activity, for example, OSHA's revision of its national consensus standards.[65]

A number of strategies have been suggested and tried in order to facilitate standards development. Dr. Corn referred to better coordination between the legal effort from the Office of the Solicitor and the scientific effort from OSHA's health standards office.[66] He recommended a "generic" approach to regulation, under which "OSHA could not wait for development of standards for each agent to regulate each of [the] specific provisions for that agent." Thus, he suggested, standards would be issued for labeling of compounds, for medical surveillance of employees, and for monitoring of the work environment.[67] An early attempt in generic regulation was the so-called Standards Completion Project, undertaken jointly by OSHA and NIOSH. This effort, begun in 1974, would have "fleshed out" the limited health standards for 400 toxic substances issued in 1971 which contained only permissible exposure limits.[68] The comprehensive standards were to include requirements for air monitoring, medical surveillance, record keeping, work practice, and others. While proposals were issued for two groups of toxic substances in 1975,[69] no

---

may comprise as much as 20% of total cancer mortality. DEPARTMENT OF HEALTH EDUCATION AND WELFARE, ESTIMATES OF THE FRACTION OF CANCER IN THE UNITED STATES RELATED TO OCCUPATIONAL FACTORS (1978). This estimate was regarded as not having any validity by Doll and Peto, who arrived at an estimate of approximately 4% with an acceptable range from 2 to 10%. *The Causes of Cancer: Quantitative Estimates of Avoidable Risks of Cancer in United States Today,* 66 J. NAT'L CANCER INST. 1191 (1981). The Office of Technology Assessment observed that "almost every estimate [of occupationally caused cancer] fits comfortably in the range of 10 ± 5 percent." ASSESSMENT OF TECHNOLOGIES FOR DETERMINING CANCER, *supra* at 88. See also D.P. McCaffrey, OSHA AND THE POLITICS OF HEALTH REGULATION 21–24 (1982).

[62]See §6(g), 29 U.S.C. §655(g).

[63]702 F.2d 1150 (D.C. Cir. 1983); this case is discussed in Chapter 6, Section C.

[64]For example, OSHA's commercial diving standard, 29 C.F.R. §§1910.401–.441 (1983), was "targeted for review" by the Task Force; see Department of Labor Office of the Secretary, Semiannual Agenda of Regulations, 48 FED. REG. 18,166 (1983); and in November 1982, OSHA issued a final rule exempting educational/scientific diving from the standards requirements, 47 FED. REG. 53,357 (1982). For a discussion of the role of the Task Force in the revisions of OSHA's commercial diving standards, see *Office of Management and Budget Control of OSHA Rulemaking: Hearings Before a Subcomm. of the House Comm. on Government Operations,* 97th Cong., 2d Sess. 197–219 (1982) (statement of Robert J. Pleasure, Associate General Counsel, Carpenters International Union); *id.* at 321–22 (statement of Christopher DeMuth, Executive Director, Presidential Task Force on Regulatory Relief). For a discussion of Exec. Order No. 12,291, see Chapter 10.

[65]See discussion in Chapter 2.

[66]Report from Assistant Secretary of OSHA Morton Corn to the Secretary of Labor 24 (Jan. 12, 1976) (status report on OSHA), reprinted in 6 OSHR 1094–1102 (1977).

[67]*Id.* at 25.

[68]These limits now appear in 29 C.F.R. §1910.1000 (1983).

[69]40 FED. REG. 20,202 (1975) (ketones); 40 FED. REG. 47,262 (1975) (alkyl benzene, cyclohexane, ketones, ozone).

final standard was ever issued and the project was eventually abandoned.[70]

A successful generic regulatory effort was OSHA's regulation on access to employee exposure and medical records, which requires that employers provide access to employees and to OSHA to existing employer-maintained exposure and medical records relevant to employees exposed to a broad range of toxic substances and harmful physical agents.[71] And in 1981, OSHA issued a proposed generic hazard identification (labeling) standard; the standard was withdrawn early in 1981, and reissued with changes in 1983.[72]

A major initiative by OSHA to provide a framework for its regulation of carcinogens in a "timely and efficient manner" was its issuance of the Carcinogens Policy in 1980. As explained more fully in the preamble, the purpose of the Policy was to avoid the reargument of the same policy issues in each substance-specific rulemaking proceeding. Instead, under the Policy, OSHA issued, with public participation, policy determinations on the regulation of carcinogens which could be questioned later in only limited circumstances.

Carcinogens Policy Final Rule

45 Fed. Reg. 5002, 5011, 5013–14 (1980)

This proposal marks a departure from OSHA's usual pattern of a substance-by-substance approach in setting health standards concerning exposures to potential occupational carcinogens. For the reasons set forth herein, OSHA believes that such an approach is not only proper and necessary but in fact compelling because of its "experience gained under this and other health and safety laws". (The Act, §6(b)(5).) Moreover, OSHA believes that by establishing a rational and predictable policy concerning the regulation of exposures to occupational carcinogens, employers, employees, the general public, the scientific community and public interest groups will not only benefit from improved employee health, but also experience greater efficiencies in their own activities.*

Employees are exposed to many substances on a daily basis. Obviously, most of these substances are not carcinogenic. Yet, some may be. OSHA believes that this general policy and procedure will facilitate the sifting

*While OSHA will not elaborate upon the lengthy time consuming "case-by-case" rulemakings experienced by other agencies under other statutes, an excellent discussion of the same problems appears in McGarity, *Substantive and Procedural Discretion in Administrative Resolution of Science Policy Questions: Regulating Carcinogens in EPA and OSHA,* 67 GEORGETOWN LAW JOURNAL 724 (February 1979). Also see Berger and Riskin, *Economic and Technological Feasibility in Regulating Toxic Substances Under the Occupational Safety and Health Act,* 7 ECOLOGY LAW QUARTERLY 285 (1978). The scientific commentators have also recommended a generic, rather than a case-by-case approach, to regulating carcinogens for the same reasons. See NATIONAL ACADEMY OF SCIENCES—NATIONAL RESEARCH COUNCIL, DECISIONMAKING FOR REGULATING CHEMICALS IN THE ENVIRONMENT 33 (1975).

[70]NIOSH, however, has issued the comprehensive requirements in a nonregulatory form. For a discussion of the standards completion project, see HOUSE COMM. ON GOVERNMENT OPERATION, CHEMICAL DANGERS IN THE WORKPLACE, H.R. REP. NO. 1688, 94th Cong., 2d Sess. 19 (1976).

[71]29 C.F.R. §1910.20 (1983).

[72]48 FED. REG. 53,260 (1983) (to be codified at 29 C.F.R. §1910.1200). The access rule is discussed in Chapter 5, Section F, and the labeling rule is discussed in Chapter 18, Section F.

through the evidence concerning substances which may be imputed to be potential carcinogens and the application of uniform, predictable criteria in order to assess whether such substances should indeed be treated as carcinogenic. Without such a system and appropriate criteria, OSHA believes that this task cannot be accomplished in a timely and efficient manner. With an appropriate system and criteria, worker health will be protected efficiently without rediscussing or relitigating, time and time again, the same issues and without unnecessarily draining limited industry, union, public interest, scientific and government resources. As a result of this policy, OSHA believes that these limited resources will focus on the most substantial issues in OSHA rulemaking proceedings, thus leading to more compact and useful Records in a shorter period of time. In addition, this policy will result in continuity of approach, even in the face of changes of policymakers. This new approach is intended to ameliorate certain administrative problems in standard setting. One significant aspect of past rulemakings concerning carcinogens was the enormous time period between OSHA's initial *Federal Register* notices and final judicial actions. * * *

OSHA, therefore, like other administrative agencies in the past, has determined that it is necessary at this time, in the fulfillment of its statutory objectives, to reshape the size and content of its rulemaking proceedings at least insofar as potential carcinogens are concerned. This set of regulations finalized today incorporates policy determinations concerning how and when chemical or physical substances should be identified, classified, and consequently regulated, as posing a carcinogenic risk to humans.

It is OSHA's intention to limit the issues in the subsequent 6(b) rulemakings on individual substances, to topics and issues specific to those individual substances. The rehearing of the validity of this classification system and most of the other general policy determinations as described in this document, including the procedural structure intended to be followed, are not to be the subject of those individual rulemakings. However, other than the foreclosed policy determination that employees shall be exposed to Category I* potential carcinogens at the lowest level feasible and that the level should be reached primarily through the use of engineering and work practice controls, hardly any scientific or policy issue is really foreclosed from discussion if certain minimum threshold criteria are met in the subsequent rulemakings. * * *

The Agency believes that this "generic" form of standard setting will not in any way deny "due process", the statutory procedural requirements set forth in §6(b) of the Act or in 5 U.S.C. 553, *et seq.*, since every opportunity for notice, comment, and public participation in extensive hearings pursuant to those provisions has been afforded. Moreover, the model regulatory standards we believe are flexible. As pointed out below, with certain limited exceptions, the model standards contain provisions that are no more than guidelines for OSHA and the public in conducting future regulatory activity and are not meant, as was perceived by some of the participants in this rulemaking, to be rigid, to involve irrebuttable presumptions or to provide the prescribed protection in the most costly manner conceivable. The opposite is our intent. * * *

OSHA has legal authority pursuant to the Act to follow the generic approach to rulemaking. Sections 6(b), 3(8), 2(b)(3) & (9) and 8(g)(2) provide for the promulgation of occupational health standards and this includes methods reasonably necessary and appropriate for such standards and regulations the Secretary deems necessary to carry out his responsibilities. The Secretary believes that the generic approach to rulemaking, establishing general policies in this proceeding and specific requirements in subsequent

*[Author's note—For a definition of Category I carcinogens, see 29 C.F.R. §1990.112(a) (1983). The classification is based on the quality of evidence indicating carcinogenicity.]

individual rulemakings is a necessary and reasonable approach to regulating occupational carcinogens in view of the large number of such substances which must be considered and regulated and the weaknesses of the past substance-by-substance approach.

Among other things, the Carcinogens Policy provided procedures for the identification, classification, and regulation of potential occupational carcinogens[73] and two "model standards" for emergency temporary and permanent standards for carcinogens.[74] The Carcinogens Policy, however, has never been implemented in a rulemaking proceeding. The Policy was challenged in several courts of appeals, and complex litigation involving questions of venue and jurisdiction between courts of appeals and district courts developed.[75] The litigation is still pending as of early 1984, and in the meantime, OSHA has undertaken a reevaluation of the Policy. Pursuant to this reassessment, OSHA has stayed those portions of the Policy which required the Agency to establish and publish candidate and priority lists for carcinogens.[76]

As an alternative to the present regulatory process, which has often been criticized as "slow," "cumbersome," and "excessively adversarial," suggestions have been made to substitute a form of negotiation between the interested parties, which would culminate in an agreement that becomes the basis for a rule.[77] In 1975, a serious attempt was made by former Secretary of Labor John Dunlop to avoid the traditional rulemaking model in developing a health standard governing coke oven emissions and instead to arrange for negotiations between the steel companies and unions in an effort to reach consensus. This effort failed, and Dunlop's approach was greeted with considerable hostility.[78] In any event, it seems clear that any

[73]These procedures are codified at 29 C.F.R. pt. 1990 (1983).

[74]45 FED. REG. 5001, 5217–357 (1980), discussed in Section C, this chapter.

[75]See *American Petroleum Inst. v. OSHA*, 8 OSHC 2025 (5th Cir. 1980).

[76]48 FED. REG. 241 (1983). The publication of these lists is the initial step in the process of regulating carcinogens under the Policy. For further discussion of "generic" rulemaking, see Administrative Conference of the United States, Federal Regulations of Cancer-Causing Chemicals, 1 C.F.R. §305.82–5 ¶V(1) (1983). The Conference stated that "generic" rules may be used in "appropriate" circumstances "for summary administrative resolution of recurrent issues," so long as the reexamination of scientific conclusions respecting the carcinogenicity of specific substances is not foreclosed. See also R.A. Merrill, REPORT TO THE ADMINISTRATIVE CONFERENCE OF THE U.S., FEDERAL REGULATIONS OF CANCER-CAUSING CHEMICALS, Apr. 1, 1982 Draft).

[77]See Administrative Conference of the United States, Procedures for Negotiating Proposed Regulations, 1 C.F.R. §305.82–4 (1983); Harter, *Negotiating Regulations: A Cure for the Malaise*, 71 GEO. L.J. 1 (1982); Note, *Rethinking Regulation: Negotiation as an Alternative to Traditional Rulemaking*, 94 HARV. L. REV. 1871 (1981).

[78]S. Kelman, REGULATING AMERICA, REGULATING SWEDEN: A COMPARATIVE STUDY OF OCCUPATIONAL SAFETY AND HEALTH POLICY 135 (1981). The attempt to "negotiate" the coke oven emission standard and the reasons for the failure of the attempt are discussed in detail by Professor Kelman in his PhD dissertation for the Department of Government of Harvard University. See Regulating Job Safety and Health: A Comparison of the U.S. Occupational Safety and Health Administration and the Swedish Worker Protection Board at 230–55. The coke oven emission standard was eventually issued by OSHA, 41 FED. REG. 46,742 (1976), and upheld by the Court of Appeals for the Third Circuit, *American Iron & Steel Inst. v. OSHA*, 577 F.2d 825 (3d Cir. 1978), *cert. dismissed*, 448 U.S. 917 (1980). S. Kelman, REGULATING AMERICA,

significant initiative toward negotiation standards would have to be brought about through amendments to the Act, a prospect which appears to be highly unlikely at this time.[79]

C. Substantive Provisions of Standards

1. Safety Standards

No standard format exists for OSHA safety standards. Because of the wide variety of hazards dealt with in the safety area, the provisions of individual standards differ greatly. An example of a recent OSHA safety standard is "Servicing Multi-Piece Rim Wheels."[80] The standard deals with the hazard of an employee being struck by a wheel component thrown from an inflated wheel during an unintended separation. The standard contains the following major provisions:

- A statement of the scope of the standard;
- A set of definitions of terms used in the standard;
- Provisions on employee training;
- Requirements for tire servicing equipment;
- Requirements for wheel component acceptability;
- A statement of safe operating procedures;
- Two appendices, one a warning chart to be shown to employees, and the other, instructions on obtaining safety charts from the National Highway Traffic Safety Administration.[81]

Safety standards are preceded by a preamble, which explains the reasons for, and the contents of, the standard. The preamble to the

supra 9–17, compares the American "adversarial" system of standard-promulgations with the Swedish process, which Kelman states reflects "different" values. *Id.* at 118. See also U.S. DEP'T OF LABOR, SWEDISH-AMERICAN CONFERENCE ON CHEMICAL HAZARDS IN THE WORK ENVIRONMENT 130–45 (1980).

[79]See Harter, *supra* note 77 at 110–12. OSHA has utilized, however, a tripartite negotiation approach in *enforcing* the arsenic standard, 29 C.F.R. §1910.1018 (1983), which, like most health standards, requires feasible engineering controls to meet the permissible exposure limits, *id.* at §1910.1018(g). In 1982, the American Smelting and Refining Company (ASARCO), the United Steelworkers of America, and OSHA signed agreements applicable to four ASARCO plants, which are to be in force until the mid-1980s, and indicate the engineering controls and work practices to be instituted by ASARCO and the schedule for their installation. See 12 OSHR 517 (1982). Assistant Secretary Auchter stated that the negotiated plans "fulfill the promise of protection for workers" while at the same time "maintaining the efficiency and competitiveness" of ASARCO's smelting and refining operations. *Id.* A similar agreement was signed on January 30, 1984 by ASARCO, the Steelworkers, and OSHA on compliance with the lead standard. 13 OSHR 947 (1984). OSHA recently stated that in connection with planned expedited rulemaking for benzene, it intended "to employ the services of a neutral mediator to assist interested persons in reaching agreement on joint recommendations to OSHA for the proposal." 13 OSHR 155, 539, 556 (1983).

[80]45 FED. REG. 6706 (1980); 29 C.F.R. §1910.177 (1983). This standard was subsequently amended to cover also servicing of single-piece wheel rims, 49 FED. REG. 4338 (1984) (to be codified at 29 C.F.R. §1910.177). See also Guarding of Low Pitched Roof Perimeters During the Performances of Built-Up Roofing Work, 45 FED. REG. 75,618 (1980); 29 C.F.R. §1926.500(g) & app. A (1983). OSHA's ongoing effort to revise §6(a), 29 U.S.C. §655(a), standards is discussed in Chapter 2.

[81]Standards that adopt a "performance" approach, see, e.g., Revised Fire Protection Standard, 29 C.F.R. §§1910.155–.165 (1983), contain appendices with "nonmandatory guidelines to assist employers in complying" with the standards' requirements, e.g., *id.* at apps. A–D. Performance standards are discussed in Chapter 2.

multipiece rim wheel standard contains the following major elements:

- A background section explaining how multipiece rim wheels are used, and describing the history of the regulation, the hazards involved, and the available accident data.
- A summary and explanation of the specific provisions of the standard and the major issues in the proceeding.
- A regulatory assessment, as required by Executive Order No. 12,044, issued by President Carter,[82] and a conclusion by OSHA that the standard is economically and technologically feasible.

2. Health Standards

Certain basic provisions have appeared in all OSHA health standards, beginning with the asbestos standard issued in 1972, even though the content of these provisions has varied somewhat, and recent standards, such as lead, have contained novel provisions. The following list of elements in health standards is based on the discussion in the preamble to the "Carcinogens Policy" regarding "model standards" for carcinogens.[83] These provisions have generally been included in standards regulating all toxic substances as well as carcinogens. Where appropriate, reference will be made to specific provisions in the vinyl chloride standard.[84]

(a) *Scope and application.* This section defines the coverage of the standard, stating if any industries or operations are exempt. The vinyl chloride standard applies to all vinyl chloride and polyvinyl chloride operations, but specifically exempts the handling and use of fabricated products made of polyvinyl chloride (subsection (a)).

(b) *Definitions.* This section defines the important terms utilized in the standard (subsection (b) of vinyl chloride standard).

(c) *Permissible exposure limit (PEL).* This is the basic requirement of a standard, setting forth the concentration level above which an employer may not expose his employees to the regulated substance. The PEL is usually stated in terms of an eight-hour time-weighted average (TWA). The PEL for vinyl chloride is one part per million (ppm), averaged over an eight-hour work period. In some cases, a ceiling level is established, barring employee exposure to concentrations of the substance for shorter periods of time. The ceiling level is higher than the eight-hour TWA; for example, in vinyl chloride, a ceiling of 5 ppm averaged over a 15-minute period was set. OSHA's policy has been to require employers to meet the PEL and ceiling level primarily by means of engineering controls.[85] Finally,

---

[82]43 FED. REG. 12,661 (1978). In implementing this order, the Department of Labor published economic identification criteria. 44 FED. REG. 5570 (1979).

[83]45 FED. REG. 5002, 5217–35 (1980).

[84]The vinyl chloride standard is reprinted in part in Section B, this chapter.

[85]See discussion *infra* this section.

some standards set an "action level," lower than the eight-hour TWA, at which level of exposure certain compliance obligations, such as monitoring and medical surveillance, but not engineering controls, are imposed. The action level of vinyl chloride, set forth in the definition section, is 0.5 ppm averaged over an eight-hour day (subsections (c) and (b)(1)).[86]

(d) *Exposure monitoring.* This section defines the employer's obligations to monitor the workplace atmosphere for the substance regulated. It contains requirements for the frequency and manner[87] of monitoring and, as provided by §8(c)(3) of the Act, requires that employees have reasonable opportunity to observe the monitoring. Under the vinyl chloride standard, monitoring must be conducted monthly if any employee is exposed (without regard to the use of respirators) above the PEL; and quarterly, if the exposure is below the PEL but above the action level (subsection (d)).

(e) *Regulated areas.* This provision, which does not appear in all health standards, requires the identification of certain areas in the plant where the toxic substance is present and limits access to authorized persons in those areas (subsection (e) of vinyl chloride standard).

(f) *Methods of compliance.* This section defines the steps an employer must take to reach the PEL (subsection (b) of vinyl chloride standard). It has been established OSHA policy that feasible engineering controls are the primary method for reaching the PEL; respiratory protection may be used only if engineering controls are not feasible, or until they are instituted, or in emergencies. OSHA usually allows the employer to choose the particular engineering controls to be used in reaching the PEL; the only exception was the coke oven emissions standard, which specified the engineering controls that must be implemented.[88] The basis for OSHA policy on engineering controls, work practices, administrative controls, and respirators as methods of compliance was explained in the preamble to OSHA's final Carcinogens Policy.

Carcinogens Policy Final Rule

45 Fed. Reg. 5002, 5223–24 (1980)

***

3. *The Reasons for the Policy.* Engineering controls with appropriate work practices are unquestionably the best method for effective and reliable

[86]See Proposed Cotton Dust Standard, 48 FED. REG. 26,962, 26,970–71 (1983) (proposing to amend the standard to include an action level because it would encourage employers to lower exposures below the action level; because it would substantially increase the cost-effectiveness of the standard; and because action levels have been successful in the past).

[87]See, e.g., Cotton Dust, 29 C.F.R. §1910.1043(d) (1983) (requiring that measurements of airborne cotton dust be done by a "vertical elutriator cotton dust sampler" or by a "method of equivalent accuracy and precision" that meets certain specified requirements). See also Ethylene Oxide Proposed Rule, 48 FED. REG. 17,284 (1983) (posing the question whether there are "serious limitations as to the accuracy or precision of the available sampling techniques").

[88]41 FED. REG. 46,742 (1976).

control of employee exposures to cancer causing agents. Engineering controls can act on the source of the emission and eliminate or reduce employee exposure without reliance on the employee to take self-protective action. Engineering controls encompass material substitution, process or equipment redesign, process or equipment sealing, enclosure, or isolation, local exhaust ventilation, and employee isolation (e.g., a standby pulpit, as opposed to personal protective equipment). Once engineering controls are implemented, the employee is permanently protected, subject only, in some cases, to periodic preventive maintenance.

***

"Work practice controls" or "work practices" accomplish the same results as engineering controls, but rely upon employees to repeatedly perform certain activities in a specified manner so that airborne concentrations are eliminated or reduced. Work practices may involve simple instructions to employees to keep lids on containers, to clean up spills immediately, or to observe required hygiene practices. Good work practices are often required in conjunction with engineering controls: for example, where employees perform an operation under an exhaust hood, they must perform their work in such a way as to maximize the efficiency of the ventilation equipment.

Work practices do not include administrative controls. The term administrative controls refers mainly to employee rotation, so that for any one employee his exposure is lower or his work hours are reduced. Although OSHA's recently promulgated lead standard included administrative controls in its definition of work practices, the preamble to that standard (43 FR 52990) made it clear that since administrative controls, by definition, expose more employees to the contaminant, they are unacceptable when the contaminant is one for which "no effect" levels are unknown, such as carcinogens. *Ibid.*

Work practices also act on the source of the emission but rely upon employee behavior, which in turn relies upon supervision, motivation and education to make them effective. For this reason work practices are not as desirable a method as engineering controls. But because the two methods often must be employed together to make either one effective and because they are the only methods that act to eliminate or reduce the hazard at its source, they have been given equal status in the compliance priorities of the final cancer policy.

4. *The Record as a Whole Supports the Policy.* The rulemaking record in the "cancer policy" generated substantial support for the use of engineering and work practice controls as the preferred means of preventing employee exposure, both from government and industry witnesses. * * *

As has been pointed out in these documents, respirators are the least satisfactory means of control because of difficulties inherent in their design and use. Respirators are capable of providing good protection only if they are properly selected for the types and concentrations of airborne contaminants present, properly fitted and refitted to the employee, worn by the employee, and replaced when they have ceased to provide protection. While it is theoretically possible for all of these conditions to be met, it is more often the case that they are not. Consequently, the protection of employees by respirators is not very effective and is therefore permitted only in certain specified circumstances. For example, proper facial fit is essential, but due to variation in individual facial dimensions, and the limited range of facepiece configurations, such fit is difficult to achieve. Often the work involved is strenuous and the increased breathing resistance of the respirator reduces their acceptability to employees. Safety problems presented by respirators must also be considered. Respirators limit vision. Speech is also limited. Voice transmission through a respirator can be difficult, annoying and fatiguing. Movement of the jaw in speaking also causes leakage. Communication may make the difference between a safe, efficient operation, on the one hand, and confusion and panic, especially in difficult and dangerous jobs, on the other hand. Also skin irritation can result from wearing a respirator in hot, humid conditions and such irritation can cause considerable distress and disrupt work schedules. The extent of this problem is even greater where the toxic substance is a skin irritant and becomes trapped between the respirator facepiece and the skin.

It is clear therefore that respirators cannot be considered as the primary means of employee health protection.

In February 1983, OSHA issued an advance notice of proposed rulemaking, stating its intention to reexamine its policy giving priority to engineering controls as the primary means to meet exposure limits.[89] The issue arises largely because engineering controls are substantially more expensive than other methods of compliance.

(g) *Respirators*. This section applies where respirators are an acceptable means of compliance. It includes requirements for the selection and use of respirators. Respirators ordinarily must be approved by NIOSH and used and maintained in accordance with the OSHA respirator standard. 29 C.F.R. §1910.134 (1983). The vinyl chloride standard contains an additional provision making the wearing of respirators voluntary for one year for concentrations less than 25 ppm. Otherwise, employers must assure that employees wear respirators in accordance with the standard (subsection (g)).

(h) *Work practices*. Health standards frequently require employers to engage in certain work practices to limit employee exposure; these requirements apply without regard to whether the PEL is exceeded. Subsections (h) and (i) of the vinyl chloride standard contain specific requirements for hazardous operations and emergency situations.

(i) *Protective clothing*. These provisions require the employer to provide, and assure, the use of protective clothing, such as gloves, coveralls, and masks. In some standards, the employer is expressly required to pay for the protective clothing.

(j) *Housekeeping*. This section is designed to minimize accumulations of the toxic substance; because vinyl chloride operations are usually in a "closed system," no housekeeping provision was included.

(k) *Hygiene facilities*. These provisions regulate such matters as change rooms, showers, lavatories, and the consumption of food and beverages where the toxic substance is present.[90]

(l) *Signs and labels*. Signs are required to warn employees of the presence of toxic substances in certain plant areas. Labels must be

[89]47 FED. REG. 187 (1982). This notice is discussed in Chapter 10, Section F, in connection with cost effectiveness and the Supreme Court's *Cotton Dust* decision. See also Proposed Asbestos Standard, 49 FED. REG. 14,116, 14,124 (1984) (proposing to permit any feasible combination of engineering controls, work practices, and personal protective equipment to reduce exposures from the current PEL to the new PEL).

[90]The vinyl chloride standard does not contain specific provisions governing protective clothing, housekeeping, and hygiene facilities. For examples of these provisions, see the OSHA lead standard, issued in 1978, 29 C.F.R. §1910.1025(g)–(i) (1983).

placed on certain products to alert employees of their hazardous nature (subsection (l) of vinyl chloride standard). These provisions are required by §6(b)(7) of the Act.[91]

(m) *Medical surveillance*. In accordance with the requirements of §6(b)(7) of the Act,[92] health standards require that medical examinations be provided to employees at no cost. These examinations are for the purposes of identifying employees who are ill so that medical attention can be provided; identifying employees who are at increased risk from continued exposure to the substance, including those suffering from early and reversible stages of disease, so that proper precautions, including transfer from exposure, can be taken; and for research purposes to determine the causes of occupational illness (subsection (k) of the vinyl chloride standard).[93]

(n) *Employee training*. These provisions obligate employers to train employees in the recognition of the hazard involved, and in the understanding of the precautions to be taken and the provisions of the standard. Training material is frequently prepared by OSHA and either attached as an appendix to the standard, as in the case of the lead standard, or otherwise distributed (subsection (j) of the vinyl chloride standard).

(o) *Record keeping*. Employers are required to keep pertinent records of their activities under the standard and to make them available to OSHA and to employees and their representatives. The most important of these records are of medical surveillance and exposure monitoring. OSHA recently promulgated a generic rule requiring access to certain employer exposure monitoring and medical records (subsection (m) of the vinyl chloride standard).[94]

(p) *Effective date*. The vinyl chloride standard was published October 4, 1974, and was scheduled to become effective January 1, 1975. This effective date was postponed by the court of appeals, because of the litigation, until 60 days after the date of its order affirming the standard.[95] In addition to the effective date for the overall standard, which is normally soon after the standard is published, OSHA often has later effective dates for specific provisions within the standard. This is particularly applicable to the requirements for engineering controls, which are often burdensome to employers. For example, the lead standard contained a schedule for various industries to reach the permissible exposure limits by means of engineering controls; the time frames vary from the effective date of the entire standard to 10 years for the primary lead production industry.[96]

[91]29 U.S.C. §655(b)(7).
[92]*Id.*
[93]For a fuller discussion of the evolution of OSHA standards' provisions on medical surveillance, see Chapter 5.
[94]This rule is discussed in Chapter 18, Section F.
[95]*Society of the Plastics Indus. v. OSHA*, 509 F.2d 1301, 1311 (2d Cir. 1975).
[96]29 C.F.R. §1910.1025(e)(1) (1983). OSHA has stayed the standard's requirements that the primary and secondary smelting industries and the battery industry submit compliance plans for the implementation of engineering controls. 47 FED. REG. 54,433 (1982). Although the Act, §6(b)(4), 29 U.S.C. §655(b)(4), provides that a standard's effective date may be delayed no longer than 90 days, the legislative history indicates that this provision was not intended to bar

In summarizing these standard elements, OSHA in the Carcinogens Policy stated that some provisions relate directly to control of the exposure level. These include provisions relating to the PEL, monitoring, and methods of compliance. Other provisions, such as medical surveillance, and signs and labels, the Agency said, are designed to give employees information they need to enable them to protect themselves and to "make decisions about their jobs and their futures."[97]

One of the most striking developments in preambles to health standards is their length. The preamble to the asbestos standard encompassed two *Federal Register* pages. The main portion of the lead preamble was 55 *Federal Register* pages, with a later published elaboration called "Attachments to Preamble," which was 155 *Federal Register* pages. Not all preambles are as long as the lead document, but all are considerably longer than the asbestos *Federal Register* notice. This development reflects the increased scope and complexity of the proceedings and the amount of evidence received. It also is in response to the insistence of reviewing courts that OSHA state the reasons for its determinations, instead of OSHA's attorneys doing so subsequently in their court briefs.[98]

the possibility that "a particular standard may provide for graduated requirements to take effect progressively on specific dates even though the intervals between the effective dates of such graduated requirements may exceed 90 days." 116 CONG. REC. 42,206 (1970) (statement of Rep. Steiger).

[97]45 FED. REG. 5002, 5218 (1980).

[98]See discussion in Chapter 8, Section F.

4

Emergency Temporary Standards

Under §6(c), OSHA is authorized to issue an emergency temporary standard (ETS), without engaging in rulemaking, if it determines that "employees are exposed to grave danger from exposure to substances or agents determined to be toxic or physically harmful or from new hazards" and that the ETS is "necessary" to protect employees from the danger.[1] The Act provides that the standard takes immediate effect and remains effective until superseded by a "permanent" standard issued under §6(b); the Act further requires that the final permanent standard be issued no later than six months after publication of the ETS.

OSHA has issued nine ETSs since the Act became effective. The emergency standards covering vinyl chloride, dibromo-3-chloropropane (DBCP), and the first ETS on asbestos were not challenged in court. They continued in effect for the entire six-month period and were ultimately superseded by permanent standards. The ETS for acrylonitrile was challenged, but the requested stay was denied by the Court of Appeals for the Sixth Circuit,[2] and a permanent standard was later issued after rulemaking. Four ETSs—covering benzene, diving, pesticides, and 14 carcinogens—were stayed or vacated by the courts of appeals; permanent standards on diving and 13 of the 14 carcinogens are presently in effect. In March 1984, the second asbestos ETS was indefinitely stayed by the Court of Appeals for the Fifth Circuit.

The most difficult issue surrounding the issuance of ETSs is whether a "grave danger" exists so as to justify OSHA's acting, without public participation, to impose legally enforceable obligations on employers.[3] In the case of OSHA's ETS on pesticides, the Court of

[1] 29 U.S.C. §655(c) (1976).

[2] *Vistron v. OSHA*, 6 OSHC 1483 (6th Cir. 1978).

[3] OSHA, of course, has discretion to solicit public comment or to hold an informal public hearing before issuing an ETS. OSHA did so before issuing the vinyl chloride ETS. See discussion in Chapter 3, Section B.

Appeals for the Fifth Circuit strongly disagreed with OSHA's conclusions that a grave danger existed and vacated OSHA's ETS.

The initial request for an ETS on pesticides was made by the Migrant Legal Action Program and other public interest groups representing farmworkers. The petition, filed in 1972, requested several actions by OSHA to provide effective job protection for these farmworkers.[4]

Petition of Migrant Legal Action Program Inc., et al. for Promulgation of Standards Relating to Pesticides

***

Although farmworkers are in need of many protections which could be provided by the Occupational Safety and Health Administration, their most pressing need is for protection against pesticides. We feel that the danger of pesticides to farmworkers is such that an emergency temporary standard is required. Section 6(c)(1) of the Act requires temporary emergency standards to be passed if employees are exposed to grave danger from exposure to substances or agents determined to be toxic or physically harmful or from new hazards, and emergency standards are necessary to protect employees from such danger. 29 U.S.C. 655(c)(1). Both the legislative history of the Act and the facts which have been publicized about the dangers of pesticides to farmworkers show that the danger from pesticides is the type of danger contemplated by the framers of the Act as requiring a temporary standard.

There has been ample evidence of the toxic and physically harmful nature of pesticides. The correlation between the symptoms of pesticide poisoning and the use of pesticides in working and living areas has been so amply documented that it can no longer be doubted. Scientists have found that the levels of chlorinated hydrocarbon pesticides in the human tissues of those exposed to these pesticides were significantly higher than in those persons not exposed to these pesticides and that the cholinesterase levels in those persons exposed to organophosphate pesticides were significantly lower than in those persons not exposed to these pesticides. Many other studies have documented the correlation between actual contact with pesticides and symptoms of pesticide poisoning among farmworkers.

It has not only been found that contact with pesticides is physically harmful but it has also been found that the extent of the harm, especially to farmworkers, is very great. The Senate Labor Committee Report on the Act stated "that an estimated 800 persons are killed each year as a result of improper use of such pesticides, and another 80,000 injured." However, there are many more cases of pesticide poisoning which go unreported each year. As Dr. Bernard C. Conley, Secretary of the Committee on Pesticides and the Committee on Toxicology of the American Medical Association has stated, "The incidence of injury from pesticides has been extremely difficult to determine. Statistical data on nonfatal and fatal poisoning have been scanty and misleading, and this has contributed to the misconception that pesticide injuries are infrequent or rare." Another expert in this area, Victoria Trasko of the Occupational Health Program stated that "the occupational poisonings identified * * * in 15 different states" were "definitely an understatement of the real occurrence of poisonings in these states."

There are several logical explanations for this dearth of reliable data on pesticide poisonings. The first is that many farmworkers do not have access

[4]Petition of Migrant Legal Action Program Inc., *et al.*, for Promulgation of Standards Relating to Pesticides 13–16 (1972); a copy of petition is available in author's files.

to medical care and as a result, there is no record of their illnesses. Farmworkers access to medical care is limited by their lack of funds, exclusion from federal programs such as medicare and medicaid, and the ineffectiveness of the Migrant Health Program. A second explanation lies in the fact that "farmworkers are generally unaware of the dangers inherent in their jobs due to pesticides and the nature and extent of how pesticides effect their physical and mental health." The following statement from Dr. Milby, Chief of the Bureau of Occupational Health and Environmental Epidemiology, California Department of Public Health, illustrates why farmworkers are often unaware that they have been poisoned.

> Common symptoms (of pesticide poisoning) include nausea, vomiting, malaise, sweating, diarrhea, and weakness. Since these are the same symptoms which characterize such everyday illnesses as "flu," the individual frequently does not suspect that he is in fact experiencing the effect of cumulative exposure to organophosphate pesticide residues. He is likely to stay home for a day or two, and the overt symptoms usually disappear, since cholinesterase loss of the magnitude we are discussing here is reversed relatively rapidly.
>
> Our research leads us to suspect, furthermore, that there is probably an even more common form of "adverse effect"—one which is even slower to develop and even subtler in its manifestations. The individual may not feel nauseated, dizzy, etc.; he may not lose any time from work at all. But he may have a little more difficulty getting to sleep at night. His appetite may be a little less hearty than it was. And there may be a gradual impairment of his hand-eye coordination and other neuromuscular functions, so that he is able to pick 10 percent fewer oranges or peaches, say, than formerly. The affected individual may not notice the difference at all; or, if he notices, will most likely shrug and ascribe it to "getting older."

This statement also illustrates why many physicians, unfamiliar with pesticide poisonings may not diagnose the true cause of the symptoms presented. Finally, the pesticide poisoning centers, whose function is to record the incidence of pesticide poisonings are usually inaccessible to the farmworker.

The second condition stated in the section of the Act describing temporary emergency standards refers to the grave danger emanating from new hazards. Again, the Senate Labor Committee Report illustrates that pesticides were intended to come under the emergency standards. The Report describes pesticides as one of "the technological advances and new processes in American industry that have brought numerous new hazards to the work place."

The most serious new hazard will be presented by the substitution of methyl parathion and other organophosphates for DDT which can no longer be used as of December 31, 1972. Although the organophosphates are less persistent in the environment, they present a greater hazard to workers coming into contact with them. The Environmental Protection Agency, in cancelling the use of DDT, has recognized this new hazard and has accordingly set aside a period of time in which workers are to be educated to use the organophosphates. However, we feel that more protection than that which can be provided through education, must be provided for the workers who will increasingly come into contact with these dangerous substances.

The third condition—that such an emergency standard is necessary to protect employees from such danger—is also referred to in the Senate Labor Committee Report. It states that "despite the unmistakable danger that these substances present, no effective controls presently exist over their safe use and no effective protections exist against toxic exposure of farmworkers or others in the rural populace." The Federal Insecticide, Fungicide and Rodenticide Act (FIFRA) of 1947 is the only legislation which deals with pesticides. However, the protections in the registration process created by this act are for the consumer and not for the farm worker. Although tolerance levels for the purposes of consuming foods exposed to pesticides have been developed, no such tolerance levels have been developed for farmworker exposure to pesticides. Although FIFRA will soon be replaced by new legislation, the new legislation does not yet require and is not likely to require registration criteria for safe worker re-entry times, compensation for medical care, posting of treated property and other protections for farmworkers.

***

OSHA issued an ETS on pesticides on May 1, 1973, a month after the Farmworkers brought an action in the U.S. District Court of the District of Columbia to compel the issuance of an ETS.[5] The ETS applied to 21 organophosphate pesticides, and imposed requirements for warning employees of pesticide hazards, field reentry times, sanitation and medical services, and first-aid. The preamble to the standard set forth in conclusory terms the Agency's basis for its finding of "grave danger."

Emergency Temporary Standard for Exposure to Organophosphorous Pesticides

38 Fed. Reg. 10,715, 10,715–16 (1973)

***

During the last several years, general awareness of, and concern with, the hazards presented by the use of pesticides have been increasingly felt and expressed. The legislative history of the Williams-Steiger Occupational Safety and Health Act of 1970 itself demonstrates this awareness. The pertinent Senate report states the following:

> Pesticides, herbicides and fungicides used in the agricultural industry have increasingly become recognized as a particular source of hazard to large numbers of farmers and farmworkers. One of the major classifications of agricultural chemicals—the organophosphates—has a chemical similarity to commonly used agents of chemical and biological warfare, and exposure, depending on degree, causes headache, fever, nausea, convulsions, long term psychological effects, or death. Another group—the chlorinated hydrocarbons—are stored in fatty tissues of the body, and have been identified as causing mutations, sterilization, and death.
>
> While the full extent of the effect that such chemicals have had upon those working in agriculture is totally unknown, an official of the Department of Health, Education, and Welfare stated, during hearings of the Migratory Labor Subcommittee, that an estimated 800 persons are killed each year as a result of improper use of such pesticides, and another 80,000 injured. Despite the unmistakable danger that these

---

[5] *Thomas v. Brennan,* No. 73-502 (D.D.C. 1973) (dismissed voluntarily after the ETS on pesticides was issued).

substances present, no effective controls presently exist over their safe use and no effective protections against toxic exposure of farmworkers or others in the rural populace.
(S. REP. No. 91-1282, 91st Cong., 2d Sess., 3–4 (1970).)

For some of the background of this statement, see also hearings on S. 2193 and S. 2788 before the Subcommittee on Labor and Public Welfare, 1st and 2d Sess. pt. 1, at 736, 737 (1970). In the fall of 1971, the Council on Environmental Quality was asked to study the feasibility of regulating, under the Occupational Safety and Health Act of 1970, the exposure of agricultural employees to organophosphorous pesticides. In his environmental message of February 8, 1972, to the Congress, the President stated that measures to protect agricultural workers from adverse exposure to pesticides were essential to a sound national pesticide policy and that he was directing the Departments of Labor and of Health, Education, and Welfare to develop standards under the Occupational Safety and Health Act of 1970 to protect agricultural workers from pesticide poisoning. By letter dated February 1972, the President requested the Secretary of Labor to develop such standards in cooperation with the Department of Health, Education, and Welfare.

In response to the congressional purpose manifested in the legislative history of the act, and to the directive of the President, plans were immediately developed to investigate and evaluate the dangers connected with the use of pesticides. In June 1972, the Standards Advisory Committee on Agriculture was appointed for the purpose, among others, to submit recommendations regarding a standard on pesticides. The committee appointed a special subcommittee to study exclusively the hazards of pesticide poisoning. The subcommittee and the full committee had several meetings, open to the public, at which the hazards of pesticides were discussed. At these meetings members of the public and experts from Federal agencies, such as the Department of Agriculture and the Environmental Protection Agency, participated.

In September 1972, the Migrant Legal Action Program, Inc., et al., filed a petition with the Occupational Safety and Health Administration, setting forth data and arguments for the promulgation of an emergency standard on pesticides. During this time, available literature on pesticide poisoning was reviewed by the staff of the Occupational Safety and Health Administration.

The information presented to us, and obtained by us, makes it clear that the pesticides listed in the standard set out below are highly harmful. The National Institute for Occupational Safety and Health confirms the inherent toxicity of the organophosphorous insecticides and believes that protection for workers exposed to these agricultural chemicals is greatly needed and that the concept of reentry intervals provides an effective means to furnish such protection. The Federal Task Group on Occupational Exposures to Pesticides has identified apples, peaches, grapes, tobacco, oranges, lemons, and grapefruits as crops which pose the greatest hazards to field workers' health, on the basis of field worker contact in cultivation and harvesting, and organophosphorous compounds as pesticides needing priority attention, because they can be cumulative and historically have caused serious health problems.

Accordingly, it is hereby found that: (1) The pesticides listed in the standard below are highly toxic; (2) that premature exposure to them would pose a grave danger; (3) that farm workers have in the past growing seasons, are now, and are expected to be in the near future exposed to these pesticides; and (4) that the standard set out below, based on the recommendations of the Standards Advisory Committee on Agriculture and suggested field reentry safety intervals from the Environmental Protection Agency, is necessary to regulate such exposure so as to protect the workers from the danger.

***

Several days after the ETS was promulgated, Dr. F. S. Arant, Chairman of the Subcommittee on Pesticides of OSHA's Agriculture Advisory Committee, wrote to Assistant Secretary John H. Stender, strongly disagreeing with the Agency finding on the need for the emergency temporary standard and resigning from his position. His letter, dated May 9, 1973, stated in part as follows:

> I was shocked to learn of the Emergency Temporary Standard for Exposure to Organophosphorous Pesticides, published in the Federal Register, May 1, 1973. The Advisory Subcommittee on Pesticides worked many hours and conferred with leading scientists throughout the United States during the past 9 months assisting OSHA in developing a realistic reentry standard to protect employees from hazards from certain organophosphorous insecticides. We considered many controversial aspects of the problems associated with exposure to pesticides and encountered some that were so controversial that agreement was difficult to reach. However, there was no disagreement in the Subcommittee regarding the absence of any need for an emergency standard. Its enactment has created unnecessary problems.
>
> In the preamble to the emergency standard, an official of the Department of Health, Education and Welfare was quoted as estimating 800 persons are killed and another 80,000 injured each year with pesticides. We would like to see the proof of this "estimate." A diligent search by the Subcommittee revealed that relatively few deaths in the United States could be attributed to pesticides and those that did occur resulted primarily from persons drinking the chemical through accident or suicide. The number of pesticide related deaths was much smaller than for household chemicals and non-prescription drugs. The emergency standard deals with employees reentering treated orchards or fields. The subcommittee was unable to find a single authentic record of a fatality resulting from a person entering or working a field treated with a pesticide.
>
> Workmen have become ill from exposure to certain organophosphorous insecticides, principally parathion on citrus and peaches in arid areas of one state (California). Many of these workers had been in contact with treated foliage in harvest operations for long periods of time, some six or seven days a week for 10 to 14 weeks or longer. The state took prompt action and the situation is now under control. A survey of pesticide safety specialists at all the Land-Grant universities revealed that no problems had arisen from workmen entering pesticide treated fields in a majority of the states and only minor problems in others.
>
> The Subcommittee on Pesticides was unanimous in its recommendation that no emergency existed and that there was no justification for emergency standards. This recommendation was made verbally at several Subcommittee meetings and adopted unanimously as a written resolution at the Iowa City meetings on December 19 and 20, 1972. The

resolution was also adopted unanimously by the entire Advisory Committee on Occupational Safety and Health in Agriculture at the same meeting. Despite the recommendations of these advisory committees, the emergency temporary standard was issued and this is a standard without recourse. No hearings are permitted.[6]

In response to strong public opposition, including serious criticism from members of Congress, OSHA first suspended the effective date and then amended the standard, narrowing its scope and limiting its requirements.[7] The Court of Appeals for the Fifth Circuit stayed the ETS and on January 9, 1974, issued its decision vacating the ETS in its entirety. The court found that no grave danger existed justifying the issuance of an emergency standard.

Florida Peach Growers Association v. Department of Labor

489 F.2d 120 (5th Cir. 1974)

RONEY, Circuit Judge:
***

THE MERITS

We find no substantial evidence in the record considered as a whole to support the determination by the Secretary that emergency temporary standards were necessary within the demands of section 6(c). There is an abundance of evidence that emergency standards are not necessary. The investigative groups convened by the Government to study the problem of occupational exposure to pesticides—the Pesticides Subcommittee, the Advisory Committee, and the Task Group*—all firmly concluded that no emergency existed and that there was no justification for use of an emergency temporary standard. Although these findings by his own investigators do not preclude the Secretary from issuing an emergency standard, they indicate the strength of the evidence contrary to his determination.

Extraordinary power is delivered to the Secretary under the emergency provisions of the Occupational Safety and Health Act. That power should be delicately exercised, and only in those emergency situations which require

*[Author's note—The Task Group was established in 1971, as a federal interagency task force by the Council on Environmental Quality, and directed to evaluate the occupational health dangers of pesticides. The chronology of events during 1973–74 relating to OSHA's attempt to regulate pesticide reentry times is set forth in COMPTROLLER GENERAL OF THE UNITED STATES, REPORT TO THE CONGRESS, MWD-75-55, EMERGENCY TEMPORARY STANDARDS ON ORGANOPHOSPHOROUS PESTICIDES app. at 2–7 (1975).]

[6]Letter from Dr. F.S. Arant, Chairman of the Subcomm. on Pesticides of OSHA's Agriculture Advisory Comm. to John H. Stender, Assistant Secretary of OSHA (May 9, 1973) (available in author's files).

[7]38 FED. REG. 15,729 (1973) (suspending effective date); 38 FED. REG. 17,214 (1973) (amending standard). The Migrant Legal Action Program again brought suit challenging OSHA's suspension of the ETS without an opportunity for notice and comment. *Villaneuva v. Brennan,* No. 73-1225 (D.D.C. 1973). The district court denied the requested injunction and ultimately the Court of Appeals for the Fifth Circuit upheld OSHA's authority to amend an ETS without rulemaking. *Florida Peach Growers Ass'n v. Department of Labor,* 489 F.2d 120, 127 (5th Cir. 1974).

it.[16] The promulgation of any standard will depend upon a balance between the protection afforded by the requirement and the effect upon economic and market conditions in the industry. As articulated by the Chairman of the Subcommittee on Pesticides, who resigned in "shock" upon finding that the recommended standards were issued on emergency basis: "It is essential that employees be protected against exposure to highly toxic materials, but this should be done without eliminating the agriculture enterprise and the associated jobs."

The need for a serious emergency upon which to ground temporary standards is reflected in the words of the Act which require the Secretary to determine "(A) that employees are exposed to grave danger from exposure to substances or agents determined to be toxic or physically harmful or from new hazards, and (B) that such emergency standard is necessary to protect employees from such danger." The Act requires determination of danger *from* exposure to harmful substances, not just a danger of exposure; and, not exposure to just a danger, but to a *grave* danger; and, not the necessity of just a temporary standard, but that an emergency standard is necessary.

The reasons published by the Secretary with the standards do not themselves evidence a factual need for emergency standards. The record supports the need for some standards, but not emergency standards. The Secretary's findings were that:

> (1) The pesticides listed in the standard below are highly toxic; (2) that premature exposure to them would pose a grave danger; (3) that farm workers have in the past growing seasons, are now, and are expected to be in the near future exposed to these pesticides; and (4) that the standard set out below, based on the recommendations of the Standards Advisory Committee on Agriculture and suggested field reentry safety intervals from the Environmental Protection Agency, is necessary to regulate such exposure so as to protect the workers from the danger.

38 FED. REG. 10716 (1973).

The able Government brief indicates the existence of an emergency because of the ban on the use of DDT, which in turn increases the usage of

[16]That Congress intended a carefully restricted use of the emergency temporary standard is reflected in the legislative history of both the Senate and the House. As pointed out by the Senate Labor and Public Welfare Committee:

> *Emergency Standards.*—Because of the obvious need for quick response to new health and safety findings, section 6(c) mandates the Secretary to promulgate temporary emergency standards if he finds that such a standard is needed to protect employees who are being exposed to grave dangers from potentially toxic materials or harmful physical agents, or from new hazards for which no applicable standard has been promulgated. Upon publication of such an emergency temporary standard, the Secretary must begin a regular standard-setting procedure for such hazard, which proceeding must be completed within six months.
>
> [S. REP. No. 1282, 91st Cong., 2d Sess. 7 (1970)] 1970 U.S. CODE CONG. & ADMIN. NEWS at 5184.

In recommending the Senate-House Conference Report for final adoption, Congressman Steiger of Wisconsin, a House Member of the Conference Committee reported:

> TEMPORARY EMERGENCY STANDARDS
>
> Section 6(c) provides for the expedited procedure for promulgating emergency standards in accordance with the desires of this House. It is intended that this procedure not be utilized to circumvent the regular standard-setting procedures. It should be used only for those limited situations where such emergency standards are necessary because employees are exposed to grave danger from substances or agents determined to be toxic or physically harmful agents, or new hazards. Section 6(c)(1)(B) makes it clear that it is also necessary that the Secretary determine that such a standard is necessary to protect employees from such danger. Thus, where the state of the art is incomplete or in some way lacking so that a determination cannot be made as to whether such a standard will protect employees, in such instances the Secretary would utilize the regular standard-setting procedures as provided in subsection 6(b) and not the emergency standard-setting authority.
>
> [116 CONG. REC. 42,206 (1970).]

the pesticides listed in the standard, and because of the approach of the 1973 growing season with seasonal peaks of exposure in the months of June, July, August, September, and October. The published reasons and the record give us no indication that the Secretary made a before-the-fact finding that an emergency situation existed on this basis. The Growers point out that the use of DDT on the crops covered by the standard was banned before the Secretary ever initiated this proceeding in 1971, and that ample time existed for promulgation of a standard under section 6(b) to become effective for the 1973 growing season. The listed pesticides have been in long time use.

The Secretary's heavy reliance on the toxicity *per se* of the organophosphorus pesticides to support the allegedly grave danger to agricultural employees through residue exposure is unjustified. In the published standards the Secretary seems to rely on the Senate report "that an estimated 800 persons are killed each year as a result of improper use of pesticides, and another 80,000 injured." This report refers to *improper* use, and encompasses all pesticides.

The Growers concede the high toxicity of these pesticides in the laboratory,[17] but assert that it has little bearing on the hazard to which the emergency standard is addressed, the occupational exposure of agricultural workers to organophosphorus residues on foliage. The Secretary points to statistics on deaths and injuries caused by pesticides. The statistics cited include deaths of all kinds, accidental ingestion by children, industrial accidents, accidents involving spraymen and their helpers, and suicides. The Growers argue that there is a substantial and significant difference in the toxicity *per se* of an organophosphorous pesticide and that of any of its residues that may remain on sprayed foliage, because the organophosphates start to decompose rapidly immediately upon application. They, therefore, challenge the Secretary's citation of sources cataloging the symptoms of severe organophosphate poisoning, because they describe the afflictions of person[s] having primary contact with pure or active organophosphates. They refer us instead to record documents detailing the generally mild nature of the relatively few cases of illness reported by crop workers exposed solely to residues. The studies of the occasional outbreaks of organophosphate poisoning among farmworkers exposed to residues indicate the nature and degree of danger. From time to time a group of workers will experience nausea, excessive salivation and perspiration, blurred vision, abdominal cramps, vomiting, and diarrhea, in approximately that sequence. There is substantial evidence that farmworkers occupationally exposed to organophosphate residues on foliage may experience headache, fatigue, and vertigo. These are not grave illnesses, however, and do not support a determination of a grave danger. A relatively small number of workers experience these difficulties, and it has been going on during the last several years thus failing to qualify for emergency measures. No deaths have been conclusively attributed to exposure to residues. The Subcommittee on Pesticides reported it was unable to find a single authentic record of a fatality resulting from a person entering or working in a field treated with a pesticide. The Secretary has pointed to no evidence to the contrary.

Statistics of "acute poisoning" of agricultural workers fail to distinguish between farmworkers exposed only to residues and persons responsible for applying the pesticides. Further, the statistics do not meaningfully break down injuries by type of pesticides. Thus, the Secretary would support his finding of danger with statistics of deaths and injuries due to causes

[17]Toxicity is the inherent capacity of a substance to produce injury or death. Hazard is a function of toxicity and exposure and is the probability that injury will result from the use of a substance in a given formulation, quantity, or manner. If taken in sufficient quantities, even water will cause death.

which the standard he promulgated will not correct. The ultimate picture in this record is one of only a few farmworkers made ill by organophosphate residues relative to the mass of agricultural workers in contact with treated foliage.

The Secretary seeks to justify his reliance on these statistics with evidence that agricultural injury statistics seriously underreport the incidence. He argues that the statistics he used are the best available. Granting that it is so, little is proven. Some of the very reasons why the statistics understate the true extent of the problem reinforce the Growers' reading of the evidence. That farmworkers do not report some illness because the symptoms are relatively minor merely supports the conclusion that the situation is not sufficiently grave to warrant use of subsection (c) procedure. Similarly, the suggestion that physicians may fail to diagnose pesticide poisoning as such because it mimics the flu is not persuasive of any grave danger.

We reject any suggestion that deaths must occur before health and safety standards may be adopted. Nevertheless, the danger of incurable, permanent, or fatal consequences to workers, as opposed to easily curable and fleeting effects on their health, becomes important in the consideration of the necessity for emergency measures to meet a grave danger.

In sum, considering the record as a whole, the Secretary has not shown by substantial evidence that agricultural workers are exposed to a grave danger from exposure to organophosphorous pesticide residues on treated plants that must necessarily be protected by an emergency temporary standard. We, of course, do not intimate any opinion as to the feasibility of the published standard to furnish adequate protection, were an emergency shown, nor does our determination in any way reflect consideration of the propriety of a permanent standard promulgated under subsection (b) of the Act.

All petitions for review, except for that in No. 73-2690, are granted and the Emergency Temporary Standard is determined to be invalid and is vacated.

OSHA never issued a permanent standard regulating field reentry times for pesticides. A proposed permanent standard was published,[8] and public hearings were held under §6(b). Sharp disagreement arose, however, between the Environmental Protection Agency and OSHA on whether OSHA or EPA, under the Federal Insecticide, Fungicide, and Rodenticide Act (FIFRA),[9] had authority to regulate pesticide reentry times. OSHA eventually conceded EPA's primary authority and EPA issued worker protection standards for agricultural pesticides.[10] The Farmworkers again brought suit against OSHA to compel it to regulate pesticide reentry times. The Court of Appeals for the District of Columbia Circuit held that OSHA was preempted under §4(b)(1) of the Act[11] from regulating pesticides because of EPA's exercise of authority in issuing a standard.[12]

[8]38 FED. REG. 17,245 (1973).
[9]7 U.S.C. §§136–136y (1982).
[10]39 FED. REG. 16,888 (1974).
[11]29 U.S.C. §653(b)(1).
[12]*Organized Migrants in Community Action v. Brennan*, 520 F.2d 1161 (D.C. Cir. 1975). For a discussion of §4(b)(1) of the Act, 29 U.S.C. §653(b)(1), see Chapter 15, Section B. OSHA's

The Fifth Circuit Court of Appeals' decision in *Florida Peach Growers* was based primarily on its conclusion that the record failed to show that agricultural workers were exposed to the danger of "incurable, permanent, or fatal consequences * * * as opposed to easily curable and fleeting effects on their health."[13] Two other considerations, however, appear also to have played a part in the court's decision. OSHA had argued that an "emergency" existed in view of the government ban on DDT, increasing the use of the regulated pesticides, and the approach of the growing season with "seasonal peaks of exposure" between June and October. The court rejected this argument on the grounds that DDT was banned before 1971 and there would have been "ample time" for OSHA to proceed with rulemaking to regulate the pesticides before the 1973 growing season.[14] The court was apparently also strongly influenced by the conclusions of the subcommittee on pesticides of the OSHA advisory committee and the interagency task group that there was no need for an emergency standard.[15] It is likely that the relatively short OSHA preamble articulating the reasons for its finding of grave danger was also significant. To be sure, the court of appeals did not vacate the standard on the grounds that the statement of reasons was inadequate under §6(e).[16] However, the preamble was conclusory, relying largely on other conclusory statements in the Senate Labor Committee Report of 1970. OSHA's difficulty in sustaining its burden in court was al-

unsuccessful attempt to regulate field reentry times for pesticide exposure has been discussed in several articles; e.g., Weissman, *Pesticides, Farmworkers and OSHA, Res Ipsa Loquitur*, 27 GEO. REV. L. & PUB. INTEREST 1 (1975); Note, *Interpreting OSHA's Preemption Clause: Farmworkers as a Case Study,* 128 U. PA. L. REV. 1509 (1980); Comment, *Farmworkers in Jeopardy: OSHA, EPA and the Pesticide Hazard,* 5 ECOLOGY L.Q. 69 (1975).

In August 1981, the Public Citizen Health Research Group and the American Federation of State, County, and Municipal Employees (AFSCME) petitioned OSHA to issue an ETS for another pesticide, ethylene oxide, lowering the PEL from 50 parts per million to 1 ppm. This petition was based on recently available evidence, both in human and animals, of the carcinogenicity of ethylene oxide at low levels. OSHA rejected the request for an ETS and published an advance notice of proposed rulemaking on ethylene oxide, 47 FED. REG. 3566 (1982). OSHA concluded that the existing record did not support the required finding of "grave danger" to employees to warrant the issuance of an ETS. The petitioners filed suit in the U.S. District Court for the District of Columbia, asking the court to require OSHA to issue the ETS. Among other arguments, in response, OSHA claimed that its authority to regulate ethylene oxide was preempted under §4(b)(1) by EPA's authority to regulate pesticides under the Federal Insecticide, Fungicide, and Rodenticide Act (FIFRA), 7 U.S.C. §§136–136y (1982). On January 5, 1983, the district court ordered OSHA to issue an ETS on ethylene oxide within 20 days. The court rejected OSHA's preemption argument and concluded that the record before the Agency presented a "solid and certain foundation showing that workers are subjected to grave health dangers from exposure to ethylene oxide at levels within the currently permissible range." *Public Citizen Health Research Group v. Auchter,* 554 F. Supp. 242, 251 (D.D.C. 1983). The Court of Appeals for the District of Columbia Circuit on March 15, 1983 reversed the District Court order that OSHA issue an emergency standard. *Public Citizen Health Research Group v. Auchter,* 702 F.2d 1150 (D.C. Cir. 1983). However, in light of the "significant risk of grave danger to human life," the court of appeals ordered OSHA to issue a proposal on ethylene oxide within 30 days and said that it "expects" promulgation of a final rule within a year's time. *Id.* at 1159. The court of appeals agreed with the district court that OSHA was not disabled from regulating ethylene oxide because of the exercise of EPA authority. 702 F.2d at 1156. The court of appeals decision is reprinted and discussed in Chapter 6, Section C.

[13]489 F.2d at 132.

[14]*Id.* at 130–31.

[15]*Id.* at 129–31.

[16]29 U.S.C. §655(e). But cf. *Dry Color Mfrs. Ass'n v. Department of Labor,* 486 F.2d 98 (3d Cir. 1973), reprinted in Chapter 8, Section F.

most certainly increased by its failure to articulate fully, as it later did in its "able" brief, the factual basis for the grave danger findings.

In vacating the pesticide ETS, the Fifth Circuit Court of Appeals strongly emphasized the "extraordinary power" that was given to OSHA under §6(c)[17] to promulgate standards without rulemaking, insisting that the power be "delicately exercised" only in "emergency" situations that require it.[18] The same theme was sounded by the Court of Appeals for the Third Circuit in *Dry Color Manufacturers Association v. Department of Labor.*[19] The court there set aside OSHA's ETS on ethyleneimine (EI) and 3,3′-dichlorobenzedine (DCB) on the grounds that the Agency had failed adequately to state its reasons for issuing the emergency standard.[20] While not basing its decision on the grounds that the standard did not meet the substantive criteria of §6(c), the court expressed "doubts" as to whether there was substantial evidence supporting OSHA's findings that EI and DCB are carcinogenic and, therefore, that a grave danger existed. The court said that "while the Act does not require an absolute certainty as to the deleterious effect of a substance on man, an emergency temporary standard must be supported by evidence that shows more than some possibility that a substance may cause cancer in man."[21] After quoting the excerpt from the Report of the Senate Committee on Labor and Public Welfare on ETSs[22] referring to the "obvious need for quick response to new health and safety findings," the court in *Dry Color* went on to say:

> This language, however, should not be read to mean that a showing of mere speculative possibility that a substance is harmful to man is sufficient to call into effect the summary procedure of subsection 6(c). It is clear from the Act that Congress considered that the ordinary process of rulemaking would be that provided for in subsection 6(b), dealing with permanent standards; emergency temporary standards should be considered an unusual response to exceptional circumstances. The courts should not permit temporary emergency standards to be used as a technique for avoiding the procedural safeguards of public comment and hearing required by subsection 6(b). Especially where the effects of a substance are in sharp dispute, the promulgation of standards under subsection 6(b) is preferable since the procedure for permanent standards is specifically designed to bring out the relevant facts.[23]

Thus, the Third Circuit Court of Appeals emphasized that emergency standards are an "unusual response to exceptional circumstances" and should not be used to circumvent the procedural safeguards of §6(b) rulemaking.

---

[17]29 U.S.C. §655(c).

[18]489 F.2d at 129.

[19]486 F.2d 98, 104 (3d Cir. 1973). The substantive issues in this case are discussed in Chapter 6, Section A.

[20]*Id.* at 106.

[21]*Id.* at 104.

[22]486 F.2d at 104 n.9a. See also *Florida Peach Growers Ass'n v. Department of Labor,* 489 F.2d at 130 n.16 (citing same excerpt), reprinted *supra* in this chapter.

[23]486 F.2d at 104 n.9a.

OSHA's ETS on diving was stayed by the Court of Appeals for the Fifth Circuit in *Taylor Diving & Salvage Co. v. Department of Labor*[24], and was later withdrawn by the Agency.[25] The ETS on benzene was also stayed by the Court of Appeals for the Fifth Circuit and never went into effect.[26] In issuing subsequent emergency temporary standards, OSHA sought to take into account the courts' dissatisfaction with OSHA's grave danger findings in the *Florida Peach Growers* and *Dry Color* proceedings. In 1977, OSHA issued an ETS covering occupational exposure to dibromo-3-chloropropane (DBCP). The preamble to the standard contained an extensive discussion of the factual data supporting OSHA's finding that a grave danger existed to employees.

Emergency Temporary Standard for Exposure to 1,2-Dibromo-3-Chloropropane (DBCP)

42 Fed. Reg. 45,536, 45,538–39 (1977)

***

III. REASONS FOR ISSUANCE OF AN EMERGENCY TEMPORARY STANDARD

The Assistant Secretary finds that exposure to DBCP poses a grave danger to humans. Specifically, human exposure to DBCP can cause oligospermia and aspermia—commonly known as sterility. Moreover, accumulated animal studies of good design have established that DBCP is clearly a carcinogen in animals, and must therefore be viewed as posing a risk of cancer to humans.

The data on sterility in workers involved in the manufacture and formulation of DBCP have only recently been collected. In fact, investigations into this ominous loss of reproductive capacity in these men are continuing. However, current data conclusively and reliably show that the deleterious effect on the testes is significant even at extremely low levels of exposure to DBCP. The preliminary investigations, which principally involve testing for sperm counts, have been interpreted by OSHA to conclusively establish that DBCP causes sterility. This evidence of a new hazard to employees is further substantiated by earlier evidence that DBCP causes sterility in animals, even though the dosages were higher. The Assistant Secretary, in light of this new and convincing evidence, concludes that occupational exposure to DBCP causes sterility.

In addition, recent animal studies have demonstrated the carcinogenicity of DBCP. Olsen, et al. in 1973, Powers, et al. in 1975, and Weisburger, et al. in 1977, have produced squamous cell carcinomas of the stomach in a high percentage of the exposed population of animals. In light of this recent evidence and the overwhelming scientific support for the view that substances shown to cause cancer in animals must be treated as posing a carcinogenic risk to man, the Assistant Secretary must treat DBCP as posing a cancer risk in humans, especially since it has been shown to have detrimental biological effect in man. The best available scientific evidence indicates

[24] 537 F.2d 819 (5th Cir. 1976) (petitioner had good prospects of prevailing on the merits, and danger of irreparable harm from ETS existed).
[25] 41 FED. REG. 48,950 (1976).
[26] The ETS on benzene is discussed in Chapter 7, Section A.

that no safe level for exposure to a carcinogen, including DBCP, can be established at present. OSHA has considered this question of a safe level in previous rulemaking proceedings (see preambles to carcinogen standards (39 FR 3758), vinyl chloride standard (39 FR 35892), and coke oven emissions standard (41 FR 46742)) and has concluded that, as a prudent policy matter, in the absence of a demonstrated safe level or threshold for a particular carcinogen, it must be assumed that none exists. This view is consistent with a considerable body of scientific opinion holding that when dealing with a carcinogen, no safe level can be determined for any given population. For example, the National Cancer Institute's Ad Hoc Committee on the Evaluation of Low Levels of Environmental Chemical Carcinogens (1970) states:

> No level of exposure to a chemical carcinogen should be considered toxicologically insignificant for man. For carcinogenic agents, a "safe level for man" cannot be established by application of our present knowledge. (NCI, 1970, p. 1.)

And NIOSH has taken the position that in regulating cancer-causing substances "* * * it is not possible at present to determine a safe exposure level for carcinogens." (Rev. Arsenic Crit. Doc. 1975.)

In conclusion, the Assistant Secretary determines that exposure of employees to a known sterility-inducing and cancer-causing substance in the workplace environment is a "grave danger" within the meaning of section 6(c)(1)(A) of the Act.

That cancer and substances which cause cancer pose a grave danger within the meaning of section 6(c)(1)(A) needs little supportive discussion. The nature of the cancer hazard differs from many other types of toxicity. Employees exposed to carcinogens risk incurable, irreversible and even fatal consequences. No symptomatic evidence of the development of the cancer may be apparent to the employee during a long latency period of 10–30 years. A single exposure episode may be sufficient to cause cancer. These factors, which establish the grave danger posed by exposure to carcinogens, also lead inexorably to the conclusion that it is necessary to provide immediate protection for employees through the issuance of an emergency temporary standard, within the meaning of section 6(c)(1)(B) of the Act.

Substances which cause involuntary sterility must also be viewed as posing a grave danger within the meaning of section 6(c)(1)(A) of the Act. Sterility in man, while perhaps not physically painful, nor equivalent to the loss of life itself, is a serious impairment of a vital and necessary bodily function—the ability to reproduce. Furthermore, sterility can have devastating psychological effects.

The need for immediate regulation to limit exposure to DBCP is apparent from the above discussion. Both as a carcinogen and as a quick-acting sterilant (some employees who became sterile had been exposed for only a few months), DBCP can pose its life-threatening danger in a very brief period of exposure. Without this emergency temporary standard, employees would continue to be exposed to this threat during the period of time necessary to complete the normal rulemaking process. Issuance of this ETS is therefore necessary to protect employees from this grave danger.

***

In justifying the ETS for DBCP, the preamble emphasizes several key points:

(1) DBCP poses a "grave danger" to employees since it causes sterility in humans and, based on accumulated animal studies of "good design," it may be viewed as posing a risk of cancer to humans.

(2) The data on DBCP causing sterility has "only recently been collected," and "current data conclusively and reliably show" the deleterious effect of DBCP even at low levels of exposure.

(3) There is a need for an emergency standard because DBCP can pose its life-threatening danger in a brief period of exposure, "a simple exposure episode may be sufficient to cause cancer" and "some employees who became sterile had been exposed [to DBCP] for only a few months."

The emergency temporary standard for DBCP was not challenged in court and was eventually superseded by a permanent standard.[27]

The reasoning utilized by OSHA in support of the ETS for DBCP would suggest the issuance of emergency standards for other workplace carcinogens. Indeed, OSHA reached this conclusion in its proposed Carcinogens Policy, where it required OSHA to issue an ETS for all Category I carcinogens.[28] In the final Carcinogens Policy OSHA reiterated its policy conclusion that carcinogenic substances cause a "grave danger" to employees, relying in part on the "widely accepted" view that "exposure to carcinogens for very short periods can begin the carcinogenic process." OSHA deleted the requirement that an ETS be issued for these carcinogens, however, and, provided instead that OSHA *may* initiate the proceeding by issuing an emergency standard. The preamble stated:

> * * * OSHA has determined that the question of whether a particular emergency standard is necessary to protect employees from such grave danger, the second prong of the statute's test, is more appropriately left for determination in each case. OSHA believes it should retain the flexibility not to issue Emergency Temporary Standards where, for instance, the limits of short-term exposure reduction already have been reached voluntarily. In such a case, although OSHA's application of the priority factors listed in §1990.132(b) and otherwise, might signal regulatory activity leading to a permanent standard, there may be no compelling need for an emergency standard. In other cases, the level of the agency's resources including compliance, legal and technical personnel,

---

[27]43 FED. REG. 11,514 (1978). A significant difference between the compliance provisions in an ETS and those in permanent standards is that, under an ETS, the employer is not required to reach the permissible exposure limit primarily by engineering controls. See, e.g., ETS for Acrylonitrile, 43 FED. REG. 2586, 2595, 2601 (1978) (requiring compliance "by any practical combination of engineering controls, work practices and personal protective equipment"). This approach of ETSs is based on the fact that emergency standards are effective for a relatively short period of time, six months, and, therefore, costly engineering controls should not be required until rulemaking takes place and a final standard issued. The production of DBCP in the United States was discontinued after promulgation of the OSHA standard. See also Infante & Tsongas, *Occupational Reproductive Hazards: Necessary Steps to Prevention,* 4 AM. J. INDUS. MED. 383, 386 (1983). Drs. Infante and Tsongas quote a letter from the National Peach Council to Assistant Secretary Bingham suggesting, among other things, that if "possible sterility" from DBCP is the problem "couldn't workers who were old enough that they no longer wanted to have children accept such positions [working with DBCP] voluntarily?" *Id.* at 385.

[28]42 FED. REG. 54,148, 54,168 (1977). Category I carcinogens are defined in 29 C.F.R. §1990.112(a) (1983). The classification is based on the quality of the evidence indicating carcinogenicity.

at a given time, may suggest that employee health may be more effectively protected by concentrating those resources in work on permanent standards.[29]

Despite OSHA's mixed success in court with ETSs, the Comptroller General in 1977 suggested that OSHA utilize its §6(c) authority more "aggressively."

Report of the Comptroller General[30]

***

CONCLUSION

OSHA has made limited use of its authority to (1) issue emergency temporary standards to protect employees from grave dangers and (2) promptly provide needed protection that can be supported by available evidence. The gravity of the dangers posed by toxic substances dictates that OSHA use its authority more aggressively. Millions of workers are exposed to substances that can cause cancer and other serious or irreversible diseases.

OSHA needs to establish criteria on the conditions under which emergency temporary standards should be issued. The criteria should define grave danger and the evidence needed to support a determination that a grave danger exists. The definition should make it clear that direct evidence of fatalities attributable to the workplace is not necessary.

OSHA's announced intent to issue emergency temporary standards on confirmed carcinogens would, if carried out, be a significant step toward establishing the needed criteria. Additional criteria are needed for substances which, although noncarcinogenic, pose grave dangers to workers.

The criteria for issuing emergency temporary standards should apply both to toxic substances not covered by standards and toxic substances covered by inadequate standards. The question of industry's general ability to comply with a standard within 6 months should not deter efforts to protect workers from a grave danger.

Reluctance to issue emergency temporary standards has been influenced by the position that such action should not be taken unless a permanent standard can be issued in 6 months to supersede the temporary standard. Although the act requires that a permanent standard be issued within 6 months after a temporary standard, it does not say whether the temporary standard expires if this requirement is not met. OSHA's interpretation that the standard expires after 6 months is not consistent with the basic intent of protecting workers from grave danger.

Regardless of whether evidence is adequate to support emergency action, OSHA should promptly issue standards based on available evidence,

---

[29]45 FED. REG. 5002, 5215–16 (1980). Compare statements by Assistant Secretary Auchter in 1983: "What happens when we issue an emergency temporary standard? We've got six months to come out with a permanent rule. Virtually all of our other standard activity stops. I have to weigh in my mind what that is going to do to the rest of the work that is going on in the agency." N.Y. Times, Apr. 18, 1983 at A15, col. 3, 4. Mr. Auchter also noted that OSHA had lost four of the five ETSs that went to court "because we had not shown the emergency need." He said: "We in the Government have to make decisions that keep us as far as possible beyond litigation." *Id.* But see D.D. McCaffrey, OSHA AND THE POLITICS OF HEALTH REGULATION 96 (1982) (*"It was not true that OSHA had been unsuccessful in defending ETSs in court* * * *. Rather, the high political and resource costs of court battles—successful or unsuccessful—made OSHA reluctant to issue the ETS." (emphasis in original)). On Nov. 4, 1983, Assistant Secretary Auchter issued an ETS on asbestos. See discussion *infra* this chapter.

[30]COMPTROLLER GENERAL OF THE UNITED STATES, REPORT TO THE CONGRESS, DELAYS IN SETTING WORKPLACE STANDARDS FOR CANCER-CAUSING AND OTHER DANGEROUS SUBSTANCES (HRD-77-71) 43–44 (1977).

even if the standards cannot include all protective measures that would be desirable if more or better data were available. The act provides that any standard may be revoked or modified as additional evidence is obtained.

RECOMMENDATIONS TO THE SECRETARY OF LABOR

The Secretary should require OSHA to take the following actions to implement section 6(c) of the act:

—Define grave danger to include exposure of workers to a toxic substance or harmful agent which has resulted or can result in incurable, irreversible, or fatal harm to health.
—Issue emergency temporary standards in all cases where they are needed to protect employees from grave danger, including any such dangers posed by toxic substances or harmful agents covered by inadequate standards.
—Require that emergency temporary standards remain in effect until superseded by a permanent standard.

The Secretary should also require OSHA to promptly issue emergency temporary or permanent standards on toxic substances to require needed protection that can be supported by available evidence and to revise and add to such standards as more and better evidence becomes available.

The Comptroller General's suggestion that an emergency temporary standard could remain in effect beyond the six-month period is not supported by the legislative history. The bill reported by the House Labor Committee provided that emergency temporary standards would remain in effect for six months, unless rulemaking on a permanent standard was commenced, in which case the standard would remain in effect until the termination of the rulemaking proceeding. Congressman Steiger criticized this language on the grounds that ETSs could remain "in effect for six months or until replaced by permanent standards" and thus for an "inordinately long period of time." He indicated that his substitute language would avoid this result by providing that ETSs would remain in effect until replaced by a permanent standard "which the [Standards] Board[31] is required to issue within six months after the emergency temporary standards are issued."[32] The Steiger substitute language was adopted by the Conference Committee and became §6(c). In light of this history, it would be extremely difficult to argue that under §6(c) an ETS can remain effective for a period longer than six months even though the time limitation is not stated explicitly. In addition, there would be considerable policy objection to a view that permitted ETSs to remain in effect indefinitely, without public participation.[33]

[31]The Steiger substitute assigned the authority to issue standards to a Standards Board rather than OSHA. See Chapter 1.

[32]116 CONG. REC. 38,372 (1970).

[33]Compare *Performance of the Occupational Safety and Health Administration: Hearings Before the Subcomm. of the House Comm. on Government Operations,* 95th Cong., 1st Sess. 7–8 (1977) (statement of Gregory Ahart, Director, Human Resources Div., GAO) (ETS can extend past six-month period) with, *id.* at 92 (statement of Dr. Bingham, Assistant Secretary of OSHA) (debatable whether ETS can extend past six-month limit).

The Comptroller General also suggested that OSHA "establish criteria on the conditions under which emergency temporary standards should be issued." No formal criteria have been issued by the Agency, although the full preambles to the ETSs for diving, benzene, acrylonitrile, and DBCP, as well as the preamble to the Carcinogens Policy, contain statements of OSHA policy on emergency standards. As indicated, contrary to the recommendation of the Comptroller General, after 1977 OSHA has not used its ETS authority more "aggressively." No additional ETSs were issued by Assistant Secretary Bingham after acrylonitrile in January 1978. In October 1981, the United States Automobile Workers of America and other unions petitioned OSHA for an ETS on formaldehyde. The petition was based on experimental evidence in animal studies showing the substance to be a carcinogen.

UAW Petition for ETS on Formaldehyde[34]

***

In support of this request we bring the following points to your attention:

1. Formaldehyde exposure causes nasal cancer in rats, and probably in mice, at exposure levels comparable to those found in many work environments. Two independent studies, conducted respectively by Battelle Columbus Laboratory for the Chemical Industry Institute of Toxicology and by New York University, have confirmed decisively that formaldehyde vapor induces squamous cell carcinoma in the nasal cavity of rats.

2. Additional supporting evidence lends weight to the CIIT and NYU findings. Short-term testing of formaldehyde *in vitro* shows it to produce gene mutations and chromosome aberrations in several test systems. Formaldehyde exerts these effects by itself and can also interact with other mutagens such as X-rays and ultra-violet light.

3. At least three prestigious scientific panels have reviewed this evidence and concluded that formaldehyde should be treated as a human carcinogen:

a. In November, 1980 the Federal Panel on Formaldehyde under the auspices of the National Toxicology Program concluded that "formaldehyde poses a cancer risk to humans."

b. In February, 1981 the Environmental Cancer Information Unit of the Mt. Sinai School of Medicine reported to the American Cancer Society that "formaldehyde is a carcinogen in rats and, data suggests, in mice, at exposure levels comparable to those found in some home and work environments. These findings indicate that effective controls should be initiated to reduce or eliminate human exposures to formaldehyde."

c. In a Current Intelligence Bulletin dated 4/15/81 the National Institute [for] Occupational Safety and Health concluded that "formaldehyde has induced a rare form of nasal cancer in both Fischer 344 rats and in B6CF1 mice. * * * Although humans and animals may differ in their susceptibility to specific chemical compounds, any substance that produces cancer in experimental animals should be considered a cancer risk to humans."

[34] Petition of UAW for ETS on Formaldehyde 3–5, *UAW v. Donovan,* No. 82-2401 (D.D.C. 1982).

4. Formaldehyde is a major industrial chemical with numerous workplace uses. U.S. production of aqueous formaldehyde was 6.4 billion pounds in 1978. During 1972 to 1974 NIOSH estimated that 1.6 million workers in more than 60 industrial categories were exposed to formaldehyde. Worker formaldehyde exposures frequently occur at air levels comparable to those leading to cancer in rats.

5. The current OSHA standard for workplace formaldehyde exposure is 3 PPM as an eight hour time weighted average with a 5 PPM acceptable ceiling and a 10 PPM maximum peak. NIOSH reviewed this standard in 1976, found it inadequate to afford protection against even the irritant effects of formaldehyde and recommended a new standard of 1 PPM for any 30-minute sampling. On the basis of the new evidence of carcinogenic potential described above, NIOSH further revised its recommendation on 4/15/81 as follows:

> "Safe levels of exposure to carcinogens have not been demonstrated, but the probability of developing cancer should be reduced by decreasing exposure. An estimate of the extent of the cancer risk to workers exposed to various levels of formaldehyde at or below the current 3 PPM standard has not yet been determined. In the interim NIOSH recommends that, as a prudent Public Health measure, engineering controls and strict work practices be employed to reduce occupational exposure to the lowest feasible limit."

Based on the information presented in this petition and supporting documents we believe there is an immediate and compelling need to take strong regulatory action. Those of our members who are currently exposed to formaldehyde face a substantial risk of irreparable harm simply by going to work each day. We therefore urge a prompt and affirmative response to this proposal.

On January 29, 1982, Assistant Secretary Auchter rejected the formaldehyde petition. His letter said in part:

> * * * I believe that emergency temporary standards are appropriate only in response to extraordinary conditions which result in the exposure of employees to a grave danger during the course of their employment. In addition, to justify such a unique remedy where broad public participation is virtually eliminated, the danger to which employees are exposed must generally be even more clearly demonstrated than in a normal §6(b) rulemaking proceeding.
>
> The primary evidence upon which your petition stands are two new animal studies performed by the Chemical Industry Institute of Toxicology and New York University. The animal data indicates that formaldehyde causes nasal tumors in test animals at relatively high levels of exposure. However, in the studies these results are statistically significant only at exposure levels of approximately 15 ppm, substantially above the current permissible exposure limit (PEL) for formaldehyde of 3 ppm. Risk assessments that OSHA has performed on the animal data suggest that the risk from formaldehyde exposure at the PEL, when compared with other occupational risks, is not sufficient to warrant a finding of "grave danger" and resulting emergency action.
>
> In all ETS situations, OSHA must not only sustain the burden of proving that the ETS is justified by a grave danger, which we do not believe exists here, but also that the emergency action is necessary to protect employees. Information which you submitted along with your

petition, such as the NIOSH monitoring results, indicates that exposure levels are generally below the current PEL of 3 ppm. This information is corroborated by OSHA inspection data. In addition, other OSHA standards, such as those found at 29 CFR 1910.132 and 1910.133 require that employees use protective equipment and eye and face protection when any chemical, including formaldehyde, is used in such a manner that it is capable of injuring or impairing employees. These standards will help provide additional protection to employees exposed to liquid formaldehyde. Since OSHA already has a standard for formaldehyde which requires employee exposure to airborne formaldehyde to be kept quite low, and there are a number of other OSHA standards which will contribute to the protection of formaldehyde-exposed employees in certain circumstances, the data you submitted would not enable us to make a showing that the emergency action you request is necessary.

While the evidence contained in your petition cannot, standing alone, support emergency action, we have recently been informed of a few cases of nasal tumors in workers. It is possible that these instances may be related to formaldehyde exposure. This information, however, is of an extremely preliminary nature and at this point we do not have sufficient details about it to support emergency action. We are investigating these cases further and will reconsider our decision if the results of our further investigation warrant action.[35]

In August 1982, the UAW brought suit in the U.S. District Court for the District of Columbia claiming that OSHA's denial of its petition was arbitrary and capricious, and asking the court to issue an order requiring OSHA to issue an ETS "significantly reducing permissible levels of workplace exposure to formaldehyde."[36]

A major regulatory action took place in early November 1983 when OSHA issued an emergency temporary standard reducing the PEL for asbestos to 0.5 fibers per cubic centimeter (f/cc).[37] This action is best understood in the context of OSHA's lengthy prior experience in regulating asbestos hazards.

OSHA's original standard on asbestos, issued in 1971 under §6(a), contained a PEL of 12 f/cc. The standard was amended in June 1972 when OSHA issued a permanent asbestos standard providing for a 2 f/cc limit after four years and an immediate effective date for a 5 f/cc limit, with a 10 f/cc ceiling.[38] The permanent standard was challenged in court by the Industrial Union Department and upheld

[35]Letter from Assistant Secretary Thorne Auchter to Howard Young, Director of UAW (Jan. 29, 1982) (on file in *UAW v. Donovan,* No. 82-2401 (D.D.C. 1982)).

[36]UAW Complaint at ¶1, *UAW v. Donovan,* No. 82-2401 (D.D.C.). See also *Gulf S. Insulation v. U.S. Consumer Product Safety Comm'n,* 701 F.2d 1137 (5th Cir. 1983) (holding that the record developed by CPSC did not contain the substantial evidence necessary to support its ban on urea-formaldehyde foam insulation). The United States decided not to file a petition for certiorari from this decision. N.Y. Times, Aug. 26, 1983 at 1, col. 5. The issue of whether OSHA should take stricter regulatory action regarding formaldehyde was the underlying issue in a hearing held by the Subcommittee on Investigations and Oversight of the House of Representatives Committee on Science and Technology in July 1981. *Proposed Firing of Dr. P. Infante by OSHA: A Case Study in Science and Regulation: Hearing Before the Subcomm. on Investigations and Oversight of the House Comm. on Science and Technology,* 97th Cong., 1st Sess. (1981). Dr. Infante is presently the head of OSHA's Office of Standards Review.

[37]48 FED. REG. 51,086 (1983).

[38]29 C.F.R. §1910.1001 (1983). The permanent standard had been preceded by an ETS, issued in December 1971. 36 FED. REG. 23,207 (1971). The ETS was not challenged in court.

in large measure by the Court of Appeals for the District of Columbia Circuit in *Industrial Union Department v. Hodgson.*[39] The court of appeals remanded to OSHA for reconsideration of the issue of whether the effective date of the 2 f/cc standard should be accelerated for certain industries. OSHA did not act on the remand of the issue of the standard's effective date, however, and the issue became moot in 1976 when the lower limit went into effect.[40]

On October 9, 1975, OSHA proposed a more stringent asbestos standard with a PEL of 0.5 f/cc and a ceiling level of 5 f/cc.[41] This standard was predicated on "new evidence" showing that asbestos causes various types of cancer and mesothelioma, and on OSHA's policy of setting the PEL for carcinogens at the "lowest feasible level."[42] No hearing was ever scheduled on the 1975 proposal.[43]

In its lengthy preamble to the 1983 ETS,[44] OSHA made it clear that the standard was being based on information and analyses which postdated the 1975 proposal and was therefore a "new regulatory initiative."[45] OSHA's summary findings and "rationale for the ETS" are stated in the following two paragraphs:

> OSHA has determined that prevailing conditions involving worker exposure to airborne asbestos dust justify the promulgation of an emergency temporary standard. OSHA estimates that approximately 375,000 workers are exposed to asbestos at various levels ranging from a high value of 20 f/cc to below 0.5 f/cc. OSHA has estimated that under current exposure conditions asbestos exposed workers face an extraordinarily high risk of contracting asbestos-related cancer whether the risk is computed over a working lifetime of exposure or for exposure periods as short as 6 months. The average excess cancer risks for all workers exposed above 0.5 f/cc using available exposure data and relying on the risk assessment are estimated as approximately 196 excess cancer deaths per 1000 workers for 45 years of exposure, 139 deaths per 1000 workers for 20 years, 10 per 1000 workers for 1 year, and 6 per 1000 workers for 6 months of exposure.
>
> OSHA believes that risks of these magnitudes, taking into account all relevant considerations such as total numbers of workers at risk and

[39]499 F.2d 467 (D.C. Cir. 1974). Portions of this decision are reprinted and discussed in Chapters 3 and 5.

[40]The court also remanded to OSHA the issue of whether the 3-year period provided in the standard for retention of monitoring records was sufficiently long. OSHA did not act on the issue until shortly before the expiration of the 3-year retention period, when it amended the standard to provide for a 20-year retention period. 41 FED. REG. 11,504 (1976). The access portion of the asbestos standard was amended in 1980 to conform to the "generic" access standard, 45 FED. REG. 35,212, 35,281 (1980); see the discussion in Chapter 5, Section F.

[41]40 FED. REG. 47,652 (1975).

[42]For a discussion of the "lowest feasible level" policy, see Chapter 9.

[43]In the 1983 ETS, OSHA noted that the "lowest feasible level" policy, upon which the 1975 proposal was based, was rejected by the Supreme Court in its decision vacating the benzene standard. *Industrial Union Dep't v. American Petroleum Inst.*, 448 U.S. 607 (1980), cited in 48 FED. REG. at 51,088. The 1975 proposal expressly excluded application to the construction industry. OSHA indicated, however, its intention to publish a separate proposal for the construction industry. 40 FED. REG. at 47,657. This was never done.

[44]The preamble comprised 53 *Federal Register* pages, including extensive health data and quantitative risk assessments upon which the Agency findings were based. The preamble to the ETS on asbestos issued in 1971 comprised *one* paragraph. See discussion in Chapter 8, Section F.

[45]48 FED. REG. at 51,088. The preamble also indicated that OSHA had consulted with the Construction Advisory Committee for Occupational Safety and Health and following its recommendation, the ETS is also applicable to the construction industry. *Id.*

quality of supporting data, constitute an emergency situation which requires immediate response by the agency.[46]

OSHA's more detailed rationale for the ETS sets forth its conclusions in terms of the specific requirements of §6(c) that employees are exposed to a "grave danger" and that the ETS is "necessary to protect employees from such danger" and takes into account the case law on OSHA's authority to issue emergency standards.

Emergency Standard for Asbestos

48 Fed. Reg. 51,086, 51,088–98 (1983)

A. GRAVE DANGER

OSHA has determined that the risk to workers from exposures to asbestos at conditions that exist in the workplace pose a grave danger of death from cancer and of severe disability from the lung disease, asbestosis. In making a "grave danger" determination, the severity of the disease produced by exposure to the regulated substance and the magnitude of the predicted risks of disease must be considered. In addition, the Supreme Court has suggested that a determination of "grave danger" indicates a situation where the risk is more than "significant" [*Industrial Union Dep't v. American Petroleum Institute,* 448 U.S. 607, 641 n.45].

OSHA has applied that analytic approach endorsed by the Supreme Court for "significant risk" determinations in evaluating the gravity of the danger faced by asbestos-exposed workers. The Supreme Court gave some general guidance as to the process to be followed. It recognized that while the Agency must support its finding that a certain level of risk exists with substantial evidence it also recognized that its determination that a particular level of risk is "significant" will be based largely on policy considerations [*IUD v. API,* 448 U.S. 607, 656, n.62].

OSHA believes, therefore, that its determinations regarding the magnitude of the risk faced by employees should, to the extent possible, rely on quantitative expressions of that risk, utilizing the best available data.

The Court stated that the significant risk determination required by the OSH Act is "not a mathematical straitjacket," and "OSHA is not required to support its finding that a significant risk exists with anything approaching scientific certainty. * * * A reviewing court [is] to give OSHA some leeway where its findings must be made on the frontiers of scientific knowledge [and that] * * * the Agency is free to use conservative assumptions in interpreting the data with respect to carcinogens, risking error on the side of overprotection rather than underprotection" [448 U.S. at 655, 656].

In the case of asbestos, the data available are of unusual breadth and high quality. However, because risk assessment itself involves many uncertainties, OSHA made certain assumptions in its analysis and evaluation of these data. In assessing the risk for asbestos-exposed workers, OSHA has attempted to use realistic assumptions, although the court stated that the Agency was free to use "conservative assumptions" in interpreting data. OSHA, in many cases, has indicated where different assumptions may produce different results. In addition OSHA cautions that because the risk figures finally derived are the products of a process which, as the Supreme Court acknowledged, is "on the frontiers of science," they should be viewed

[46] *Id.*

as approximations of the degree of risk faced by asbestos-exposed workers and not as precise fixed predictions of the number of workers who will actually develop disease.

OSHA has evaluated the kinds of dangers presented by asbestos exposure, the quantification of those dangers under present asbestos exposure conditions, the quality of the data on which risk estimates are based, a comparison of asbestos risks to other occupational risks, and relevant policy and legal considerations in concluding that workers are exposed to a grave danger from asbestos.

1. *Nature of the Diseases.* As stated above, the nature of the disease associated with exposure to a toxic substance is one of the most important elements OSHA evaluates in determining whether a grave danger exists. This factor was discussed in *Florida Peach Growers Association, Inc. v United States Department of Labor, supra.* The court, in overruling OSHA's organophosphate pesticide ETS, observed:

> We reject any suggestion that deaths must occur before health and safety standards may be adopted. Nevertheless, *the danger of incurable, permanent, or fatal consequences to workers, as opposed to easily curable and fleeting effects on their health, becomes important in the consideration of the necessity for emergency measures to meet a grave danger.* 489 F 2nd at 132 (emphasis added)

OSHA is aware of no instances in which exposure to a toxic substance has more clearly demonstrated detrimental health effects on humans than has asbestos exposure. The diseases caused by asbestos exposures are in large part life-threatening or disabling. Among these diseases are lung cancer, cancer of the mesothelial lining of the pleura and peritoneum, and asbestosis. In addition, workers exposed to asbestos are at increased risk of gastrointestinal cancer, as shown by epidemiologic studies. Although colorectal cancer may be curable if detected in an early stage, other gastrointestinal cancers are usually fatal. OSHA also believes that asbestos might induce cancers at other sites, which are also often fatal.

Of these, lung cancer constitutes the greatest health risk for American asbestos workers and has accounted for more than half of excess mortality in some occupational cohorts. About 90% of lung cancer patients die within 5 years of diagnosis. Mesothelioma is an incurable cancer which is usually fatal within a year after diagnosis. It is epidemiologically linked to asbestos exposure, and occurs very rarely, if at all, in persons never exposed to asbestos. Asbestosis, a type of pulmonary fibrosis, is usually non-reversible, its advanced stages are disabling, and can be fatal. OSHA concludes that all these diseases are very serious, and that the excess mortality from such severe diseases must be considered an important factor for making a grave danger determination.

2. *Degree of Risk of Developing Dangerous Disease.* OSHA based its calculations of extent of risk faced by workers under current exposure conditions primarily on the results of a quantitative analysis which derived numerical estimates of cancer risk at various cumulative exposures corresponding to levels at which workers are exposed.

Although 2 f/cc is the current PEL for asbestos exposure, actual exposure conditions vary widely, mostly by industry segment. As explained later in this document and as set forth in Table 1 [omitted], average ambient exposure levels in various industries include high exposure levels such as 20 f/cc in drywall removal, renovation and demolition; 5 f/cc in shipbuilding and repair; midrange exposure levels such as 2 f/cc in secondary fabricating of cement sheet, packing and gaskets and paper products and rebuilding and refacing brakes; 1.5 f/cc for dry processing of textiles; and lower exposure levels such as 0.5 f/cc and 0.2 f/cc in the manufacture of floor tile.

Because OSHA is required to consider the actual danger faced by workers in assessing whether exposure to a substance presents a "grave danger", OSHA looked at the risk of developing disease not only at the 2 f/cc permissible level but at all exposure levels which workers currently face. * * *

***

These calculations show that the risks of asbestos-related disease are alarmingly high at current occupational exposure levels. For example, an estimated total cancer risk of 265 excess deaths per 1000 workers exists for workers exposed for a 45-year lifetime at 10 f/cc, a level which currently exists on some construction sites. At 5 f/cc, the exposure levels which are considered average in shipbuilding and repair, the risk of developing asbestos-related cancer for a 45-year exposure period is 149 excess deaths per 1000 workers. At the current permissible level of 2 f/cc which also represents actual exposure levels in such industries as secondary fabricating of cement sheet, packing gaskets and paper products and rebuilding and refacing brakes, risk is estimated as 64 excess cancer deaths per 1000 workers for a 45-year exposure period.

These risks remain very high when the period of exposure for which calculations are done is shortened to 20 years, which OSHA believes is another appropriate point for examination. The period of 20 years is the approximate midpoint between 1 year and 45 years of exposure; also many workers receive 20 years of exposure. Counterpart risk calculations using a 20-year exposure period are: for workers exposed to 10 f/cc, 140 excess cancer deaths per 1000 workers; for exposures to 5 f/cc, 105 excess cancer deaths per 1000 workers and for exposures to 2 f/cc, 44 excess cancer deaths per 1000 workers.

OSHA also estimated risks of developing cancer for a one year period of exposure at various levels to which employees are exposed. The counterpart risks for exposures to 10 f/cc for one year are: 15 excess cancer deaths per 1000 workers; to 5 f/cc, 7 excess cancer deaths per 1000 workers and to 2 f/cc, 3 excess cancer deaths per 1000 workers.

Even at current workplace exposure levels which are less than the current PEL, extraordinarily high risks of disease exist. At 0.5 f/cc, 17 excess cancer deaths per 1000 workers are predicted for a 45-year lifetime exposure, and 11 excess cancer deaths per 1000 workers for a 20-year exposure period.

* * * OSHA has considered that the additional and independent risk of developing asbestosis increases the danger faced by exposed workers and underscores the gravity of the health threats to employees posed by asbestos.

3. *Quality of Data on Which Risk Estimates are Based.* The underlying data upon which the quantitative risk assessments for asbestos are based are high quality epidemiologic studies, conducted in occupational environments. OSHA emphasizes that the data bases for asbestos are of unusual quality and size. Unlike most potential occupational carcinogens, asbestos has been studied often and thoroughly for evaluation of its effects on occupational populations.

In deriving these quantitative estimates for cancer risk, OSHA utilized eleven studies for the calculation of the lung cancer risk, four of which were also used to calculate the mesothelioma risk. Investigations involved "cohort" studies where the frequencies of various types of cancers in workers exposed to asbestos were compared to those in "control" groups not exposed to asbestos or to those of general populations such as U.S. males. Studies of such design are able to provide direct estimates of excess risk.

***

No extrapolation from animal data to human data is necessary in order to show carcinogenicity of asbestos. For most substances, OSHA must infer human health effects, such as carcinogenicity, from animal data.

***

4. *Comparative Analysis.* Insight into the magnitude of the risk associated with asbestos exposure can be gained by reviewing other occupational risks. OSHA believes it is instructive to compare asbestos risks with other workplace hazards agreed on as presenting an unusually high degree of hazard, where the data are considered both available and reliable.

***

Therefore, the excess cancer risk at 2 f/cc for asbestos workers is estimated as more than twice as high as the maximum permitted radiation cancer risk and about 25 times higher than the estimated cancer risk of 95 percent of the workers exposed to radiation. At existing conditions, asbestos workers' excess cancer risks are estimated to be 85 times higher than the cancer risk faced by 95 percent of the workers exposed to radiation. The risk of asbestosis further increases the significance of the risk from asbestos exposure.

At 0.5 f/cc, OSHA estimates that 17 excess cancer deaths will occur in 1000 workers exposed 45 years. This risk is approximately 7 times higher than the cancer risk faced by 95 percent of the workers exposed to radiation. OSHA finds that these comparative risks strongly support OSHA's finding that workers exposed to air concentrations above 0.5 f/cc are far above the point of significant risk and are at grave danger of dying from cancer.

5. *Conclusion.* OSHA's finding of "grave danger" is based on evidentiary and policy considerations. OSHA's determination that the magnitude of the estimated risk to exposed workers is alarmingly high constitutes the major component of the "grave danger" finding. The overall extraordinary degree of risk, the extent that very high risk is found in many asbestos using industries, and the unusually high quality of the data utilized to make these assessments present a very strong evidentiary basis for a "grave danger" finding. Just as importantly, the unique gravity of asbestos-caused diseases, in particular cancer, such as mesothelioma which is linked almost exclusively to asbestos exposure, strongly supports OSHA's finding of grave danger. Also OSHA's comparison of the risk of asbestos-related disease to other industrial risks underscores the extraordinarily high risk estimated for asbestos exposure. OSHA has also noted the concerns of workers about current workplace conditions and the numerous petitions for an ETS from unions representing many exposed workers. Finally OSHA has relied on its experience in evaluating and regulating workplace hazards in recognizing the extraordinary degree of risk currently faced by asbestos workers and in determining that such risk constitutes a grave danger to those workers.

B. NEED FOR AN ETS

OSHA has determined that this ETS is necessary to protect employees from grave danger, the second prong of the Act's test of OSHA's exercise of its ETS authority (Section 6(c) of the Act). As explained in detail, the effect of this ETS is to save many lives which would otherwise be lost to asbestos-related disease if current working conditions were not changed. OSHA believes that employees can be adequately protected against this grave danger only by issuing an ETS. This is because no other Agency action and no other foreseeable event would result in sufficiently reduced asbestos exposures that would alleviate the grave danger. Further, the provisions of the ETS are tailored to effect the necessary exposure reductions expeditiously.

1. *Lives Saved by Issuing an ETS.* OSHA has estimated the number of deaths avoided as a result of an ETS which would reduce the PEL to 0.5 f/cc. For cancer only, based on continuing exposures under currently existing conditions for 6 months, the potential number of lives saved is estimated as

approximately 210. Based on continuing exposures at currently existing conditions for 1 year, the potential number of lives saved is estimated at approximately 426. Also, OSHA has estimated that the promulgation of an ETS setting a 0.5 f/cc PEL may avoid 5725 cancer deaths assuming 20 years exposure to asbestos of the current workforce at current conditions and 7815 cancer deaths assuming 45 years exposure.

OSHA is aware, of course, that Section 6(c) of the Act limits the effective time of an ETS to 6 months, and OSHA concludes that a grave danger exists and an ETS is necessary even if OSHA focuses exclusively on this six month period. However, the Agency believes it is appropriate to calculate benefits deriving from an ETS using lifetime risks from 20 and 45 years of exposure to the PEL of 0.5 f/cc established by the ETS. Although the ETS expires within 6 months, Section 6(c) requires that rulemaking on a permanent standard also be completed within 6 months, so that there will be no gap in protection for exposed employees. In OSHA's experience and judgment, complying with this statutory directive and completing rulemaking for a permanent standard within 6 months of an ETS has and can be done.

OSHA also believes, based on its experience, that it is very likely that the PEL established after 6(b) rulemaking will be no higher than 0.5 f/cc, the ETS limit. Therefore, OSHA believes that the ETS will result in a reduced lifetime worker exposures of 0.5 f/cc or lower for 20 or 45 years, and that the benefits derived from these exposure reductions for these time periods are appropriately attributed to OSHA's promulgation of this emergency standard.

***

3. *No Other Agency Action is Adequate to Protect Against the Grave Danger.* * * *

***

(b) OSHA rejected relying on merely beginning Section 6(b) rulemaking proceedings to revise the standard to reduce the PEL as an inadequate response to the grave danger faced by asbestos-exposed workers. Beginning rulemaking proceedings results in no immediate workplace changes. Employees would still continue to be exposed to those conditions which define a grave danger for at least the pendency of the rulemaking. In OSHA's experience, completing 6(b) rulemakings not initiated by an ETS concerning hazardous substances can take many years. For example, the coke oven emission standard took approximately 3½ years, the lead standard, more than 6 years and the cotton dust standard, more than 4 years. These periods do not include any of the additional delays in the effective dates of OSHA standards that were due to judicially imposed stays, which have resulted in delays lasting several years. Under the most favorable circumstances, however, OSHA believes that it is possible that a section 6(b) rulemaking limited to the issues raised herein might be completed in approximately one year, absent an ETS.

As shown above, the estimated risks of developing asbestos-related cancer due to exposure for one year under current conditions, are still extraordinarily high. The additional risks of developing asbestosis due to one year's exposure under current conditions, although quantified with less certainty, are also more than significant. OSHA also believes that the risks of six months exposure, approximated by taking over half of the one year risks under current conditions, also are unacceptably high. OSHA emphasizes as stated above, that OSHA's experience shows that without an ETS, proceedings leading to a permanent health standard are unlikely to be completed within a six month period. The explanation of OSHA's capability to produce a standard within 6 months of an ETS lies in the urgency generated by OSHA's finding of a grave danger, the existence of a specific statutory deadline to complete a rulemaking within 6 months and the need to prevent a gap in protection between the expiration of the ETS and the imposition of the permanent standard for a substance already determined to present a grave danger.

***

A number of elements of this rationale for the ETS are particularly noteworthy.

1. OSHA relied only on data from studies on humans to establish the carcinogenicity of asbestos; no extrapolation from animal data was necessary, as is the case for "most substances" regulated by OSHA.[47] OSHA also said that the data bases for asbestos regulation are of "unusual quality and size."[48]

2. OSHA performed quantitative risk assessments on the basis of this "high quality" data in order to estimate the excess cancer risk to employees at various levels of exposure. The Agency then compared that level of risk with other occupational risks and concluded that there was an "overall extraordinary degree of risk" to asbestos workers.[49] This "alarmingly" high risk constituted the main component of the "grave danger" finding. The preamble noted that in its *Benzene* decision,[50] the Supreme Court said that OSHA must make a finding of "significant risk" for standards issued after rulemaking,[51] but for emergency standards, the Agency must meet a higher standard of proof, namely of "grave" risk.

3. With respect to the second aspect of the statutory requirement—that the ETS is "necessary"—OSHA made specific findings on the number of lives that would be saved by issuing an ETS. Although OSHA had construed §6(c) to mean that an ETS may remain in effect for no longer than six months,[52] it nonetheless calculated the number of lives to be saved by the ETS on the basis of lifetime risks from 20 to 45 years of exposure. The Agency's rationale was that (a) the Act requires a permanent standard to be promulgated within six months; (b) completion of this rulemaking "has and can be done"; and (c) it is likely that the PEL in the permanent standard will be no higher than 0.5 f/cc, the level in the ETS. Accordingly, under OSHA's reasoning, benefits to be derived from the permanent standard—that is, lifetime employee exposures to asbestos—"are appropriately attributed to OSHA's promulgation of this emergency standard."[53]

This rationale is more understandable in the context of OSHA's explanation elsewhere in the preamble of why the Agency did not undertake §6(b) rulemaking as a response to the "grave danger" from asbestos. OSHA pointed to the long delays involved in rulemaking on

[47]48 FED. REG. at 51,090.
[48]*Id.*
[49]*Id.* at 51,091.
[50]448 U.S. 607, 641 n.45 (1980).
[51]See discussion in Chapter 9.
[52]See discussion *infra* this chapter.
[53]48 FED. REG. at 51,091.

toxic substances, often lasting several years, and said that without an ETS, proceedings for a permanent standard are "unlikely" to be completed within a six-month period. The preamble states:

> The explanation of OSHA's capability to produce a standard within 6 months of an ETS lies in the urgency generated by OSHA's finding of a grave danger, the existence of a specific statutory deadline to complete a rulemaking within 6 months and the need to prevent a gap in protection between the expiration of the ETS and the imposition of the permanent standard for a substance already determined to pose a grave danger.[54]

In other words, the justification for an ETS, at least in part, was that as a practical matter it is the only procedure OSHA could utilize which would bring about *long-term employee protection* from asbestos within a reasonable period of time.

OSHA's emergency standard on asbestos was quickly challenged in the Court of Appeals for the Fifth Circuit[55] by a trade association, representing 47 employers, and individual employers who mine, manufacture, fabricate, distribute, and use asbestos or asbestos products. The petition for review was soon followed by a motion for a stay of the ETS. In requesting the stay, the petitioners argued that the ETS would produce "virtually no worker health benefits" because existing levels of exposure already average "near" the 0.5 f/cc level established by the ETS. In the "few situations" where higher exposures persist, according to the petitioners, OSHA's "new enforcement program" under the existing standard would be sufficiently protective. On the other hand, the petitioners argued, OSHA's "sensational proclamation" in the ETS that a "health emergency" exists was likely to cause irreparable injury to the petitioners' businesses by cutting "deeply" into the sales of asbestos products.[56]

OSHA opposed the stay. However, on November 23, 1983, a panel of the Court of the Appeals for the Fifth Circuit[57] granted the stay.[58]

On March 7, 1984, the Court of Appeals for the Fifth Circuit decided that the asbestos ETS was invalid, finding that the record in the proceeding did not establish that the risk the ETS sought to eliminate was "grave" or that the ETS was "necessary" to protect employees from that danger.

---

[54]*Id.* at 51,098. OSHA separately concluded that the risks of exposure for six months were also "unacceptably high." *Id.*

[55]*Asbestos Information Ass'n/North America v. OSHA,* (Cases Nos. 83-4687, 4688, 4689, 4711, filed Nov. 2, 1983 and Nov. 15, 1983). An official of the AFL-CIO stated that it was "disappointed and dismayed" by the ETS because the new level was "five times higher than what the unions had requested" and because the standard would allow employers to use respirators rather than engineering controls to meet the requirements of the standard. AFL-CIO NEWS, Nov. 7, 1983, at 4, cols. 4–5. On the issue of engineering controls in emergency standards, see discussion *supra* note 27.

[56]Memorandum in Support of Petitioners' Motion for Stay Pending Judicial Review at 35–44.

[57]Chief Judge Charles Clark and Circuit Judge E. Grady Jolly, with Circuit Judge Alvin B. Rubin dissenting.

[58]13 OSHR 691 (1983).

Asbestos Information Association v. OSHA

727 F.2d 415, 11 OSHC 1817 (5th Cir. 1984)

JOLLY, Circuit Judge:
***

II.

***

No new data or discovery leads OSHA to invoke its extraordinary ETS powers and lower the asbestos PEL. Rather, OSHA bases its conclusion that a grave danger exists on quantitative risk assessments, which are mathematical extrapolations, of the likelihood of contracting an asbestos-related disease at various levels of exposure to asbestos particles. The risk assessment, which OSHA completed in July of 1983, and a meeting a few months earlier between the Assistant Secretary of Labor and a recognized expert in the asbestos epidemiology field, heightened OSHA's awareness of the asbestos situation and precipitated the ETS.

OSHA calculated the likelihood of developing lung cancer, mesothelioma, and gastrointestinal cancer due to contact with ambient asbestos fibers at different exposure levels. By applying its calculations to an estimated working population exposed to asbestos, OSHA claims that 210 lives eventually can be saved from cancer by lowering the PEL to 0.5 f/cc for six months. 48 FED. REG. at 51,086. These figures include deaths that will occur at OSHA's estimated current actual exposure levels and include employees working in environments where the density of ambient asbestos particles is 20 f/cc, ten times the current PEL. Even if, however, OSHA removes from the computation those employees who do not enjoy the benefit of the current 2.0 f/cc PEL because it is not enforced in their work place, and counts only those employees who are exposed to ambient asbestos between the levels of 2.0 f/cc and 0.5 f/cc, OSHA estimates it can save 80 lives by lowering the PEL for six months.[8]

OSHA calculated the number of lives saved by first deriving a mortality rate, which is the number of excess deaths[9] because of exposure to ambient asbestos particles at different levels.[10] It then multiplied the number of

[8]OSHA's own data, however, indicates that the actual number of asbestos-related cancer deaths prevented by a 0.5 f/cc standard would be approximately 40 for six months. Approximately 71% of the benefits of OSHA's ETS accrue in the drywall construction industry where demolition and other activities generate large amounts of ambient asbestos particles. OSHA specifically estimated it could save 157 lives in that industry by lowering the PEL for six months, assuming the entire industry currently complies with the 2 f/cc standard. 48 FED. REG. at 51,097 (Table 4). Its estimate of employee exposure, however, indicates that of all 51,621 employees estimated to be working in that industry, 38,666 or approximately 75% currently are exposed to only 0.2 f/cc. *Id.* at 51,094 (Table 1). Consequently, these workers would not benefit by lowering the standard to 0.5 f/cc. Reducing OSHA's calculations by 75% indicates that approximately 14 drywall construction workers will benefit from the ETS. Stated differently, approximately 43 of the 57 workers already are exposed to levels below that which the ETS would permit.

[9]Excess deaths are those that would not occur in a control group not exposed to asbestos. 48 FED. REG. at 51,101. For mesothelioma, the number of excess deaths projected due to asbestos exposure also is the total number of deaths, because mesothelioma almost never occurs in someone not exposed to asbestos. *Id.* at 51,089.

[10]Because of the latency period for most asbestos-related cancers, OSHA had no observations for asbestos-related deaths under the current 2 f/cc PEL. OSHA calculated a mortality rate for the 2 f/cc PEL by estimating the cumulative number of particles to which a worker will be exposed over varying lengths of time at 2.0 f/cc, and then applying the mortality rate to higher exposures over a shorter period of time. For example, a worker exposed to 0.2 f/cc for ten years would be as likely to develop cancer as a worker exposed to 0.4 f/cc for five years. The validity of this calculation depends on the assumption accepted by most, but not all, researchers that the likelihood of developing asbestos-related cancer varies directly with the number of fibers inhaled, and is not independently affected by intensity or duration of exposure.

workers currently exposed at those levels by the mortality rate. Finally, to obtain a projected number of lives saved, it subtracted the number of deaths that it estimates will continue to occur even at the new PEL from the number of deaths likely to occur at the higher levels of exposure, which resulted in 210 deaths for six months exposure. 48 FED. REG. at 51,095–51,097 and Chart 4.

The underlying data base from which OSHA derived its mortality rates consists of eleven epidemiological studies which OSHA felt contained sufficient data to allow computation of quantitative risk assessments for lung cancer. Four of these studies OSHA decided also provide sufficient data to compute risk assessments for mesothelioma. These eleven studies observe a total of approximately 53,000 workers in several countries and in a wide range of occupations. The studies include insulation workers, production workers, maintenance employees, textile workers, miners, and millers. They observe a variety of exposure levels, and include workers exposed to the three commonly occurring asbestos fiber types. 48 FED. REG. at 51,101–51,105.

***

III.

***

In reviewing the ETS, we also must remain aware that the plain wording of the statute limits us to assessing the harm likely to accrue, or the grave danger that the ETS may alleviate, during the six-month period that is the life of the standard. OSHA urges us to assess the harm likely to accrue over at least a year, even though the ETS expires six months from its promulgation. At oral argument OSHA said that even if the ETS lapsed before OSHA promulgated a permanent regulation, the benefits of the ETS likely would continue because employers will have expended the resources to comply with the new lower standard and would have no incentive to revert to old practices. These post hoc rationalizations cannot be accepted as basis for our review; first, because the ETS statute does not contemplate the Secretary's allowing an ETS to lapse before he promulgates a permanent standard, and second, because to assume that employers will not revert to less exacting standards is pure speculation. The opposite is equally plausible, especially given that OSHA allows compliance with the ETS through methods as simple as wetting floors or wearing respirators. See 48 FED. REG. at 51,086.

In its November 4 publication OSHA partially justified its decision to issue an ETS on the fact that notice-and-comment rulemaking often takes several years to complete, excluding possible subsequent postponements of the effective date caused by court-ordered stays pending judicial review. 50 FED. REG. at 51,089. OSHA apparently would have us assess benefits in this light. We cannot do so. As noted earlier, OSHA concedes that it can complete rulemaking within one year. Additionally, as its legislative history makes clear, the ETS statute is not to be used merely as an interim relief measure, but treated as an extraordinary power to be used only in "limited situations" in which a grave danger exists, and then, to be "delicately exercised." *Public Citizen Health Research Group v. Auchter,* 702 F.2d at 1150 (D.C. Cir. 1983). See also *Taylor Diving & Salvage v. Department of Labor,* 537 F.2d 819, 820–21 (5th Cir. 1976); *Florida Peach Growers,* 489 F.2d at 129; *Dry Color Manufacturers' Ass'n,* 486 F.2d at 104 n.9a (3d Cir. 1973). The Agency cannot use its ETS powers as a stop-gap measure. This would allow it to displace its clear obligations to promulgate rules after public notice and opportunity for comment in any case, not just in those in which an ETS is necessary to avert grave danger. See 29 U.S.C. §655(b).

IV.

A.

The AIA urges us to hold that OSHA must have new information before it promulgates an ETS. An "emergency" cannot exist, it argues, when the agency has known for years that asbestos constitutes a serious health risk and, in fact, has had all the data it uses to support its November 4 action at hand, but nevertheless failed to act on it. Although new information may be a sound basis for an ETS, we decline to hold that OSHA cannot issue an ETS in its absence. As OSHA admits, the Agency's failure to act may be evidence that a situation is not a true emergency, but we agree with OSHA that failure to act does not conclusively establish that a situation is not an emergency.

The ETS statute itself, allowing the Secretary to promulgate an ETS in response to "grave danger * * * or * * * new hazards," precludes our imposing a "new information" requirement on OSHA. Additionally, to impose such a requirement would imprudently circumscribe the Secretary's ability to act in response to serious situations. If exposure to 2.0 f/cc of asbestos fibers creates a grave danger, to hold that because OSHA did not act previously it cannot do so now only compounds the consequences of the Agency's failure to act.

OSHA should, of course, offer some explanation of its timing in promulgating an ETS, especially when, as here, for years it has known of the serious health risk the regulated substance poses, and has possessed, albeit in unrefined form, the substantive data forming the basis for the ETS. In this case OSHA says it acted in response to new awareness of the danger of asbestos and in response to extrapolated data that did not become available until July of 1983, four months before it promulgated the ETS. We are not prepared to say that such heightened awareness cannot justify the Secretary's action.

Additionally, even if adequately explained, an ETS must, on balance, produce a benefit the costs of which are not unreasonable. The protection afforded to workers should outweigh the economic consequences to the regulated industry. *American Petroleum Institute v. OSHA,* 581 F.2d 493, 502–03 (5th Cir. 1978), *aff'd sub nom. Industrial Union Department v. American Petroleum Institute,* 448 U.S. 607 (1980); *Florida Peach Growers,* 489 F.2d at 130.[18] OSHA conducted a benefits analysis prior to promulgating the ETS, and concluded that the cost of compliance with the lower PEL is reasonable compared to total industry sales volume. Further, OSHA concluded that the costs are fairly distributed, because the industries in which asbestos-related risks currently are greatest are the industries that must spend the most to comply with the lower standard.[19] In no industry does the compliance cost exceed 7.2 cents per dollar of sales, and in most industries, the cost of compliance is less than one cent per dollar. 48 FED. REG. at 51,136–51,137 and Tables 17 and 18. We cannot say that the cost of compliance is unreasonable if the ETS in fact alleviates a grave danger.

***

[18]Although in this case the agency conducted a formal cost-benefit analysis, we do not imply that the Occupational Safety and Health Act requires the agency to do so before it promulgates an ETS. Indeed, in true "emergency" situations, that the agency would have time to conduct such an analysis is unlikely. The *American Petroleum Institute* and *Florida Peach Growers* cases require only that in reviewing whether the agency's action was reasonable under the circumstances, we analyze the anticipated benefit of the ETS in light of its probable consequences.

[19]For the six months the ETS remains in effect, the average cost per worker of compliance is $708. The construction industry cost per worker is $973, the highest of any industry segment. The automotive aftermarket industry has the lowest cost per worker at $251. 48 FED. REG. at 51,137, Table 17. The estimated employee exposure in the automotive aftermarket currently is less than 0.1 f/cc for the vast majority of employees. *Id.* at 51,093, Table 1.

B.

The ETS statute requires that the Secretary issue an ETS only after he finds substantial evidence indicating both that a "grave danger" exists and that an emergency standard is "necessary" to protect workers from such danger. Thus, the gravity and necessity requirements lie at the center of proper invocation of the ETS powers. No one doubts that asbestos is a gravely dangerous product. The gravity we are concerned with, however, is not of the product itself, but of six months exposure to it at 0.5 f/cc, as compared with six months exposure at 2.0 f/cc. Our inquiry, then, is a narrow one, and requires us to evaluate both the nature of the consequences of exposure, and also the number of workers likely to suffer those consequences.

***

OSHA claims that by permanently lowering the present 2.0 f/cc PEL to 0.5 f/cc, it will save sixty-four lives per one thousand workers over a working lifetime of forty-five years. See 48 FED. REG. at 51,100. Over six months, this works out to eighty lives out of an estimated worker population of 375,399.[21] 48 FED. REG. at 51,094–51,095. As the Supreme Court has noted, the determination of what constitutes a risk worthy of Agency action is a policy consideration that belongs, in the first instance to the Agency. *Industrial Workers Union,* 448 U.S. at 656 n.62. "Some risks are plainly acceptable and others are plainly unacceptable." *Id.* at 655. The Secretary determined that eighty lives at risk is a grave danger. We are not prepared to say it is not.

***

* * * OSHA has made the *number* of deaths avoided—at least 80—the basis for its rulemaking. Yet it is apparent from an examination of the record that the actual number of lives saved is uncertain, and is likely to be substantially less than 80.[22] Both the gravity of the risk as defined by OSHA and the necessity of an ETS to protect against it are therefore questionable.

Additionally, although risk assessment analysis is an extremely useful tool, especially when used to project lifetime consequences of exposure, the results of its application to a small slice of time are speculative because the underlying database projects only long-term risks. Epidemiologists generally study only the consequences of long-term exposure to asbestos. Indeed, OSHA concedes some unreliability and uncertainty to be inherent in risk assessment generally. Applying the risk assessment process to a period of six months, one-ninetieth of OSHA's estimated working lifetime, only magnifies those inherent uncertainties.

By holding as we do in this case, however, we do not intimate at all that risk-assessment analysis is inappropriate evidence on which to base any standard, temporary or permanent. We say no more than that evidence based on risk-assessment analysis is precisely the type of data that may be more uncritically accepted after public scrutiny, through notice-and-comment rulemaking, especially when the conclusions it suggests are controversial or subject to different interpretations.

C.

Even assuming that OSHA's projected benefits would accrue from the ETS, however, we hold that OSHA's action must fail for another reason. The

[21]In the November 4 publication, OSHA stated that 210 lives will be saved over six months. The Agency, however, concedes that this figure is inflated because it includes those lives that OSHA could save by enforcing its current 2.0 f/cc standard. See *supra* note 8 and accompanying text.

[22]See *supra* note 8.

Agency has not proved that the ETS, OSHA's most dramatic weapon in its enforcement arsenal, is "necessary" to achieve the projected benefits.

As OSHA concedes, the probable practical effect of the ETS, which allows compliance through "any feasible combination of engineering controls, work practices, and personal protective equipment and devices," would be that employers would require employees to wear respirators. Current regulations already require employers to outfit workers with respirators that can provide up to one hundred-fold protection. 29 C.F.R. §§1910.1001(c)(2)(iii), (d)(2)(ii), and (d)(2)(iii). Yet OSHA did not include in its calculations the effect of enforcing the current standard by requiring employers in the drywall construction and demolition industry to furnish these respirators. Counsel for OSHA informed the court at oral argument that the Secretary considers the regulation requiring construction and demolition workers to wear respirators to be unenforceable absent actual monitoring to show that ambient asbestos particles are so far above the permissible limit that respirators are necessary to bring the employees' exposure within the PEL of 2.0 f/cc. Fear of a successful judicial challenge to enforcement of OSHA's permanent standard regarding respirator use hardly justifies resort to the most dramatic weapon in OSHA's enforcement arsenal. Thus, lacking a satisfactory explanation why the ETS is a necessary means to achieve the added saving obtainable by application of the current regulations, we must assume that OSHA's claimed benefit should be discounted by some additional, uncertain amount.

OSHA also attempts to justify the ETS by emphasizing that the ETS does more to protect workers' health than simply lowering the asbestos fiber PEL. An ETS, however, is not necessary to achieve these ancillary benefits. The ETS requires employers to educate employees concerning the risks of asbestos exposure and the proper steps necessary to minimize exposure. While education is a worthy objective, OSHA could achieve it without invoking its extraordinary ETS power. Indeed, current regulations provide for worker training and education. *Id.* at §1926.21. Similarly, OSHA supports its action by arguing that it plans to increase enforcement efforts, with the aim of encouraging greater compliance with the new standard than it estimates currently exists under the present standard. Increasing enforcement is another worthy objective; but it likewise cannot justify use of the ETS power, especially when, as in this case, much of the claimed benefit could be obtained simply by enforcing the current standard.

***

Thus, despite OSHA's extensive use of risk assessments to establish grave danger from asbestos, the Agency for the fourth time was unable to persuade the Court of Appeals for the Fifth Circuit that its emergency power under §6(c) was validly exercised. The court emphasized its view that OSHA's power was not to be used "merely as an interim relief measure"[59] but was an "extraordinary power to be used only in 'limited situations' in which a grave danger exists, and then, to be 'delicately exercised.'" In support of these propositions, the Court cited not only its own ETS decisions but also *Public Citizen Health Research Group v. Auchter,* [60] in which the Court of Appeals

[59]The court expressly rejected OSHA's argument that it could rely on long-range benefits from lowering the PEL to justify the ETS. 11 OSHC at 1821. See 48 FED. REG. at 51,091, 51,098, and the discussion *supra*.

[60]702 F.2d 1150 (D.C. Cir. 1983). See discussion in Chapter 6, Section C.

for the District of Columbia reversed the District Court order that OSHA be required to issue an ETS on ethylene oxide. The Court of Appeals for the Fifth Circuit here accepted OSHA's view that the ETS authority may be utilized even where "new information" does not become available to the Agency,[61] but went on to disagree with OSHA's calculation that the lowered PEL would save 80 lives in 6 months. The court stated that no more than 40 lives would be saved,[62] and since OSHA calculations on the number of lives that would be saved were incorrect, the court concluded that the Agency's findings on "gravity" and "necessity" were "questionable."

The court's discussion of the use of risk assessments as a basis for ETSs is also noteworthy. The court questioned particularly the reliability of risk assessments in determining risks over a short period, and referred to OSHA's concession that there was "some unreliability and uncertainty * * * inherent in risk assessments generally." In addition, the court said that risk assessment data is "precisely" that type of information that should undergo public comment before it is accepted by an agency as a basis for regulation. It is indeed ironic that in its decision on the *Benzene* standard, the court of appeals refused to accept OSHA's argument that risk assessments were unreliable and insisted that the Agency show "measurable" benefits; in this case, where OSHA used risk assessments, the court embraced the unreliability of risk assessments in concluding that they cannot be relied on as a basis for an ETS.[63]

On April 10, 1984, OSHA proposed a revision of its existing asbestos standard. The notice proposed two alternative permissible exposure limits: .2 fibers/cc and .5 fibers/cc (the PEL in the 1983 ETS). The proposed standard, which included other changes in the existing standard, would apply to all industries covered by the Act, including the maritime and construction industries. 49 FED. REG. 14,116 (1984).

---

[61]11 OSHC at 1822. The court's statement that the benzene ETS was set aside for lack of substantial evidence, 11 OSHC at 1822 n.16, appears to be in error. The court's citation is to its decision vacating the "permanent" benzene standard; the ETS was stayed, became involved in litigation over proper venue for the proceeding, and was later withdrawn. No decision on the merits was issued. See discussion in Chapter 7.

[62]11 OSHC at 1819 n.8.

[63]The court's statements that the ETS "must, on balance, produce a benefit the costs of which are not unreasonable," and the statement that the "protection afforded to workers should outweigh the economic consequences to the regulated industry," 11 OSHC at 1822–23, are difficult to understand in light of the Supreme Court decision in *American Textile Mfrs. Inst. v. Donovan,* 452 U.S. 490 (1981). The court of appeals, however, did not vacate the standard because of cost factors.

5

Medical Surveillance: Evolution of OSHA Policy

A. Background

This chapter traces the development of OSHA policy respecting one of the important elements of health standards: the requirement that medical examinations be made available, without cost, to employees exposed to occupational hazards.

In addition to the more general criteria for OSHA health standards contained in §6(b)(5),[1] §6(b)(7)[2] requires that certain specific provisions be included in these standards. These include provisions on labeling or other forms of warning to apprise employees of the hazards to which they are exposed; provisions on protective equipment and "control or technological procedures" to be used in connection with hazards; and provisions for the monitoring or measuring of employee exposure levels. Section 6(b)(7) further provides: "Where appropriate, any such standards shall prescribe the type and frequency of medical examinations or other tests which shall be made available, by the employer or at his cost, to employees exposed to such hazards in order to most effectively determine whether the health of such employees is adversely affected by such exposure." If the examinations are "in the nature of research," they may be furnished at the expense of the Secretary of Health and Human Services (HHS). This section further states that the "results of such examinations or tests" shall be furnished "only" to the Secretaries of Labor and HHS "and at the request of the employee, to his physician." All OSHA health standards issued under §6(b), commencing with the asbestos standard issued in 1972,[3] have included the required medical surveillance provisions. And, while the broad outline of medical surveillance programs has remained essentially the same since

[1] 29 U.S.C. §655(b)(5) (1976).
[2] *Id.* at §655(b)(7).
[3] 37 Fed. Reg. 11,318, 11,322 (1972), *amended by* 45 Fed. Reg. 35,281 (1980) (codified at 29 C.F.R. §1910.1001(j) (1983)).

OSHA's asbestos standard, there have been significant changes in the details of these requirements. The formulating of medical surveillance provisions has frequently confronted OSHA with sensitive legal and policy issues.[4]

The medical surveillance program has two primary purposes: (1) to give the employee notice of any adverse health effects that he or she may have suffered, so that proper medical attention may be obtained and proper precautions, such as removal from exposure, taken; and (2) to provide OSHA and NIOSH with data for purposes of research into the relationship of toxic substances and occupational illnesses. In addition, employee medical records often provide OSHA with information necessary for enforcement of the Act.[5]

The basic elements of a medical surveillance program in a health standard were set forth by OSHA in its discussion of a model carcinogen standard in the Agency's Carcinogens Policy issued in 1980.[6]

Carcinogens Policy Final Standard

45 Fed. Reg. 5001, 5230–31 (1980)

***

The model standards require that the medical surveillance program provide each covered employee with an opportunity for medical examination. As noted above, the authority and requirement for this provision are found in section 6(b)(7) of the Act.

All examinations and procedures are required to be performed by or under the supervision of a licensed physician and provided without cost to the employee. While the physician will usually be selected by the employer, the standard does not so mandate, leaving the employer free to institute alternative procedures such as joint selection with the employee or selection by the employee. Clearly, a licensed physician is the appropriate person to be conducting a medical examination. However, certain parts of the required examination (e.g., taking a history) do not necessarily require a physician's expertise and may be conducted by another person under the supervision of the physician. As noted above, the Congress has mandated, by reason of section 6(b)(7) of the Act, that medical examinations and procedures required by OSHA standards be provided at no cost to the employee.

The model standards provide that a work history, medical history and medical examination be performed. The purpose of this requirement is to establish a baseline health condition against which changes in an employee's health may be compared in the detection of changes in physical condition.

The model standards leave blank the specific medical protocols for the examination. The various tests that comprise the medical examination to be prescribed in the individual rulemaking are determined by the specific toxicology and effects of each substance. The tests that are designed to be used in an initial assessment of an employee's health and to detect changes in

[4]See, e.g., Vinyl Chloride Standard, 39 FED. REG. 35,890, 35,895 (1974), reprinted in Chapter 3, Section B.

[5]See Access to Employee Exposure and Medical Records, 45 FED. REG. 35,212, 35,226–27 (1980) (discussion of the various ways employee medical records are important to OSHA in the performance of its statutory responsibilities in the compliance and standards-setting areas).

[6]This language on medical surveillance programs also appears in the proposal. 42 FED. REG. 54,148, 54,178 (1977).

health which may occur are to be required as appropriate. Of course, if additional tests and procedures are appropriate for a given substance, they may be added in the specific rulemaking covering that substance.

The model standards require that the employer provide the physician with certain specified information. This information includes a copy of the regulation, a description of the affected employee's duties as they relate to the employee's exposure, the results of the employee's exposure measurement, if any personal protective equipment is used or is to be used, and information from previous medical examinations of the affected employee to the extent that they are not readily available to the physician. The purpose in making this information available to the physician is to aid in the evaluation of the employee's fitness to wear personal protective equipment when required.* * *

The employer is required to obtain a written opinion from the examining physician containing: the physician's opinion as to whether the employee has any detected medical conditions which would place the employee at increased risk of material impairment of health from exposure to the toxic substance; the results of the medical tests performed; any recommended limitations upon the employee's exposure to the toxic substance and upon the use of protective clothing and equipment such as respirators; and a statement that the employee has been orally informed by the physician of any medical conditions which require further examination or treatment. This written opinion must be relevant to specific findings or diagnoses unrelated to occupational exposure, and a copy of the opinion must be provided to the affected employee. The purpose in requiring the examining physician to supply the employer with a written opinion containing the above mentioned analyses is to provide the employer with a medical basis to aid in his determination of the employee's initial placement and continuing capability to use protective clothing and equipment. Requiring that the opinion be in written form will serve as an objective check that employers have actually had the benefit of the information in making these determinations. Likewise, the requirement that the employee be provided with a copy of the physician's written opinion will assure that the employee is informed of the results of the medical examination and may take any appropriate action. The purpose in requiring that specific findings or diagnoses unrelated to occupational exposure not be included in the written opinion is to encourage employees to submit to a medical examination by removing the fear that employers may find out information about their physical condition that has no relation to occupational exposures.

Additional important issues that arise in medical surveillance programs in OSHA health standards are as follows:

1. Employees must be given "an opportunity" to take a medical examination; OSHA generally has not required that employees take the examination.[7]

2. The examinations are given at the expense of the employer. The employer must pay for the examination itself, and although the standards have not dealt explicitly with the issue of pay, they have generally been understood to mean that the employer also pays the employee for his time spent taking the examination if it is given

[7]See discussion in Section B, this chapter.

during normal working hours. The arsenic standard goes further, stating that the examination be given "without cost to the employee, without loss of pay, at a reasonable time and place,"[8] with the preamble explaining that the employer "is obligated to pay for the time spent taking the medical examination if it is taken outside normal working hours."[9]

3. The standard defines the group of employees who are to take the medical examination. In some cases, the requirement applies to all exposed employees; under some standards, it applies only to employees exposed above the action level.[10]

4. The medical surveillance provision states how frequently the examination is to be given. In the asbestos standard, examinations must be given before placement, annually thereafter, and on termination of employment.[11] Under some standards, the frequency of examination depends on the age of the employee,[12] the length of time the employee had been exposed to the regulated substance,[13] or the extent to which the results of earlier examinations show that the employee is suffering from adverse medical conditions.[14]

5. The standard normally describes the minimum protocol for the medical examination. In its standard for 14 carcinogens, OSHA did not include detailed criteria for the contents of the medical examinations. This portion of the standard was challenged by the Oil and Chemical Workers Union, and the Court of Appeals for the Third Circuit remanded the case to OSHA for further action on this issue. The court said "[i]t is important that the worker be assured of the benefits of medical procedures that are presently available and that the industry be advised of what is expected of it."[15]

6. The standards also deal with the issue of access to the results of the medical examinations by the employer, the employee, and employee representatives, and the action, if any, that must be taken by the employer if the examination shows an adverse medical condition.[16]

---

[8]29 C.F.R. §1910.1018(n) (1983).

[9]43 FED. REG. 19,584, 19,621 (1978). On February 14, 1984, the Court of Appeals for the Ninth Circuit held that a company violated the arsenic standard by failing to compensate employees for the time and costs connected with taking medical exams during nonworking hours. *Phelps Dodge Corp. v. OSHRC*, 11 OSHC 1769 (9th Cir. 1984). See also the Ethylene Oxide Proposal, 48 FED. REG. 17,284, 17,311 (1983), providing that the physical must be provided to the employee "without loss of pay at a reasonable time and place." The preamble to the standard explains that these provisions are designed "to assure that they [the medical examinations] are taken." *Id.* at 17,305.

[10]E.g., Benzene, 43 FED. REG. 5918, 5965 (1978). The plurality opinion of the Supreme Court in the *Benzene* decision was critical of OSHA's exclusion of employees exposed below the action level from the medical surveillance requirements. *Industrial Union Dep't v. American Petroleum Inst.*, 448 U.S. 607, 650, 658 & n.66 (1980).

[11]29 C.F.R. §1910.1001(j) (1982).

[12]E.g., Coke Oven Emissions, 29 C.F.R. §1910.1029(j)(3) (1983).

[13]*Id.*

[14]E.g., Lead, 29 C.F.R. §1910.1025(j)(2) (1983) (relating to biological monitoring).

[15]*Synthetic Organic Chem. Mfrs. Ass'n v. Brennan*, 506 F.2d 385, 391 (3d Cir. 1974), *cert. denied sub nom. Oil, Chem. & Atomic Workers Internat'l Union v. Dunlop*, 423 U.S. 992 (1975). The question of genetic screening and monitoring of employees is discussed in Section E, this chapter.

[16]These issues are discussed in Section D, this chapter.

7. The medical surveillance provisions also require employers to maintain the examination records for a specified period of time. Under the asbestos standard, for example, this period was not less than 20 years. The Industrial Union Department challenged this period as being too short, but the provision was upheld by the court.[17]

B. Are the Examinations Voluntary for the Employee?

Section 6(b)(7) provides that the examination shall "be made available" to exposed employees[18]; this language implies that the examinations are voluntary and the employee may refuse to take the examination.[19] The asbestos standard does not deal explicitly with this issue, although it suggests that the examinations are voluntary.[20] The standard for 14 carcinogens, issued in 1974, is ambiguous on this issue.[21] The vinyl chloride standard, however, also promulgated in 1974, made it clear that the examinations were voluntary, stating that the medical surveillance program "shall provide each such employee with an opportunity for examinations and tests in accordance with this paragraph."[22] The coke oven emission standard, as well as all later standards (other than diving[23]), also made clear that employees are not required to undergo medical examinations.[24] In order to assure that the employee refusal to take an examination would be "informed," however, the coke oven emission standard required that the employer inform the employee of the health consequences of a refusal to take the examination and obtain a signed statement from the employee stating that he or she has been informed by the employer of the consequences of not taking the examination and understood those consequences.[25]

The issue of whether medical examinations should be voluntary was discussed most fully in the OSHA lead standard, in the context of the medical removal protection (MRP) provisions.[26] The major purpose of medical removal protection was to assure full employee participation in the medical surveillance program by eliminating employee concern that they would suffer loss of pay if they were

[17]*Industrial Union Dep't v. Hodgson,* 499 F.2d 467, 486 (D.C. Cir. 1974). This provision in the asbestos standard appears at 29 C.F.R. §1910.1001(i) (1983).

[18]29 U.S.C. §655(b)(7).

[19]This discussion relates only to the requirements of the OSHA standard and does not deal with any work rules governing medical examinations which may be promulgated by the employer, consistent with the requirements of other statutes, and collective bargaining agreements.

[20]The court of appeals so understood the provision. *Industrial Union Dep't v. Hodgson,* 499 F.2d at 485 n.45, reprinted in Chapter 6, Section A. The pertinent language is that the employer shall "provide or make available" the examinations. The import of the word "provide," in contrast to "make available," is not clear.

[21]39 FED. REG. 3756, 3762 (1974). The standard states that the medical examination "shall be provided."

[22]39 FED. REG. 35,890, 35,897 (1974) (codified at 29 C.F.R. §1910.1017(k) (1983)).

[23]See discussion *infra* this section.

[24]E.g., 29 C.F.R. §1910.1029(j)(1)(iii) (1983).

[25]*Id.*

[26]Medical removal protection is discussed in Section D, this chapter.

transferred from their jobs if the physical examination disclosed an adverse medical condition. OSHA concluded that this purpose could *not* be served by making the medical examinations mandatory. The lead preamble states:

> 5. *Alternatives to MRP considered by OSHA*. Before deciding to include MRP in the final lead standard, OSHA considered and rejected several possible alternatives. Mandating that employers compel all employees to participate in medical surveillance offered under the standard was rejected in part due to the fact that this step could not possibly assure the voluntary and meaningful worker participation upon which success of the standard's medical surveillance program depends. Mere participation is not an end in and of itself. For example, no degree of compulsion can prevent workers from obtaining and misusing chelating agents so as to yield apparently low blood lead level results. No degree of compulsion can force workers to reveal subtle, subjective symptoms of lead poisoning which a physician needs to know as part of an adequate medical history.
>
> In addition, OSHA declined to mandate worker participation in medical surveillance due to the substantial personal privacy and religious concerns involved in health care matters. Governmental coercion in this sensitive area would prove counterproductive to the goal of meaningful worker participation. Finally, the foregoing arguments against mandatory participation arise irrespective of whether or not MRP benefits are provided to removed workers. Thus, mandatory worker participation with MRP is no more satisfactory an alternative than mandatory worker participation without MRP.[27]

The only standard in which OSHA made the medical surveillance program mandatory was diving, issued in 1977.[28] The reasons given by OSHA in the preamble to the standard were that "diving is basically a high-stress occupation performed under difficult environmental conditions, and that the safety of the diver and other dive team members can depend on the health of the individual diver."[29] The standard further provided that the employer must then decide whether the employee was medically fit to perform assigned tasks, consistent with the medical opinion.[30]

Thus, as a general matter, OSHA has not required employees to take medical examinations. This policy decision has not elicited significant opposition from either employee or employer representatives. From the point of employees, their rights are protected if the examination is made available; mandating the examination causes employee concern that the results of the examination would be used as a basis for adverse personnel action against the employee.[31] There

[27]43 FED. REG. 52,952, 52,973–74 (1978).
[28]42 FED. REG. 37,650 (1977).
[29]*Id.* at 37,656–57.
[30]*Id.* at 37,657. The diving standard also contained provisions for three-physician review of medical determinations; these were held invalid by the court of appeals, which vacated the entire medical surveillance program in the diving standard. See discussion in Section C, this chapter. Although the preamble did not expressly relate the mandatory nature of the examinations to the availability of three-physician review, it would appear that the employee's right to appeal an "adverse" medical opinion to a second and third physician would have minimized somewhat the coercive impact on employees of mandatory medical examinations.
[31]See discussion in Section D, this chapter.

are also privacy interests which would militate against employees advocating mandatory examinations. Employers have not opposed voluntary examinations, most likely because mandatory examinations could prove hard to enforce, and, to the extent the employer wished to require physical examinations, he or she could impose them, relying on management work rules.

C. Who Selects the Physician?

Section 6(b)(7) provides that medical examinations shall be made available "by the employer or at his cost."[32] The question whether the employer should choose the physician and the related issue of whether the employer should have a right to obtain the medical records of an employee were important issues in the proceedings leading to the asbestos standard. In the final standard, OSHA determined that the employer should have the option of choosing the physician and should have access to the results of the medical examination. The preamble explains this determination briefly as follows:

> One question which has been raised goes to whether the employer or the employee should be allowed to choose the examining physician. The standard gives the option to the employer. Since some employers already have a medical examination program in operation, and, also, have medical departments with some expertise in the diagnosis of asbestos-related diseases, it seems more reasonable to permit them to utilize the present programs and expertise, than to permit an employee to choose a private general practitioner.[33]

The Industrial Union Department (IUD), among other parties, challenged this result in the court of appeals, arguing that employer access to the results of the examination could well prejudice the employee's job status. The court, however, upheld OSHA's determination.

Industrial Union Department v. Hodgson

499 F.2d 467 (D.C. Cir. 1974)

McGOWAN, Circuit Judge:

***

The standards require that all employers provide a medical examination (1) when an employee is first assigned to a job involving exposure to asbestos, (2) when he leaves that job, and (3) annually during his employment in the position. Petitioners do not attack this timetable or the substantive requirements governing the nature of the examination, but they do assert that the examinations should be given by a physician of the employee's choice, and that the results should be given to the employer only if the employee decides to do so.

Petitioners argue that the standards violate the principle of physician/

[32] 29 U.S.C. §655(b)(7).

[33] 37 FED. REG. 11,318, 11,319 (1972).

patient confidentiality because they would allow the physical examinations to be conducted by company doctors and would make the results available to the employer. Confidentiality is necessary, they argue, to avert the possibility that, in hiring and discharging employees, employers will discriminate against those with symptoms of asbestos-related disease or prior histories of exposure to asbestos dust. The Secretary recognized this potential problem, and stated that uses of the records would be scrutinized carefully. However, he did not consider the possibility of such abuse sufficient to outweigh the opposing considerations.

The standards require the employer to take into consideration the result of an employee's most recent physical examination in making assignments to jobs requiring the use of respirators. An employee who cannot safely perform such a job is to be reassigned without loss of seniority or wages. The Secretary reasoned that the salutary purposes of this provision could not be fulfilled if employers were denied access to the medical records.[45]

Since the results are to be made available to the employer, allowing the employer to select the physician has the advantages of both convenience and efficiency. Some employers already maintain an industrial medical staff skilled in dealing with asbestos-related medical problems, and the requirements of the standards as promulgated appear likely to encourage the development of additional staff of this sort. The employee would then benefit from the expertise associated with specialization, and the state of medical knowledge concerning these diseases should likewise be improved. All of these factors, identified in the record, operate to make the Secretary's decision on this point a non-arbitrary one.

***

The question of who chooses the examining physician is related to a central controversy in the medical surveillance program. A major purpose of the medical surveillance program is to allow protective action to be taken if the employee shows symptoms of disease; this action might include, as the court of appeals pointed out, transfer of the employee from his or her job to avoid or minimize continued exposure to the toxic substance. While this result is medically necessary, employees have become concerned that this transfer would lead to loss of jobs or removal to lower-paying positions. The Act does not directly provide protection against discharge or layoffs in those circumstances,[34] and workers' compensation remedies are often inadequate. In an attempt to eliminate or minimize this adverse impact, IUD argued in the *Asbestos* case that the employer should not choose the physician and should not have access to the results of the medical examination so that the employer would not be in a position to rely on these results to fire or transfer the employee. The issue was decided in court against the IUD, however, and all health standards have followed the *Asbestos* decision on this issue. Employee groups have generally opposed mandatory medical examinations, favored

[45]The ultimate choice remains in the hands of the employees in any event since the examinations provided by the employer are entirely voluntary insofar as employees are concerned.

[34]Section 11(c) of the Act would not appear generally to prohibit discrimination in employment conditions because of the employee's medical condition. 29 U.S.C. §660(c).

three-physician review, and opposed mandatory removal of employees from jobs where examinations show them at excess risk.[35] Through these positions, the unions have sought to minimize the possibility that the employer would take adverse and prejudicial personnel action against employees with symptoms of illness or histories of exposure to the toxic substance, arguing in part that the medical surveillance program's effectiveness would be undermined if employees were reluctant to participate because of the possibility of adverse economic consequences. This is not to say that the unions opposed medical examinations or removals; on the contrary, they vigorously supported them. However, they contended that OSHA standards imposing these as requirements would not adequately serve their health purpose, unless they also provide for the protection of employee wages and other benefits. The issue was resolved to the satisfaction of the unions when OSHA adopted medical removal protection, which permitted a fully effective medical surveillance by largely avoiding wage loss to employees.

OSHA's diving standard included for the first time provisions for three-physician review. In substance, an employee who is dissatisfied with the medical opinion of the physician selected by the employer may obtain a second examination, at the expense of the employer, from a physician of the employee's choice. If the second opinion differs from the first, the two physicians select a third physician, who gives a binding opinion. The preamble to the diving standard explained the three-physician review.

Final Standard for Diving Operations

42 Fed. Reg. 37,650, 37,658 (1977)

***

The employer's decision on diving assignments must be consistent with medical opinion. Therefore, the function of the physician's medical report is to serve as a basis of the employer's determination. If the physician's opinion is that an employee is medically fit, the employer should be able to rely on that opinion and assign the employee to any task for which the employee is otherwise qualified. On the other hand, if the physician recommends a restriction or limitation on the employee's hyperbaric exposure, OSHA recognizes that both the employer and employee are put in a difficult position by the standard's requirement that employees who are medically unfit, as determined by the employer based on a mandatory medical examination, not be permitted further hyperbaric exposure. By its nature, diving demands that employees whose assignments require hyperbaric exposure be medically fit. It is recognized that certain medical conditions may be incompatible with diving; persons with these conditions who continue to be exposed jeopardize not only their own lives but may risk the lives of others as well. OSHA must also be cognizant of the employees' countervailing rights to be protected in their choice of occupation. The agency must endeavor not to create, through a health and safety standard, a situation which restricts

[35]See Section D, this chapter.

entry into a profession or allows employees to be dismissed for a cause which is less than substantial.

The proposal provided a procedure to be used if an adverse medical opinion, based on certain mandatory disqualifying conditions, led to an employer's determination to withdraw an employee from further hyperbaric exposure. This decision gave the employee the right to obtain a second opinion from a physician chosen by the employee. If the two medical opinions rendered were in disagreement, the proposal's procedure would have required that a binding third opinion by a physician agreed upon by the first two physicians be obtained.

***

The authority of OSHA to require the three-physician review was challenged in the court of appeals by diving contractors. The Court of Appeals for the Fifth Circuit found the provision beyond OSHA's statutory authority and vacated the entire medical surveillance program in the diving standard.

Taylor Diving & Salvage Co. v. Department of Labor

599 F.2d 622 (5th Cir. 1979)

JAMES C. HILL, Circuit Judge:

***

I.

The petitioners first claim that the Secretary of Labor exceeded his statutory authority in promulgating §1910.411, the medical requirements provisions of the standard. Basically, the petitioners object to §1910.411 because it gives an employee the right to appeal an initial determination of medical unfitness for diving to a second, and possibly third, doctor until the first examining doctor's evaluation is either concurred in by the second doctor or overruled by the third. Also, the employer is required to pay for all medical examinations accorded under §1910.411.

***

We look first to the language of the Act, 29 U.S.C.A. §§651–678, and its legislative history, [1970] *U.S. Code Cong. & Admin. News*, p. 5177, and we see that Congress created OSHA for the sole purpose of protecting the health and safety of workers and improving physical working conditions on employment premises. *Brennan v. OSHRC*, 488 F.2d 337, 338 (5th Cir. 1973); see especially 29 U.S.C.A. §651. As we noted earlier, the Secretary is authorized to promulgate only regulations which are reasonably necessary or appropriate to achieve that Congressional goal.

In promulgating §1910.411, however, OSHA's choice of medical examination procedures was controlled by its "cognizan[ce] of the employees' countervailing rights to be protected in their choice of occupation." 42 *Fed. Reg*. 37,650, 37,658 (1977). In its effort "not to create, through a health and safety standard, a situation which restricts entry into a profession or allows employees to be dismissed for a cause which is less than substantial," OSHA sought to achieve a "balance between the need for a mandatory medical examination and the employee's right to a thorough medical assessment."

Id. In effect, §1910.411 imposes upon employers a mandatory job security provision. OSHA is simply not authorized to regulate job security as it has done here.

The medical examination procedures established by OSHA provide that the fitness evaluation of the third doctor in the examining process controls the medical fitness determination of the employer. Thus, even though the employer may genuinely question the medical fitness of a particular employee, due to the first doctor's conclusion of medical unfitness, the employer's determination of medical fitness is conclusively determined by the third doctor's evaluation, which may be that the employee is medically fit. Furthermore, the employer has no control over the third doctor's fitness standards, so that the employer is prevented from setting higher health standards for employees than the secondary examining doctors choose to set.

While we commend OSHA for trying to avoid doing harm with regulations, we must condemn its efforts here because §1910.411 invalidly binds employers in their determinations of employees' medical fitness. "When adopting OSHA, Congress deliberately sought to achieve job safety while maintaining proper employee-employer relations." *Marshall v. Daniel Construction Co.,* 563 F.2d 707, 716 (5th Cir. 1977), *cert. denied,* 439 U.S. 880 (1978).

The Secretary argues that consideration of and accounting for the economic impact of regulations is permissible when promulgating occupational safety and health standards. We agree. Indeed, a regulation enacted in the single-minded pursuit of absolutely risk free workplaces regardless of cost or limitation of job opportunities might well be struck down; it would certainly be ill-advised. This Court has repudiated OSHA's use of such tunnel vision in an earlier opinion. See *American Petroleum Institute,* 581 F.2d at 502–03. Had OSHA limited the employment of commercial divers by prohibiting the hiring of those who fail to pass a medical examination and then, mindful of unfairly prohibiting a banned employee from obtaining employment, provided for a challenge of the initial determination without making the results of that challenge binding on the employer's employment decision, we would be faced with a different question on appeal. OSHA may arguably provide for a threshold determination of medical fitness, but OSHA may not impose a ceiling on the medical fitness standards used by employers in hiring divers. To countenance the Secretary's regulation here would countenance OSHA's imposing not only minimum standards of safety, but also *maximum* standards which employers would be allowed to impose upon themselves. No such Congressional intent is to be found in the Act. Section 1910.411 is thus vacated.[5]

***

OSHA revisited the issue of three-physician review in the lead standard which was promulgated in 1978. The final lead standard included both provisions for multiple-physician review[36] (as it was called there) and medical removal protection.[37] Ultimately the District of Columbia Circuit Court of Appeals upheld the multiple-physi-

[5]Our conclusion that employers' maximum safety standards may not be limited by OSHA does not, of course, mean that an employer's maximum safety standards are never subject to outside limitation. The unwarranted exclusion from employment of applicants or former employees might well be limited through labor-management relations or some other appropriate avenue. We merely find that OSHA is an inappropriate authority for such regulation.

[36]43 FED. REG. 52,952, 52,998–53,001 (1978).
[37]*Id.* at 52,972–77.

cian review provision, distinguishing *Taylor Diving* and reasoning that OSHA's rationale for multiple-physician review in the lead standard departed substantially from the rationale stated in the diving standard.[38]

D. Employee Removal Requirements and Medical Removal Protection

OSHA has made significant changes since 1972 in its policy regarding the action an employer must take when employees are shown by medical examinations to be at excess risk from continued exposure to a toxic substance. The asbestos standard provided that the employer shall receive a copy of the results of the medical examination, but contained no requirement that employees would be removed from asbestos exposure if they were shown to have any symptoms of asbestos disease. The standard, however, contained the following provision:

> (*c*) No employee shall be assigned to tasks requiring the use of respirators if, based upon his most recent examination, an examining physician determines that the employee will be unable to function normally wearing a respirator, or that the safety or health of the employee or other employees will be impaired by his use of a respirator. Such employee shall be rotated to another job or given the opportunity to transfer to a different position whose duties he is able to perform with the same employer, in the same geographical area and with the same seniority, status, and rate of pay he had just prior to such transfer, if such a different position is available.[39]

The preamble contained no explanation of this limited removal and income protection provision, and it was not challenged in court. In discussing the issue of medical surveillance, the District of Columbia Circuit Court of Appeals referred in passing to the "salutary purposes of this provision."[40]

The standard on 14 carcinogens provided that the physician conducting the medical examination shall "furnish the employer a statement of the employee's suitability for employment in the specific exposure."[41] The provision contained no requirement for removal of employees, nor a provision similar to the asbestos standard for the transfer of employees who were unable to wear respirators for medical reasons.

In the vinyl chloride standard, issued in 1974, OSHA for the first time *required* removal of employees from exposure who were at higher risk. The standard provided that a statement of each employ-

[38]*United Steelworkers v. Marshall,* 647 F.2d 1189, 1239 n.76 (D.C. Cir. 1980), *cert. denied sub nom. Lead Indus. Ass'n v. Donovan,* 453 U.S. 913 (1981).

[39]37 FED. REG. 11,318, 11,321 (1972) (codified at 29 C.F.R. §1910.1001(d)(2)(iv)(c) (1983)).

[40]*Industrial Union Dep't v. Hodgson,* 499 F.2d 467, 485 (D.C. Cir. 1974), reprinted in Section C, this chapter.

[41]See e.g., 4-Nitrobiphenyl, 39 FED. REG. 3756, 3762 (1974) (codified at 29 C.F.R. §1910.1003(g)(2)(iii) (1983)).

ee's "suitability for continued exposure to vinyl chloride, including use of protective equipment and respirators," shall be obtained from the physician. The standard further stated that "[i]f any employee's health would be materially impaired by continued exposure, such employee shall be withdrawn from possible contact with vinyl chloride."[42]

The validity of the removal provision in the vinyl chloride standard was not litigated in court. However, the mandatory removal requirements, which were not coupled with any protection of the wages and benefits of employees who were transferred, was a matter of serious concern to employee representatives. The issue culminated in the development of the coke oven emission standard, with the United Steelworkers of America and the Oil, Chemical, and Atomic Workers Union particularly pressing for income protection for employees who were removed from their jobs.[43] OSHA dealt with the question at some length in its preamble to the coke oven emission standard. The Agency indicated its sympathy with the union concerns, but felt that the record was deficient as a basis for a "rate retention" requirement, as it was then called. At the same time, concluding that mandatory removal was closely tied to rate retention, OSHA decided that neither should be addressed in the coke oven emission standard and deleted the mandatory removal provisions from the standard.

Final Standard for Coke Oven Emissions

41 Fed. Reg. 46,742, 46,780 (1976)

***

The proposed standard included a provision prohibiting the exposure of an employee to coke oven emissions if the employee would be placed at increased risk of material impairment to his or her health from such exposure. Under the proposal, this determination could be based on the physician's written opinion. The proposal did not include any provision requiring the transfer of that employee to another job, nor did it include the Advisory Committee recommendation that any removal from exposure "shall not result in loss of earnings or seniority status to the affected employee." These provisions have been referred to collectively as rate retention.

In this proceeding, representatives of unions indicated their great concern regarding any requirement for the mandatory removal of employees because of increased risk in the absence of a rate retention right for employees so removed. The major argument presented was that the absence of a rate retention provision would constitute a major disincentive to employees to submit to physical examinations because they would fear that an adverse medical opinion could result in loss of employment. As a result, the purpose of the medical surveillance requirements would be subverted and early de-

[42]39 FED. REG. 35,890, 35,897–98 (1974) (codified at 29 C.F.R. §1910.1017(k)(4) & (5) (1983)).

[43]An example of the criticism of OSHA's coke oven emission proposal, which did not include wage protection, is the statement of Steven Wodka of the Oil, Chemical, & Atomic Workers Union: "OSHA is fooling itself if it thinks that workers will submit to medical exams when the results may cost them their job, their seniority, or present pay rate." Quoted in D.P. McCaffery, OSHA AND THE POLITICS OF HEALTH REGULATION 89 (1982).

tection of illness would, too often, not occur. It was also suggested that the absence of a rate retention provision creates a dilemma antithetical to the purposes of the Act—namely, the employee's need to choose between continuing to work but risking his life by continuing to do so, and protecting his health, but losing his job.

This dilemma was articulated by Dr. Eula Bingham, Chairperson of the Advisory Committee, who said: "It is to me an impossible situation for a worker to be afraid to take a physical examination because he is going to lose the job that he uses to feed his family. It is unbelievable."*

The record contains testimony regarding cases where employees were reluctant to take physical examinations because of their fear that they will lose their jobs or be transferred to lower paying jobs, and cases where employees were in fact transferred to lower paying jobs or laid off because of the results of medical examinations.

The Agency agrees that the approach taken in the proposed standard confronts the employee with a difficult choice and we are sympathetic to the concerns reflected in the unions' position on this issue. However, we believe that the present record does not contain sufficient evidence on the propriety, scope and implications of a rate retention requirement so as to constitute an adequate basis for the incorporation of such a provision in the standard.

***

While we are not providing for rate retention in the standard, we are convinced that further exploration of this issue is necessary in order to deal in considerably more depth with the numerous issues raised by such a provision. It is therefore our intention to conduct prompt further study, through an advisory committee or other means, of the need and implications of rate retention as an aspect of an OSHA health standard. On the basis of this study, the Agency will take further action under the Act, as appropriate, regarding rate retention.

In the meantime, we have also determined to modify the proposed standard to delete the mandatory removal provision. In our view, the issue of mandatory removal is closely related to the issue of rate retention and neither should be addressed in the present standard. The Agency's further study of rate retention will also involve consideration of the mandatory removal question.

The approach taken in the coke oven emission standard—no rate retention and no mandatory removal of employees—was continued by OSHA in the standards on benzene,[44] arsenic,[45] and cotton dust.[46] In the cotton dust standard, however, OSHA adopted a provision for removal of employees who were unable to wear respirators for medical reasons, which was similar to the early asbestos clause. The provision in the cotton dust standard also provided for income retention. The preamble to the cotton dust standard contained a brief justification for the provision.

*[Author's note—This expression of view by Dr. Bingham, who later became Assistant Secretary, was, in part, the basis for an argument by the Lead Industrial Association in the *Lead* case that she had prejudged the medical removal protection issue. See Chapter 8, Section B.]

[44] 43 FED. REG. 5918, 5958 (1978).
[45] 43 FED. REG. 19,584, 19,621–22 (1978).
[46] 43 FED. REG. 27,350, 27,392 (1978).

Final Standard for Cotton Dust

43 Fed. Reg. 27,350, 27,387 (1978)

***

Due to the high incidence of diminished pulmonary function among existing cotton dust workers, OSHA expects that some employees will be found to be incapable of wearing a single use or other form of negative pressure respirator. As a result, the medical surveillance provisions of the standard include a requirement that an examining physician determine an employee's ability to wear respirators. In situations where a negative pressure respirator cannot be worn, the examining physician must determine the employee's ability to wear a powered air purifying respirator (PAPR).*

OSHA expects that in a limited set of circumstances, certain employees will be incapable of even wearing a PAPR. An employee under such circumstances, in the absence of some opportunity to transfer to a position where a respirator need not be worn, might very well be discharged, or otherwise sustain economic loss, due to his or her inability to wear a respirator. OSHA views such a result as exceedingly harsh. It is manifestly unfair that employees who are unable to wear respirators suffer loss of jobs or other economic detriment because their employers have not yet achieved compliance with the engineering control requirements of the standard, but are relying instead on the interim and less effective device of respirators. During the multi-year period prior to the full implementation of engineering controls, this risk of adverse economic impact on employees is a real one. Accordingly, we conclude that where the employer is relying upon respirators as the means of limiting worker exposure, his or her compliance obligation must include the obligation to assure that employees who are unable to wear respirators not suffer any economic detriment. * * *

It is OSHA's judgment that this restricted form of job transfer without loss of earnings or other employment rights or benefits should only have to be invoked in the most unusual of circumstances, but, the protection it affords should greatly increase the success of the standard's respiratory protection provisions. As noted elsewhere in the preamble, the broad questions of mandatory medical removal of cotton dust workers, and medical protection, are not addressed by this final cotton dust standard, but will be discussed in the forthcoming inorganic lead standard. It is appropriate to note that the limited transfer option contained in this final cotton dust standard is adopted solely in response to the special role of respiratory protection in this standard, and is not meant to reflect the agency's evolving perception of the overall problem of medical removal protection.

***

The American Textile Manufacturers Institute challenged the provision in the District of Columbia Circuit Court of Appeals.[47] The court, in upholding these requirements, stated:

*[Author's note—Both negative pressure and PAPRs are respirators which mechanically filter airborne contaminants. PAPRs use blowers or compressors to pass contaminated air through the filter element. In the negative pressure respirator, the wearer's inhalation produces the pressure to draw the air through the filter.]

[47] *AFL-CIO v. Marshall*, 617 F.2d 636 (D.C. Cir. 1979).

> The textile petitioners argue that this provision exceeds OSHA's statutory authority. We find to the contrary that this provision falls within the agency's authority to include in its standards such "practices, means, methods, operations, or processes, [that are] reasonably necessary or appropriate to provide safe or healthful employment." [§3(8) of the Act, 29 U.S.C. §652(8)] OSHA determined that a limited number of employees cannot be protected even temporarily by respirators. [§6(b)(5) of the Act, 29 U.S.C. §655(b)(5)] Absent the medical transfer and wage guarantee provisions, during the interim compliance period these employees would be forced to suffer continuing exposure to impermissibly high dust levels or else risk disadvantageous transfers, or even risk losing their jobs. Given these alternatives, such employees may refrain from disclosing actual health impairments from the dust exposure. OSHA is authorized to guard against such problems. Its mandate is to assure that "*no* employee will suffer material impairment of health or functional capacity" on the job. In light of the Act's purposes, OSHA sought to avoid these problems and properly placed the cost of employee protection on the employer.
>
> Employers will be burdened for only a short time by the medical transfer and wage guarantee provisions. Once employers achieve the requisite exposure levels, no employees, to be protected, will need respirators or job transfers. In the meantime, the provisions provide a reasonable incentive for employers to speed compliance efforts. They also sensibly allocate the transfer decision to doctors surveying the individual workers. We find, therefore, that the medical transfer and wage guarantee provisions fall within OSHA's authority to adopt methods reasonably necessary to ensure safe working conditions.[48]

The Supreme Court later vacated the provision on the grounds that OSHA had improperly justified the provision as necessary "to minimize any adverse economic impact on the employee by virtue of the inability to wear a respirator" rather than as an approach designed to contribute to increased health protection.[49] The argument in the OSHA brief justifying the provision on health grounds was noted by the Supreme Court, which said that this argument "very well" may have merit, but it refused to accept the "*post hoc* rationalizations" as a basis for agency action.[50]

In the lead standard, OSHA included a comprehensive medical removal protection (MRP) provision. The MRP program and its justification were discussed at length in the preamble to the standard.[51]

---

[48]617 F.2d at 674–75.

[49]*American Textile Mfrs. Inst. v. Donovan,* 452 U.S. 490, 538 (1981).

[50]*Id.* at 539. The Supreme Court decision was handed down after OSHA had published the lead standard containing medical removal protection and the court of appeals had upheld its validity. *Infra* this section. In proposing changes to the cotton dust standard in 1983, OSHA stated that it did not intend to include a wage retention provision. It explained that in view of the substantial progress of industry towards compliance, the provision was "unnecessary." OSHA said further that it did not believe that it should become involved in determining wages and terms of employment, "an area traditionally reserved to employers' personnel practices and the collective bargaining process, unless there is a compelling health reason." In the case of cotton dust, OSHA did not find the compelling reasons "at this time." Proposed Cotton Dust Standard, 48 FED. REG. 26,962, 26,973–74 (1983).

[51]An even more extensive rationale for the MRP provision was contained in the attachments to the preamble, 43 FED. REG. 54,354, 54,440–73 (1978).

Final Standard for Lead

43 Fed. Reg. 52,952, 52,973–74 (1978)

***

C. MEDICAL REMOVAL PROTECTION

1. *Introduction.* The final standard includes provisions entitled Medical Removal Protection. Medical Removal Protection, or MRP, is a protective, preventive health mechanism integrated with the medical surveillance provisions of the final standard. MRP provides temporary medical removals for workers discovered through medical surveillance to be at risk of sustaining material impairment to health from continued exposure to lead. MRP also provides temporary economic protection for those removed. * * *

2. *Importance of temporary medical removals.* A central element of MRP is the temporary medical removal of workers at risk of sustaining material impairment to health from continued exposure to lead. This preventive health mechanism is especially well suited to the lead standard due to the reversible character of the early stages of lead diseases, and to the relative ease with which a worker's body may be biologically monitored for the presence of harmful quantities of lead. Temporary medical removal protects worker health both by severely limiting subsequent occupational exposure to lead, and by enabling a worker's body to naturally excrete previously absorbed lead which has accumulated in various tissues.

Temporary medical removal is an indispensable part of the lead standard for two significant reasons. Little margin for safety is provided by the final standard's 50 ug/m^3 permissible exposure limit, thus it is highly likely that some small fraction of workers (much less than 6 percent) will not be adequately protected even if an employer complies with all other provisions of the standard. Temporary medical removal will be the only means of protecting these workers. Many years will be needed for some segments of the lead industry to completely engineer out excessive plant air lead emissions. During this time heavy reliance will have to be placed on respiratory protection—a frequently inadequate means of worker protection. Again, temporary medical removal is essential for those inadequately protected. * * *

3. *MRP as a means of effectuating the medical surveillance sections of the lead standard.* Temporary medical removals depend on voluntary and meaningful worker participation in the standard's medical surveillance program. Medical surveillance, a major element of the Act's integrated approach to preventive health, can only function as intended where workers (1) voluntarily seek medical attention when they feel ill, (2) fully cooperate with examining physicians to facilitate accurate medical diagnoses, and (3) refrain from efforts to conceal their true health status. No one can coerce these qualities of worker participation—they will occur only where no major disincentives to meaningful worker participation exist. Absent these qualities of worker participation, medical surveillance cannot serve to identify those workers who need temporary medical removals, and consequently the overall protection offered by the lead standard will be diminished.

Participation in medical surveillance offered under the lead standard will sometimes prompt the temporary medical removal of a worker. Absent some countervailing requirement, removal could easily take the form of a transfer to a lower paying job, a temporary lay off, or even a permanent termination. The possibility of these consequences of a medical removal present a dramatic and painful dilemma to many workers exposed to inorganic lead. A worker could fully participate in the medical surveillance program and risk losing his or her livelihood, or resist participating in a meaningful fashion and thereby lose the many benefits that medical surveillance and temporary medical removals can provide. Convincing evidence presented during the lead proceeding established that many workers will either refuse or resist meaningful participation in medical surveillance unless economic protection is provided.

***

4. *MRP as a means of allocating the costs of temporary medical removals.* Temporary medical removal is fundamentally a protective, control mechanism, as is the elimination of air lead emissions through the use of engineering controls. The use of a temporary removal carries the possibility of dislocation costs to an employer through the temporary loss of a trained and experienced employee. And, a removed worker might easily lose substantial earnings or other rights or benefits by virtue of the removal. These costs are a direct result of the use of temporary medical removal as a means of protecting worker health. MRP is meant to place these costs of worker protection directly on the lead industry rather than on the shoulders of individual workers unfortunate enough to be at risk of sustaining material impairment to health due to occupational exposure to lead. The costs of protecting worker health are appropriate costs of doing business since employers under the Act have the primary obligation to provide safe and healthful places of employment.

One beneficial side-effect of MRP will be its role as an economic incentive for employers to comply with the final standard. * * *

***

This preamble discussion makes it clear that the MRP provision is an integral part of the medical surveillance program and thus is designed to protect employee health. The provision, therefore, would probably not have been legally vulnerable under the Supreme Court decision vacating the analogous but more limited income protection provision in the cotton dust standard.[52] In the lead standard, OSHA concluded that the removal of high-risk employees from continued exposure to lead was essential; since the detection of those high-risk employees would be much more difficult if employee cooperation in medical surveillance were undermined, medical removal protection was necessary to eliminate a major reason for employee nonparticipation in medical surveillance. Earlier in the coke oven emission standard, OSHA attempted to encourage employee participation in medical surveillance by eliminating the mandatory removal requirements. In the lead standard, however, OSHA rejected this alternative, emphasizing that the preventive health mechanism of removal is "especially well suited to the lead standard due to the reversible character of the early stages of lead disease, and to the relative ease with which a worker's body may be biologically monitored for the presence of harmful quantities of lead."[53]

In refusing to adopt the alternative of providing for removal at the election of the employee, OSHA said that "far too often" employ-

---

[52]See *supra* this section.
[53]43 FED. REG. at 52,972.

ees would choose not to be removed, and this would be "inconsistent with the preventive purposes of the Act."[54]

OSHA also adopted multiple-physician review in the lead standard, relying again on the argument that the provisions were necessary to insure the success of the medical surveillance program.

Final Standard for Lead

43 Fed. Reg. 52,952, 52,998–99 (1978)

***

OSHA's first reason for the provisions of a physician review opportunity is to strengthen and broaden the basis for medical determinations made under the standard in situations where a worker questions the results of an initial medical examination or consultation. The education and training provisions of the lead standard should assure that workers become knowledgeable in the nature and symptoms of the numerous lead-related diseases. Thus, when a worker disputes the results of an initial medical examination or consultation conducted by an employer-retained physician, adequate justification will exist for seeking a second medical opinion.

***

OSHA's second reason for the provision of a physician review opportunity is to assure employee confidence in the soundness of the medical determinations made pursuant to the standard. Considerable evidence in the lead record documents the fact that workers question the objectivity of some employer-retained physicians. * * * The standard's ability to prevent material impairment to worker health and functional capacity—particularly with respect to reproductive health, and the health of the long term lead worker—will significantly depend on workers trusting and confiding in examining physicians. OSHA adopted the multiple physician review mechanism as a means of providing workers with an opportunity to obtain independent review of the determinations of physicians they do not trust. More importantly, use of this review mechanism should serve to engender worker trust and confidence in the employer-retained physician where merited. * * *

* * * The discussion concerning and the inclusion of this mechanism, however, is not implicit criticism of the general medical community. Based on the lead record, OSHA has no cause to conclude that a majority of employer-retained physicians are not sincerely devoted to worker protection. Even worker representatives most critical of some "company doctors" agree that there are many competent and concerned corporate physicians. The multiple physician review opportunity contained in the final standard addresses problems presented by a minority of physicians. * * *

***

There remained the serious question whether the Act authorized the innovative provision of medical removal protection. Recognizing the importance of the issue, in the lead preamble itself, OSHA concluded that MRP "flows directly from and is fully consistent with the Act's express language" and was not contrary to the mandate of

[54] *Id.* at 52,974.

§4(b)(4)[55] or national labor policy.[56] In the litigation in the court of appeals, OSHA's authority to require MRP was one of the major issues. The Lead Industries Association (LIA) argued that MRP (which LIA referred to as "earnings protection") was inconsistent with the language and legislative history of the Act.[57]

Consolidated Brief of Industry Petitioners and Intervenors, United Steelworkers v. Marshall

* * * If in fact workers refuse to participate in medical programs because of their concern about economic repercussions, that concern arises only because the benefits available under existing workmen's compensation programs are not sufficient. As Grover Wrenn, OSHA's Director of Standards Development, pointed out during the hearings,

> If employees in fact did not lose pay as a result of medical removal, because of the existence of these [workmen's] compensation programs, no disincentive should exist to participate in medical surveillance programs.

In short, the earnings protection element of OSHA's proposal is designed to correct what OSHA perceives to be the inadequacies of existing workmen's compensation programs and to do so by superseding those programs and expanding employers' responsibilities. Both the legislative history and the language of the Act itself establish beyond doubt that Congress did not intend OSHA to have any such authority.

Section 4(b)(4) of the Act provides that "nothing in this Act"—and that "nothing" includes OSHA's authority under Section 6(b)(7) to require medical examinations

> shall be construed to *supersede or in any manner affect any workmen's compensation law* * * *. 29 U.S.C. §653(b)(4).

Nevertheless, as OSHA's own expert witness, Peter Barth, testified at the hearings, the rate retention component of MRP completely "replaces" coverage under existing workmen's compensation laws.

In no state is a worker entitled to workmen's compensation as a result of illness or disease due to occupational exposure to lead if he has not suffered a diminution of wages. As Barth explained, "the worker has to demonstrate some earning loss. It is one of the conditions to getting into the compensation system." Under existing workmen's compensation programs, the maximum that a lead worker can receive under applicable compensation programs is two-thirds of his regular salary. In no jurisdiction does a worker who is entitled to receive workmen's compensation receive 100 percent of his salary.

However, *under OSHA's MRP earnings protection requirement, no lead worker—whether he is removed from exposure because he is actually ill or merely because he has an "elevated" blood-lead level—can suffer any diminution of wages.* Indeed, the entire purpose of the earnings protection element of MRP is to "assure that a removed worker suffers neither economic loss nor loss of employment opportunities due to the removal." Since a lead worker with MRP will receive 100 percent (not two-thirds) of his pay and will not suffer any diminution of wages, he will not be entitled to workmen's

[55]Sec. 4(b)(4) of the Act, 29 U.S.C. §653(b)(4), is quoted, in relevant part, *infra* this section.
[56]43 FED. REG. at 52,976–77.
[57]Consolidated Brief of Industry Petitioners and Intervenors at 102–05.

compensation. Consequently, his rate retention under the Standard will have effectively "replace[d]" workmen's compensation and will have "affect[ed] and indeed * * * enlarge[d] the * * * duties and liabilities, which [the employer] otherwise would have had under state compensation programs."

This is the very impact which the Act forbids.

***

The Court of Appeals for the District of Columbia Circuit by a 2–1 vote, rejected the LIA arguments on the invalidity of medical removal protection.[58] Chief Judge Wright, writing for the majority, first ruled that, based on the Act's language and legislative history, "MRP * * * appears to lie well within the range of OSHA's powers."[59] He then rejected LIA's argument that since Congress included job removal and earnings protection in the Federal Coal Mine Safety Act of 1969,[60] its failure to do so in the OSH Act, or to expressly give OSHA the power to impose MRP, shows that Congress intended that OSHA *not* have authority to include MRP in its standards. He said: "Congress may well have avoided all mention of medical removal protection in the OSH Act simply because it thought that mandating such a specific program was inappropriate in a statute so much broader and so much more dependent on agency implementation than the Coal Act."[61] Chief Judge Wright then rejected LIA's arguments based on §4(b)(4),[62] with Judge MacKinnon vigorously dissenting on this issue.

United Steelworkers v. Marshall

647 F.2d 1189 (D.C. Cir. 1980)

WRIGHT, Chief Judge:

***

* * * LIA argues vigorously that MRP contravenes [§4(b)(4)] in both purpose and effect. * * * [The] evidence shows little more than that LIA was able to put its ideas into the mouths of OSHA witnesses during cross-examination, and in any event we do not construe Section 4(b)(4) as being concerned with agency motive. LIA's argument on the effect of MRP does, however, deserve serious attention.

Under workmen's compensation law, a worker suffering disablement of a designated type can recover, depending on the state, up to two thirds of his lost wages. These laws presume that if a disabled worker could recover all his lost wages, he would have no incentive to return to work. LIA contends that under MRP a worker would never seek worker's compensation, nor would he ever become medically or financially eligible for it. A worker re-

---

[58]*United Steelworkers v. Marshall,* 647 F.2d 1189, 1228–38 (D.C. Cir. 1980).

[59]*Id.* at 1232.

[60]30 U.S.C. §843(b)(2)–(3) (1976).

[61]647 F.2d at 1232. The principle relied on by LIA is known as *expressio unius est exclusio alterius.* Compare the treatment of this issue by the Supreme Court in *Whirlpool Corp. v. Marshall,* 445 U.S. 1, 13–21 (1980), reprinted in Chapter 18, Section B.

[62]29 U.S.C. §653(b)(4).

moved for such a disability would be guaranteed all the earnings rights of his high-exposure job; suffering no loss of wages, he would be entitled to no wage replacement under worker's compensation. Moreover, under MRP most workers would be removed because of high blood-lead levels—before they exhibit clinical symptoms of disablement. Thus they would enjoy their guaranteed salary before they became disabled in the eyes of workmen's compensation law. LIA concludes that for workers vulnerable to lead disease MRP would "supersede" and "affect" worker's compensation law to the point of wholly replacing it.

***

To resolve this issue we first recognize that Section 4(b)(4) is vague and ambiguous on its face. We must seek then the best, not a perfect, reading. Thus LIA's argument under Section 4(b)(4) really shows that no literal reading of it is possible, since any health standard that reduces the number of workers who become disabled will of course "affect" and even "supersede" worker's compensation by ensuring that those workers never seek or obtain workmen's compensation benefits.

Applied to those workers who, thanks to MRP, never become disabled, the argument certainly proves too much. Nevertheless, LIA's argument, if sharpened, remains plausible, since it can still sensibly apply to a small and special group of workers: those who are removed under MRP pursuant to a final medical determination that they are already disabled, and for whom no low-exposure job of equal salary is available.

***

* * * [W]e must take seriously LIA's argument with respect to the special group of workers we have described, [but] we do not think it proves MRP to violate Section 4(b)(4). First, this special group of workers is probably very small, and will become progressively smaller. As the new PEL lowers lead exposure throughout industry, and as older workers with high accumulations of lead retire, fewer and fewer workers will require removal because of manifest disablement from lead. Moreover, workers remaining in this group will still have an incentive to file worker's compensation claims. First, worker's compensation laws universally reimburse workers for the medical expenses of their disablement, while MRP does not. Second, MRP benefits can only last slightly more than 18 months, whereas worker's compensation may replace lost wages for longer periods or even indefinitely. 82 AM. JUR. 2D *Workmen's Compensation* §382 (1976).

The question remains, then, what *does* Section 4(b)(4) mean, if it *does not* mean that OSHA is barred from creating medical removal protection? We see two plausible meanings. First, as courts have already held, Section 4(b)(4) bars workers from asserting a private cause of action against employers under OSHA standards. *Jeter v. St. Regis Paper Co.,* 507 F.2d 973 (5th Cir. 1975); *Byrd v. Fieldcrest Mills, Inc.,* 496 F.2d 1323 (4th Cir. 1974). Second, when a worker actually asserts a claim under workmen's compensation law or some other state law, Section 4(b)(4) intends that neither the worker nor the party against whom the claim is made can assert that any OSHA regulation or the OSH Act itself *preempts* any element of the state law. For example, where OSHA protects a worker against a form of disablement not compensable under state law, the worker cannot obtain state relief for that disablement. Conversely, where state law covers a wider range of disablements than OSHA aims to prevent, an employer cannot escape liability under state law for a disablement not covered by OSHA. In short, OSHA cannot *legally* preempt state compensation law, even if it *practically* preempts it in some situations.

We conclude that though MRP may indeed have a great practical effect on workmen's compensation claims, it leaves the state schemes wholly intact as a *legal* matter, and so does not violate Section 4(b)(4).

MACKINNON, Circuit Judge (dissenting):

***

OSHA and the majority are reading far more authority into this Act than Congress ever intended it to have. The general impression gleaned from these provisions is that Congress was referring to the regulation of conditions *within the workplace*. The Act uses such terms as "working conditions","places of employment", and "control procedures * * * used *in connection with* [the hazards to which the employee is exposed]". The latter authorization is stated in the same phrase with "protective clothing", indicating that the control techniques are intended for application within the working environment. Congress never even hinted that the Secretary, by this language, could demand the medical removal program, which has nothing to do with providing safe "working *conditions*". In fact the MRP imposes no conditions whatsoever on the working place.[6]

***

However, to culminate the review of congressional intent which is adverse to the MRP, 29 U.S.C. §653(b)(4) must be read. This section of the OSH Act precludes the promulgation of such a far reaching MRP as OSHA seeks to establish for the lead industries. Section 653(b)(4) proscribes anything in the OSH Act from "*supersed[ing] or in any manner affect[ing] any workmen's compensation law.* * * *" *Id.* (Emphasis added). Since an employee may be removed before he is overtly ill or incapable of performing his job under the lead standard, he receives *greater protection* under it than under any state workmen's compensation law in existence. In addition, even though he may be disabled because of lead poisoning, he will be removed under the authority of new regulations with *full compensation*. Since most workmen's compensation statutes require a diminution of wages before they become effective, there is no way that lead will ever again be the cause of disabling conditions warranting the receipt of workmen's compensation. Finally, most workmen's compensation laws only provide benefits which amount to a portion of the employee's wages, whereas the lead industries regulation, at its maximum, mandates full seniority and the payment of full wages for eighteen months. For these reasons, the medical removal protection system in the lead standard clearly affects and supersedes workmen's compensation laws in violation of §653(b)(4).*

The court of appeals, without dissent, also upheld the multiple-physician review provisions.[63] Judge Wright concluded that the

[6]Although a bare reading of the statute is sufficient to arrive at this conclusion, the legislative history of the OSH Act discussed by the majority also evidences the lack of congressional intent to authorize this regulation. Congress had discussed a form of "strike with pay" in the Daniels Amendment, and rejected it. H.R. Rep. No. 91-1291, 91st Cong., 2d Sess. 30 (1970). This Amendment was distinguishable from the MRP, but in many respects its effect on the employer, i.e. having to pay for laid off employees, is comparable. And Congress refused to authorize this.

In addition, weight must be given to the fact that only one year prior to the enactment of the OSH Act, Congress enacted the Federal Coal Mine Health and Safety Act of 1969, 30 U.S.C. §801 *et seq.*, which *did* expressly authorize a removal program. 30 U.S.C. §843(b)(3). With Congress' full knowledge of this, and rejection of the strike with pay provision, it is difficult to conclude that congressional silence on the benefits issue connotes implicit authorization of the financial and seniority benefits provided in the MRP system.

*[Author's note—But see Note, *The Validity of Medical Removal Protection in OSHA's Lead Standard,* 59 TEX. L. REV. 1461, 1480 (1981) (Medical removal protection works to "complement rather than preempt" state workers' compensation schemes and other industries should view MRP as a "model for future efforts to confront the problems of occupational disease."). See also generally OFFICE OF TECHNOLOGY ASSESSMENT, THE ROLE OF GENETIC TESTING IN THE PREVENTION OF OCCUPATIONAL DISEASE 122–23 (1983) (possible application of MRP to job removals resulting from genetic testing).]

[63]*United Steelworkers v. Marshall,* 647 F.2d at 1239.

mechanism "furthers OSHA's legitimate goals both directly and indirectly"; (1) by decreasing the chances of a single erroneous diagnosis becoming the final medical determination, and (2) by enhancing worker confidence in examining physicians and thus enhancing their health.[64] The court also distinguished *Taylor Diving,* in which the Court of Appeals for the Fifth Circuit held invalid the three-physician review in the diving standard. Chief Judge Wright said:

> LIA cites one recent decision holding an OSHA multiple physician review scheme to lie beyond OSHA's statutory power, *Taylor Diving & Salvage Co. v. U.S. Dep't of Labor,* 599 F.2d 622 (5th Cir. 1979), but that case is wholly distinguishable from the present one. *Taylor* addressed an OSHA standard for commercial divers which included a three-tiered physician review scheme to determine whether divers are medically fit to tolerate hyperbaric conditions. If a company-appointed physician found an employee unfit, the employee could procure a second opinion at the company's expense and, if the first two physicians disagreed, a third doctor, chosen by the first two and paid by the company, would make the final decision. As construed by the court and apparently as presented by the agency, the diving scheme was designed deliberately to protect employee job security and to eliminate barriers to entry into the profession. *Id.* at 625. The court held that protecting workers' choice of occupation was not reasonably related to OSHA's statutory mission. Moreover, since the scheme allowed employee-appointed physicians to overrule an employer's decision that a worker was unfit, it impermissibly set a ceiling on medical fitness standards established by employers. OSHA, the court held, was created to set minimum, not maximum, health standards. *Id.*
>
> The diving standard is thoroughly inapposite to the lead standard. First, OSHA has justified the multiple physician review mechanism in the lead standard as a device to enhance worker health; the agency neither designed nor intended the mechanism to protect access to or security in employment. Second, though a worker who wanted to remain on his job could theoretically obtain a second or third opinion to overturn the original decision of a company doctor to remove him, the lead standard's guarantee of medical removal benefits will leave such an employee little practical incentive to do so. Finally, the provision allowing employers to remove employees—with guaranteed benefits—on more conservative medical criteria than those imposed by the standard, §1910.1025(k)(2)(vii), ensures that the lead standard does not impose maximum health standards.[65]

E. Genetic Testing

The role of genetic testing in the working environment is becoming an increasingly important and controversial issue. The subject was considered at length in a recent study by the Office of Technology Assessment (OTA) published in 1983.[66] The scope of the problem, and of its study, was stated by OTA as follows:

> Genetic testing, as used in the workplace, encompasses two types of techniques. Genetic screening involves examining individuals for cer-

[64] *Id.*

[65] *Id.* at 1239 n.76.

[66] OFFICE OF TECHNOLOGY ASSESSMENT, THE ROLE OF GENETIC TESTING IN THE PREVENTION OF OCCUPATIONAL DISEASE (1983).

tain inherited genetic traits. Genetic monitoring involves examining individuals periodically for environmentally induced changes in the genetic material of certain cells in their bodies. The assumption underlying both types of procedures is that the traits or changes may predispose the individuals to occupational diseases. (Changes in the germ cells egg and sperm could result in birth defects in offspring but such reproductive effects are not part of this study.)

Although this technology is still in its infancy, it has the potential to play a role in the prevention of occupational diseases. It is technologically and economically impossible to lower the level of exposure to hazardous agents to zero. However, if individuals or groups who were predisposed to specific types of occupational illness could be identified, other preventive measures could be specifically directed at those persons. This is the promise of genetic testing. At the same time, however, the technology has potential drawbacks and problems. For example, the ability of the techniques to identify people who are predisposed to occupational illness has not been demonstrated. In addition, some people are concerned that its use could result in workers being unfairly excluded from jobs or in attention being directed away from efforts to reduce workplace hazards.[67]

Since genetic testing may result in refusals to hire, discharge, or transfer to lower-paying jobs, these programs raise not only scientific and ethical[68] questions, but legal questions as well. The major legal questions relating to genetic testing arise under Title VII of the Civil Rights Act of 1964[69] and the Rehabilitation Act of 1973.[70] In addition, genetic testing of employees has raised questions under the OSH Act. In 1980 it was suggested widely that the OSHA standards on 14 carcinogens *require* employers to conduct genetic screening as part of the medical surveillance program.[71] Assistant Secretary Bingham announced at that time that it was not OSHA's intention to require genetic testing or the exclusion from employment of otherwise qualified employees because of the results of the testing.[72] In a proposed standard issued in April 1983 on ethylene oxide, however, OSHA expressly posed the question whether genetic screening and chromosome analysis, among other tests, should be provided as part of the routine physical examination.[73]

On the issue of OSHA authority respecting genetic testing, the OTA study concluded:

The OSH Act and regulations thereunder neither prohibit nor require genetic testing. However, the Secretary of Labor has broad au-

---

[67]*Id.* at 5.

[68]For a discussion of the ethical questions, see *Genetic Screening in the Workplace: Hearing Before the Subcomm. on Investigation and Oversight of the House Comm. on Science and Technology,* 97th Cong., 2d Sess. 75 (1982) (testimony of Dr. Thomas Murray, Inst. of Society, Ethics, and Life Sciences, Hastings Center).

[69]42 U.S.C. §§2000e–2000e-17 (1976 & Supp. V 1981) (current version).

[70]29 U.S.C. §§701–796i (1976 & Supp. V 1981). For a discussion of these issues, see Note, *Genetic Testing in Employment: Employee Protection or Threat,* 15 SUFFOLK U.L. REV. 1187 (1981).

[71]See, e.g., Ethyleneimine Standard, 29 C.F.R. §1910.1012(g)(1) (1982) (requiring that the medical examination include "the personal history of the employee, family, and occupational background, including genetic and environmental factors").

[72]Bingham, Letter to the Editor, N.Y. Times, Mar. 22, 1980 at 20, col. 5.

[73]48 FED. REG. 17,284, 17,285 (1983).

> thority to regulate employer medical procedures as long as the regulation is related to worker health and meets the feasibility and significant risk requirements. Therefore, the Secretary could require genetic testing in its various forms, if the techniques were shown to be reliable and reasonably predictive of future illness. The Secretary also could regulate the use of genetic testing, but only to the extent that the regulation was related to employee health. The act grants no authority over rights or conditions of employment per se and no authority to protect applicants for employment from discrimination.[74]

Indeed, it has also been suggested that OSHA is the "most appropriate" agency to address the problems of workers who are identified as a result of genetic testing as having become hypersusceptible after beginning a job that exposes them to carcinogens; and that if a "significant risk" is found, OSHA could have the authority to mandate removal of hypersusceptible workers and to provide a wage compensation scheme, similar to MRP in the lead standard, "to ensure the employee compliance necessary to the success of the medical surveillance and removal program."[75]

F. Access to Medical Records

Two statutory provisions are particularly relevant to the right of Agency and employee access to employee medical records. Section 6(b)(7),[76] setting forth criteria for health standards, requires employee medical examinations and provides that the "results" of these examinations "be furnished only" to OSHA, NIOSH, and, "at the request of the employee, to his physician." Section 8(c)(1)[77] authorizes the Secretary of Labor to require each employer to "make, keep, and make available" to OSHA and NIOSH records "regarding his activities relating to this Act" as are necessary "for developing information regarding the causes and prevention of occupational accidents and illness." In addition, §8(g)(2)[78] provides general rulemaking authority to OSHA.

The asbestos standard required the retention of medical records for a period of at least 20 years,[79] and provided access to these medical records to OSHA, NIOSH, "authorized physicians and medical consultants of either of them," to the physician of an employee or former employee, and to the employer.[80] Substantially the same access provisions were contained in the 14 carcinogen, vinyl chloride,

[74]OFFICE OF TECHNOLOGY ASSESSMENT, THE ROLE OF GENETIC TESTING, *supra* note 66 at 12.

[75]Note, *Occupationally Induced Cancer Susceptibility: Regulating the Risk,* 96 HARV. L. REV. 697, 710–14 (1983). See also *Genetic Screening of Workers: Hearing Before the Subcomm. on Investigations and Oversight of the House Comm. on Science and Technology,* 97th Cong., 2d Sess. 97–128 (1982); Goodrich, *Are Your Genes Right for Your Job,* 3 CAL. LAW. 25 (1983); Rothstein, *Employee Selection Based on Susceptibility to Occupational Illness,* 81 MICH. L. REV. 1379 (1983).

[76]29 U.S.C. §655(b)(7).

[77]*Id.* at §657(c)(1).

[78]*Id.* at §657(g)(2).

[79]The District of Columbia Circuit Court of Appeals rejected a union challenge that this period was insufficient. *Industrial Union Dep't v. Hodgson,* 499 F.2d 467, 486 (D.C. Cir. 1974).

[80]37 FED. REG. 11,318, 11,322 (1972).

and coke oven emission standards.[81] In the preamble to the coke oven standard, OSHA explained that access to medical records was not provided directly to employees, but only to their physicians because of "the special character and sensitivity of the components of a medical record."[82]

In the benzene standard, however, OSHA also provided for access to medical records to any "other individual designated by the affected employee or former employee."[83] While noting that medical records "may contain a wide range of personal and medical information deemed confidential or private," the preamble did not otherwise explain the broader access provision, except that it mentions that the standard limits access to information related to benzene exposure.[84] In any event, in the benzene standard, OSHA, for the first time, allowed access to medical records to the employee himself or herself, or to any individual designated by the employee, rather than only to the physician designated by the employee.[85]

Later in 1978, OSHA issued the lead standard,[86] which contained detailed provisions on access to medical records. Specifically, the standard provided for access to all medical records to OSHA, NIOSH, and to employees or former employees or "to a physician or any other individual" designated by an employee or former employee.[87] In addition, the standard stated that "biological monitoring and medical removal records" shall be available both to employees and former employees "or their authorized representatives."[88] Before the court of appeals, LIA first challenged the unrestricted access to records granted to OSHA and NIOSH. The court unanimously rejected this challenge.

United Steelworkers v. Marshall

647 F.2d 1189 (D.C. Cir. 1980)

WRIGHT, Chief Judge:

***

We have little trouble rejecting LIA's challenge to the unrestricted access to records the standard grants to OSHA and NIOSH. The statute itself unquestionably permits and even requires such access. 29 U.S.C. §§655(b)(7), 657(c)(1) (1976). And, contrary to LIA's contention, the Supreme Court's decision in *Whalen v. Roe*, 429 U.S. 589 (1977), does not invalidate the government's access to medical records as a violation of employees'

---

[81]See, e.g., Coke Oven Emissions, 41 FED. REG. 46,742, 46,789 (1976) (codified at 29 C.F.R. §1910.1029(m)(3)(ii) (1983)). Under this standard, the employer was required to maintain medical records for 40 years or the duration of employment plus 20 years, whichever is longer. 29 C.F.R. §1910.1029(m)(3)(iii) (1983).

[82]41 FED. REG. at 46,783.

[83]43 FED. REG. 5918, 5967 (1978).

[84]*Id.* at 5962.

[85]This, of course, is in addition to the access provided to OSHA, NIOSH, and the employer, as already discussed.

[86]43 FED. REG. 52,952 (1978).

[87]*Id.* at 53,014.

[88]*Id.* The standard also contained provisions on access to employee exposure monitoring records. On exposure records, see §8(c)(3), 29 U.S.C. §657(c)(3).

constitutional rights. In holding that the New York system for recording the identities of people taking prescribed addictive drugs did not violate these people's "zone of privacy," the *Whalen* Court did rely in part on the detailed protections against unwarranted disclosure contained in the statute creating the system. *Id.* at 601, 605. But the Court stated explicitly that it was expressing no opinion on the constitutionality of any system of government collection of confidential information that did not contain similar protection. *Id.* at 605–606. *Whalen,* indeed, *supports* OSHA's position here, since the Court clearly rejected the notion that required disclosure of private medical information to public health agencies or other bodies charged with the public welfare violated the Constitution, *id.* at 602, and refused, in the absence of concrete proof, to entertain speculation that such bodies will engage in unwarranted public disclosure of such information, *id,* at 601.

LIA finally argues that even if the government is entitled to *some* private information, OSHA has not shown why the government needs *unlimited* access to *identifiable* medical records. We think, however, that OSHA was reasonable in finding that to ensure compliance with the standard it needs not only general information on employee health, but also specific information on individual workers to ensure that no single exposed employee suffers any illegal threat from lead. Moreover, to obtain express permission for disclosure from each of millions of workers would create an unthinkable administrative burden and risk the health of workers who for unfortunate but understandable reasons might fear any disclosure of their health records. The rule on government access thus meets the demand of *Whalen* that such required disclosure be a reasonable exercise of government responsibility over public welfare, *id.* at 597–598, and the requirement under the OSH Act that any OSHA program be reasonably related to the general goal of preventing occupational lead disease, 29 U.S.C. §652(8) (1976).

***

LIA also challenged the provision granting access to certain records to "authorized" representatives of employees. The court said that the meaning of the standard was not altogether clear on this matter, but it construed the provision as giving an "independent right of access" of these records to officials of a union representing employees, without requiring employee consent. The court, however, agreed with OSHA's interpretation that the standard gave unions access only to a limited class of medical removal records.[89] So construed, the court of appeals held that the union access did not infringe on privacy interests or constitutional protections. The court made it clear, however, that if the standard allows "the unions, without the employee's permission, to examine the intimate results of physician examinations—information to which the employee and the government have rightful access—it may violate the statute and the Constitution."[90]

[89]These records include the name and social security number of the employee; the dates of the employee removal and his or her return to the job; an explanation of how the removal was being accomplished; and a statement whether the removal was because of an elevated blood level. See 43 FED. REG. at 53,013 (codified at 29 C.F.R. §1910.1025(n)(3)(ii) (1983)).

[90]*United Steelworkers v. Marshall,* 647 F.2d at 1243.

The court's view on access to employee medical records may be summarized as follows: OSHA and NIOSH are entitled to unrestricted access to employee medical records without first having to obtain employee consent; employees (and former employees), their physicians, and their specifically designated representatives are entitled to the employees' own medical records; but unions, although representing employees for purposes of collective bargaining, cannot obtain confidential medical records, that is, those containing intimate results of physician's examinations without the consent of the employee involved.

Many of the same issues were involved in the making of OSHA's rule, "Access to Employee Exposure and Medical Records,"[91] which was issued shortly before the court of appeals decision in the *Lead* case. In the past, OSHA had provided for employee access to medical records (and exposure records) only as part of standards regulating specific toxic substances.[92] The new access rule adopted a "generic" approach: it provided for "employee, designated representative, and OSHA access to employer-maintained exposure and medical records relevant to employees exposed to toxic substances and harmful physical agents."[93] The rule also provided for preservation of these records by employers for 40 years, or the duration of employment plus 20 years, whichever is longer. The preamble to the access rule discusses the major legal and policy issues that had been raised in the proceeding.[94]

Final Rules on Access to Employee Exposure and Medical Records

45 Fed. Reg. 35,212, 35,217–18 (1980)

***

In promulgating this final standard, OSHA not only determined that broad worker and designated representative access to records would serve important occupational health purposes, but also carefully evaluated the various arguments which were raised for and against OSHA's proposed means for achieving these objectives. Most employer participants, for example, opposed the standard's access principles on a variety of grounds. Concerns were voiced about potential harm to employees and the occupational physician-patient relationship should direct employee and designated representative access be provided to medical records. Arguments were made that direct access could prove to be genuinely harmful in certain cases; thus the employer's physician should retain discretion as to what information is released, and to whom. On the basis of the record and its own judgments, OSHA concluded that reliance on physician discretion to disclose informa-

[91] 45 FED. REG. 35,212 (1980) (codified at 29 C.F.R. §1910.20 (1983)). The more neutral term "rule" has been used because of the controversy over whether the "rule" was a "standard" or "regulation" for purposes of court review. See discussion *infra* this section.

[92] See, e.g., Vinyl Chloride Standard, reprinted in Chapter 3, Section B.

[93] 45 FED. REG. 35,212 (1980).

[94] See also Note, *Occupational Health Risks and the Worker's Rights to Know,* 90 YALE L.J. 1792 (1981) (notes that a disproportionate share of health risks are borne by industrial workers and argues that it would be more equitable if these risks were knowingly encountered and workers received adequate compensation for resulting illnesses).

tion to an employee, or reliance solely on physician-to-physician transfers of medical records, were inadequate responses to the needs for direct worker access.

Unrestricted patient access to medical records has been a major public policy issue during the past decade, and the trend throughout the nation both on the State and Federal level has been to provide direct patient access to medical records. The final standard's access principles draw support from the recommendations of such bodies as the 1976 Privacy Protection Study Commission, the National Commission on the Confidentiality of Health Records, and the American Medical Records Association. It is OSHA's judgment that there is no basis for anticipating harm to an employee from direct access to medical information as provided in the final standard. The rare occasion where direct access may possibly be harmful can adequately be dealt with by the standard's provisions concerning physician consultations and access through a designated representative.

Employee and designated representative access to both medical and exposure records were also opposed on the grounds that these records would often be misinterpreted and misunderstood. Statements were made that most employees are incapable of understanding the highly technical language, abbreviated shorthand and illegible writing that is often found in medical records. The likelihood of misinterpretation leading to unnecessary anxiety or inappropriate action was therefore considered great. OSHA agrees that professional evaluation and interpretation will often be important, but does not agree that the possibility of occasional misunderstanding should enable employers to deny access to either medical or exposure records. The solution reflected in the final rule is to provide full worker access to the records, while at the same time encouraging the employer to offer whatever professional interpretation the employer feels is necessary. The worker, however, retains the right to personally evaluate the record, and have independent analysis conducted by professionals and non-professionals alike. If a worker is incapable of understanding something in a medical or exposure record, OSHA expects that the worker will naturally seek the assistance of someone more knowledgeable whom he or she trusts.

Arguments were also made that provisions allowing designated representative access to medical records would seriously interfere with the physician-patient relationship due to the open-ended nature of the proposal's written consent process. Objections to the proposal focused on potential invasions of an employee's privacy expectations from unrestrained third party access to identifiable medical records. To preclude unrestrained third party access, the final standard conditions the access of designated representatives to employee medical records upon the specific written consent of the employee. The elements of specific written consent are a reflection of recommendations in the record, including those of the Privacy Commission and the American Medical Records Association.

As to who can be a designated representative, the final standard embodies the view that once specific written consent is obtained, no additional restrictions are needed, such as limiting access to physicians or industrial hygienists. Since it is the employee's right to have access to the complete record, the employee should, by the consent procedure, control the conditions of disclosure and redisclosure.

Arguments were also made that broadened access to medical records would inevitably impair the creation, expansion and effectiveness of occupational medical programs. It is OSHA's judgment that these predictions are exaggerated, since no concrete evidence was presented which indicated that the standard would have a negative impact on corporate efforts to provide occupational health programs. Corporate witnesses stated that, in fact, there would likely be no reduction in their occupational medical efforts. In addition, the record supports the view that direct patient access to medical records actually promotes the therapeutic relationship between physician and patient.

The most significant issue posed by OSHA access to employee records concerns a potential clash with the right of privacy vis-a-vis employee medical records. Employee medical records subject to this standard may sometimes contain intimate details concerning a person's life; e.g., disease experience, psychiatric disorders, venereal disease, abortion, alcohol or drug abuse, sexual preferences, family medical problems, etc. OSHA access to complete medical records was felt by numerous participants to raise the threat of misuse since harmful medical information could be disseminated in a way that would adversely affect the employee. It was also argued that governmental access to personal medical information threatens to destroy expectations of confidentiality and severely impair medical programs.

There was universal agreement that if OSHA obtained access to employee medical records, this access should be accompanied by stringent internal agency procedures to preclude abuse of personally identifiable medical information. Provided these procedures were established, many participants, including all union and employee participants, endorsed unconsented OSHA access to employee medical records for occupational safety and health purposes.

Having considered the record, and analyzed the legal issues involved, the agency decided to include a requirement in the final standard providing for unconsented OSHA access to employee medical records. As recognized by numerous participants, acquiring consent is not always a realistic possibility. Large numbers of people may be involved, emergency situations may not permit delay, the residences of terminated employees may be unknown, or an absence of consent may decrease the statistical validity of a study. In so providing for unconsented agency access, OSHA also agrees with the views of several participants that strong public health considerations must weigh heavily when balanced against the privacy interest in precluding unconsented access.

Even though OSHA has legal authority to seek unconsented access to identifiable medical records, the agency concluded that this authority should, as a matter of sound public policy, be exercised with great care. To protect the employee's privacy interests, OSHA recognizes the need for stringent internal procedures to: (1) limit the circumstances in which identifiable medical records are examined or obtained by OSHA personnel, and (2) control use of identifiable medical information once in the agency's possession. In order to effectuate these decisions, the agency is simultaneously promulgating detailed procedural regulations governing all aspects of OSHA examination and use of personally identifiable medical records.*

***

The OSHA access rule was challenged in several courts by both employer and union groups, and litigation over the validity of the rule is pending in the Courts of Appeals for the District of Columbia Circuit and the Fifth Circuit. A preliminary issue in this litigation was whether the access rule was a "standard" and therefore subject to the exclusive review jurisdiction of the courts of appeals under §6(f),[95] or a "regulation" and therefore reviewable in the district

*[Author's note—The internal OSHA procedures on medical records appear at 29 C.F.R. pt. 1913 (1983)].

[95]29 U.S.C. §655(f).

courts. The District Court of the Western District of Louisiana first held that the access provisions were a standard and dismissed the suit brought by an employer association for lack of subject matter jurisdiction.[96] The Fifth Circuit Court of Appeals reversed on the jurisdictional question and remanded the case for a ruling by the district court on the validity of the rule.[97] The court of appeals found that the record access rule "fits neatly within the language and history of section 8"[98] (which deals with regulations) because "it is among the more general class of enforcement and detection regulations contemplated by Congress in Section 8."[99]

The district court decided on November 5, 1982, that the rule was valid in all respects.[100] The court rejected the employers' arguments that the access rule contravenes Fourth Amendment rights of employers, privacy rights of employees, employers' rights under the Trade Secrets Act,[101] and interfered with the exclusive jurisdiction of the National Labor Relations Board. Meanwhile, on July 13, 1982, OSHA proposed a number of modifications to the access rule,[102] the most important of which related to the definition of "toxic substance," the rights of unions representing employees to exposure data, and the protection of trade secrets in exposure records.[103]

---

[96] *Louisiana Chem. Ass'n v. Bingham,* 496 F. Supp. 1188 (W.D. La. 1980).
[97] *Louisiana Chem. Ass'n v. Bingham,* 657 F.2d 777 (5th Cir. 1981).
[98] 29 U.S.C. §657.
[99] 657 F.2d at 783.
[100] *Louisiana Chem. Ass'n v. Bingham,* 550 F. Supp. 1136 (W.D. La. 1982).
[101] 18 U.S.C. §1905 (Supp. V 1981).
[102] 47 FED. REG. 30,420 (1982).
[103] The issue of government access to identifiable health records in the face of confidentiality contentions has also arisen in the context of health hazard evaluations conducted by NIOSH under §22(a), 29 U.S.C. §671(a). See, e.g., *United States v. Westinghouse Elec. Corp.,* 638 F.2d 570 (3d Cir. 1980) (strong public interest in NIOSH research and investigation justifies "minimal" privacy intrusion surrounding employee medical records; court does not require individual employee consent before records are released, but employees are entitled to prior notice and opportunity to object); *General Motors Corp. v. Director of NIOSH,* 636 F.2d 163 (6th Cir. 1980), *cert. denied,* 454 U.S. 877 (1981). (NIOSH subpoena for records enforced, with remand to district court to formulate "security provisions" to ensure "proper" disposition of the records). These cases, as well as the entire issue, are discussed in Comment, *Employee Medical Records and the Constitutional Right to Privacy,* 38 WASH. & LEE L. REV. 1267 (1981).

PART II

OSHA Standards and the Courts

6

OSHA and the Courts: An Uneasy Partnership

A. Court Review of OSHA Standards

OSHA's authority to issue legally enforceable occupational safety and health standards is broad. To be sure, the Act, in the "Findings and Purpose" section and in §6(b)(5),[1] contains policy guidance on the contents of the OSHA standards that emphasizes the overriding protective goal of standards, subject to the constraints of feasibility. In addition, there are the requirements, contained in §6(b)(7),[2] that certain specific provisions, such as those mandating employee medical examinations,[3] be included in health standards. But within this framework, OSHA retains considerable flexibility and discretion to decide the stringency, and thus the cost, of standards.[4]

As part of the legislative compromise leading to the enactment of OSHA, Congress subjected the Agency's standard-setting authority to various constraints. Among the most important of these are the following:

1. The rulemaking procedures prescribed under §6 include a notice of proposed rulemaking, public comment, a hearing, if requested, and a statement of reasons upon publication of the final standard. These procedures allow for the expression of views by in-

[1]29 U.S.C. §655(b)(5) (1976).

[2]29 U.S.C. §655(b)(7).

[3]See discussion in Chapter 5.

[4]A number of commentators have discussed, and some have criticized, the broad discretion assigned by Congress to administrative agencies under regulatory statutes. See, e.g., Stewart, *The Reformation of American Administrative Law,* 88 Harv. L. Rev. 1669, 1693–97 (1975) (Stewart emphasizes the "serious institutional constraints" on Congress' ability to specify regulatory policy in meaningful detail.). The Supreme Court has not held a federal law unconstitutional on the grounds of overbroad delegation of legislative authority since the 1930s. See *A.L.A. Schechter Poultry Corp. v. United States,* 295 U.S. 495 (1935); *Panama Ref. Co. v. Ryan,* 293 U.S. 388 (1935). But cf. *American Textile Mfrs. Inst. v. Donovan,* 452 U.S. 490, 543 (1981) (Rehnquist, J., dissenting); *Industrial Union Dep't v. American Petroleum Inst.,* 448 U.S. 607, 671 (1980) (Rehnquist, J., concurring in judgment) (in enacting §6(b)(5), Congress unconstitutionally delegated to the executive branch the authority to make the "hard policy choices" that are properly the task of the legislative branch). Excerpts from these latter two decisions are reprinted in Chapters 9 and 10.

terested persons and are designed to bring about more reasoned decision making.[5]

2. Standards issued by OSHA may be challenged in a U.S. court of appeals under §6(f) "by any person who may be adversely affected by the standard."[6] In these challenges, the standard will be affirmed only if supported by "substantial evidence." The remainder of this chapter deals with the scope of judicial review of OSHA standards.

3. Citations issued under OSHA standards may be contested by an employer before the independent Occupational Safety and Health Review Commission. The applicability of the standard to the particular facts is determined in these proceedings; and in some situations, the Review Commission and the courts may entertain challenges to the underlying standard.[7]

4. Employers may apply for temporary or permanent variances from the requirements of standards by following certain prescribed procedures in the Act and in OSHA regulations.[8]

In addition to these statutory limitations, other constraints on OSHA regulatory activity exist through the general operation of "checks and balances" in our governmental system. The President and his advisors have the authority, and the responsibility, to assure that executive agencies act in accordance with national policy. In the regulatory context, a series of executive orders have specified procedures that must be followed by agencies before issuing rules with a major impact on the economy.[9] The Congress, too, has continued to exercise scrutiny over rulemaking activity of agencies such as OSHA, in part through the enactment of statutes prescribing that administrative agencies follow certain steps before issuing final rules[10] and through the appropriations process and the congressional oversight functions.[11]

[5]See Chapter 8 for a discussion of the procedural issues in OSHA rulemaking.

[6]29 U.S.C. §655(f). In *Fire Equip. Mfrs. Ass'n v. Marshall*, 679 F.2d 679 (7th Cir. 1982), *cert. denied*, 103 S. Ct. 728 (1983). The OSHA revision of the fire protection standard (29 C.F.R. §§1910.155–.165 (1983)) was challenged on various procedural and substantive grounds by an association representing manufacturers of fire-fighting equipment and individual manufacturers. The court of appeals dismissed the petition for review for lack of standing, finding that these manufacturers had not been "adversely affected" by the standard. More specifically, the court concluded that the manufacturers were not the "most effective advocate" of employee interests in safety; and that neither the potential loss of profits as a result of the standard nor the competitive disadvantage resulting from the unequal concern for safety among employers are within the protected "zone of interests" of the Act. 679 F.2d at 681–83. Cf. R.W. Crandall, *The Twilight of Regulation*, 18 Brookings Bull., Win.-Spr. 1982 at 1, 5 (suggests that regulatory reform has not been successful, in part, because some business firms, such as control equipment manufacturers and suppliers, benefit financially from the rules and therefore have opposed modifications that would reduce the control technology required).

[7]See Chapter 15, Section C for a discussion of challenges to the validity of standards in enforcement cases.

[8]The Act's provisions dealing with variances are §§6(b)(6), 6(d), and 16 (national defense variance). 29 U.S.C. §§655(b)(6), 655(d), 665. OSHA's regulations on variances appear in 29 C.F.R. pt. 1905 (1983). See discussion in Chapter 15, Section F.

[9]There have been suggestions for an expansion of the presidential role in regulation; see Cutler and Johnson, *Regulation and the Political Process*, 84 Yale L.J. 1395 (1975). See also Comment, *Capitalizing on a Congressional Void: Executive Order No. 12,291*, 31 Am. U.L. Rev. 613 (1982) (discussing President Reagan's extension of presidential supervisory authority over regulatory activity).

[10]Two notable examples are the National Environmental Policy Act (NEPA), 42 U.S.C. §§4321–61 (1976 & Supp. V 1981), and the Regulatory Flexibility Act, 5 U.S.C. §§601–12 (1982).

[11]For a discussion of OSHA and the Congress, see Chapter 21.

OSHA has been largely, but not uniformly, successful in upholding its standards in the federal courts. Its most significant setback was the vacating of the benzene standard by the Supreme Court. On the other hand, the Supreme Court upheld the cotton dust standard insofar as it applied to the textile industry, and various courts of appeals have affirmed, either totally or in most important respects, the standards for asbestos, carcinogens, vinyl chloride, coke oven emissions, and lead. Most OSHA safety standards have not been challenged; those challenged, with the exception of the lavatory standard, which was vacated by the Court of Appeals for the Second Circuit, have been upheld. The OSHA record in defending emergency temporary standards has been far less successful.[12]

A list of court decisions in the major OSHA standards cases follows:

- *Associated Industries of New York State v. Department of Labor,* 487 F.2d 342 (2d Cir. 1973) (vacating lavatory standard);
- *Dry Color Manufacturers' Association v. Department of Labor,* 486 F.2d 98 (3d Cir. 1973) (vacating carcinogen emergency temporary standard);
- *Florida Peach Growers Association v. Department of Labor,* 489 F.2d 120 (5th Cir. 1974) (vacating pesticide emergency temporary standard);
- *Industrial Union Department v. Hodgson,* 499 F.2d 467 (D.C. Cir. 1974) (affirming asbestos standard);
- *Synthetic Organic Chemical Manufacturers Association v. Brennan,* 503 F.2d 1155 (3d Cir. 1974), *cert. denied,* 420 U.S. 973 (1975) (affirming standard for ethyleneimine) (*SOCMA* I);
- *Synthetic Organic Chemical Manufacturers Association v. Brennan,* 506 F.2d 385 (3d Cir. 1974), *cert. denied sub nom. Oil, Chemical and Atomic Workers International Union v. Dunlop,* 423 U.S. 830 (1975) (vacating MOCA standard) (*SOCMA* II);
- *Society of the Plastics Industry v. OSHA,* 509 F.2d 1301 (2d Cir.), *cert. denied sub nom. Firestone Plastics Co. v. Department of Labor,* 421 U.S. 992 (1975) (affirming vinyl chloride standard);
- *AFL-CIO v. Brennan,* 530 F.2d 109 (3d Cir. 1975) (remanding to OSHA an amendment to no-hands-in-die standard);
- *American Iron & Steel Institute v. OSHA,* 577 F.2d 825 (3d Cir. 1978), *cert. dismissed,* 448 U.S. 917 (1980) (affirming coke oven standard);
- *American Petroleum Institute v. OSHA,* 581 F.2d 493 (5th Cir. 1978) (vacating benzene standard), *aff'd sub nom. Industrial Union Department v. American Petroleum Institute,* 448 U.S. 607 (1980);
- *AFL-CIO v. Marshall,* 617 F.2d 636 (D.C. Cir. 1979) (uphold-

[12]See discussion in Chapter 4.

ing cotton dust standard), *aff'd sub nom. American Textile Manufacturers Institute v. Donovan,* 452 U.S. 490 (1981);

- *Texas Independent Ginners Association v. Marshall,* 630 F.2d 398 (5th Cir. 1980) (vacating cotton ginning standard);
- *United Steelworkers of America v. Marshall,* 647 F.2d 1189 (D.C. Cir. 1980) (lead standard affirmed in part, remanded in part), *cert. denied sub nom. Lead Industries Association v. Donovan,* 453 U.S. 913 (1981);
- *Asbestos Information Association v. OSHA,* 727 F.2d 415 (5th Cir. 1984) (vacating OSHA's 1983 asbestos emergency standard).

The decisions of the courts, in addition to their obvious importance in ruling on the validity of standards, have also had a major impact on OSHA's standards development process. A number of the changes in the procedural aspects of OSHA standards rulemaking have resulted in large measure from court rulings. Notable among them are OSHA's publication of extensive preambles that contain a detailed analysis of the record and a discussion of the alternatives that were rejected and OSHA's practice of presenting its own expert testimony at hearings. Substantively, the courts have established binding principles on OSHA regulatory activity, particularly in the health standard area. The Supreme Court decision in the *Benzene* case[13] imposed the requirement that OSHA regulate only where it finds a "significant risk." In the *Cotton Dust* case,[14] the Supreme Court defined the limits of the statutory term "feasibility," prohibiting OSHA's use of cost-benefit analysis in issuing standards for toxic substances. The courts have also generally upheld the Agency approach to err on the side of employee protection in imposing regulatory obligations.

The courts of appeals have reached varying results in OSHA standards cases, with the District of Columbia and the Second and Third Circuits most often acting favorably towards the Agency, and the Court of Appeals for the Fifth Circuit often setting the standards aside.[15] This uneven pattern suggests strongly that the view of a particular court of appeals on its proper role on review may be an important factor in determining the ultimate result in a case. Thus, a critical threshold question in proceedings to review OSHA standards is the scope of the court's authority to review the Agency's determinations. This is often referred to as the "standard of review." Traditionally, courts have applied the more searching "substantial evidence" test to decisions made by an agency after adjudications or formal rulemaking[16] and the "arbitrary and capricious" test to Agency rules issued after informal rulemaking proceedings.[17] The

[13]*Industrial Union Dep't v. American Petroleum Inst.,* 448 U.S. 607 (1980).
[14]*American Textile Mfrs. Inst. v. Donovan,* 452 U.S. 490 (1981).
[15]The fact that the chances for a particular party to prevail in an OSHA standards review case depends in some measure on the forum in which the case is heard has led to the "race to the courthouse." See Chapter 7.
[16]Administrative Procedure Act, 5 U.S.C. §556 (1982).
[17]*Id.* at §§553, 706.

proper application of these general principles to OSHA standards was a major issue confronting the Court of Appeals for the District of Columbia Circuit in reviewing OSHA's first standard issued after rulemaking, the standard regulating asbestos hazards.[18]

Industrial Union Department v. Hodgson

499 F.2d 467 (D.C. Cir. 1974)

McGOWAN, Circuit Judge:
***

II

OSHA is a self-contained statute in the sense that it does not depend upon reference to the Administrative Procedure Act for specification of the procedures to be followed. It prescribes that the process of promulgating a standard is to be initiated by the publication of a proposed rule. Interested persons are given a period of 30 days thereafter within which to submit written data or comments. Within this period any interested person may submit written objections, and may request a public hearing thereon. In such event, the Secretary shall publish a notice specifying the particular standard involved and stating the time and place of the hearing. Within 60 days after the completion of such hearing, the Secretary shall make his decision. Judicial review by the courts of appeals is provided.[11]

This procedure is characteristic of the informal rulemaking contemplated by Section 4 of the APA, 5 U.S.C. §553, and it was so understood by the Congress. By regulation, however, the Secretary, although describing it as "legislative in type," has provided that the oral hearing called for in the statute shall contain some elements normally associated with the adjudicatory or formal rulemaking model. As indicated in the text of the regulations, set forth in the margin,[12] the Secretary apparently concluded that this was necessary because of the necessity of having a record to which the statutorily mandated substantial evidence test could be meaningfully applied by a reviewing court. The only controversy we have in this case as to

[11]29 U.S.C. §655(f) reads in relevant part: "The determinations of the Secretary shall be conclusive if supported by substantial evidence in the record considered as a whole."

[12]In 29 C.F.R. §1911.15 ("Nature of Hearing"), the Secretary stated in relevant part:

"(a)(2) Section 6(b)(3) provides an opportunity for a hearing on objections to proposed rule making, and section 6(f) provides in connection with the judicial review of standards, that determinations of the Secretary shall be conclusive if supported by substantial evidence in the record as a whole. Although these sections are not read as requiring a rule making proceeding within the meaning of the last sentence of 5 U.S.C. 553(c) requiring the application of the formal requirements of 5 U.S.C. 556 and 557, they do suggest a Congressional expectation that the rule making would be on the basis of a record to which a substantial evidence test, where pertinent, may be applied in the event an informal hearing is held.

"(3) The oral hearing shall be legislative in type. However, fairness may require an opportunity for cross-examination on crucial issues. The presiding officer is empowered to permit cross-examination under such circumstances. * * *

"(b) Although any hearing shall be informal and legislative in type, this part is intended to provide more than the bare essentials of informal rule making under 5 U.S.C. 553. The additional requirements are the following:

"(1) The presiding officer shall be a hearing examiner appointed under 5 U.S.C. 3105.

"(2) The presiding officer shall provide an opportunity for cross-examination on crucial issues.

"(3) The hearing shall be reported verbatim, and a transcript shall be available to any interested person on such terms as the presiding officer may provide."

[18]For a discussion of OSHA's ETS on asbestos issued in 1983, see Chapter 4.

the procedural requirements of the statute is not with respect to the manner in which the rulemaking was done by the Secretary, but as to the reach of the substantial evidence test in the course of judicial review.

The substantial evidence test has customarily been directed to adjudicatory proceedings or formal rulemaking. The hybrid nature of OSHA in this respect can be explained historically, if not logically, as a legislative compromise. The Conference Report reflects that the Senate bill called for informal rulemaking, but the House version specified formal rulemaking and substantial evidence review. The House receded on the procedure for promulgating standards, but the substantial evidence standard of review was adopted.

One question generated by this anomalous combination is whether the determinations in question here are of the kind to which substantial evidence review can appropriately be applied. The Government in its argument suggested that a proper accommodation could be effected by construing the statute to require substantial evidence review of factual determinations, while weighing the inferences of policy drawn from those facts in terms of their freedom from arbitrariness or irrationality. We do not believe this approach would affect the rigorousness of our review to the extent the Government seems to suppose, or that petitioners purport to fear. The analysis may, however, be useful for the purpose of clarifying the diverse nature of the judicial task imposed upon us by a statute like OSHA.

Another problem arising from substantial evidence review of informal proceedings concerns the adequacy of the record to permit meaningful performance of the required review. Although this issue has not been directly raised in argument, it underlies much of the controversy concerning the sufficiency of the evidence to support various specific determinations of the Secretary. Thus some explication of the procedural implications of the prescribed substantial evidence standard of review should help to clarify our resolution of the particular substantive issues presented by this petition.

Faced with the fact that his determinations were commanded by Congress to be reviewed under a substantial evidence standard, the Secretary did voluntarily move his procedures significantly towards the formal model. He directed that (1) a qualified hearing examiner should preside over the oral hearing, (2) cross-examination should be permitted, and (3) a verbatim transcript made. The total record in this case was in part created under the conditions that obtain in a formal proceeding. In substantial remaining part, however, it consists of a melange of written statements, letters, reports, and similar materials received outside the bounds of the oral hearing and untested by anything approaching the adversary process.

Thus, in some degree the record approaches the form of one customarily conceived of as appropriate for substantial evidence review. In other respects, it does not. On a record of this mixed nature, when the facts underlying the Secretary's determinations are susceptible of being found in the usual sense, that must be done, and the reviewing court will weigh them by the substantial evidence standard. But, in a statute like OSHA where the decision making vested in the Secretary is legislative in character, there are areas where explicit factual findings are not possible, and the act of decision is essentially a prediction based upon pure legislative judgment, as when a Congressman decides to vote for or against a particular bill.

OSHA sets forth general policy objectives and establishes the basic procedural framework for the promulgation of standards, but the formulation of specific substantive provisions is left largely to the Secretary. The Secretary's task thus contains "elements of both a legislative policy determination and an adjudicative resolution of disputed facts." *Mobil Oil Corp. v. FPC*, 483 F.2d 1238, 1257 (D.C. Cir. 1973). Although in practice these elements may so intertwine as to be virtually inseparable, they are conceptually distinct and can only be regarded as such by a reviewing court.

From extensive and often conflicting evidence, the Secretary in this case made numerous factual determinations. With respect to some of those questions, the evidence was such that the task consisted primarily of evaluating the data and drawing conclusions from it. The court can review that data in the record and determine whether it reflects substantial support for the Secretary's findings. But some of the questions involved in the promulgation of these standards are on the frontiers of scientific knowledge, and consequently as to them insufficient data is presently available to make a fully informed factual determination. Decision making must in that circumstance depend to a greater extent upon policy judgments and less upon purely factual analysis. Thus, in addition to currently unresolved factual issues, the formulation of standards involves choices that by their nature require basic policy determinations rather than resolution of factual controversies. Judicial review of inherently legislative decisions of this sort is obviously an undertaking of different dimensions.

For example, in this case the evidence indicated that reliable data is not currently available with respect to the precisely predictable health effects of various levels of exposure to asbestos dust; nevertheless, the Secretary was obligated to establish some specific level as the maximum permissible exposure. After considering all the conflicting evidence, the Secretary explained his decision to adopt, over strong employer objection, a relatively low limit in terms of the severe health consequences which could result from overexposure. Inasmuch as the protection of the health of employees is the overriding concern of OSHA, this choice is doubtless sound, but it rests in the final analysis on an essentially legislative policy judgment, rather than a factual determination, concerning the relative risks of underprotection as compared to overprotection.

Regardless of the manner in which the task of judicial review is articulated, policy choices of this sort are not susceptible to the same type of verification or refutation by reference to the record as are some factual questions. Consequently, the court's approach must necessarily be different no matter how the standards of review are labeled. That does not mean that such decisions escape exacting scrutiny, for, as this court has stated in a similar context:

> This exercise need be no less searching and strict in its weighing of whether the agency has performed in accordance with the Congressional purposes, but, because it is addressed to different materials, it inevitably varies from the adjudicatory model. The paramount objective is to see whether the agency, given an essentially legislative task to perform, has carried it out in a manner calculated to negate the dangers of arbitrariness and irrationality in the formulation of rules for general application in the future.

Automotive Parts & Accessories Association v. Boyd, 407 F.2d 330, 338 (1968).

We do not understand Congress to have in this instance nullified this approach for all purposes by directing substantial evidence review. As noted above, that provision is important as an indication of how we should approach certain kinds of questions and what kind of record we should demand of the Secretary. But it is surely not to be taken as a direction by Congress that we treat the Secretary's decision making under OSHA as something different from what it is, namely, the exercise of delegated power to make within certain limits decisions that Congress normally makes itself, and by processes, as the courts have long recognized and accepted, peculiar to itself. A due respect for the boundaries between the legislative and the judicial function dictates that we approach our reviewing task with a flexibility informed and shaped by sensitivity to the diverse origins of the determinations that enter into a legislative judgment.

What we are entitled to at all events is a careful identification by the Secretary, when his proposed standards are challenged, of the reasons why he chooses to follow one course rather than another. Where that choice purports to be based on the existence of certain determinable facts, the Secretary must, in form as well as substance, find those facts from evidence in the record. By the same token, when the Secretary is obliged to make policy judgments where no factual certainties exist or where facts alone do not provide the answer, he should so state and go on to identify the considerations he found persuasive.

Judge Friendly concluded his ruminations in *Associated Industries* with an expression of doubt as to "whether judicial review of legislative standards resulting from informal rule-making will ultimately prove to be feasible."* That is certainly a serious and substantial question. Whether it can eventually be answered affirmatively must depend in large measure upon the care and good sense with which both the delegatee of what is essentially legislative power and the reviewing court go about their respective duties. In the case of OSHA, the Secretary has wisely acted by regulation to go beyond the minimum requirements of the statute and to expand his capacity to find facts by providing an evidentiary hearing in which cross-examination is available. We think it equally the part of wisdom and restraint on our part to show a comparable flexibility, and to be always mindful that at least some legislative judgments cannot be anchored securely, and solely in demonstrable fact. Such a principle, far from being destructive of the Congressional purpose to provide judicial review, seems to us within the Congressional contemplation as essential to its preservation.

This landmark decision has provided a framework for all subsequent discussions of the scope of court review of OSHA standards.[19] In deciding that the substantial evidence test should be applied flexibly to OSHA determinations, the District of Columbia Circuit relied on two major considerations: (1) many of the issues confronting OSHA are on the "frontiers of scientific knowledge requiring" basic policy determinations rather than factual decision"; these policy decisions, the court said, cannot be judged by a "factual yardstick"; and (2) since "the protection of the health of employees is the overriding concern of OSHA," its decision to resolve doubts in favor of employee health is "doubtless sound." Thus, while the court of appeals insisted that its review must be calculated to "negate the dangers of arbitrari-

*[Author's note—In *Associated Indus. v. Department of Labor,* 487 F.2d 342 (2d Cir. 1973), Judge Friendly, writing for the court, set aside OSHA's amendments to its lavatory standard, originally adopted as a national consensus standard in 1971, because of the inadequacy of the statement of reasons and the lack of "substantial evidence" supporting OSHA's determination. In the course of his opinion, Judge Friendly offered views on the meaning of the substantial evidence test in the Act. OSHA had argued that, in light of the legislative history, the court should apply a "less severe" standard of review than the substantial evidence test. While rejecting the OSHA argument, the court suggested that the controversy was "semantic to some degree," and because the two tests—substantial evidence and arbitrary and capricious—"tend to converge," the issue lacks "dispositional importance." *Id.* at 347–50.]

[19]The subsequent history of OSHA's regulation of asbestos hazards is discussed in Chapter 4.

ness and irrationality," it enunciated a relatively deferential approach to OSHA's standards designed to protect employees. The principles stated by the court in the *Asbestos* case have been adopted in substance by all courts of appeals dealing with OSHA standards, except for the Fifth Circuit Court of Appeals.

The statutory substantial evidence test, when applied to OSHA policy determinations, was a more flexible instrument of review than it had been when utilized in the setting of adjudication and formal rulemaking. The arbitrary and capricious test, traditionally applied to informal rulemaking, has also undergone an opposite transformation, from a highly deferential standard to one encompassing more searching and careful review. This development, which received a great impetus from the Supreme Court decision in *Citizens to Preserve Overton Park, Inc. v. Volpe,*[20] and which paralleled the evolution of hybrid rulemaking procedures[21] has been traced by commentators to the huge growth in the 1970s of the scope and impact of government regulation and a growing distrust on the part of substantial segments of the public in the activities of federal regulatory agencies.[22] As a result of these and other developments, the two tests for review—substantial evidence and arbitrary and capricious—became very similar and, as has been observed in several decisions, for practical purposes normally lead to the same result.[23]

B. Adjudicating "Science-Policy" Issues

The court in *Industrial Union Department v. Hodgson* referred to "legislative-policy" issues, on the frontiers of scientific knowledge, which often confront OSHA in the development of standards. Examples of these "science-policy" questions, as they have sometimes been called,[24] are whether a substance shown to be carcinogenic in animals should be regulated as causing cancer in humans, and the level of exposure at which a carcinogen should be regulated.[25] OSHA determinations of carcinogenicity for the most part have been based on

[20]401 U.S. 402 (1971).

[21]See discussion in Chapter 3, Section A.

[22]See, e.g., Delong, *Informal Rulemaking and the Integration of Law and Policy,* 65 VA. L. REV. 257, 262–89 (1979). See also SENATE COMM. ON THE JUDICIARY, THE REGULATORY REFORM ACT, S. REP. No. 284, 97th Cong., 1st Sess. 50–52, 93–106 (1981).

[23]See *Associated Indus. v. Department of Labor,* 487 F.2d 342, 349–50 (2d Cir. 1973) (stating that the "two criteria [for review] tend to converge."). See also *State Farm Mut. Auto. Ins. Co. v. Department of Trans.,* 680 F.2d 206, 228–29 (D.C. Cir. 1982) *vacated and remanded sub nom. Motor Vehicle Mfrs. Ass'n v. State Farm Mut. Auto. Ins. Co.,* 51 USLW 4953 (1983); *Ethyl Corp. v. EPA,* 541 F.2d 1, 37 n.79 (D.C. Cir. 1976).

[24]For a discussion of science-policy questions, see McGarity, *Substantive and Procedural Discretion in Administrative Resolution of Science Policy Questions: Regulating Carcinogens in EPA and OSHA,* 67 GEO. L.J. 729, 731–47 (1979).

[25]Analogous issues arise with respect to noncarcinogenic toxic substances. For example, in issuing a standard on cotton dust, OSHA confronted the questions whether the substance causes byssinosis in workers, and the level at which it should be regulated. See Chapter 10. Carcinogenic substances have consistently been OSHA's highest regulatory priority. See generally preamble to Carcinogens Policy, 45 FED. REG. 5002 (1980), and discussion in Chapter 3, Section B.

two types of scientific data: epidemiological studies and animal studies, as discussed in the following report:[26]

Merrill, *Federal Regulation of Cancer-Causing Chemicals*

***

2. EPIDEMIOLOGICAL STUDIES

***

b. *Standard approaches.* Epidemiologists follow two basic approaches in investigating the connection between environmental insults and cancer: cohort studies and case-control studies. Cohort studies, which may be either prospective or retrospective, compare the health histories of groups of individuals who differ in their exposure to the chemical under study. For example, a retrospective cohort study might compare the health histories of workers in a shipbuilding facility that used asbestos insulation to the histories of other workers who had no exposure to asbestos. To the extent possible, workers would be subdivided according to their level of exposure to asbestos. The incidence of lung cancer among workers in each group would be compared to identify associations between the level of asbestos exposure and the occurrence of disease. A cohort study can convincingly attribute the differences in cancer rate to asbestos exposure only if all of the cohorts are well matched with respect to other variables, e.g., sex, occupation, age, and exposure to other carcinogens such as cigarette smoke, that may influence the incidence of the disease. This ideal is very difficult to achieve.

The second common format is the case-control study. Such a study begins with the identification of individuals who suffer from a disease. These "cases" are then compared with matched "controls," i.e., individuals who do not have the disease, to determine whether there are notable differences in exposure to suspect agents. As in cohort studies, the objective of the case-control approach is to minimize the effect of confounding variables so that any differences in cancer rates can be correlated with chemical exposure.

The reliability of the conclusions from either cohort or case-control studies depends on the closeness of the match between the compared groups, the accuracy of exposure and disease incidence data, and the statistical limits of detecting differences in disease incidences between groups. The case-control format has particular appeal for the study of possible chemical carcinogens because the expected incidence of most cancers is low. The probative weight of findings from such a study depends on whether the investigator has identified a control population that closely matches the cases. Selecting matched controls requires informed guesses about the factors that could affect disease incidence, to which individual exposure may often be impossible to verify.

[26]R. A. Merrill, FEDERAL REGULATION OF CANCER-CAUSING CHEMICALS, REPORT TO THE ADMINISTRATIVE CONFERENCE OF THE UNITED STATES, 47–67, April 1, 1982 (Draft). See also CONGRESS OF THE UNITED STATES, OFFICE OF TECHNOLOGY ASSESSMENT, ASSESSMENT OF TECHNOLOGIES FOR DETERMINING CANCER RISKS FROM THE ENVIRONMENT 122–62 (1981); OSHA's Carcinogens Policy, 45 FED. REG. 5002 *passim* (1980); *id.* at 5035–60 (the use of epidemiological data); *id.* at 5060–5201 (animal studies). OSHA based its standard on coke oven emissions, for example, on both epidemiological and animal studies showing carcinogenic effects. 41 FED. REG. 46,742, 46,744–51 (1976). The arsenic standard was set, in the absence of animal data showing carcinogenicity, solely on the basis of epidemiological studies, 43 FED. REG. 19,584, 19,586–97 (1978). See also the Supplemental Statement of Reasons for Arsenic Standard, 48 FED. REG. 1864, 1890–91 (1983) (most animal studies on arsenic have obtained negative results; "some" have obtained positive results; lack of "good" animal studies does not detract from conclusion that arsenic is a human carcinogen).

Epidemiologists confront three main problems: detection of disease frequencies; recordation of exposure histories; and discernment of linkages between them. The North American data base lags well behind modern needs to describe the health effects of environmental agents. The decentralized, predominantly private U.S. health care system does facilitate uniform data gathering as readily as do state systems. The classical uncertainties of disease recording—inaccuracies in mortality and morbidity classification—are thus exacerbated by the lack of a uniform national reporting system. Data gathering is also impeded by the mobility of the U.S. population, which produces frequent changes in dietary and exposure patterns and discontinuities in medical records. The probable costs of more accurate and comprehensive health records, furthermore, are very high, particularly if this requires a large fraction of deaths to be followed by autopsy. Finally, the number of unidentified carcinogens to which individuals are exposed in their daily lives practically defies attempts to control for their effects.

c. *Power and limitations of epidemiological studies.* Roughly two dozen chemicals have been identified as carcinogens in epidemiological studies. The most appealing feature of such studies is the obvious one: they involve human populations. Using data from human experience eliminates the uncertainties that accompany extrapolation of animal test results to humans. Epidemiological studies can also incorporate, albeit at substantial expense, very large populations, which are not practical in animal experiments. A large population extends the lower range of detectable effects and enhances the statistical significance of observed associations.

***

Accordingly, while epidemiological investigations have illuminated the role of environmental factors in some human diseases, they have played a limited role in identifying health effects of specific chemicals. Their expense alone would preclude routine use of epidemiological studies as a screen for carcinogenic effects. Moreover, epidemiologists can evaluate only those substances to which humans are already exposed; ethical constraints generally preclude nontherapeutic experiments with human subjects. In consequence, the data used in epidemiological studies are often poor. Epidemiologists do not conduct "experiments"; instead, they take human circumstances as they find them and attempt to select from recorded data the information required for a well-designed study of carcinogenicity. Most available data, e.g., from hospital or employer records or death certificates, are at best an imperfect approximation of actual experience. Their deficiencies are particularly evident in cancer studies, which often try to reconstruct conditions that existed many years earlier at the time of initial exposure to a chemical.

***

3. ANIMAL EXPERIMENTS

a. *Background.* * * *

Toxicologists agree that, although not ideal, the best practical test for identifying causes of human cancer is the long-term bioassay in rodents. Though their details still occasion scientific debate, these tests generally follow a well-accepted pattern: A small population of animals (e.g., rarely more than 400), which have been maintained under similar conditions in a controlled laboratory setting, is divided into groups equally distributed by sex, e.g., 50 animals of each sex per group. One and usually more of these groups is regularly exposed to the test chemical for the major part of their lifespan. A control group of similar size is left untreated. All animals are carefully observed for the appearance of tumors during the study and, at the conclusion, are sacrificed and examined pathologically. The incidence(s) of tumors observed in the exposed group(s) is compared to the incidence of tumors in the controls. Because the groups are otherwise theoretically identical, any increase in tumor incidence in the exposed animals presumably can be attributed to the test chemical.

Although animal bioassays are straightforward in concept, they are difficult to design, execute and evaluate. The standard rodent bioassay requires three to five years to complete and demands coordinated efforts by scientists and technicians who possess skills ranging from routine animal care to pathological diagnosis. The cost of evaluating one chemical in a single rodent species now substantially exceeds $500,000. Each element in the design of an experiment calls for scientific judgment, which later may provide the basis for argument when a regulatory agency seeks to interpret the results. The following discussion highlights common sources of disagreement.

The critical design parameters of animal carcinogenicity tests are the species used, population size, and the dose levels of the test chemical. Elaborate though they are, current tests reflect major compromises of the theoretical ideal on each point.

b. *Choice of species.* * * *

c. *Dose levels.* The practical limits on the size of animal experiments have led to the use of high test doses, which are necessary to produce a detectable incidence of tumors. Experimenters customarily use the highest dose the animals can tolerate without experiencing other unacceptable adverse health effects. Higher doses may overload the animals' system and distort their metabolism of the test chemical; even the "maximum tolerated dose" (MTD) may produce other adverse effects in the animals. More important, high test doses complicate interpretation because they generally exceed, sometimes by several orders of magnitude, the levels at which humans are likely to be exposed to the test chemical. Despite these difficulties, toxicologists generally support the use of high experimental doses to determine whether a substance is capable of causing cancer. This practice has, however, often occasioned public misunderstanding of test results and sometimes evoked barbed criticism of agency reliance on animal data.

A desired product of any animal bioassay is a positive correlation between increasing dose and response which supports the inference that the chemical caused the response. * * *

A "no effect" level is the dose level of a chemical, known to be toxic at high doses, at which—and thus presumably below which—adverse effects are not observed. No-effect levels can be determined for many non-carcinogens and are commonly used by regulators in setting exposure levels that are presumed safe. But routine animal bioassays are not capable of demonstrating a no-effect level for a carcinogen. First, their insensitivity makes them incapable of ruling out the possibility of tumors in a small percentage of exposed individuals. In addition, assumptions about the mechanism of carcinogenesis are consistent with the occurrence of tumors at extremely low doses. Thus there currently is no practical method for determining no-effect levels for carcinogens experimentally. The most any negative bioassay can convincingly demonstrate is the maximum risk of cancer that a chemical may pose, i.e., the incidence that could escape detection.

***

For the first time, in 1974, the question of how to treat animal carcinogens faced OSHA in setting a standard for 14 carcinogens.[27] Although there was no epidemiological data for several of these substances showing that they cause cancer in humans, OSHA decided to treat them as human carcinogens, based on animal studies.

Health Standards for Carcinogens

39 Fed. Reg. 3,756, 3,757–58 (1974)

***

Ethyleneimine. The carcinogenic potential of ethyleneimine (EI) has been confirmed by a study conducted by Walpole in 1954 involving rats and one sponsored by the National Cancer Institute involving mice. In the first study, animals developed injection site sarcomas which the investigators attributed to the direct action of Ethyleneimine, and in the second study 80 percent of the animals developed tumors, including more than one-half with hepatomas (which the investigators stated had "malignant potentiality") and almost three-quarters with pulmonary tumors. Although high doses of EI were administered, the investigators stated there was no way to predict whether man would be more or less susceptible to tumor induction by EI.

The case for the carcinogenicity of EI, then, rests on the extrapolation to humans of the findings in two separate, controlled animal studies. This position is compatible with that of NIOSH concerning the prior demonstration of carcinogenicity in at least two animal studies.

A major question of occupational carcinogenesis relates to the extrapolation of results of animal experimentation to humans. The basis of numerous objections to the proposals is that, even assuming the validity of animal experiments, such do not furnish sufficient evidence that the substances involved are carcinogenic to humans. Extrapolation of results obtained by animal experimentation is alleged to be vitiated by several considerations: (a) that certain cancers are specific only to some species; (b) that the conditions of animal experiments are out of proportion to, and not consistent with, conditions prevailing in industrial exposure; and (c) that no cancers have yet been detected in humans exposed to the substances. For those substances whose metabolism is understood, and is similar in both animals and man, the fact that they induce cancers in animals warrants the expectation that they will induce cancers in men. This applies to the substances which cause urinary bladder cancers in animals acting, not directly, but indirectly through the mediation of metabolites formed both in experimental animals and in exposed workers. This is also true of those substances which apparently require no metabolic alteration but attack a particular biologic system (e.g., respiratory tract, alimentary canal) which is similar in both animals and humans.

The objections raise the much broader issue of human exposure to a chemical which is only known to have caused cancers in experimental animals.

It is important to note that some opponents of the regulation of such chemicals do not advocate treating them as if they were harmless with respect to carcinogenic potential. Several employers, for instance insist that such substances must be treated with "care" or "respect," while also insist-

[27]The question of the extrapolation of carcinogenic effects—whether found in epidemological studies or in animal experiments—from high-dose levels to low-dose levels is discussed more fully in relation to the Supreme Court *Benzene* decision, *Industrial Union Dep't v. American Petroleum Inst.*, 448 U.S. 607 (1980), in Chapter 9.

ing that they call for significantly less protection than those substances known to be human carcinogens.

We think it improper to afford less protection to workers when exposed to substances found to be carcinogenic only in experimental animals. Once the carcinogenicity of a substance has been demonstrated in animal experiments, the practical regulatory alternatives are to consider them either non-carcinogenic or carcinogenic to humans, until evidence to the contrary is produced. The first alternative would logically require, not relaxed controls on exposure, but exclusion from regulation. The other alternative logically leads to the treatment of a substance as if it was known to be carcinogenic in man.

We agree with the Director of NIOSH, and the report of the Ad Hoc Committee on the Evaluation of Low Levels of Environmental Chemical Carcinogens to the Surgeon General, U.S. Public Health Service, April 22, 1970, that the second alternative is the responsible and correct one. * * *

In the face of scientific uncertainty, OSHA decided, as a matter of policy, that the only "responsible and correct" course of action was to regulate animal carcinogens as carcinogens in humans; otherwise, the Agency said, the substance would escape regulation altogether, an obviously unacceptable result. The standard for ethyleneimine (EI) was challenged in court and, predictably, a major issue was whether the substance was a human carcinogen. The Court of Appeals for the Third Circuit upheld OSHA's carcinogenicity findings, applying a standard of review substantially the same as the District of Columbia Circuit Court of Appeals in the *Asbestos* case.

Synthetic Organic Chemical Manufacturers Association v. Brennan (SOCMA I)

503 F.2d 1155 (3d Cir. 1974)

STALEY, Circuit Judge:

***

This case is a good illustration of the difficulty of attempting to measure a legislative policy decision against a factual yardstick. OSHA's position with respect to EI is bottomed on an extrapolation from data gathered in two animal studies. The first is a study carried out in 1954 by Walpole which concluded that EI may be regarded as carcinogenic in rats and mice. The Innes study, the second upon which OSHA relied, concluded that mice under experimental exposure to EI developed tumors with "malignant potentiality." The extrapolation from the data gathered in these rodent experiments to humans is justified by the Report of the Ad Hoc Committee on the Evaluation of Low Levels of Environmental Chemical Carcinogens to the Surgeon General. This report states:

> "Any substance which is shown conclusively to cause tumors in animals should be considered carcinogenic and therefore a potential cancer hazard for man."

If the issue to be reviewed were merely whether EI was carcinogenic in rats and mice, we believe that we could point to the Walpole and Innes studies and safely conclude that the Secretary's determination of animal carcinogenicity was supported by substantial evidence. But the extrapolation of that determination from animals to humans is not really a factual matter.

***

It seems to us that what the Secretary has done in extrapolating from animal studies to humans is to make a legal rather than a factual determination. He has said in effect that if carcinogenicity in two animal species is established, as a matter of law §§6(a) and 6(b)(5) require that they be treated as carcinogenic in man. This is in the nature of a recommendation for prudent legislative action.

***

It seems, then, that judicial review of a §6 standard properly includes at least the following:

(1) determining whether the Secretary's notice of proposed rule making adequately informed interested persons of the action taken;
(2) determining whether the Secretary's promulgation adequately sets forth reasons for his action;
(3) determining whether the statement of reasons reflects consideration of factors relevant under the statute;
(4) determining whether presently available alternatives were at least considered; and
(5) if the Secretary's determination is based in whole or in part on factual matters subject to evidentiary development, whether substantial evidence in the record as a whole supports the determination.

***

The court in the *Synthetic Organic* case said that the issue whether EI is a human carcinogen was a "legal" rather than "factual" determination. This articulation apparently means that, as in the review of legal determination, the court would not require the Agency to adduce "substantial evidence" in the traditional sense in support of its finding of carcinogenicity and would accord a significant measure of deference to the views of the Agency charged with administering the statute.[28]

In reviewing OSHA's vinyl chloride standard, the Court of Appeals for the Third Circuit was confronted with the issue whether the Agency had properly imposed a 1 part per million (ppm) standard when the experimental evidence then available showed that animal cancer was induced at levels no lower than 50 ppm. (There was also

[28]See *E.I. du Pont de Nemours & Co. v. Train,* 430 U.S. 112, 135 n.25 (1977). The resolution by the Third Circuit Court of Appeals of the "mouse to man" question in the *Synthetic Organic* case appears inconsistent with its decision in *Dry Color Mfrs. Ass'n v. Department of Labor,* 486 F.2d 98 (3d Cir. 1973). The court there agreed that extrapolations from animal experiments may "in appropriate cases" be used to establish "a sufficient probability of harm to man"; but, in setting aside the carcinogen emergency temporary standard, the court said that there must be "more than some possibility that a substance may cause cancer in man." *Id.* at 104; see Chapter 8, Section F. The *Dry Color* case, however, involved an emergency standard, promulgated after the "summary procedures" of §6(c), 29 U.S.C. §655(c). The court stated that "especially where the effects of a substance are in sharp dispute," normal rulemaking should be used "to bring out the relevant facts." 486 F.2d at 105 n.9a. On the other hand, in the *Synthetic Organic* case, a rulemaking record had been developed, and the Third Circuit, therefore, did not act inconsistently in upholding the human carcinogenicity finding. Cf. McGarity, *supra* note 24, at 803. See generally the discussion in Chapter 4 on emergency temporary standards.

evidence of cancer in humans as a result of exposure to vinyl chloride but the level of exposure was not known.) The court upheld the 1 ppm level, saying:

> As in *Industrial Union Department, AFL-CIO v. Hodgson,* the ultimate facts here in dispute are "on the frontiers of scientific knowledge," and, though the factual finger points, it does not conclude. Under the command of OSHA, it remains the duty of the Secretary to act to protect the workingman, and to act even in circumstances where existing methodology or research is deficient. The Secretary, in extrapolating the MCA [Manufacturing Chemists Association] study's finding from mouse to man, has chosen to reduce the permissible level to the lowest detectable one. We find no error in this respect.[29]

The fullest articulation of views on the scope for review of OSHA standards was made by Senior Circuit Judge Bazelon for the Court of Appeals for the District of Columbia Circuit in the case in which the cotton dust standard was upheld as it applied to a number of industries.

AFL-CIO v. Marshall

617 F.2d 636 (D.C. Cir. 1979)

BAZELON, Senior Circuit Judge:
***

II. SCOPE OF REVIEW

The Occupational Safety and Health Act of 1970 is one of a number of recent Congressional statutes that designates the stringent "substantial evidence" test for judicial review of notice-and-comment rulemaking. The explicit language of the Act, its legislative history, and its application by the courts confirm that regulations promulgated under the Act are to be upheld on review if supported by "substantial evidence on the record considered as a whole."

The substantial evidence test provides for more rigorous scrutiny than the usual "arbitrary and capricious" test applicable to informal rulemaking. Although Congress required this more rigorous judicial review, it nevertheless delegated unusually broad discretionary authority to regulate against possible harms. We have already resolved this seeming anomaly in *Industrial Union Dep't v. Hodgson,* 499 F.2d 467 (D.C. Cir. 1974). There we concluded that the reviewing court's task under the Act is to provide a careful check on the agency's determinations without substituting its judgment for

[29]*Society of the Plastics Indus., Inc. v. OSHA,* 509 F.2d 1301, 1308 (2d Cir. 1975). The court in this case also enunciated the doctrine of "technology forcing" in deciding that the standard was "feasible." See Chapter 10. In concluding that OSHA was not "restricted to the status quo" in deciding feasibility issues, 509 F.2d at 1309, the court was, as in its view on the permissible exposure level, *id.* at 1308, deferring to the Agency's decision to act "even in circumstances where existing methodology or research is deficient." Thus, feasibility issues may also be on the "frontiers of * * * knowledge," and therefore the "substantial evidence" test could not be applied in its strict adjudicatory sense. See *id.* at 1304. See also *American Textile Mfrs. Inst. v. Donovan,* 452 U.S. 490, 528 n.52, 529 n.54 (1981) (examining OSHA's economic feasibility findings for the cotton dust standard); *American Iron & Steel Inst. v. OSHA,* 577 F.2d 825 (3d Cir. 1978) (upholding in substantial part the OSHA coke oven emission standard).

that of the agency. Congress apparently created an "uneasy partnership" between the agency and the reviewing court[49] to check extravagant exercises of the agency's authority to regulate risk. Our role in this partnership is to ensure that the regulations resulted from a process of reasoned decisionmaking consistent with the agency's mandate from Congress. By statute, this process must include notice to interested parties of issues presented in the proposed rule. The agency must also provide opportunities for these parties to offer contrary evidence and arguments.

OSHA adopted additional procedures to improve its decisionmaking process. A qualified hearing examiner must preside at oral hearings on proposed standards. A verbatim transcript of the hearing is required, and cross-examination is permitted. These procedures, which were followed in this case, transform OSHA's action into "hybrid" rulemaking, and produce a record more susceptible to rigorous judicial review than the more usual informal rulemaking record.

The tasks of this reviewing court are thus to ensure that the agency has (1) acted within the scope of its authority;[57] (2) followed the procedures required by statute and by its own regulations;[58] (3) explicated the bases for its decision; (4) adduced substantial evidence in the record to support its determinations.

The meaning of "substantial evidence" in this context is problematic. Factual proof about particular health risks may not be substantial in the traditional sense simply because the medical and scientific communities do not yet completely understand the nature of threatening diseases. To protect workers from material health impairments, OSHA must rely on predictions of possible future events and extrapolations from limited data. It may have to fill gaps in knowledge with policy considerations. Congress recognized this problem by authorizing the agency to promulgate rules on the basis of the "best available evidence."[60] OSHA's mandate necessarily requires it to act—even if information is incomplete—when the best available evidence indicates a serious threat to the health of workers. Thus, a court entrusted with the rigorous "substantial evidence" review must examine not only OSHA's factual support, but also the "judgment calls" and reasoning that contribute to its final decision. Otherwise, an agency's claim of ignorance would clothe it with unreviewable discretion.

Therefore, the reviewing court must examine both factual evidence and the agency's policy considerations set forth in the record. To facilitate this review of the record, the agency must pinpoint the factual evidence and the

---

[49]*Assoc. Indus. v. Dep't of Labor* at 354. This partnership is uncomfortable for the courts chiefly because the record produced by informal rulemaking is not easily suited to close judicial scrutiny. *See Florida Growers Ass'n v. Dep't of Labor,* 489 F.2d 120, 129 [1 OSHC 1472] (5th Cir. 1974). The record generally is a compendium of letters, studies, reports, and statements, untested by the adversary process. Reflecting the legislative nature of informal rulemaking, the record often does not even display the full range of considerations before the agency when the decision was made. Pedersen, *Formal Records and Informal Rulemaking,* 85 YALE L.J. 38, 62 (1975). Furthermore, the decision necessarily rests on policy considerations authorized by the agency's mandate. Judicial review of such considerations cannot be identical to review of factual determinations.

[57]*Citizens to Preserve Overton Park v. Volpe,* 401 U.S. 402, 415 (1971); *City of Chicago v. FPC,* 458 F.2d 731, 745 (D.C. Cir. 1971); *Automotive Parts & Accessories Ass'n v. Boyd,* 407 F.2d 330, 343 (D.C. Cir. 1968); 5 U.S.C. §706(a)(C)(1976).

[58]*Citizens to Preserve Overton Park v. Volpe,* 401 U.S. 402, 417 (1971); *Friends of the Earth v. United States Atomic Energy Comm'n,* 485 F.2d 1031, 1033 (D.C. Cir. 1973); 5 U.S.C. §706(2)(D) (1976).

[60]29 U.S.C. §655(b)(5) (1976). The best available evidence in an area of changing technology and incomplete scientific data may leave gaps in knowledge that require policy judgments in constructing the health and safety standard. In recent statutes Congress has defined "evidence" to mean "any matter in the rulemaking record." E.g., Toxic Substances Control Act, 15 U.S.C. §2618(c)(1)(B) (1976).

policy considerations upon which it relied. This requires explication of the assumptions underlying predictions or extrapolations,[64] and of the basis for its resolution of conflicts and ambiguities.[65] In enforcing these requirements, the court does not reach out to resolve controversies over technical data.[66] Instead, it seeks to ensure public accountability. Explicit explanation for the basis of the agency's decision not only facilitates proper judicial review but also provides the opportunity for effective peer review, legislative oversight, and public education. This requirement is in the best interest of everyone, including the decisionmakers themselves. If the decisionmaking process is open and candid, it will inspire more confidence in those who are affected. Further, by opening the process to public scrutiny and criticism, we reduce the risk that important information will be overlooked or ignored. Instructed by these ends, this court on review will "combine supervision with restraint."[67]

***

Judge Bazelon emphasized the policy nature of much of OSHA decision making and stated that the court should not resolve "controversies over technical data." He concluded that on review the court must "combine supervision with restraint." This approach on review is substantially the same as explicated by Judge McGowan in the *Asbestos* case four years earlier.

In one critical respect, the later decisions elaborated upon a comment in the *Asbestos* case regarding the importance of OSHA's adherence to the procedures promoting full public participation in the development of standards. In the *Asbestos* case, the court said that, although the Agency would not be required to provide strict factual

---

[64]With regard to environmental rulemaking, this court held that

Where a statute is precautionary in nature, the evidence difficult to come by, uncertain, or conflicting because it is on the frontiers of scientific knowledge, the regulations designed to protect the public health, and the decision that of an expert administrator, we will not demand rigorous step-by-step proof of cause and effect. Such proof may be impossible to obtain if the precautionary purpose of the statute is to be served. * * * The Administrator may apply his expertise to draw conclusion from suspected, but not completely substantiated, relationships between facts, from trends among facts, from probative preliminary data not yet certifiable as 'fact,' and the like.

Ethyl Corp. v. EPA, 541 F.2d 1, 28 (D.C. Cir. 1976) (citations omitted). *See Amoco Oil Co. v. EPA*, 501 F.2d 722, 740–41 (D.C. Cir. 1974) (reasons and explanations, but not findings, required for predictions used in risk regulation).

[65][W]hen the Secretary is obliged to make policy judgments where no factual certainties exist or where facts alone do not provide the answer, he should so state and go on to identify the considerations he found persuasive.

499 F.2d at 476.

[66]*See Ethyl Corp. v. EPA*, 541 F.2d 1, 67 (D.C. Cir. 1976) (Bazelon, C.J., and McGowan, J., concurring); *Internat'l Harvester Co. v. Ruckelshaus*, 478 F.2d 615, 652 (D.C. Cir. 1973) (Bazelon, C.J., concurring in result). From their stance outside of both scientific and political debates, courts can help to ensure that decisionmakers articulate the basis for their decisions. But once courts step beyond that role and endeavor to judge the merits of competing expert views, they leave the terrain they know. In so doing, the judiciary may mislead the public into believing it provides an expert check on decisions that in fact it does not fully comprehend.

Thus, a court applying the substantial evidence test to a numerical standard considers "whether the agency's numbers are within a 'zone of reasonableness,' not whether its numbers are precisely right." *Hercules, Inc. v. EPA*, 598 F.2d 91, 107 (D.C. Cir. 1978). Similarly, the possibility of drawing two inconsistent conclusions from the evidence does not prevent an administrative agency's finding from being supported by substantial evidence. *Environmental Defense Fund v. EPA*, 510 F.2d 1292, 1298 (D.C. Cir. 1975); *accord Bayside Enterprises, Inc. v. NLRB*, 429 U.S. 248, 302 (1976).

[67]*Public Serv. Comm'n v. FPC*, 511 F.2d 338, 355 (D.C. Cir. 1975).

support for its policy resolutions, the court was entitled to a "careful identification" by OSHA "of the reasons why [it] chooses to follow one course rather than another."[30] This theme was underscored by the Third Circuit Court of Appeals in the *Synthetic Organic* case; of the five criteria for review which the court listed, three were basically procedural in nature (adequacy of notice, adequacy of statement of reasons, and considerations of presently available alternatives). And in the *Cotton Dust* case, the District of Columbia Circuit Court of Appeals emphasized the importance of an "explicit explanation for the basis of the agency's decision" and said that "if the decision making process is open and candid, it will inspire more confidence in those who are affected."[31]

For a number of years, there was a debate between Judge Bazelon, on the one hand, and Judges Leventhal and Wright, on the other (all on the District of Columbia Circuit Court of Appeals) concerning the proper role of courts of appeals in reviewing Agency policy rules. Judges Leventhal and Wright insisted that the courts take a "hard look" at both the substantive and procedural aspects of Agency decisions to satisfy themselves that "reasoned discretion" has been exercised.[32] Judge Bazelon, on the other hand, expressed the view that the courts were not sufficiently versed in technical matters to examine the substantive aspects of these rules in detail; rather, he emphasized, the courts should insist on Agency compliance with procedural requirements so as to provide a "framework for reasoned decision-making."[33] The Bazelon approach was dealt a severe blow in the Supreme Court decision in *Vermont Yankee Nuclear Power Corp. v. Natural Resources Defense Council, Inc.*,[34] which rejected his view that the courts of appeals, in the interest of what was considered better decision making, could require agencies to follow procedures more formal than those prescribed in the Administrative Procedure Act. By 1979, in the *Cotton Dust* case, Judge Bazelon balanced substantive and procedural review (combining "supervision with restraint") and seemed to have satisfied the previously unconvinced Chief Judge Wright. For in reviewing the OSHA lead standard and upholding it in most major respects, Chief Judge Wright, writing an opinion for a 2–1 majority of the court of appeals, fully subscribed to the views of Judge Bazelon on the scope of review as defined in the *Cotton Dust* case.[35]

After several years it became apparent that in promulgating health standards, OSHA was repeatedly being confronted with the same science-policy questions. The relitigation of these issues was a significant factor, OSHA concluded, in the delays that OSHA was

[30] *Industrial Union Dep't*, 499 F.2d at 475.
[31] *AFL-CIO v. Marshall*, 617 F.2d at 652.
[32] See, e.g., Wright, *The Courts and the Rulemaking Process: The Limits of Judicial Review*, 59 CORNELL L. REV. 375 (1974).
[33] See Bazelon, *Risk and Responsibility*, 205 SCIENCE 277 (1979). The debate between Judges Bazelon, Leventhal, and Wright is discussed in McGarity, *supra* note 24 at 796–808.
[34] 435 U.S. 519 (1978).
[35] *United Steelworkers v. Marshall*, 647 F.2d 1189, 1206–07 (D.C. Cir. 1980).

experiencing in the regulation of workplace carcinogens. In order to facilitate the issuance of standards governing carcinogens, OSHA in 1980 issued its Carcinogens Policy.[36] Among other things, this Policy included a number of policy determinations, reached after full participation by the public, which, the Agency said, would be binding in all future carcinogen rulemaking, except in limited circumstances. One of OSHA's main policy determinations was that "positive results obtained in one or more experimental studies conducted in one or more mammalian species will be used to establish the qualitative inference of carcinogenic hazard to workers." This determination, of course, was consistent with OSHA's conclusions in a number of earlier rulemakings, such as the rulemaking on 14 carcinogens. Arguments that animal data should not be relied on to establish human carcinogenicity under the Carcinogens Policy would be considered by OSHA in rulemaking on individual substances only if the data offered meet strict criteria for admissibility.[37] The scientific and policy basis for OSHA's determinations on this issue is set forth at length in the preamble to the Carcinogens Policy.[38]

C. Proposals for Reform

It has been frequently argued that scientific and technical disputes "fall outside the limits of judicial competence" and therefore the courts should have a greatly reduced role in resolving these issues.[39] Suggestions for change have been of several types. Some would seek to improve the competence of the courts to decide technical issues; others would provide alternative procedures to relieve the courts of much of the decision-making responsibility. The proposal for the establishment of a science court is one of the best known.[40]

The Science Court Experiment: An Interim Report

There are many cases in which technical experts disagree on scientific facts that are relevant to important public decisions. Nuclear power, disturbances to the ozone layer, and food additives are recent examples. As a result, there is a pressing need to find better methods for resolving factual disputes to provide a sounder basis for public decisions. We accordingly propose a series of experiments to develop adversary proceedings and test their value in resolving technical disputes over questions of scientific fact. One such approach is embodied in a proposed Science Court that is to be concerned solely with questions of scientific fact. It will leave social value questions—the ultimate policy decision—to the normal decision making appara-

[36] 45 FED. REG. 5002.
[37] 45 FED. REG. at 5287 (codified at 29 C.F.R. §1990.144(c) (1983)).
[38] 45 FED. REG. at 5067–5263 (1980). The Carcinogens Policy has not been implemented and is presently undergoing reconsideration. See Chapter 3, Section B.
[39] Jasanoff & Nelkin, *Science, Technology and the Limits of Judicial Competence,* 68 A.B.A. J. 1094 (1982).
[40] Task Force of the Presidential Advisory Groups on Anticipated Advances in Science & Technology, *The Science Court Experiment: An Interim Report,* 193 SCIENCE 653 (1976).

tus of our society, namely, the executive, legislative, and judicial branches of government as well as popular referenda. Similar proposals have been made by several authors, and those which have come to the attention of the Task Force are listed in the bibliography.

In many of the technical controversies that are conducted in public, technical claims are made but not challenged or answered directly. Instead, the opponents make other technical claims, and the escalating process generates enormous confusion in the minds of the public. One purpose of the Science Court is to create a situation in which the adversaries direct their best arguments at each other and at a panel of sophisticated scientific judges rather than at the general public. The disputants themselves are in the best position to display the strengths of their own views and to probe the weak points of opposing positions. In turn, scientifically sophisticated outsiders are best able to juxtapose the opposing arguments, determine whether there are genuine or only apparent disagreements, and suggest further studies which may resolve the differences.

We have no illusions that this procedure will arrive at the truth, which is elusive and tends to change from year to year. But we do expect to be able to describe the current state of technical knowledge and to obtain statements founded on that knowledge, which will provide defensible, credible, technical bases for urgent policy decisions.

The basic mechanism proposed here is an adversary hearing, open to the public, governed by a disinterested referee, in which expert proponents of the opposing scientific positions argue their case before a panel of scientist/judges. The judges themselves will be established experts in areas adjacent to the dispute. They will not be drawn from researchers working in the area of dispute, nor will they include anyone with an organizational affiliation or personal bias that would clearly predispose him or her toward one side or the other. After the evidence has been presented, questioned, and defended, the panel of judges will prepare a report on the dispute, noting points on which the advocates agree and reaching judgments on disputed statements of fact. They may also suggest specific research projects to clarify points that remain unsettled.

The Science Court is directed at reducing the extension of authority beyond competence, which was Pascal's definition of tyranny. It will stand in opposition to efforts to impose the value systems of scientific advisers on other people. As previously stated, the Science Court will be strictly limited to providing the best available judgments about matters of scientific fact. It is so constructed in the belief that more broadly based institutions should apply societal values and develop public policies in the areas to which the facts are relevant.

***

RESULTS OF THE PROCEEDING

The primary results to be expected are a series of factual statements which will be arrived at in two ways. First there will be the statements of fact made by the case managers and not challenged by their opponents.* A second group of results will be the opinions of the judges regarding statements that were challenged. Some or most of these statements of fact will be qualified with statements about probable validity or margins of error. An important secondary consequence will be the lines drawn between areas where scientific knowledge exists and where it does not exist. Since important knowledge that is lacking will be pointed out, judgments of the science

*[Author's note—The "case managers" are the individuals selected to represent the various "sides" of a pending issue. The procedures that would be used by the Science Court are described in 193 SCIENCE at 653–55.]

court will suggest areas where new research should be stimulated. In almost all cases the boundary between knowledge and ignorance will continuously shift, and revisions to take account of new knowledge may have to be made frequently when issues of great national importance are at stake.

It bears repeating that the Science Court will stop at a statement of the facts and will not make value-laden recommendations.

A related proposal was made in 1980 by the American Industrial Health Council (AIHC) in the rulemaking on OSHA's Carcinogens Policy.[41] AIHC, which represented various employers and employer associations, recommended the creation of a science panel which would make "scientific judgments" about chronic health risks to humans associated with industrial activity.

Post-Hearing Brief for the American Industrial Health Council

Introduction. The purpose of this document is to summarize AIHC's views on the role of an independent Science Panel and the proper constitution and functioning of the Panel in making essentially scientific judgments about qualitative and quantitative aspects of human chronic health risks associated with industrial activity. AIHC advocates use of a panel of eminent scientists to perform these functions separate from the regulatory agencies. AIHC proposes that the Science Panel perform risk estimations on a case-by-case basis for the benefit of all federal regulatory agencies and that, largely separate from, but utilizing, such risk estimations, the agencies then make regulatory or social policy decisions.

The need for soundness of major regulatory programs requires that they be based on the best science. The purpose of this proposal is to provide the best assurance that the science upon which the agencies rely will achieve scientific expertise and freedom from conflict of interest. The proposal is intended to attain the level of excellence achieved by the National Academy of Science or the equivalent.

Summary. The essence of AIHC's proposal for achieving a more cohesive approach to the development of federal carcinogen and other chronic health control policies is to recognize that the determination of whether a material is likely to cause cancer, or to induce other adverse chronic health effects involves scientific rather than regulatory judgments. AIHC advocates that in the development of carcinogen and other federal chronic health control policies scientific determinations should be made separate from regulatory considerations and that such determinations, assessing the most probable human risk, should be made by the best scientists available following a review of all relevant data. These determinations should be made by a Panel of eminent scientists located centrally somewhere within government or elsewhere as appropriate but separate from the regulatory agencies whose actions would be affected by the determinations. These determinations would be limited to scientific issues and would not intrude upon the regulatory responsibilities of the individual agencies involved. AIHC recognizes that these regulatory responsibilities, quite properly, do differ from one agency to another.

[41]Appendix C, Post-Hearing for American Industrial Health Council OSHA Carcinogens Policy (Docket No. H-090) (1980). This document is on file in the OSHA Docket Office, Washington, D.C.

Panelists would be selected on the basis of their eminence and would convene periodically to assess materials submitted for their consideration by agencies on a case-by-case basis. The Panel would have its own staff.

The strength of the Panel's determinations would depend upon the eminence of the scientists, the soundness of their judgments, and the objectivity of their review process.

***

In issuing the Carcinogens Policy, OSHA expressly rejected the proposed science panel.[42] OSHA said that such a system "would be inefficient in its division of responsibilities, would be subject to unreasonable delays, and would not ensure the protection of workers."[43] OSHA further analogized the AIHC science panel to the "independent" standards board considered, and rejected, by Congress in the legislative development of OSHA.[44] The preamble to the Carcinogens Policy stated:

> In this regard, OSHA agrees with the observations of the AFL-CIO in its post-hearing brief, that some of the concerns expressed in the congressional debate apply equally to the AIHC recommended panel. These were: that the Board was not accountable to any one person, that many other regulatory agencies had combined standard-setting and enforcement authority and that the key need was for the technical personnel at the operating level of the agency.[45]

Judge Bazelon, in 1970, while supporting a science court in theory, was troubled by many of its implications.[46]

Bazelon, *Coping with Technology Through the Legal Process*

***

While I thus support the goals of the Science Court proposal, I find some of its features troubling. First, I fear that a lengthy adversary proceeding, limited solely to factual issues, might well exaggerate the importance of those issues, and might tend to diminish the importance of the underlying value choices. A factual decision by a Science Court, surrounded by all the mystique of both science and the law, might well have enormous, and unwarranted, political impact.

Moreover, it is not entirely clear to me that all disputes among experts either could or should be "resolved." Experts usually disagree not so much about the objectively verifiable facts, but about the inferences that can be drawn from those facts. And they disagree precisely because it is impossible to say with certainty which of those inferences are "correct."

Consider, for example, the discussion of the possibility of catastrophic accidents at nuclear reactors. Most experts agree that the likelihood of such an accident is pretty low, but they disagree about just how low. And there is no experiment we can conduct to determine which experts are right—it is

---

[42] 45 FED. REG. at 5002, 5203–04.

[43] *Id.*

[44] *Id.* at 5204. See Chapter 1 for the congressional debate over an independent standards-setting board.

[45] 45 FED. REG. at 5204.

[46] Bazelon, *Coping with Technology Through the Legal Process,* 62 CORNELL L. REV. 817, 827–28 (1977).

just not practical to construct 1,000 reactors and monitor them for 10,000 years, to determine how safe reactors really are. Physicist Alvin Weinberg has called such questions "trans-scientific";[33] and while it *is* appropriate for scientists to address these questions—they do involve facts rather than values—the scientific method simply cannot provide definitive answers. A Science Court might choose not to address such issues at all; in that case, its usefulness would be limited to a fairly small category of controversial scientific issues. If it *did* try to resolve such essentially unresolvable issues, however, it is hard to see how it could do much more than affix a "seal of approval" to a majority or "establishment" point of view—and that might simply discourage dissent.

Finally, I fear that the kind of adjudicatory procedures contemplated by the Science Court idea might prove to be very time-consuming. Of course, no one is fonder of adjudicatory procedures than judges—but we are acutely conscious of just how costly they can be.

Having expressed all these misgivings, I must reiterate that I fully support the goals of the Science Court idea; and I think that Dr. Arthur Kantrowitz and the other backers of the proposal have performed a great public service, simply by forcing us all to think and talk about these issues.

The practicality of the science court proposal may be judged by examining the manner in which the issues posed in the standard for ethyleneimine (EI) might be resolved. There was reliable experimental evidence that EI caused cancer in rodents, but no experimental data as to humans. If the issue whether EI was an animal carcinogen alone were presented to a science court, there would presumably be little or no dissent that EI is an animal carcinogen. However, since the crucial issue facing OSHA in the rulemaking proceeding was whether animal data should be extrapolated to humans, the determination of the science court that EI was an animal carcinogen would have only marginal significance for regulatory purposes.

The issue of extrapolation from animals to humans, on the other hand, is not a factual or scientific question but rather a science-policy issue, not resolvable by reference to facts alone. While scientists undoubtedly have views on this issue as well as numerous other policy issues, these value-laden questions of policy, it would seem, should be determined not by scientists but by responsible government officials. Submission of these issues to a science court would only mislead the public into believing that they were being decided in an objective manner because of the scientific background of the court members; but in fact, because of the nature of the issues, they are being resolved on policy grounds—that is, without reliance on scientific expertise. To be sure, it was not suggested that "social value" questions be submitted to a science court. But, aside from the difficulty in distinguishing between "factual" and "social value" questions, the problem, as Judge Bazelon noted, is that, accepting that limitation, a science court's "usefulness would be limited to a fairly small category of controversial scientific issues"; in the case of OSHA's EI standard,

[33]Weinberg, [*Science and Trans-Science,* 10 MINERVA 209] at 219 (1972).

the factual issue would not have been central to the rulemaking.[47]

One of the underlying reasons for objection to the science court proposal is skepticism as to the objectivity of scientists. As Judge Bazelon said in another connection, "the very concept of objectivity embodied in the word disinterested is now discredited."[48] As a result, if a science court were established, the focus of the policy controversy would shift, with interested parties trying to make certain that scientists favorable to their views were selected as judges. The same considerations would seem to have been uppermost in the mind of OSHA in rejecting AIHC's science panel proposal on the grounds that regulatory policy decisions should be made by "accountable" political officials rather than by an "expert" body operating under the mantle of "disinterestedness."[49]

Another more modest, but no less controversial, proposal was made by Judge Leventhal. He suggested that "scientific aides" be hired by appellate courts to advise the court in cases involving difficult technical issues.[50] This proposal was criticized by Judge Bazelon, saying that "[i]n highly controversial areas, where the experts disagree, it would be dangerous indeed to allow one expert with one point of view to have special access to the Judge's ear."[51] This is yet another example of the doubt held by the public on whether scientists can separate their "scientific" views from their "policy" views.

D. Suits to Compel OSHA to Issue Standards

The relationship between OSHA and the courts discussed to this point has been in the context of judicial control over Agency *action;*

[47]See also Talbott, *"Science Court": A Possible Way to Obtain Scientific Certainty for Decisions Based on Scientific "Fact"?* 8 ENVTL. L. 827 (1978) (the stress on technical aspects of regulatory decision making not well founded since "real issues" tend to be those of a social nature).

[48]Bazelon, *Risk and Responsibility,* 205 SCIENCE 277 (1979).

[49]See also OFFICE OF TECHNOLOGY ASSESSMENT, *supra* note 26, at 192–96 (discussing several alternative "locations" for federal carcinogenic risk-assessment activities). See also Martin, *The Proposed Science Court,* 75 MICH. L. REV. 1058 (1977) (critical discussion of the science court proposal).

[50]Leventhal, *Environmental Decision-Making and the Role of the Courts,* 122 U. PA. L. REV. 509, 550 (1974).

[51]Bazelon, *Coping with Technology Through the Legal Process,* 62 CORNELL L. REV. 817, 838 (1977). For other proposals on judicial handling of science and technology questions, see Jasanoff & Nelkin, *supra* note 39 at 1098–99. See also the works cited by McGarity, *supra* note 24, at 729 n.1. Patricia M. Wald, Judge of the District of Columbia Circuit Court of Appeals, suggested that after oral argument in appeals from administrative agency decisions, "the dynamics of court-counsel exchange should be as fluid and flexible as the difficulty of the case demands." Wald, *Judicial Review of Complex Administrative Agency Decisions,* 462 ANNALS 72, 84–86 (1982).

A major effort has been undertaken during the last several years to make significant changes in the scope of court review of administrative agency determinations. The so-called Bumpers Amendments and its progeny would, in various ways, limit judicial deference to agency expertise and reverse the presumption of validity that attaches to agency rules and regulations. E.g., SENATE COMM. ON THE JUDICIARY, THE REGULATORY REFORM ACT, S. REP. No. 284, 97th Cong., 1st Sess. 164–73 (1981). See also Recommendation No. 81–2, 1 C.F.R. §305.81–2 (1983) (Although the Administrative Conference of the United States opposed various versions of the Bumpers Amendments, it has favored more limited changes in the judicial review provisions of the A.P.A.). But see SENATE COMM. ON GOVERNMENT AFFAIRS, THE REGULATORY REFORM ACT, S. REP. No. 305, 97th Cong., 1st Sess. 84–92 (1981); Woodward & Levin, *In Defense of Deference: Judicial Review of Agency Action,* 31 AD. L. REV. 329 (1979).

that is, the question facing the court was whether to uphold an issued standard as being adequately supported by the record. Another issue has arisen in the OSHA regulatory context: whether the court, on petition from interested parties, can and should compel the Agency to institute, and complete, rulemaking proceedings on a particular standard on the grounds that the Agency's failure to act was arbitrary and capricious. This issue has been litigated, extensively, in connection with the OSHA field sanitation standard. The suit was brought in December 1978 by a migrant worker group, the Migrant Legal Action Program, and about nine years later, after two court of appeals decisions, the litigation was settled. As of early 1984, however, no final field sanitation standard had yet been issued.[52]

National Congress of Hispanic American Citizens (El Congreso) v. Marshall (Hispanic II)

626 F.2d 882 (D.C. Cir. 1979)

LEVENTHAL, Circuit Judge:
***

I. BACKGROUND

A. *The History of the Litigation:* In December, 1973, the National Congress of Hispanic American Citizens (El Congreso) filed this action to compel the Secretary of Labor to promulgate safety and health standards for the agricultural industry, specifically identifying standards governing field sanitation, farm safety equipment, roll-over tractor protection, personal protective equipment, nuisance dust and noise.[2] El Congreso argued that, by not issuing these standards within a designated time after he had begun action on them, the Secretary had abused his discretion and had unlawfully withheld and unreasonably delayed agency action in violation of §6(b)(1)–(4) of the Occupational Safety and Health Act of 1970. The district court, holding that mandatory time frames are triggered once the Secretary begins action on a standard, ordered the Secretary to publish a final farm machinery guarding standard, and to proceed, according to the time frames of §6(b), with publishing final protective equipment and field sanitation standards.[4] Since the district court believed that any departure from the statutory timetable violated the Act, it did not examine the Secretary's criteria for setting the priorities that led to delay in issuing the particular standards in dispute, it did not reach the question of whether that delay constituted an abuse of discretion, and it made no finding regarding the relative need for a field sanitation standard or any other standard.

On appeal, we reversed stating, in our 1977 opinion, that the manda-

[2]El Congreso had petitioned the Occupational Safety and Health Administration (OSHA) in September, 1972, for the promulgation of these standards. Subsequently, the agency promulgated a roll-over tractor protection standard, 29 C.F.R. 1928.51 (1978), 40 FED. REG. 18,254 (April 25, 1975), and a farm machinery guarding standard, 29 C.F.R. 1928.57 (1978), 41 FED. REG. 10,190 (March 9, 1976).

[4]*National Congress of Hispanic American Citizens v. Dunlop,* 425 F. Supp. 900, 903 (1975).

[52]A detailed history of the proceedings and the litigation appears in Advance Notice of Proposed Rulemaking on Field Sanitation, 48 FED. REG. 8493, 8494–95 (1983). See also *infra* note 57.

tory language of the Act did not negate the "implicit acknowledgement that traditional agency discretion to alter priorities and defer action due to legitimate statutory considerations was preserved."[5] We pointed out that the Secretary may "rationally order priorities and reallocate his resources at any rulemaking stage" so long as "his discretion is honestly and fairly exercised." We ordered the trial court to require the Secretary to file a report on the situation with regard to each proposed standard, including timetables for their development. We stated that if the district court was satisfied with the sincerity of the Secretary's effort, it should hold the case for further reports; but that if it was not so satisfied, it should act accordingly.

On December 26, 1978, in a memorandum opinion and final order, the district court concluded its consideration upon remand with respect to the standard for field sanitation. Finding that "the Secretary [had] not established any criteria which would enable the Court to determine that the agency [had] acted in a rational manner," the district judge concluded that the Secretary's refusal to complete a field sanitation standard was inconsistent with the requirements of the Act and the mandate of this court, and that, therefore, the Secretary "must be directed to complete development of a field sanitation standard * * * as soon as possible."

The Secretary appealed and sought and obtained from us a stay of the district court order. *Sua sponte,* we ordered expedited treatment.

***

II. ANALYSIS

On several important issues, the parties agree. The Secretary does not deny that in the case of the field sanitation standard he has departed from the timetables set forth in 29 U.S.C. §655(b). El Congreso does not contest the authority of the Secretary to set priorities for OSHA and to allocate the resources available to him. Finally, since the decision of our court in *Hispanic* I, it has been clear that the timetables set forth in 29 U.S.C. §655(b) are not etched in stone, that they do not so circumscribe the discretion of the Secretary that his failure to act within their limits is, by itself, an abuse by him of his discretionary powers. So long as his action is rational in the context of the statute, and is taken in good faith, the Secretary has authority to delay development of a standard at any stage as priorities demand. The issue before us, therefore, is whether the Secretary's action in the instant case was rational and taken in good faith.

A. WAS THE SECRETARY'S ACTION RATIONAL?

The Secretary has delayed action on the field sanitation standards because he has given priority to the development of other standards. El Congreso argues that this setting of priorities was irrational. The district court agreed, concluding that "the Secretary [had] not established any criteria which would enable the Court to determine that the agency [had] acted in a rational manner." We disagree.

On this review, a court examines agency action to determine whether it was "arbitrary, capricious, an abuse of discretion, or otherwise not in accordance with law."

> To make [such a] finding the court must consider whether the decision was based on a consideration of the relevant factors and whether there has been a clear error of judgment. Although this inquiry into the facts

[5] *National Congress of Hispanic American Citizens v. Usery,* 180 U.S.App.D.C. 337, 341, 554 F.2d 1196, 1200 (1977). [Author's note—Prior case is *Hispanic* I; present case is *Hispanic* II.]

> is to be searching and careful, the ultimate standard of review is a narrow one. The court is not empowered to substitute its judgment for that of the agency.[33]

Here, the Secretary, both in his initial presentation to the district court and in his later responses to interrogatories, has outlined the criteria which he used in setting his priorities. He has explained at length how the application of those criteria produced the set of priorities he developed. The criteria selected by the Secretary adequately reflect the purposes and provisions of the statute, and are rational within that context.[36] Though other relevant criteria could be imagined, and though even given the Secretary's criteria a different set of priorities could be developed, it is not the function of a reviewing court to substitute its judgment for that of the Secretary, where the Secretary has reasonably exercised his discretion. In our view, the Secretary has acted reasonably in this instance, and the district court developed "its own view of appropriate priorities for standards development" and this was error.

The district court's "own view of appropriate priorities" disregarded, without warrant, material findings made by the agency. For example, the Secretary specifically concluded that rulemaking concerning the field sanitation standard would be quite lengthy, involving the allocation of substantial resources (approximately 3600 man-hours). The Secretary also concluded that the greatest hazards to agricultural employees had already been remedied, and that other industries merited allocation of the available resources (the accident rate in the agricultural industry was fifth among eight major groupings). The district court emphasized almost exclusively the number of employees to be benefited, ignoring numerous other criteria considered by the Secretary such as the nature and the severity of the hazard exposure.

This court is of the view that greater respect is due the Secretary's judgment that promulgation of a cancer policy, a lead standard, an anhydrous ammonia standard and the like, merited higher priority than a field sanitation standard. With its broader perspective, and access to a broad range of undertakings, and not merely the program before the court, the agency has a better capacity than the court to make the comparative judgments involved in determining priorities and allocating resources. The district court impermissibly substituted its judgment for that of the agency.

At another level, El Congreso indicts the Secretary's *modus operandi* in approaching problems of occupational health and safety. It argues that "the Department is dissipating its energy by undertaking a myriad of entirely new ventures [in a] 'shot gun' approach [which] means that hundreds of standards are simultaneously 'being developed' but that few are ever finalized." Such an approach, El Congreso argues, is so irrational a use of agency resources as to be an abuse of power. We cannot agree.

The Occupational Health and Safety Act represents an attempt by Congress to address wide-ranging, serious, and complicated problems. The Secretary already has promulgated standards addressing significant problems in a number of areas,[39] including some covering the most serious hazards in

[33]*Citizens to Preserve Overton Park v. Volpe,* 401 U.S. 402, 416 (1970) (citations omitted).

[36]Giving priority to the most severe hazards, for example, comports with the Act's provisions for expedited rulemaking to deal with grave dangers. 29 U.S.C. §655(c) (1977). The other criteria listed by the Secretary also find support in the statute. See, e.g., 29 U.S.C. §§655(b)(1), 655(b)(5), 655(g), 656, 669, and 671 [1976].

[39]See, e.g., 43 FED. REG. 27962 (June 27, 1978, amending 43 FED. REG. 5918, Feb. 10, 1978) (benzene); 43 FED. REG. 19584 (May 5, 1978) (inorganic arsenic); 43 FED. REG. 52952 (Nov. 14, 1978) (lead); 43 FED. REG. 27350 (June 19, 1978) (cotton dust) as amended, 43 FED. REG. 28473 (June 30, 1978), 43 FED. REG. 35032 (August 8, 1978), 43 FED. REG. 56893 (December 5, 1978); 43 FED. REG. 27418 (June 23, 1978) (cotton dust in cotton gins); 43 FED. REG. 11514 (March 17, 1978) (dibromochloropropane); 43 FED. REG. 45762 (Oct. 3, 1978) (acrylonitrile).

the agricultural industry.[40] The development of additional standards is proceeding. The Act clearly envisaged such a broad scope approach to the problem, with the informed discretion of the Secretary guiding decisions regarding the number of projects to be undertaken at one time. The 18-month plan submitted by the Secretary is a reasonable exercise of that discretion.

B. DID THE SECRETARY ACT IN GOOD FAITH?

***

The Secretary's claim of a good faith effort to discharge his statutory obligations is supported by the fact that he has already promulgated standards addressing many significant hazards, including standards regulating the most hazardous aspects of the agricultural industry. In so doing, he has demonstrated a willingness to face strong political opposition by promulgating standards which were strenuously, even bitterly, opposed by those to be regulated. Nothing in the record would lead us to conclude at this time that the Secretary has been insincere in setting priorities or in presenting his timetable to the district court.

C. THE ADEQUACY OF THE SECRETARY'S SUBMISSION

Our opinion in *Hispanic* I made clear that the timetable of 29 U.S.C. §655(d)(1)–(4) is not mandatory. But it also made clear that El Congreso was entitled to *some* timetable for the development of a field sanitation standard. Where the Secretary deems a problem significant enough to warrant initiation of the standard setting process, the Act requires that he have a plan to shepherd through the development of the standard—that he take pains, regardless of the press of other priorities, to ensure that the standard is not inadvertently lost in the process.

It is not enough for the Secretary merely to state that the standard will not be issued over the next 18 months. If other priorities preclude promulgation of a field sanitation standard within that frame, then the Secretary must provide a timetable—at least for the standard in question—which covers a larger period.

Upon remand to the district court, the Secretary should be granted leeway to reconsider the timetable submitted on January 22, 1979, since it was developed without input from the official charged with responsibility for this area. In constructing the timetable, the Secretary need not be constrained, as he would have been under the district court order, to rearrange priorities that were rationally set. But, the Secretary must give due regard to the principle, presumed in the timetable of 29 U.S.C. §655(d)(1)–(4) and developed here, that once the process of developing a standard begins, a good faith effort must be made to complete it.[44] It is for the district court to review the timetable submitted—not with regard to the Secretary's setting of priorities, which we have held in this instance to be a rational exercise of his discretionary powers, but with regard to the narrower concerns we have just delineated.

***

[40]See, e.g., 29 C.F.R. 1928.51 (1978) (tractor rollover protection); 29 CFR 1928.57 (1978) (guarding of farm machinery); 29 CFR 1928.21(a)(5) (1978) (cotton dust in cotton gins).

[44]Our previous opinion did not require, and the timetable submitted by the Secretary as a result of this opinion will not require, that the Secretary commit himself irrevocably to developing a standard by a date certain, or within a given time frame. Intervening circumstances may require a readjustment of priorities, and a concomitant modification of the timetable. Good faith adjustments of this sort are proper.

In response to the court decision, Assistant Secretary Bingham on June 2, 1980, filed a timetable for the development of a field sanitation standard.[53] The timetable provided a "reasonable estimation" that a final standard would be issued in 44 months. In anticipation of reaction that the timetable was unduly lengthy and therefore not in "good faith," Dr. Bingham attached the chronology of several recent OSHA standards, including: employee access to medical and exposure records—almost five years from appointment of advisory committee to final standard; lead standard—almost six years from initial collection of information to anticipated publication of final rule; fire protection—six years from collection of information to anticipated publication of final rule.[54] The point presumably being made by Dr. Bingham was that if the field sanitation timetable was unduly long, so were other OSHA standards proceedings, and OSHA may be tardy but not in bad faith. District Judge June Green refused to accept the argument or the timetable and conducted a five-day trial in December 1980 on whether OSHA's timetable was in good faith. Meanwhile, in March 1981, NIOSH submitted a report to OSHA on field sanitation concluding that "a standard for field sanitation could and should be supported on the basis of the well-known and long documented sanitary requirements of public health practice and the need for equalizing the working conditions with other occupational groups." In August 1981, the new OSHA Assistant Secretary, Thorne G. Auchter, filed a revised timetable, estimating a period of 58–63 months before the final field sanitation standard would be issued. In stating that no work could be done on the field sanitation standard for two years, the court document filed by OSHA said the following:[55]

Timetable for Issuing Field Sanitation Standard

***

In setting health standard priorities, the Secretary has considered: the number of workers exposed to the hazard; the severity of such hazards; the existence of research relevant to hazard identification and methods of hazard control; formal recommendations of the National Institute for Occupational Safety and Health (NIOSH); petitions for standards on the hazards; and court decisions and other factors affecting the enforceability of the standard. These criteria were explicitly approved by the D.C. Circuit Court of Appeals in this case. [*Hispanic* II], 626 F.2d at 889.

***

In accord with the criteria stated above, OSHA has identified a number of activities it will be taking with respect to many toxic substances which take priority over the development of a field sanitation standard. These high priority activities preclude development of the field sanitation standard for at least 2 years.

***

[53]The timetable of Assistant Secretary Bingham is on file in *National Congress of Hispanic Am. Citizens v. Donovan,* No. 2142–73 (D.D.C. 1981).

[54]Final rules have since been published for lead, 43 FED. REG. 52,952 (1978), and fire protection, 45 FED. REG. 60,656 (1980).

[55]On file in *National Congress of Hispanic Am. Citizens v. Donovan,* No. 2142–73 (D.D.C. 1981).

The majority of OSHA's health standards staff (13 project officers and seven supporting professionals) will be required to devote their full time to OSHA's highest priorities: the development of standards for arsenic; asbestos; access to employee exposure and medical records; noise; cancer policy; lead; cotton dust; labeling; respirators; and the relationship of personal protective equipment and engineering controls.

Several of these standards address serious hazards affecting extremely large numbers of employees: access to employee exposure and medical records; noise; cancer policy; labeling; respirators; and the relationship of personal protective equipment and engineering controls. The remaining four address life threatening hazards and large numbers of employees: arsenic; asbestos; lead and cotton dust. These high priority standards also impose large costs on employers and OSHA will review them with the view of reducing unnecessary burdens in accord with the President's directive that regulatory burdens be reduced. These high priority standards also may impose substantial economic impact on small business and OSHA has decided to review them in order to minimize such impact. OSHA also has considered that these high priority standards are, in most cases, major rulemaking projects to which OSHA has been committed for several years.

The remainder of OSHA's health standards staff (3 project officers and 4 supporting professionals) will assist in the development of the highest priority standards detailed above, and will also examine other hazards which OSHA believes merit consideration at this time. This second group includes: ethylene oxide; cadmium; anesthetic gases; chromium; nickel; MBOCA; and radiofrequency radiation. * * *

***

In sum, the field sanitation standard could only be worked on during the next two years at the expense of OSHA's not undertaking necessary work on higher priority projects. In this respect OSHA has considered that many of the illnesses complained of by farmworkers were not severe and did not require farmworkers to miss significant periods of work. OSHA also has considered that there is presently no method for quantifying the risks posed by inadequate field sanitation. Therefore OSHA has considered its high priorities in comparison to field sanitation and found that development of a field sanitation standard must be deferred for the next two years to enable OSHA to complete necessary work on its high priorities. * * *

On October 30, 1981, District Judge Green issued yet another decision in this proceeding, holding OSHA was in "bad faith" in submitting the new timetable. She expressed the view that OSHA's timetables "belong in Alice in Wonderland tales: each step forward brings us two steps backward." Judge Green said further that the existence of other OSHA work does not "of itself justify relegating a simple standard of field sanitation to the dust bin." Accordingly, she ordered the Secretary of Labor "to make his best, good faith efforts" to issue a field sanitation standard within the next 18 months.[56]

On June 10, 1982, the Migrant Legal Action Program and OSHA entered into a settlement agreement, concluding this case. The key paragraphs in the agreement are the following:

3. *Development of the Field Sanitation Standard*—Once the attached Order is entered, Defendants shall immediately begin development of the field sanitation standard. Defendants shall make a good

[56]*National Congress of Hispanic Am. Citizens v. Donovan,* No. 2142–73 (D.D.C. 1981).

faith effort to publish a proposed standard in the *Federal Register* within 15–18 months after the entry of the attached Order and to complete the standard within 31 months after that Order. Assuming that a proposed field sanitation standard is published, Defendants agree to hold a hearing on the proposal within 20–23 months of the entry of the attached Order.

If it appears to Defendants that the 15–18 month proposal date, the 20–23 month hearing date or the 31-month final promulgation date cannot be met, one of the named Defendants, or their successors, shall immediately file with the Court, with copies to plaintiffs' counsel, an affidavit which demonstrates why the proposal, hearing or final promulgation date cannot be met, together with the agency's current best estimate of the date on which the task at issue will be completed. During the 30 days following the submission of Defendants' affidavit Plaintiffs shall have the right to object to any such delay by requesting the Court to compel compliance with the proposal date, the hearing date, or the final promulgation date set out in this paragraph. In the event Plaintiffs seek such an order, Defendants shall have the burden of demonstrating that the proposed deviation from the timeframes set out in this paragraph reflect a "good faith effort" to complete the field sanitation standard, *National Congress of Hispanic American Citizens v. Marshall,* 626 F.2d 882, 890 (D.C. Cir. 1979), subject to appellate review.[57]

It is apparent that a petitioner faces substantial difficulties in seeking to compel an agency such as OSHA to issue a standard through court action. This is due in part to the practicalities of the matter; the court, even if agreeing with the petitioner, would be requiring the Agency to undertake a complex, lengthy rulemaking process. Even if the Agency acted in total good faith, there is considerable doubt whether the process could be completed successfully because of a court order, if the Agency thought that the standard was not entitled to a high priority and that it did not have the resources to accomplish the task. In addition, the applicable legal principles are not favorable to the petitioner in these cases. As Judge Leventhal pointed out, "[w]ith its broader perspective, and access to a broad range of undertakings, and not merely the program before the court, the agency has a better capacity than the court to make the comparative judgment involved in determining priorities and allocating resources."[58] This would clearly be the case where OSHA determined

[57]Settlement Agreement at 2–3, *National Congress of Hispanic Am. Citizens v. Donovan,* No. 2142–73 (D.D.C. 1981). On March 1, 1983, almost 10 years after suit was filed in this proceeding, OSHA issued an advance notice announcing that it was considering proposing a new field sanitation standard and soliciting "quantitative and qualitative data, expert opinion comments, and information regarding field sanitation." 48 FED. REG. 8493 (1983). On March 1, 1984, OSHA issued a proposed field sanitation standard, 49 FED. REG. 7589 (1984). Previously, migrant workers' representative had asked Judge Green to take action against OSHA for having missed its agreed-upon deadline. See 13 OSHR 993–94; 1011–12 (1984).

[58]*Hispanic* II, 626 F.2d at 889. Accord *WWHT, Inc. v. FCC,* 656 F.2d 807, 818–19 (D.C. Cir. 1981) ("It is only in the rarest and most compelling circumstances that" courts should "overturn an agency judgment not to institute rulemaking."). See also *State Farm Mut. Auto. Ins. Co. v. Department of Transp.,* 680 F.2d 206, 228–29 (D.C. Cir. 1982). In *Hispanic* I, referred to in the decision in *Hispanic* II, *supra* this section, the court of appeals refused to compel the Agency to comply strictly with the timetables set forth in §6(b)(1)–(4), 29 U.S.C. §655(b)(1)–(4). OSHA's experience with these mandatory time frames is discussed in Tomlinson, REPORT ON THE EXPERIENCE OF VARIOUS AGENCIES WITH STATUTORY TIME LIMITS APPLICABLE TO LICENSING OR CLEARANCE FUNCTIONS AND TO RULEMAKING, ADMINISTRATIVE CONFERENCE OF THE UNITED STATES, RECOMMENDATIONS AND REPORTS 119, 195–211 (1978). Tomlinson concluded that statutory deadlines were "unduly rigid" and that their disadvantages should "discourage their use by Congress." *Id.* at 122. See also Recommendation of Administrative Conference of the United States, Time Limits on Agency Actions, 1 C.F.R. §305.78-3 (1983).

that it was giving priority to the regulation of substances causing cancer over regulation of conditions causing dysentery and heat exhaustion, which the Agency viewed as lesser hazards. The issue becomes more difficult when OSHA gives a high priority to the revision and simplification of safety standards (as OSHA did in September 1977) or to the revision of health standards as part of its efforts for "regulatory reform" (as it did in August 1981). But even in that more difficult context the logic of the court of appeals decision would seem to lead to the conclusion that, so long as the Agency was acting in good faith, the court should not second-guess its determination of priorities. It should be emphasized that the issue is being discussed in the legal context of whether the court can or should oversee OSHA's priority-setting for standards; whether, under a statute designed to assure all workers safe and healthful working conditions, OSHA, as a policy matter, acted properly in not regulating field sanitation hazards for nine years is a much different question.

An issue not expressly litigated in this case is the significance of congressional pressure on OSHA not to issue a field sanitation standard. The OSHA proposal on field sanitation was issued on April 27, 1976.[59] Less than a month later, during the debate on the Department of Labor appropriations bill, OSHA was subjected to vitriolic criticism in large part because of the field sanitation proposal. One Congressman called OSHA "the mandator of the privy on the plains."[60] This criticism led to the enactment of an appropriations rider, renewed every year since then, exempting small farms from OSHA jurisdiction.[61] Although there was no evidence in the court proceeding that OSHA's discontinuing the field sanitation proceeding was the result of political pressure, this presumably was the belief of some members of the public. Judge Leventhal dealt with the issue indirectly in saying that OSHA in the past had "demonstrated a willingness to face strong political opposition" by promulgating certain standards; and, therefore, nothing in the record leads to the conclusion that OSHA's timetable was "insincere."[62] As a hypothetical matter, however, if it were admitted that OSHA ended the rulemaking because of its "sincere" fear of debilitating congressional amendments, it is arguable that this "political" basis for setting priorities in some circumstances would be in "good faith" under the principles in *Hispanic* II.

Despite the considerable obstacles to the success of a court suit to compel the issuance of a standard, these suits have been brought on a number of occasions against OSHA, and in some cases the mere filing of the suit may have been a substantial factor in causing the Agency to act, or, at least, in bringing about action more quickly. One example was the suit to compel OSHA to issue an emergency temporary standard on pesticides; it was followed quickly by the issuance of an

[59]41 FED. REG. 17,576.
[60]122 CONG. REC. 20,367 (1976) (statement of Rep. Skubitz).
[61]This and other riders to OSHA appropriations bills are discussed in Chapter 21, Section C.
[62]*Hispanic* II, 626 F.2d at 890.

ETS.[63] The suit brought by the Textile Workers Union in the District Court of the District of Columbia to compel OSHA to issue a standard on cotton dust was a factor in the issuance of the standard despite an effort by some White House officials to delay the standard.[64] Finally, in 1976 a suit was brought by a member of Congress and the Public Citizen Health Research Group to force OSHA to issue a labeling standard. The suit was dismissed by the district court,[65] but a proposed standard was eventually issued under the administration of Dr. Bingham.[66]

The effort to compel OSHA to issue a field sanitation standard did not bring about prompt Agency action and, indeed, has not yet succeeded in obtaining the standard. On the other hand, the District of Columbia Circuit Court of Appeals reacted more sympathetically to a suit by the Public Citizen Health Research Group to force the issuance of an ethylene oxide standard. While reversing the decision of the federal district judge which would have required OSHA to issue an emergency temporary standard on ethylene oxide within 20 days,[67] the court of appeals ordered OSHA to issue an ethylene oxide *proposal* within 30 days[68] and directed OSHA to proceed on a "priority, expedited basis and to issue a permanent standard as promptly as possible, well in advance of the current latter part of 1984 estimate."[69]

Public Citizen Health Research Group v. Auchter

702 F.2d 1150 (D.C. Cir. 1983)

Before: ROBINSON, Chief Judge, WILKEY and GINSBURG, Circuit Judges.

Opinion Per Curiam.

Per Curiam: In this expedited appeal, Assistant Secretary of Labor Thorne G. Auchter challenges a January 5, 1983, district court order directing the Occupational Safety and Health Administration (OSHA) to issue by January 25, 1983, an emergency temporary standard (ETS) regulating workplace exposure to ethylene oxide (EtO), a synthetic organic chemical.[2] Used, *inter alia,* as a sterilizing agent, fumigant, pesticide, and industrial chemical additive, EtO is critically employed as a sterilant in the health

[2]On January 20, 1983, this court issued an administrative stay of the district court's order. On February 3, because of the urgent need for full appellate review, we expedited the case on our own motion and heard oral argument on February 16, 1983.

[63]See Chapter 4.

[64]See Chapter 11.

[65]*Public Citizen Health Research Group v. Marshall,* 485 F. Supp. 845 (D.D.C. 1980).

[66]46 FED. REG. 4412 (1981). For a discussion of the history of the labeling standard, which is now known as the hazard communication standard, see Chapter 18, Section F. In August 1982, various groups filed suit to compel issuance of an emergency standard on formaldehyde. See Chapter 4.

[67]*Public Citizen Health Research Group v. Auchter,* 554 F. Supp. 242 (D.D.C. 1983). Judge Barrington Parker concluded that the record before the Agency "represented a solid and certain foundation that workers are subjected to grave health dangers from exposure to ethylene oxide at levels within the currently permissible range" and, therefore, OSHA's decision not to issue an ETS was an "abuse of discretion." *Id.* at 251.

[68]The proposal was issued on April 21, 1983. 48 FED. REG. 17,284. It proposed a permissible exposure limit of 1 part per million of ethylene oxide.

[69]702 F.2d 1150, 1159 (D.C. Cir. 1983).

care and medical products industries. The exposure of workers who work near sterilizing equipment in the health care industry is the principal focus of this case.[3]

The current OSHA standard for EtO has been in effect for well over a decade; it allows a permissible exposure limit in the workplace of 50 parts per million (ppm) averaged over an eight hour workday ("time-weighted average" or TWA). 29 C.F.R. §1910.1000.[4] That level, information now available reveals, poses a serious risk of causing chromosomal abnormalities and cancer in exposed animals and humans.

In August 1981, Public Citizen Health Research Group ("Public Citizen") petitioned OSHA to issue an emergency temporary standard reducing the permissible exposure limit of EtO to a TWA of 1 ppm. Public Citizen invoked a provision of the Occupational Safety and Health Act, 29 U.S.C. §§651–78 (1976 and Supp. IV 1980) (OSH Act or "Act"), that authorizes OSHA to adopt an immediately effective temporary standard upon determining that failure to take emergency action would subject employees to "grave danger" from exposure to a toxic substance. 29 U.S.C. §655(c). In September 1981, OSHA denied the petition on the ground that "on the available evidence * * * current conditions do not constitute an emergency situation." The evidence did persuade the Assistant Secretary, however, that "the current OSHA standard of 50 ppm may not be sufficiently protective"; he therefore stated he would proceed with normal non-emergency rulemaking to revise the standard.

Four months later, in January 1982, OSHA published an "Advance Notice of Proposed Rulemaking." 47 FED. REG. 3566 (Jan. 26, 1982).[7] The Assistant Secretary projects issuance of an actual Notice of Proposed Rulemaking in June 1983, and a final regulation sometime in the fall of 1984.

Shortly before OSHA denied its petition, Public Citizen commenced this civil action to obtain an order commanding the issuance of an emergency standard. On cross-motions for summary judgment, the district court concluded that "[t]he record before the agency presented a solid and certain foundation showing that workers are subjected to grave health dangers from exposure to ethylene oxide within the currently permissible range." App. 141. OSHA had abused its discretion, the district court held, by denying the Public Citizen petition and, instead, proceeding on a course "which insured the continuing existence of the challenged standard." *Id.* Accordingly, the district court ordered OSHA to "promulgate within 20 days * * * an appropriate emergency temporary standard addressing worker exposure to ethylene oxide."

This difficult case, which we must decide under pressing circumstances, has two novel aspects. First, OSHA has in the past, on its own initiative, issued emergency standards which were promptly brought to court for review by industry complainants. See *Florida Peach Growers Association, Inc. v. United States Department of Labor,* 489 F.2d 120 (5th Cir. 1974); *Dry Color Manufacturers' Association, Inc. v. United States Department of Labor,* 486 F.2d 98 (3d Cir. 1973) (both vacating emergency temporary standards). We are unaware, as was the district court, App. 134, of any prior case in which a court was invited to review OSHA's denial of a petition to issue an

[3]Self-regulation in other affected industries has reduced EtO exposure to levels at which petitioners concede a diminished need, if any, for emergency action.

[4]The existing standard was the Walsh-Healey federal standard in effect at the time the Occupational Safety and Health Act was passed in 1970. OSHA adopted the standard under the Act's provision for adoption without rulemaking of "established federal standard[s]." 29 U.S.C. §§655(a), 652(10). The Walsh-Healey standard "preceded the recognition of the carcinogenic potential of EtO and was based only on consideration of acute and chronic nonmalignant health effects." 47 FED. REG. 3566 (Jan. 26, 1982).

[7]This Notice simply solicited information "to determine whether it is necessary and appropriate to pursue further regulatory activity regarding occupational exposure to EtO." 47 FED. REG. 3571 (Jan. 26, 1982).

emergency standard. Second, in declining to replace the current standard on an emergency basis, and in justifying its position in court, OSHA did not rely on any government action. Rather, it relied heavily on voluntary efforts of employers, "alerted and responsive to the new health data concerning [EtO]," to "lower[] their in-house allowable exposure limits to a fraction of the OSHA standard." OSHA Brief at 22, 23.

The Assistant Secretary has emphasized the extraordinary authority and large measure of discretion 29 U.S.C. §655(c) vests in him to determine whether an "emergency standard is necessary to protect employees from [grave] danger." While it is a close question, our review of the record indicates that, in ordering an emergency standard, the most drastic measure in the Agency's standard-setting arsenal, the district court impermissibly substituted its evaluation for that of OSHA. Nonetheless, we fully agree with the district court that "OSHA has embarked upon the least responsive course short of inaction." Beyond question, despite the efforts of "[m]any companies," OSHA Brief at 23, the record shows a significant risk that some workers, and the children they will hereafter conceive, are subject to grave danger from the employees' exposure to EtO. We therefore hold that OSHA must expedite the rulemaking in which it is now engaged.[9]

Congress has instructed OSHA, in determining the priority for establishing standards, to "give due regard to the urgency of the need for mandatory * * * health standards for particular * * * workplaces." 29 U.S.C. §655(g). Further, the Administrative Procedure Act directs an agency "to conclude [within a reasonable time] a matter presented to it." 5 U.S.C. §555(b). Complementing that provision, this court has authority, pursuant to 5 U.S.C. §706(1), to compel agency action unreasonably delayed. There is an obvious need, apparent to OSHA, for an EtO standard that reflects, as the current standard does not, the mutagenic and carcinogenic potential of the chemical. Because it is also plain that industry efforts fall a considerable distance from taking every worker exposed to EtO out of the grave danger zone, a more than three-year span from Public Citizen's petition to projected final regulation is not tolerable. OSHA's failure to date to issue even a notice of proposed rulemaking—some eighteen months after announcing its intention to commence rulemaking—is in our judgment, and in light of the risk to current and future lives, agency action unreasonably delayed.

To assure that OSHA will give due regard to the need, urgent for some workers, for a new EtO standard, and to prevent undue protraction in OSHA's conclusion of this matter, we direct the Assistant Secretary to issue a notice of proposed rulemaking within thirty days of the date of this decision and to proceed expeditiously thereafter toward issuance of a permanent standard for EtO.[12]

I. BACKGROUND

It is undisputed that evidence now available shows EtO to pose serious qualitative health risks unappreciated when OSHA adopted the current standard. The chemical is both mutagenic and carcinogenic in animals and humans. One uncontradicted study finds significant chromosomal aberrations in workers chemically exposed to 36 ppm; another study shows similar results for concentrations described, more generally, as within the current 50 ppm standard. In addition, dose responsive relationships between expo-

[9]As the district court observed, staff scientists in OSHA's Directorate of Health Standards had recommended either the issuance of an emergency standard or expeditious rulemaking.

[12]While we do not set a day certain for a final rule, in view of the significant risk of grave danger to which some workers and the children they may conceive are exposed, and the time OSHA has already devoted to EtO, we expect the Agency to bring this matter to a close within a year, well in advance of the current, perhaps September 1984, estimate.

sure to EtO and fertility in rats have been observed. 47 FED. REG. 3568 (Jan. 26, 1982).

OSHA concedes that scientific reports show a "statistically significant" increase in cancer in animals exposed at 33 ppm and above.[13] The National Institute [for] Occupational Safety and Health (NIOSH) determined from the prior reports that "exposures as low as 10 ppm increased the proportion of female rats with * * * leukemia." App. 954. OSHA has acknowledged this dose responsive relationship at 10 ppm. 47 FED. REG. 3568 (Jan. 26, 1982).

The two principal human studies reported a "significant excess" of both leukemia and stomach cancer from exposure to EtO, OSHA Brief at 15; see also 47 *Fed. Reg.* 3568; the same cancers appeared in animals exposed to EtO. The human studies indicated a 15-fold increase in leukemia at 20 ± 10 ppm and a statistically significant increase in total cancer mortality at exposures averaging well within the current 50 ppm standard. 47 FED. REG. 3568.[16]

The evidence is more problematic on the quantitative risk posed by EtO, i.e., the extent to which workers are actually exposed to the serious qualitative risks of the chemical. NIOSH estimates that 75,000 health care workers are regularly exposed to EtO and another 25,000 "casually" or intermittently exposed. But uncontradicted evidence in the record indicates that many hospitals have voluntarily improved internal procedures or modified their sterilizers to reduce exposure levels to 10 ppm and below. Other evidence presented to the district court, however, suggests that a not insignificant number of workers are still regularly exposed to levels much higher than 10 ppm, but within the current 50 ppm standard. A 1977 study by the American Hospital Association (AHA) reported that although 24 of the 27 hospitals studied had chronic daily exposure levels of 10 ppm and below, the remaining three had chronic exposures of 50 ppm and above. A 1981 MITRE Corporation study, incorporating the AHA 1977 study results, found of 65 hospitals examined, 24 did not even monitor EtO levels.

Evaluating this evidence, the Assistant Secretary acknowledged that "there is important new health data on the carcinogenic risks of EtO," but determined that at current levels of exposure EtO did not present a danger sufficiently grave to require issuance of an ETS. The Agency initially stated:

> [t]he limited data currently available to OSHA * * * indicate that most hospitals have reduced exposures below 10 ppm TWA for sterilizing operations. The data also suggest that the number of employees at any given hospital who may be involved in using EtO as a sterilant is very small, and that their exposure is intermittent.

During litigation, OSHA slightly qualified or clarified this assessment and estimated the "average" hospital exposure level *at* (neither above nor below) 10 ppm.

Public Citizen disputes OSHA's findings as unsupported by the record. Further, Public Citizen asserts that even if the "average" exposure level is indeed 10 ppm and even if the risk of harm at that level is not "grave," the record still shows many workers exposed to levels over 10 ppm, yet below the current standard of 50 ppm, levels at which the risk of harm is grave.

[13]* * * OSHA defines "statistically significant" to mean "that the events observed are extremely unlikely to have occurred as a result of chance. In other words, it is highly probable that ethylene oxide is responsible for the elevated animal cancer rates in this study." OSHA Brief at 14 n.15.

[16]The Assistant Secretary contends, however, that these studies are unreliable because of methodological errors; the exposed populations were too small, subjects may have been exposed to other carcinogenic substances, and exposure levels were not actually known. OSHA Brief at 15–16, 50 & n.54.

II. DECISION

A. STANDARD OF REVIEW

The legislative history of 29 U.S.C. §655(c) makes clear that emergency standards are to be used only in "limited situations." Correspondingly, courts have observed that authority to adopt such standards entails "[e]xtraordinary power," *Florida Peach Growers Association, Inc. v. United States Department of Labor,* 489 F.2d 120, 129 (5th Cir. 1974), that should be "delicately exercised," *id.,* and used only as "an unusual response to exceptional circumstances." *Dry Color Manufacturers' Association, Inc. v. United States Department of Labor,* 486 F.2d 98, 104 n.9a (3d Cir. 1973); see also *Taylor Diving & Salvage v. United States Department of Labor,* 537 F.2d 819, 820–21 (5th Cir. 1976).

In deciding whether to issue an ETS, OSHA must make both factual and policy judgments on the basis of information that may be incomplete. The Supreme Court, in the context of permanent regulation, has noted that threshold determinations by OSHA of whether a sufficient risk of harm exists to justify regulation "must necessarily be based on considerations of policy as well as empirically verifiable facts." *Industrial Union Department v. American Petroleum Institute,* 448 U.S. 607, 655 n.62 (1980) (*"Benzene").* While OSHA determinations "which are essentially legislative and rooted in inferences from complex scientific and factual data," *United Steelworkers of America v. Marshall,* 647 F.2d 1189, 1206 (D.C. Cir. 1980), *cert. denied sub nom. Lead Industries Association, Inc. v. Donovan,* 453 U.S. 913 (1981), are entitled to great deference, our review of the Assistant Secretary's refusal to issue an ETS for EtO exposure must take into account the mandatory language of 29 U.S.C. §655(c) and the fact that "the interests at stake are not merely economic interests in a license or a rate structure, but personal interests in life and health." *Wellford v. Ruckelshaus,* 439 F.2d 598, 601 (D.C. Cir. 1971); see *Nader v. Nuclear Regulatory Commission,* 513 F.2d 1045, 1047 (D.C. Cir. 1975).

B. MERITS

1. OSHA's finding of no "grave danger" is rational. The district court, after reviewing the available evidence, concluded that "the agency's decision resulted from a clear error of judgment." Responding to OSHA's position that there was no emergency because private institutions had voluntarily reduced EtO exposure to levels OSHA considered "safe," the district court declared "[s]uch actions do nothing to protect those workers *who are still exposed* to EtO at hazardous levels or those employed at institutions which may at any time elect to withdraw the voluntary restrictions." (emphasis added). The district court thus cited *actual* exposure risks and *not,* as OSHA would restate the court's position, simply a future prospect should employers relax their vigilance.

In light of the mixed fact/policy judgment Congress empowered OSHA to make on uncertain evidence, we cannot say, as the district court did, that the decision not to issue an ETS lacked support in the record. The Agency's estimates of a 10 ppm average exposure level and a statistically lower risk of harm at that level may rest on impressionistic information. However, nothing offered to the district court contradicts those estimates. Given the substantial evidence of voluntary reduction in exposure levels and the predominantly lower levels reported in the available studies, a 10 ppm "average" appears to us a reasonable assumption. Similarly, the absence of clear scientific evidence of the degree of harm from exposure at levels 10 ppm and below counsels deference to the Assistant Secretary's assessment.

The difficult question in this case is whether the Assistant Secretary's reliance on an *average* exposure level in finding no grave danger was irrational or an abuse of discretion.[23] We are not positioned to say that the expert Agency acted impermissibly in this regard. From the statistically small sampling and the paucity of current data concerning actual EtO exposure levels, we are unable to venture even a guess as to existing exposure patterns over the average. Under the circumstances, we must defer to the Assistant Secretary's determination that the available evidence is inadequate to show the existence of "grave danger" necessitating an emergency standard.

All we say today is that in the absence of a more complete record as to actual exposure levels, we are hesitant to *compel* the Assistant Secretary to grant extraordinary relief. We express no opinion as to whether the same record would support *voluntary* issuance by OSHA of an emergency standard.

2. The significant risk of grave danger necessitates expedited rulemaking.

Though we disagree with the district court that a finding of grave danger is *compelled* on this record such that the Assistant Secretary should be *required* to exercise his "extraordinary power" to direct immediate relief in the form of an ETS, our review would be less than "thorough" and "probing"[24] if we simply overturned the judgment below. Ample evidence in the record indicates a significant risk that some workers, who are actually being exposed to levels of EtO greater than the 10 ppm "average" (yet within the 50 ppm standard), currently encounter a potentially grave danger to both their health and the health of their progeny. In face of this evidence, the Assistant Secretary's unaccounted-for delay in issuing a Notice of Proposed Rulemaking and his refusal to assign to the EtO rulemaking any priority status constitute agency action "unreasonably delayed." 5 U.S.C. §706(1).

In its September 1981 denial of Public Citizen's petition for an ETS, OSHA conceded that its current 50 ppm standard "may not be sufficiently protective," and stated its "intent[] to proceed with rulemaking under §6(b) of the OSH Act [29 U.S.C. §655(b)] to lower the permissible exposure limit and to incorporate appropriate provisions for monitoring, medical surveillance and the like." App. 5b. The Assistant Secretary encouraged Public Citizen "to provide information in response to OSHA's advance notice of proposed rulemaking, and to participate in the full rulemaking proceedings on EtO." *Id.* Despite this announced intent in September 1981 to proceed with a rulemaking, *no notice of proposed rulemaking has yet been issued.*

Instead, the Assistant Secretary issued in January 1982, four months after committing OSHA to a rulemaking, merely an "Advance Notice of Proposed Rulemaking". See 47 FED. REG. 3566 (Jan. 26, 1982). The comment period on this "advance notice" ended March 31, 1982. Almost a year has passed without commencement of the rulemaking. The Assistant Secretary now estimates that a notice of proposed rulemaking will issue in June 1983, with promulgation of a final rule sometime in the fall of 1984.

Three years from announced intent to regulate to final rule is simply

[23]An easy question to resolve, however, is the Assistant Secretary's assertion that "there is a serious question as to OSHA's jurisdiction over hospital employees engaged in EtO sterilization activities," because of EPA's regulation of the chemical under the pesticide statute (the Federal Insecticide, Fungicide, and Rodenticide Act, 7 U.S.C. §§136–136y). OSHA, as the district court pointed out, see App. 140, has dealt with exposure to EtO for over a decade and has committed itself to eventual replacement of its dated standard. We agree entirely with the district court's conclusion that OSHA is not disabled from issuing an EtO standard in "areas—such as the health care industry—where EPA has apparently exercised minimal, if any, regulatory authority in an overlapping manner."

[24]See *Citizens to Preserve Overton Park v. Volpe,* 401 U.S. 402, 415 (1971) (though agency decision is subject to "presumption of regularity," that presumption should not "shield [agency] action from a thorough, probing, in-depth review").

too long given the significant risk of grave danger EtO poses to the lives of current workers and the lives and well-being of their offspring. Delays that might be altogether reasonable in the sphere of economic regulation are less tolerable when human lives are at stake. See, e.g., *Blankenship v. Secretary of Health, Education, and Welfare,* 587 F.2d 329, 334 (6th Cir. 1978); *Environmental Defense Fund, Inc. v. Hardin,* 428 F.2d 1093, 1099 (D.C. Cir. 1970).[26] This is particularly true when the very purpose of the governing Act is to protect those lives. 29 U.S.C. §651(b).

We would hesitate to require the Assistant Secretary to expedite the EtO rulemaking if such a command would seriously disrupt other rulemakings of higher or competing priority. But we do not confront such a case. Prior to oral argument, we asked the parties to address the status of other ongoing OSHA rulemakings and the feasibility of expediting the EtO proceedings in the absence of an ETS. OSHA informed us that the Agency is currently engaged in three proceedings that it believes would be disturbed by speedier EtO rulemaking: (1) rulemaking requiring chemical manufacturers to assess and communicate to employees the hazards posed by chemicals they produce; (2) a reassessment of the current regulations governing worker exposure to asbestos; and (3) rulemaking governing exposure to ethylene dibromide.

The hazard communication rulemaking has been on the agency's docket since 1977, see 42 FED. REG. 5372 (Jan. 28, 1977); by OSHA's own admission, development of the standard is nearly complete with a final rule expected by July 1983. 47 FED. REG. 48549 (Oct. 28, 1982). In addition to the fact that OSHA's work on the rule is substantially done, we note that this standard relates solely to information employers must give to workers concerning workplace hazards; it does not substantively regulate the workplace environment. The asbestos docket has been open since 1975, and OSHA estimates a Notice of Proposed Rulemaking may be issued in March 1984.[27] Ethylene dibromide, a chemical with both carcinogenic and mutagenic risks, is appropriately on a somewhat faster track. An Advance Notice of Proposed Rulemaking issued in December 1981, and the Notice of Proposed Rulemaking is scheduled for May 1983. OSHA concedes, however, that virtually all exposure to ethylene dibromide takes place in California, which has recently issued an emergency standard setting the permissible exposure limit at .130 ppm. See 46 FED. REG. 61672 (Dec. 18, 1981); OSHA Brief at 57 n.60. None of these proceedings appears to approach in urgency the need for prompt issuance of a new EtO exposure standard, and OSHA has provided us with no reasoned explanation why it has protracted the EtO rulemaking despite the documented risks to workers' lives and the lives of children they may conceive.

The court is not disarmed in these circumstances. OSHA is obliged by its governing Act to "give due regard to the urgency of the need" for a new EtO standard in setting priorities, and the record before us strongly indicates that the Agency has not done so.[29] OSHA, we are convinced, is not proceeding to conclude this matter "within a reasonable time." 5 U.S.C. §555(b).[30] We therefore act to "compel * * * action" OSHA has "unreasonably delayed." 5 U.S.C. §706(1).

---

[26]The risk to human life need not be a certainty to justify expedition. As in *Hardin,* "if petitioners are right" in their claim that EtO presents a serious hazard for a significant number of workers, then "even a temporary refusal [to act] results in irreparable injury on a massive scale." 428 F.2d at 1099.

[27]Unlike EtO and ethylene dibromide, asbestos is not now suspected of being mutagenic as well as carcinogenic.

[29]We recall here, as the district court reported, that OSHA's Directorate of Health Standards had recommended either issuance of an ETS or expeditious rulemaking to address the hazard posed by EtO.

[30]The reasonableness of the delay must be judged "in the context of the statute" which authorizes the agency's action. *National Congress of Hispanic Am. Citizens v. Marshall,* 626

CONCLUSION

We cannot "compel solutions where none exist," *American Broadcasting Co. v. FCC,* 191 F.2d 492, 501 (D.C. Cir. 1951), but we "must act to make certain that what can be done is done." *Id.* To that end, we order the Assistant Secretary to issue a notice of proposed rulemaking within 30 days of the date of this decision. Although we dictate no fixed date for issuance of a final rule, we do direct OSHA to proceed on a priority, expedited basis and to issue a permanent standard as promptly as possible, well in advance of the current, latter part of 1984 estimate. Under the circumstances presented here, i.e., the significant risk of grave danger to human life, and the time OSHA has already devoted to EtO, we expect promulgation of a final rule within a year's time.

Several significant aspects of this decision should be noted, particularly in light of the earlier decision of the Court of Appeals for the District of Columbia Circuit in the *Field Sanitation* case. In the *Ethylene Oxide* case the panel was plainly dissatisfied with the Agency's regulatory response to the new information on the chemical. However, it was also reluctant to order OSHA to issue an ETS, as the district judge had done. The court of appeals' reference to the ETSs which had been vacated by other courts of appeals strongly suggests that the court here recognized the legal vulnerability of an ETS, if challenged; indeed, the uncertainty of the Agency succeeding in vindicating the ETS in court could be increased where the standard was issued under legal compulsion. Ultimately, it may be suggested, the court of appeals' view was that prompt protection of employees from ethylene oxide was necessary, but, in the long run, more effective protection would be afforded by means of a §6(b) rulemaking rather than through an ETS.[70] The court of appeals, however, without any noticeable hesitation, required OSHA to proceed expeditiously under §6(b). The prior decision in the *Field Sanitation* case, which reflected, it would seem, a more tolerant attitude to OSHA discretion, was no precedential barrier.[71] In *Hispanic* II, the court of appeals insisted that the district court give "greater respect" to OSHA judgment,[72] and on the record before it decided that OSHA's decision was not in bad faith or irrational. In *Ethylene Oxide,* on the other hand, the court of appeals agreed with the district court that OSHA had embarked "on the least responsive course short of inaction";[73] in effect,

F.2d 882, 888 (D.C. Cir. 1979). In the context of the OSH Act, designed to protect workers' health, the Assistant Secretary's protracted course in face of potentially grave health risks cannot be characterized as reasonable.

[70]Compare OSHA's statement in its Carcinogens Policy in 1980, 45 FED. REG. at 5215–16, quoted in Chapter 4, that in some cases "employee health may be more effectively protected by concentrating those [compliance, legal, and technical] resources in work on permanent standards." But see discussion of emergency standard for asbestos issued in 1983 in Chapter 4.

[71]*Hispanic* II was cited once by the court of appeals in the ethylene oxide case, 702 F.2d at 1158, n.30, and not for the purpose of distinguishing the case.

[72]626 F.2d at 889.

[73]702 F.2d at 1153.

that the Agency's timetable that would lead to a Fall 1984 promulgation of a permanent standard was, if not in bad faith, irrational. The major distinguishing feature from *Hispanic* II was that field sanitation hazards do not cause cancer and the field sanitation standard was competing for priority with standards for carcinogens and other highly toxic substances.[74] Finally, in the case of ethylene oxide, dramatic new evidence had become available of the chemical's harmful effects. In any event, it is probable that the court of appeals would ultimately have required OSHA to issue a field sanitation proposal within some reasonable time, if the case had not been settled in the meantime.

[74] *Id.* at 1158.

7

"The Race to the Courthouse": Forum-Shopping in Review of OSHA Standards

A. When Does the Race Begin?

The marked difference in the attitude among some of the courts of appeals in the review of OSHA standards has led the parties challenging standards to undertake major efforts to locate the standards review proceedings in forums sympathetic to their position. In cases involving the review of determinations of federal administrative agencies, the agency is required to file the administrative record in the court in which the first petition in point of time was filed.[1] The application of this provision in OSHA standards cases has led to bitterly fought and complex litigation in which several major issues have arisen. First, when was the challenged standard promulgated? This issue is critical since petitions for review may be filed only *after* the administrative action has taken place. Second, it must be determined which petition was first filed. Finally, the issue arises whether the proceeding should be transferred from the court of first filing to some other court of appeals "for the convenience of the parties in the interest of justice," as provided in the last sentence of 28 U.S.C. §2112(a).

The first case involving a "race to the courthouse" resulted from OSHA's promulgation of an emergency temporary standard for benzene in April 1977. The complex sequence of events preceding the litigation is described in detail in Judge Leventhal's main opinion for the Court of Appeals for the District of Columbia Circuit. While there was no majority opinion, the court, with Judge Fahy dissenting, decided to transfer the proceeding to the Fifth Circuit Court of Appeals.

[1]28 U.S.C. §2112(a). The applicable provision is quoted in *Industrial Union Dep't v. Bingham,* 570 F.2d 965, 968 n.3 (D.C. Cir. 1977), *infra* this section.

Industrial Union Department v. Bingham

570 F.2d 965 (D.C. Cir. 1977)

LEVENTHAL, Circuit Judge:

These consolidated proceedings for review of a standard promulgated by the Occupational Safety and Health Administration (OSHA) are now before us on a motion for transfer to the Fifth Circuit pursuant to 28 U.S.C. §2112(a).

I.

Disposition of this motion requires rather detailed exposition of the procedural history of these cases.

On April 28, 1977, representatives of approximately twelve organizations were invited to a meeting the following morning in the office of Dr. Eula Bingham, Assistant Secretary of Labor (OSHA). The meeting began shortly after 9:30 a.m. and concluded about 10:00 a.m. At the meeting, Dr. Bingham signed an emergency temporary standard regulating occupational exposure to benzene. She described the standard to those present, and answered their questions about it. The evidence is in conflict whether copies of the text of the standard were made available to those present.

Those invited to the meeting represented a cross-section of organizations with an interest in the benzene standards, including: the American Petroleum Institute, the Chamber of Commerce of the United States, the Manufacturing Chemists Association, the National Association of Manufacturers, the Rubber Manufacturers of America, the Organization Resources Counselors, Inc., the Industrial Union Department of the AFL-CIO (hereafter simply "AFL-CIO"), the United Rubber Workers of America, the United Steelworkers of America, the United Auto Workers, and the Health Research Group.

The representative of the AFL-CIO immediately after leaving the meeting with Dr. Bingham prepared a petition for review of the standard. The petition was received in the office of the Clerk of this court at 10:08 a.m. on April 29. It was assigned docket number 77-1395.

At 10:30 a.m., April 29, Dr. Bingham and F. Ray Marshall, Secretary of Labor, held a press conference at which they discussed the issuance of the benzene standard. At the conclusion of the press conference, copies of the text of the standard and a statement describing it were made available to those in attendance. The printed statement indicated on its face that it was for release at 10:30 a.m.

The standard was filed with the *Federal Register* at 10:55 a.m. on April 29. It was published in the *Federal Register* for May 3, 1977.

On May 10, the American Petroleum Institute (API), the National Petroleum Refiners Association, and ten oil companies (hereafter collectively the "petroleum industry petitioners") filed a petition for review of the benzene standard with the Court of Appeals for the Fifth Circuit in New Orleans, Louisiana. On May 20, the day before the standard was to take effect, Judge Morgan of that court granted an interim stay of its effectiveness. On June 7, a three-judge division of that court transferred that proceeding to this circuit pursuant to 28 U.S.C. §2112(a).[3] They extended the stay pending the action of this court. The stay remains in effect.

---

[3]The pertinent portions of this statute read:

If proceedings have been instituted in two or more courts of appeals with respect to the same order the agency, board, commission, or officer concerned shall file the record in that one of such courts in which a proceeding with respect to such order was first instituted. The other courts in which such proceedings are pending shall thereupon transfer them to the court of appeals in which the record has been filed. For the convenience of the parties in the interest of justice such court may thereafter transfer all the proceedings with respect to such order to any other court of appeals.

The petition filed by the American Petroleum Institute, et al., was assigned docket number 77-1516 in this court and was consolidated with the review proceeding filed by the AFL-CIO. The petroleum industry petitioners have now moved for retransfer to the Fifth Circuit. A number of other motions are also pending, but in view of our decision to grant the transfer, we will leave them for determination by the Fifth Circuit.

II

Initially, we must consider our jurisdiction over the AFL-CIO's petition. Section 2112(a) provides for the transfer of multiple review proceedings to the circuit in which a petition was first filed, but if the AFL-CIO's petition is invalid, our jurisdiction was not validly invoked and we are not the court of first filing even though the AFL-CIO's petition here was earlier in time than the petition in the Fifth Circuit. The issue is whether the AFL-CIO filed prematurely, since it filed before the press conference at which the benzene standard was made known to the general public. We have concluded that disclosure to the general public is not necessary to make agency action ripe for judicial review.

In *Saturn Airways, Inc. v. CAB,* 155 U.S. App. D.C. 151, 153, 476 F.2d 907, 909 (1973), we held that a petition was not premature when "at the time of filing it was clear both that the Board had taken what it deemed official action and that the substance of that action had been communicated to the public in some detail." The petitions at issue in *Saturn* had been filed after the agency adopted regulations and issued a press release describing the regulations "in some detail," (*id.*) but before the text of the regulations had been released. In the instant case, the AFL-CIO's petition was filed before the agency either held a press conference or issued a press release. Nevertheless, it was filed after the agency "had taken what it deemed official action" (*id.*)—Dr. Bingham's signing the standard—and after the substance of that action had been communicated to a fairly representative cross-section of the interested public.

***

[We are not] concerned here with secret agency action, in which the agency's decision is not communicated to anyone outside the agency. The Supreme Court has made clear that "[k]nowledge of the substance must to some extent be made manifest" in order for the agency's decision to be "issued," for substantive rights to be defined, and for the time for seeking review to begin to run. *Skelly Oil Co. v. Phillips Petroleum Co.,* 339 U.S. 667, 676 (1950). A valid petition for review could not have been filed before the press conference if Dr. Bingham had signed the standard alone in her office, or only in the presence of an aide, and not held the invitational briefing.

We recognize that advance disclosure of agency action to selected individuals or groups is likely to present difficulties. No matter how conscientious and careful an agency is in selecting those to be invited to a briefing, it is likely that some interested parties will not be represented. It might be concluded that a regulation or standard is issued for purposes of judicial review only when its text is made available to the public at large, on the ground that such a rule would insure a fair and even start for the race to the courthouse. But so interpreting the statutes would not bar the agency from discussing proposed or anticipated action with invited groups in advance of the vote or signature which makes it official agency action. Those participating in such discussions would have an advantage in preparing petitions for review and filing them quickly once the public announcement was made. It is not part of our judicial function, nor do we have any inclination, to dictate to agencies how they may or may not promulgate their actions. Agencies are vested with considerable discretion in such matters. *Skelly Oil*

Co. v. Phillips Petroleum Co., 339 U.S. 667, 676 (1950). We should give deference to the agency's choice, if it is reasonable. *Udall v. Tallman,* 380 U.S. 1, 16–17 (1965); *Brennan v. Occupational Safety & Health Comm'n,* 513 F.2d 553, 554 (10th Cir. 1975).

We would have a substantially different case if OSHA had invited, for example, only representatives of organized labor, or only representatives of industry, to the briefing by Dr. Bingham. Nothing in the record contravenes the prima facie impression that OSHA made a good-faith effort to invite representatives of a cross-section of the groups with an interest in the benzene standards.[8]

***

It bears emphasis that the entire dispute now before the court could have been avoided had OSHA established and followed regulations defining the process of issuing and promulgating standards. In recent years, we have often seen multiple review petitions filed within hours or minutes of a major agency action. We deplore such hasty filing. But while it exists as a fact or a potential, it appears incumbent upon agencies to make an effort to deal with the problem. * * *

The existence of such a regulation, making clear exactly what constitutes the issuance or promulgation of a standard, and providing that from and after that event all interested persons may have access to the text of the standard, would permit more expeditious review of OSHA actions without the lengthy and complex wrangling over jurisdiction and venue which we have seen in this case. * * *

In the context of this case, we hold that we do not lack jurisdiction over a petition for review filed after an agency takes action and the petitioner becomes aware of the substance of that action, even if the action is disclosed to the general public only after the petition has been filed. A person with actual notice is bound by an agency action, even if the act which imparts constructive notice to others—filing with the Federal Register for publication—has not yet occurred. We have jurisdiction over the petition filed by the AFL-CIO, and we are the court of first filing within the meaning of 28 U.S.C. §2112(a).

III

This does not end our inquiry, however. The last sentence of section 2112(a) provides that "[f]or the convenience of the parties in the interest of justice such court [of first filing] may thereafter transfer all the proceedings with respect to such [agency] order to any other court of appeals." For the reasons expressed in this section, I would vote to exercise our discretion to transfer these consolidated proceedings to the Fifth Circuit in the interest of justice.

***

As a simplifying rule of administration, I am of the view that in general a case should be transferred to the circuit in which the first petition was filed subsequent to the disclosure of the agency decision to the general public. This would be subject to the possibility that in any particular case the court would be of the view that the interest of justice was furthered by retention or review in another circuit.

Applying that standard in the instant case, I am of the view that there is no dominating reason in the interest of justice for retention of the proceeding in the District of Columbia Circuit, that the movants have established that they were not given an opportunity to make any determination of the

[8]There is no intention to suggest or condone, for example, ex parte contacts between quasi-judicial regulatory bodies and litigants, or *sub rosa* "leaks" of impending agency decisions to particularly favored persons or interests.

convenient forum after the decision was made known to the public, and that this opportunity was denied them because of the advance disclosure of the benzene standard to an invited group and the action of the AFL-CIO in filing its petition for review prior to disclosure to the public.[17]

The motion for transfer is granted, and the other pending motions in these cases will be transmitted to the Fifth Circuit for its consideration.

Two other opinions were handed down in this case. Judge Wilkey agreed that the proceeding should be transferred to the Fifth Circuit Court of Appeals, thus forming a majority with Judge Leventhal. His rationale was different, however; in his view, the standard was not issued until its substance was communicated to the general public at the news conference held after the Industrial Union Department (IUD) petition was filed, and, therefore, the union petition was prematurely filed.[2] Judge Fahy dissented. He agreed with Judge Leventhal that the standard was issued when signed. However, unlike Judge Leventhal, he did not believe that the interests of justice required a transfer of the case to the Fifth Circuit Court of Appeals.[3]

By the time the court decided this case, the issue of the validity of the benzene ETS had already become moot. The standard was issued on April 29, 1977, and had been stayed immediately by the Fifth Circuit Court of Appeals. Since under the Act an ETS could remain in effect only for a six-month period, and the court decision was not issued until October 17, OSHA decided that further litigation to uphold the standard would serve no useful purpose. OHSA therefore abandoned the ETS.[4]

In his opinion in *Industrial Union Department v. Bingham,* Judge Leventhal suggested that OSHA promulgate a regulation making clear what constitutes the issuance of a standard. This, he said, "would permit more expeditious review of OSHA actions without the lengthy and complex wrangling over jurisdiction and venue," which occurred in that case.[5] Accepting this suggestion, OSHA shortly thereafter promulgated a regulation providing that a standard is deemed issued for purpose of judicial review when it is "officially filed in the Office of the Federal Register."[6] The procedure at

[17]The petroleum industry petitioners make the additional argument in favor of transfer that their disagreement with the OSHA standard is far more substantial than the AFL-CIO's. The AFL-CIO, which initially urged OSHA to promulgate a benzene standard, is, they urge, not "genuinely aggrieved" by the standard actually issued.

I do not rely on this argument in voting to transfer these cases. There is nothing which indicates that the AFL-CIO petitioned, alleging a mere technical, cosmetic, or non-substantive grievance, in order to give this court jurisdiction over petitions which might be filed by others, or in order to collusively assist OSHA in defending its standard. Absent any such evidence, I decline to examine the relative merits of each petitioner's substantive claims against the agency, merely to decide the proper venue. * * *

[2]570 F.2d at 973.

[3]*Id.* at 976.

[4]OSHA's further efforts to regulate benzene hazards are discussed in Chapter 9.

[5]570 F.2d at 970–71.

[6]42 FED. REG. 65,166 (1977) (amending 29 C.F.R. §§1911.12, 1911.18).

the Office of the Federal Register is as follows: the Agency document embodying the rule or regulation is brought to the Office of the Federal Register; it is reviewed by the staff and, if acceptable for publication, the document is "officially filed," that is, date-stamped, placed in a public place, and made generally available.[7] The document is then published in the *Federal Register,* which is distributed to the public.

While the new OSHA regulation made litigation such as *Industrial Union Department v. Bingham* less likely, issues continued to arise on the question of when a standard was promulgated. An example was the race to the courthouse involving the cotton dust standard.

The chronology of events was as follows: OSHA's cotton dust standard was signed by Assistant Secretary Bingham and brought to the Office of the Federal Register at 9:00 a.m. on June 19, 1978. While the document was being reviewed by the staff, a petition was filed by the American Textile Manufacturers Institute (ATMI) in the Court of Appeals for the Fourth Circuit at 11:00 a.m. The standard was "officially filed," that is, date-stamped by the Federal Register office at 11:53 a.m. Two minutes later the Industrial Union Department (IUD) filed a petition for review in the District of Columbia Circuit Court of Appeals. ATMI argued that the standard was "filed" under the OSHA regulation when it was brought to the Federal Register office at 9:00 a.m.; it relied on the fact that the OSHA regulation states that the issuance occurs when the document is "filed in," and not "filed by," the Register. The IUD argued that the document was not "officially filed" until 11:53 a.m. and, therefore, its petition was the earlier one. In an unpublished opinion, the Court of Appeals for the Fourth Circuit agreed with the IUD and transferred the case to the District of Columbia Circuit Court,[8] which, in turn, refused to retransfer the case for the "convenience of the parties in the interest of justice"[9] and, after staying the entire standard, decided the case on the merits, upholding its validity, except as to one industry.[10]

The venue controversy in the *Cotton Dust* case may have been avoided had OSHA chosen the approach in defining the time of issuance of standards adopted by EPA under the Clean Air Act.[11] The EPA regulation provided that a standard is deemed to be issued two weeks after it is published in the *Federal Register.*[12] The preamble explained the reasons for this approach.

---

[7]The document is normally filed 8:30 a.m. on the day before publication, which is several days after the document is presented to the Office of the Federal Register, but, at the request of the Agency, it may be filed earlier.

[8]*American Textile Mfrs. Inst. v. Bingham,* No. 78-1378 (4th Cir. 1978). See 8 OSHR 568 (1978).

[9]*AFL-CIO v. Marshall,* 1978 OSHD ¶23,088 (D.C. Cir.) (per curiam).

[10]*AFL-CIO v. Marshall,* 617 F.2d 636 (D.C. Cir. 1979), *aff'd sub nom. American Textile Mfrs. Inst. v. Donovan,* 452 U.S. 490 (1981). The substantive aspects of the *Cotton Dust* case are discussed in Chapter 10.

[11]42 U.S.C. §§7401–7642 (Supp. V 1981) (revising and reclassifying 42 U.S.C. §§1857–58a (1976)).

[12]The proposal had provided that the standard would be deemed issued *one* week after publication.

Races to the Courthouse Final Rule

45 Fed. Reg. 26,046 (1980)

***

C. UNCERTAINTY AS TO WHEN THE RACE BEGINS

There is substantial uncertainty over precisely what event triggers the start of the race. It could be when the Administrator signs a rule; when he announces it at a press conference (or in a press release); when any party finds out he signed the rule; when the document arrives at the Office of the Federal Register ("OFR"); when the document is made officially available for public inspection at the OFR; when the document actually appears in the *Federal Register*; or at some other time specified by the Agency.

Such uncertainty has often resulted in expensive time-consuming and silly legal wrangling. Attorneys obsessed with winning the race have gone to great lengths to learn, often from "inside" information, when a rule will be signed or delivered to the OFR. They will also, in an effort to cover all bets, file many separate lawsuits within a few days under different theories as to when the race begins. Such multiple filings may then cause months of delay while attorneys file bundles of paper in various courts arguing over when the race began and who won it.

III. EPA'S PROPOSAL AND ITS RATIONALE

***

EPA's "one week 1:00 p.m. deferral" proposal was designed to curtail the "protracted procedural disputes" associated with racing and to accomplish several other purposes. First, EPA sought to assure fairness in the racing process by providing every party with advance notice of the time of promulgation. Under EPA's proposal, no party could benefit by inside information about when a rule would be signed, delivered, or filed.

Second, EPA wanted to make racing simple and inexpensive. Under EPA's proposal, promulgation occurs at a definitely ascertainable time which is not dependent upon the occurrence of a physical act. Experience of other agencies has shown that a race triggered by a physical event (such as filing a document at the OFR or stamping a document with an agency seal) prompts some of the most bizarre behavior in the entire sport of racing. Lawyers and their agents set up "human chains" utilizing walkie-talkies, sophisticated chronometers, hand signals, and diversionary tactics. These chains establish a communications link between the locus of the triggering event and a court clerk's office so that with a skilled racing team, a lawsuit can be filed within seconds of promulgation. Moreover, in anticipation of a rumored impending promulgation, racing teams often "camp out" for days at the locus of the triggering event, causing disruptive congestion at federal offices. Under EPA's proposal, no party could benefit from expensive, sophisticated human chains. Any party of modest means can simply appear at his favorite court at the appointed hour and file his petition.

Third, EPA wanted to curtail the lengthy litigation delays often associated with racing. Under EPA's proposal, the question of what event triggered the race would no longer be an issue. Nor should the question of who won the race be an issue; presumably, all racers will file at exactly the same time.

Fourth, EPA wanted to provide all interested parties with sufficient time to read a rule before the race begins. In some circumstances, a party might want to know what EPA has done before deciding to sue EPA.

B. "Convenience of the Parties in the Interest of Justice"

Under its regulation on the issuance of standards, OSHA adopted the practice of advising the public at what time the newly issued standard would be filed with the *Federal Register*. This practice assured that all members of the public would have the same information on the filing of standards and, thus, that any "race" would be entirely fair. However, this practice, like the EPA regulation, virtually assured a tie in any race since all persons were then in a position to file a petition for review in the circuit of their choice at precisely the same instant. This occurred, for example, after the issuance of OSHA's lead standard in 1978, when "simultaneous" petitions were filed in the Courts of Appeals for the Third and Fifth Circuits. The Court of Appeals for the Third Circuit unanimously decided that the proceeding should be transferred to the Court of Appeals for the District of Columbia Circuit.

United Steelworkers v. Marshall

592 F.2d 693 (3d Cir. 1979)

GIBBONS, Circuit Judge:

This case presents the problem of the appropriate court to review agency action when petitions for review have been filed in this court and in the [C]ourt of [A]ppeals for the Fifth Circuit. We conclude that the agency action in question should be reviewed by the Court of Appeals for the District of Columbia Circuit.

On November 9, 1978, the Occupational Safety and Health Administration of the United States Department of Labor (OSHA) held separate briefings for labor union representatives and industry representatives at which it disclosed the substantive contents of a proposed occupational health standard regulating employee exposure to lead, a highly toxic substance. The United Steelworkers of America, AFL-CIO-CLC (Steelworkers) had participated in the rulemaking proceedings in which the proposed standard was developed, representing members exposed to lead in industrial processes. The Lead Industries Association, Inc. (LIA), a trade association of manufacturers smelting or using lead, and Chloride Incorporated, a manufacturer, had also participated in those proceedings. Counsel for the Steelworkers had in the rulemaking proceedings sought a standard requiring a lower exposure level than OSHA was about to propose, while counsel for LIA and Chloride, Inc. had sought a standard less stringent in several respects. LIA is the chief spokesman for all employers in the primary lead industry. Both sides of the dispute over the standard were aware that the proposed standard would be filed in the Office of the Federal Register on November 13, 1978, and both sides anticipated filing a petition for review pursuant to §6(f) of the Occupational Safety and Health Act, 29 U.S.C. §655(f). Each side, therefore, had a representative at the Office of the Federal Register when it opened for business on November 13, 1978. Aware of the provisions of 28 U.S.C. §2112(a), each side had prepared in advance a petition for review for filing in the court of appeals of its choice. Counsel for LIA and Chloride chose the Fifth Circuit, while counsel for the Steelworkers chose the Third Circuit. When at 8:45 a.m. the OSHA standard was stamped filed at the Office of the Federal Register, counsel for the Steelworkers, having arranged for an open telephone line to a public telephone near the entrance to the Third Circuit Clerk's Office, instructed its representative in Philadelphia to file the Steelworkers' petition for review. The Third Circuit Clerk's office marked the petition filed at 8:45 a.m., eastern standard time. Counsel for LIA and Chloride, Inc., having arranged for an open line to a telephone near the entrance of the Fifth Circuit Clerk's office in New Orleans, instructed its representative there to file their petition for review. It was marked filed at 7:45 a.m., central standard time. Thus, the court records of the respective clerks' offices reflect a dead heat in the race to the courthouse.

The governing statute provides that "[i]f proceedings have been instituted in two or more courts of appeals with respect to the same order the * * * commission * * * concerned shall file the record in that one of such courts in which a proceeding with respect to such order was first instituted." 28 U.S.C. §2112(a). On November 29, 1978 the Solicitor of Labor informed the Chief Judges of both courts by letter that in the Labor Department's view there is no court of first filing. The letter called to the courts' attention the additional provision in 28 U.S.C. §2112(a) that "[f]or the convenience of the parties in the interest of justice such court may thereafter transfer all the proceedings with respect to such order to any other court of appeals," and suggested consultation between the respective courts as to which of the two should make a determination as to venue for the review of the lead standard.

The petitions for review were referred to panels in each court. Following the procedure adopted by the District of Columbia Circuit in *American Public Gas Ass'n v. FPC*, 555 F.2d 852 (D. C. 1976) this court conferred with the Fifth Circuit panel which on December 14, 1978 filed a per curiam opinion on the LIA petition for review as follows:

> Having conferred with the judges of the Third Circuit, and by agreement with them, we defer proceedings pending a decision by the Third Circuit Court of Appeals designating the forum to consider and decide this matter.

On December 20, 1978 we heard argument on the question of the proper forum, during which counsel for LIA and Chloride, Inc. advanced several arguments in support of its contention that the case should be heard in the Fifth Circuit.

The first contention of LIA is that, despite the fact that the court records suggest a simultaneous filing in both courts, in fact the filing in New Orleans occurred some ten seconds prior to that in Philadelphia. In support of that contention LIA filed in the Fifth Circuit affidavits by the persons who comprised its chain of communication between the Office of the Federal Register and the Clerk's Office in New Orleans tending to suggest that their timepieces were more accurate and their communications were more rapid than those of their Steelworker rivals. Counsel for the Steelworkers filed opposing affidavits tending to suggest that at best the race produced no more than a dead heat. Unlike race tracks, however, courts are not equipped with photoelectric timers, and we decline the invitation to speculate which nose would show as first in a photo finish.

The duty of determining who was first to file falls, under the express provisions of 28 U.S.C. §2112(a), upon the agency whose proceedings are under review. Prior to 1958, under many statutes providing for judicial review in the courts of appeals, when petitions were filed in several jurisdictions it was possible for the agency, by filing the record in one of them, to select its forum. One purpose of §2112(a) was to eliminate that possibility by substituting a mechanical first filing rule. What has resulted in practice has been the substitution of forum shopping by counsel for petitioners by the exercise of technological ingenuity in achieving first filing. When counsel succeeds in obtaining an earlier time stamp from the Clerk of one court the agency under review must file there. Where ingenuity has produced times of filing recorded as simultaneous, but one petitioner challenges the accuracy of the record time, three constructions of §2112(a) are possible. We could hold that the agency under review must hold a hearing to determine who filed first. This would reintroduce the problem of an interested agency making a decision as to the forum. We could hold that one of the reviewing courts could, when the fact of simultaneous filing was disputed, hold a hearing. But Courts of Appeals lack the ready means of holding hearings on disputed factual matters, and such a course would inevitably cause delays in the reviewing process. Finally, we could, and do, adopt a rule that in the absence of extraordinary circumstances, the official notations of time of filing are conclusive. When those notations show a simultaneous filing the agency should proceed, as did the Labor Department here, to notify both courts, who by agreement will determine which one will determine venue "[f]or the convenience of the parties in the interest of justice." 28 U.S.C. §2112(a).

LIA also contends that "the convenience of the parties in the interest of justice" mandates transfer to the Fifth Circuit. In support of this contention it urges that the statute gave LIA a choice of forum, and that where it did not quite succeed in effectuating that choice because the race to the courthouse resulted in a tie, the tie should be broken by weighing the relative aggrievement of the two petitioners. The Steelworkers, according to LIA, achieved in the standard as promulgated by OSHA most of the employee protection they sought, whereas the standard is significantly more stringent than the industry thinks necessary. Since LIA members suffer greater aggrievement, the argument continues, "the interest of justice" demands that its choice of a forum prevail. Other courts have limited the inquiry into the respective merits of the petitions to a determination whether the petitioning party's claim of aggrievement is so frivolous or insubstantial as to undercut the assumption of a good faith petition for review. E.g., *Ball v. NLRB,* 299 F.2d 683, 687 (4th Cir.), *cert. denied,* 369 U.S. 838 (1962); *UAW v. NLRB,* 373 F.2d 671, 673–74 (D.C. Cir. 1967); *Public Service Comm'n v. FPC,* 472 F.2d 1270, 1272 (D.C. Cir. 1972). The Steelworkers petition meets that minimal threshold, for the Union has at all times advocated a more stringent standard than that adopted by OSHA. Any more refined inquiry would require that the court considering a venue matter take into account the relative merits of the substantive positions being advanced. Certainly the reference to "the interest of justice" in §2112(a) was not intended to require such a preliminary examination of the merits. Rather the entire clause is directed at a balancing of competing claims of convenience in the prompt disposition of the petitions.

LIA suggests that we should take into account the Steelworkers' motivation in filing in the Third Circuit. That motivation purportedly is a desire to avoid application to the OSHA lead standard of the cost/benefit criteria announced in *American Petroleum Institute v. OSHA*, 581 F.2d 494 (5th Cir. 1978). LIA candidly concedes, however, that its motivation in filing in the Fifth Circuit was the presumed benefit its members would derive from that precedent. We have not been told why LIA's motivation for seeking what it suspects will be a favorable forum should be less rigorously scrutinized than that of the Steelworkers. Were we to consider such motivation to be relevant to the venue determination we would have to conclude that the scales are evenly balanced. But more fundamentally, we think it would be improper, in making a venue determination "[f]or the convenience of parties in the interest of justice", to take into account factors bearing on the merits of the agency action under review rather than factors bearing on convenience in carrying on the litigation. We decline to do so. Whether the interpretation of the statute announced by one court of appeals differs from that of another, and which interpretation is more just are not matters properly relevant to a §2112(a) transfer decision.

***

Finally LIA urges that a large number of employers aggrieved by the lead standard, including some who have moved to intervene in this review proceeding, have plants located in the Fifth Circuit. The Steelworkers counter this factor by observing that it has fifteen thousand members working in plants in the Third Circuit which will be covered by the standard. Neither the location of the plants nor the location of the workers bears, however, on the issue of relative convenience for the review of an OSHA standard adopted pursuant to §6 of the Act. Review is confined to the agency record. Thus neither the officials of employers nor the affected employees will appear in court. The only significant convenience factor which affects petitioners seeking review of rulemaking on an agency record is the convenience of counsel who will brief and argue the petitions.

Aside from that factor, three possible institutional considerations occur to us. One is the relative expertise of a given court of appeals in the area of law under review. We think that Congress, by providing for review in the circuit wherein the petitioner resides or has his principal place of business, 29 U.S.C. §655(f), has implicitly determined that OSHA litigation should not be concentrated in a single "expert" court. Thus we think it would be improper to speculate that any circuit court of appeals is more expert in OSHA matters than another. Accord, *Public Service Comm'n v. FPC, supra,* 472 F.2d at 1272–73. A second institutional consideration is the relative state of the dockets of those courts to which the case might be relegated. But since no court which is likely to be convenient to counsel has an uncrowded docket, that consideration is essentially neutral. A third institutional consideration is the desirability of concentrating litigation over closely related issues in the same forum so as to avoid duplication of judicial effort. In her November 29, 1978 letter the Solicitor of Labor informed us that on November 21, 1978 LIA filed in the District of Columbia Circuit a petition to review ambient air quality standards for lead issued pursuant to the Clean Air Act by the Environmental Protection Agency (EPA). 43 FED. REG. 46246 et seq. (October 5, 1978). Under that statute, in contrast to the Occupational Safety and Health Act, judicial review of ambient air standards may be brought only in the Court of Appeals for the District of Columbia. 42 U.S.C. §7607(b)(1). Thus there is no possibility of transferring the review proceedings dealing with air quality standards for lead to this or any other court. Since it seems likely that the District of Columbia litigation will involve issues with respect to the health need for limiting exposure to lead similar to those raised in the LIA and Steelworkers petitions pending here and in the Fifth Circuit there appears to be a strong institutional interest in having these petitions considered in the District of Columbia forum.

That institutional interest would not in our view suffice to justify a serious imposition of inconvenience upon those counsel [sic] most closely involved in the litigation. But in this instance the District of Columbia is obviously a convenient forum. The agency and its counsel are located there. The law firm which represents the Steelworkers, and which was the leading participant in the agency proceedings on the side urging more stringent standards is located in Washington. The counsel for LIA, who was the leading participant in the agency proceedings on the side urging less stringent standards, is located in New York City. As between Washington and Philadelphia there is no measurable difference in convenience. For us the institutional interest in having one court consider air standards for lead issued by both federal agencies issuing such standards then, is decisive.

***

As indicated in the court of appeals decision, the unsuccessful threshold argument of the Lead Industries Association was that LIA's petition was filed "some ten seconds" prior to the Union petition, despite the identical time stamp. In support, LIA filed an affidavit in court, which described the factual basis for its argument in greater detail than stated in Circuit Judge Gibbons' opinion for the court.[13]

Affidavit for Lead Industries Association

1. We are employees of Washington Experts, Ltd., a service enterprise which was retained by Lead Industries Association, Inc., through its counsel, to determine when OSHA's new lead standard was officially filed and to assist in the filing of a petition for review from that standard.

2. On Monday, November 13, 1978, your deponents, together with three other employees of Washington Experts, Ltd., established a telephone hook-up between a public telephone near the Federal Register office and a public telephone just outside the Clerk's office in the federal courthouse in New Orleans. The telephone line between our employees in Washington and petitioners' counsel in New Orleans was kept open from approximately 8:00 a.m. EST until after the petition for review in this proceeding had been filed. In addition, prior to the events described below, readings of the Federal Register's running time-stamp clock and its wall clock were taken (with the Federal Register's permission) and were synchronized with the watches worn by those doing the monitoring and with the watches worn by petitioners' counsel in New Orleans.

3. The standard was placed on the filing table for public view (and hence was "officially filed") at 8:45:20 EST according to the wall clock and 8:44:39 EST according to the running time-stamp clock. Signals were immediately relayed by our monitors to your deponent * * * in the public telephone booth in Washington, and she, in turn, instructed petitioners' attorneys in New Orleans to file the petition. The moment of filing in New Orleans was given by telephone to us in Washington. Having previously synchronized our own watches with the Federal Register clocks, we were able to determine that the time lapse between the promulgation of the standard in Washington and the filing of the petition for review in New Orleans was ten seconds.

4. Later during the morning of November 13, 1978 (but before learning that the United Steelworkers of America had filed a petition for review in the Court of Appeals for the Third Circuit) we prepared a memorandum summarizing the times of filing. * * *

---

[13]Affidavit filed by Lead Industries Association in *United Steelworkers v. Marshall,* 592 F.2d 693 (3d Cir. 1979).

5. We have been advised by petitioners' counsel that the Steelworkers' petition in the Third Circuit carries a handwritten notation indicating a filing time of 8:45 EST on November 13. Although the Steelworkers may have filed their petition within a minute of the official filing of the lead standard in Washington, D.C., the events which occurred in Washington confirm that the Steelworkers' petition in the Third Circuit was filed after petitioners' papers in this appeal had already been submitted to the Clerk:

(a) Immediately prior to the official filing of the lead standard in Washington, D.C., there was only one person other than representatives of Washington Experts, Ltd. who was monitoring the Federal Register office. He was not sitting at the filing table (as were we), and from his chair across the room could not see what was placed on the table.

(b) The unidentified person (who, we assume, was acting on behalf of the Steelworkers) had a walky-talky in his briefcase but he did not remove the walky-talky to communicate the words "Go! Go!" (presumably to someone in the *Washington* area, not in Philadelphia) until after our monitor had given the signal to [your deponent] and had returned to the Federal Register filing room to verify the time. In other words, it was *our* post-signal activities which first alerted him to the fact that the standard had been filed, and even then he did not bother to look at the document on the table before he communicated into his walky-talky. Nor did he return to the office to verify what the document was or to determine the time of filing.

(c) Based upon the personal observations of your deponent * * * and other Washington Experts' employees (all of whom had the unidentifiable person in view during the entire relevant period), we estimate that he did not even start to communicate with his radio device until some ten seconds or more after we had already given the predesignated signal to petitioners' counsel in New Orleans.

For the foregoing reasons we believe that the Steelworkers' petition was filed subsequent to the petition in this Court.

The Steelworkers also submitted an affidavit purporting to show that its petition was filed promptly at 8:45 a.m. The Court of Appeals for the Third Circuit, however, refused to become involved in this issue of whether *in fact* one party had filed its petition earlier, saying that it would "decline the invitation to speculate which nose would show as first in a photo finish." The court's further refusal to weigh the "relative aggrievement" of the two petitioners is in accord with the views of Judge Leventhal as stated in *Industrial Union Department v. Bingham.*[14]

In its ultimate decision on the merits in the *Lead* proceeding, the Court of Appeals for the District of Columbia Circuit, upholding the standard for the most part, disposed of the Steelworkers' substantive arguments summarily, but dealt with the LIA arguments at great length.[15] This disposition, however, in no way undermines the thrust

[14]570 F.2d at 973 n.17; see Section A, this chapter.

[15]*United Steelworkers v. Marshall,* 647 F.2d 1189, 1277–1310 (D.C. Cir. 1980) (lead standard aff'd in part, remanded in part), *cert. denied sub nom. Lead Indus. Ass'n v. Donovan,* 453 U.S. 913 (1981).

of the court's rationale in the earlier case; specifically, that determining the "relative aggrievement" would involve a more "refined" inquiry, which, the court believed, would be inappropriate in the "preliminary" litigation on the issue of jurisdiction.[16]

C. Suggestions for Change

Races to the courthouse have been widely described as unseemly spectacles, wasting the resources of the courts and of the parties, and reflecting some discredit on the legal profession. The Administrative Conference of the United States has addressed this problem,[17] and, based on its recommendation, regulatory reform legislation passed in 1982 by the U.S. Senate contains provisions intended to eliminate at least certain aspects of the race.[18]

Regulatory Reform Act, S. 1080

Senate Committee on Governmental Affairs
S. Rep. No. 305, 97th Cong., 1st Sess. 92–94 (1981)

***

SECTION 6—VENUE

Section 6 would revise 28 U.S.C. §2112(a) to eliminate unseemly "races" to preferred courts of appeals by parties seeking review of agency actions. These increasingly frequent spectacles arise from section 2112(a)'s present requirement that an agency file the record on appeal with the court in which the first petition to review the action is received. Since the venue of a review proceeding affecting multiple parties might lie in any of several circuits, and since in proceedings subject to section 2112 it is difficult to demonstrate that the convenience of the parties or the interests of justice favor transfer to another circuit, the court where the record is filed is likely to be the court which decides the case. Inasmuch as some affected parties, or their lawyers, believe that certain circuits are likely to be more receptive than others to their arguments, section 2112(a) has precipitated some quite refined and costly racing schemes.

[16]See also Haworth, *Modest Proposals to Smooth the Track for the Race to the Court House,* 48 Geo. Wash. L. Rev. 211, 225–26 (1980) (suggesting that 28 U.S.C. §2112 should be amended to require that a party must be "substantially aggrieved" for venue determinations).

[17]Eliminating or Simplifying the "Race to the Courthouse" in Appeals from Agency Action, 1 C.F.R. §305.80–5 (1982). See also McGarity, *Multi-Party Forum Shopping for Appellate Review of Administrative Action,* 129 U. Pa. L. Rev. 302 (1980) (an earlier version of this article was prepared as a report to the Administrative Conference of the United States).

[18]See also Regulatory Reform Act of 1982, House Comm. on the Judiciary, H.R. Rep. No. 435, 97th Cong., 2d Sess. 80–84 (1982) (adopting the Administrative Conference's recommendation to randomly select the court of review and rejecting proposals to amend 5 U.S.C. §704 to shift general review of agency actions from the district courts to the courts of appeals and to amend 28 U.S.C. §2112 to limit the right of individuals to sue federal agencies in the District of Columbia Circuit). Cf. *The Wheel of Fortune: Federal Licensing by Lottery,* 35 Ad. L. Rev. 33 (1983) (symposium on the Federal Communications Commission's proposed use of a random selection system for the granting of licenses); McGarity & Schroeder, *Risk-Oriented Employment Screening,* 59 Tex. L. Rev. 999, 1023–24 n.109 (1981) ("a lottery for allocating kidney dialysis machines might be the most efficient way to make that decision, given the difficulty of specifying and applying less arbitrary criteria").

The Committee does not believe that Congress intended or foresaw these unfortunate consequences in 1958, when it amended section 2112(a) to cure a weakness in earlier statutes that effectively allowed Federal agencies themselves to select the circuit for review. "Races to the courthouse" are wasteful, and they are likely to detract from the public's perception of the Federal courts as impartial dispensers of justice. The races—often decided by seconds or fractions of sections—produce no economic benefit, yet can cost private participants tens of thousands of dollars and waste the time of courts and agencies in contests over venue: Piecemeal reforms within the framework of the current section 2112(a) appear unlikely to have much impact.

S. 1080, which substantially embodies a proposal for statutory revision recommended by the Administrative Conference of the United States, will eliminate incentives to race by giving all parties who file within a ten-day period an equal chance to have the case assigned initially to their preferred forum. Its random selection approach would eliminate racing with fewer side effects on judicial review than other proposals for legislation.

Under the revised section 2112(a), a party wishing to qualify for the random selection procedure will have to meet two conditions. First, it must either file first or bring a review proceeding within ten days after the institution of the first proceeding with respect to an order. Second, it would have to notify the agency in writing of the filing, to ensure a quick, accurate determination of eligible parties. If the agency receives notices indicating that proceedings have been instituted in two or more courts of appeals within ten days of each other, it must promptly identify these courts to the Administrative Office of the U.S. Courts. That body would then choose, pursuant to a system of random selection that it would devise for this purpose, the court in which the record shall be filed. The scheme envisions that the Administrative Office would have only the ministerial duty of devising and administering a random selective scheme; it would not be authorized to make decisions of a judicial nature such as interpreting the statute or determining which court ultimately should hear a case. As the bill suggests, there would be one entry per circuit in which proceedings are pending, rather than one per petitioner or proceeding.

Under S. 1080's approach, no random selection will be required if a second proceeding is commenced more than ten days after the initial one. If proceedings are instituted in only one circuit, or if the agency receives notice of proceedings pending in only one circuit, the pool would contain only one entry. Thus, a party who files for review on the first possible day but does not notify the agency in writing of its action might not qualify for the random selection procedure if other parties file and give notice within ten days.

Under this proposal, the court chosen at random will then take jurisdiction over all review proceedings dealing with the same order, subject to its existing power to transfer for the convenience of the parties in the interest of justice. Section 6 will not change existing standards for transfer. S. 1080 would not cover cases where venue is specified by statute to lie in one particular circuit, or cases filed in the district courts. If the court in which the record is filed determines that it lacks jurisdiction or venue is improperly laid but that jurisdiction and venue may be proper in another circuit, the court would notify the Administrative Office of the U.S. Courts of that fact, and the Administrative Office then would choose from among the remaining courts in which petitions have been filed according to the same random method.

Section 6 does not affect existing standards for granting stays pending review. However, the revised section 2112(a) will place a time limit on stays issued prior to the lottery. Unless extended by the court selected, such stays would expire automatically 15 days after the Administrative Office has made its selection. The Committee believes this limitation should minimize forum shopping for temporary stays and ensure that judicial comity does not prevent the court ultimately chosen by lottery from lifting a stay.

***

This legislation would not eliminate all aspects of forum shopping. On the one hand, it would normally remove the incentive for parties to file the first petition. But, under the legislation, the court chosen at random would still have authority to transfer to another circuit for the convenience of the parties in the interest of justice. Thus, parties would still have an opportunity to argue that some other circuit is more appropriate, and wasteful threshold litigation could ensue. The alternative would be to bar any further transfer from the court chosen by random selection. While this alternative would avoid threshold litigation over venue, it would also eliminate the flexibility that allows the courts to deal with cases in which justice and convenience clearly point to some other court.[19] Professor Thomas McGarity of the University of Texas School of Law has stated that the random selection system does not preclude the possibility of forum shopping to avoid a particular circuit, a problem, he says, which "seems ultimately unsolvable."[20] Thus, as sometimes happens, if a party or group of parties is most interested in *avoiding* a particular circuit rather than in *locating* the appeal in a particular circuit, under the random selection procedures, the party could substantially improve its chances of success by filing petitions in all circuits other than its disfavored one.

Other proposals to reform the law governing review of agency decisions have been advanced. Notable among these are establishment of exclusive venue for review in a particular court, as is the case for certain standards under the Clean Air Act,[21] and the application of judicial "expertise" as a criterion for discretionary transfer.[22] One of the significant problems with a single court for review, apart from the political controversy that would surround any attempt to determine which court of appeals that would be, is that it "would completely deny the Supreme Court the percolating and signaling advan-

---

[19]See McGarity *supra* note 17, at 373 (suggesting that those cases would be "rare").

[20]*Id.* at 375.

[21]42 U.S.C. §7607(b) (Supp. V 1981).

[22]Professor Mark Rothstein advocates an amendment to 28 U.S.C. §2112(a) providing that when petitions for review are filed in two or more circuits, venue would lie in the D.C. Circuit. Rothstein, *OSHA After Ten Years: A Review and Some Proposed Reforms,* 34 VAND. L. REV. 71, 91–92 (1981). Although Rothstein said this would be the "simplest" solution, it might be one of the most controversial and least likely to be enacted. A strong argument that more weight should be given to considerations of the expertise of the circuits in determining venue is presented by Haworth, *supra* note 16 at 225–26. Cf. Administrative Conference Recommendation No. 82–3, 1 C.F.R. §305.82–3 (1983) (recommending Congressional review of statutes, such as the Clean Air Act, which provide for exclusive court of appeals review in the District of Columbia Circuit to determine if the need for authoritative determinations on nationally applicable requirements outweighs the benefits of providing litigants the choice of forum).

tages of multi-circuit review"; and that "over time the relationship between the two institutions [the Agency and the court with exclusive jurisdiction] would either blossom or sour to such an extent that the court would be either too hesitant or too eager to reverse agency decisions."[23]

The race to the courthouse, in the OSHA context, had developed because of the significant divergence of views on their review responsibilities between the Fifth Circuit Court of Appeals, on the one hand, and the District of Columbia, Second, and Third Circuits on the other. As a result, however, of the Supreme Court decisions in the *Benzene* and *Cotton Dust* cases setting forth principles for the substantive content of OSHA standards, these differences in approach may be less marked in the future and, therefore, even apart from legislative reform, the race may become less significant. At the same time, arguably, a particular circuit's view in reviewing an agency's action goes beyond the question of the formal rule of law being applied; rather, it raises the far broader question of whether the court is in sympathy with the agency's regulatory goals. And, so long as the parties perceive that certain courts of appeals are more (or less) sympathetic to the agency's regulatory efforts, the race will continue.[24]

[23]McGarity, *supra* note 17 at 356–60. For further elaboration of the proposal to base venue on expertise, see Note, *Venue for Judicial Review of Administrative Actions: A New Approach,* 93 HARV. L. REV. 1735 (1980).

[24]The race to the courthouse becomes more complex where one or more challenges to the agency action are filed in a U.S. district court in addition to those filed in the courts of appeals. See *American Petroleum Inst. v. OSHA,* 8 OSHC 2025 (5th Cir. 1980) (validity of OSHA's Carcinogens Policy).

8

Procedural Issues in Standards Rulemaking

A. Background

While the courts of appeals have differed in defining the scope of their responsibility to review the substance of OSHA standards, all courts have agreed on the principle that OSHA must strictly adhere to the procedural requirements of the Act and the OSHA regulations in rulemaking proceedings.[1] One of the major procedural themes was sounded in the two earliest standards review cases. The courts in *Associated Industries v. Department of Labor*[2] and *Industrial Union Department v. Hodgson*[3] both emphasized the importance of OSHA's articulation of the reasons for its standards. "What we are entitled to in all events," said Judge McGowan in *Industrial Union Department*, "is a careful identification by the Secretary, when his proposed standards are challenged, of the reasons why he chooses to follow one course rather than another."[4] In the *Synthetic Organic Chemical* case, the Court of Appeals for the Third Circuit, in affirming OSHA's standard on ethyleneimine, articulated its responsibility for review as including a determination of whether the proposal "adequately informed interested persons of the action taken" and whether the promulgation "adequately" sets forth the reasons for the action.[5] And in the *Cotton Dust* case, Judge Bazelon, noting the "uneasy partnership" between the Agency and the reviewing court, stated that the court's role in the partnership "is to ensure that the regulations resulted from a process of reasoned decisionmaking."[6] He asserted, particularly, that the decision-making process must include, by statute,

[1]The statutory procedural requirements are contained in §6, 29 U.S.C. §655 (1976); the OSHA rules of procedure for rulemakings appear in 29 C.F.R. pt. 1911 (1983). The requirements of the Administrative Procedure Act, 5 U.S.C. §§551–59, 701–06 (1976 & Supp. V 1981), are also relevant.

[2]487 F.2d 342 (2d Cir. 1973).

[3]499 F.2d 467 (D.C. Cir. 1974).

[4]*Id.* at 475.

[5]*Synthetic Organic Chem. Mfrs. Ass'n v. Brennan,* 503 F.2d 1155, 1160 (3d Cir. 1974), *cert. denied,* 420 U.S. 973 (1975); see Chapter 6, Section A.

[6]*AFL-CIO v. Marshall,* 617 F.2d 636, 649–50 (D.C. Cir. 1979), *aff'd sub nom. American Textile Mfrs. Inst. v. Donovan,* 452 U.S. 490 (1981); see Chapter 6, Section A.

"notice to interested parties of issues presented in the proposed rule," and that the Agency must also provide "opportunities for these parties to offer contrary evidence and arguments." Finally, Judge Bazelon defined the task of the reviewing court as including two procedural aspects: ensuring that the Agency has followed the procedures required by statute and by its own regulations, and that it has explicated the basis for its decision.[7]

This chapter deals with the major procedural objections that have been raised to OSHA standards. These are: (1) bias of the decision maker;[8] (2) inadequacy of notice; (3) *ex parte* communications and lack of separation of functions; and (4) inadequacy of the statement of reasons. All but the last of these issues were involved in the litigation in the District of Columbia Circuit Court of Appeals on the validity of OSHA's lead standard,[9] and the lengthy court of appeals' opinion in that case provides a convenient framework for discussion.

The lead standard was proposed by OSHA on October 3, 1975.[10] The proposal contained a permissible exposure limit of 100 micrograms of lead per cubic meter of air, with requirements for engineering controls, environmental monitoring, and employee medical surveillance. Hearings were held in March, April, and May 1977 and, after OSHA reopened the record on the issue of medical removal protection,[11] additional hearings were held in November and December 1977. The final standard was issued in November 1978.[12] It provided for a permissible exposure limit of 50 micrograms, to be achieved primarily by engineering controls. The final standard also contained medical removal protection requirements, as well as various other provisions normally included in OSHA health standards. The standard was challenged by both employer and employee groups, and a "race to the courthouse" ensued. The proceeding was ultimately transferred to the District of Columbia Circuit Court of Appeals.[13]

The court of appeals upheld the lead standard in major respects, with Judge MacKinnon dissenting on a number of key issues. The majority, in an opinion by Chief Judge Skelly Wright, rejected various procedural challenges; held that the standard met the "significant risk" test stated by the Supreme Court in the *Benzene* case;[14] and, as to 10 industries, the standard was economically and technologically feasible. As to a number of other industries, the court remanded the case to OSHA for additional feasibility findings.[15]

---

[7]*Id.* at 650.

[8]This issue is not strictly procedural, but it has been included in this chapter for reasons of editorial convenience.

[9]*United Steelworkers v. Marshall,* 647 F.2d 1189 (D.C. Cir. 1980), *cert. denied sub nom. Lead Indus. Ass'n v. Donovan,* 453 U.S. 913 (1981).

[10]40 FED. REG. 45,934 (1975).

[11]See Chapter 5, Section D.

[12]43 FED. REG. 52,952 (1978).

[13]*United Steelworkers v. Marshall,* 592 F.2d 693 (3d Cir. 1979), discussed in Chapter 7.

[14]*Industrial Union Dep't v. American Petroleum Inst.,* 448 U.S. 607 (1980), discussed in Chapter 9.

[15]After remand, OSHA held an additional hearing and issued a Supplemental Statement of Reasons, finding the lead standard feasible for most of the industries that were the subject of the remand. 46 FED. REG. 6134 (1981). OSHA subsequently stayed the effective date of the

B. Bias of the Decision Maker

The first procedural issue discussed by the court of appeals in the lead proceeding is the alleged bias of the person who made the ultimate decision on the lead standard, Dr. Eula Bingham, the Assistant Secretary for OSHA. The Lead Industries Association (LIA) argued that she prejudged the issue of medical removal protection—a critical issue in the standard—and therefore the standard should be vacated.

United Steelworkers v. Marshall

647 F.2d 1189 (D.C. Cir. 1980)

J. SKELLY WRIGHT, Chief Judge:
***

A. BIAS OF THE DECISIONMAKER

LIA urges us to vacate the entire lead standard because, in its view, the official who ultimately set the standard, Assistant Secretary of Labor Eula Bingham, had prejudged the essential issues in the rulemaking proceeding. For proof of this allegedly fatal bias, LIA points to a speech Bingham delivered on November 3, 1978 to a United Steelworkers of America conference on occupational exposure to lead.

Bingham's speech began innocuously, if dramatically ("Brothers and Sisters"), by noting her concern for workers and by recognizing how much OSHA depended on their unique perspective when it gathered information in setting safety standards. But after asserting that she and Secretary of Labor Marshall were "determined" to have a lead standard, Bingham proceeded to suggest her predisposition on important issues. As to the medical removal protection provision (MRP):

> I think that there may be some apprehension because Assistant Secretaries in the past have not always understood, or have not known how to spell the words medical removal protection, or rate retention * * *. Well, I learned to spell those words a long time ago on the Coke Oven Advisory Committee, and if you want to know how I feel about it, you need only to look up my comments during those Committee hearings. As far as I'm concerned, it is impossible to have a Lead Standard without it. * * *

Appendix to Lodged Documents (ALD) 3. As to the dangers of lead:

> * * * I can tell you about a plant within 300 miles of the city where workers are told to go to the hospital from work and receive therapy that would drag out poison and precious metals. And then they're sent back to be poisoned again. I bet I could go down to the hospitals of this

supplemental statement and on December 11, 1981, issued a Revised Supplemental Statement of Reasons, amending the final lead standard in several respects. 46 FED. REG. 60,758 (1981). In addition, in 1982 OSHA stayed the requirement that three industries (primary and secondary smelting and battery manufacturing) submit plans for implementing engineering controls. 47 FED. REG. 54,433 (1982). Union challenges to OSHA's action are still pending before the District of Columbia Circuit Court of Appeals in early 1984. See also discussion in Section C of this chapter.

city and find a worker that is undergoing kidney dialysis, and I'll bet you a dinner that some of those workers have been in lead plants.

Id. at 4. As to economic feasibility:

> I have told some people that I have never aspired to be an economist, but I tell you I can smell a phony issue when I see one. And to say that safety and health regulations are inflationary is phony. * * *
>
> * * * I don't understand a society such as ours who is not willing to pay a dollar more for a battery to insure that workers do not have to pay for that battery with their lives.

Id. at 5. The speech went on to urge workers "to control their own destiny" by educating themselves about the lead problem, and ended by calling for political support in the imminent congressional elections for candidates sympathetic to OSHA's goals. *Id.* at 9.

Were it our task to assess the wisdom and propriety of an administrator's public conduct, we might well admonish Dr. Bingham for this speech. She served her agency poorly by making statements so susceptible to an inference of bias, especially statements to a group so passionately involved in the proceedings. But our task is rather to measure her conduct against the legal standards for determining whether an official is so biased as to be incapable of finding facts and setting policy on the basis of the objective record before her. Moreover, we must bear in mind that this particular speech, though delivered five days before the Secretary of Labor signed the final standard and ten days before he released it, came 30 days after Bingham had effectively made her own decision on the standard and ten days after she had approved the final language.

An administrative official is presumed to be objective and "capable of judging a particular controversy fairly on the basis of its own circumstances." *United States v. Morgan,* 313 U.S. 409, 421 (1941). Whether the official is engaged in adjudication or rulemaking, mere proof that she has taken a public position, or has expressed strong views, or holds an underlying philosophy with respect to an issue in dispute cannot overcome that presumption. *Hortonville Joint School District No. 1 v. Hortonville Educ. Ass'n,* 426 U.S. 482, 493 (1976); *United States v. Morgan, supra,* 313 U.S. at 421. Nor is that presumption overcome when the official's alleged predisposition derives from her participation in earlier proceedings on the same issue. *FTC v. Cement Institute,* 333 U.S. 683, 702–703 (1948). To disqualify administrators because of opinions they expressed or developed in earlier proceedings would mean that "experience acquired from their work * * * would be a handicap instead of an advantage." *Id.* at 702.

When Congress creates an agency with an express mission—in OSHA's case, to protect workers' health and safety—the agency officials will almost inevitably form views on the best means of carrying out that mission. The subjective partiality of an official of such an agency does not invalidate a proceeding that the agency conducts in good faith. *Lead Industries Ass'n, Inc. v. EPA,* 647 F.2d 1130, 1179 (D.C. Cir. 1980); *Carolina Environmental Study Group v. United States,* 510 F.2d 796, 801 (D.C. Cir. 1975).

This court has indeed required disqualification of an agency adjudicator when his public statements about pending cases revealed he "'has in some measure adjudged the facts as well as the law of a particular case in advance of hearing it.'" *Cinderella Career & Finishing Schools, Inc. v. FTC,* 425 F.2d 583, 591 (D.C. Cir. 1970), *quoting Gilligan Will & Co. v. SEC,* 267 F.2d 461, 469 (2d Cir.), *cert. denied,* 361 U.S. 896 (1959); see *Texaco, Inc. v. FTC,* 336 F.2d 754, 760 (D.C. Cir. 1964), *vacated and remanded per curiam on other grounds,* 381 U.S. 739 (1965). And, although these cases involved adjudication, we could perhaps logically apply them to hybrid rulemaking proceedings like the present one in which the factual predicates of final rules are subject to review under the substantial evidence test.

So applied, however, these cases would lead us to vacate the lead standard only if Dr. Bingham had demonstrably made up her mind about important and specific factual questions and was impervious to contrary evidence. This test would be hard enough for petitioners to meet. But in *Ass'n of Nat'l Advertisers, Inc. v. FTC,* 627 F.2d 1151 (D.C. Cir. 1979), handed down after oral argument in the present case, we raised an even higher barrier to claims of bias in rulemaking proceedings. We stressed there the difference between the essentially "legislative" factfinding of a rulemaker and the trial-type factfinding of an adjudicator, and thus held that the *Cinderella* test was inappropriate. We concluded that an agency official must be disqualified from rulemaking "only when there has been a clear and convincing showing that [she] has an unalterably closed mind on matters critical to the disposition of the proceeding." 627 F.2d at 1195.

The relevant statute in *Ass'n of Nat'l Advertisers, Inc. v. FTC, supra,* Section 18 of the Federal Trade Commission Act, 15 U.S.C. §57a (1976), like the OSH Act, creates procedures more formal than the minimal ones required for informal rulemaking by 5 U.S.C. §553 (1976). We held, however, that even in such hybrid rulemaking the findings of fact so intertwine with the policies that emerge from them that we could not, as we could in *Cinderella,* "cleave law from fact" in deciding whether the official had prejudged factual issues. 627 F.2d at 1168.

Dr. Bingham's general expression of solidarity with the Steelworkers was legally harmless. Her call for support for congressional candidates sympathetic to her agency's mission did not bear on any specific issues in the case, and is probably the sort of political activity we simply must accept from a political appointee. Thus her bias, if any, shows up in her remarks about MRP, the dangers of lead poisoning, and the inflationary effect of the lead standard.

Had she made these remarks before the rulemaking began or while OSHA was receiving public comments, we might still have had to strain precedent to find grounds for disqualification. Her remarks on MRP do not bear on any specific factual issues, but rather reveal a general predisposition on a matter of policy, of the sort held legally harmless in *FTC v. Cement Institute, supra,* and *Ass'n of Nat'l Advertisers, Inc. v. FTC, supra.* Her remarks about endangered workers do bear on a factual question, but only very generally; they reveal no prejudgment on the precise and complex factual issues in the case, such as the exact blood-lead level at which disease develops. Finally, although the speech does allude specifically to the cost of the standard to the battery industry, Dr. Bingham's expression of disbelief in the inflationary effect of the standard is really part of a general rhetorical flourish about the danger of undervaluing worker health.

In any event, the fact remains that Dr. Bingham delivered the speech *after* she had decided on the standard and *after* the record had been closed. We can thus infer bias only if we construe her remarks retroactively. There may be cases warranting such judicial mindreading, but they would have to involve far more explicit and detailed statements by the allegedly biased person. The only language of predisposition in Bingham's speech that we can plausibly read retroactively is that on MRP, and her statement on that subject falls within the category of views derived from administrative experience to which the Supreme Court referred in *FTC v. Cement Institute, supra,* 333 U.S. at 702. Thus, Bingham's speech simply does not reveal prejudgment with sufficient specificity to prove bias under the *Cinderella* standard, and, all the more so, does not constitute the "clear and convincing" evidence demanded by *Ass'n of Nat'l Advertisers, Inc. v. FTC, supra.* Judicial review of rulemaking, unlike the ABA Canon of Ethics, does not attack the mere appearance of impropriety. Bingham's speech, however unfortunate, does not prove the proceedings unfair.

***

In deciding that the speech by Dr. Bingham did not disqualify her from acting as decision maker in the lead proceeding, the court's majority relied heavily on its earlier decision in *Association of National Advertisers v. FTC*. That case involved statements made by FTC Chairman Pertschuk that advertising on children's television had adverse effects on children. The Court of Appeals for the District of Columbia Circuit in that case stated more fully its view of the critical differences between the role of officials in adjudicatory agencies and that of an official with the responsibility to issue policy rules.

Association of National Advertisers v. FTC

627 F.2d 1151 (D.C. Cir. 1979)

TAMM, Circuit Judge:

***

B

We never intended the *Cinderella* rule to apply to a rulemaking procedure such as the one under review. The *Cinderella* rule disqualifies a decisionmaker if "'a disinterested observer may conclude that [he] has in some measure adjudged the facts as well as the law of a particular case in advance of hearing it.'" 425 F.2d at 591 (quoting *Gilligan, Will & Co. v. SEC,* 267 F.2d at 469). As we already have noted, legislative facts adduced in rulemaking partake of agency expertise, prediction, and risk assessment. In *Cinderella,* the court was able to cleave fact from law in deciding whether Chairman Dixon had prejudged particular factual issues. In the rulemaking context, however, the factual component of the policy decision is not easily assessed in terms of an empirically verifiable condition. Rulemaking involves the kind of issues "where a month of experience will be worth a year of hearings."[37] Application of *Cinderella*'s strict law-fact dichotomy would necessarily limit the ability of administrators to discuss policy questions.

The legitimate functions of a policymaker, unlike an adjudicator, demand interchange and discussion about important issues. We must not impose judicial roles upon administrators when they perform functions very different from those of judges. As Professor Glen O. Robinson, a former member of the Federal Communications Commission, has commented:

> Although members of agencies such as the FCC certainly do perform significant judicial functions in deciding individual cases, they perform even more tasks of a legislative or an executive character. When the FCC, for example, promulgated regulations barring common ownership of local newspapers and broadcast stations, it performed a legislative task, pure and simple. In reaching the decision, the Commission was

[37] *American Airlines, Inc. v. CAB,* 359 F.2d 624, 633 (D.C. Cir. 1966) (*en banc*).

> neither bound by, nor expected to conform to, the confining procedures or standards of a court. Why then should the decisionmakers be stamped from a judicial cast? Insofar as the agency is delegated broad legislative powers and responsibilities, would it not be at least as appropriate to measure agency members against standards used to evaluate legislators? Such standards would place agency members on a better standing with respect to judges and would create an entirely new frame of reference for assessing agency performance. The supremacy of carefully reasoned principle—the supposed ideal of judicial decision—necessarily would yield to the dictates of political compromise and expediency, which are the accepted hallmarks of legislative action. Correspondingly, the standard for evaluating the composition of the agencies would shift from an emphasis on professional training to an emphasis on representativeness.

Robinson, *The Federal Communications Commission: An Essay on Regulatory Watchdogs,* 64 VA. L. REV. 169, 185–86 (1978) (footnotes omitted).

The *Cinderella* view of a neutral and detached adjudicator is simply an inapposite role model for an administrator who must translate broad statutory commands into concrete social policies. If an agency official is to be effective he must engage in debate and discussion about the policy matters before him. [As] this court has recognized before, "informal contacts between agencies and the public are the 'bread and butter' of the process of administration." *Home Box Office, Inc. v. FCC,* 567 F.2d 9, 57 (D.C. Cir.) (per curiam), *cert. denied,* 434 U.S. 829 (1977).

***

Judge MacKinnon dissented in the *Association of National Advertisers* case; he would have found that a rulemaking official was disqualified if there was a showing "by a preponderance of the evidence that the decisionmaker could not participate fairly in the formulation of the rule because of a substantial bias or prejudgment on any critical fact that must be resolved * * *."[16] Judge MacKinnon thus would permit disqualification on a "preponderance of the evidence" rather than, as found by the majority, a "clear and convincing showing." Further, Judge MacKinnon would require only "substantial bias or prejudgment" while the majority insisted on an "unalterably closed mind." Under either test, however, disqualification of a rulemaking official would be rare; significantly, Judge MacKinnon, who dissented on a number of major issues of appeal in the *Lead* case, was in agreement with the majority that Dr. Bingham should not be disqualified.

While the court in the *Lead* proceeding refused to disqualify Dr. Bingham as a legal matter, the opinion of the court said that it "might well admonish Dr. Bingham for this speech. She served her agency poorly by making statements so susceptible to an inference of bias, especially statements to a group so passionately involved in the proceedings."[17]

---

[16]627 F.2d at 1197.
[17]647 F.2d at 1208.

In his study on the disqualification issue,[18] Professor Peter Strauss states that a "proper test" for disqualifying rulemaking officials must not interfere with the necessary elements of policy-making activity or "chill important aspects of agency public life." He concludes that "neither acts nor beliefs producing the decision to initiate a rule, the formulation of a proposal, or the conduct of policy discussion with Congress, the regulated industry, or the public should be regarded as disqualifying." He further said that "only prejudgment as to 'fact' is even arguably disqualifying," and, in the case of hybrid rulemaking, these would be the "facts" as to which full adversary procedures may be required. Finally, he said "one might well prefer to leave correction of indecorous behavior entirely to the political realm—to congressional reaction, loss of public acceptance, and collegial censure."[19]

Professor Strauss applied his conclusions specifically to the Dr. Bingham speech involved in the *Lead* case and found it legally not disqualifying, but objectionable in other respects.[20]

Strauss, *Disqualification of Decisional Officials in Rulemaking*

***

One may test these conclusions against Assistant Secretary Bingham's passionate speech to the Steelworkers. Here, unlike *ANA* [*Association of National Advertisers*], the question of disqualification was raised only after the rulemaking was complete, on the basis of a speech given at a time when, inevitably, the intellectual process of reaching judgment would have been complete. The remarks she made indicate strong commitments on a variety of matters, ranging from the political to the technical. Undoubtedly, too, Dr. Bingham brought from her role in private life as a specialist in workplace health hazards strongly held views about the needs for worker safety; she had been selected precisely because of her background and attitude.

The largest element of the objections, as with Chairman Pertschuk, appears to have been to Dr. Bingham's choice of forum and manner. The speech was openly political, given as the most hotly objected of Chairman Pertschuk's remarks had been, to a constituency on the opposite side of the issue. From an industrialist's perspective, it must seem very raw, and would equally have seemed so had the speech been given after adoption of the standard. But at this level of generality, raw as it is, no requirement of detachment can be inferred from the statute or otherwise. In describing her trips through "the palace guard," Dr. Bingham described existing political processes of control that are an inherent part of rulemaking as currently understood; repudiation of those processes, if warranted, should not be accomplished through the back door of a disqualification standard. Solicitation of political support is equally part of an Assistant Secretary's job as now understood. Remarks about inflationary impact or attitudes toward the willingness of society to bear the added costs of worker protection, however strongly felt or misguided, are far too general to suggest prejudgment; they

[18]Strauss, *Disqualification of Decisional Officers in Rulemaking,* 80 COLUM. L. REV. 990 (1980).
[19]*Id.* at 1048–49.
[20]*Id.* at 1047–48.

reflect the sorts of broad political attitudes that all experienced persons bring to public office.

Dr. Bingham's remarks about medical removal protection similarly fail to suggest prejudgments of fact. "Medical removal protection" is a regulatory measure requiring companies to maintain workers removed from a work environment for health reasons at their former rate of pay. OSHA's authority to impose such a measure and the need for it were hotly contested, but neither issue suggests disputes of fact as to which cross-examination or the ability to decide solely on the basis of a record are requisite. Even if, as appears, Dr. Bingham came to office with unshakeable views on these matters, that would be no occasion to disqualify. Whether she was wrong on the question of authority will of course be resolved by the courts: the disposition to impose a certain measure within her authority, if known, would be a valid reason only for making or withholding her appointment.

In short, the speech—fiery as it was—appears not to have reflected prejudgments on particular contested issues of specific fact. Its fire may well have cast discredit on the result in some quarters, contributed to a sense of unfairness or social malaise unfortunate in a society that depends largely on good will and voluntary compliance for adherence to law. The proper correctives, however, are political, not judicial; a judicial corrective would of necessity extend requirements of objectivity beyond their reach even in analogous judicial circumstances and correspondingly impair the necessary political controls of the regulatory system.

***

The Administrative Conference of the United States adopted a recommendation entitled "Decisional Officials' Participation in Rulemaking Proceedings,"[21] based on Professor Strauss' report. In fashioning rules for disqualification of officials in rulemaking, the Administrative Conference sought to distinguish between prejudgments of particular "adjudicative" or "specific" facts—which would be disqualifying—and prejudgments on policy.[22] The District of Columbia Circuit Court of Appeals seems to reject this distinction; the court emphasized the difficulty in attempting to "cleave fact from law" and stated its disqualification test in terms of an "unalterable closed mind" on "matters critical to the disposition of the proceeding."[23] The difficulties implicit in the Administrative Conference approach may be seen in the context of Dr. Bingham's comments on medical removal protection. Her statements, in the words of the court of appeals, "reveal a general predisposition on a matter of policy."[24]

Accordingly, under both the court and the Conference tests, she should not have been disqualified. But under the Administrative Conference approach, it is difficult to conceive of what Dr. Bingham could have said on medical removal protection that would have constituted disqualifying prejudgment. If the Assistant Secretary stated

---

[21]1 C.F.R. §305.80–4 (1983).

[22]*Id.* at §305.80–4(c).

[23]*United Steelworkers v. Marshall,* 647 F.2d at 1210.

[24]*Association of Nat'l Advertisers,* 627 F.2d at 1195, *quoted in United Steelworkers v. Marshall,* 647 F.2d at 1209.

a view on the number of employees with high blood-lead levels at a particular plant, this might be considered a prejudgment on an "adjudicative fact"; but because of the relative remoteness of this fact to the ultimate determination on medical removal protection, this prejudgment would not pertain to "facts which will be materially at issue in the proceeding." The court test, therefore, would seem more practical; without attempting to distinguish between "facts" and "policy," it would disqualify the official if the prejudgment was of sufficient magnitude ("unalterably closed mind") on a sufficiently critical matter.

C. Notice Requirements

The Administrative Procedure Act requires the Agency to publish "either the terms or substance of the proposed rule or a description of the subjects and issues involved."[25] OSHA, on the other hand, is required to initiate a standards rulemaking proceeding by publishing a proposed rule.[26] A mere statement of the "subjects and issues involved," arguably, is insufficient. The Attorney General's Manual on the Administrative Procedure Act states that the notice must be "sufficiently informative to assure interested persons an opportunity to participate intelligently in the rulemaking process."[27] The adequacy of OSHA's notice of proposed rulemaking was another major issue in the litigation involving the validity of the OSHA lead standard.

United Steelworkers v. Marshall

647 F.2d 1189 (D.C. Cir. 1980)

J. SKELLY WRIGHT, Chief Judge:
***

D. NOTICE OF RULEMAKING

The industry's most serious procedural attack on the lead standard goes to the sufficiency of the original notice of proposed rulemaking. The notice issue illustrates as well as any other that the rulemaking to set the lead standard was something less than a masterpiece of administrative procedure. Our task, however, is only to see whether the agency has complied with the law, and though the notice of rulemaking could well have been clearer and more specific, it meets the demands of that ubiquitous term of art in administrative law—"adequacy."

The OSH Act itself simply requires the Secretary to publish a proposed rule in the *Federal Register*, 29 U.S.C. §655(b)(2) (1976), but implicitly incorporates the general requirement for informal rulemaking in 5 U.S.C. §553(b)(3) (1976): notice of "the terms or substance of the proposed rule or a

[25] 5 U.S.C. §553(b) (1982).
[26] 29 U.S.C. §655(b)(2).
[27] U.S. DEP'T OF JUSTICE, ATTORNEY GENERAL'S MANUAL ON THE ADMINISTRATIVE PROCEDURE ACT 29–30 (1947).

description of the subjects and issues involved." The agency must "fairly apprise interested persons" of the nature of the rulemaking, *American Iron & Steel Institute v. EPA,* 568 F.2d 284, 293 (3d Cir. 1977), but a final rule may properly differ from a proposed rule—and indeed must so differ—when the record evidence warrants the change. "A contrary rule would lead to the absurdity that in rule-making under the APA the agency can learn from the comments on its proposals only at the peril of starting a new procedural round of commentary." *International Harvester Co. v. Ruckelshaus,* 478 F.2d 615, 632 n.51 (D.C. Cir. 1973). Where the change between proposed and final rule is important, the question for the court is whether the final rule is a "logical outgrowth" of the rulemaking proceeding. *South Terminal Corp. v. EPA,* 504 F.2d 646, 659 (1st Cir. 1974). The courts have described the notice requirement with other verbal formulas, but general principles only take us so far. We must proceed to compare carefully the specific language of the proposal with that of the final rule, in light of the evidence adduced at the hearings.

LIA and other industry parties stress [a serious difference] between the proposed and final lead standard: the proposal set a PEL of 100 ug/m^3 while the final standard's PEL is 50 ug/m^3. * * *

1. *Notice of the PEL.* The difference between a PEL of 100 ug/m^3 and one of 50 ug/m^3 is obviously substantial. Though OSHA finds the latter PEL feasible, it cannot deny that the change in the final rule greatly increases the number of employees affected by the standard, as well as the standard's economic and technological demands on industry. Nevertheless, the published explanation accompanying the proposed rule does in fact give notice that OSHA might set a PEL lower than 100 ug/m^3.

OSHA first listed the major issues raised by the rulemaking, including:

> 1. Whether the proposed permissible exposure limit to lead should be 100 ug/m^3: and whether this level incorporates an appropriate margin of safety;
>
> 2. Whether subclinical effects of exposure should be considered in establishing a standard for occupational exposure to * * * lead;
>
> ***
>
> 8. To what extent are there groups with increased susceptibility to lead in the working population, such as women of childbearing age; and should such increased susceptibility, if it exists, be considered in establishing a standard for occupational exposure to * * * lead[.]

40 FED. REG. 45934/1-2 (1975). The question of "appropriate margin of safety" raised the possibility that OSHA might find the 100 ug/m^3 PEL not safe enough. The issue of subclinical effects, in the context of the published proposal as a whole, also portended a lower PEL. There is scientific consensus that people rarely suffer actual clinical symptoms of lead intoxication at blood-lead levels below 80 micrograms of lead per 100 grams of whole blood (80 ug/100g), which would probably correlate with air-lead levels over 100 ug/m^3. Nevertheless, OSHA noted that there were studies showing such subclinical effects of lead as inhibition of the important enzyme ALA at a blood level of 40 ug/100g, *id.* at 45936/1, which might correlate with an air level as low as 50 ug/m. The proposal noted, in fact, that "it is not known with certainty at what level this enzyme inhibition becomes clinically important," *id.* at 45935/1, and, most important, clearly stated:

> In any event, the question of both clinical and subclinical effects should be fully discussed in comments submitted as well as at the hearing, if one is held, *and might necessitate a different permissible exposure limit in the final standard than that proposed.*

Id. (emphasis added).

Finally, notice of the issue of the special susceptibility of women of child-bearing age also should have alerted the parties that OSHA might lower the proposed PEL. The proposal noted there was evidence that fetuses and children, as well as people already suffering such conditions as anemia or renal insufficiency, needed their blood-lead levels kept as low as 30 ug/100g, and plainly stated that OSHA believed its statutory mandate required it to protect these groups. *Id.* at 45936/3. Indeed, the supplemental notice announcing the public hearing stated that comments received from both industry and health groups since the original proposal had revealed that the 100 ug/m^3 standard "is actually inadequate to protect a developing fetus." 42 FED. REG. 810/1 (1977).

Obviously, OSHA would have served the parties far better had it listed in the proposal two or more alternative PEL's and invited comments on each, even if it had stated a preference for one of them.[41] But the language of the proposal contains enough suggestions of the possibility of a lower PEL to meet the test of "adequate" notice. Moreover, although at this point we need not consider "substantial evidence," the rulemaking produced enough evidence on such issues as subclinical effects of lead to make the lower PEL a "logical outgrowth" of the proceeding.

***

MACKINNON, Circuit Judge (dissenting):

***

Here the agency "proposed [a] permissible exposure limit [of] 100 ug/m^3 * * *." It never expressly stated it was considering or might consider the 50 ug/m^3 standard that it eventually promulgated. Very little evidence was submitted by the industry petitioners *or the agency* on any level but the 100 ug/m^3, or higher. *OSHA concedes that no evidence whatsoever was introduced on the economic feasibility of complying with the 50 ug/m^3 level.* Because of this, and after noting the thousands of pages introduced by all the parties, supporting both a stringent or lenient rule, and finding no evidence on the 50 ug/m^3 level, it is a plain absurdity to conclude that the parties were sufficiently informed to permit their meaningful participation in discussing the possibility that the proposed permissible exposure limit of 100 ug/m^3 would be reduced 50 percent.

A review of the vague statements in the notice of proposed rulemaking also points out that the 50 ug/m^3 PEL is not a "logical outgrowth" of the proposed rulemaking at 100 ug/m^3. Whether a 100 ug/m^3 PEL would provide an appropriate margin of safety, considering the uncertainty of OSHA's scientific models and medical evidence, is a very open-ended question. Further, whether subclinical effects should be considered is also misleading since subclinical effects undoubtedly can appear at the 100 ug/m^3 level in addition to lower exposure levels. This does not lead to a logical outgrowth that the proposed PEL should be cut in half. Finally, the extra consideration given to women of childbearing age most logically leads to the conclusion that special precautions, in the form of respirators or alternative work assignments, might be taken for these susceptible individuals, *not* that a definite level at 50 ug/m^3 would be set.

However, one could accept the most liberal interpretations of these statements in the notice of proposed rulemaking *had they indeed produced any significant amount of evidence* at the 50 ug/m^3 level. But no such evi-

[41]OSHA might also have issued a supplemental notice of rulemaking on this change, as it in fact did to invite comments on medical removal, 42 FED. REG. 46547/1 (1977), and other issues, *id.* at 809. Since OSHA hired DBA to prepare a posthearing report on the feasibility of the lower PEL in November 1977, months before the hearing record was closed, see ALD 20–26, OSHA obviously anticipated the change in time to issue a new notice. We note, however, that the supplemental notice in which OSHA announced the public hearing in the rulemaking, 42 FED. REG. 810/1 (1977), did announce that comments received by that time had shown that the 100 ug/m^3 PEL would not protect fetuses.

dence evolved. That no such evidence was offered by anyone during the extended hearings, especially considering the intensity of the participants' adversarial positions, constitutes the best support for concluding that the 50 ug/m^3 level was *not* a "logical outgrowth" of the Notice of the Rulemaking proposal of a "100 ug/m^3 * * * level."

Finally, OSHA and the majority rely on a feeble and totally irresponsible rationale for upholding the adequacy of this notice. They assert and conclude that since the LIA and other industry petitioners took the position throughout the proceeding that the proposed 100 ug/m^3 level was infeasible, then, *a fortiori,* they would likewise have claimed the 50 ug/m^3 level was infeasible. Thus, why bother hearing their comments and evidence on the stricter level, for it would only be the same "in greater volume or [stated] more vociferously." *BASF Wyandotte v. Costle,* 598 F.2d at 644.

This reasoning presumes that the evidence of feasibility for 50 and 100 ug/m^3 is interchangeable, and there is no way the decisionmaker could reach a different conclusion on the two. This presumes too much. One level may require an assortment of conventional engineering controls, while the 50 ug/m^3 level may demand total rebuilding of the workplace or dislocation of the entire industry. The cost may also increase disproportionately. Moreover, without any evidence of the cost and difficulties of complying with the 50 ug/m^3, how would the decisionmaker have the vaguest idea of its feasibility? Taking the majority's reasoning to its logical, albeit extreme, position one would have to conclude that OSHA could have reduced the PEL to 10 ug/m^3. By the agency's contention, the fact that OSHA may arrive at a lower PEL is allegedly forewarned by the notice of proposed rulemaking, and any evidence that the protestants would have introduced would have been the same; infeasibility would still be argued only more "vociferously." It is obviously absurd to compare the probative effect of evidence of infeasibility for 100, 50, or 10 ug/m^3 and state that the only difference is in the vocal force of the advocacy.

Therefore, since reasonable notice was not given, nor evidence received as to the 50 ug/m^3 level, and the majority's defense of such slipshod practice is wholly illogical, I would remand the case to OSHA for a rulemaking based on proper notice to the interested parties and the public. Thereafter, substantial *public* evidence would be required to support the agency's conclusion.

***

The court also decided that, while OSHA could have given "clearer notice," it had done a "legally adequate job" in framing its proposal regarding respirators as a means of compliance;[28] Judge MacKinnon agreed.

In deciding that the ultimate 50-microgram level was the "logical outgrowth" of the proposed 100-microgram level, the majority relied on the fact that the proposal contained "enough suggestions" of the lower PEL to meet the test of adequate notice. At the same time, the majority indicated some dissatisfaction with OSHA's performance, saying that the Agency would have "served the parties better" by listing alternative PELs and inviting comments on each. A more substantial complaint about the adequacy of the notice on the PEL in

[28]647 F.2d at 1223–25.

the *Lead* proceeding was voiced by the Committee on the Judiciary of the U.S. Senate in reporting the Regulatory Reform Act in 1981.[29]

Regulatory Reform Act, S.1080
Senate Committee on the Judiciary

S. Rep. No. 284, 97th Cong., 1st Sess. 120–21 (1981)

***

The Committee has determined that a successful public comment process requires fair and complete notice to the public of the subject of the rulemaking and of the agency's approach to that subject. When the agency's approach changes greatly in the course of the rule making, the public may no longer be in a position to offer valuable comment. Accordingly, paragraph (c)(2) is intended to require agencies to keep interested members of the public fairly and contemporaneously apprised of the material issues being considered by the agency and of the potential substance of rules under consideration. Sometimes an initial notice of proposed rulemaking will identify all of the significant issues, identify all persons who may be covered by the rule, and indicate the potential substance of a final rule, including alternatives that are under consideration even though not reflected in the proposed language of the rule. This paragraph is intended to encourage agencies to do so by not requiring further notices where alternatives being considered were fully described in the initial notice of proposed rule making. In *Sierra Club v. Costle,* [657 F.2d 298, 353–54] (D.C. Cir. 1981), for example, the court found that EPA had announced in the preamble to the proposed rule that it was still considering a range of alternatives on various issues and expressly requested public comment on the alternatives.

In other cases, additional issues and alternative proposals will be identified in comments on the proposed rule. Where these comments have been made available in a public file and an opportunity for response thereto allowed, the Committee does not intend that this paragraph should require the agency to publish a separate notice of the issues raised. If the agency decides to make substantial changes in the proposed rule, however, such as by adopting a significantly different approach recommended in comments or subsequently developed by the agency, or by changing the scope of the rule to cover a significantly broader or different category of persons, a notice describing these changes and providing the necessary supporting information and analysis would be required. E.g., *American Iron and Steel Institute v. EPA,* 568 F.2d 284, 292 (3d Cir. 1977) (holding that the agency "failed to give adequate notice that [its rule] would govern the specialty steel industry). The notice required by this subsection need not be in the form of a reproposal, but should be designed to provide sufficient opportunity for comment by affected persons.

The objective of this paragraph is to preclude agencies from making drastic departures from proposals or from the foundation for rule making developed by the agency which could not have been reasonably anticipated by interested persons. E.g., *United Steelworkers of America v. Marshall,* 647 F.2d 1189 (D.C. Cir. 1980). Such revisions would require prior notice under this section and reasonable opportunity to comment. Where a change goes to

[29]The bill reported by the committee, and passed by the Senate, would have required a new notice of proposed rulemaking "whenever the provisions of a final rule that an agency plans to adopt are so different from the provisions of the proposed rule that the original notice of proposed rule making did not fairly apprise the public of the issues ultimately to be resolved in the rule making or the substance of the rule." See 128 CONG. REC. S2713 (daily ed. Mar. 24, 1982).

the heart of a rule and represents a substantial departure from the proposal commented on by the public commenters should be informed of that change and allowed to comment on it.

***

The Senate Judiciary Committee sought to preclude an administrative agency from making "drastic departures" from an initial proposal "which could not have been reasonably anticipated by interested persons." The committee apparently felt that, notwithstanding the hints in the lead proposal, the public could not "reasonably" anticipate a 50-microgram limit. OSHA has occasionally reopened the record for additional public comment when sufficient new data have come to its attention after the close of the record or when it is contemplating a major departure from the original proposal for other reasons.[30] Indeed, in the lead proceeding itself, OSHA issued a supplemental notice on the issue of medical removal protection because the issue had not been raised in the original lead proposal.[31] To be sure, medical removal protection was a *new* issue; the public would not have had any reason to believe that the MRP question would be decided in the final standard and therefore would have submitted no comment on the issue. Arguably, however, since the *issue* of permissible exposure limit had been raised in the proposal, the Agency was justified in changing the level in the final standard from 100 to 50 micrograms. But this reasoning would seem to justify virtually any change in the PEL. Even the court of appeals majority did not accept this argument, looking at the definition of "issue" far more narrowly. In the court's view, the issue was the 100-microgram (or 50-microgram) limit, and not the PEL generally. The court of appeals, therefore, was compelled to conclude that the 50-microgram limit was actually raised in the proposal. Judge MacKinnon, in dissent, rejected even this narrow rationale, refusing to rely on "vague statements" in the notice. The Senate Judiciary Committee did not focus so much on the definition of issue, but rather emphasized the practical consideration: is the Agency considering a "drastic departure" from the proposal?

In publishing proposals on permissible exposure limits, OSHA has followed, on at least one occasion, the suggestion to list alternatives. OSHA listed several possible PELs and invited comment on each in its proposed standard on acrylonitrile (AN).[32]

---

[30]For example, in connection with the inorganic arsenic standard, OSHA reopened the record at least twice after the proposed rule was published. Notice, Opportunity to Comment on New Information, 41 FED. REG. 48,746 (1976); Notice, Additional Issue, Sputum Cytology, to be Considered at Informal Public Hearing, 41 FED. REG. 29,425 (1976).

[31]42 FED. REG. 46,547 (1977). The D.C. Circuit Court of Appeals referred to this notice, 647 F.2d at 1222 n.41.

[32]The proposed permanent standard for acrylonitrile was issued simultaneously with an emergency temporary standard for the substance, 43 FED. REG. 2600 (1978). See also proposed asbestos standard, 49 FED. REG. 14,116 (1984) proposing alternative PELs of .2 fibers/cc and .5 fibers/cc.

Proposed Acrylonitrile Standard

43 Fed. Reg. 2608, 2610 (1978)

***

3. *Paragraph (c).—Permissible exposure limits (PEL).* The permissible exposure limits set forth in the proposal are more stringent than those in the ETS. The proposal provides three alternative sets of permissible exposure limits for consideration; 2 ppm as an eight (8)-hour time-weighted average (TWA), with a ceiling limit of 10 ppm measured over any 15-minute period during the workshift (2/10 proposal); 1 ppm TWA with a 5 ppm ceiling (1/5 proposal); and 0.2 ppm TWA with a 1 ppm ceiling (0.2/1 proposal). The 2/10 proposal is clearly more stringent than the ETS, which also has exposure limits of 2/10, because of the greater flexibility available under the ETS as to methods of compliance, as discussed below.

***

By including several sets of alternative permissible exposure limits in the proposal, OSHA acknowledges that there is much data and information yet to be gathered as to what constitutes the lowest feasible level of exposure to AN in the affected industries. It should be noted that although OSHA has expressly proposed three alternative sets of permissible exposure limits, the PEL in the final rule will be the lowest feasible levels, based upon the entire record of the proceeding, and may differ from the proposed levels.

***

OSHA welcomes any and all data available as to the appropriate permissible exposure limits for AN, the feasibility of the proposed exposure limits and any other exposure limits which should be considered by OSHA, as well as other related issues.

This approach, while adequately apprising the public of the possibility that the Agency may adopt one of several levels, may be deficient in another respect. By proposing a range of widely divergent levels, the Agency, arguably, was making it difficult for members of the public to determine to which level they should direct their comments, and thus adversely affected interested persons' opportunity to participate intelligently in the proceeding. In responding to a health standard proposal, employers may wish to conduct a survey of the feasibility of a particular level; if OSHA were to propose a level for lead somewhere between 25 and 250 micrograms, the employers would be hard-pressed in commissioning a study and preparing for the hearing.[33] As Judge MacKinnon said in his dissent, the evidence on the feasibility of different levels of exposure is not "interchangeable."[34] Accordingly, the Agency, in order to provide fair notice, must steer a difficult course between an impermissibly narrow notice, which does not alert the public to all the alternatives they must

[33]The acrylonitrile proposal suggested levels from 0.2–2.0 ppm.
[34]647 F.2d at 1316.

address, and an unduly broad notice, which does not adequately focus on the issues being actively considered by the Agency.[35]

The question of the adequacy of OSHA's notice of proposed rulemaking in a standards proceeding was raised and decided in several additional cases.

- *Synthetic Organic Chemical Manufacturers Association v. Brennan,*[36] in which the court set aside the rules governing laboratories under OSHA's ethyleneimine standard because of inadequacy of notice.
- *American Iron & Steel Institute v. OSHA,*[37] where the court held that OSHA had failed to properly advise the public that the coke oven standard would also apply to construction employers. Because of the unique features of the exposure of construction employees and of the compliance obligations of their employers, the court concluded that on such a "matter of vital importance," more specific notice was required.
- *AFL-CIO v. Marshall,*[38] in which the court upheld OSHA's pulmonary function tables, which were part of the medical surveillance provision in the cotton dust standard. The court rejected the employers' notice argument on the ground that the "general concept" of these tables was "thoroughly discussed" during the rulemaking and the tables finally adopted were "a logical outgrowth" of the testimony at the hearing.
- *Taylor Diving & Salvage Co. v. Department of Labor,*[39] where the court held that the provision in the diving standard on employee access to records was the "logical outgrowth" of the proposal.
- *Daniel International Corp. v. OSHRC,*[40] where the court upheld an amendment to OSHA's temporary flooring standard against an inadequacy-of-notice challenge.
- *Louisiana Chemical Association v. Bingham,*[41] where the court rejected the argument that the definition of "toxic substance and harmful physical agent" in the final records access rule was an im-

[35]See Rochvarg, *Adequacy of Notice of Rulemaking Under the Federal Administrative Procedure Act—When Should a Second Round of Notice and Comment be Provided,* 31 AM. U.L. REV. 1 (1981). Professor Rochvarg offers extensive comment on the court of appeals' *Lead* case in discussing two separate tests for determining the adequacy of notice: the "logical outgrowth" approach and the "harmless error" approach. One of Professor Rochvarg's suggestions is that in its notice of proposed rulemaking the Agency include statements on the "general area to be regulated," specific proposed rules, and either "alternative" rules or a statement that "additional rules consistent with the general purpose of the rulemaking proceeding and complementary to the specific proposed rules are under consideration." *Id.* at 16–17. However, at least in the OSHA context, a notice written along the lines proposed by Professor Rochvarg would presumably allow OSHA to issue a PEL far different from the specific PEL proposed, on the grounds that it was "consistent" with the rulemaking and "complementary" to the specific proposal. And yet, as a practical matter, this type of notice would impose a heavy burden on the public in having to comment on a broad range of PELs.

[36]503 F.2d 1115 (3d Cir. 1974).

[37]577 F.2d 825 (3d Cir. 1978), *cert. dismissed,* 448 U.S. 917 (1980).

[38]617 F.2d 636, 675 (D.C. Cir. 1979).

[39]599 F.2d 622, 626 (5th Cir. 1979).

[40]656 F.2d 925 (4th Cir. 1981). This case also involved the issue of whether procedural objections to the manner in which a standard was promulgated may be raised in a proceeding to enforce the standard. See discussion in Chapter 15, Section C.

[41]550 F. Supp. 1136 (W.D. La. 1982).

permissible "radical expansion" from the meaning of the term as used in the proposal.[42]

A separate issue is whether, under the Administrative Procedure Act (APA),[43] the Agency action requires any notice and comment. Under the APA, no notice and comment is required for "interpretative rules, general statements of policy, or rules of agency organization, procedure or practice," or where the agency finds "good cause" that public proceedings are "impracticable, unnecessary, or contrary to the public interest."[44] Thus, for example, in *Chamber of Commerce v. OSHA,*[45] the question was whether OSHA's walkaround pay "rule" was an interpretative rule, exempt from notice and comment requirements. The applicability of the APA's notice and comment requirements[46] has also arisen when OSHA has sought to suspend the effective date of a standard that had already been issued.[47]

In 1981, based on the provisions of Executive Order No. 12,291,[48] OSHA twice postponed the effective date of the hearing conservation amendment, without affording notice and comment. The amendment, issued in January 1981 and to be effective 90 days later,[49] had previously been challenged on substantive grounds by various employers under §6(f).[50] OSHA's postponements of the effective date of the amendment[51] were challenged by the AFL-CIO primarily on the grounds that no opportunity for notice and comment had been given.[52] OSHA argued in response that the postponement "was essentially an administrative stay of the amendment in order to obviate judicial review" and thus was aptly analogized to a stay under the APA, which permits an agency "to postpone the effective date of any action taken by it" pending judicial review where it "finds that jus-

[42]*Id.* at 1147–48. See also *Chlorine Inst. v. OSHA,* 613 F.2d 120 (5th Cir.), *cert. denied,* 449 U.S. 826 (1980). There, OSHA, in 1978, modified its chlorine standard, 29 C.F.R. §1910.1000, to provide a ceiling limit rather than an eight-hour time-weighted average. (The ceiling limit is considerably more stringent from employers' point of view; see Chapter 3, Section C.) OSHA argued that the original time-weighted average promulgated in 1971 was a clerical error and that its correction did not, as the petitioners claimed, require notice and comment. The court, after noting that the "case abounds with examples of bureaucratic ineptness," concluded that it was "difficult * * * but not impossible to believe OSHA took seven years to uncover an error such as this," and upheld OSHA's modification.

[43]5 U.S.C. §§551–59 (1982).

[44]*Id.* at §553(b).

[45]636 F.2d 464 (D.C. Cir. 1980) (holding that the rule was not "interpretative" and, therefore, required notice and comment), discussed in Chapter 16, Section B.

[46]It has been generally assumed that if the Agency action does not require notice and comment under the APA, it is likewise exempt from the §6(b) requirements for notice and comment.

[47]Cf., e.g., *Council of the S. Mountains, Inc. v. Donovan,* 653 F.2d 573 (D.C. Cir. 1981) (affirming on narrow grounds Mine Safety and Health Administration's action in postponing, without notice and comment, effective date of standard requiring coal operators to equip all underground miners with self-contained self-rescuers).

[48]Sec. 7, 46 FED. REG. 13,193 (1981). The order, among other things, required that, in some circumstances, agencies postpone the effective date of regulations which had not yet become effective to permit review of the regulations by the new Administration.

[49]46 FED. REG. 4078 (1981).

[50]29 U.S.C. §655(f). As of early 1984, some of these proceedings were still pending in the Fourth Circuit Court of Appeals. See 12 OSHR 1080 (1983).

[51]46 FED. REG. 21,365, 28,845, 39,137 (1981).

[52]*AFL-CIO v. Donovan,* No. 81-1719 (D.C. Cir. filed July 2, 1981).

tice so requires."[53] The proceedings were later transferred to the Court of Appeals for the Fourth Circuit on the motion of OSHA since the §6(f) proceedings on the noise amendment were already pending in that circuit. The suit was later withdrawn by the AFL-CIO after OSHA solicited comment on the changes to the hearing conservation amendment and permitted portions to go into effect.[54] A revised hearing conservation amendment was published in March 1983.[55]

A similar challenge was filed by the United Steelworkers of America, joined by the United Automobile Workers, when OSHA "administratively stayed" a provision in the lead standard requiring employers in the primary and secondary lead smelting industries to complete the mandated compliance plans.[56] The Agency had twice stayed this provision without notice and comment.[57] The third stay, issued on December 3, 1982, followed notice and comment proceedings. The Agency explained that since it was in the process of reconsidering the lead standard, the stay was required because it would prevent "wasteful expenditures without adversely affecting worker health."[58] The unions argued that the first two stays, issued without notice and comment, were procedurally defective and that the comment on the third stay was insufficient to cure the defect.[59] They further argued that the stays constitute a modification of the lead standard and cannot "survive the close scrutiny" that OSHA modification must be subjected to.[60] The case is still pending before the Court of Appeals for the District of Columbia Circuit. In January 1984, OSHA advised the court that the Agency had decided not to propose major revisions of the lead standard; that a tripartite effort was being made among industry, unions, and the Agency to resolve issues surrounding compliance with the engineering control requirements, and accordingly the Agency would propose to vacate the stay as soon as clearance was obtained.[61]

D. *Ex Parte* Contacts: The Public

The propriety of *ex parte* communications—that is, off-the-record communications between interested members of the public and Agency officials during informal rulemaking proceedings—has been

[53] 5 U.S.C. §705 (1982).
[54] 46 FED. REG. 42,622 (1981).
[55] 48 FED. REG. 9738 (1983).
[56] *United Steelworkers v. Auchter,* No. 83-1022 (D.C. Cir. filed Sept. 20, 1982).
[57] 47 FED. REG. 26,557, 40,410 (1982).
[58] 47 FED. REG. 54,433 (1982).
[59] Brief of Petitioners at 13 n.15, *United Steelworkers v. Auchter,* No. 83-1022 (D.C. Cir. filed Sept. 20, 1982).
[60] *Id.* at 8. For a full discussion of the issue of the legality of postponement of effective dates without notice and comment, see *National Resources Defense Council v. EPA,* 683 F.2d 752 (3d Cir. 1982) (holding that Exec. Order No. 12,291 requirements did not constitute "good cause" for EPA's postponement of the effective date of certain environmental regulations, and ordering the EPA to reinstate the amendments retroactively).
[61] Supplemental Brief of Department of Labor at 1–5, *United Steelworkers v. Auchter,* No. 83-1022 (D.C. Cir. filed Sept. 20, 1982).

considered by the District of Columbia Circuit Court of Appeals in a number of major decisions. The first of these was *Home Box Office, Inc. v. FCC.*[62] The substance of this decision was summarized by Professor Nathanson as follows:

> The gist of that opinion is that once a notice of proposed rulemaking is issued pursuant to the Administrative Procedure Act (5 U.S.C. §553) there should be no oral communications between interested persons and officials of the agency dealing with the merits of the proposed rules outside the confines of public meetings or hearings conducted in accordance with public notices, and no written communications dealing with the merits of the proposed rules that are not placed in the public file or record of the rulemaking proceedings in accordance with established procedures designed to make them available for examination and rebuttal by interested persons. The rationale of the opinion is that the existence of such ex parte communications in rulemaking proceedings is inconsistent with adequate judicial review on "the full administrative record" in accordance with *Citizens to Preserve Overton Park, Inc. v. Volpe,* 401 U.S. 402 (1971), and also inconsistent with "fundamental notions of fairness implicit in due process", citing *Sangamon Valley Television Corp. v. United States,* 269 F.2d 221 (D.C. Cir. 1959).[63]

In a subsequent decision, the broad ruling of the *Home Box Office* court was limited to rulemakings involving the grant of a valuable privilege—such as a license.[64] In the recent case *Sierra Club v. Costle,*[65] the court stated the view that the *Home Box Office* rule was of "questionable utility" and should not be applied to informal rulemaking "of the general policymaking sort" involved under the Clean Air Act. The policy rationale for its view restricting the applicability of the *Home Box Office* rule was stated by the court in the *Sierra Club* decision, as follows:

> Under our system of government, the very legitimacy of general policymaking performed by unelected administrators depends in no small part upon the openness, accessibility, and amenability of these officials to the needs and ideas of the public from whom their ultimate authority derives, and upon whom their commands must fall. As judges we are insulated from these pressures because of the nature of the judicial process in which we participate; but we must refrain from the easy temptation to look askance at all face-to-face lobbying efforts, regardless of the forum in which they occur, merely because we see them as inappropriate in the judicial context. Furthermore, the importance to effective regulation of continuing contact with a regulated industry, other affected groups, and the public cannot be underestimated. Informal contacts may enable the agency to win needed support for its program, reduce future enforcement requirements by helping those regulated to anticipate and shape their plans for the future, and spur the provision of information which the agency needs. * * *[66]

[62]567 F.2d 9 (D.C. Cir.), *cert. denied,* 434 U.S. 829 (1977).
[63]Nathanson, *Report to the Select Committee on* Ex Parte *Communications in Informal Rulemaking Proceedings,* 30 AD. L. REV. 377 (1978).
[64]*Action for Children's Television v. FCC,* 564 F.2d 458 (D.C. Cir. 1977).
[65]657 F.2d 298 (D.C. Cir. 1981).
[66]657 F.2d at 400–01.

At the same time, the court in *Sierra Club* strongly suggested that certain documents received outside of the record—those documents of "central relevance to the rulemaking" as well as those which produce "significant new information"—should be placed in the docket as soon as possible after they became available.[67]

In the *Lead* litigation, LIA relied on *Home Box Office* and related cases in arguing that OSHA staff attorneys acted as advocates in the rulemaking proceeding, and therefore, the Assistant Secretary, by consulting with these advocates at the decisional stage, engaged in unlawful *ex parte* contact with one of the adverse sides in the rulemaking. The court, assuming that the OSHA attorneys were analogous to advocates for a particular position, rejected this contention, holding that there was no legal bar to a staff advocate advising a decision maker in the setting of policy rules as part of the "deliberative" process. The court held further that the *Home Box Office* line of cases applied only to contacts between *outside parties* and the Agency, and that neither the holding nor the reasoning of these cases extends to off-the-record contacts with agency staff.[68] In a similar vein, the court held that OSHA consultants in the lead proceeding were the "functional equivalent of agency staff," and therefore their written communications with the Agency decision makers that were not part of the record were not unlawful *ex parte* contacts.[69]

E. *Ex Parte* Contacts: The Executive Department

A related question is whether the substance of discussions between agency staff and officials in the White House and the Office of Management and Budget (OMB) regarding issues in a pending informal rulemaking must be disclosed as part of the rulemaking record. The question has become far more significant under Executive Order No. 12,291, which authorizes broad review by the Office of Manage-

[67]*Id.* at 397. Accord Administrative Conference Recommendation No. 77-3, Ex Parte Communications in Informal Rulemaking Proceedings, 1 C.F.R. §305.77-3 (1983). See also Pederson, *Formal Records and Informal Rulemaking,* 85 YALE L.J. 38 (1975) (an influential article on the development of the "record" in informal proceedings); Note, *Due Process and* Ex Parte *Contacts in Informal Rulemaking,* 89 YALE L.J. 194 (1979).

[68]647 F.2d at 1213. See also *Hercules, Inc. v. EPA*, 598 F.2d 91, 119–28 (D.C. Cir. 1978); M. Asimow, *When the Curtain Falls: Separation of Functions in Federal Administrative Agencies,* 81 COLUM. L. REV. 759 (1981).

[69]The court held that the consultants based their reports on record evidence and did not introduce new factual material into the deliberative process; if that had been the case, a new opportunity for comment would have been required. See also *Lead Industries Ass'n v. OSHA,* 610 F.2d 70 (2d Cir. 1979), where the court held that the consultants' reports in the lead rulemaking were not subject to disclosure under the Freedom of Information Act. The court in the D.C. Circuit *Lead* case relied on Judge Friendly's opinion in the Freedom of Information case, 647 F.2d at 1220. Department of Labor regulations on disclosure of information under the Freedom of Information Act appear at 29 C.F.R. §§70.1–.77 (1983). More specific OSHA instructions are contained in the Field Operations Manual, ch. XIV, OSHR [Reference File 77:4701–:4702]. See also *Chemical Mfrs. Ass'n v. OSHA,* 9 OSHC 1105 (D.D.C. 1980) (rejecting employer request for disclosure of consultant reports on hazard communication standard, relying on exemption 5 under FOIA covering interagency memoranda).

ment and Budget of agency rulemaking.[70] In *Sierra Club v. Costle,* the court refused to impose this requirement.

Sierra Club v. Costle

657 F.2d 298 (D.C. Cir. 1981)

WALD, Circuit Judge:

***

The court recognizes the basic need of the President and his White House staff to monitor the consistency of executive agency regulations with Administration policy. He and his White House advisers surely must be briefed fully and frequently about rules in the making, and their contributions to policymaking considered. The executive power under our Constitution, after all, is not shared—it rests exclusively with the President. The idea of a "plural executive," or a President with a council of state, was considered and rejected by the Constitutional Convention. Instead the Founders chose to risk the potential for tyranny inherent in placing power in one person, in order to gain the advantages of accountability fixed on a single source. To ensure the President's control and supervision over the Executive Branch, the Constitution—and its judicial gloss—vests him with the powers of appointment and removal, the power to demand written opinions from executive officers, and the right to invoke executive privilege to protect consultative privacy. In the particular case of EPA, Presidential authority is clear since it has never been considered an "independent agency," but always part of the Executive Branch.

The authority of the President to control and supervise executive policymaking is derived from the Constitution; the desirability of such control is demonstrable from the practical realities of administrative rulemaking. Regulations such as those involved here demand a careful weighing of cost, environmental, and energy considerations. They also have broad implications for national economic policy. Our form of government simply could not function effectively or rationally if key executive policymakers were isolated from each other and from the Chief Executive. Single mission agencies do not always have the answers to complex regulatory problems. An overworked administrator exposed on a 24-hour basis to a dedicated but zealous staff needs to know the arguments and ideas of policymakers in other agencies as well as in the White House.

***

The purposes of full-record review which underlie the need for disclosing ex parte conversations in some settings do not require that courts know the details of every White House contact, including a Presidential one, in this informal rulemaking setting. After all, any rule issued here with or without White House assistance must have the requisite *factual support* in the rulemaking record, and under this particular statute the Administrator may not base the rule in whole or in part on any *"information or data"* which is not in the record, no matter what the source. The courts will monitor all this, but they need not be omniscient to perform their role effectively. Of course, it is always possible that undisclosed Presidential prodding may direct an outcome that *is* factually based on the record, but different from the outcome that would have obtained in the absence of Presidential in-

[70]See Comment, *Capitalizing on a Congressional Void: Executive Order No. 12,291,* 31 AM. U.L. REV. 613, 631–37 (1982). For a discussion of the implementation of Exec. Order No. 12,291 during its first year, see Scalia, *Regulation—The First Year, Regulatory Review and Management,* REGULATION, Jan.–Feb. 1982 at 19–21.

volvement. In such a case, it would be true that the political process did affect the outcome in a way the courts could not police. But we do not believe that Congress intended that the courts convert informal rulemaking into a rarified technocratic process, unaffected by political considerations or the presence of Presidential power. In sum, we find that the existence of intra-Executive Branch meetings during the post-comment period, and the failure to docket one such meeting involving the President, violated neither the procedures mandated by the Clean Air Act nor due process.

***

The *ex parte* issue as applied to discussions between the Agency and White House officials has not been litigated directly in an OSHA case. After the close of the record in the cotton dust rulemaking, but before the issuance of the standard, there were well-publicized discussions between the President, Secretary of Labor Marshall, Assistant Secretary Bingham, and their staffs, as a result of which changes were made in the standard;[71] the issue was not raised in court, however. And, under Executive Order 12,291, there have been extensive exchanges between OSHA and officials in OMB and the White House on issues relating to the development of standards. A subcommittee of the House Committee on Government Operations held hearings in 1982 on the question of OMB involvement in OSHA standards, and in December 1983, the full committee issued a report, entitled, "OMB Interference with OSHA Rulemaking."[72] Focusing particularly on OSHA's hazard communication and diving rulemakings, a majority of the Committee concluded that Executive Order 12,291 had been "used as a back door, unpublicized, channel of access to the highest levels of political authority in the Administration for industry alone." The majority also expressed "concern" about the "great deference" that OSHA officials have paid to OMB and the White House on rulemaking issues. The majority of the Committee recommended, among other things, that the agency with responsibility over the program should be overruled by OMB only where "there is overwhelming evidence to do so," and that the evidence should be part of the public record.[73]

In dealing with this issue, the Administrative Conference of the United States recommended that there be no requirement that the contents of conversations between presidential advisers and agency officials be placed in the rulemaking file except where the communications contain "material factual information (as distinct from indications of governmental policy) pertaining to or affecting the proposed rule" or the communication originated from "persons outside

[71]See Verkuil, *Jawboning Administrative Agencies:* Ex Parte *Contacts by the White House,* 80 COLUM L. REV. 943, 945 (1980). See also N.Y. Times, May 24, 1978, at D1, col. 6; Washington Post, June 8, 1978, at A9, Col. 1; and discussion in Chapter 10.

[72]*Office of Management and Budget Control of OSHA Rulemaking: Hearings Before a Subcomm. of the Home Comm. on Government Operations,* 97th Cong., 2d Sess. (1982).

[73]H.R. REP. No. 98, 98th Cong., 1st Sess. 11, 13 (1983). Twelve members of the Committee dissented from the majority report in two separate statements of views, *id.* at 15 and 24, and Cong. Frank submitted a separate statement of concurring views.

the government."[74] Senate committees reporting regulatory reform legislation in 1981 rejected proposals that would have required that all oral and written communications between the executive office and the President and the agencies with respect to a rule be recorded and placed in the rulemaking file.[75] On the other hand, the American Bar Association, Commission on Law and the Economy, recommended that memoranda exchanged between an agency and the President and his staff be made part of the public record; that the fact that a discussion was held also be placed in the record, but that the substance of the discussion remain private—in order to encourage a "full and frank exchange of opinions and advice."[76]

F. Statement of Reasons

Under §6(e) of the Act, when promulgating a standard, OSHA must "include a statement of the reasons for such action."[77] The OSHA regulations go further and mandate that the statement accompanying the issuance of a standard "show the significant issues which have been faced and * * * articulate the rationale for their solution."[78] The importance of a full explanation by OSHA of the reasons for its determinations in the standards rulemaking has been emphasized by the courts in virtually all decisions reviewing OSHA standards. Indeed, as Judge McGowan emphasized in *Industrial Union Department v. Hodgson,*[79] which upheld for the most part OSHA's asbestos standard, the need for a statement of reasons is even more pressing when the Agency is "obliged to make policy judgments where no factual certainties exist or where facts alone do not provide the answer." In these circumstances, Judge McGowan said, the Agency "should so state and go on to identify the considerations [it] found persuasive."[80]

Two early OSHA standards, the lavatory standard and the emergency temporary standard on 14 carcinogens, both issued in 1973, were set aside by the courts because of the inadequacy of OSHA's statement of reasons. *Associated Industries v. Department of Labor* involved the lavatory standard;[81] OSHA had amended its §6(a) sanitation standard by significantly reducing the number of lavatories required for office employees, but retaining the original requirements for industrial establishments, despite employer objections that the latter were too stringent. The court, in an opinion by Judge

[74]Administrative Conference of the United States, Recommendation No. 80-6, Intergovernmental Communications in Informal Rulemaking, 1 C.F.R. §305.80-6 (1983).
[75]See SENATE COMM. ON GOVERNMENTAL AFFAIRS, REGULATORY REFORM ACT, S. REP. NO. 305, 97th Cong., 1st Sess. 67–72 (1981). See also generally Verkuil, *supra* note 71.
[76]ABA COMM'N ON LAW AND THE ECONOMY, FEDERAL REGULATION: ROADS TO REFORM 81 (1979).
[77]29 U.S.C. §655(e).
[78]29 C.F.R. §1911.18(b) (1983).
[79]499 F.2d 467 (D.C. Cir. 1974).
[80]*Id.* at 476.
[81]487 F.2d 342 (2d Cir. 1973).

Friendly, vacated the industrial portion of the lavatory standard, first, because of lack of substantial evidence, and second, because of inadequacy of the statement of reasons. He said that "[i]n a case where the proposed standard under OSHA has been opposed on grounds as substantial as those presented here, the Department has the burden of offering *some* reasoned explanation."[82] And in a major case, the Court of Appeals for the Third Circuit vacated OSHA's emergency temporary standard applying to 3,3′-dichlorobenzedine (DCB) and ethyleneimine (EI) because of deficiencies in the statement of reasons.

Dry Color Manufacturers' Association v. Department of Labor

486 F.2d 98 (3d Cir. 1973)

VAN DUSEN, Circuit Judge:

***

The Emergency Temporary Standard in question was originally published by the Occupational Safety and Health Administration ("OSHA") on May 3, 1973, after about one year of consulting with the National Institute for Occupational Safety and Health ("NIOSH") and receiving data and commentary from other interested groups. The Emergency Temporary Standard was based on the following findings, which are stated in the preamble of the standard and constitute the entire statement of reasons for its issuance:

> On the basis of all the relevant information before us, it is hereby found that: (1) the 14 carcinogens listed in the emergency temporary standard are toxic and physically harmful; (2) that exposure to any of the 14 substances poses a grave danger to employees; (3) that employees are presently being exposed to the substances; and (4) that the emergency temporary standard set forth below is necessary to protect the employees from such exposure.

38 FED. REG. 10929.

The standard goes on to prescribe plant operating procedures and equipment, work practices and procedures for preventing exposure of employees to the 14 chemicals found to cause cancer, together with a timetable of effective dates. * * *

***

II.

Subsection 6(e) of the Act specifically mandates that "[w]henever the Secretary promulgates any standard, * * * he shall include a statement of the reasons for such action, which shall be published in the *Federal Register*." 29 U.S.C. §655(e)

This requirement is designed to serve several general functions: it provides an internal check on arbitrary agency action by insuring that prior to taking action an agency can clearly articulate the reasons for its decision; it makes possible informed public criticism of a decision by making known its underlying rationale; and it facilitates judicial review of agency action by providing an important part of the record of the decision. Furthermore, the statement-of-reasons requirement of subsection 6(e) complements the re-

---

[82] *Id.* at 354.

quirement of subsection 6(c)(1) that certain factual findings be made by the Secretary of Labor prior to promulgation of an emergency temporary standard. Congress contemplated that emergency temporary standards would be developed and issued without the benefit of ordinary standard-setting procedures involving public comment and open hearings in the interest of permitting rapid action to meet emergencies. But, by the language of subsection 6(e), it clearly refused to eliminate the general requirement of articulation of the reasons for such action as an essential safeguard to emergency temporary standard-setting.

The question before us is whether the statement of reasons included in the preamble to the Emergency Temporary Standard in question is adequate to satisfy subsection 6(e). There are no cases applying this provision of the Act to guide our decision. However, it is an increasingly accepted principle of administrative law, even when there is no statutory requirement for a statement of reasons, that "the basis for * * * decision should appear clearly on the record, not in conclusory terms but in sufficient detail to permit prompt and effectual review." Furthermore, the informal or expeditious nature of the administrative procedure leading to an action or decision should not induce courts to relent in their demand for an adequate statement of reasons. Thus, in *Kennecott Copper Corp. v. Environmental Protection Agency,* 462 F.2d 846 (D.C. Cir. 1972), the court remanded the record to the Director of the Environmental Protection Agency "to supply an implementing statement that will enlighten the court as to the basis on which he reached the * * * [particular] standard [he issued]." *Id.* at 850.

The only statement of reasons offered by OSHA and contained in the preamble to the Emergency Temporary Standard consists of the finding that the 14 chemicals listed in the standard are carcinogens and the conclusion, reciting the language of subsection 6(c)(1) of the Act, that the conditions necessary for the issuance of an emergency temporary standard have been met. We find this statement of reasons inadequate. In the context of a voluminous factual record, such a conclusory statement of reasons places too great a burden on interested persons to determine and challenge the basis for the standard, and makes possible in any subsequent judicial review the use of *post hoc* rationalizations that do not necessarily reflect the reasoning of the agency at the time the standard was issued.

In particular, we find the statement of reasons here insufficient in two respects. First, it fails to set forth the basis for its finding that the 14 chemicals listed in the standard are carcinogens. To satisfy subsection 6(e), the statement of reasons should indicate which data in the record is being principally relied on and why that data suffices to show that the substances covered by the standard are harmful and pose a grave danger of exposure to employees. This could have been accomplished here by a brief statement in the May 3, 1973, notice that certain scientific data (citing the record documents) showed the DCB and EI produced cancer in rodents and supported the conclusion that they were therefore carcinogenic in man. Second, the statement of reasons failed to offer any explanation as to why this particular standard is "necessary to protect the employees from such exposure." We do not mean to say that every procedure must be justified as to every substance, type of use or production technique. But we do read subsection 6(e) as requiring at least a general explanation as to why the procedures prescribed were chosen in light of the recommendations of scientific experts and other governmental bodies, the types of industrial practices with these chemicals, and the alternative kinds of regulations considered by OSHA.

Because of these deficiencies in the statement of reasons, the Emergency Temporary Standard will be vacated and remanded to OSHA as to DCB and EI. This court has been asked to review the adequacy of the statement of reasons only with respect to DCB and EI, and hence no relief is requested in the proceedings now before the court on these petitions with respect to the other 12 chemicals listed in the standard. However, if OSHA decides to issue another emergency temporary standard on carcinogenic chemicals dealing with DCB and EI, rather than await issuance of the permanent standard to be made on or before November 3, 1973, it should on remand issue a statement of reasons that satisfies the requirement of subsection 6(e) as to these chemicals. However, we do not contemplate that this decision will interfere with the current proceedings before OSHA to promulgate a permanent standard for carcinogens.

***

In this case, the court stated three principal reasons for the need for an adequate statement of reasons by the Agency for its final action:

(1) acts as an internal check on arbitrary Agency action; it ensures that before taking action the Agency can articulate to itself the reasons for decision;

(2) makes possible informed public criticism of a decision by making known the Agency's underlying rationales;

(3) facilitates judicial review of Agency action.

The policy reasons for court insistence that the Agency explicate the basis for its decisions was also articulated by Judge Bazelon in the *Cotton Dust* case:

> Explicit explanation for the basis of the agency's decision not only facilitates proper judicial review but also provides the opportunity for effective peer review, legislative oversight, and public education. The requirement is in the best interest of everyone, including the decision-makers themselves. If the decision-making process is open and candid, it will inspire more confidence in those who are affected * * *.[83]

In the *Dry Color* case, the court of appeals also made it clear that the "need for expedition" in the case of ETSs was no "excuse" for the lack of an adequate statement of reasons.[84] Indeed, the court concluded that in issuing an ETS, OSHA is required by §6(c)(1) to make "certain factual findings";[85] this parallels the requirement for a "statement of reasons" under §6(e) for standards issued after rulemaking.[86]

The preamble to OSHA's ETS on pesticides[87] also contained a conclusory statement of reasons, similar to the one published for DCB and EI. The Court of Appeals for the Fifth Circuit, however, vacated the standard on substantive grounds;[88] it cited the *Dry Color*

---

[83]*AFL-CIO v. Marshall,* 617 F.2d 636, 651–52 (D.C. Cir. 1979), reprinted in Chapter 6, Section A.

[84]486 F.2d at 106 n.12.

[85]*Id.* at 106.

[86]*Id.* Cf. *AFL-CIO v. Brennan,* 530 F.2d 109, 116 (3d Cir. 1975) (holding that §6(b)(8), 29 U.S.C. §655(b)(8), requires that promulgation of a permanent standard that modifies an existing national consensus standard must be accompanied by a more particularized statement of reasons as to why the modified standard will better effectuate the purposes of OSHA).

[87]Discussed in Chapter 4.

[88]*Florida Peach Growers Ass'n v. Department of Labor,* 489 F.2d 120, 129–32 (5th Cir. 1974).

case[89] but did not rely directly on the inadequacy of the statement of OSHA's reasons in reaching its decision.

Largely in response to the court decisions in *Associated Industries* and *Dry Color,* OSHA began to prepare far more extensive preambles to its proposed and final standards.[90] While no recent OHSA standard has been set aside in its entirety because of the inadequacy of the statement of reasons, the Supreme Court, in reviewing the OSHA cotton dust standard, held that OSHA had not presented a legally sufficient explanation for the income-protection provisions in the standard. The Supreme Court refused to accept the arguments in OSHA's brief to the Court as a substitute for the Agency's own statement of the basis for its action.[91] In addition, in the *Cotton Dust* decision, the court of appeals' determination that substantial evidence did not support the Agency's finding that the standard was feasible for the cotton-seed oil industry was tied to a finding of an inadequate statement of reasons. The court said that "the record does not sufficiently establish [the standard's] economic feasibility" and also that "the agency's position is too unclear to permit us to complete our reviewing function."[92] Similarly, in its *Lead* decision, the Court of Appeals for the District of Columbia Circuit remanded the record to the Agency on the issue of the technological feasibility of the standard for nonferrous foundries (among other industries) saying:

> Given its reasonable effort at specificity in presenting and analyzing record evidence in the primary and secondary smelters, we were willing to show considerable deference to OSHA's conclusions for those industries. On the bare, unsupported explanation offered here, we cannot do the same for the nonferrous foundries.[93]

In sum, a court determination that no substantial evidence supports a standard may be made even though a full statement of reasons exists; on the other hand, a finding of an inadequate statement of reasons may well mean not only that the Agency failed to state the basis for its decision, but also that no basis in the record exists.

G. Use of Advisory Committees

Under §§6 and 7 of the Act,[94] OSHA may, in its discretion, establish an advisory committee to make recommendations on the development of a standard. According to the statute, the advisory committee first makes its recommendations to OSHA, then OSHA publishes

[89]*Id.* at 124.

[90]See, e.g., Vinyl Chloride Standard, 39 FED. REG. 35,890 (1974), reprinted in part in Chapter 3, Section B.

[91]*American Textile Mfrs. Inst. v. Donovan,* 452 U.S. 490, 539 (1981), discussed in Chapter 10.

[92]*AFL-CIO v. Marshall,* 617 F.2d 636, 669, 672–73 (D.C. Cir. 1979). See also *id.* at 670. ("The agency's position on economic feasibility * * * is neither clear nor adequately supported by the record.")

[93]*United Steelworkers v. Marshall,* 647 F.2d 1189, 1294 (D.C. Cir. 1980).

[94]29 U.S.C. §§655, 656.

a proposal. In *Synthetic Organic Chemical Manufacturers Association v. OSHA,*[95] the court set aside OSHA's final standard on the carcinogen MOCA [4,4′ methylene bis (2-chloroaniline)] because, in its view, the proper sequence was not followed since the proposed standard was published before the advisory committee recommendations were received by OSHA. OSHA argued that special circumstances were involved because it had earlier issued an ETS for MOCA; it was therefore required by §6(c) to publish the proposal immediately without awaiting the advisory committee's recommendations. The court rejected the argument, insisting on strict adherence to the statutory procedure.

On the basis of the legislative history of the OSH Act, OSHA's 1971 regulations on the issuance of standards provides that OSHA "shall" consult with the Construction Safety and Health Advisory Committee "in the formulation of" a rule to promulgate, modify, or revoke a construction standard.[96] The Construction Safety and Health Advisory Committee had been originally established under the Construction Safety Act,[97] a predecessor to the OSH Act. In *National Constructors Association v. Marshall,*[98] the court remanded a construction standard on "ground-fault circuit interrupters" because OSHA had failed to consult with the construction advisory committee a second time prior to issuing a standard that was significantly modified after the original consultation with the committee. The court rejected the argument that further consultation with the committee was unnecessary because the modified standard was a "logical outgrowth" of the earlier version.[99] After remand, OSHA conducted further consultations with the committee and the standard was ultimately upheld by the court.[100]

---

[95] 506 F.2d 385 (3d Cir. 1974), *cert. denied sub nom. Oil, Chemical & Atomic Workers v. Dunlop*, 423 U.S. 830 (1975).

[96] 29 C.F.R. §1911.10(a) (1983).

[97] 40 U.S.C. §333(e) (1976).

[98] 581 F.2d 960 (D.C. Cir. 1978).

[99] *Id.* at 970–71.

[100] The standard appears at 29 C.F.R. §1926.400(h) (1983).

9

The *Benzene* Decision: OSHA and Risk Assessments

A. Early Extrapolations

A key issue in OSHA's regulation of toxic chemicals, particularly carcinogens, is the establishment of a permissible exposure limit (PEL). OSHA must first decide whether the chemical being regulated causes carcinogenic (or other toxic) effects in humans. These determinations are usually made on the basis of epidemiological or animal studies.[1] OSHA is then confronted with the difficult question regarding substances determined to be carcinogenic: at what specific point should the permissible exposure limit be set where the data, whether from animal experiments or from epidemiological studies, show carcinogenic risk only at higher levels, or unknown levels, of exposure. To assist in making these determinations, scientists and regulatory officials have used quantitative risk assessments (or risk analysis), defined by OSHA recently as a "statistical technique used [to] estimate risk at levels outside the range of observed exposure levels."[2]

Merrill, *Federal Regulation of Cancer-Causing Chemicals*[3]

***

E. QUANTITATIVE ASSESSMENT OF RISK

1. BACKGROUND.

Some of the most intensely debated issues in health regulation surround the efforts to develop quantitative estimates of the risk posed by spe-

[1]See Chapter 6, Section A. The scientific view, which has been accepted by OSHA (and other agencies) and by the courts, is that animal evidence may permissibly be extrapolated from "mouse to man." *Society of the Plastics Indus. v. OSHA,* 509 F.2d 1301 at 1308 (2d Cir.), *cert. denied sub nom. Firestone Plastics Co. v. Dept. of Labor,* 421 U.S. 992 (1975). This chapter will focus on carcinogenic risk, although analogous issues arise respecting other toxic effects.

[2]Notice of Limited Reopening of Arsenic Rulemaking Record, 47 FED. REG. 15,358, 15,361 (1982). See *infra* this chapter for a discussion of OSHA regulation of arsenic. For a discussion of quantitative risk assessment, see McGarity, *Substantive and Procedural Discretion in Administrative Resolution of Science Policy Questions: Regulating Carcinogens in EPA and OSHA,* 67 GEO. L.J. 729, 733–36 (1979).

[3]Merrill, FEDERAL REGULATION OF CANCER-CAUSING CHEMICALS, REPORT TO THE ADMINISTRATIVE CONFERENCE, 83–92, April 1, 1982 (Draft).

cific substances. A variety of incentives have encouraged regulators to devise methods for quantifying risk. Quantification permits differentiation among chemicals that are potential candidates for regulation and, thus, aids in setting priorities. Numerical depiction of the risks of particular chemicals can be used to illustrate the health benefits obtainable from controlling human exposures. Perhaps most important, and certainly most controversial, quantitative risk assessment facilitates analysis of the costs and benefits of alternative control strategies.

Many observers have criticized quantification, particularly if used to set exposure limits for carcinogens. They call attention to the wide disparities in the risks predicted by different models. They stress, as well, the lack of empirical verification of any of the models and the uncertainties inherent in extrapolating from animals to humans. Supporters of quantitative risk assessment acknowledge these deficiencies, but defend its use as a means of structuring decisions that inevitably involve substantial uncertainties. They further contend that intelligent evaluation of efforts to reduce human exposure to carcinogens requires some method for estimating the magnitude of the health consequences of alternative control measures.

Broadly defined, quantitative risk assessment is a method or process of calculating the estimated likelihood that a particular exposure to a substance will cause cancer. While the mathematical models available to perform such a calculation vary in both their operation and outcomes, they start from the common assumption that the likelihood of harm from exposure to a toxic substance is a function of two variables: the potency of the substance and the dose, the latter characterized by the level, route, and duration of exposure. In this respect, the assumptions underlying quantitative risk assessment for carcinogens do not in principle differ from those historically relied on in evaluating other types of toxic effects.

Government agencies have used quantitative techniques in regulating hazardous materials for several decades. Like the contemporary extrapolation models for carcinogens, these techniques assumed that the capacity of a substance to harm an organism was a function of the dose; lower doses were likely to be safer than higher ones. Traditional regulation of non-carcinogens relied on so-called "safety factors," which were applied in the following way. On the premise that many cells had to be damaged before an organism sustained permanent harm, toxicologists conducted tests to determine the highest dosage that produced no adverse effects in experimental animals. The regulator would divide this "no observed effect" level by a predetermined number, i.e., a safety factor, to arrive at a dose predicted to be without risk for humans. For direct food additives, for example, FDA ordinarily used a safety factor of 100. While there were attempts to justify these safety factors biologically, most toxicologists would concede that the concept was simply an attempt to account for the possibility that humans might be more sensitive than the species tested, and that the precise numbers used were selected for reasons of convenience.

It should be stressed that this approach to quantification was not designed to estimate the magnitude of the risk associated with exposure to a chemical. Rather, the approach rested on the assumption that toxic materials had thresholds. The "no observed effect level" was an approximation of this threshold for the species tested; the safety factor accounted for the possibility that humans, being more heterogeneous, might also be more vulnerable. The permitted exposure level derived by this method was accepted, probably literally as well as operationally, as "safe." Although regulatory agencies occasionally used safety factors in regulating carcinogens, it was recognized that this approach was inconsistent with the no threshold hypothesis that dominates current thinking.

***

2. ESTIMATING CANCER RISKS.

Assessment of the magnitude of the human cancer risk presented by a carcinogen requires two types of information and a method for relating them. The first is information about the incidence of cancer produced by different doses of substance, i.e., a dose-response curve. The second type of information needed is information about the exposures humans are encountering or are likely to encounter. There are perhaps half a dozen mathematical extrapolation models that can be used to relate these two types of information. Some such model is necessary because the information about potency and exposure [is] almost invariably derived from different sources.

Human epidemiological studies often, perhaps usually, are not sensitive enough to detect the effects of prevailing levels of exposure to most substances that cause cancer. In the majority of cases, therefore, agencies must make decisions on the basis of data derived from experiments with laboratory animals. But standard animal bioassays, for both economic and scientific reasons, * * * employ doses that proportionally are much higher than those to which humans are exposed. Some method is needed to translate these data into realistic estimates of the magnitude of the risk to humans.

Accordingly, quantitative risk assessment usually involves two extrapolations, both of which introduce uncertainties. The first extrapolation is from laboratory animals to humans; the second is from the high doses observed to cause cancer in animals to likely human exposure levels below the range for which data are available. Actually, where both extrapolations are being performed, there is a third source of uncertainty. For we are asked to assume not only that magnitude of risk is a function of dose and that animals are reliable qualitative predictors of potential human hazard, but also that the dose-response experience of animals is likely to replicate that of man. The premise of the first extrapolation—that an agent which induces cancer in other mammals presents a potential risk for man—is well-accepted. There is also considerable evidence that, within species, cancer incidence generally increases with dose. Physiological differences between experimental animals and man, however, weaken confidence in quantitative extrapolation between the species.

If a regulator desires to quantify the risk associated with human exposure to animal carcinogens, however, this extrapolation is unavoidable. To perform this extrapolation, some adjustment must be made for the obvious disparity in the sizes of the two species to posit biologically comparable human doses. This adjustment is reflected in a so-called "scaling factor" by which animal dose-response data are translated into ostensibly equivalent human doses. Several different approaches have been advanced as biologically rational. The scaling factors currently popular with regulatory agencies make comparisons on the basis of dose per unit of surface area, e.g., micrograms per square centimeter, or on the basis of relative dietary concentrations. A third scaling factor, originally quite popular, bases comparisons on the relative body weights of the two species. The current preference for the other approaches reflects official conservatism; the use of relative surface area has the effect of increasing estimates of human risk because the disparity in surface area between rodents and humans is relatively less than the disparity in their weights. What little empirical evidence exists, however, suggests that all four of the common scaling factors may modestly overstate human risk.

The attempt to estimate the effects of low doses of a carcinogen from its observed effects at high doses is more controversial than the qualitative extrapolation from animals to man. To make this extrapolation with confidence an agency would need, but almost never has, knowledge of the underlying mechanism by which the substance causes cancer. Moreover, in most cases there are no data from animal exposures at doses corresponding to the human exposure levels that are of regulatory interest. Thus an agency must have some way of estimating risks at doses for which no experimental data exist. In the absence of data, quantitative risk assessment requires that an agency select some biostatistical model for estimating low dose effects. Several mathematical models have been formulated for this purpose. While the models vary widely in their biological rationales, all of them purport to describe the relationship between recorded doses of a carcinogen and the observed incidence of tumors, and then to predict the probable relationship at lower doses.

Most extrapolation models make assumptions about the stochastic properties of the interactions between carcinogens and target organisms. One procedure relies on the classic dose-response relationship derived from pharmacologic studies, which is an S shaped curve, indicating that some organisms in an exposed population are extremely sensitive to toxic agents and others are highly resistant. This relationship led Mantel and Bryan to hypothesize that the dose-response function of a carcinogen can be approximated by a probit curve in the low dose range. Another model starts from the premise that no effect can occur when the dose of a carcinogen is zero and estimates the effects of low exposures by a linear extrapolation from observed positive doses to zero. A third, the so-called "one hit" model, assumes that a single "hit" of a carcinogen can produce a positive response. Any molecule of the substance is regarded as having the same probability of causing the reaction; the cumulative probability of cancer is thus a function of cumulative exposure. The linear model predicts direct straight-line proportionality between dose and effect until the dose decreases to zero. The one-hit model generally produces higher estimates of risk at low doses. None of the models that have gained popularity, however, suggests a threshold dose below which a carcinogen poses no risk of cancer.

Because several extrapolation models enjoy respectable support, a regulator is faced with a difficult choice at the outset. One approach would be to select an extrapolation model on the basis of correspondence with the data in the observable, i.e., high-dose, range. If one model fit[s] the observable data better than others, it might be favored as the best predictor of effects at low doses. Choosing among extrapolation models based on the closeness of "fit" to the available data is rarely possible, however, because the data usually fit several models equally well. The inadequacy of closeness of fit as a criterion for selecting an extrapolation model is illustrated by the results of applying several models to the data derived from a unique animal experiment performed by the National Center for Toxicological Research. This so-called "megamouse study" used 24,000 rodents to measure the effects of 2-acetyl-aminofluorine (2-AAF), a known carcinogen, administered at doses an order of magnitude lower than those customarily used in animal bioassays. The final data from the study were statistically fit to the five best-known mathematical models to determine whether "closeness of fit" would provide a basis for preference. While two models could be eliminated, each of the other three was equally compatible with the observed data.

The choice of an extrapolation model can significantly affect risk estimates. While the familiar models often fail to reveal differences at high doses, they often predict sharply different risks in the low-dose range. This disagreement among the models is illustrated numerically and graphically in Tables [III and IV], which are based on the NCTR study.

Because it is usually not feasible to prefer one extrapolation model based on its compatibility with observed data, regulatory agencies have relied on other criteria. For a time some espoused the linear model, which

TABLE III

RANGE OF ESTIMATE "VIRTUALLY SAFE" DOSES*

Extrapolation Model	Virtually Safe Doses, ppb Liver Neoplasm	Bladder Neoplasm
Probit	250.	33000.
Mantel-Bryan	11.	220.
Multi-Stage	.41	8.8
One-Hit	.36	.80
Linear	.39	.80

*"Virtually safe" defined as a risk of tumor no greater than 1 in 1 million over a lifetime (1/10–6), extrapolated from a study in which mice were administered 2-AAF in the diet for 24 months. Estimates are based on 99 percent confidence level.

assumes that the incidence of cancer will be directly proportional to dose at low exposures. One reason for preferring this model is its so-called "conservatism," i.e., the model is less likely than others to underestimate the numbers of cancers expected as the result of exposure to low doses of a carcinogen. If this were the sole concern, an agency would always select the model that produced the highest estimated risk of cancer so that regulation could prevent the hypothesized "worst case." Regulators, however, have generally tended to seek one extrapolation procedure that can be followed in evaluating all carcinogens. This is not essential; it has been suggested that regulators should use several different models to display the range of potential risk associated with exposure to a carcinogen. In the absence of a powerful desire for flexibility, however, consistency and predictability argue for the selection of a single model so that risk estimates for different substances are readily comparable. This approach facilitates ranking of carcinogens. The agency with the most experience in quantitative risk estimation for carcinogens, EPA, currently favors the multi-stage model, which is claimed to best represent the mechanism of carcinogenesis. Other agencies are likely to follow EPA's lead.

OSHA first confronted the question of determining carcinogenic risks at low levels of exposure in issuing the asbestos standard in 1972.[4] OSHA stated in the preamble that, although there is clear evidence that asbestos causes asbestosis, lung cancer, and mesothelioma, "we do not have, in general, accurate measures of the levels of exposure occurring 20 or 30 years ago, which have given rise to these consequences." The Agency decided, however, that the two-fiber level (two fibers of asbestos per cubic centimeter of air (fibers/cc)) should be adopted, rejecting arguments for higher permissible exposure lev-

[4] 37 FED. REG. 11,318 (1972).

TABLE IV

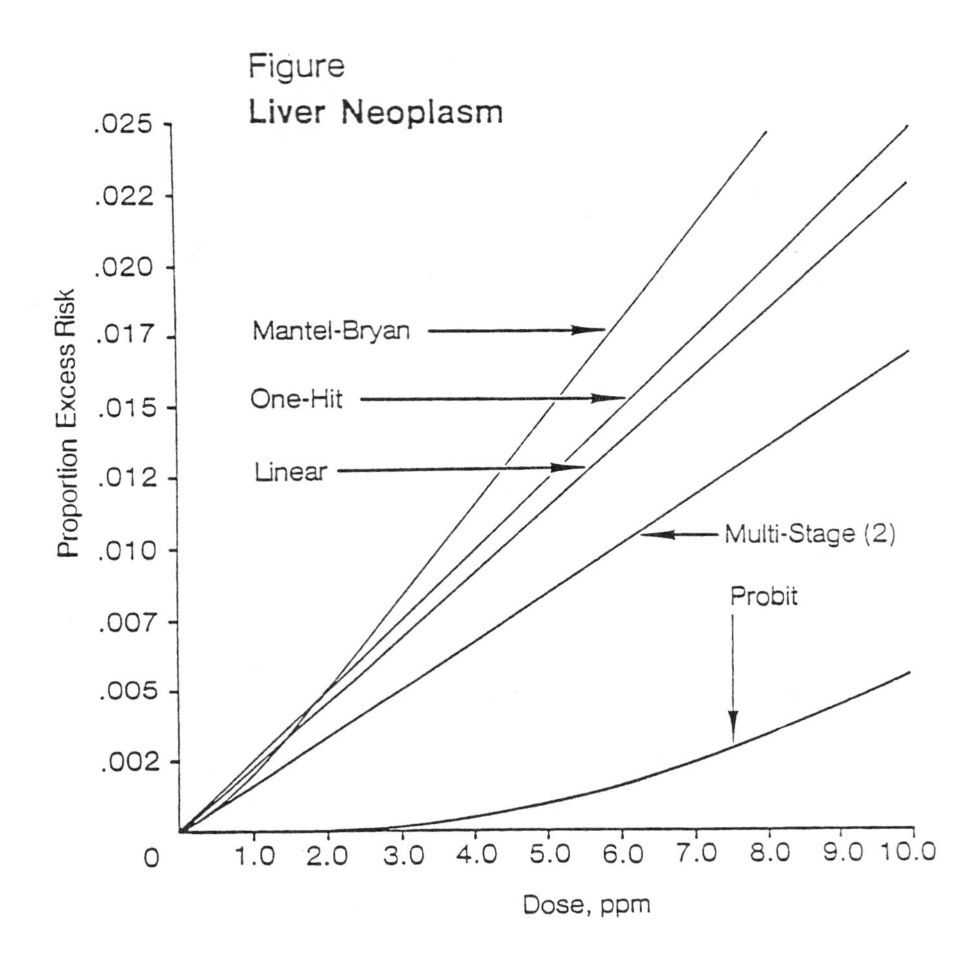

els, but it delayed the effective date of the new standard for four years because of feasibility considerations.[5] OSHA concluded that in view of the "grave consequences" of exposure, asbestos must be regulated, "on the basis of the best evidence available now, even though it may not be as good as scientifically desirable," and that the "conflict in the medical evidence is resolved in favor of the health of employees."[6] Essentially the same conclusion was reached by OSHA in regulating 14 carcinogens[7] and vinyl chloride.[8] These stand-

[5]On the feasibility issue in asbestos, see Chapter 10, Section A.
[6]37 FED. REG. 11,318 (1972).
[7]39 FED. REG. 3756, 3758 (1974).
[8]39 FED. REG. 35,890, 35,892–93 (1974).

ards were upheld by the courts of appeals in relevant respects.[9]

In issuing a standard regulating coke oven emissions in 1976, OSHA for the first time expressly articulated its policy of regulating carcinogens to the "lowest feasible level."[10] OSHA said in the coke oven emission preamble that "from the point of view of choosing a safe level of exposure, the permissible exposure limit should be set at zero." Because of feasibility considerations, however, OSHA said that the PEL "must be set at the lowest level feasible."[11] Although conceding that such determination involves "a measure of subjective judgment," it is justified "by the nature of the hazard being dealt with" and the intent of the Act, as expressed in §6(b)(5).[12] OSHA set the permissible exposure limit at 150 micrograms per cubic meter (ug/m^3).[13]

B. The *Benzene* Proceeding

OSHA issued a permanent standard for benzene in February 1978.[14] The standard lowered the previous permissible exposure limit, adopted in 1971 under §6(a),[15] from 10 parts per million (ppm) to 1 ppm, with a ceiling of 5 ppm during any 15-minute period.[16] While there were no animal data on benzene and, in general, the epidemiological data showed humans contracting leukemia at concentrations significantly higher than the prior level of 10 ppm,[17] the Agency nonetheless adopted the stringent 1 ppm level based on the "lowest feasible level" policy. OSHA stated that the only conclusion possibly to be derived at that time regarding the dose-response relationship for benzene was that higher exposures carry a greater risk; and since a determination of a precise level of benzene that presents no hazard cannot be made, the question "whether a 'safe' level of exposure to benzene exists cannot be answered."[18] Prudent health

[9]*Synthetic Organic Chem. Mfrs. Ass'n. v. Brennan,* 503 F.2d 1155 (3d Cir. 1974), *cert. denied,* 420 U.S. 973 (1975) (ethyleneimine standard affirmed); *Society of the Plastics Indus. v. OSHA,* 509 F.2d 1301 (2d Cir. 1975) (vinyl chloride standard affirmed).

[10]41 FED. REG. 46,742 (1976).

[11]*Id.* at 46,755.

[12]29 U.S.C. §655(b)(5)(1976).

[13]41 FED. REG. at 46,755. While OSHA in its coke oven preamble articulated its rationale in terms of the "lowest feasible level" policy, at the hearing, it requested Dr. Charles Land of the National Cancer Institute to estimate the excess lung cancer mortality risk to coke oven workers at various levels of exposure. The risk analysis was performed, both on the basis of the linear and the quadratic models, and OSHA concluded that Dr. Land's analysis "supported" the conclusions that there was no exposure level that could be considered safe. *Id.* at 46,753–55. OSHA's use of the risk assessment performed by Dr. Land is discussed in D.D. Briggs & L.B. Lave, *Regulating Coke Oven Emissions,* in QUANTITATIVE RISK ASSESSMENT IN REGULATION 135–50 (L.B. Lave ed. 1982).

[14]The permanent standard was preceded by an emergency temporary standard, 42 FED. REG. 22,516 (1977), which was stayed by the Fifth Circuit Court of Appeals and never went into effect. See *Industrial Union Dep't v. Bingham,* 570 F.2d 965, 968 (D.C. Cir. 1977), and discussion in Chapter 4.

[15]29 C.F.R. §1910.1000 Table Z-2 (1983).

[16]43 FED. REG. 5918 (1978).

[17]*American Petroleum Inst. v. OSHA,* 581 F.2d 493, 498 n.13 (5th Cir. 1978), *aff'd sub nom. Industrial Union Dep't v. American Petroleum Inst.,* 448 U.S. 607 (1980).

[18]43 FED. REG. at 5931, 5946.

policy, according to OSHA, therefore requires limiting exposure to the maximum extent feasible.

Final Benzene Standard

43 Fed. Reg. 5918, 5946–47 (1978)

***

The issue of the levels at which cancer is induced by chemical agents and whether or not there is a "threshold" has been a major issue in every OSHA rulemaking concerning the regulation of occupational carcinogens (See preambles to Carcinogen standard (39 FR 3758); Vinyl Chloride (39 FR 35892); Coke Oven Emissions (41 FR 46742)). The benzene hearing was no exception. As in the case of arsenic, the lack of an unequivocal animal model requires that the Secretary's decision rely primarily on data obtained from human evidence. However, the epidemiologic method is by its very nature, a retrospective view of the evidence, i.e. findings of a recent excess of mortality among workers may relate to initial exposures occurring as much as 20 years or more previously and which most certainly correlate to exposures at higher concentrations than present levels. Because of these variables, it is extremely difficult to derive definite conclusions as to the quantitative health risk to the workers at exposures near 1 ppm. Studies which report negative findings, such as those studies conducted by Stallones, Tabershaw-Cooper and Thorpe suffer from the disadvantage of not being able to clearly define an exposed cohort, i.e. the identification of a group of workers actually exposed to benzene and to what levels they were, in fact, exposed. OSHA recognizes that it is extremely difficult to reconstruct and define employee exposures retrospectively. Therefore, given the uncertainty of the definition of exposure and the potential for dilution of mortality excess among those actually exposed to benzene and other methodological deficiencies, OSHA is reluctant to place substantial reliance upon these negative reports. Thus, there is little definitive information available pertaining to the leukemogenic risk of an adequate size cohort of workers exposed to benzene less than 10 ppm and who have been followed for an adequate amount of time. Furthermore, it is OSHA's view following a careful review of the record that, at the present time, it is impossible to derive any conclusions regarding dose-response relationships for benzene beyond the general observation that higher exposure levels carry a greater risk than do lower exposure levels. What is apparent however, is that a decrease in exposure level and/or duration will result in a decreased risk of leukemia.

***

It is, therefore, clear from an examination of the record that a determination of a precise level of benzene exposure which presents no hazard cannot be made and that the corollary question of whether a "safe" level of exposure to benzene exists cannot be answered. The agency is aware of and has examined scientific opinion and data submitted by industry that thresholds for carcinogens may exist. However, prudent public health policy requires a conservative course of action until such evidence is of a definitive nature.

In its conclusionary document the International Workshop also stated: "[t]he workshop discussed but could not agree on whether there was a concentration below which there would be no leukemogenic effect clearly attributable to benzene." Kraybill testified that because there was not enough data at the lower part of the dose-effect curve, he was unable to estimate a "safe" level. As early as 1939, in presenting data on the effects of chronic benzene exposure including leukemia, Hunter stated: "It is doubtful whether any concentration of benzene greater than zero is safe over a long period of time." In 1974, the MAK-Werte Working Group stated "* * * [O]n the proven carcinogenic (leukemogenic) effects of benzene and the lack of quantitative measuring data in the low concentration ranges, it is not possible at this time to establish an MAC (maximum allowable concentration) which might be regarded to be without danger."

***

* * * In OSHA's view, the demonstration of cancer induction in humans at a particular level is not in the regulatory context, a prerequisite to a determination that a substance represents a cancer hazard for humans at that level. Relying upon a substantial body of scientific opinion, OSHA has concluded that, when dealing with a carcinogen, no safe level exists for any given population. For example, the National Cancer Institute's Ad Hoc committee on the Evaluation of Low Levels of Environmental Carcinogens (1970) states:

> no level exposure to a chemical carcinogen should be considered toxicologically insignificant for man. For carcinogenic agents, a "safe level for man cannot be established by application of our present knowledge."

Thus, OSHA takes the position that in promulgating a health standard for a life threatening hazard such as benzene-induced leukemia, the exposure limits be established on a conservative basis so that worker health is properly safeguarded.

From the point of view of choosing a safe level of exposure, therefore, the permissible exposure limit should be set at zero. However, based on the evidence in the record, it is OSHA's judgment that a zero standard for exposure to benzene is not technologically feasible. In fact, it is clear that certain quantities of benzene are present in the ambient environment as a result of natural phenomena and as artifacts of human activity. While a permissible exposure limit equal to zero plus background would represent the lowest level theoretically possible, OSHA believes that the record shows such an approach is not feasible. Even if such a number could be determined, achieving a standard of zero plus background would require that exposure to benzene be effectively zero so as not to increase employee exposure above background levels. There has been no evidence presented that convinces OSHA that such a complete elimination of benzene in all industries can be achieved by existing or future technology. In determining the appropriate permissible level of employee exposure to benzene, OSHA relies in part on the record of this proceeding and in part on policy considerations which lead the Agency to conclude that, in dealing with a carcinogen or other toxic substance for which no safe level of exposure has been demonstrated, the permissible exposure limit must be set at the lowest level feasible. Such a determination involves a measure of subjective judgment which OSHA believes is justified by the nature of the hazard being dealt with, and the intent of the Act. Section 6(b)5 provides that the standards for toxic substances shall be feasible. * * *

OSHA has determined that 1 ppm TWA, 5 ppm ceiling for 15 minutes is the level which most adequately assures, to the extent feasible, the protection of workers exposed to benzene.

The permanent standard on benzene was challenged in court principally by the American Petroleum Institute (API). The Fifth

Circuit Court of Appeals vacated the standard, holding that, in issuing standards, OSHA must show by substantial evidence that the "measurable" benefits of the standard bear a "reasonable relationship" to the costs.

American Petroleum Institute v. OSHA

581 F.2d 493 (5th Cir. 1978)

CLARK, Circuit Judge:
***

As a result of its toxicity, benzene's history has been one of regulation. In 1946, the American Conference of Governmental Industrial Hygienists recommended a threshold limit value for benzene exposure of 100 ppm. This value was reduced to 50 ppm in 1947, to 35 ppm in 1948, to 25 ppm in 1963, and to 10 ppm in 1974. The American National Standards Institute adopted a threshold limit value of 10 ppm in 1969, which OSHA adopted in 1971 without rulemaking under the authority of 29 U.S.C.A. §655(a). This standard, codified at 29 C.F.R. §1910.1000 Table Z-2 (1977) and still in effect, was based on the nonmalignant toxic effects of benzene exposure and not on any possible leukemia hazard.

Widely scattered through the benzene literature are studies suggesting a link between benzene exposure and leukemia, a usually fatal cancer of the blood-forming organs. During the 1970's several additional studies reported a statistically significant increased risk of leukemia among workers occupationally exposed to high levels of benzene and concluded benzene was a leukemogen.[13] As a result of this new evidence, OSHA began procedures which culminated with the present proposal, among other things, to reduce the permissible exposure level from 10 ppm to 1 ppm.

In January 1977 OSHA issued voluntary Guidelines for Control of Occupational Exposure to Benzene recommending exposure not to exceed an eight-hour time-weighted average of 1 ppm. An Emergency Temporary Standard for Occupational Exposure to Benzene also providing for a reduction in the permissible exposure limit to 1 ppm was issued in May 1977, but this standard never went into effect because of judicial challenges. The proposed permanent benzene standard, which was based on OSHA's determination that the available scientific evidence established that employee expo-

---

[13]One such study was reported in 1975 by Dr. Enrico Vigliani. In 1963, Dr. Vigliani participated in a study of workers exposed to resins, inks, varnishes, and glues containing various amounts of benzene, and found a risk of leukemia among these workers twenty times greater than that for the general population. Toluene was substituted for benzene in 1964 in one of the industries studied, and the 1975 study showed no new cases of leukemia among workers in that industry.

A second study was reported in 1972 by Dr. Muzaffer Aksoy, a hematologist who testified at the rulemaking hearing. In this study Dr. Aksoy reported four leukemia deaths among Turkish shoemakers resulting from their exposure to benzene concentrations in excess of 150 ppm for periods ranging from six to fourteen years, and at the hearing he estimated that the incidence of leukemia among the population he studied was twice what would have been expected for the population as a whole. Dr. Aksoy also noted a decline in leukemia cases after other solvents were substituted for benzene.

The study most heavily relied upon by OSHA was one reported by Dr. Peter Infante of the National Institute for Occupational Safety and Health, a body created to conduct research and recommend occupational safety and health standards. See 29 U.S.C.A. §§669–71. Dr. Infante studied workers exposed to benzene in the production of Pliofilm at Goodyear's Akron and St. Mary's plants between 1940 and 1949 and found among them a five-fold increased risk of dying of leukemia when compared to two control groups. No specific exposure level during the period covered by the study was established, but testimony at the hearing indicated that exposure was probably around 100 ppm during most of the period studied with occasional exposure levels as high as several hundred parts per million.

sure to benzene presents a leukemia hazard and that exposure therefore should be limited to the lowest feasible level, was published on May 27, 1977. This proposal provided for a reduction in the permissible exposure limit from 10 ppm to 1 ppm and established requirements relating to dermal and eye contact, exposure monitoring, medical surveillance, methods of compliance, labeling, and recordkeeping. Public hearings were held July 19 through August 10, 1977, at which 95 witnesses testified. In addition, numerous exhibits and documents were submitted to OSHA as part of the rulemaking record. The resulting permanent benzene standard was promulgated on February 3 and published on February 10, 1978, with a March 13, 1978 effective date.

***

IV.

OSHA justifies the reduction of the permissible exposure limit for benzene from 10 ppm to 1 ppm by coupling two factual findings, which it contends are supported by substantial evidence in the record, with a regulatory policy which OSHA contends is required by its mandate to protect workers. The factual findings are that benzene causes leukemia and that there presently exists no known safe level for benzene exposure. The regulatory policy is to limit employee exposure to carcinogens to the lowest feasible level.

* * * The petitioners argue that by defining an "occupational safety and health standard" as one requiring conditions "reasonably necessary" to provide safe or healthful places of employment, 29 U.S.C.A. §652(8), Congress recognized that safety and health resources are not unlimited and required OSHA somewhere in its decisionmaking process to (1) attempt to determine the extent to which its standards will benefit workers, and (2) decide whether the projected benefits justify the costs of compliance with the standard. Only if all standards are subjected to such assessment, argue the petitioners, can OSHA assure maximum benefit from the finite amount industry can expend on safety and health and thus carry out Congress' overriding policy "to assure so far as possible every working man and woman in the Nation safe and healthful working conditions." 29 U.S.C.A. §651(b). Since OSHA has not made a valid determination that reducing the permissible exposure level of benzene from 10 ppm to 1 ppm is reasonably necessary to protect workers from a risk of leukemia, the producers ask us to set that part of the standard aside.

***

We are not persuaded by OSHA's argument that this standard should be upheld since the lack of knowledge concerning the effects of exposure to benzene at low levels makes an estimate of benefits expected from reducing the permissible exposure level impossible.[23] The statute requires all conditions imposed by a standard to be reasonably necessary to provide safe or healthful employment, and it requires decisions to be based on "the best

[23]Although OSHA asserts that risk quantification at low exposure levels and therefore estimates of expected benefits from the standard cannot presently be made, OSHA has provided us with a Preliminary Report on Population Risk to Ambient Benzene Exposures, recently released by the Environmental Protection Agency, which attempts to extrapolate from the results of the Infante study a determination of the risk of leukemia to the general population at the exposure level of 1 part per billion. In addition, the petitioners introduced at the rulemaking proceeding a preliminary risk assessment for occupational exposure to benzene at 10 ppm and 1 ppm based on the studies at higher exposure levels relied on by OSHA. Finally, OSHA's economic consultant testified that it could perform a cost-effectiveness analysis for the benzene standard, an analysis which would have included some kind of risk quantification. Although OSHA's assertion that present knowledge is insufficient to construct a valid dose-response curve for benzene may be correct, the record reflects that preliminary assessments are now being made and that valid extrapolations will be possible as more is known about the effects of past exposure at higher levels.

available evidence," "research, demonstrations, experiments, and such other information as may be appropriate," "the latest scientific data in the field," and "experience gained under this and other health and safety laws." By requiring the consideration of such kinds of information, Congress provided that OSHA regulate on the basis of knowledge rather than on the unknown. But see *Society of Plastics Industry, Inc. v. OSHA,* 509 F.2d 1301, 1308 (2d Cir. 1975). Until OSHA can provide substantial evidence that the benefits to be achieved by reducing the permissible exposure limit from 10 ppm to 1 ppm bear a reasonable relationship to the costs imposed by the reduction, it cannot show that the standard is reasonably necessary to provide safe or healthful workplaces.

This does not mean that OSHA must wait until deaths occur as a result of exposure at levels below 10 ppm before it may validly promulgate a standard reducing the permissible exposure limit. See *Florida Peach Growers Association, Inc. v. United States Department of Labor,* 489 F.2d 120, 132 (5th Cir. 1974). Nevertheless, OSHA must have some factual basis for an estimate of expected benefits before it can determine that a one-half billion dollar standard is reasonably necessary. For example, when studies of the effects of human exposure to benzene at higher concentration levels in the past are sufficient to enable a dose-response curve to be charted that can reasonably be projected to the lower exposure levels, or when studies of the effects of animal exposure to benzene are sufficient to make projections of the risks involved with exposure at low levels, then OSHA will be able to make rough but educated estimates of the extent of benefits expected from reducing the permissible exposure level from 10 ppm to 1 ppm. Until such estimates are possible, OSHA does not have sufficient information to determine that a standard such as the one under review which it can only say might protect some worker from a leukemia risk is reasonably necessary.

We will not attempt to reconcile our decision with the cases from other circuits which uphold other standards regulating exposure to carcinogens. See *Industrial Union Department, AFL-CIO v. Hodgson,* 162 U.S. App. D.C. 331, 499 F.2d 467 (1974) (asbestos dust standard); *Society of Plastics Industry, Inc. v. OSHA,* 509 F.2d 1301 (2d Cir. 1975) (vinyl chloride standard); *American Iron & Steel Institute et al. v. OSHA*, 577 F.2d 825, No. 76-2358 *et al.* (3d Cir., filed March 28, 1978) (coke oven emission standard). Those opinions did not address what Congress meant by requiring the conditions imposed by standards to be reasonably necessary to provide safe or healthful places of employment. In this circuit, under [*Aqua Slide 'n' Dive Corp. v. Consumer Product Safety Commission,* 569 F.2d 831 (5th Cir. 1978)], substantial evidence must support a finding that those conditions are reasonably necessary, a showing that OSHA has not made. In addition, those cases were decided on their own records. Without critical analysis of what was established in those proceedings, we hold in today's case that Congress intended for OSHA to regulate on the basis of more knowledge and fewer assumptions than this record reflects.

OSHA's failure to provide an estimate of expected benefits for reducing the permissible exposure limit, supported by substantial evidence, makes it impossible to assess the reasonableness of the relationship between expected costs and benefits. This failure means that the required support is lacking to show reasonable necessity for the standard promulgated. Consequently, the reduction of the permissible exposure limit from 10 ppm to 1 ppm and all other parts of the standard geared to the 1 ppm level must be set aside.

***

The Court of Appeals for the Fifth Circuit vacated OSHA's benzene standard on two grounds: (1) the Agency had failed to provide an estimate, supported by substantial evidence, of the expected benefits for reducing the permissible exposure limit; and (2) it therefore necessarily failed to assess the reasonableness of the relationship between expected costs and benefits. The latter issue—the balancing of costs and benefits in OSHA standards-setting activity—will be discussed in Chapter 10 in connection with the litigation involving OSHA's cotton dust standard.

The court's analysis is noteworthy for several reasons. In overruling the OSHA standard, the court departed substantially from the view of other courts of appeals regarding the scope of review of Agency policy-making decisions. As has already been noted, the District of Columbia, Second, and Third Circuit Courts of Appeals had adopted a relatively deferential attitude in reviewing OSHA standards, particularly where the issues were legislative in nature, "on the frontiers of scientific knowledge." These earlier decisions were based in large measure on the overriding policy consideration, derived from the OSH Act, that uncertainties should be resolved in favor of the safety and health of employees. The Fifth Circuit, however, without attempting to distinguish these earlier opinions, insisted that OSHA "regulate on the basis of knowledge rather than on the unknown."[19] Indeed, the thrust of this decision was that OSHA must await the availability of better data before undertaking further regulation. The court said that "[*u*]*ntil* OSHA can provide substantial evidence that the benefits to be achieved by reducing the permissible exposure limit from 10 ppm to 1 ppm bear a reasonable relationship to the costs imposed by the reduction, it cannot show that the standard is reasonably necessary to provide safe or healthful workplaces."[20]

One of the crucial issues in the benzene rulemaking and litigation was whether it was possible to construct a dose-response curve for benzene showing the relationship between different exposure levels and the risk from cancer, that is, perform a quantitative risk assessment. In the preamble to the benzene standard, published in 1978, OSHA concluded on the basis of a "careful review" of the record that it was "impossible to derive any conclusions regarding dose-response relationships for benzene."[21] The Court of Appeals for the Fifth Circuit, on the other hand, stated that "preliminary assessments are now being made and that valid extrapolations will be possible as more is known."[22] In reaching this conclusion, the court re-

[19]581 F.2d at 504.

[20]*Id.* (emphasis added). The court also set aside OSHA's prohibition on dermal contact with benzene on the ground that OSHA relied on "dated, inconclusive data when modern experimental methods can quickly and efficiently provide reliable information." This, the court said, contravenes the statutory requirement in §6(b)(5), 29 U.S.C. §655(b)(5), that OSHA act on the "best available evidence." *Id.* at 507. On the meaning of the phrase "best available evidence," the Fifth Circuit Court of Appeals disagreed with the District of Columbia Circuit Court of Appeals. In reviewing OSHA's cotton dust standard, the District of Columbia Circuit Court of Appeals said that "[OSHA] may have to fill gaps in knowledge with policy considerations. Congress recognized this problem by authorizing the agency to promulgate rules on the basis of the 'best available evidence'." *AFL-CIO v. Marshall,* 617 F.2d 636, 651 (D.C. Cir. 1979). Thus, in the view of the District of Columbia Circuit Court of Appeals, this statutory language was not a constraint on OSHA authority but rather a mandate to act in the face of uncertainty.

[21]43 FED. REG. at 5946.

[22]581 F.2d at 504 n.23.

lied in part on a Preliminary Report on Population Risk to Ambient Benzene Exposures, released by EPA, which was based on data from the Infante epidemiological study introduced in the OSHA record.[23] While the EPA report was not in the rulemaking record, it was attached by OSHA to its brief in an effort to persuade the court of appeals that another federal agency had found an excess of 60 to 80 leukemia cases at ambient air levels of 1 part per *billion;* in other words, that the leukemia risk was substantial at low levels of exposure. It is indeed ironic that this report, introduced by OSHA after the close of the record, was used by the court to disprove OSHA's contention that quantitative risk assessments could not be performed for benzene.[24]

The Supreme Court granted OSHA and the Industrial Union Department's petitions for certiorari from the decision by the Fifth Circuit Court of Appeals vacating the benzene standards. Since the available epidemiological studies showing benzene to be a leukemogen were based on exposure levels significantly higher than the 1 ppm PEL adopted by OSHA,[25] a major issue in the litigation was whether the statute required OSHA to provide an estimate of "measurable" benefits that would result from the reduction of the permissible exposure limit to 1 ppm; this would almost certainly involve performing a quantitative risk assessment. The industry respondents urged the Supreme Court to affirm the appellate court's decision on this issue, arguing that OSHA's finding that the benefits from the modified standard would be "appreciable"[26] was insufficient. Their brief to the Supreme Court contained a "Summary of Argument," which stated in part:

> In short, there are compelling reasons why the decision below should be affirmed. The court below was asked to uphold an enormously costly standard based upon nothing more than a scanty *Federal Register* notice that waffles back and forth between a poorly articulated interpretation of the Act and a completely unsupported "appreciable" benefits "determination." This is precisely the type of arbitrary action that Congress intended to prevent when it subjected OSHA standards to the "substantial evidence" standard of review. If OSHA can rely upon a completely unsupported assumption to proclaim that the benefits of its standard "are likely to be appreciable" when the uncontradicted evidence shows the contrary to be true, then much of the Act, including the "substantial evidence" test, will have been drained of its intended meaning.[27]

Supporting its position that OSHA incorrectly concluded that risk assessments could not be reliably performed, the industry petitioners relied heavily on the testimony of Dr. Richard Wilson who

[23]*Id.* at 498 n.13.

[24]The court also relied on a preliminary risk assessment done by an industry expert (Dr. Richard Wilson of Harvard University) which had been introduced in the rulemaking proceeding. *Id.* at 504 n.23.

[25]*Id.* at 498 n.13.

[26]See benzene standard, 43 FED. REG. 5918, 5941 (1978).

[27]Brief for Respondents at 30, *Industrial Union Dep't v. American Petroleum Inst.,* 448 U.S. 607 (1980).

testified that he had prepared an "upper bound" risk assessment for benzene which demonstrated that, at most, the OSHA 1 ppm level would avoid two cases of leukemia every six years. The court of appeals had in part relied on the Wilson testimony, referring to it as a "preliminary risk assessment,"[28] to show that, contrary to the OSHA assertions, quantitative risk assessments could be performed. OSHA, as part of its reply brief to the Supreme Court, submitted a statement entitled "Risk Assessment Uncertainty," which sought to show that based on the same data used by Dr. Wilson, but employing "different yet reasonable assumptions," the Agency would obtain a result showing that the new standard would save 800 times as many lives as estimated by Dr. Wilson. This, OSHA claimed, illustrated "the wide range of uncertainty presented by attempts to quantify the health risk for benzene."

Reply Brief for the Federal Parties in Industrial Union Department v. American Petroleum Institute[29]

RISK ASSESSMENT UNCERTAINTY

The limitations of estimating the number of lives at risk due to long-term, low-level exposure to benzene are illustrated by the Wilson and the EPA studies. Although the two calculations were based on the same mathematical method of extrapolation, they employed a number of different assumptions and yielded markedly different results.

The procedure for extrapolating risks was presented in mathematical form in the EPA's report and in graphical form in Wilson's report. The general principle is that the adverse health effects observed in one or more groups of workers exposed to benzene at various levels can be used to predict the risk to other workers exposed to benzene (now and in the future) at much lower levels. To do so requires explicit assumptions about the relationships between leukemia incidence, intensity of exposure, duration of exposure, and other relevant factors such as age, latent period, etc.

***

As we show below, each calculation incorporated, explicitly or implicitly, a number of other assumptions, each of which introduces a further wide range of uncertainty into the final estimates of risk. Some of these uncertainties result from lack of scientific understanding of the mechanisms of carcinogenesis, whereas others result from specific deficiencies or gaps in the data on the effects of exposure to benzene. Although EPA pointed out the wide range of uncertainty in their estimates, Dr. Wilson did not do so, leaving an entirely spurious impression of precision.

***

9. *The total range of uncertainty.* The use of the fragmentary existing data on excess cancer incidence in workers formerly exposed to benzene to estimate lifetime risks to workers currently exposed to different levels thus involves a number of assumptions, explicit and implicit, each of which introduces a substantial range of uncertainty into the calculations. Because risk extrapolation is a sequential procedure in which each step utilizes the results of the preceding steps, these uncertainties expand as the calculation progresses and combine multiplicatively to yield an extremely wide range of uncertainty in the final estimates.

---

[28]581 F.2d at 504 n.23.

[29]*Industrial Union Dep't v. American Petroleum Inst.*, 448 U.S. 607, Appendix A (1980).

It is difficult to estimate the full magnitude of this range of uncertainty, because information is lacking about the consequences of many of the assumptions. However, as shown above, there is at least an 80-fold variation between the dose-response data in various epidemiologic studies; in addition, there may be uncertainties ranging up to ten-fold resulting from lack of knowledge of the duration of exposure in these studies and up to three-fold resulting from lack of knowledge of current exposure. Since the uncertainties combine multiplicatively, these factors alone introduce uncertainties into the final estimates of risk that may range from one to 2,400. Additional uncertainties of unknown magnitude are introduced by the other factors discussed above, including incomplete follow-up of the reference populations, errors in recording causes of death, fluctuations in exposure, extrapolation of partial-lifetime to full-time risks, possible risks from other diseases, uncertainties in the form of the dose-response and duration-response relationships, age, sex, and synergisms with other risk factors.

So many factors are involved that the range between the lowest and highest reasonable estimates of risks may range into the tens of thousands, if not millions. This range of uncertainty cannot be ignored. Although EPA recognized that its own estimates of general population risks could for these reasons "easily be in error by several orders of magnitude", Dr. Wilson presented only a single estimate of risks and did not discuss uncertainties except to claim that his estimate was conservative and pessimistic. The analysis above shows, to the contrary, that his estimate of risk could easily be too low by many orders of magnitude.

The following "risk assessment," using assumptions different from those used by Dr. Wilson (yet wholly reasonable) illustrates the multiplicative effect of these uncertainties.

1. Rather than reject Dr. Infante's findings, as Dr. Wilson did, our counterexample accepts Dr. Infante's finding that there were 5.6 excess leukemias among the 748 workers whose vital status was studied through 1975. We also assume that the average level of benzene to which workers were actually exposed was 10 ppm.

2. Next, we adjust the number of excess leukemias found by Dr. Infante to take into account uncertainties in his data: (i) we assume that each employee in his cohort was exposed for five years; (ii) Dr. Infante reported 5.6 excess deaths through 1975 when over 600 of the cohort were still alive; due to the long latency period, we assume that benzene-induced leukemias will continue to emerge as the cohort ages; we also assume that Dr. Infante missed some leukemias among the 25% of the cohort whose vital status was unknown in 1975, so that a total of 17 excess leukemias will result over the entire lifetime of the 748 members of the cohort. These adjustments mean that approximately 23 of every 1000 employees exposed for five years at 10 ppm will be expected to die of leukemia as a result of occupational exposure.

3. If we assume that there are 30,000 job positions in the United States now exposed at 10 ppm and that job turnover is such that these positions are filled anew every five years (as we assume was true in the Infante study) then 690 leukemias would eventually result for every five years of exposure at 10 ppm. If there are 60,000 job positions, then there would eventually be 1380 leukemias for every five years of occupational exposure. Dr. Wilson predicted 2 cancers every six years (or roughly 1.7 every five years). Thus, simply by making different yet reasonable assumptions, we obtain a result about 800 times greater than did Dr. Wilson—illustrating the wide range of uncertainty presented by attempts to quantify the health risk for benzene. (These numbers, moreover, do not include death due to nonmalignant diseases caused by benzene.)

***

C. The Supreme Court Decision

On July 2, 1980, the Supreme Court vacated the OSHA benzene standard.[30] The Court was sharply divided, with five different opinions being handed down. A four-Justice plurality, consisting of the Chief Justice, and Justices Stewart, Powell, and Stevens, in an opinion by Justice Stevens, concluded that OSHA must find as a threshold matter that the toxic substance at issue created a "significant" health risk in the workplace and that the lower standard would eliminate or reduce that risk. This, the plurality held, OSHA had failed to do for benzene. In view of this conclusion, the plurality found it unnecessary to reach the question whether OSHA must also balance costs and benefits in issuing standards. Justice Powell largely concurred in the plurality opinion but would also have required OSHA to do a cost-benefit analysis. Chief Justice Burger filed a brief concurrence. Justice Rehnquist would also have vacated the standard, but on the grounds that the first sentence of §6(b)(5) was an unconstitutional delegation of legislative power to OSHA. Thus, there was a majority of five Court members voting to vacate the standard. Justice Marshall, in a vigorous opinion for himself and Justices Brennan, White, and Blackmun, would have affirmed the standard.

Industrial Union Department v. American Petroleum Institute

448 U.S. 607 (1980)

MR. JUSTICE STEVENS announced the judgment of the Court and delivered an opinion in which THE CHIEF JUSTICE and MR. JUSTICE STEWART join * * * [MR. JUSTICE POWELL concurred in part].

***

II

The critical issue at this point in the litigation is whether the Court of Appeals was correct in refusing to enforce the 1 ppm exposure limit on the ground that it was not supported by appropriate findings.

Any discussion of the 1 ppm exposure limit must, of course, begin with the Agency's rationale for imposing that limit. The written explanation of the standard fills 184 pages of the printed appendix. Much of it is devoted to a discussion of the voluminous evidence of the adverse effects of exposure to benzene at levels of concentration well above 10 ppm. This discussion demonstrates that there is ample justification for regulating occupational exposure to benzene and that the prior limit of 10 ppm, with a ceiling of 25 ppm (or a peak of 50 ppm) was reasonable. It does not, however, provide direct support for the Agency's conclusion that the limit should be reduced from 10 ppm to 1 ppm.

The evidence in the administrative record of adverse effects of benzene exposure at 10 ppm is sketchy at best. * * *

[30] *Industrial Union Dep't. v. American Petroleum Inst.*, 448 U.S. 607 (1980).

***

With respect to leukemia, evidence of an increased risk (i.e., a risk greater than that borne by the general population) due to benzene exposures at or below 10 ppm was even sketchier. Once OSHA acknowledged that the NIOSH study it had relied upon in promulgating the emergency standard did not support its earlier view that benzene had been shown to cause leukemia at concentrations below 25 ppm, there was only one study that provided any evidence of such an increased risk. That study, conducted by the Dow Chemical Co., uncovered three leukemia deaths, versus 0.2 expected deaths, out of a population of 594 workers; it appeared that the three workers had never been exposed to more than 2 to 9 ppm of benzene. The authors of the study, however, concluded that it could not be viewed as proof of a relationship between low-level benzene exposure and leukemia because all three workers had probably been occupationally exposed to a number of other potentially carcinogenic chemicals at other points in their careers and because no leukemia deaths had been uncovered among workers who had been exposed to much higher levels of benzene. In its explanation of the permanent standard, OSHA stated that the possibility that these three leukemias had been caused by benzene exposure could not be ruled out and that the study, although not evidence of an increased risk of leukemia at 10 ppm, was therefore "consistent with the findings of many studies that there is an excess leukemia risk among benzene exposed employees." 43 FED. REG., at 5928. The Agency made no finding that the Dow study, any other empirical evidence or any opinion testimony demonstrated that exposure to benzene at or below the 10 ppm level had ever in fact caused leukemia. See 581 F.2d, at 503, where the Court of Appeals noted that OSHA was "unable to point to any empirical evidence documenting a leukemia risk at 10 ppm * * *."

In the end OSHA's rationale for lowering the permissible exposure limit to 1 ppm was based, not on any finding that leukemia has ever been caused by exposure to 10 ppm of benzene and that it will *not* be caused by exposure to 1 ppm, but rather on a series of assumptions indicating that some leukemias might result from exposure to 10 ppm and that the number of cases might be reduced by reducing the exposure level to 1 ppm. In reaching that result, the Agency first unequivocally concluded that benzene is a human carcinogen.[36] Second, it concluded that industry had failed to prove that there is a safe threshold level of exposure to benzene below which no excess leukemia cases would occur. In reaching this conclusion OSHA rejected industry contentions that certain epidemiological studies indicating no excess risk of leukemia among workers exposed at levels below 10 ppm were sufficient to establish that the threshold level of safe exposure was at or above 10 ppm.[37] It also rejected an industry witness' testimony that a dose-response curve could be constructed on the basis of the reported epidemiological studies and that this curve indicated that reducing the permissi-

[36]"The evidence in the record conclusively establishes that benzene is a human carcinogen. The determination of benzene's leukemogenicity is derived from the evaluation of all the evidence in totality and is not based on any one particular study. OSHA recognizes, as indicated above that individual reports vary considerably in quality, and that some investigations have significant methodological deficiencies. While recognizing the strengths and weaknesses in individual studies, OSHA nevertheless concludes that the benzene record as a whole clearly establishes a causal relationship between benzene and leukemia." 43 FED. REG., at 5931.

[37]In rejecting these studies, OSHA stated that: "Although the epidemiological method can provide strong evidence of a causal relationship between exposure and disease in the case of positive findings, it is by its very nature relatively crude and an insensitive measure." After noting a number of specific ways in which such studies are often defective, the Agency stated that it is "* * * OSHA's policy when evaluating negative studies, to hold them to a higher standard of methodological accuracy." 43 FED. REG., at 5931–5932. Viewing the industry studies in this light, OSHA concluded that each of them had sufficient methodological defects to make them unreliable indicators of the safety of low-level exposures to benzene.

ble exposure limit from 10 to 1 ppm would prevent at most one leukemia and one other cancer death every six years.[38]

Third, the Agency applied its standard policy with respect to carcinogens, concluding that, in the absence of definitive proof of a safe level, it must be assumed that *any* level above zero presents *some* increased risk of cancer.[40] As the Government points out in its brief, there are a number of scientists and public health specialists who subscribe to this view, theorizing that a susceptible person may contract cancer from the absorption of even one molecule of a carcinogen like benzene. Brief for Federal Parties, at 18–19.

Fourth, the Agency reiterated its view of the Act, stating that it was required by §6(b)(5) to set the standard either at the level that has been demonstrated to be safe or at the lowest level feasible, whichever is higher. If no safe level is established, as in this case, the Secretary's interpretation of the statute automatically leads to the selection of an exposure limit that is the lowest feasible.[42] Because of benzene's importance to the economy, no one has ever suggested that it would be feasible to eliminate its use entirely, or to try to limit exposures to the small amounts that are omnipresent. Rather, the Agency selected 1 ppm as a workable exposure level, * * * and then determined that compliance with that level was technologically feasible and that "the economic impact of * * * [compliance] will not be such as to threaten the financial welfare of the affected firms or the general economy." 43 FED. REG., at 5939. It therefore held that 1 ppm was the minimum feasible exposure level within the meaning of §6(b)(5) of the Act.

Finally, although the Agency did not refer in its discussion of the pertinent legal authority to any duty to identify the anticipated benefits of the new standard, it did conclude that some benefits were likely to result from reducing the exposure limit from 10 ppm to 1 ppm. This conclusion was based, again, not on evidence, but rather on the assumption that the risk of leukemia will decrease as exposure levels decrease. Although the Agency had found it impossible to construct a dose-response curve that would predict with any accuracy the number of leukemias that could be expected to result from exposures at 10 ppm, at 1 ppm, or at any intermediate level, it nevertheless "determined that the benefits of the proposed standard are likely to be appreciable."[43] 43 FED. REG., at 5941. In light of the Agency's disavowal of any ability to determine the numbers of employees likely to be

---

[38]OSHA rejected this testimony in part because it believed the exposure data in the epidemiological studies to be inadequate to formulate a dose-response curve. It also indicated that even if the testimony was accepted—indeed as long as there was any increase in the risk of cancer—the agency was under an obligation to "select the level of exposure which is most protective of exposed employees." 43 FED. REG., at 5941.

[40]"As stated above, the positive studies on benzene demonstrate the causal relationship of benzene to the induction of leukemia. Although these studies, for the most part involve high exposure levels, it is OSHA's view that once the carcinogenicity of a substance has been established qualitatively, any exposure must be considered to be attended by risk when considering any given population. OSHA therefore believes that occupational exposure to benzene at low levels poses a carcinogenic risk to workers." 43 FED. REG., at 5932.

[42]"There is no doubt that benzene is a carcinogen and must, for the protection and safety of workers, be regulated as such. Given the inability to demonstrate a threshold or establish a safe level, it is appropriate that OSHA prescribe that the permissible exposure to benzene be reduced to the lowest level feasible." 43 FED. REG., at 5932.

[43]At an earlier point in its explanation, OSHA stated:

"There is general agreement that benzene exposure causes leukemia as well as other fatal diseases of the bloodforming organs. In spite of the certainty of this conclusion, there does not exist an adequate scientific basis for establishing the quantitative dose response relationship between exposure to benzene and the induction of leukemia and other blood diseases. The uncertainty in both the actual magnitude of expected deaths and in the theory of extrapolation from existing data to the OSHA exposure levels places the estimation of benefits on 'the frontiers of scientific knowledge.' While the actual estimation of the number of cancers to be prevented is highly uncertain, the evidence indicates that the number may be appreciable. There is general agreement that even in the absence of the ability to establish a 'threshold' or 'safe'

adversely affected by exposures of 10 ppm, the Court of Appeals held this finding to be unsupported by the record. 581 F.2d, at 503.[44]

It is noteworthy that at no point in its lengthy explanation did the Agency quote or even cite §3(8) of the Act. It made no finding that any of the provisions of the new standard were "reasonably necessary or appropriate to provide safe or healthful employment and places of employment." Nor did it allude to the possibility that any such finding might have been appropriate.

III

Our resolution of the issues in this case turns, to a large extent, on the meaning of and the relationship between §3(8), which defines a health and safety standard as a standard that is "reasonably necessary and appropriate to provide safe or healthful employment," §6(b)(5), which directs the Secretary in promulgating a health and safety standard for toxic materials to "set the standard which most adequately assures, to the extent feasible, on the basis of the best available evidence, that no employee will suffer material impairment of health or functional capacity * * *."

In the Government's view, §3(8)'s definition of the term "standard" has no legal significance or at best merely requires that a standard not be totally irrational. It takes the position that §6(b)(5) is controlling and that it requires OSHA to promulgate a standard that either gives an absolute assurance of safety for each and every worker or that reduces exposures to the lowest level feasible. The Government interprets "feasible" as meaning technologically achievable at a cost that would not impair the viability of the industries subject to the regulation. The respondent industry representatives, on the other hand, argue that the Court of Appeals was correct in holding that the "reasonably necessary and appropriate" language of §3(8), along with the feasibility requirement of §6(b)(5), requires the Agency to quantify both the costs and the benefits of a proposed rule and to conclude that they are roughly commensurate.

In our view, it is not necessary to decide whether either the Government or industry is entirely correct. For we think it is clear that §3(8) does apply to all permanent standards promulgated under the Act and that it requires the Secretary, before issuing any standard, to determine that it is reasonably necessary and appropriate to remedy a significant risk of material health impairment. Only after the Secretary has made the threshold determination that such a risk exists with respect to a toxic substance, would it be necessary to decide whether §6(b)(5) requires him to select the most protective standard he can consistent with economic and technological feasibility, or whether, as respondents argue, the benefits of the regulation must be commensurate with the costs of its implementation. Because the Secretary did not make the required threshold finding in this case, we have no occasion to determine whether costs must be weighed against benefits in an appropriate case.

***

B

By empowering the Secretary to promulgate standards that are "reasonably necessary or appropriate to provide safe or healthful employment and places of employment," the Act implies that, before promulgating any

level for benzene and other carcinogens, a dose response relationship is likely to exist; that is, exposure to higher doses carries with it a higher risk of cancer, and conversely, exposure to lower levels is accompanied by a reduced risk, even though a precise quantitative relationship cannot be established." 43 FED. REG., at 5940.

[44]The court did, however, hold that the Agency's other conclusions—that there is *some* risk of leukemia at 10 ppm and that the risk would decrease by decreasing the exposure limit to 1 ppm—were supported by substantial evidence. 581 F.2d, at 503.

standard, the Secretary must make a finding that the workplaces in question are not safe. But "safe" is not the equivalent of "risk-free." There are many activities that we engage in every day—such as driving a car or even breathing city air—that entail some risk of accident or material health impairment; nevertheless, few people would consider these activities "unsafe." Similarly, a workplace can hardly be considered "unsafe" unless it threatens the workers with a significant risk of harm.

Therefore, before he can promulgate *any* permanent health or safety standard, the Secretary is required to make a threshold finding that a place of employment is unsafe—in the sense that significant risks are present and can be eliminated or lessened by a change in practices. This requirement applies to permanent standards promulgated pursuant to §6(b)(5), as well as to other types of permanent standards. For there is no reason why §3(8)'s definition of a standard should not be deemed incorporated by reference into §6(b)(5). The standards promulgated pursuant to §6(b)(5) are just one species of the genus of standards governed by the basic requirement. That section repeatedly uses the term "standard" without suggesting any exception from, or qualification of, the general definition; on the contrary, it directs the Secretary to select "*the* standard"—that is to say, one of various possible alternatives that satisfy the basic definition in §3(8)—that is most protective. Moreover, requiring the Secretary to make a threshold finding of significant risk is consistent with the scope of the regulatory power granted to him by §6(b)(5), which empowers the Secretary to promulgate standards, not for chemicals and physical agents generally, but for "*toxic* chemicals" and "*harmful* physical agents."

***

In the absence of a clear mandate in the Act, it is unreasonable to assume that Congress intended to give the Secretary the unprecedented power over American industry that would result from the Government's view of §§3(8) and 6(b)(5), coupled with OSHA's cancer policy. Expert testimony that a substance is probably a human carcinogen—either because it has caused cancer in animals or because individuals have contracted cancer following extremely high exposures—would justify the conclusion that the substance poses some risk of serious harm no matter how minute the exposure and no matter how many experts testified that they regarded the risk as insignificant. That conclusion would in turn justify pervasive regulation limited only by the constraint of feasibility. In light of the fact that there are literally thousands of substances used in the workplace that have been identified as carcinogens or suspect carcinogens, the Government's theory would give OSHA power to impose enormous costs that might produce little, if any, discernible benefit.[51]

If the Government were correct in arguing that neither §3(8) nor §6(b)(5) requires that the risk from a toxic substance be quantified sufficiently to enable the Secretary to characterize it as significant in an under-

[51]OSHA's proposed generic cancer policy, 42 FED. REG. 54148 (Oct. 4, 1977), indicates that this possibility is not merely hypothetical. Under its proposal, whenever there is a certain quantum of proof—either from animal experiments, or, less frequently, from epidemiological studies—that a substance causes cancer at any exposure level, an emergency temporary standard would be promulgated immediately, requiring employers to provide monitoring and medical examinations and to reduce exposures to the lowest feasible level. A proposed rule would then be issued along the same lines, with objecting employers effectively foreclosed from presenting evidence that there is little or no risk associated with current exposure levels. 42 FED. REG., at 54154–54155, 54184.

The scope of the proposed regulation is indicated by the fact that NIOSH has published a list of 2,415 potential occupational carcinogens, NIOSH Suspected Carcinogens: A Subfile of the NIOSH Registry of Toxic Effects of Chemical Substances, HEW Pub. No. 77-149 (Dec. 1976). OSHA has tentatively concluded that 269 of these substances have been proved to be carcinogens and therefore should be subject to full regulation. See OSHA Press Release, USDL 78-625 (July 14, 1978).

standable way, the statute would make such a "sweeping delegation of legislative power" that it might be unconstitutional under the Court's reasoning in *Schechter Poultry Corp. v. United States,* 295 U.S. 495, 539, and *Panama Refining Co. v. Ryan,* 293 U.S. 388. A construction of the statute that avoids this kind of open-ended grant should certainly be favored.

C

The legislative history also supports the conclusion that Congress was concerned, not with absolute safety, but with the elimination of significant harm. The examples of industrial hazards referred to in the committee hearings and debates all involved situations in which the risk was unquestionably significant. For example, the Senate Committee on Labor and Public Welfare noted that byssinosis, a disabling lung disease caused by breathing cotton dust, affected as many as 30% of the workers in carding or spinning rooms in some American cotton mills and that as many as 100,000 active or retired workers were then suffering from the disease. It also noted that statistics indicated that 20,000 out of 50,000 workers who had performed insulation work were likely to die of asbestosis, lung cancer or mesothelyioma as a result of breathing asbestos fibers. Another example given of an occupational health hazard that would be controlled by the Act was betanaphthylamine, a "chemical so toxic that any exposure at all is likely to cause the development of bladder cancer over a period of years." S. Rep. No. 91-1282, at 3–4 (91st Cong., 2d Sess.).

Moreover, Congress specifically amended §6(b)(5) to make it perfectly clear that it does not require the Secretary to promulgate standards that would assure an absolutely risk-free workplace. Section 6(b)(5) of the initial Committee bill provided that

> The Secretary in promulgating standards under this subsection, shall set the standard which most adequately and feasibly assures, on the basis of the best available evidence, that no employee will suffer *any* impairment of health or functional capacity, or diminished life expectancy even if such employee has regular exposure to the hazard dealt with by such standard for the period of his working life. (Emphasis supplied.) S. 2193, 91st Cong., 2d Sess., at 40.

On the floor of the Senate, Senator Dominick questioned the wisdom of this provision, stating:

> How in the world are we ever going to live up to that? What are we going to do about a place in Florida where mosquitoes are getting at the employee—perish the thought that there may be mosquitoes in Florida? But there are black flies in Minnesota and Wisconsin. Are we going to say that if employees get bitten by those for the rest of their lives they will not have been done any harm at all? Probably they will not be, but do we know? [116 CONG. REC. 36,522 (1970).]

He then offered an amendment deleting the entire subsection. After discussions with the sponsors of the Committee bill, Senator Dominick revised his amendment. Instead of deleting the first sentence of §6(b)(5) entirely, his new amendment limited the application of that subsection to toxic materials and harmful physical agents and changed "any" impairment of health to "material" impairment. In discussing this change, Senator Dominick noted that the Committee's bill read as if a standard had to "assure that, no matter what anybody was doing, the standard would protect him for the rest of his life against any foreseeable hazard." Such an "unrealistic standard," he stated, had not been intended by the sponsors of the bill. Rather, he explained that the intention of the bill, as implemented by the amendment, was to require the Secretary

> * * * to use his best efforts to promulgate the best available standards, and in so doing, * * * he should take into account that anyone working in toxic agents and physical agents which might be harmful may be subjected to such conditions for the rest of his working life, so that we can get at something which might not be toxic now, if he works in it a short time, but if he works in it the rest of his life might be very dangerous; and we want to make sure that such things are taken into consideration in establishing standards." [116 CONG. REC. 37,623 (1970).]

Senator Williams, one of the sponsors of the Committee bill, agreed with the interpretation, and the amendment was adopted.

In its reply brief the Government argues that the Dominick amendment simply means that the Secretary is not required to eliminate threats of insignificant harm; it argues that §6(b)(5) still requires the Secretary to set standards that ensure that not even one employee will be subject to any risk of serious harm—no matter how small that risk may be. This interpretation is at odds with Congress' express recognition of the futility of trying to make all workplaces totally risk-free. Moreover, not even OSHA follows this interpretation of §6(b)(5) to its logical conclusion. Thus, if OSHA is correct that the only no-risk level for leukemia due to benzene exposure is zero and if its interpretation of §6(b)(5) is correct, OSHA should have set the exposure limit as close to zero as feasible. But OSHA did not go about its task in that way. Rather, it began with a 1 ppm level, selected at least in part to ensure that employers would not be required to eliminate benzene concentrations that were little greater than the so-called "background" exposures experienced by the population at large. Then, despite suggestions by some labor unions that it was feasible for at least some industries to reduce exposures to well below 1 ppm, OSHA decided to apply the same limit to all, largely as a matter of administrative convenience. 43 FED. REG. 5947 (1978).

OSHA also deviated from its own interpretation of §6(b)(5) in adopting an action level of 0.5 ppm below which monitoring and medical examinations are not required. In light of OSHA's cancer policy, it must have assumed that some employees would be at risk because of exposures below 0.5 ppm. These employees would thus presumably benefit from medical examinations, which might uncover any benzene-related problems. OSHA's consultant advised the Agency that it was technologically and economically feasible to require that such examinations be provided. Nevertheless, OSHA adopted an action level, largely because the insignificant benefits of giving such examinations and performing the necessary monitoring did not justify the substantial cost.

OSHA's concessions to practicality in beginning with a 1 ppm exposure limit and using an action level concept implicitly adopt an interpretation of the statute as not requiring regulation of insignificant risks.[58] It is entirely consistent with this interpretation to hold that the Act also requires the Agency to limit its endeavors in the standard-setting area to eliminating significant risks of harm.

Finally, with respect to the legislative history, it is important to note that Congress repeatedly expressed its concern about allowing the Secretary to have too much power over American industry. Thus, Congress refused to give the Secretary the power to shut down plants unilaterally because of an imminent danger, see *Whirlpool Corp. v. Marshall,* 445 U.S. 1

[58]The Government also states that it is OSHA's policy to attempt to quantify benefits wherever possible. While this is certainly a reasonable position, it is not consistent with OSHA's own view of its duty under §6(b)(5). In light of the inconsistencies in OSHA's position and the legislative history of the Act, we decline to defer to the Agency's interpretation.

(1980), and narrowly circumscribed the Secretary's power to issue temporary emergency standards.[59] This effort by Congress to limit the Secretary's power is not consistent with a view that the mere possibility that some employee somewhere in the country may confront some risk of cancer is a sufficient basis for the exercise of the Secretary's power to require the expenditure of hundreds of millions of dollars to minimize that risk.

D

Given the conclusion that the Act empowers the Secretary to promulgate health and safety standards only where a significant risk of harm exists, the critical issue becomes how to define and allocate the burden of proving the significance of the risk in a case such as this, where scientific knowledge is imperfect and the precise quantification of risks is therefore impossible. The Agency's position is that there is substantial evidence in the record to support its conclusion that there is no absolutely safe level for a carcinogen and that, therefore, the burden is properly on industry to prove, apparently beyond a shadow of a doubt, that there *is* a safe level for benzene exposure. The Agency argues that, because of the uncertainties in this area, any other approach would render it helpless, forcing it to wait for the leukemia deaths that it believes are likely to occur before taking any regulatory action.

We disagree. As we read the statute, the burden was on the Agency to show, on the basis of substantial evidence, that it is at least more likely than not that long-term exposure to 10 ppm of benzene presents a significant risk of material health impairment. Ordinarily, it is the proponent of a rule or order who has the burden of proof in administrative proceedings. See 5 U.S.C. §556(d). In some cases involving toxic substances, Congress has shifted the burden of proving that a particular substance is safe onto the party opposing the proposed rule. The fact that Congress did not follow this course in enacting OSHA indicates that it intended the Agency to bear the normal burden of establishing the need for a proposed standard.

In this case OSHA did not even attempt to carry its burden of proof. The closest it came to making a finding that benzene presented a significant risk of harm in the workplace was its statement that the benefits to be derived from lowering the permissible exposure level from 10 to 1 ppm were "likely" to be "appreciable." The Court of Appeals held that this finding was not supported by substantial evidence. Of greater importance, even if it were supported by substantial evidence, such a finding would not be sufficient to satisfy the Agency's obligations under the Act.

***

Contrary to the Government's contentions, imposing a burden on the Agency of demonstrating a significant risk of harm will not strip it of its ability to regulate carcinogens, nor will it require the Agency to wait for deaths to occur before taking any action. First, the requirement that a "significant" risk be identified is not a mathematical straitjacket. It is the Agency's responsibility to determine, in the first instance, what it considers to be a "significant" risk. Some risks are plainly acceptable and others are plainly

[59]In *Florida Peach Growers Assn., Inc. v. Dept. of Labor,* 489 F.2d 120, 130, and n.16 (CA5 1974), the court noted that Congress intended to restrict the use of emergency standards, which are promulgated without any notice or hearing. It held that, in promulgating an emergency standard, OSHA must find not only a danger *of* exposure or even some danger *from* exposure, but a grave danger from exposure necessitating emergency action. Accord, *Dry Color Mfrs. Assn., Inc. v. Dept. of Labor,* 486 F.2d 98, 100 (CA3 1973) (an emergency standard must be supported by something more than a possibility that a substance may cause cancer in man).

Congress also carefully circumscribed the Secretary's enforcement powers by creating a new, independent board to handle appeals from citations issued by the Secretary for noncompliance with health and safety standards. See 29 U.S.C. §§659–661.

unacceptable. If, for example, the odds are one in a billion that a person will die from cancer by taking a drink of chlorinated water, the risk clearly could not be considered significant. On the other hand, if the odds are one in a thousand that regular inhalation of gasoline vapors that are two percent benzene will be fatal, a reasonable person might well consider the risk significant and take appropriate steps to decrease or eliminate it. Although the Agency has no duty to calculate the exact probability of harm, it does have an obligation to find that a significant risk is present before it can characterize a place of employment as "unsafe."

Second, OSHA is not required to support its finding that a significant risk exists with anything approaching scientific certainty. Although the Agency's findings must be supported by substantial evidence, 29 U.S.C. §655(f), §6(b)(5) specifically allows the Secretary to regulate on the basis of the "best available evidence." As several courts of appeals have held, this provision requires a reviewing court to give OSHA some leeway where its findings must be made on the frontiers of scientific knowledge. See *Industrial Union Dept., AFL-CIO v. Hodgson,* 162 U.S. App. D.C. 331, 340, 499 F.2d 467, 476 (1974); *Society of the Plastics Industry, Inc. v. OSHA,* 509 F.2d 1301, 1308 (CA2 1975), *cert. denied,* 421 U.S. 992. Thus, so long as they are supported by a body of reputable scientific thought, the Agency is free to use conservative assumptions in interpreting the data with respect to carcinogens, risking error on the side of over-protection rather than under-protection.

Finally, the record in this case and OSHA's own rulings on other carcinogens indicate that there are a number of ways in which the Agency can make a rational judgment about the relative significance of the risks associated with exposure to a particular carcinogen.[64]

***

E

Because our review of this case has involved a more detailed examination of the record than is customary, it must be emphasized that we have neither made any factual determinations of our own, nor have we rejected

[64]For example, in the coke oven emissions standard, OSHA had calculated that 21,000 exposed coke oven workers had an annual excess mortality of over 200 and that the proposed standard might well eliminate the risk entirely. 41 FED. REG. 46742, 46750 (Oct. 22, 1976), upheld in *American Iron & Steel Inst. v. OSHA,* 577 F.2d 825 (CA3 1978), [*cert. dismissed,* 448 U.S. 917 (1980)]. In hearings on the coke oven emissions standard, the Council on Wage and Price Stability estimated that 8 to 35 lives would be saved each year, out of an estimated population of 14,000 workers, as a result of the proposed standard. Although noting that the range of benefits would vary depending on the assumptions used, OSHA did not make a finding as to whether its own staff estimate or CWPS's was correct, on the ground that it was not required to quantify the expected benefits of the standard or to weigh those benefits against the projected costs.

In other proceedings, the Agency has had a good deal of data from animal experiments on which it could base a conclusion on the significance of the risk. For example, the record on the vinyl chloride standard indicated that a significant number of animals had developed tumors of the liver, lung and skin when they were exposed to 50 ppm of vinyl chloride over a period of 11 months. One hundred out of 200 animals died during that period. 39 FED. REG. 35890, 35891 (Oct. 4, 1974). Similarly, in a 1974 standard regulating 14 carcinogens, OSHA found that one of the substances had caused lung cancer in mice or rats at 1 ppm and even 0.1 ppm, while another had caused tumors in 80% of the animals subjected to high doses. 39 FED. REG. 3756, 3757 (Jan. 29, 1974), upheld in *Synthetic Organic Chemical Mfrs. Assn. v. Brennan,* 503 F.2d 1155 (CA3 1974), *cert. denied,* 420 U.S. 973, and 506 F.2d 385 (CA3 1974), *cert denied,* 423 U.S. 830.

In this case the Agency did not have the benefit of animal studies, because scientists have been unable as yet to induce leukemia in experimental animals as a result of benzene exposure. It did, however, have a fair amount of epidemiological evidence, including both positive and negative studies. Although the Agency stated that this evidence was insufficient to construct a precise correlation between exposure levels and cancer risks, it would at least be helpful in determining whether it is more likely than not that there is a significant risk at 10 ppm.

any factual findings made by the Secretary. We express no opinion on what factual findings this record might support, either on the basis of empirical evidence or on the basis of expert testimony; nor do we express any opinion on the more difficult question of what factual determinations would warrant a conclusion that significant risks are present which make promulgation of a new standard reasonably necessary or appropriate. The standard must, of course, be supported by the findings actually made by the Secretary, not merely by findings that we believe he might have made.

In this case the record makes it perfectly clear that the Secretary relied squarely on a special policy for carcinogens that imposed the burden on industry of proving the existence of a safe level of exposure, thereby avoiding the Secretary's threshold responsibility of establishing the need for more stringent standards. In so interpreting his statutory authority, the Secretary exceeded his power.

***

MR. JUSTICE POWELL, concurring in part and in the judgment.

***

Although I regard the question as close, I do not disagree with the plurality's view that OSHA has failed, on this record, to carry its burden of proof on the threshold issues summarized above. But even if one assumes that OSHA properly met this burden, * * * I conclude that the statute also requires the agency to determine that the economic effects of its standard bear a reasonable relationship to the expected benefits. An occupational health standard is neither "reasonably necessary" nor "feasible," as required by statute, if it calls for expenditures wholly disproportionate to the expected health and safety benefits.

***

MR. JUSTICE REHNQUIST, concurring in the judgment.

***

If we are ever to reshoulder the burden of ensuring that Congress itself make the critical policy decisions, this is surely the case in which to do it. It is difficult to imagine a more obvious example of Congress simply avoiding a choice which was both fundamental for purposes of the statute and yet politically so divisive that the necessary decision or compromise was difficult, if not impossible, to hammer out in the legislative forge. Far from detracting from the substantive authority of Congress, a declaration that the first sentence of §6(b)(5) of the OSHA constitutes an invalid delegation to the Secretary of Labor would preserve the authority of Congress. If Congress wishes to legislate in an area which it has not previously sought to enter, it will in today's political world undoubtedly run into opposition no matter how the legislation is formulated. But that is the very essence of legislative authority under our system. It is the hard choices, and not the filling in of the blanks, which must be made by the elected representatives of the people. When fundamental policy decisions underlying important legislation about to be enacted are to be made, the buck stops with Congress and the President insofar as he exercises his constitutional role in the legislative process.

I would invalidate the first sentence of §6(b)(5) of the Occupational Safety and Health Act of 1970 as it applies to any toxic substance or harmful physical agent for which a safe level, that is a level at which "no employee will suffer material impairment of health or functional capacity even if such employee has regular exposure to [that hazard] for the period of his working life[,]" is, according to the Secretary, unknown or otherwise "infeasible." Absent further congressional action, the Secretary would then have to choose, when acting pursuant to §6(b)(5), between setting a safe standard or setting no standard at all. Accordingly, for the reasons stated above, I concur in the judgment of the Court affirming the judgment of the Court of Appeals.

MR. JUSTICE MARSHALL, with whom MR. JUSTICE BRENNAN, MR. JUSTICE WHITE, and MR. JUSTICE BLACKMUN join, dissenting.

In cases of statutory construction, this Court's authority is limited. If the statutory language and legislative intent are plain, the judicial inquiry is at an end. Under our jurisprudence, it is presumed that ill-considered or unwise legislation will be corrected through the democratic process; a court is not permitted to distort a statute's meaning in order to make it conform with the Justices' own views of sound social policy. See *TVA v. Hill,* 437 U.S. 153 (1978).

Today's decision flagrantly disregards these restrictions on judicial authority. The plurality ignores the plain meaning of the Occupational Safety and Health Act of 1970 in order to bring the authority of the Secretary of Labor in line with the plurality's own views of proper regulatory policy. The unfortunate consequence is that the Federal Government's efforts to protect American workers from cancer and other crippling diseases may be substantially impaired.

* * * In this case the Secretary of Labor found, on the basis of substantial evidence, that (1) exposure to benzene creates a risk of cancer, chromosomal damage, and a variety of nonmalignant but potentially fatal blood disorders, even at the level of 1 ppm; (2) no safe level of exposure has been shown; (3) benefits in the form of saved lives would be derived from the permanent standard; (4) the number of lives that would be saved could turn out to be either substantial or relatively small; (5) under the present state of scientific knowledge, it is impossible to calculate even in a rough way the number of lives that would be saved, at least without making assumptions that would appear absurd to much of the medical community; and (6) the standard would not materially harm the financial condition of the covered industries. The Court does not set aside any of these findings. Thus, it could not be plainer that the Secretary's decision was fully in accord with his statutory mandate "most adequately [to] assure[] * * * that no employee will suffer material impairment of health or functional capacity * * * "

The plurality's conclusion to the contrary is based on its interpretation of 29 U.S.C. §652(8), which defines an occupational safety and health standard as one "which requires conditions * * * reasonably necessary or appropriate to provide safe or healthful employment * * *." According to the plurality, a standard is not "reasonably necessary or appropriate" unless the Secretary is able to show that it is "at least more likely than not" that the risk he seeks to regulate is a "significant" one. Nothing in the statute's language or legislative history, however, indicates that the "reasonably necessary or appropriate" language should be given this meaning. Indeed, both demonstrate that the plurality's standard bears no connection with the acts or intentions of Congress and is based only on the plurality's solicitude for the welfare of regulated industries. And the plurality used this standard to evaluate not the agency's decision in this case, but a strawman of its own creation.

Unlike the plurality, I do not purport to know whether the actions taken by Congress and its delegates to ensure occupational safety represent sound or unsound regulatory policy. The critical problem in cases like the one at bar is scientific uncertainty. While science has determined that exposure to benzene at levels above 1 ppm creates a definite risk of health impairment, the magnitude of the risk cannot be quantified at the present time. The risk at issue has hardly been shown to be insignificant; indeed, future research may reveal that the risk is in fact considerable. But the existing evidence may frequently be inadequate to enable the Secretary to make the threshold finding of "significance" that the Court requires today. If so, the consequence of the plurality's approach would be to subject American workers to a continuing risk of cancer and other fatal diseases, and to render the Federal Government powerless to take protective action on their behalf. Such an approach would place the burden of medical uncertainty squarely on the shoulders of the American worker, the intended beneficiary of the Occupational Safety and Health Act. It is fortunate indeed that at least a majority of the Justices reject the view that the Secretary is prevented from taking regulatory action when the magnitude of a health risk cannot be quantified on the basis of current techniques. * * *

Because today's holding has no basis in the Act, and because the Court has no authority to impose its own regulatory policies on the Nation, I dissent.

***

II

The plurality's discussion of the record in this case is both extraordinarily arrogant and extraordinarily unfair. It is arrogant because the plurality presumes to make its own factual findings with respect to a variety of disputed issues relating to carcinogen regulation. * * * It should not be necessary to remind the Members of this Court that they were not appointed to undertake independent review of adequately supported scientific findings made by a technically expert agency.[9]

***

The plurality is obviously more interested in the consequences of its decision than in discerning the intention of Congress. But since the language and legislative history of the Act are plain, there is no need for conjecture about the effects of today's decision. "It is not for us to speculate, much less act, on whether Congress would have altered its stance had the specific events of this case been anticipated." *TVA v. Hill,* 437 U.S., at 185. I do not pretend to know whether the test the plurality erects today is, as a matter of policy, preferable to that created by Congress and its delegates: the area is too fraught with scientific uncertainty, and too dependent on considerations of policy, for a court to be able to determine whether it is desirable to require identification of a "significant" risk before allowing an administrative agency to take regulatory action. But in light of the tenor of the plurality opinion, it is necessary to point out that the question is not one-sided, and that Congress' decision to authorize the Secretary to promulgate the regulation at issue here was a reasonable one.

***

In these circumstances it seems clear that the Secretary found a risk that is "significant" in the sense that the word is normally used. There was some direct evidence of chromosomal damage, nonmalignant blood disorders, and leukemia at exposures at or near 10 ppm and below. In addition, expert after expert testified that the recorded effects of benzene exposure at higher levels justified an inference that an exposure level above 1 ppm was dangerous. The plurality's extraordinarily searching scrutiny of this factual

[9] I do not, of course, suggest that it is appropriate for a federal court reviewing agency action blindly to defer to the agency's findings of fact and determinations of policy. Under *Citizens to Preserve Overton Park, Inc. v. Volpe,* 401 U.S. 402, 416 (1971), courts must undertake a "searching and careful" judicial inquiry into these factors. Such an inquiry is designed to require the agency to take a "'hard look,'" *Kleppe v. Sierra Club,* 427 U.S. 390, 410 (1976) (citation omitted), by considering the proper factors and weighing them in a reasonable manner. There is also room for especially rigorous judicial scrutiny of agency decisions under a rationale akin to that offered in *United States v. Carolene Products, Inc.,* 304 U.S. 144, 152, n.4 (1938). See *Environmental Defense Fund v. Ruckelshaus,* 142 U.S. App. D.C. 74, 439 F.2d 584 (1971).

I see no basis, however, for the approach taken by the plurality today, which amounts to nearly *de novo* review of questions of fact and of regulatory policy on behalf of institutions that are by no means unable to protect themselves in the political process. Such review is especially inappropriate when the factual questions at issue are ones about which the Court cannot reasonably be expected to have expertise.

record reveals no basis for a conclusion that quantification is, on the basis of "the best available evidence," possible at the present time. If the Secretary decided to wait until definitive information was available, American workers would be subjected for the indefinite future to a possibly substantial risk of benzene-induced leukemia and other illnesses. It is unsurprising, at least to me, that he concluded that the statute authorized him to take regulatory action now.

***

In recent years there has been increasing recognition that the products of technological development may have harmful effects whose incidence and severity cannot be predicted with certainty. The responsibility to regulate such products has fallen to administrative agencies. Their task is not an enviable one. Frequently no clear causal link can be established between the regulated substance and the harm to be averted. Risks of harm are often uncertain, but inaction has considerable costs of its own. The agency must decide whether to take regulatory action against possibly substantial risks or to wait until more definitive information becomes available—a judgment which by its very nature cannot be based solely on determinations of fact.

Those delegations, in turn, have been made on the understanding that judicial review would be available to ensure that the agency's determinations are supported by substantial evidence and that its actions do not exceed the limits set by Congress. In the Occupational Safety and Health Act, Congress expressed confidence that the courts would carry out this important responsibility. But in this case the plurality has far exceeded its authority. The plurality's "threshold finding" requirement is nowhere to be found in the Act and is antithetical to its basic purposes. "The fundamental policy questions appropriately resolved in Congress * * * are *not* subject to re-examination in the federal courts under the guise of judicial review of agency action." *Vermont Yankee Nuclear Power Corp. v. NRDC,* 435 U.S. 519, 558 (1978) (emphasis in original). Surely this is no less true of the decision to ensure safety for the American worker than the decision to proceed with nuclear power. See *ibid.*

Because the approach taken by the plurality is so plainly irreconcilable with the Court's proper institutional role, I am certain that it will not stand the test of time. In all likelihood, today's decision will come to be regarded as an extreme reaction to a regulatory scheme that, as the Members of the plurality perceived it, imposed an unduly harsh burden on regulated industries. But as the Constitution "does not enact Mr. Herbert Spencer's Social Statics," *Lochner v. New York,* 198 U.S. 45, 75 (1905) (Holmes, J., dissenting), so the responsibility to scrutinize federal administrative action does not authorize this Court to strike its own balance between the costs and benefits of occupational safety standards. I am confident that the approach taken by the plurality today, like that in *Lochner* itself, will eventually be abandoned, and that the representative branches of government will once again be allowed to determine the level of safety and health protection to be accorded to the American worker.

Several aspects of this highly significant decision should be emphasized.[31] First, the plurality expressly struck down the OSHA policy of regulating carcinogens to the "lowest feasible level." The Court

[31]The majority of the Court also set aside OSHA's ban on dermal exposure to benzene on the grounds that the Agency failed to make the requisite finding that the prohibition was "reasonably necessary and appropriate" to remove a significant risk of harm from such contact. The Court, therefore, found it unnecessary to reach the question, decided by the court of appeals, of whether OSHA was required by the "best available evidence" requirement to obtain more information on the hazards from dermal exposure. 448 U.S. at 662. Justice Marshall dissented on this issue as well, *id.* at 722 n.35.

held that the policy was based on "assumptions" rather than evidence and it impermissibly shifted the burden of proof from the Agency to the regulated industry to show that there was a safe level of exposure. Shortly after the Supreme Court decision, OSHA amended its Carcinogens Policy to delete its "lowest feasible level" policy in conformance with the *Benzene* decision.[32]

Justice Stevens' plurality opinion also expressed serious concern over the regulatory implications of the government's "lowest feasible level" policy. He noted the possibility of "pervasive regulation" by OSHA subject only to feasibility constraints, "no matter how minute the exposure," and said it would be "unreasonable to assume that Congress intended to give the Secretary the unprecedented power over industry that would result from the Government view."[33] He stated further that Congress' repeated concerns over OSHA having "too much power over American industry" was not consistent with the OSHA position that "the mere possibility that some employee somewhere in the country may confront some risk of concern is a sufficient basis for the exercise of the Secretary's power to require the expenditure of hundreds of millions of dollars to minimize that risk."[34]

Second, the principal holding of the *Benzene* plurality was that §3(8)[35] requires an Agency showing, by substantial evidence, that "at least more likely than not" the new standard will eliminate or reduce a "significant risk" in the workplace. The Court declined to defer to OSHA's contrary interpretation of the Act, partly because of the "inconsistencies in OSHA's position." The Court relied in this regard particularly on the exemption of gasoline station employees from the requirements of the standard despite the fact that they were regularly exposed to the inhalation of gasoline vapors;[36] on OSHA's setting the same limit for all industries, "largely as a matter of administrative convenience," although there was a suggestion that in some industries a lower level would be feasible;[37] and on OSHA's establishing an action level of 0.5 ppm for medical surveillance requirements, despite the record showing that such examinations would be technologically and economically feasible and all exposed employees would benefit from them.[38] Thus, in the Court's view, OSHA's "concessions to practicality" undermined its arguments favoring a highly protective standard.

---

[32]46 FED. REG. 4889 (1981).

[33]448 U.S. at 645.

[34]*Id.* at 651–52. Although the plurality did not reach the cost-benefit issue, this and other remarks in Justice Stevens' opinion suggest that cost-benefit considerations may have played some part in its thinking. However, Justice Stevens later joined the majority in the *Cotton Dust* decision in barring cost-benefit under §6(b)(5). *American Textile Mfrs. Inst. v. Donovan,* 452 U.S. 490 (1981), reprinted in Chapter 10.

[35]29 U.S.C. §652(8).

[36]See 448 U.S. at 628, 655.

[37]*Id.* at 650. The Supreme Court's view is similar to the conclusion of the District of Columbia Circuit Court of Appeals in the *Asbestos* case, which disagreed with OSHA's decision to delay the lower PEL for *all* industries. See Chapter 6, Section A.

[38]448 U.S. at 650. The plurality also questioned OSHA's establishing an action level for air monitoring requirements. *Id.*

Finally, the plurality said that the risk from a substance must be "quantified sufficiently to enable the Secretary to characterize it as significant in an understandable way."[39] The plurality made clear that there were a "number of ways" by means of which the Agency could make a "rational judgment" on the degree of risk from a carcinogen at low levels. It noted OSHA's use of epidemiological data as a basis for estimating the degree of risk in the coke oven emission standard and the similar use of animal experimental data in regulating vinyl chloride and 14 carcinogens.[40]

And, significantly, the plurality made clear that the Supreme Court, by requiring a significant risk finding, did not intend to "strip [OSHA] of its ability to regulate carcinogens," nor will the Court's decision "require the Agency to wait for deaths to occur before taking any action." First, the plurality said, the significant risk requirement is not a "mathematical straitjacket"; the determination that a particular level of risk is "significant" will be "based largely on policy considerations," presumably with considerable court deference to the Agency decision.[41] Secondly, the Court said that OSHA need not support its finding "with anything approaching scientific certainty" and that it can utilize "conservative assumptions" in interpreting data, "risking error on the side of overprotection rather than underprotection."[42] With particular regard to benzene, the opinion said, while a "precise correlation between exposure levels and cancer risks may not be possible," on the basis of the available epidemiological evidence, it would "at least be helpful" in determining if there was a significant risk.[43] The plurality of the Court would thus give the Agency considerable flexibility in performing risk assessments and determining that a risk was significant; however, the Court made it clear that it would not permit OSHA to avoid the process entirely by relying on a "policy" that makes attempts at quantitative estimation completely unnecessary.

The impact of the Supreme Court decision on OSHA regulatory policy would appear to be less constraining than that of the Fifth Circuit opinion. The Supreme Court plurality left open the possibility that it would permit OSHA to reduce the level of benzene to 1 ppm if the requisite evidence of significant risk was adduced. The Fifth Circuit, on the other hand, seemed to accept OSHA's view that reliable risk assessments could not be done, but reached the conclusion that regulation would have to wait until better data were available.[44] Thus, consistent with the Supreme Court's opinion and on the basis of the same record, OSHA could utilize "conservative assumptions" and "[risk] error on the side of overprotection" in constructing a dose-response curve showing benzene to be a significant risk. Indeed,

[39]*Id.* at 646.
[40]*Id.* at 656–57 and n.64.
[41]*Id.* at 655 n.62.
[42]*Id.* at 656.
[43]*Id.* at 657 n.64.
[44]*American Petroleum Inst. v. OSHA*, 581 F.2d at 504, and n.23.

OSHA did precisely that in its reply brief analyzing the Professor Wilson testimony. OSHA's basic argument that risk assessments could not be done reliably may have been undercut by the analysis in its reply brief of Dr. Wilson's testimony—in purporting to show that the Wilson risk analysis was not reliable, OSHA showed the opposite, namely, that risk assessments could be, and were being, done. The Supreme Court took OSHA at its word, insisting that OSHA engage in risk quantification, but allowing OSHA a degree of flexibility in making "conservative"—that is, protective—assumptions in performing the quantifications.

It was on this point that the dissenters strongly disagreed. The plurality approach, said Justice Marshall, would force OSHA "to deceive the public by acting on the basis of assumptions that must be considered too speculative to support any realistic assessment of the relevant risk".[45] Stated somewhat differently, the dissenting view was that the quantification of risk through risk assessments was no less a policy determination by the Agency than the "lowest feasible level" policy which was rejected by the plurality.

The reaction to the Supreme Court decision in *Benzene* was strong and mixed. Dr. Lester B. Lave, writing in 1982 on the issue of quantitative risk assessment in regulation, concluded that the *Benzene* case is a "classic illustration of the cost of being unprepared."[46] He suggested that "more careful analysis initially could have produced better scientific evidence" and a regulation that would have been sustained. He stated that "[t]his case shows the unfortunate consequences of a decision to base a health standard on political grounds without careful analysis of the scientific support for such action."[47] Dr. Lave's general conclusion from the benzene and other regulatory case studies is that "quantitative risk assessment could have improved estimation of risks associated with each substance and could have helped the agency to design better standards."[48]

A highly critical view of the result was expressed by Professor Neil J. Sullivan of Baruch College, City University of New York.[49]

Sullivan, *The Benzene Decision: A Contribution to Regulatory Confusion*

***

A review of the decision in the benzene case leads to three important conclusions. The first is that the Court has confused the development of

[45]448 U.S. at 716 (Marshall, J., dissenting).
[46]INTRODUCTION TO QUANTITATIVE RISK ASSESSMENT IN REGULATION 18 (L.B. Lave ed. 1982).
[47]*Id.*
[48]*Id.* at 21. In 1981, Dr. Lave, testifying in favor of regulatory reform legislation, stated that Justice Rehnquist's dissent in the *Cotton Dust* Supreme Court decision was an "insightful statement as to the fundamental problem with current regulation: Congress has not decided what it wants." *Regulatory Reform Legislation of 1981: Hearings Before the Senate Comm. on Governmental Affairs,* 97th Cong., 1st Sess. 256 (1981).
[49]Sullivan, *The Benzene Decision: A Contribution to Regulatory Confusion,* 33 AD. L. REV. 351 at 363–65 (1981).

policy making for OSHA and similar regulatory agencies. The assessment of risks, costs and benefits in the effort to minimize health hazards in the workplace is an acutely complex exercise. Any appropriate policy should rest on the most capable judgment of health specialists balanced by the considerations of economists and others who can estimate the impact of particular regulatory schemes.

What the Court has done by its decision in *IUD v. API* is to subordinate the expertise of agency and private sector specialists to the opinions of various federal judges. Only after those complex scientific and economic arguments have been filtered through the presentations of attorneys will final policies be established. Since the interest of those attorneys is successful litigation, the development of carcinogen and toxic substance policies rests on the machinations of the judicial process rather than on the judgment of those who are best trained to know how to treat the hazards of our workplaces.

Agencies, workers and business executives cannot respond effectively to regulatory standards because no such standards exist. The opinions of individual judges will vary on the merits of any particular case; and, as the debate between Judges Bazelon and Leventhal indicates, even the method of deciding complex regulatory issues is open to dispute. Agencies can adopt policies which they hope the courts will find palatable, and companies can decide to comply or to resist those policies. But neither government nor business can be confident in its actions since the ultimate decision rests on the opinion of a magistrate.

The second conclusion is that the benzene decision simply places too much authority for regulatory policies in the hands of the judiciary. The problem is not that a particular standard has been approved or rejected by the courts. Rather, the scope of the courts' consideration is troubling. The benzene regulation was not held to be invalid on a narrow constitutional issue or because it clearly contradicted the mandate of the enabling act. The plurality's opinion rest[s] on the inexpert conclusion that OSHA had failed to demonstrate the significance of risk to benzene exposure.

The opinions of experts will, of course, conflict on such issues as the significance of risks, but the court offers no solace that the judicial opinion is superior to the bureaucratic one. Certainly one cannot argue that the judgment of the court is more democratic, since judges are insulated from even such indirect popular controls as the budgetary process. Nor can one argue that the Supreme Court has articulated a policy which will offer clear guidance to all lower courts in their rulings of regulatory standards. Not even the majority of the Burger Court itself could agree about the basis for its conclusion to invalidate the benzene standard. The prospect for coherent review in the various appellate circuits is indeed dim.

One must conclude that the Court has assumed authority without constraint. The absence of predictability which is inherent in the benzene decision renders the courts policymakers whose expertise is suspect and whose reasoning is haphazard. Government regulation has been criticized in recent years for being arbitrary and unrelated to the realities of the business community. The benzene decision may relieve those who dreaded compliance with OSHA's regulation, but it can hardly encourage those who are concerned about the more serious implications of arbitrary government policymaking.

The final conclusion about this case concerns the appropriate object for censure. One can fault the courts for venturing into a realm of decision-making where they have little of value to offer, but the real source of this dismal prospect is Congress. The benzene decision represents one of a number of consequences of legislation which is little more than a thinly veiled compromise among political interests.

The fact that the Burger Court could reach such a remarkable number of distinct conclusions about OSHA's responsibility in rulemaking suggests that legislative intent is unacceptably vague. The vagaries of case-by-case policymaking in the federal courts can be greatly minimized if Congress exercises its appropriate authority and legislates in a clear manner what it wants OSHA to consider in the promulgation of its standards.

The rulemaking authority of regulatory agencies is a legislative function. It necessarily means tradeoffs among affected parties and the promotion and retarding of the interests of numerous groups. Such considerations should be resolved in a political forum, the legislative arena, rather than in the courts. Challenges to regulations turn on matters of political and economic interest much more than on strained considerations of constitutional rights.

The appropriate recourse for a review of agency regulations is for the courts to limit their review to constitutional issues and for Congress to entertain political objections to standards. For example, a regulation could take effect unless the majority of each House votes affirmatively to invalidate the regulation. The objection to the benzene regulation would thus be reviewed in a more appropriate forum, and the legislative intent concerning such concepts as the significance of a hazard could be clarified.

After such reform, regulations would remain complex; interests would still be greatly affected and political repercussions would remain high. An abundance of interesting contests would thus remain for our entertainment and illumination. But authority and responsibility in government would be more appropriately allocated than it currently is, and the judicial abetting of legislative irresponsibility and administrative futility would be checked.

Even those who would agree with Professor Sullivan's critique of the Supreme Court decision in the *Benzene* case might be troubled with his proposed remedy, i.e., the legislative veto. Apart from its constitutional invalidity, the legislative veto has often been criticized on both theoretical and practical grounds.[50]

A more favorable view of the Supreme Court decision was expressed by Richard M. Cooper, formerly chief counsel of the Food and Drug Administration.[51]

Cooper, *The Benzene Ruling—What the Court Decided*

Only the Supreme Court could have done it so stylishly.

In its most important toxic substances decision to date, the court recently overturned a new standard developed by OSHA to reduce workers' exposure to benzene, a human carcinogen. The case offers a fascinating view of the interactions of the Supreme Court, Congress and a regulatory agency. And it presents in sharp focus a conflict between the court's duty to carry out the expressed intent of Congress and its duty to decide in favor of sensible public policy.

[50]For a discussion of the legislative veto, see Chapter 21. The Supreme Court held legislative veto unconstitutional in *Immigration & Naturalization Service v. Chadha*, 51 USLW 4907 (1983). For another critical view of the *Benzene* decision, see Shaw & Wolfe, *A Legal and Ethical Critique of Using Cost-Benefit Analysis in Public Law*, 19 HOUS. L. REV. 899, 924–26 (1982) (arguing that *Benzene* was "essentially" a cost-benefit decision, and that it is "morally repugnant" and an "impossible task" for OSHA to be expected to put a price tag on benefits and show that the price is reasonable in comparison to costs, as required by the *Benzene* decision). See also Note, *Regulating Toxic Substances Under the Occupational Safety and Health Act: A Reconsideration of the 'Feasibility' Provisions*, 33 SYRACUSE L. REV. 887 (1982).

[51]Cooper, *The Benzene Ruling—What the Court Decided*, Washington Post, Aug. 27, 1980, at A15, col. 1. Reprinted with permission.

The decision did what the four dissenters said it did: it took away from the labor movement a political victory it had won when the Occupational Safety and Health Act was enacted in 1970. The statute directed OSHA, when regulating a toxic substance in the work place, to set an exposure standard that would ensure, to the extent feasible, "that no employee will suffer material impairment of health" even if he is exposed to the substance during his entire working life. Strong language. OSHA was trying to carry it out.

But four justices held that OSHA erred by failing to make a specific finding that exposure to benzene under the previous standard presented a "significant risk." The term "significant risk" did not appear in the statute. In a magisterial feat of magic, the plurality found the requirement in a definitional provision.

***

If you disregard the language of the statute, there is much to be said for the plurality's position, and it should not set back the cause of health protection in the work place. Under OSHA's view, an industry had to be forced to the brink of disaster, if necessary, to reduce the perhaps negligible risk from a single toxic substance; the industry would then have few or no resources left to reduce the risks from other toxic substances. Even if OSHA's reading of the statute were correct, it couldn't work over time. Eventually OSHA would have had to alter its course.

The plurality's requirement of a finding of "significant risk" should not present serious problems—even where data are sketchy. The plurality did not dispute the principle used by all the health regulatory agencies that any exposure to a carcinogen presents some risk. It merely held that under the Occupational Safety and Health Act only significant risks are to be regulated by standards. Even in the heartland of cancer protection—the world of the Delaney Clause—the U.S. Circuit Court of Appeals, in an opinion by the late Judge Harold Leventhal, recently reached an analogous conclusion about an important class of food additives.

The plurality made it clear that OSHA should decide in the first instance what is a "significant" risk and that the criteria for that judgment do not have to be quantitative and do not have to be based on scientific certainty. OSHA can rely on scientifically reputable opinion and conservative assumptions, and need not await evidence of actual human disease. Even the new benzene standard may be sustainable if the record supports a finding of "significant risk" under the plurality's broad view of that concept, an issue the court left open. The court also left for another day the question of whether OSHA must weigh the health benefits from a standard against the costs of complying with it.

Sometimes a loss is liberating. There is a limit to the amount of resources even our society will devote to protection of health and safety. OSHA has been freed from a policy that over time might have driven it to require vast misallocations of those limited resources. The result may well be more effective health regulation in the work place.

Whether it was appropriate for the court to have distorted the statute as it did is a question that will attract jurisprudential controversy for years to come. It was a bold step even for a Supreme Court.

Both commentators seem to agree that the Supreme Court plurality took substantial liberties with the plain terms of the Act in making its decision, which was based in large measure on the Court's policy predilections. While Cooper finds the decision "liberating," Sullivan objects to the Supreme Court's substituting its judgment for that of the expert agency on a policy matter. In his view, the courts should limit themselves to "constitutional" issues, and Congress should "entertain political objections to standards." Whether increased congressional scrutiny of rulemaking is on the horizon, either through more specific delegations of authority[52] or through some type of legislative veto, is far from clear. But, in the meantime, so long as agencies such as OSHA are charged with rulemaking responsibilities involving legislative-policy type determinations, and courts are assigned the task of reviewing those decisions, it seems inevitable that the courts will continue to make policy judgments in performing their judicial responsibilities.[53]

Several years passed before OSHA again acted to regulate benzene. On July 8, 1983, OSHA published a notice in the *Federal Register* asking for data, developed since 1977, on, among other things, the health effects of benzene, estimates of risk presented by the substance, and current exposure levels. OSHA noted that the Agency itself had performed two quantitative cancer risk assessments on benzene and that a third had been performed under contract. These, the Agency stated, concluded that the estimated excess leukemia risk from lifetime exposure to benzene at 10 ppm ranged from 14 to 207 deaths per 1000 workers. OSHA indicated, however, that it had decided to deny a petition filed by various unions for an emergency temporary standard on benzene and instead to undertake "expedited consideration" of a new standard.[54]

D. The Post-*Benzene* Era: OSHA Use of Risk Assessments

The Supreme Court's significant risk test has been applied to two OSHA standards cases since the *Benzene* decision—the *Cotton Dust* and *Lead* cases. The Court of Appeals for the District of Columbia Circuit, on August 15, 1980, upheld the OSHA 50-microgram limit for lead exposures in major industries, concluding that OSHA had "clearly" met the Supreme Court's significant risk test.

United Steelworkers v. Marshall

647 F.2d 1189 (D.C. Cir. 1980)

J. SKELLY WRIGHT, Chief Judge:

***

For the plurality, the fatal flaw in the benzene standard was that OSHA had relied on evidence of benzene's carcinogenic effects at extremely high exposure levels, and then simply assumed that lowering the PEL as far

---

[52]Cf. 448 U.S. at 671–88 (Rehnquist, J., concurring in judgment).

[53]There have been, of course, other comments on the Supreme Court *Benzene* decision. See, e.g., Comment, *The Significant Risk Requirement in OSHA Regulation of Carcinogens,* 33 STAN. L. REV. 551 (1981); Comment, *Billion Dollar Benzene Blunder,* 16 TULSA L.J. 252 (1980). On the more general issue of the role of courts in setting social policy, see Friendly, *The Courts and Social Policy: Substance and Procedure,* 33 U. MIAMI L. REV. 21, 38 (1978) (deploring courts' substituting their views for those of agencies on social policy issues).

[54]48 FED. REG. 31,412 (1983).

as possible would yield "appreciable" benefits. OSHA had produced no reliable empirical evidence of benzene-caused leukemia at levels of 10 ppm or below. Moreover, OSHA had not fashioned—and indeed insisted that it could not fashion—a dose-response curve that would have enabled it to infer the likely risk of leukemia at low benzene levels from the established evidence at higher levels. * * *

***

In creating the new standard, OSHA has clearly met the Section 3(8) threshold test of proving "significant harm" described by the *American Petroleum Institute* plurality. OSHA nowhere relied on categorical assumptions about the effects of lead poisoning; indeed, since the lead standard does not rest on the carcinogenic effects of lead, OSHA did not even have available to it any general policy dictating that there is no safe level of lead. Nor did OSHA rest on evidence of the dangers of lead at very high exposure levels and then simply infer that those dangers would decrease as the PEL was lowered. Rather, OSHA amassed voluminous evidence of the specific harmful effects of lead at particular blood-lead levels, and correlated these blood-lead levels with air-lead levels. By this means OSHA was able to describe the actual harmful effects of lead on a worker population at both the current PEL and the new PEL. In its proof of significant harm from lead at the current PEL and its careful measurement of the likely reduction, in that harm at the new PEL, the lead standard stands in marked contrast to the benzene standard struck down by the Supreme Court.

***

* * * Thus OSHA found that at the current PEL the vast majority of workers would have blood-lead levels above [40 micrograms per 100 grams], 40 ug/100g, the point at which, according to the evidence, dangerous subclinical effects of lead are likely to occur. By contrast, at the new PEL the great majority of workers will have blood-lead levels below the 40 ug/100g danger point, and virtually none will have blood-lead levels over 60 ug/100g. At the current PEL, then, the vast majority of workers face a significant harm from lead, and at the PEL that harm disappears for most workers. Of course, at the new PEL, a substantial minority of workers may have blood-lead levels above 40 ug/100g, and other workers are likely to have blood-lead levels above the 30 ug/100g mark at which lead may have serious subclinical effects on the reproductive system. But the lead rulemaking, unlike the benzene rulemaking—as characterized by the *American Petroleum Institute* plurality—did not aim at an absolutely risk-free workplace. Rather, it focused on the serious subclinical effects of lead, and reduced the risk of those serious effects as far as it could within the limits of feasibility. OSHA therefore has carried its burden under Section 3(8).

The Supreme Court upheld OSHA's cotton dust standard in 1981, concluding on the basis of the epidemiological study conducted by Dr. James Merchant that it was "difficult to imagine what else the agency could do to comply" with the significant risk requirements in the *Benzene* decision.[55]

The *Benzene* decision also had an impact on pending litigation involving the OSHA arsenic standard. Arsenic had been shown to be a carcinogen, and in May 1978 OSHA promulgated a standard for arsenic, lowering the permissible level in the §6(a) standard from 500

[55]*American Textile Mfrs. Inst. v. Donovan,* 452 U.S. 490, 506 n.25 (1981). The relevant portion of the *Cotton Dust* decision appears in Chapter 10.

to 10 micrograms per cubic meter (ug/m^3).[56] Although several reliable epidemiological studies on arsenic had been conducted, OSHA applied its lowest feasible level theory, concluding that there was no adequate scientific basis to determining a quantitative dose-response relationship at the lower levels of exposure.[57] The challenge to the validity of the arsenic standard was pending in the Court of Appeals for the Ninth Circuit when the *Benzene* case was decided by the Supreme Court. OSHA asked the court of appeals to remand the standard to the Agency for further rulemaking on the significant risk issue; at the same time, it asked the court to allow the standard to remain in effect, except to the extent it had already been stayed. In support, OSHA submitted three quantitative risk assessments, incorporating dose-response curves based on studies in the record which demonstrated, OSHA said, that the standard would probably ultimately meet the significant risk test for arsenic. The court granted OSHA's request, and on April 9, 1982, OSHA reopened the record in the arsenic proceeding on the issue of significant risk. On the basis of the record and three separate quantitative risk assessments, OSHA issued a supplemental statement of reasons concluding that arsenic presents a significant risk to employees at the prior 500-microgram level and that the 10-microgram level is necessary to significantly reduce the health risk.[58]

Inorganic Arsenic Supplemental Statement of Reasons for Final Rule

48 Fed. Reg. 1864, 1865–67 (1983)

***

I. SUMMARY OF OSHA'S ANALYSIS

OSHA's overall analytic approach for setting worker health standards is a four-step process consistent with recent court interpretations of the Occupational Safety and Health Act and rational, objective, policy formulation. In the first step, risk assessments are performed where possible and considered with other relevant factors to determine whether the substance to be regulated poses a significant risk to workers. Then in the second step, OSHA considers which, if any, of the proposed standards being considered for that substance will substantially reduce the risk. In the third step, OSHA looks at the best available data to set the most protective exposure limit necessary to reduce significant risk that is both technologically and economically feasible. In the fourth and final step, OSHA considers the most cost-effective way to achieve the objective.

The Ninth Circuit's remand provided that OSHA consider the issues presented by the first two steps and the elements of the third step dealing with risk issues. This final *Federal Register* document directly addresses those matters. A cooperative evaluation by technical experts from OSHA,

[56] 43 FED. REG. 19,584 (1978).

[57] *Id.* at 19,600.

[58] The litigation under §6(f) in which the validity of the arsenic standard has been challenged is still pending in the Court of Appeals for the Ninth Circuit. *ASARCO Inc. v. OSHA*, Nos. 78-1959, 2764, 3038.

the smelter companies and the United Steelworkers of America, which is not part of this record, gives additional consideration to the final steps.

It is appropriate to consider a number of different factors in arriving at a determination of significant risk with respect to inorganic arsenic. The Supreme Court gave some general guidance as to the process to be followed. It indicated that the Secretary is to make the initial determination of the existence of a significant risk, but recognized that "while the Agency must support its finding that a certain level of risk exists with substantial evidence, we recognize that its determination that a particular level of risk is 'significant' will be based largely on policy considerations." (*IUD v. API,* 448 U.S. 655, 656, n.62). In order for such a policy judgment to have a rational foundation, it is appropriate to consider such factors as quality of the underlying data, reasonableness of the risk assessment, statistical significance of the findings, the type of risk presented and the significance of the numerical risk relative to other risk factors.

In the April 9, 1982 (47 FR 15358) document which opened the issue of significant risk, OSHA pointed out that there were a number of high quality epidemiology studies such as Lee and Fraumeni, Pinto and Enterline, Ott et al., and Hill and Faning which strongly associated exposure to inorganic arsenic with excess risk of lung cancer among workers in both smelters and chemical plants. Many of these studies demonstrated a good dose response relationship and provided a good basis for risk assessment. Several demonstrated measured excess risk below the prior 500 ug/m^3 exposure limit. For example, the Lee and Fraumeni study indicated that for long term exposure to inorganic arsenic a 445 to 567% excess risk (334 to 425 excess cases per 1000 exposed employees) of lung cancer exists at 580 ug/m^3 and a 150 to 210% excess risk (112 to 158 excess cases per 1000 exposed employees) exists at 290 ug/m^3.

During the notice and comment period OSHA received published versions of additional studies, including Lee-Feldstein, Enterline and Marsh, Higgins et al., Mabuchi et al., and Lubin et al., which continued to strongly associate exposure to inorganic arsenic with excess risk of lung cancer. Recently, the International Agency for Research on Cancer, the World Health Organization-Arsenic Working Group and the Chemical Manufacturers Association have also judged inorganic arsenic to be a human carcinogen. The new data is of high quality and confirms OSHA's earlier conclusion that inorganic arsenic is strongly associated with excess risk of lung cancer. The new data also includes measured excess risk below 500 ug/m^3. For example, the Lee-Feldstein study covering 8045 employees and including 4448 low exposure workers whose average exposure was 290 ug/m^3, and whose mortality was observed over a 39 year period, indicates a 131% excess risk for those low exposure employees. Also, Enterline and Marsh observed a 168% excess risk for employees exposed to an average of 49 ug/m^3 (estimated from urinary level of 163 ug/l) for 10–19 years.

In the April 9th document, OSHA pointed out that a number of the epidemiology studies provided a good basis for risk assessment because of their high quality and because of the availability of quantitative estimates of exposure. OSHA presented three risk assessments and reached the preliminary conclusion that they presented reasonable estimates of risk, with OSHA selecting as most reasonable estimates ranging from a 500–620% excess risk (375 to 465 excess cases of lung cancer [per] 1000 exposed employees) for a working lifetime of exposure at 500 ug/m^3 to a 10–14% excess risk (7 to 10 excess cases per 1000) at 10 ug/m^3. These estimates were based on a linear model. OSHA also presented estimates based on a quadratic model, but new analysis indicate[s] that the data strongly supports a linear model in the case of inorganic arsenic.

Additional data were submitted which strongly support estimates of risk in this range. Dr. Crump submitted risk assessments based on the new epidemiologic studies which were in this range and which demonstrated good fits between the data and the linear model. Dr. Radford submitted an estimate of risk which was somewhat higher and Dr. Enterline an estimate which was somewhat lower. The National Institute [for] Occupational Safety and Health agreed with OSHA's estimates and approach.

Based on the earlier data and the data submitted in response to the April 9th document, OSHA concludes that the range of reasonable estimates of risk from a working lifetime of exposure to inorganic arsenic are from 148 to 767 excess deaths from lung cancer per 1000 exposed employees at 500 ug/m^3 to 2.2 to 29 excess lung cancer deaths per 1000 exposed employees at 10 ug/m^3. The OSHA preferred estimates within that range are approximately 400 excess deaths per 1000 at 500 ug/m^3, 40 excess deaths per 1000 at 50 ug/m^3 and 8 excess deaths per 1000 at 10 ug/m^3.

Consultants in Environmental and Occupational Health (CEOH) presented an alternate analysis principally based on the results of the study by Higgins et al. Employees with average exposures between 100 and 500 ug/m^3, including those who had peak exposures over 500 ug/m^3, had statistically significant increased respiratory cancer mortality. However, they found that employees whose ceiling exposures never exceeded 500 ug/m^3 had SMR's between 116 and 129 (16 to 29% excess risk), which were not statistically significant (Method I analysis). CEOH, therefore, suggested that 500 ug/m^3 was a practical threshold and there would be little excess risk for employees with no peak exposures over that limit.

This hypothesis is not nearly as strongly supported as the estimates of risk OSHA has presented. First, the OSHA estimates are based on a generally accepted model, with a biologic basis, which fits well a substantial body of high quality data. Second, both the Lee and Fraumeni, and Lee-Feldstein studies of the entire Anaconda cohort (not just 22%) demonstrated a statistically significant excess risk (from 86% to 213%) for low exposure employees who did not have any peak exposures over approximately 500 ug/m^3. This result directly contradicts the ceiling hypothesis. Third, the Higgins data was based on a 22% sampling of Anaconda employees, resulting in very low statistical power. The study only had a 16–37% chance of detecting a 50% excess risk, if it actually existed. Fourth, the employees actually had an excess risk (under Higgins Method I analysis) which was not very different (116–129 SMR) from OSHA's estimate (150 SMR) for employees with their relatively low average exposure. Fifth, the ceiling hypothesis has only been preliminarily tested at one location and before the possibility would develop of general acceptance in the scientific community, there would need to be supportive results in a number of locations.

Based on measured data in the record of excess risk below 500 ug/m^3 and estimates from the risk assessments summarized above and discussed in depth below indicating approximately 400 excess cases per 1000 exposed employees at 500 ug/m^3, OSHA concludes that a significant risk is presented by inorganic arsenic at the prior 500 ug/m^3 limit and that a lower exposure limit is needed. The Ninth Circuit has already agreed with this conclusion stating "it is undisputed that exposure to inorganic arsenic at the level of 500 ug/m^3 * * * poses a significant health risk * * *" (ASARCO et al. v. OSHA, No. 78-1959. Memorandum, April 7, 1981, p. 3).

OSHA also concludes, based on the estimates from the risk assessments and the dose-response demonstrated in many of the epidemiology studies, that a 10 ug/m^3 exposure limit, the lowest level feasible, together with the industrial hygiene provisions in the arsenic standard are necessary and appropriate to significantly reduce the health risk. These requirements will very substantially reduce the risk, by approximately 98%, and will protect employees principally in the nonferrous metal smelting, automobile and arsenical chemical industries.

Finally OSHA concludes that the new inorganic arsenic standard setting exposures at 10 ug/m^3 does not reduce the risk of the exposure to inorganic arsenic below the level of significance. The level of risk from working a lifetime of exposure at 10 ug/m^3 is estimated at approximately 8 excess lung cancer deaths per 1000 employees. OSHA believes that this level of risk does not appear to be insignificant. It is below risk levels in high risk occupations but it is above risk levels in occupations with average levels of risk. The OSHA Act was enacted in order to reduce significant risk insofar as feasible. It should be noted that the Supreme Court stated as to a 1 in 1000 level of risk of fatality that "a reasonable person might well consider the risk significant and take appropriate steps to decrease or eliminate it" (*IUD v. API,* 448 U.S. 655). OSHA believes the risk assessments and significant risk analysis support the retention of the 10 ug/m^3 level.

By achieving the 10 ug/m^3 limit, industry will have taken reasonable steps to protect their employees from the risks of arsenic. Substantial progress has already been made. Separate from this notice, and not part of this record, OSHA proposed to the affected smelter companies and the United Steelworkers of America, which represents smelter workers, a cooperative assessment by technical experts representing the three sectors to evaluate control methodology to protect employees while maintaining the efficiency of the smelting industry. This suggestion was accepted by USWA, ASARCO, and Kennecott. Agreements carrying out this proposal have been signed by OSHA, ASARCO and the United Steelworkers for 5 ASARCO facilities, and OSHA believes those parties have made exceptional progress in protecting exposed employees.

***

Although in 1978 OSHA had concluded that there was no "adequate scientific basis" for doing a risk assessment for arsenic, four years later, in its notice reopening the arsenic record[59] on the basis of the same studies, OSHA conducted a risk analysis, concluding that "reasonable confidence can be placed in the estimates of the risk presented." In estimating risk, OSHA used both the linear and quadratic models, showing that the risk was "significant" under either and explaining why the linear model was "preferable." These conclusions were reaffirmed when OSHA issued a supplement statement of reasons for the final arsenic standard. Whether OSHA could have similarly prepared a risk assessment for benzene based on the evidence in the record is a difficult question. OSHA had concluded that the "quality" of the arsenic studies was "high"; it is questionable whether the benzene studies were of similar high quality and would have provided a risk analysis in which "reasonable confidence" could be placed. In light of Justice Stevens' statements, however, that

[59] 47 FED. REG. 15,358, 15,365 (1982). Additional epidemiological studies were placed in the record and relied on by OSHA in the reopened arsenic proceeding. 48 FED. REG. at 1866. See also proposed standard for ethylene oxide, 48 FED. REG. 17,284 (1983). In proposing a permissible exposure limit of one part per million, OSHA performed several quantitative risk analyses, and concluded, preliminarily, that at the current exposure limit the estimates of risk are "well above the one per thousand guideline suggested by the Supreme Court in the *Benzene* decision as representing 'significant risk'"; and that the significant risk was not eliminated at a 1 ppm PEL with a projected 1 to 2 excess cases of cancer per 1000 exposed workers. 48 FED. REG. at 17,292–96. OSHA also made extensive use of quantitative risk assessments in its emergency standard on asbestos issued in 1983, 48 FED. REG. 51,086, 51,122–32 (1983).

OSHA is not required to support its significant risk findings "with anything approaching scientific certainty,"[60] it is at least arguable that a benzene quantitative risk analysis, perhaps based on more "conservative" assumptions than the one developed in OSHA's reply brief, would have been approved by the Supreme Court.[61]

Arguments have been advanced after the *Benzene* decision that OSHA must establish "significant risk of harm" in proving violations of the general duty clause and certain standards. In *Kelly Springfield Tire Company v. Donovan,*[62] the significant risk argument was rejected in a general duty case.

---

[60]448 U.S. at 656.

[61]As indicated in Section C of this chapter, in 1983 OSHA announced that it had performed risk assessments showing an excess risk from leukemia from lifetime exposure to benzene at 10 ppm ranging from 14 to 207 deaths per 1000 workers. 48 FED. REG. 31,412 (1983). See also White, Infante & Chu, *A Quantitative Estimate of Leukemia Mortality Associated with Occupational Exposure to Benzene,* 2 RISK ANALYSIS 195 (1982) (on the basis of essentially the same data that were in the OSHA rulemaking record in benzene, and using the "one-hit model," see discussion in Section A, this chapter, the study concluded that at 10 ppm excess leukemia deaths would be between 44 and 152 per thousand workers and at 1 ppm between 5 and 16 per thousand workers).

There has been considerable debate and literature on the subject of quantitative risk assessments. See e.g., *Comparative Risk Assessment: Hearings Before the Subcomm. on Science, Research and Technology of the House Comm. on Science and Technology,* 96th Cong., 2d Sess. (1980); OFFICE OF TECHNOLOGY ASSESSMENT, CONGRESS OF THE U.S., ASSESSMENT OF TECHNOLOGIES FOR DETERMINING CANCER RISKS FROM THE ENVIRONMENT 157–72 (1981); McGarity, *supra* note 2, at 733–36 and nn.17–27. See also Exec. Order No. 12,291 §2(d)(1), 46 FED. REG. 13,193, 13,194 (1981) (requires that agencies intending to issue "major" rules prepare "regulatory impact statements," which shall include, among other things, "a description of the potential benefits of the rules including any beneficial effects that cannot be quantified in monetary terms"); Comment, *Capitalizing on a Congressional Void: Executive Order No. 12,291,* 31 AM. U.L. REV. 613, 617 n.28 (1982). Cf. Regulatory Reform Act, S.1080, 97th Cong., 2d Sess. §4(a) (1982) (passed by the Senate; 128 CONG. REC. S2713, daily ed. Mar. 24, 1982). In March 1983, the National Research Council's Commission on Life Sciences issued a report entitled: *Risk Assessment in the Federal Government: Managing the Process.* The report made three recommendations: (1) regulatory agencies should explicitly distinguish between "scientific findings and policy judgments embodied in risk assessments" and the "political, economic and technical considerations that influence the design and choice of regulatory strategies"; (2) "uniform guidelines" that "would structure the interpretation of scientific and technical information relevant to the assessment of health risks" should be developed and followed; and (3) Congress should create an independent scientific board called the Board of Risk Assessment to perform various functions in the area of risk assessment. COMM. ON THE INSTITUTIONAL MEANS FOR ASSESSMENT OF RISKS TO PUBLIC HEALTH, NATIONAL RESEARCH COUNCIL, RISK ASSESSMENT IN THE FEDERAL GOVERNMENT: MANAGING THE PROCESS 5–8 (1983).

[62]11 OSHC 1889 (5th Cir. 1984).

10

The *Cotton Dust* Decision and Cost-Benefit Analysis

A. Early Interpretations of "Feasibility"

Under §6(b)(5), OSHA, in setting standards dealing with toxic materials or harmful physical agents, must assure that "no employee will suffer material impairment of health" even if the employee is exposed to the toxic substance for the period of his or her working life.[1] At the same time, §6(b)(5) places an important constraint on this requirement: it must be achieved "to the extent feasible." The interpretation of the "feasibility" requirements for OSHA health standards has been one of the most controversial legal and policy issues in the history of the Agency.

OSHA was first confronted with the feasibility issue in issuing an asbestos standard in 1972.[2] On the basis of the medical data available, OSHA determined that a permissible exposure limit (PEL) of two fibers of asbestos per cubic centimeter of air (fibers/cc) was needed to protect employees from asbestosis and mesothelioma. OSHA also found, however, that "many work operations" involving asbestos will "meet varying degrees of difficulty" in complying with the standard. More specifically, the preamble stated that, "[i]n some plants, extensive redesign and relocation of equipment may be needed."[3] OSHA concluded that a four-year delay in the two-fiber effective date would give all employers "reasonable time to comply," and that no harm to employees was "reasonably expected" to result during the transitional period.[4] For the interim period, OSHA immediately reduced the permissible level to five fibers.[5]

The asbestos standard was challenged in the Court of Appeals for the District of Columbia Circuit by both the Industrial Union

[1] 29 U.S.C. §655(b)(5) (1976).
[2] 37 Fed. Reg. 11,318 (1972).
[3] *Id.* at 11,319.
[4] *Id.*
[5] *Id.* at 11,318–19.

Department of the AFL-CIO and the Health Research Group, a public interest group.[6] They objected particularly to the delay in the effective date of the lower PEL. OSHA argued that the term "feasibility" in §6(b)(5) includes both technological and economic factors,[7] and that an immediate effective date for the lower PEL was infeasible. The court of appeals' decision upheld OSHA's view on the meaning of feasibility, but remanded the case to OSHA, in part, for further development of the record on the ability of different industries to comply with the lower PEL in a shorter time frame. In his opinion for the court, Judge McGowan discussed in detail the meaning of the feasibility limitation in §6(b)(5).

Industrial Union Department v. Hodgson

499 F.2d 467 (D.C. Cir. 1974)

MCGOWAN, Circuit Judge:

***

III

* * * [T]he statutory authority for the promulgation of standards reads in relevant part:

> The Secretary * * * shall set the standard which most adequately assures, *to the extent feasible,* on the basis of the best available evidence, that no employee will suffer material impairment of health or functional capacity. * * * 29 U.S.C. §655(b)(5) (emphasis supplied).

The standards as promulgated retain the concentration level specified by the temporary emergency standard until 1976 when a lower permanent standard becomes effective. The Secretary explained his decision to delay, two years longer than the period suggested by NIOSH, implementation of the tougher standard as "necessary to allow employers to make the needed changes for coming into compliance," and petitioners argue that the Secretary improperly considered economic factors in reaching this conclusion. We conclude that the factors entering into the Secretary's conclusion could properly include problems of economic feasibility.

There can be no question that OSHA represents a decision to require safeguards for the health of employees even if such measures substantially increase production costs. This is not, however, the same thing as saying that Congress intended to require immediate implementation of all protective measures technologically achievable without regard for their economic impact. To the contrary, it would comport with common usage to say that a standard that is prohibitively expensive is not "feasible."[22] Senator Javits,

[22]A discussion of some of the costs of dust control is found in Hills, *Economics of Dust Control,* 132 ANNALS OF THE NEW YORK ACADEMY OF SCIENCES 322 (1965), App. at 442–54. Several industry representatives testified in detail concerning the cost of attempting to meet the standards. Cf. *H & H Tire Co. v. United States Dept. of Transportation,* 471 F.2d 350 (7th Cir. 1972); *Chrysler Corporation v. Department of Transportation,* 472 F.2d 659 (6th Cir. 1972). These cases support the proposition that "practical" as employed in the Automobile Safety Act of 1966, 15 U.S.C. §1392(a), includes economic considerations, but the legislative history of that statute, unlike the history of OSHA, is more explicit on that point.

[6]*Industrial Union Dep't v. Hodgson,* 499 F.2d 467 (D.C. Cir. 1974).

[7]Although no distinction was explicitly made in the preamble between technological and economic feasibility, the language of the preamble seems to relate only to technological factors.

author of the amendment that added the phrase in question to the Act, explained it in these terms:

> As a result of this amendment the Secretary, in setting standards, is expressly required to consider feasibility of proposed standards. This is an improvement over the Daniels bill, which might be interpreted to require absolute health and safety in all cases, regardless of feasibility, and the Administration bill, which contains no criteria for standards at all.

S. REP. No. 91-1282, 91st Cong., 2d Sess., p. 58.

The thrust of these remarks would seem to be that practical considerations can temper protective requirements. Congress does not appear to have intended to protect employees by putting their employers out of business—either by requiring protective devices unavailable under existing technology or by making financial viability generally impossible.

This qualification is not intended to provide a route by which recalcitrant employers or industries may avoid the reforms contemplated by the Act. Standards may be economically feasible even though, from the standpoint of employers, they are financially burdensome and affect profit margins adversely. Nor does the concept of economic feasibility necessarily guarantee the continued existence of individual employers. It would appear to be consistent with the purposes of the Act to envisage the economic demise of an employer who has lagged behind the rest of the industry in protecting the health and safety of employees and is consequently financially unable to comply with new standards as quickly as other employers.[23] As the effect becomes more widespread within an industry, the problem of economic feasibility becomes more pressing. For example, if the standard requires changes that only a few leading firms could quickly achieve, delay might be necessary to avoid increasing the concentration of that industry. Similarly, if the competitive structure or posture of the industry would be otherwise adversely affected—perhaps rendered unable to compete with imports or with substitute products—the Secretary could properly consider that factor. These tentative examples are offered not to illustrate concrete instances of economic unfeasibility but rather to suggest the complex elements that may be relevant to such a determination.[25]

With the aid of the foregoing analytic background of the procedural and substantive provisions of the Act, we turn to the specific objections raised by petitioners to the standards.

1. EFFECTIVE DATE FOR THE TWO FIBER STANDARD.

The most important aspect of setting the standards was the determination of an acceptable dust concentration level. Under the emergency standards, the eight hour time-weighted average airborne concentration of asbestos dust had been limited to five fibers greater than five microns in length

[23]Temporary variances may be obtained when timely compliance is technologically impossible.

[25]Since technological progress is here linked to objectives other than the traditional competitive, profit-oriented concerns of industry, accommodation of both sets of values will sometimes involve novel economic problems. *International Harvester Co. v. Ruckelshaus,* 478 F.2d 615 (D.C. Cir. 1973), illustrates some of these problems in the context of the automobile emissions standards of the Clean Air Act. 42 U.S.C. §§1857 *et seq.* In the highly concentrated automobile industry the court deemed it likely that, by virtue of their size and importance to the economy, any one of the three major companies could obtain a relaxation of the automobile emissions standards if it could not meet them. If this occurred after other manufacturers had prepared to comply with the standard, the technological laggard would enjoy a competitive advantage because installation of the control devices renders the vehicle less efficient to operate. This circumstance justified insuring that the standards could be met by all major producers before they became effective.

per milliliter of air (hereinafter "the five fiber standard"). 36 F.R. 23207, 23208. A principal issue at the hearings on the permanent standards was whether the standard should remain at five fibers or be lowered to two. Proponents of standards ranging from zero to 12 fibers appeared, and it is fair to say that the evidence did not establish any one position as clearly correct. The Secretary decided to resolve this doubt in favor of greater protection of the health of employees, and established the two fiber standard recommended by NIOSH and his Advisory Committee as the level ultimately to be achieved.

Industry representatives testified that they simply could not reduce concentrations to the two fiber level in the foreseeable future. In the course of formulating its proposal, NIOSH had undertaken a limited analysis of industry's capacity to comply and had recommended delaying the effective date of the two-fiber standard for two years, i.e., July 1, 1974. The Secretary decided to retain the five fiber standard for approximately four years (July 1, 1976) before requiring the reduction to two fibers, in order to give employers time to prepare for the lower limit. Petitioners assert that the four year delay permitted by the Secretary is too long because (1) the health of employees is endangered thereby, and (2) employers do not need that much time.

***

A. HEALTH HAZARDS OCCASIONED BY THE DELAY.

***

B. INDUSTRIAL COMPLIANCE CAPABILITY.

The evidence indicates that significant inter-industry, as well as intra-industry, differences exist concerning the time needed by employers to meet a two fiber standard. Within particular industries the concentration levels at some plants, usually newer ones, are much lower than at others. More importantly, some industries could implement a two fiber standard more quickly than others. The Director of NIOSH recommended that the Secretary require compliance sooner than 1976 where possible, and that he prohibit degradation of workplaces with concentrations currently below the limits.

Despite this recommendation and the evidence of these differences, the Secretary issued a single uniform effective date for all employers in all industries. He explained this decision as follows:

> It is concluded there should be one minimum standard of exposure to asbestos applicable to all workplaces exposed to any kind of mixture of kinds of asbestos. Reasons of practical administration preclude a variety of standards for different kinds of asbestos and of workplaces.

We cannot say on this record that an attempt to assign differing effective dates to employers *within* an industry based on the time needed by each employer to alter his plant would be practicable. However, insofar as inter-industry differences are concerned, those reasons of practical administration are neither explained nor readily apparent.

***

It may be that the task of devising categories and classifying employers by industry would be unmanageable in view of the many diverse uses of asbestos. However, there is no evidence to that effect in the record, and it is not for the court to guess at the Secretary's reasoning or to supply justifications for his action. We have noted his cryptic reference to "reasons of practical administration," but, insofar as inter-industry distinctions are concerned, those reasons are not self-evident. Therefore, we remand this aspect of the standards to the Secretary for clarification or reconsideration.

This opinion has provided a framework for all subsequent Agency and court discussions of the issue of economic feasibility in the context of OSHA standards.[8] The Court of Appeals for the Third Circuit quoted extensively from, and agreed with, this decision in its opinion in *AFL-CIO v. Brennan*.[9] The Third Circuit used the phrase "massive economic dislocation" to describe the type of economic impact which properly could be taken into account by OSHA.[10] In the *Asbestos* case, the D.C. Circuit Court of Appeals referred to a standard that would make "financial viability generally impossible,"[11] but the import of the two formulations would seem to be the same. The Third Circuit Court of Appeals emphasized the practical difficulties in enforcing a standard that was economically infeasible. The court said that

> [a]n economically impossible standard would in all likelihood prove unenforceable, inducing employers faced with going out of business to evade rather than comply with the regulation * * * . The burden of enforcing a regulation uniformly ignored by a majority of industry members would prove overwhelming.[12]

The same test for economic feasibility was applied by the District of Columbia Circuit Court of Appeals in reviewing the OSHA *Cotton Dust* standard.[13]

In 1974, OSHA issued the standard for vinyl chloride,[14] and OSHA was once again confronted with the question of feasibility. The Agency found that vinyl chloride was a carcinogen and promulgated a standard setting a PEL of one part per million (ppm).[15] In doing so, OSHA recognized that this level could not be achieved in the near future by all covered establishments; it said, however, that

[8]The insistence that OSHA seek additional data on the compliance capabilities of various industries would appear to be inconsistent with the court's later view, stated in the *Cotton Dust* case, that OSHA standards may be promulgated on the basis of the "best available evidence" (§6(b)(5)). *AFL-CIO v. Marshall,* 617 F.2d 636 (D.C. Cir. 1979), *aff'd sub nom. American Textile Mfrs. Inst. v. Donovan,* 452 U.S. 490 (1981). A distinction could be drawn, however, between the *Asbestos* case, where OSHA was required to obtain additional data to justify the delaying of a more protective standard—a diminution of protection—and the *Cotton Dust* case, where OSHA's promulgation of a protective standard—an increase in protection—was upheld even though the available scientific evidence did not provide full factual support for OSHA's determination of the permissible exposure limit.

[9]530 F.2d 109 (3d Cir. 1975). This case involved the validity of the OSHA modification of the no-hands-in-dies standard. 29 C.F.R. §1910.217(d)(1) (1983). The Agency action was based in part on economic considerations and was challenged by the AFL-CIO. Although agreeing that the cost of compliance was a proper consideration in its determination, the court found that the OSHA statement of reasons was inadequate and remanded for further amplification, as required by §6(b)(8), 29 U.S.C. §655(b)(8) (1976). See discussion, Chapter 2.

[10]530 F.2d at 123.

[11]499 F.2d at 478.

[12]530 F.2d at 123.

[13]*AFL-CIO v. Marshall,* 617 F.2d 636, 655 (D.C. Cir. 1979), discussed in Section B, this chapter.

[14]39 FED. REG. 35,890 (1974).

[15]*Id.* at 35,892.

they would "in time" be able to reach this level "for most job classifications most of the time."[16] This would require, according to OSHA, "some new technology and work practices" and the adoption of technology presently used in other industries.[17] The standard was challenged by industry in the Court of Appeals for the Second Circuit, and the court upheld the standard in full, dealing at length with the issue of technological feasibility.

Society of the Plastics Industry, Inc. v. OSHA

509 F.2d 1301 (2d Cir. 1975)

MR. JUSTICE CLARK:

***

* * * [P]etitioners strongly urge that the Secretary breached his statutory mandate to insure that the standard selected is a "feasible" one. Relying on the so-called Snell Report,[4] petitioners claim that VCM [vinyl chloride monomer] and PVC [polyvinyl chloride] manufacturers will never be able to reduce levels of exposure to 1 ppm through engineering means. They point to the conclusion reached by the Snell Report that:

> The costs of compliance increase rapidly with decreasing VCM target levels and represent significant engineering uncertainty or infeasibility beyond 10 ppm ceiling and 2–5 ppm TWA for the VCM industry and 15–25 ppm ceiling and 10–15 ppm TWA for the PVC industry.

According to the report, "[b]ased on the industry surveys and Snell's independent assessments of the state-of-the-art of the technology," the standard price of VCM would only rise from 7.41¢/lb. at present to 7.69¢/lb. at a target level of 2–5 ppm TWA, but would supposedly soar to 12.71¢/lb. at the "no-detectable" level.

In his statement of reasons in support of the standard, the Assistant Secretary acknowledged the industry contention and the Snell conclusion about the infeasibility of the 1 ppm level, but noted that: "Labor union spokesmen and the Health Research Group, Inc., however, have suggested that such a level is attainable." The Assistant Secretary went on to say:

> Since there is no actual evidence that any of the VC or PVC manufacturers have already attained a 1 ppm level or in fact instituted all available engineering and work practice controls, any estimate as to the lowest feasible level attainable must necessarily involve subjective judgment. Likewise, the projections of industry, labor, and others concerning feasibility are essentially conjectural. Indeed, as Firestone has suggested, it is not possible to accurately predict the degree of improvement to be obtained from engineering changes until such changes are actually implemented.
>
> We agree that the PVC and VC establishments will not be able to attain a 1 ppm TWA level for all job classifications in the near future. We do believe, however, that they will, in time, be able to attain levels

[4]This report, "Economic Impact Studies of the Effects of Proposed OSHA Standards for Vinyl Chloride," was prepared in September of 1974 by Foster D. Snell, Inc., an independent consultant, at the request of the Secretary of Labor and estimates the cost to the industry of complying with various exposure levels between 50 ppm and 0–1 ppm. Plant and industry visits were the principal means of information gathering.

[16]*Id.*

[17]*Id.* The vinyl chloride rulemaking is discussed in Chapter 3, Section B.

of 1 ppm TWA for most job classifications most of the time. It is apparent that reaching such levels may require some new technology and work practices. It may also be necessary to utilize technology presently used in other industries. In any event the VC and PVC industries have already made great strides in reducing exposure levels. For example, B.F. Goodrich testified that it has reduced average exposure levels in several PVC plants from 35–40 ppm early this year to 12–13 ppm at the time of the hearing. We are confident that industry will continue to do so. [39 FED. REG. at 35892.]

We cannot agree with petitioners that the standard is so clearly impossible of attainment. It appears that they simply need more faith in their own technological potentialities, since the record reveals that, despite similar predictions of impossibility regarding the emergency 50 ppm standard, vast improvements were made in a matter of weeks, and a variety of useful engineering and work practice controls have yet to be instituted. In the area of safety, we wish to emphasize, the Secretary is not restricted by the status quo. He may raise standards which require improvements in existing technologies or which require the development of new technology, and he is not limited to issuing standards based solely on devices already fully developed. Cf. *Chrysler Corp. v. Dept. of Transportation,* 472 F.2d 659, 673 (6th Cir. 1972); *Natural Resources Defense Council, Inc. v. E.P.A.,* 489 F.2d 390, 401 (5th Cir. 1974).

There is much testimony in the record, especially in the Snell Report, indicating that VCM concentration can be easily pinpointed and largely corrected. For example, many of the companies engaged in PVC manufacture still perform the cleaning of batch reactors by opening the vessel and having the worker physically enter it. In chipping off the accretion from the walls of the vessel, the worker is thus exposed to a high concentration of VCM. Yet other, less hazardous methods are currently available and in use, in which the vessels are cleaned by machinery, emulsions or simply water under high pressure. Other sources of exposure are encountered in filling tank cars, measuring, testing and repairing pipe joints or other connections. The Snell Report indicates that much of this may be alleviated.

But whether it can or not, the Secretary's compliance scheme does not rest only on engineering and work practice controls. He does mandate that the industry use such technology to the extent feasible, but, more importantly, he requires that, in addition, respiratory protection be used if engineering means cannot bring the VCM level down to the permissible limit.

To be sure, respirators have their drawbacks. These problems were detailed at the hearings and recognized by the Assistant Secretary. Self-contained and air-hose type breathing equipment is bulky, expensive and infeasible for full-time use, as well as potentially hazardous in terms of tripping, restricted mobility, and over-exhaustion of workers. But the fact remains that they effectively eliminate exposure to VCM, and they are already being used by some PVC companies in the cleaning process and at other points in production with good success.

Like the industry's claims about the impossibility of achieving compliance through technological means, petitioners' claims of dire consequences from the requirement of respiratory protection are exaggerated. It does not appear that full-time use of respirators is necessary, and the Snell Report points this out. Furthermore, lightweight, inexpensive cartridge or cannister-type respirators, which can effectively filter out VCM at low levels, are now available and acceptable. Contrary to petitioners' assertions, the Snell Report indicates that a variety of respirators are reasonably available.

***

The view of the Court of Appeals for the Second Circuit was reaffirmed by the Court of Appeals for the Third Circuit in its decision in *AFL-CIO v. Brennan.*[18] The court there said that OSHA would not be justified "in dismissing an alternative to a proposed health and safety standard as infeasible when the necessary technology looms on the horizon."[19] The court affirmed OSHA's finding that compliance with the existing no-hands-in-dies standard was not technologically feasible where the Agency had considered not only "existing technological capabilities" but also "imminent advances in the art."[20]

As discussed in the court opinion in *Society of the Plastics Industry,* OSHA contracted with an economic consultant to prepare a report on the estimated cost to industry in complying with the various PELs under consideration. This was the first time a formal economic analysis was used in OSHA standards rulemaking. In subsequent rulemakings OSHA routinely arranged for economic analyses, and these were required in the case of "major" actions under Executive Orders issued by Presidents Ford, Carter, and Reagan.[21] Although the court focused on the technological problems of compliance with the PEL, it also referred to the cost impact of the standard, implicitly holding that it was economically feasible. Indeed, the vinyl chloride standard preamble and litigation demonstrate that the two issues—technological and economic feasibility—are closely tied together. As the PEL becomes lower and the technological difficulties of compliance increase, the cost also increases, often at a considerable rate. Thus, most standards could be viewed as being technologically feasible, if enough money were to be spent to develop and implement the

---

[18]*Supra* note 9.

[19]530 F.2d at 121.

[20]*Id.* at 122. The principle of technology-forcing was reaffirmed by the Court of Appeals for the Third Circuit in the *Coke Oven Emission* case, *American Iron and Steel Inst. v. OSHA,* 577 F.2d 825, 833–34 (3d Cir. 1978), *cert. dismissed,* 448 U.S. 917 (1980), and by the Court of Appeals for the District of Columbia Circuit in the *Cotton Dust* case, *AFL-CIO v. Marshall,* 617 F.2d 636, 658–59 (D.C. Cir. 1979), *aff'd sub nom. American Textile Mfrs. Inst. v. Donovan,* 452 U.S. 490 (1981), and in the *Lead* decision, *United Steelworkers v. Marshall,* 647 F.2d 1189, 1264–65 (D.C. Cir. 1980), *cert. denied sub nom. Lead Industries Ass'n v. Donovan,* 453 U.S. 913 (1981).

[21]Exec. Order No. 11,821, 3 C.F.R. 926 (1971–75 Comp.), amended by Exec. Order No. 11,949, 3 C.F.R. 161 (1977 Comp.), was issued by President Ford, and required federal agencies to consider the inflationary impact of major legislative proposals, regulations, and rules. The Office of Management and Budget and the Council on Wage and Price Stability (COWPS), monitored federal agency activity under this order. In 1978, President Carter issued Exec. Order No. 12,044, 3 C.F.R. 152 (1978 Comp.), requiring a regulatory analysis for all major federal regulations. The analysis was required to state the problem addressed, discuss alternatives, and explain the reasons why a particular regulatory strategy was selected. The Regulatory Analysis Review Group (RARG) was informally established by President Carter to review regulatory analyses prepared by agencies; RARG was staffed by COWPS and the Council of Economic Advisors. For a discussion of these executive orders, see CONGRESSIONAL QUARTERLY INC., REGULATION: PROCESS AND POLITICS 61–71 (1982); COUNCIL ON WAGE AND PRICE STABILITY, BENEFIT-COST ANALYSIS OF SOCIAL REGULATION (J. Miller III & B. Yandle, eds. 1979); Baram, *Cost Benefit Analysis: An Inadequate Basis for Health, Safety and Environmental Regulatory Decisionmaking,* 8 ECOLOGY L.Q. 473, 502–15 (1980); Roberts & Kossek, *Implementation of Economic Impact Analysis: The Lessons of OSHA,* 83 W. VA. L. REV. 449 (1981). Exec. Order No. 12,291, 3 C.F.R. 127 (1982), issued by President Reagan, is discussed *infra* this chapter.

necessary technology. This means that a determination on technological feasibility usually involves economic feasibility considerations as well. It is possible, of course, that a standard could be deemed technologically infeasible regardless of the amount spent to develop the technology, although this probably would be infrequent in light of the "technology-forcing" principles adopted in the *Vinyl Chloride* case.[22]

In the *Vinyl Chloride* decision, the court found that the standard was feasible, relying in part on the fact that OSHA permitted the use of respirators if engineering controls proved infeasible. This reasoning was criticized by Chief Judge Wright of the D.C. Circuit Court of Appeals in his opinion upholding, in major respects, the OSHA lead standard.[23] He referred to the "potential circularity of the scheme."[24] "If employers need only use those engineering and work practice controls which prove feasible, * * * [h]ow can such a standard ever be infeasible?"[25] In an effort to resolve this apparent circularity, Chief Judge Wright said:

> When affected parties petition for pre-enforcement review of an OSHA standard, the standard must pass a preliminary test of general feasibility. First, within the limits of the best available evidence, and subject to the court's search for substantial evidence, OSHA must prove a reasonable possibility that the typical firm will be able to develop and install engineering and work practice controls that can meet the PEL in most of its operations. OSHA can do so by pointing to technology that is either already in use or has been conceived and is reasonably capable of experimental refinement and distribution within the standard's deadlines. The effect of such proof is to establish a presumption that industry can meet the PEL without relying on respirators, a presumption which firms will have to overcome to obtain relief in any secondary inquiry into feasibility in any of the proceedings we discuss below. Insufficient proof of technological feasibility for a few isolated operations within an industry, or even OSHA's concession that respirators will be necessary in a few such operations, will not undermine this general presumption in favor of feasibility. Rather, in such operations firms will remain responsible for installing engineering and work practice controls to the extent feasible, and for using them to reduce lead exposure

[22]In 1978, in the cotton dust standard OSHA found that the permissible exposure limit was technologically feasible, concluding that in both the textile and nontextile industries, "modification and adaptation of existing dust control systems should bring the vast majority of workplaces into compliance." 43 FED. REG. 27,350, 27,362 (1978). This finding was affirmed by the Court of Appeals for the District of Columbia Circuit, *AFL-CIO v. Marshall,* 617 F.2d 636, 656–59 (D.C. Cir. 1979). The preamble to the lead standard also contains an extensive discussion of technological feasibility; there OSHA discusses technology-forcing requirements in relation to the time period allowed for compliance and the cost of compliance. Attachments to the Preamble, Occupational Exposure to Lead, 43 FED. REG. 54,354, 54,476 (1978). See generally Berger and Riskin, *Economic and Technological Feasibility in Regulating Toxic Substances Under the Occupational Safety and Health Act,* 7 ECOLOGY L.Q. 285 (1978); and Latin, *The Feasibility of Occupational Health Standards: An Essay on Legal Decisionmaking Under Uncertainty,* 78 NW. U.L. REV. 583 (1983) (discusses the burden on OSHA in demonstrating the feasibility of standards, and concludes that OSHA must make a "more persuasive showing of technological feasibility than of economic feasibility"). For a discussion of technological feasibility in the *enforcement* of OSHA standards, see Chapter 15, Section D.

[23]*United Steelworkers v. Marshall,* 647 F.2d 1189, 1267–69 (D.C. Cir. 1980), *cert. denied, sub nom. Lead Indus. Ass'n v. Donovan,* 453 U.S. 913 (1981).

[24]647 F.2d at 1269.

[25]*Id.*

as far as these controls can do so. In any proceeding to obtain relief from an impractical standard for such operations, however, the insufficient proof or conceded lack of proof will *reduce* the strength of the presumption a firm will have to overcome in justifying its use of respirators.[26]

B. Cost Benefit: The Coke Oven Emissions and Cotton Dust Proceedings

The issue of a standard's feasibility also arose during the rulemaking on OSHA's coke oven emission standard.[27] OSHA published a proposed standard for coke oven emissions in 1975[28] after receiving recommendations from an advisory committee it had appointed in 1974.[29] Two economic studies on the feasibility of the standard were prepared for OSHA by D.B. Associates, and steel industry representatives also submitted various cost estimates. The Council on Wage and Price Stability, acting under the authority of President Ford's Executive Order No. 11,821 requiring inflationary impact statements,[30] participated in the proceedings, and urged OSHA to use cost-benefit analysis in its decisionmaking process.[31] In the preamble to the final coke oven emissions standard, OSHA unequivocally rejected the cost-benefit approach.

Final Standard on Coke Oven Emissions
41 Fed. Reg. 46,742, 46,750–51 (1976)

***

C. *Benefits.* It is clear that the overriding purpose of the Act is to protect employee safety and health even if such protection results in the expenditure of large sums of money, increased production costs or reduced profit margins. On the other hand, the Act is not intended to impose unnecessary or inappropriate financial or other burdens upon affected employers.

In an effort to assist OSHA in its decision-making process, CWPS suggested that OSHA utilize cost-benefit analysis. That is, benefits of the coke oven standard would be quantified in dollars and measured against the dollar costs of implementing the standard. Cost-benefit analysis is a common method for making economic decisions. In recent years, some economists have sought to apply this analysis to the value of human life and the cost of health care. However, there is no consensus as to an appropriate methodology to arrive at dollar values for benefits.

---

[26]*Id.* at 1272.

[27]See 41 FED. REG. 46,742 (1976).

[28]40 FED. REG. 32,268 (1975).

[29]See 41 FED. REG. 46,742 (1976). In 1973, NIOSH recommended criteria for a coke oven emission standard.

[30]3 C.F.R. 926 (1971–75 Comp.). See also Baram, *supra* note 21.

[31]Statement on Behalf of the Council on Wage and Price Stability, Coke Oven Emission Standard Docket No. H-017A (1976) (copy available in OSHA Docket Office, Washington, D.C.). In its testimony, the Council asks: "Could the $160 million to $1.3 billion that might save at most 35.4 [coke oven] workers per year be better spent on cancer research that has the potential of preventing hundreds of thousands of cancer deaths per year?" *Id.* at 12. See also Transcript of Coke Oven Emission Hearing at 4570–83, Docket No. H-017A (May 11, 1976) (testimony of James C. Miller III & John Morrall III).

There are insuperable obstacles to any attempt to estimate accurately and to reduce to dollar terms the value of any health regulation. To begin with, since life and health are neither bought nor sold in our society, any estimate as to dollar values must necessarily be speculative. Yet, such an estimate requires unambiguous determinations of preventable mortality and morbidity and accepted standards of dollar values of life, illness, pain, and grief of those directly and indirectly affected. Indeed as CWPS suggested, dollar values for benefits require a subjective judgment as to social utility or disutility. CWPS suggested a general approach to estimating benefits as the social benefits to society as a whole, including the individuals who comprise it and reflecting any net reduction in their disutilities. For reasons that are discussed more fully below, OSHA believes that there are so many difficulties involved in attempting to assign a dollar value to the benefits of the standard that such figures would not provide a meaningful indication of the true value of the standard.

Industry representatives recognized the seriousness of the health problem and took no position on the number of lives that would be saved by the proposed regulation, and no witnesses were willing to equate dollars with lives. The primary issues then, as raised by CWPS, were (1) whether a dollar value for benefits should or indeed could be established; (2) the elements to be included and evaluated as benefits, and (3) the calculation of preventable mortality.

***

Even if a meaningful estimate of reduced mortality could be established, we do not believe that there is an adequate methodology to quantify the value of a life. Various methodologies were suggested in the IIS, but none was viewed as satisfactory.

One method commonly used in analysis of programs involving health care or disease control is often referred to as the "human capital" approach. The "human capital" method derives a minimum monetary value of human life based on the value of an individual's future earnings which would be lost as a result of premature death. Such calculations are occasionally supplemented by the "suggestion that auxiliary calculations be made in order to take account of the suffering of the victim, his loss of utility from ceasing to be alive, and/or of the bereavement of his family". Others, such as Dorothy Rice, extend this concept by "totaling the amount that is spent on medical care and the value of earnings foregone as a result of disability or death" to obtain a minimum value of human life.

Use of the human capital approach is qualified by its reliance on the arguable assumption that the sum of foregone wages (or foregone wages plus medical care costs) is the best estimate of the value placed on human life by society. Use of this method is further handicapped because it implies, for example, that retired persons (who are no longer "earning") are worthless, and that men are worth more than women (because the average earnings of men are higher than the average earnings of women).

Another method, somewhat similar to the first, is sometimes called the "net output" approach. The value of an individual's life under this method is found by "calculating the present discounted value of the losses over time accruing to others as a result of the death of a particular individual". Use of this method requires acceptance of the attitude that what is most important to society is simply the resultant net loss or gain following the death of one or more of its members. If accepted, the approach implies that the death of any person whose earning power or productivity is negative (such as a retired person regardless of his or her ownership of property), represents a net benefit to society. The method has no regard for the feelings of the potential victim or his family, restricting itself only to the interests of the surviving members of society as a whole.

A third method advocated by many for use in benefit assessment approaches the problem from a "social" aspect and bases the value of life on the amounts invested by government in social programs aimed at reducing the number of deaths. Renal dialysis for persons with kidney failure (the costs of which range from $15,000 to $25,000 per patient per year) is just one example of the free medical care available under a government-sponsored program (Social Security). Under this benefits analysis approach, the costs involved in the program imply that society places a value on life substantially higher than the sum of the wages these persons would earn over their working lifetimes.

While some have also suggested that an implicit value of human life could be derived from decisions on amounts spent in other programs to prevent mortality, Mishan notes that such values may properly differ among programs. He also notes that no democratic voting process is involved directly in such program decisions and, even if that were the case, an independent economic criterion for the value of life would be required for rational decisions. Some have felt that such an independent value could be derived from examination of wage rates paid in hazardous occupations. However, this would assume that workers have perfect knowledge of the nature of the hazards, and this would be more likely in obvious exposures than in the case of exposure to occupational carcinogens which have a long latency period so that the time of death is remote from the initial exposure. The time difference also introduces questions on whether the future benefits of reduced mortality should be discounted to arrive at some present value, but there is substantial disagreement among economists on the use of discounting in estimating the value of a life to be saved in future years. Finally, even if the value of life could somehow be assessed, there appears no way to value a difference between the slow and painful process of dying from cancer as compared to other dying processes with different levels of pain and suffering.

OSHA believes that these methodologies do not adequately quantify the value of life. Accordingly, we decline to do so.

It was suggested that the cost-benefit analysis should include an estimate of the dollar benefits of the standard in relation to reduced morbidity. Again, it is not possible to precisely estimate the excess morbidity resulting from exposure to coke oven emissions, although we do know that excess morbidity does result.

CWPS testified that a previous study (of asbestos workers) indicated that the amount of excess morbidity exceeded that of excess mortality, but that the value of illness was several times less than the value associated with death, so that, in that study, equal dollar values were assigned to excess morbidity and excess mortality. They therefore proposed allowing for the value of morbidity by dividing annual costs by 2 before relating annual cost estimates to estimates of excess mortality in any cost-benefit analysis. Not only is the number of disabling illness[es] which will be prospectively avoided unknown, but, their average duration and the number of nondisabling illnesses and their duration is also unknown.

In these circumstances, we find that it is inappropriate to arbitrarily establish a dollar value on the benefits of the standard relating to anticipated declines in worker morbidity.

CWPS testified that, by relating estimates of benefits (in terms of preventable deaths or their equivalent) to estimates of the costs of compliance, it is possible to estimate the implicit cost of reduced mortality. Moreover, they testified that a decision on implementing a proposed regulation involved acceptance of such an estimate of the implicit cost as the minimum value of a life. However, for reasons noted above, we do not believe we can forecast accurately the amounts of annual reductions in mortality or morbidity that will result from the regulation, nor do we have an independent estimate or standard of the dollar value of life.

Based upon the foregoing and the record as a whole, OSHA finds that compliance with the standard (even if the higher cost estimate were used) is well within the financial capability of the coking industry. Moreover, although we cannot rationally quantify in dollars the benefits of the standard careful consideration has been given to the question of whether these substantial costs are justified in light of the hazards. OSHA concludes that these costs are necessary in order to adequately protect employees from the hazards associated with coke oven emissions.

***

After rejecting the cost-benefit approach, OSHA set a PEL for coke oven emissions at 150 micrograms per cubic meter (ug/m^3). The Agency determined that the 150-microgram level was necessary to assure that employees were protected from coke oven hazards and that achievement of this level was technologically and economically feasible.[32] OSHA did, however, do a "general" cost-benefit analysis, saying that, although the benefits could not be quantified in dollars, "these substantial costs are justified in light of the hazards * * * associated with coke oven emissions."[33] Significantly, OSHA in this preamble did not deal with the legal aspects of cost-benefit analysis. Its rejection of this approach was based solely on policy and practical grounds, that is, the "insuperable obstacles" involved in carrying out the exercise in a meaningful way. OSHA did not maintain, as it was to do later before the Supreme Court in the *Benzene* and *Cotton Dust* litigation,[34] that cost-benefit analysis was prohibited by the Act. And while the OSHA policy determination to not use cost-benefit analysis necessarily assumes that cost benefit is not required by the statute, the issue was not addressed by OSHA explicitly.

The Court of Appeals for the Third Circuit upheld in relevant respects the coke oven emissions standard,[35] affirming OSHA's conclusions on the need for the 150-microgram level and the feasibility of that level, but did not discuss the cost-benefit issues. The court also affirmed OSHA's decision to require a combination of specific engineering controls and work practices and a "performance" standard (150-microgram PEL); the court held that the use of this "double-barreled" approach was within OSHA's authority and was supported by the record.[36]

[32]41 FED. REG. at 46,752–56.

[33]*Id.* at 46,751.

[34]See *infra* this chapter.

[35]*American Iron & Steel Inst. v. OSHA,* 577 F.2d 825 (3d Cir. 1978), *cert. dismissed,* 448 U.S. 917 (1980).

[36]577 F.2d at 837–40. The court, however, ruled that the research and development requirements of the standard were invalid and unenforceable, *id.* at 838, and stated "serious reservations" as to the applicability of the standard to independent contractors in the construction industry, *id.* at 840. Industry parties petitioned for certiorari from this decision; the petition was granted by the Supreme Court, but was later withdrawn by the industry petitioners; the Court instead agreed to review the decision of the District of Columbia Circuit Court of Appeals in the *Cotton Dust* case.

In issuing the benzene standard in 1978, OSHA again followed its two-step analysis—the level necessary for employee protection as defined in §6(b)(5), as limited by feasibility considerations—in setting a PEL of one part per million (ppm).[37] The standard was challenged and was vacated by the Court of Appeals for the Fifth Circuit.[38] The court determined that OSHA must show that the "measurable benefits" expected by the standard bear a "reasonable relationship" to the costs imposed.[39] In requiring that OSHA perform a cost-benefit analysis, the court did not accept the Agency's attempt to balance the "appreciable" benefits of the standard against the costs; according to the court, this conclusion is not supported by substantial evidence since there is no estimate of what these appreciable benefits are likely to be.[40] The court of appeals' *Benzene* decision was not clear as to the manner in which OSHA is required to do a cost-benefit analysis. The court was willing to accept "rough but educated" estimates of the benefits expected, and it stated that OSHA was not required to "conduct an elaborate cost-benefit analysis."[41] The precise variables of a cost-benefit requirement, however, were not spelled out; the court said only that the benefits must bear a "reasonable relationship" to the costs.

The Supreme Court agreed to review the court of appeals' *Benzene* decision, and, in its brief to the Supreme Court, OSHA argued for the first time that the Act *prohibits* basing a standard on a cost-benefit test, relying on the statutory language and the legislative history.[42] The Agency contended initially that it could not determine "measurable" benefits or do a cost-benefit analysis because the "best available evidence" was "not good enough to enable the Secretary to determine the relationship between exposure to benzene and the risk of cancer."[43] OSHA went on, however, to argue that there were "problems" in engaging in cost-benefit analysis even when the available data were sufficient to allow a more precise prediction of the potential benefits deriving from the new standard. It said:

> Even when available data are sufficiently reliable to permit the probable health benefits of a standard to be predicted, accurately, there are problems in using cost-benefit criteria to set an exposure standard where, as here, it is possible to set a "feasible" exposure limit and reduce the risk of employees more directly. The most important difficulty in setting a higher-than-feasible exposure level using cost-benefit techniques is placing an economic value (implicit or explicit) on life, health, bereavement, pain and suffering, and other consequences of occupational disease and death. There is no generally accepted method for

---

[37] 43 FED. REG. 5,918 (1978). The benzene standard is discussed in Chapter 9; the Benzene ETS is discussed in Chapter 7.

[38] *American Petroleum Inst. v. OSHA,* 581 F.2d 493 (5th Cir. 1978), *aff'd sub nom. Industrial Union Dep't v. American Petroleum Inst.,* 448 U.S. 607 (1980).

[39] 581 F.2d at 501–05. See also the discussion in Chapter 9, Section B.

[40] *Id.* at 503.

[41] *Id.* at 504.

[42] Brief of Federal Parties at 48–72, *Industrial Union Dep't v. American Petroleum Inst.,* 448 U.S. 607 (1980).

[43] *Id.* at 61.

valuing such considerations, even implicitly. See Senate Comm. on Governmental Affairs, 95th Cong., 2d Sess., Study on Federal Regulation, Vol. VI, at XXIV (Comm. Print 1978). For a discussion of the various methods occasionally proposed to quantify the value of life and other intangibles, see 41 FED. REG. 46742, 46751 (1976) (coke-oven-emissions preamble). See generally Fried, *The Value of Life,* 82 HARV. L. REV. 1415 (1969); Rhoads and Singer, *What is Life Worth?,* 51 PUB. INTEREST 74 (Spring 1978). Second, there is no general agreement concerning whether a stream of health benefits and saved lives extending into the future should be discounted to present value in the same way that future economic values are generally discounted in awarding tort damages. Third, the Secretary has found that industry tends to overestimate the cost of implementing proposed standards, which makes it difficult accurately to weigh benefits against costs. For example, industry asserted that the Secretary's vinyl chloride standard would seriously disrupt the industry. In fact, however, the industry promptly complied, at a cost much lower than it had projected, by devising new technology to meet the standard. Vinyl chloride production has continued to increase. See Doniger, *Federal Regulation of Vinyl Chloride: A Short Course in the Law and Policy of Toxic Substances Control,* 7 ECOL. L.Q. 497 (1978).[44]

Although the Fifth Circuit had said that no "formal" cost-benefit balance was required, and that "rough but educated" benefit estimates were sufficient, OSHA argued that *any* cost-benefit technique places an "economic value (implicit or explicit) on life, health, bereavement, pain and suffering, and other consequences of occupational disease and death" and is therefore prohibited.[45] The Supreme Court affirmed the lower court decision vacating the benzene standard on significant risk grounds, but did not reach the cost-benefit issue.[46]

Meanwhile, on June 23, 1978, OSHA issued a standard regulating cotton dust hazards.[47] Relying on a number of epidemiological studies, particularly the extensive textile industry study by Dr.

[44]*Id.* at 62 n.52.

[45]*Id.* The same assumption was made by OSHA in the coke oven emissions preamble, that cost benefit requires the monetary quantification of the benefits of the standard. In OSHA litigation, cost-benefit analysis has been defined in various ways. Those who argue in support of cost-benefit analysis describe it as an informal and flexible process; OSHA, in opposition, emphasized its quantitative aspects. For a critical discussion of the Fifth Circuit decision on *Benzene,* see Comment, *Cost-Benefit Analysis for Standards Regulating Toxic Substances Under the Occupational Safety and Health Act:* American Petroleum Institute v. OSHA, 60 B.U.L. REV. 115 (1980).

[46]*Industrial Union Dep't v. American Petroleum Inst.,* 448 U.S. 607 (1980). Portions of this decision are reprinted in Chapter 9, Section C.

[47]43 FED. REG. 27,350 (1978). A suit had been brought in the U.S. District Court for the District of Columbia by the Textile Workers Union of America against the Secretary of Labor in December 1975 (No. 75-2157) to compel OSHA to commence §6(b) proceedings to revise the existing cotton dust standard. See 7 OSHR 208 (1977). After the cotton dust proposal had been published, OSHA moved to dismiss the case as moot but District Judge Corcoran denied the motion and retained jurisdiction until a final standard was issued. *Textile Workers Union v. Marshall,* 1977–1978 OSHD ¶21,914 (D.D.C. 1977). The litigation became active again in 1978 when final publication of the standard was delayed because of White House intervention. On June 7, 1978, Judge Corcoran told OSHA that he would accept "no more excuses" and that unless the standard were issued in a very short time, he would order OSHA to issue the standard. On June 9, OSHA said that the standard would be issued by June 19, and after a top-level meeting in the White House, the standard was issued on time. N.Y. Times, May 24, 1978, at D1, col. 6; Washington Post, June 8, 1978, at A9, col. 1; 8 OSHR 27–28, 59–60 (1978). See also the discussion in Chapter 8, Section E, on the legal implication of White House and OMB intervention in rulemaking proceedings.

James Merchant of NIOSH, showing the prevalence of byssinosis among workers exposed to cotton dust, OSHA established three different PELs under the standard: 200 micrograms (ug/m^3) for yarn manufacturing, 750 micrograms for slashing and weaving operations in the textile industry, and 500 micrograms for all other processes in the cotton industry and for nontextile operations.[48] OSHA found these levels to be technologically and economically feasible for the affected industries, and again refused to perform a cost-benefit analysis. The preamble states:

> Based upon the foregoing and the record as a whole, OSHA finds that compliance with the standard is well within the financial capability of the covered industries. Moreover, although the benefits of the standard cannot rationally be quantified in dollars, OSHA has given careful consideration to the question of whether these substantial costs are justified in light of the hazards of exposure to cotton dust. OSHA concludes that these costs are necessary in order to effectuate the statutory purpose of the Act and to adequately protect employees from the hazards of exposure to cotton dust.
>
> In making judgments about specific hazards, OSHA is given discretion which is essentially legislative in nature. In setting an exposure limit for a substance like cotton dust, OSHA has concluded that it is inappropriate to substitute cost benefit criteria for the legislatively determined directive of protecting all exposed employees against material impairment of health or bodily function. Where the health effectiveness of alternative approaches are extremely uncertain and likely to vary from situation to situation, OSHA believes it is appropriate to adopt the compliance strategy which provides the greatest certainty of worker protection even if the approach carries with it greater economic burdens for the affected employers.
>
> In the case of the cotton dust standard, the evidence in the record indicates that the costs of compliance are not overly burdensome to industry. Having determined that the benefits of the proposed standard are likely to be appreciable, OSHA is not obligated to carry out further exercises toward more precise calculations of benefit which would not significantly clarify the ultimate decision. Previous attempts to quantify benefits as an aid to decision making in setting health standards have not proved fruitful (41 FR 46742).[49]

Consistent with its position, OSHA did not prepare a cost-benefit analysis on cotton dust as part of the rulemaking proceeding. At the direction of the Congress,[50] however, on May 14, 1979, OSHA filed a study conducted after the rulemaking record was closed, entitled "Report to the Congress, Cotton Dust: Review of Alternative Technical Standards and Control Technologies." OSHA made it clear that

---

[48]OSHA issued a separate standard for cotton-gin operations, 43 FED. REG. 27,418 (1978): this standard did not contain a permissible exposure limit, but imposed requirements for work practices, respirator usage, medical surveillance, and recordkeeping. The Court of Appeals for the Fifth Circuit ultimately vacated the cotton-gin standard, relying on its earlier decision in the *Benzene* case in holding that OSHA had not found as a "threshold matter" that cotton dust poses a "significant health risk" in cotton gins. *Texas Indep. Ginners Ass'n v. Marshall,* 630 F.2d 398 (5th Cir. 1980).

[49]43 FED. REG. at 27,379.

[50]Departments of Labor-HEW Appropriations, Fiscal Year 1979, Pub. L. No. 95-480, 92 Stat. 1567 (1978); H.R. CONF. REP. No. 12,929 Amend. No. 109, 124 CONG. REC. 34,163 (1978) (directing OSHA to study alternatives to cotton dust standard). See 8 OSHR 565, 637 (1978).

the analysis in the study was not the basis for the standard that had been issued. The study contained what OSHA referred to as a "cost-effectiveness" analysis of the PELs for cotton dust concluding that the PEL selected in the standard was cost effective.[51]

The cotton dust standard was challenged in the Court of Appeals for the Fourth Circuit, and after preliminary litigation on the issue of proper forum,[52] the proceeding was transferred to the District of Columbia Circuit. In *AFL-CIO v. Marshall*,[53] the court of appeals affirmed the standard in most respects except as it applied to the cottonseed oil industry.[54] The court rejected the American Textile Manufacturers Institute's (ATMI) proposed alternative for a 500-microgram standard to be achieved by respirators, with medical surveillance and employee transfer requirements; it also disagreed with the contention that OSHA was not empowered to provide protection against acute byssinosis.[55] Although recognizing that the issues were difficult, the court of appeals, relying principally on *Industrial Union Department v. Hodgson*,[56] also affirmed OSHA's feasibility findings for all but one industry, accepting the Agency's estimate of the costs of compliance and its conclusion that these costs could be borne by the industry.[57] The court noted that OSHA's cost estimates were "not free from 'imprecision'," and stated:

> The very nature of economic analysis frequently imposes practical limits on the precision which reasonably can be required of the agency. This is especially the case where, as here, the industry chooses to withhold from the agency part of the data underlying the industry's cost estimates. OSHA's mandate authorizes it to promulgate standards on the basis of the "best available evidence." We find that OSHA reasonably evaluated the cost estimates before it, considered criticisms of each, and selected suitable estimates of compliance costs.[58]

In affirming the OSHA standard as it applied to most industries, the court of appeals explicitly rejected the argument that cost-benefit analysis was required in OSHA standards-setting.

[51]Compare the analysis in J. Morrall III, *Cotton Dust: An Economist's View*, in SCIENTIFIC BASIS OF HEALTH AND SAFETY REGULATION 93, 104–08 (1981). Dr. Morrall, an economist on the Council on Wage and Price Stability, concluded on the basis of what he also called "cost-effectiveness" analysis that OSHA could have provided "the same amount of benefits at a fraction of the costs if a pure performance standard had been promulgated." *Id.* at 108. In his view, a "pure performance" standard would prohibit employees from advancing to a harmful stage of byssinosis and would impose a fine on the employer for each employee case of byssinosis. For a discussion of "injury tax" proposals, see Chapter 11, Section C.

[52]See Chapter 7.

[53]617 F.2d 636 (D.C. Cir. 1979).

[54]The court found the standard infeasible in the cottonseed oil industry and remanded that portion of the standard to the Agency for further development of the record. 617 F.2d at 669–73. In June, 1983, OSHA issued a notice of proposed rulemaking on that issue, as well as on a number of others relating to cotton dust. 48 FED. REG. 26,962 (1983).

[55]ATMI argued that acute byssinosis was reversible and therefore not a "material impairment" within the meaning of §6(b)(5). The court said that the record supported OSHA's decision to implement its "dust control strategy" to protect workers from byssinosis, "which is conceded by all to cause material impairment in its chronic stage." 617 F.2d at 654 n.83.

[56]499 F.2d 467 (D.C. Cir. 1974).

[57]617 F.2d at 661–62. The court relied on the principles on feasibility originally stated in the *Asbestos* case.

[58]*Id.* at 661.

AFL-CIO v. Marshall

617 F.2d 636 (D.C. Cir. 1979)

Bazelon, Senior Circuit Judge:

***

Other statutory schemes explicitly require such particular kinds of analysis. In the Clean Air Act, for example, Congress required the Environmental Protection Agency to perform a "cost benefit analysis" before prohibiting the manufacture or sale of a fuel or fuel additive which endangers public health or welfare. Some Congressional Acts require a showing of "unreasonable risk" prior to regulation. The legislative histories of these acts have led the courts to construe this provision to require regulatory agencies to balance costs and benefits of proposed action.

In the OSH Act, in contrast, Congress itself struck the balance between costs and benefits in the mandate to the agency. Section 6(b)(5) unequivocally mandates OSHA to

> set the standard which most adequately assures, to the extent feasible, on the basis of the best available evidence, that *no employee will suffer material impairment* of health or functional capacity.

Thus Congress concluded that the benefits of health protection warranted the expense of an effective standard. In the legislative debates on the Act, Senator Yarborough who sponsored the bill responded in no uncertain terms to the claim that the proposed legislation would be too expensive:

> We are talking about people's lives, not the indifference of some cost accountants. * * * We are talking about assuring our American workers who work with deadly chemicals that when they have accumulated a few years seniority they will not have accumulated *lung congestion* and poison in their bodies, or something that will strike them down before they reach retirement age.

In contrast to the Acts for which Congress contemplated a cost-benefit requirement, the legislative history of the OSH Act contains no reference to this kind of economic analysis.

Instead, Congress determined in the OSH Act that any severe risk to employee health must be eliminated or reduced if feasible means to do so exist. This calls for a two-step analysis by the agency: (1) determining whether health impairment is threatened by the suspect substance, and (2) determining whether the selected strategy is to protect workers from this risk both technologically and economically feasible. Nothing in the statute or its legislative history requires a further determination that the costs of the standard bear a "reasonable" relationship to its benefits. Nor may this court impose additional procedural requirements. Indeed, the only authorities cited by petitioners are not binding on this circuit, and in any event, they are not persuasive in this context.

Further, cost-benefit analysis would not necessarily improve agency health and safety determinations. These techniques require the expression of costs, benefits and performance in often arbitrary, measureable terms. They may hide assumptions and qualifications in the seeming[] objectivity of numerical estimates. Especially where a policy aims to protect the health and lives of thousands of people, the difficulties in comparing widely dispersed benefits with more concentrated and calculable costs may overwhelm the advantages of such analysis.

***

The District of Columbia Circuit Court of Appeals' rationale on the cost-benefit issue is noteworthy in several important respects. First, its rejection of cost-benefit analysis was based not only on a *legal* analysis of the Act, but also on the court's conclusion that, as a technique of decision making, cost benefit would "not necessarily improve agency health and safety determinations." Secondly, the court agreed with the Fifth Circuit's view in the *Benzene* case that, to properly perform cost-benefit analysis, the benefits would have to be expressed in "measurable" terms; it disagreed with the Fifth Circuit, however, that such quantification for cost-benefit analysis would add objectivity or reliability to OSHA determinations. Finally the court, in accord with the generally accepted view, stated that "in contrast to cost-benefit analysis, cost-effectiveness is only used to compare alternatives for reaching the same goal."[59] The court concluded, however, that cost-effectiveness analysis "often raises intractable problems in obtaining the requisite measures that enable comparison."[60] OSHA, on the other hand, has consistently taken the position that cost-effectiveness analysis—achieving the requisite level of protection at the least cost—was permissible and was being used by the Agency;[61] it is not apparent why this technique, if properly used, would cause insuperable difficulties or be objectionable under the Act.[62]

C. The Supreme Court Litigation on *Cotton Dust*

The Supreme Court agreed to review the court of appeals' decision in the *Cotton Dust* case. ATMI argued before the Court that the OSHA approach to feasibility was that "the costs of a standard are irrelevant short of the point at which they threaten to ruin an entire industry."[63] It argued further that OSHA's feasibility findings were not supported by substantial evidence.[64] And, more broadly, ATMI claimed that the Act requires OSHA to make an assessment of the "significance of the risk" in light of the costs of eliminating that risk, that is, to perform a cost-benefit analysis.[65] Its brief stated:

> For one thing, engaging in such an assessment will lead OSHA to reject control strategies the costs of which are grossly disproportionate

[59]617 F.2d at 663 n.153.

[60]*Id.*

[61]In its brief to the Supreme Court in the *Cotton Dust* case, OSHA said that "the statute permits the secretary to select the least expensive means of compliance that will provide an adequate level of protection." Brief of Federal Respondent at 56, *American Textile Mfrs. Inst. v. Donovan,* 452 U.S. 490 (1981).

[62]The Supreme Court in the *Cotton Dust* case stated that OSHA "might" be required by §3(8) to adopt the least costly means to achieve the necessary protection. *American Textile Mfrs. Inst. v. Donovan,* 452 U.S. at 513 n.32, reprinted in Section C of this chapter. See also discussion of cost-effectiveness in Section D of this chapter.

[63]Brief for Petitioners at 31, *American Textile Mfrs. Inst. v. Donovan,* 452 U.S. 490 (1981).

[64]*Id.* at 27.

[65]*Id.* at 36–51.

to their benefits. It also will cause the agency to give some consideration to the impact that its standard would have on the industry's ability to address other, more significant health and safety risks in the future—including risks that OSHA anticipates will be addressed in standards that are at least in the planning stage. Finally, forcing OSHA to engage in such an assessment and to explain the determination it has reached provides a potential legislative check on what might otherwise amount to the exercise of virtually untrammeled authority. It will allow for correction, through the political process, of actions that are deemed by the Congress to be extreme, unwarranted, and inconsistent with congressional intent.[66]

In response, OSHA stated that it had construed the feasibility requirement in "a reasonable and practical fashion that provides a meaningful brake on regulation * * * . At bottom, the Secretary must determine that the industry will maintain long-term profitability and competitiveness."[67] In addition, OSHA argued that the cost-benefit test was not necessary to prevent "serious misallocation of resources"[68] and that there were a number of "serious obstacles" to the use of cost benefit in this area.[69]

OSHA, therefore, argued not only the traditional legal bases for rejecting cost-benefit analysis—through statutory language and legislative history—but also contended that any attempt to employ cost-benefit analysis would have to "overcome a number of serious obstacles." The AFL-CIO, on the other hand, although taking the same position as the government, made no argument on the desirability of cost-benefit analysis as a regulatory tool.

Brief of Federal Respondent in American Textile Manufacturers Institute v. Donovan[70]

***

For their part, petitioners would engraft upon the statute a balancing test that would require the Secretary to determine whether the benefits of a standard bear a "reasonable relationship" to its costs. Apparently in recognition of the fundamental problems with that approach, petitioners quickly add that they do not mean to suggest any need for a "rigidly formal cost-benefit calculation that places a dollar value on employee lives or health". But there is no other way to perform a meaningful cost-benefit analysis.[65] Unless there was agreement as to the dollar value of the benefits to be gained, any determination that costs and benefits are "reasonably related" would be

[65]A "formal" cost-benefit analysis may be characterized as one that seeks to weigh the costs and benefits of a particular course of action by computing and comparing the net present values of both the costs and the benefits. As applied to the setting of occupational health standards, the benefits of improved health would have to be valued in monetary terms and then compared with the costs associated with the regulation.

[66]*Id.* at 40–41.
[67]Brief for the Federal Respondent, *supra* note 61 at 48–49.
[68]*Id.* at 58.
[69]*Id.* at 59.
[70]Brief of Federal Respondent at 59–61, *American Textile Mfrs. Inst. v. Donovan,* 452 U.S. 490 (1981); see also Brief of Union Respondent *passim.*

arbitrary. Hence, placing a value on human life and freedom from suffering is an indispensable part of petitioners' program.

At all events, any attempt to employ cost-benefit analysis in this area would have to overcome a number of serious obstacles. As just noted, if costs and benefits are to be compared, they must be computed in the same units, which means that benefits must be assigned a dollar value. Even the assessment of the dollar costs of avoiding lost wages and medical expenses, which are already valued quantitatively, would be a prodigious undertaking. To be accurate, these calculations would require comprehensive data on the affected workforce, including age and earnings profiles, medical histories, postretirement experience, and premature death tables. Cost information concerning an employer's increased workers' compensation and medical insurance premiums, direct health care costs, and indirect productivity losses resulting from excess absenteeism, worker turnover, and other related inefficiencies also would have to be taken into account. See 41 FED. REG. 46751 (1976) [vinyl chloride]; 45 FED. REG. 5248 (1980) [carcinogen policy]. Even if reliable data were available for an entire industry, which is unlikely, there remain the insurmountable problems of placing a wholesale economic value on the pain, suffering, and premature deaths that will be visited upon the workforce of an industry, which is necessarily composed of persons with widely different characteristics and circumstances.[67]

Moreover, there would have to be general agreement whether a stream of health benefits and saved lives extending into the future should be ignored, discounted to present value, or otherwise calculated. A cost-benefit analysis of the cotton dust standard, for example, would surely have to consider more than the current employees in the industry. The permanent technological health improvements required by the standard would benefit untold millions of future employees as well. See *Benzene,* 448 U.S. 607, 701 n.22 (Marshall, J., dissenting).

Another defect in traditional cost-benefit analysis is that it cannot account for those considerations of equity that underlay the very concept of occupational health regulation. While the cost-benefit technique has gained some adherents as an allegedly value-free method for dealing with all manner of difficult questions, this technique originated in the context of investment capital decisions, where costs and benefits not only were already expressed in dollars, but also accrued to the same party. In occupational health regulation, however, employees bear the risk of occupational disease, while consumers and employers may acquire some benefit if the risk is not

[67]Pain and suffering attendant to group illness and injury are not easily reducible to monetary terms. 41 FED. REG. 46751 (1976); 45 FED. REG. 5248 (1980). There is no generally accepted method for valuing such considerations, even implicitly. See Senate Comm. on Governmental Affairs, 95th Cong., 2d Sess., Study on Federal Regulation, Vol. VI, at XXIV (Comm. Print 1978). Attempts have been made, however, to attach a value to life itself for this type of analysis. One such approach, the "human capital" approach, calculates the value of an individual life by computing the present value of future income as a measure of the individual's contribution to the national economy. Because this valuation system depends on earning levels, it leads to lower values on the lives of retired workers, women, and other low wage groups such as textile workers. This calculation also necessarily assumes that there is no other value to human life than producing income to maximize the gross national product. Another type of valuation has attempted to analyze wage rates for high risk jobs and to utilize the so-called "risk premium" paid for such jobs as a measure of a life. This analysis necessarily, and probably incorrectly, assumes that employees have the freedom to move to lower-risk jobs and that they have adequate knowledge to make a well-informed choice. It was for these reasons that the Secretary concluded that there was no adequate methodology to quantify the value of a life. 41 FED. REG. 46751 (1976); 45 FED. REG. 5248 (1980). These problems are necessarily compounded, of course, when the calculations purport to give some economic value to large numbers of lives. See generally Note, *Cost-Benefit Analysis for Standards Regulating Toxic Substances Under the Occupational Safety and Health Act:* American Petroleum Institute v. OSHA, 60 B.U.L. REV. 115 (1980).

curbed. See 45 FED. REG. 5249 (1980). The unequal distribution of costs and benefits in this context necessarily raises the question of this statute's purpose and its determination of what is equitable; the mere calculation of costs and benefits cannot supply that answer.[68]

***

After the briefing in the *Cotton Dust* case was complete and oral argument was heard, a new OSHA administration took office in early 1981. In a Supplemental Memorandum to the Supreme Court filed in March 1981, OSHA asked the Court to refrain from further consideration of the *Cotton Dust* cases.[71] It said:

> The Secretary believes that, as indicated in Executive Order No. 12291, an analysis of the costs and benefits of federal regulation is of vital concern to the national welfare and should be taken into account by the government in setting its priorities. See 29 U.S.C. 655(g). Consonant with the policy underlying the Executive Order, it is the Secretary's view that an assessment of the practicality of cost-benefit balancing is best achieved in the context of an actual standard such as the one concerning cotton dust and in a manner that permits public comments.
>
> ***
>
> The Secretary believes that the information, data and comments likely to be received through a public proceeding will permit him to make an informed judgment as to the feasibility and utility of cost-benefit analysis in this area and, if feasible, to produce a comprehensive and thorough cost-benefit analysis of the cotton dust standard. This experience, plus the comparative experience under other health and safety laws (see 29 U.S.C. 655(b)(5)) will enable the Secretary to decide under what circumstances it is appropriate to factor such an analysis into the setting of standards for toxic substances. In particular, because the Secretary must base standards on the "best available evidence" (*ibid.*), the usefulness *vel non* of cost-benefit analysis bears on the legality of employing that analysis in the standard-setting process.
>
> In view of the Secretary's determination to undertake comprehensive supplemental rulemaking to reconsider the cotton dust standard and the role of cost-benefit analysis under the Act, the Court may wish to refrain from further consideration of the issues now before it in the pending cases. The Secretary anticipates that, as a result of the rulemaking proceeding, extensive new information will come to light respecting such matters as the economic strength of the textile industry and the ability of that industry to implement the technology required

[68]In addition to the problems inherent in attributing a dollar value to benefits, a cost-benefit analysis depends for its accuracy on the quality of available information concerning the extent of the risk. While the agency had the benefit of the comprehensive dose-response curve derived from Dr. Merchant's data in the case of cotton dust, this may well be the exception rather than the rule. Quantitative risk assessments typically yield only crude and inexact numerical indicators of risk because large numbers of uncertainties are involved. Although there may be substantial evidence of risk as a qualitative matter, numerical figures assigned to different exposures could well be erroneous by several orders of magnitude. Thus, a great degree of uncertainty would be implicit in any cost-benefit calculation in this context. On the cost side, moreover, the agency's experience with the vinyl chloride regulation shows that predictions of anticipated expenses may be grossly inflated when compared to actual expenses. See Doniger, *Federal Regulation of Vinyl Chloride: A Short Course in the Law and Policy of Toxic Substances Control,* 7 ECOLOGY L.Q. 497 (1978).

[71]Supplemental Memorandum of Federal Respondent, *American Textile Mfrs. Inst. v. Donovan,* 452 U.S. 490 (1981).

by the standard. As a result, it is possible that some of the requirements of the standard will be modified. For example, recent reports of the improved economic health of the textile industry might provide a basis for an even more protective standard.[72]

The AFL-CIO and Amalgamated Clothing and Textile Workers Union, the union respondents, vigorously opposed the OSHA request, which was ultimately rejected by the Supreme Court without stating a reason.[73] On the merits, the Supreme Court upheld the OSHA standard as it applied to the textile industry.[74]

American Textile Manufacturers Institute v. Donovan

452 U.S. 490 (1981)

JUSTICE BRENNAN delivered the opinion of the Court.
***

The Court of Appeals upheld the Standard in all major respects. The court rejected the industry's claim that OSHA failed to consider its proposed alternative or give sufficient reasons for failing to adopt it. 617 F.2d, at 652–654. The court also held that the Standard was "reasonably necessary and appropriate" within the meaning of §3(8) of the Act, 29 U.S.C. §652(8), because of the risk of material health impairment caused by exposure to cotton dust. 617 F.2d, at 654–655, 654, n.83. Rejecting the industry position that OSHA must demonstrate that the benefits of the Standard are proportionate to its costs, the court instead agreed with OSHA's interpretation that the Standard must protect employees against material health impairment subject only to the limits of technological and economic feasibility. *Id.*, at 662–666. The court held that "Congress itself struck the balance between costs and benefits in the mandate to the agency" under §6(b)(5) of the Act, 29 U.S.C. §655(b)(5), and that OSHA is powerless to circumvent that judgment by adopting less than the most protective feasible standard. 199 U.S. App. D.C., at 81, 617 F.2d, at 663. Finally, the court held that the agency's determination of technological and economic feasibility was supported by substantial evidence in the record as a whole. *Id.*, at 73–80, 617 F.2d, at 655–662.

We affirm in part, and vacate in part.[25]

[25]The postargument motions of the several parties for leave to file supplemental memoranda are granted. We decline to adopt the suggestion of the Secretary of Labor that we should "vacate the judgment of the court of appeals and remand the case so that the record may be returned to the Secretary for further consideration and development." Supplemental Memorandum for Federal Respondent 4. We also decline to adopt the suggestion of petitioners that we should "hold these cases in abeyance and * * * remand the record to the court of appeals with an instruction that the record be remanded to the agency for further proceedings." Re-

[72]*Id.* at 2–4. OSHA made it clear that if its request were granted, the Agency intended that the new cotton dust standard, the validity of which was being litigated, would remain in effect during the reconsideration. *Id.* at 5.

[73]452 U.S. at 505 n. 25. In a later decision, *North Haven Bd. of Ed. v. Bell,* 456 U.S. 512, 522 n.12 (1982), the Supreme Court cited its decision in the *Cotton Dust* case, suggesting that it did not agree to the OSHA request because the cotton dust standard was still in effect and the change in the OSHA view did not moot the litigation.

[74]The standard, insofar as it was applicable to other industries, had been either vacated (cotton-ginning and cottonseed-oil industries); stayed by a court of appeals (waste processing, cotton batting, bedding, and upholstery); or stayed by OSHA (warehousing and classing, and knitting and hosiery). See 47 FED. REG. 5906, 5907 (1982); 48 FED. REG. 5267 (1983).

II

The principal question presented in these cases is whether the Occupational Safety and Health Act requires the Secretary, in promulgating a standard pursuant to §6(b)(5) of the Act, 29 U.S.C. §655(b)(5), to determine that the costs of the standard bear a reasonable relationship to its benefits. Relying on §§6(b)(5) and 3(8) of the Act, 29 U.S.C. §§655(b)(5) and 652(8), petitioners urge not only that OSHA must show that a standard addresses a significant risk of material health impairment, see *Industrial Union Dept. v. American Petroleum Institute,* 448 U.S., at 639 (plurality opinion), but also that OSHA must demonstrate that the reduction in risk of material health impairment is significant in light of the costs of attaining that reduction. See Brief for Petitioners in No. 79-1429, pp. 38–41.[26] Respondents on the

sponse of Petitioners to Supplemental Memorandum for Federal Respondent 4.

At oral argument, and in a letter addressed to the Court after oral argument, petitioners contended that the Secretary's recent amendment of OSHA's so-called "Cancer Policy" in light of this Court's decision in *Industrial Union Dept. v. American Petroleum Institute,* 448 U.S. 607 (1980), was relevant to the issues in the present cases. We disagree.

OSHA amended its Cancer Policy to "carry out the Court's interpretation of the Occupational Safety and Health Act of 1970 that consideration must be given to the significance of the risk in the issuance of a carcinogen standard and that OSHA must consider all relevant evidence in making these determinations." 46 FED. REG. 4889, col. 3 (1981). Previously, although lacking such evidence as dose-response data, the Secretary presumed that no safe exposure level existed for carcinogenic substances. *Industrial Union Dept. v. American Petroleum Institute, supra,* at 620, 624–625, 635–636, nn. 39 and 40 (plurality opinion). Following this Court's decision, OSHA deleted those provisions of the Cancer Policy which required the "automatic setting of the lowest feasible level" without regard to determinations of risk significance. 46 FED. REG. 4890, col. 1 (1981).

In distinct contrast with its Cancer Policy, OSHA expressly found that "exposure to cotton dust presents a significant health hazard to employees," 43 FED. REG. 27350, col. 1 (1978), and that "cotton dust produced significant health effects at low levels of exposure," *id.,* at 27358, col. 2. In addition, the agency noted that "grade ½ byssinosis and associated pulmonary function decrements are significant health effects in themselves and should be prevented in so far as possible." *Id.,* at 27354, col. 2. In making its assessment of significant risk, OSHA relied on dose-response curve data (the Merchant Study) showing that 25% of employees suffered at least Grade ½ byssinosis at a 500 ug/m^3 PEL, and that 12.7% of all employees would suffer byssinosis at the 200 ug/m^3 PEL standard. *Id.,* at 27358, cols. 2 and 3. Examining the Merchant Study in light of other studies in the record, the agency found that "the Merchant study provides a reliable assessment of health risk to cotton textile workers from cotton dust." Id., at 27357, col. 3. OSHA concluded that the "prevalence of byssinosis should be significantly reduced" by the 200 ug/m^3 PEL. *Id.,* at 27359, col. 3; see *id.,* at 27359, col. 1 ("200 ug/m^3 represents a significant reduction in the number of affected workers"). It is difficult to imagine what else the agency could do to comply with this Court's decision in *Industrial Union Dept. v. American Petroleum Institute.*

[26]Petitioners ATMI et al. express their position in several ways. They maintain that OSHA "is required to show that a reasonable relationship exists between the risk reduction benefits and the costs of its standards." Petitioners also suggest that OSHA must show that "the standard is expected to achieve a *significant reduction in* [the significant risk of material health impairment]" based on "an assessment of the costs of achieving it." Allowing that "[t]his does not mean that OSHA must engage in a rigidly formal cost-benefit calculation that places a dollar value on employee lives or health," petitioners describe the required exercise as follows: "First, OSHA must make a responsible determination of the costs and risk reduction benefits of its standard. Pursuant to the requirement of Section 6(f) of the Act, this determination must be factually supported by substantial evidence in the record. The subsequent determination whether the reduction in health risk is 'significant' (based upon the factual assessment of costs and benefits) is a judgment to be made by the agency in the first instance."

Respondent Secretary disputes petitioners' description of the exercise, claiming that any meaningful balancing must involve "placing a [dollar] value on human life and freedom from suffering," Brief for Federal Respondent, and that there is no other way but through formal cost-benefit analysis to accomplish petitioners' desired balancing, *id.* Cost-benefit analysis contemplates "systematic enumeration of all benefits and all costs, tangible and intangible, whether readily quantifiable or difficult to measure, that will accrue to all members of society if a particular project is adopted." E. Stokey & R. Zeckhauser, A PRIMER FOR POLICY ANALYSIS 134 (1978); see Commission on Natural Resources, National Research Council, DECISION MAKING FOR REGULATING CHEMICALS IN THE ENVIRONMENT 38 (1975). See generally E. Mishan, COST-BENEFIT ANALYSIS (1976); Prest & Turvey, *Cost-Benefit Analysis,* 300 ECONOMIC JOURNAL 683 (1965). Whether petitioners' or respondent's characterization is correct, we will sometimes refer to petitioners' proposed exercise as "cost-benefit analysis."

other hand contend that the Act requires OSHA to promulgate standards that eliminate or reduce such risks "to the extent such protection is technologically and economically feasible."[27] To resolve this debate, we must turn to the language, structure, and legislative history of the Act.

A

The starting point of our analysis is the language of the statute itself. * * *

The plain meaning of the word "feasible" supports respondents' interpretation of the statute. According to Webster's Third New International Dictionary of the English Language 831 (1976), "feasible" means "capable of being done, executed, or effected." Accord, the Oxford English Dictionary 116 (1933) ("Capable of being done, accomplished or carried out"); Funk & Wagnalls New "Standard" Dictionary of the English Language 903 (1957) ("That may be done, performed or effected"). Thus, §6(b)(5) directs the Secretary to issue the standard that "most adequately assures * * * that no employee will suffer material impairment of health," limited only by the extent to which this is "capable of being done." In effect then, as the Court of Appeals held, Congress itself defined the basic relationship between costs and benefits, by placing the "benefit" of worker health above all other considerations save those making attainment of this "benefit" unachievable. Any standard based on a balancing of costs and benefits by the Secretary that strikes a different balance than that struck by Congress would be inconsistent with the command set forth in §6(b)(5). Thus, cost-benefit analysis by OSHA is not required by the statute because feasibility analysis is. See *Industrial Union Dept. v. American Petroleum Institute,* 448 U.S., at 718–719 (MARSHALL, J., dissenting).

When Congress has intended that an agency engage in cost-benefit analysis, it has clearly indicated such intent on the face of the statute * * *. We therefore reject the argument that Congress required cost-benefit analysis in §6(b)(5).

B

***

Agreement with petitioners' argument that §3(8) imposes an additional and overriding requirement of cost-benefit analysis on the issuance of §6(b)(5) standards would eviscerate the "to the extent feasible" requirement. Standards would inevitably be set at the level indicated by cost-benefit analysis, and not at the level specified by §6(b)(5). For example, if cost-benefit analysis indicated a protective standard of 1000 ug/m^3 PEL, while feasibility analysis indicated a 500 ug/m^3 PEL, the agency would be forced by the cost-benefit requirement to choose the less stringent point. We cannot believe that Congress intended the general terms of §3(8) to countermand the specific feasibility requirement of §6(b)(5). Adoption of petitioners' interpretation would effectively write §6(b)(5) out of the Act. We decline to

[27]As described by the union respondents, the test for determining whether a standard promulgated to regulate a "toxic material or harmful physical agent" satisfies the Act has three parts:
"First, whether the 'place of employment is unsafe—in the sense that significant risks are present and can be eliminated or lessened by a change in practices.' [*Industrial Union Dept., supra,* at 642 (plurality opinion).] Second, whether of the possible available correctives the Secretary has selected '*the* standard * * * that is most protective.' *Ibid.* Third, whether that standard is 'feasible.'" Brief for Union Respondents 40–41.
We will sometimes refer to this test as "feasibility analysis."

render Congress' decision to include a feasibility requirement nugatory, thereby offending the well-settled rule that all parts of a statute, if possible, are to be given effect. * * *[32]

C

The legislative history of the Act, while concededly not crystal clear, provides general support for respondents' interpretation of the Act. The congressional reports and debates certainly confirm that Congress meant "feasible" and nothing else in using that term. Congress was concerned that the Act might be thought to require achievement of absolute safety, an impossible standard, and therefore insisted that health and safety goals be capable of economic and technological accomplishment. Perhaps most telling is the absence of any indication whatsoever that Congress intended OSHA to conduct its own cost-benefit analysis before promulgating a toxic material or harmful physical agent standard. The legislative history demonstrates conclusively that Congress was fully aware that the Act would impose real and substantial costs of compliance on industry, and believed that such costs were part of the cost of doing business. * * *

***

Not only does the legislative history confirm that Congress meant "feasible" rather than "cost-benefit" when it used the former term, but it also shows that Congress understood that the Act would create substantial costs for employers, yet intended to impose such costs when necessary to create a safe and healthful working environment. Congress viewed the costs of health and safety as a cost of doing business. Senator Yarborough, a cosponsor of the Williams bill, stated: "We know the costs would be put into consumer goods but that is the price we should pay for the 80 million workers in America." [116 CONG. REC. 37,345 (1970)] He asked:

> "One may well ask too expensive for whom? Is it too expensive for the company who for lack of proper safety equipment loses the services of its skilled employees? Is it too expensive for the employee who loses his hand or leg or eyesight? Is it too expensive for the widow trying to raise her children on meager allowance under workmen's compensation and social security? And what about the man—a good hardworking man—tied to a wheel chair or hospital bed for the rest of his life? That is what we are dealing with when we talk about industrial safety. * * * We are talking about people's lives, not the indifference of some cost accountants."

Id. at 37,625.

Senator Eagleton commented that "[t]he costs that will be incurred by employers in meeting the standards of health and safety to be established under this bill are, in my view, *reasonable and necessary costs of doing business." Id.* at 41,764.

[32]This is not to say that §3(8) might not require the balancing of costs and benefits for standards promulgated under provisions other than §6(b)(5) of the Act. As a plurality of this Court noted in *Industrial Union Department,* if §3(8) had no substantive content, "there would be no statutory criteria at all to guide the Secretary in promulgating either national consensus standards or permanent standards other than those dealing with toxic materials and harmful physical agents." 448 U.S. at 640 n.45. Furthermore, the mere fact that a §6(b)(5) standard is "feasible" does not mean that §3(8)'s "reasonably necessary or appropriate" language might not impose additional restraints on OSHA. For example, all §6(b)(5) standards must be addressed to "significant risks" of material health impairment. *Id.* at 642. In addition, if the use of one respirator would achieve the same reduction in health risk as the use of five, the use of five respirators was "technologically and economically feasible," and OSHA thus insisted on the use of five, then the "reasonably necessary or appropriate" limitation might come into play as an additional restriction on OSHA to choose the one-respirator standard. In this case we need not decide all the applications that §3(8) might have, either alone or together with §6(b)(5).

Other Members of Congress voiced similar views. Nowhere is there any indication that Congress contemplated a different balancing by OSHA of the benefits of worker health and safety against the costs of achieving them. Indeed Congress thought that the *financial costs* of health and safety problems in the workplace were as large or larger than the *financial costs* of eliminating these problems. In its statement of findings and declaration of purpose encompassed in the Act itself, Congress announced that "personal injuries and illnesses arising out of work situations impose a substantial burden upon, and are a hindrance to, interstate commerce in terms of lost production, wage loss, medical expenses, and disability compensation payment." 29 U.S.C. §651(a). The Senate was well aware of the magnitude of these costs:

> [T]he economic impact of industrial deaths and disability is staggering. Over $1.5 billion is wasted in lost wages, and the annual loss to the Gross National Product is estimated to be over $8 billion. Vast resources that could be available for productive use are siphoned off to pay workmen's compensation benefits and medical expenses.

S. Rep. No. 91-1282, at 2.
Senator Eagleton summarized, "Whether we, as individuals, are motivated by simple humanity or by simple economics, we can no longer permit profits to be dependent upon an unsafe or unhealthy worksite." 116 Cong. Rec. 41,764 (1970).

***

The Supreme Court then confronted the issue of whether the cotton dust standard was feasible within the meaning of §6(b)(5). The Court accepted OSHA's view that a standard is feasible if it will allow the industry to maintain "long-term profitability and competitiveness."[75] Stating that it would not reverse "substantial evidence" findings of a court of appeals unless the standard was "misapprehended or grossly misapplied," the Supreme Court affirmed the District of Columbia Circuit's conclusion that the cost of compliance with the standard, as estimated by OSHA, was "economically feasible."[76] The Supreme Court also agreed with the court of appeals' view on the difficulty confronting OSHA in making an accurate cost analysis. It said:

> The Court of Appeals observed that "the agency's underlying cost estimates are not free from imprecision," 617 F.2d at 662, but that "[t]he very nature of economic analysis frequently imposes practical limits on the precision which reasonably can be required of the agency," *id.* at 661. We suspect that this results not only from the difficulty of obtaining accurate data, but also from the inherent crudeness of estimation tools. Of necessity both the RTI and Hocutt-Thomas studies had to rely on assumptions the truth or falsity of which could wreak havoc on the validity of their final numerical cost estimates. As the official charged by Congress with the promulgation of occupational safety and health standards that protect workers "to the extent feasible," the Sec-

[75] 452 U.S. at 530 n.55.
[76] *Id.* at 529–30.

retary was obligated to subject such assumptions to careful scrutiny, and to decide how they might affect the correctness of the proferred estimates.[77]

Justice Powell did not participate in the decision. Justice Rehnquist dissented, joined by Chief Justice Burger, on the grounds that the first sentence of §6(b)(5)[78] was an unconstitutional delegation of authority to OSHA.[79] Justice Stewart dissented on the ground that OSHA's feasibility findings were not supported by substantial evidence.[80]

In its decision, the Supreme Court held that the Act does not *require* cost-benefit analysis. It did not expressly rule that cost-benefit analysis is precluded; however, the statement in the opinion that cost benefit was "inconsistent" with the language, legislative history, and underlying policies of the Act was a strong indication that the Court believes that cost-benefit analysis is prohibited. OSHA has interpreted the decision as *precluding* cost-benefit analysis.[81]

The Supreme Court decision applies only to standards covered by the first sentence of §6(b)(5), i.e., standards regulating toxic substances or harmful physical agents. The Court did not preclude the utilization of cost benefit by OSHA in setting priorities under §6(g),[82] or in taking other regulatory actions.[83]

The Court based its conclusion, as noted, on an analysis of the statutory language, the relevant legislative history, and the overriding policy of the Act favoring employee protection. Unlike the District of Columbia Circuit Court of Appeals, however, the Supreme Court expressed no view on the efficacy of cost-benefit analysis as a tool in decision making in the standards area. This is understandable in light of the Court's recognition that under other statutes Congress had expressly mandated the use of cost-benefit analysis.[84] Moreover, the Court did not attempt to resolve the issue of whether, as the government claimed, the cost-benefit exercise necessarily involves placing a dollar value on human life and suffering. The Court alludes to the conflicting positions, without deciding between them.[85] The Supreme Court did indicate, however, that the practical impact of using cost-benefit analysis would be to force OSHA "to choose a less stringent point" in setting health standards than would be set in accordance with OSHA's feasibility approach.[86] Thus, the Court concurred in what had generally been understood throughout the his-

[77]*Id.* at 529 n.54.
[78]29 U.S.C. §655(b)(5).
[79]452 U.S. at 543.
[80]*Id.* at 541. The Court determination vacating the income-protection portion of the cotton dust standard is discussed in Chapter 5, Section D.
[81]*Oversight on the Administration of the Occupational Safety and Health Act, 1981: Joint Hearings Before the Subcomm. on Investigations and General Oversight and the Subcomm. on Labor of the Senate Comm. on Labor and Human Resources,* 97th Cong., 1st Sess. 25-6 (1981) (statement of Thorne Auchter, Assistant Secretary of OSHA).
[82]29 U.S.C. §655(g).
[83]452 U.S. at 509 n.29.
[84]*Id.* at 510.
[85]*Id.* at 506 n.26.
[86]*Id.* at 513.

tory of OSHA—cost-benefit analysis was advocated to authorize less stringent OSHA standards.[87]

D. Cost Effectiveness and Cost Benefit

OSHA had accepted the principle of cost effectiveness in its brief to the Supreme Court, saying that, "the statute permits the Secretary to select the least expensive means of compliance that will provide an adequate level of protection."[88] The Supreme Court went further, saying that the "reasonably necessary and appropriate" language of §3(8) "might come into play," requiring OSHA to choose a one-respirator rather than a five-respirator standard, if the same level of protection could be achieved by both strategies.[89] It is not entirely clear why the Supreme Court chose this rather extravagant example. Since the cost-effectiveness issue has often been debated in the context of engineering controls *versus* respirators,[90] it may be that the Supreme Court sought to avoid becoming prematurely involved in that difficult issue, which was not before the Court. OSHA has indicated since the Supreme Court decision that in setting standards, it intends to "examine the cost-effectiveness of alternative approaches to the hazard in question."[91]

In February 1983, OSHA issued an advance notice of proposed rulemaking on the Carcinogens Policy, stating its intention to reexamine its policy giving engineering controls priority as the compliance method to reach permissible exposure limits.[92] The purpose of this reexamination, the notice said, was to determine if modification of the policy would permit "more cost-effective compliance strategies"; to investigate the impact of "advances in respirator design, technology, and applications"; to attempt to identify the circumstances appropriate for particular compliance strategies; and to assess workplace conditions and employee health in industries and operations employing different strategies.[93]

In 1982, OSHA had published an advance notice of proposed rulemaking on reevaluation of the cotton dust standard. In that no-

[87]In its brief to the Supreme Court, OSHA suggested that cost-benefit analysis would not always yield a less protective standard, as, for example, where "an industry's process was extremely hazardous and its product readily replaceable." Brief of Federal Respondents at 39 n.43. However, since the traditional feasibility requirement would presumably continue to be applied even if cost benefit were accepted, even a standard deemed cost beneficial could not be issued if an industry would be put out of business.

[88]Brief for the Federal Respondent, *supra* note 61 at 56.

[89]452 U.S. at 513 n.32.

[90]See *infra* this section.

[91]Statement of Assistant Secretary Auchter, *supra* note 81 at 25. Cf. Note, *Cost Benefit Analysis, Cost Effectiveness Analysis and the Cotton Dust Standard: A Matter of Life and Death,* 35 RUTGERS L. REV. 133, 153 (1982) (argues that cost-effectiveness analysis as well as cost-benefit analysis is prohibited by the Act: "In the end, when we do cost-effectiveness analysis we are left with the need to balance cost factors against reliability [of various methods of compliance, such as engineering controls or respirators] * * * . This is exactly the same balance that is required by cost-benefit analysis * * * .").

[92]48 FED. REG. 7473 (1983).

[93]*Id.* at 7474. Earlier, in May 1982, OSHA had published an advance notice seeking data on the adequacy of respiratory protection programs and equipment. 47 FED. REG. 20,803 (1982).

tice, under the topic "compliance methods-engineering controls," OSHA asked for comment on the question of whether there are more "cost-effective ways of reducing cotton dust related illnesses."[94] In its proposed standard, however, published in June 1983,[95] OSHA stated that it was adhering *"in this instance"* to its policy of requiring feasible engineering controls.[96] On other issues, OSHA did rely on cost-effectiveness analysis, for example, in proposing action levels.[97]

The issue of cost-benefit analysis has consistently elicited strong expressions of view by both proponents and opponents.[98] Its supporters argue that it can be "most useful" to the policy maker in rational decision making; its opponents have claimed, among other things, that in the health context, cost-benefit analysis "obliterates * * * moral values." The two following excerpts are typical of the opposing views.

Benefit-Cost Analysis of Social Regulation: Case Studies from the Council on Wage and Price Stability[99]

BENEFIT-COST ANALYSIS

Economists have developed a technique for evaluating programs that involve scarcity and trade-offs. It is called "benefit-cost analysis." Although much maligned by its detractors, benefit-cost analysis can be most useful to a policy maker. Basically, all it entails is weighing the benefits and costs of a proposal before action is taken. At one extreme, the analysis might be no

---

[94]47 FED. REG. 5906, 5909 (1982).

[95]48 FED. REG. 26,962 (1983).

[96]*Id.* at 26,964 (emphasis in original). The Agency gave several reasons, among them, that "the [cotton dust] industry itself has shown little interest in a performance standard [allowing the alternate use of protective equipment] at this late stage of the process." *Id.*

[97]*Id.* at 26,970–71. The notice proposed other changes in the cotton dust standard, including modifications of the exposure monitoring requirements, extensions of some compliance dates, and modifications of some definitions. In addition, the proposed standard would not cover several segments of the nontextile industry (knitting, classing, warehousing, and cottonseed processing) in which, the Agency concluded, there was no evidence of "significant risk." *Id.* at 26,968–69. A lengthy dispute between OSHA and the Office of Management and Budget over the scope of the reconsideration, particularly the issue of engineering controls, preceded this proposal. See the text of the OSHA-OMB correspondence included in the record, Docket H-025E, available in the OSHA Docket Office, Washington, D.C. See also Wall Street Journal, May 20, 1983 at 4.

[98]There has been extensive literature on cost-benefit analysis in the regulatory context. See, e.g., SUBCOMM. ON OVERSIGHT AND INVESTIGATIONS OF THE HOUSE COMM. ON INTERSTATE AND FOREIGN COMMERCE, 96TH CONG., 2D SESS., COST-BENEFIT ANALYSIS: WONDER TOOL OR MIRAGE? (Comm. Print 1980); COMM. ON PRINCIPLES OF DECISIONMAKING FOR REGULATING CHEMICALS IN THE ENVIRONMENT, NATIONAL ACADEMY OF SCIENCES, DECISIONMAKING FOR REGULATING CARCINOGENS IN THE ENVIRONMENT 38 (1975); Baram, *supra* note 21; Kelman, *Cost-Benefit Analysis: An Ethical Critique,* REGULATION Jan.–Feb. 1981, 33; Kirschten, *Can Government Place a Value on Saving a Human Life,* NAT'L J., Feb. 17, 1979 at 252; Spiller, *Cost-Benefit: The Trojan Horse of Regulatory Reform,* Legal Times of Wash., Dec. 10, 1979, at 10. See also Carcinogens Policy, 45 FED. REG. 5001, 5245 (1980) ("The Role of Economics in Setting Exposure Levels"); OFFICE OF TECHNOLOGY ASSESSMENT, CONGRESS OF THE UNITED STATES, THE IMPLICATIONS OF COST-EFFECTIVENESS ANALYSIS OF MEDICAL TECHNOLOGY app. D (Aug. 1980) (excerpt of a report prepared for OTA by the Hastings Center, Institute of Society, Ethics and the Life Sciences). The subject of cost-benefit analysis was also discussed by the Senate Judiciary Committee in reporting favorably a regulatory reform bill. S. REP. No. 284, 97th Cong., 1st Sess. 70–93 (1981). See also Fisher, *Controlling Government Regulation: Cost-Benefit Analysis Before and After the* Cotton Dust *Case,* 36 AD. L. REV. 179 (1984).

[99]BENEFIT-COST ANALYSIS OF SOCIAL REGULATION: CASE STUDIES FROM THE COUNCIL ON WAGE AND PRICE STABILITY, at 4–5 (J. Miller III & B. Yandle, eds.) © 1979 by American Enterprise Institute for Public Policy Research. Reprinted with permission.

more quantitative than the policy maker's concluding that he or she "feels" the benefits of an action outweigh the costs (or the other way around). At the other extreme are the highly sophisticated estimates of benefits and costs usually conjured up at the mention of the term "benefit-cost analysis."

Without question, estimating benefits and costs is often difficult, especially in the areas of social regulation, where the benefits may be in terms of lives saved or pain and suffering avoided. Some say this means putting a value on human pain, suffering, and death, which is not only ludicrous, but downright immoral. If anything, we would argue, the reverse is true. Since resources are limited, we cannot avoid the need to identify—and, in some way, to estimate—benefits and costs. The more compassion we have for our fellow human beings, the more important this becomes. For benefit-cost analysis is merely a shorthand summary expression of who would benefit by how much from a proposed policy change, and who would be forced to sacrifice, at what cost. Indeed, benefit-cost analysis can point the way toward increasing the degree of perfection by identifying ways to minimize costs. Whether or not benefit-cost studies are used, however, *every* policy action reveals that, in the mind of the decision maker(s), benefits do exceed costs and that the distribution of benefits and costs is socially desirable. Thus, all policy decisions—whether to adopt a new initiative, to modify it, or to accept the status quo—reveal an implicit benefit-cost analysis.

Benefit-cost analysis really has two dimensions. The first is that of aiding the decision maker in deciding whether a given proposal should or should not be adopted. That is, "Do the benefits (somehow measured) exceed the costs (also somehow measured)?" In practice, this may simply be a listing of the measurable benefits on one side of the ledger and the measurable costs on the other, and then noting the various subjective benefits and costs to be weighed in the balance. The second dimension of benefit-cost analysis has to do with analyzing alternatives. That is, are there alternative ways of securing the same social regulatory objective that impose lower costs on society, or at least distribute the costs differently? Or, for a given cost, are there alternatives that would come closer to the goal of perfection, or at least result in a different distribution of the benefits? This use of benefit-cost analysis to evaluate alternatives is sometimes called "cost-effectiveness analysis." Basically, it is nothing more than an application of the efficiency axiom: maximize production for a given total cost; or minimize total cost for a given level of production.

***

Comments of United Steelworkers of America on the Advance Notice of Proposed Rulemaking for Cotton Dust[100]

***

A. THE USE OF COST-BENEFIT ANALYSIS WHERE HUMAN LIFE AND HEALTH ARE AT RISK OBLITERATES THE MORAL VALUES WHICH SHOULD UNDERLIE DECISIONS THE SECRETARY MUST MAKE.

Should the cost of saving a human life be weighed against the value of that life so that a decision can be made on whether the life is worth saving? To almost anyone outside of government, the statement of the question carries with it an immediate negative response.

If a report were radioed in from an old man in a yacht at sea that he has run out of gas and is drifting further out to sea where he will eventually die,

[100]Docket No. 052B, May 29, 1981 (on file in OSHA Docket Office, Washington, D.C.).

we would be horrified if the Coast Guard delayed launching a rescue operation until it had estimated the cost of the gas to power the rescue boat, the wages (including overtime) which would have to be paid to the rescuers, the future earning power or value of the man in the drifting boat to society, the statistical probability of finding the old man and a multitude of other cost and benefit factors. And, if Coast Guard officers refused to launch the rescue operation because the man had a short life expectancy and the value (in dollars) of his life was less than the cost of the rescue operation, the public pressure for discharge of the Coast Guard officers would more than likely be irresistible.

We view the rescue situation involving a known person instinctively and morally. Unfortunately, when dealing with unknown persons at risk the typical Washington regulator is likely to close his mind to any consideration of moral issues, smugly telling himself that moral issues should be dealt with by religious groups—not by the government.

Such a reaction reveals an ignorance of the moral foundation of our social legislation including the Occupational Safety and Health Act.

The OSH Act authorizes intervention by government into employer-employee relations in order to raise the moral standards of the labor market. It is one of a series of such moral interventions by government in the United States, each of which has been vigorously opposed by some short-sighted employers. The first of these interventions was the prohibition of slavery, and the list includes child labor laws, Social Security, Fair Labor Standards and pension reform.

Like each of these, the Occupational Safety and Health Act is a conscious effort by our society to regulate the conditions of labor in accordance with recognized ethical principles.

The Occupational Safety and Health Act was passed "to assure safe and healthful working conditions." In Section 2 Congress declared "it to be its purpose * * * to assure so far as possible every working man and woman in the Nation safe and healthful working conditions * * * by authorizing the Secretary to set mandatory occupational safety and health standards." In the case of toxic substances the Secretary is directed to set "the standard which most adequately assures, to the extent feasible, on the basis of the best available evidence that no employee will suffer material impairment of health or functional capacity." Congress imposed a moral value in the marketplace. "Thou shalt not kill (for profit)."

The Secretary, in promulgating safety and health standards, must implement this Congressionally imposed value. But the Secretary cannot do so with cost-benefit analysis.

The "science" of cost-benefit analysis is an effort to measure, in common units of measurement, the total costs and the total benefits which may be reasonably expected to result from a given action. To "rely" on such analysis is to automatically reject the action if the units of cost exceed the units of benefit—and to take the action in the reverse case.

Although other units of measurement are conceptually possible the proponents of cost-benefit analysis for health standards have uniformly used money values to make the comparisons. Under the usual system, dollars of "cost" reflect the material worth of the capital investment needed to achieve compliance, plus that of the ongoing expenses required. Similarly, the dollars of "benefit" reflect some material valuation of the worth of the lives which would be saved, and the illnesses avoided.

Under cost-benefit analysis, to kill or not to kill becomes a question of expense, not ethics. In a triumph of materialism over morals, the ancient commandment and the Congressional purpose simply becomes inoperative at a certain level of cost. "Thou shalt not kill unless it is cheaper."

Instead of asking "Is it economically and technologically feasible to save this life?", the decision-maker who relies on cost-benefit analysis asks, "Is this life worth the cost of saving?" The decision-maker who asks the first question assumes that the life is worth saving and examines the practicalities of doing so.

Cost-benefit analysis places a monetary value on human life thereby obliterating the moral purpose which led Congress to pass the OSH Act.

***

* * * [The] principal use [of cost-benefit analysis] is as a tool for making private and public investment decisions—where the investor receives the returns. If costs are paid and benefits obtained by the *same* individual, firm, set of taxpayers, or the like, there is some merit to the use of this tool.

This is not the case with occupational health standards. Costs are paid by large groups—stockholders, consumers and others who deal with the company. The benefits of a standard are enjoyed by another group—those who otherwise must endure the pain and suffering, death and disability caused by the toxic substance.

If costs could be accurately determined, benefits properly evaluated, numbers juggled impartially, and all biases overcome, this issue of fairness would still confront the decision-makers.

Cost-benefit analysis sheds no light on the equities of the present distribution, or proposed future distributions of costs and benefits. A recent report by the Subcommittee on Oversight and Investigations of the House Interstate and Foreign Commerce Committee * * * summarized this problem in the following language:

> The essence of the equity argument is that economists have no way of making value determinations when the costs and benefits accrue to different groups within the society. One of the ways that they attempt to avoid this problem is by assuming it away, i.e., they assume that the distribution of good and bad things within the society is equitable to begin with and therefore alterations in the distribution of income or in the distribution of the negative consequences which result from the failure to regulate have no particular value to society. However, ignoring equity considerations constitutes a value judgment. As Professor Guido Calabresi of Yale points out, "the willingness of a poor man, confronting a tragic situation, to choose money rather than the tragically scarce resource [his health or safety] always represents an unquiet indictment of society's distribution of wealth."[4]
> ***

E. Executive Order No. 12,291: Quantifying Benefits

On February 17, 1981, President Reagan issued Executive Order No. 12,291, whose purpose was "to reduce the burdens of existing and future regulations, increase agency accountability for regulatory actions, provide for presidential oversight of the regulatory process, minimize duplication and conflict of regulations, and insure well reasoned regulations."[101] A significant requirement of the Order is the

[4]"Cost/Benefit Analysis-Wonder Tool or Mirage?" Report of the Subcommittee on Oversight and Investigations of the Committee on Interstate and Foreign Commerce, U.S. House of Representatives, December 1980. U.S. Government Printing Office, p. 27.

[101]46 FED. REG. 13,193 (1981) (codified at 3 C.F.R. 127 (1982)).

preparation by agencies of a preliminary and final Regulatory Impact Analysis (RIA) for each "major" rule.[102] The RIA must contain, among other things, "to the extent permitted by law," the following:

> (1) A description of the potential benefits of the rule, including any beneficial effects that cannot be quantified in monetary terms, and the identification of those likely to receive the benefits;
> (2) A description of the potential costs of the rule, including any adverse effects that cannot be quantified in monetary terms, and the identification of those likely to bear the costs;
> (3) A determination of the potential net benefits of the rule, including an evaluation of effects that cannot be quantified in monetary terms;
> (4) A description of alternative approaches that could substantially achieve the same regulatory goal at lower cost, together with an analysis of this potential benefit and costs and a brief explanation of the legal reasons why such alternatives, if proposed, could not be adopted; and
> (5) Unless covered by the description required under paragraph (4) of this subsection, an explanation of any legal reasons why the rule cannot be based on the requirements set forth in Section 2 of this Order.[103]

The Executive Order further provides that, if permitted by law, regulatory action should not be undertaken unless the "potential benefits to society for the regulations outweigh the potential costs."[104]

Under the Supreme Court *Cotton Dust* decision, as construed, OSHA is not permitted to use cost-benefit balancing in setting standards for toxic substances. Accordingly, the determination of "net benefits" of a rule cannot be, and has not been made by OSHA for toxic substance standards. The requirements in the Order for the separate description of potential costs and benefits of a proposed rule are applicable, however, to all OSHA standards activity. On June 12, 1981, the Office of Management and Budget issued a statement on the requirements of Executive Order No. 12,291, entitled "Interim Regulatory Impact Analysis Guidance."[105] With respect to benefits, the Interim Guidance stated as follows:

> The RIA should state the beneficial effects of the proposed regulatory change and its principal alternatives. It should include estimates of the present value of all potential real incremental benefits to society. Benefits that can be estimated in monetary terms should be expressed in constant dollars. Other favorable effects should be described in detail and quantified where possible. An annual discount rate of 10 percent should be used; however, where it appears desirable, other discount rates also may be used to test the sensitivity of the results. Assumptions should be stated, and the RIA should identify the data or studies on which the analysis is based.

[102] A "major" rule is a regulation "likely" to result in an annual effect on the economy of $100 million or more; a major increase in prices; or a "significant adverse effect" on competition, employment, investment, productivity, innovation, or the ability of American enterprises to compete with foreign enterprises. Exec. Order No. 12,291 §1(b), 3 C.F.R. 127 (1982).

[103] *Id.* at §3(d), 3 C.F.R. at 129.

[104] *Id.* at §§2(b), 3(d)(5), 3 C.F.R. at 128–29.

[105] Reprinted in *Hearings on the Role of OMB in Regulation Before the Subcomm. on Oversight and Investigations of the House Comm. on Energy and Commerce,* 97th Cong., 1st Sess. 360 (Comm. Print 1981).

There should be an explanation of the mechanism by which the proposed action is expected to yield the anticipated benefits.

A schedule of benefits should be included that would show the *type* of benefit, to *whom* it would accrue, and *when* it would accrue. The numbers in this table should be expressed in constant dollar terms.[106]

These interim requirements have changed OSHA practice in at least one important respect. In the past, while expressing the costs of a standard in dollar terms, OSHA has consistently refused to do so for benefits.[107] Under the Executive Order, however, benefits too must be stated in "constant dollar terms," as OSHA did in its standard on hazard communication.[108]

The types of questions that arise in calculating the dollar value of benefits were articulated in the following excerpt from a study prepared for OSHA and used by the Agency in its preliminary regulatory analysis on ethylene oxide.[109]

Economic and Environmental Impact Study of Ethylene Oxide (EtO)

***

ECONOMIC IMPACTS OF EtO-RELATED CANCER

Although OSHA does not subscribe to any method of evaluating human life, JRB has, for informational purposes, described some of the economic impacts of EtO-related cancer. This section presents these impacts to show that the adverse health effects of occupational exposure to EtO are important and quantifiable economic impacts. These impacts are not, in themselves, a full measure of the benefits of a reduced PEL for EtO, because they do not include many of the unquantifiable benefits associated with occupational cancer.

Three economic impact measures of EtO-related cancers are discussed in this section:

(1) The direct costs associated with cases of EtO-related cancer. These include the costs of diagnosis, treatment, continuing care, rehabilitation, and terminal care for cancer patients.
(2) The decline in the gross national product (GNP) caused by the loss of earnings among workers who are unable to work due to EtO-related cancer.
(3) In theoretical terms, the decline in the value of the payment a fully informed employee would demand to accept the increased risk of cancer associated with a specific job having that level of risk.

Each of these measures omits certain values and economic impacts that are important. For example, direct costs do not measure the psychic losses to

---

[106]*Id.* at 362.

[107]See, e.g., Coke Oven Emissions Standard, 41 FED. REG. 46,742, 46,750–51 (1976), reprinted in Section B of this chapter.

[108]48 FED. REG. 53,280, 53,327–30 (1983).

[109]Economic and Environmental Impact Study of Ethylene Oxide, prepared by M.B. Kent and Associates for JRB Associates, McLean, Va., 3-62–3-77 (Apr. 1983). Copy on file in OSHA Docket Office.

OSHA's preliminary regulatory analysis also sought to "monetize" the benefits of preventing ethylene-oxide-caused cancer. Preliminary Regulatory Impact and Regulatory Flexibility Assessment of the Proposed Standard for Ethylene Oxide (1982), Ch. V, at v-14–v-17. The published preamble to the proposed standard did not state the benefits in "dollar" terms, 48 FED. REG. at 17,297–98 (1983). For a discussion of the ethylene oxide proceeding, see Chapter 6, Section C.

family and friends or the productivity loss to the economy. The decline in GNP indicator does not include direct costs, and implicitly assigns a value to an individual that does not exceed his/her salary. The payment-to-accept-risk measure does not consider any costs the affected individual would not be required to pay (such as medical and other costs borne by insurance or Medicare, and welfare for his/her survivors), and additionally ignores the value that people other than the affected individual might place on his/her health and well-being.

***

DECLINE IN GROSS NATIONAL PRODUCT DUE TO FOREGONE EARNINGS

***

* * * The assumptions JRB used in this study result in an estimate of the present value over 50 years of foregone earnings for workers in the health provider industry of $33,612,911, * * *.

***

***

MEASURES OF THE VALUE OF REDUCED CANCER RISK

The purpose of this section is to determine the increase in payments that would be demanded by workers to accept a marginal increase in the risk of contracting cancer, assuming that employees were fully informed of the risks involved and operated in perfectly competitive markets. The reduction in the size of the payments that would be demanded if this risk were then reduced is also calculated; this reduction is referred to in this section as savings in the payment to accept risk. JRB also estimates willingness to pay, an alternative measure of the economic value of a reduction in the risk of incurring cancer. Using economic theory, this section will attempt to determine the shadow price that would be attached to the risk of cancer if perfect markets existed in which persons started with some baseline level of cancer risk and could be paid to accept a marginal increase in the risk of cancer. Such a perfect market does not exist. Even if it did, computation of the shadow price would still be a difficult statistical exercise.

One can, however, examine the ordinary labor market, in which a part of the wage rate can be seen as payment to accept job risk. The risk associated with a job is, however, only one of many components determining wage rate, and separating and quantifying the risk element is difficult. The values for payment to accept risk used by JRB are based on statistical studies of the labor market. However, even if the exact effect of risk on wage rate were certain, the shadow price placed on risk to life would still not represent the perfect market price. For example, there are inadequacies and uncertainties in information that distort risk premiums in wage rates, and there may be other market imperfections making wage rates different from those in a perfect market. Nevertheless, estimates of risk premiums in wage rates are the closest available approximation of payment-to-accept-risk values in a perfect free market.

Empirical evidence based on analyses of wage premiums for risk indicates that workers in occupations roughly similar to those typical of the EtO industries have been found to place a value on their lives of $1–$2 million.[1]

[1]Empirical estimates of the value of life of this order of magnitude have been obtained in the study by Robert Smith, THE OCCUPATIONAL SAFETY AND HEALTH ACT: ITS GOALS AND ACHIEVEMENTS. Washington: American Enterprise Institute, 1976; and W. Kip Viscusi, *Labor Market Valuations of Life and Limb: Empirical Estimates and Policy Implications,* in PUBLIC POLICY 26:359–386, 1978, and EMPLOYMENT HAZARDS: AN INVESTIGATION OF MARKET PLACE PERFORMANCE, Massachusetts: Harvard University Press, 1979.

This estimate is based on data from the period 1969–1970 (Viscusi 1979). JRB updated this estimate to 1982 dollars by multiplying it times the ratio of the GNP deflator for the first quarter of 1982 to the GNP deflator for the last quarter of 1969 (i.e., 203.68/88.62). This results in an estimate that $2.3–$4.6 million would be the payment demanded by a worker to accept a risk to life. To assess the sensitivity of these estimates, both the $2.3 million per case and $4.6 million per case values will be used to estimate the total value of avoidable cancer cases. Values as high as $7 million per worker have been obtained in some studies where the sample consisted of individuals particularly risk averse, as evidenced by their choice of highly safe occupations (Viscusi 1978).

The payment to accept an increase in the risk of cancer is not the only possible measure related to individual preference, assuming a perfect market. An alternative measure is termed willingness to pay. This measures what an individual would be willing to pay to avoid a given risk of cancer. This measure differs from payment to accept risk in that it implies a different assignment of implicit rights. Payment to accept risk is relevant in situations in which the individual theoretically has some control over whether or not to be exposed to a given risk and would have to be paid to accept the risk. The willingness-to-pay method is applicable in situations in which the person is already assumed to be exposed to the risk and must pay for protection to lower the risk.

In general, the payment-to-accept-risk value will be higher than the willingness-to-pay-to-avoid-risk value, although in extreme cases the two may be equal. The reason a payment to avoid risk will normally be higher than willingness to pay is that willingness to pay is inherently limited by the total assets available to the employee, while payment to accept risk is not. The payment to accept risk appears to be more relevant to job-related risk, and will receive the main emphasis in this section. However, a lower-bound estimate of willingness to pay using foregone earnings is included for sensitivity analysis purposes. This is a lower-bound estimate because a person will, in the event of death, lose all earnings he might have had. If the individual prefers to avoid risk, and the chance of death is not a certainty, he will pay more than his total earnings because there will be an additional payment to avoid risk. (The principle here is the same as that which explains why people buy insurance that has a greater cost than the expected value of its payoff.)

To calculate the total increase in payments that would be required to accept the risk of cancer, the shadow prices assigned to human lives are multiplied by the discounted number of cases avoided. To use these figures, two issues must be addressed. First, what is the relationship between a case of cancer and the loss of life? Cancer often results in death after a long interval, and before one's death one incurs substantial additional pain and financial costs. It is not at all clear *a priori* whether or not individuals would be willing to place a greater value on averting an instantaneous death than they would on averting a case of cancer; even though the death from cancer is deferred there will be an additional loss in individual welfare because of the pain and suffering involved. No attempt will be made here to resolve these issues. Rather, these value of life estimates will be used to evaluate a case of cancer.

A second major question is that of what discount rate to use. The payment-to-accept-risk values cited are for risks to life in the same year, not risks to life 15 years in the future, as necessary for this analysis. Since the question posed here is one of what the employee would be willing to pay or accept in payment, the relevant discount rate would need to be one related to the real rate of return employees could be expected to earn on their money. A 6 percent real rate of return seems high for this purpose, and even a 2 percent real rate of return might be realistic for after-tax returns on investments available to individuals of relatively modest income.[2]

The key to the argument for a lower discount rate is that the relevant question is not what discount rate is appropriate to a calculation of costs to the nation but rather, by the definition of a payment-to-accept-risk measure, what discount rate the employee might use in determining the answer to a question concerning his or her willingness to pay or to accept payment to accept risk. For this reason, present values using a discount rate of 6 percent are presented as an alternative to those using a 10 percent discount rate.*

***

OSHA's recent exercises in the hazard communication and ethylene oxide proceedings in calculating, in dollar terms, the value of benefits from a standard should be compared with OSHA's view, presented seven years before, that calculating benefits for the coke oven emissions standard presents "insuperable obstacles."[110]

[2]A discount rate of 2% was used by P. MacAvoy in the study: P. MacAvoy, The Economic Consequence of Asbestos-Related Diseases, Working Paper Number 27, January 1982.

*[Author's note—The 10% discount rate is in accord with the OMB Interim Guidance, *supra* this section. The appropriate discount rate, as can be seen, is a major problem in the quantification of benefits. See COST-BENEFIT ANALYSIS: WONDER TOOL OR MIRAGE?, *supra* note 98 at 22–24. For other discussions of quantification of benefits from safety regulation, see M. Bailey, REDUCING RISKS TO LIFE: MEASUREMENT OF THE BENEFITS (1980); Thaler & Rosen, *The Value of Saving a Life: Evidence From the Labor Market,* in HOUSEHOLD PRODUCTION AND CONSUMPTION 265 (N.E. Terlecky ed. 1976). Professor Baram critically discusses the problems with quantifying the value of human life and other "traditionally unquantifiable attributes," Baram, *supra* note 21 at 483–86.]

[110]41 FED. REG. 46,742, 46,750–51 (1976), reprinted in part *supra* pp. 304–307.

PART III

OSHA Enforcement

11

The Structure of OSHA Enforcement

A. Statutory Scheme

The purpose of the Act, as stated in the congressional findings, was "to assure so far as possible every working man and woman in the Nation safe and healthful working conditions,"[1] and it would be achieved, Congress believed, through various mechanisms; foremost among these was the enforcement of occupational safety and health standards by the Secretary of Labor. OSHA conducts surprise workplace inspections (the Act prohibits the giving of advance notice of inspections)[2] and, if violations are found, citations are issued and civil monetary penalties proposed.[3] These citations and penalties may be contested by employers before the independent Occupational Safety and Health Review Commission,[4] whose decisions are subject to court review.[5] In Congress' view, although OSHA would not have resources to inspect all covered workplaces in the country, this authority, conscientiously implemented, would provide an incentive to *all* employers to abate workplace hazards, even before inspection.

This chapter provides an introduction to OSHA enforcement; discusses the views of interested parties on whether this enforcement scheme has achieved its goals; and finally, presents alternatives or complements to the present approach. The statutory enforcement scheme was summarized by OSHA in its brief to the Supreme Court in the *Atlas Roofing* case, which involved the constitutionality under the Seventh Amendment of the statutory procedure for the imposition of civil penalties.[6]

[1]Sec. 2(b), 29 U.S.C. §651(b) (1976).
[2]Sec. 17(f), 29 U.S.C. §666(f).
[3]Sec. 9, 29 U.S.C. §658.
[4]Sec. 10(c), 29 U.S.C. §659(c).
[5]Sec. 11, 29 U.S.C. §660.
[6]Brief for the Respondents, 3–9. *Atlas Roofing Company, Inc. v. OSHRC,* 430 U.S. 442 (1977).

Brief for the Respondents in Atlas Roofing Company, Inc. v. OSHRC

***

The Occupational Safety and Health Act of 1970, 84 Stat. 1590, 29 U.S.C. 651 *et. seq.* (OSHA), was passed to alleviate the "drastic" national problem of work-related deaths and injuries with a view to assuring "so far as possible every working man and woman in the Nation safe and healthful working conditions." 29 U.S.C. 651; *National Realty and Construction Co. v. Review Commission,* 489 F.2d 1257, 1260–1261 (D.C. Cir. 1973). The Act neither alters nor addresses traditional private obligations between employers and employees relating to injuries actually caused by unsafe or unhealthy conditions. *Ibid.,* 489 F.2d at 1260, n.6; §4(b)(4), 29 U.S.C. 653(b)(4). Instead, it creates a new statutory duty to avoid maintaining such conditions, 29 U.S.C. 654(a)(1) and (2), applicable to all nongovernmental employers whose business affects commerce. The law is enforced by the Secretary of Labor through a self-contained administrative mechanism whose central features are speedy, expert and uniform resolution of contested cases by the independent Review Commission, subject to the usual appellate review. 29 U.S.C. 651(2), (3) and (10), 652(5), 657(a), 658–661. * * *

Under the Act's expedited enforcement procedures, the Secretary's inspectors are authorized to conduct safety and health inspections at reasonable times and in a reasonable manner at places of employment. 29 U.S.C. 657(a). If, upon inspection or investigation, the Secretary has cause to believe that the Act or its implementing regulations have been violated, he is empowered to issue a citation to the employer specifically describing the violation, proposing a reasonable time for its abatement, and (in his discretion) proposing a civil monetary penalty. 29 U.S.C. 658, 659.

The amount of the proposed penalty "if any" (29 U.S.C. 659(a)) depends on the severity of the hazard and the cited employer's past diligence in attempting to discover and correct it (*ibid.,* see 29 U.S.C. 666 (i) and (j)), and is designed to promote voluntary compliance by employers before any inspector arrives. 29 U.S.C. 651(1); * * *. Such proposed penalties may range from nothing for *de minimis* and nonserious violations, to not more than $1,000 for serious violations, to a maximum of $10,000 for willful or repeated violations. 29 U.S.C. 658(a), 659(a), 666(a)–(c) and (j). The Secretary may also propose a civil penalty of not more than $1,000 per day of nonabatement where subsequent inspection reveals noncompliance with a final Commission order, 29 U.S.C. 659(b), 666(d), and may seek temporary injunctions in federal district court to correct imminent dangers before administrative enforcement would result in their abatement. 29 U.S.C. 662. Finally, in cases of willful violations that cause employee death, the Secretary is authorized to refer the matter to the Justice Department for criminal prosecution, which may result in a maximum sentence of six months' imprisonment and a $10,000 fine, but unlike civil administrative enforcement entails no explicit abatement requirement. 29 U.S.C. 666(e).

If an employer wishes to contest the citation or the proposed abatement period or penalty, he must notify the Secretary within 15 working days, otherwise the citation, the abatement period and assessment as proposed become final. 29 U.S.C. 659(a). If the employer contests, the Secretary's proposed abatement order is automatically stayed and no final order directing correction of the violations is entered until proceedings before the Commission are concluded. 29 U.S.C. 659(b), 666(d). Upon the filing of an employer contest, an evidentiary hearing is held before an administrative law judge of the Occupational Safety and Health Review Commission, an agency independent of the Secretary created to "carry out adjudicatory functions under the Act" and composed of three members with six-year terms, qualified to review contested citations and assess penalties "by reason of training, education or experience." 29 U.S.C. 651(3), 659(c), 661, 666(i). At this hearing the burden is on the Secretary to establish the elements of the alleged violation and the propriety of his proposed abatement period and proposed penalty, and the judge is empowered to affirm, modify, or vacate any or all of these items, giving due consideration in his penalty assessment to "the size of the business of the employer * * *, the gravity of the violation, the good faith of the employer, and the history of previous violations." 29 U.S.C. 666(i). The judge's decision becomes the Commission's final order unless within thirty days a Commissioner directs that it be reviewed. 29 U.S.C. 659(c), 661(i); see 29 CFR 2200.90 and 2200.91.

If review is granted, the Commission's subsequent order directing abatement and the payment of any assessed penalty becomes final unless the employer timely petitions for judicial review in the appropriate court of appeals. 29 U.S.C. 660(a). The Secretary similarly may seek review of Commission orders, 29 U.S.C. 660(b), but, in either case, "[t]he findings of the Commission with respect to questions of fact, if supported by substantial evidence on the record considered as a whole, shall be conclusive." 29 U.S.C. 660(a). If the cited employer fails timely to contest a citation or to petition for court of appeals review, the Secretary is empowered to ensure abatement by seeking a summary judicial decree, enforceable by contempt, in the court of appeals. 29 U.S.C. 660(b). If the cited employer abates but fails to pay the penalty finally assessed by a Commission order, the Act provides for a civil collection action in federal district court, in which neither the fact of violation nor the propriety of the penalty assessed may be retried. 29 U.S.C. 666(k).

On March 23, 1977, the Supreme Court decided the *Atlas Roofing* case; it upheld the constitutionality of the OSHA enforcement provisions and rejected employer arguments that civil monetary penalties could not legally be assessed under the Seventh Amendment to the Constitution without a jury trial.[7]

Enforcement by OSHA through inspections, citations, and the assessment of civil penalties operates to achieve safe and healthful workplaces in two significant ways. First, *with respect to the workplace inspected,* the issuance of a legally enforceable abatement order, with the possibility of a follow-up inspection and daily penalties of up to $1,000 for nonabatement, was expected to bring about neutralization of the hazards at *that* workplace. To the extent that OSHA's selection policies succeed in identifying workplaces containing significant hazards, OSHA enforcement activity could achieve a substantial elimination of these hazards in these workplaces.

[7]*Atlas Roofing Co. v. OSHRC,* 430 U.S. 442 (1977). On the issue whether OSHA civil penalties are criminal in effect and, therefore, subject to the requirements of the Sixth Amendment, see *Atlas Roofing Co. v. OSHRC,* 518 F.2d 990 (5th Cir. 1975) (holding OSHA penalties are not penal in nature) *aff'd on other grounds,* 430 U.S. 442 (1977). See also Levin, *OSHA and the Sixth Amendment: When is a "Civil Penalty" Criminal in Effect?,* 5 HASTINGS CONST. L.Q. 1013 (1978). OSHA's enforcement regulations issued in 1971, appear at 29 C.F.R. §§1903.1–.21 (1983). Basic policy instructions from OSHA's National Office to field staff are contained in the Field Operations Manual (FOM). The FOM was first issued as the Compliance Operations Manual in 1971, was revised periodically, and completely revised in April 1983. The current version of the FOM is reprinted in OSHR [Reference File 77:2101–:5051]. The Industrial Hygiene Manual was first issued in 1976 and is reprinted in OSHR [Reference File 77:8001–:8412].

It has been universally recognized, however, that because of the pervasive coverage of OSHA (approximately 5 million workplaces) and the limited available resources (at no time more than approximately 2,500 federal and state compliance officers[8]), large numbers of workplaces would be inspected either infrequently or not at all. For OSHA to achieve more extensive elimination of workplace hazards, it was, therefore, essential that the enforcement scheme also constitute an incentive to covered employers to abate hazards whether or not an inspection had taken place. This is the second means by which OSHA enforcement protects employees. Two elements in the statutory scheme are essential to the achievement of this purpose: meaningful first instance sanctions issued following unannounced OSHA inspections. As stated by Basil Whiting, formerly Deputy Assistant Secretary for OSHA:

> First-instance sanctions under the OSH Act are intended to provide an incentive to employers to abate hazards even before the OSHA inspection takes place. Without first-instance sanctions, unconscientious employers may well await the OSHA citation and abate only in obedience to its requirements. To make this incentive meaningful, proposed penalties cannot be at a level constituting a "slap on the wrist." OSHA's full utility of its penalty authority is designed primarily to implement this Congressional purpose.[9]

[8]OSHA field inspectors are referred to as compliance safety and health officers (CSHOs) and are classified as either "safety officers" or "industrial hygienists." According to statistics from OSHA's Office of Management Data Systems, in fiscal year 1979, there were 1158 inspectors (724 safety officers and 434 hygienists); in fiscal year 1980, there were 1199 inspectors (737 safety and 462 hygienists); in fiscal year 1981, 1130 inspectors (689 safety and 441 hygienists); and in fiscal year 1982, 1004 (626 safety and 378 hygienists). This may be compared with June 1973, when OSHA had 456 safety inspectors and 68 hygienists. See Rothstein, *OSHA After Ten Years: A Review and Some Proposed Reforms,* 34 VAND L. REV. 71, 94 (1981). According to OSHA's office of State Programs, in fiscal year 1982, there were 1091 inspectors employed in the 24 states with approved plans (787 safety and 304 hygienists) and in fiscal year 1983, 1075 inspectors (757 safety and 318 hygienists).

The training of compliance personnel takes place at the OSHA Training Institute, Des Plaines, Illinois. See generally, OFFICE OF TRAINING AND EDUCATION, OSHA, NOTICE TED 1, FISCAL YEAR 1983 OSHA TRAINING INSTITUTE SCHEDULE OF COURSES AND REGISTRATION REQUIREMENTS (1983). The training schedule contains courses for newly hired compliance staff as well as advanced technical courses for experienced personnel. Among the courses offered, in addition to basic safety and health technical training, are: introduction to industrial hygiene for safety officers; safety hazard recognition for industrial hygienists; and for all CSHOs: accident investigation; civil law enforcement; guides to "voluntary compliance"; and communications and human relations. Courses are also offered at the Institute for state consultants, see Section D, this chapter, for other federal department personnel, and for employers, workers, and their representatives. From fiscal year 1972 through fiscal year 1980 (first quarter), there were 24,429 enrollments at the OSHA Training Institute, of which 9910 were federal CSHOs, 5129 were state compliance personnel, 3306 were from other federal agencies, 1060 were employee representatives and 5024 were employer representatives. *Oversight on the Administration of the Occupational Safety and Health Act: Hearings Before the Senate Comm. on Labor and Human Resources,* 96th Cong., 2d Sess. pt. 1, table 1, at 300 (1980) (statement of Deputy Assistant Secretary Basil Whiting) [hereinafter cited as *1980 Senate Oversight Hearings*].

[9]*1980 Senate Oversight Hearings, supra* note 8, pt. 1 at 285 (1980). See also Administrative Conference of the United States, Recommendation No. 79-3, Agency Assessment and Mitigation of Civil Monetary Penalties, 1 C.F.R. §305.79-3 (1983) (recommending standards for the determination of penalty amount, the initial assessment of penalties, the mitigation of penalties, and the use of evidentiary hearings); Diver, *The Assessment and Mitigation of Civil Money Penalties by Federal Administrative Agencies,* 79 COLUM. L. REV. 1435 (1979) (discusses, among other things, the administration of civil monetary penalties by the Mine Safety and Health Administration in the Labor Department). For instructions to OSHA field staff on calculation of penalties, see Field Operations Manual, ch. VI, OSHR [Reference File 77:3101–:3108].

Section 17(f)[10] furthers this policy by imposing criminal sanctions on any person who gives unauthorized advance notice of an inspection to be conducted under the Act. If an employer knew that he or she would be notified before an inspection took place, there would be less incentive to eliminate hazards prior to inspection. The statutory policy of unannounced inspections thus serves a critical purpose under the Act's enforcement scheme.[11]

B. How Effective is OSHA?

A number of factors have emerged, however, which have tended to undermine the effectiveness of OSHA enforcement as an incentive to employers to abate hazards before they are inspected. The credibility of first instance sanctions depends both on the likelihood of an OSHA inspection taking place, and on the stringency of the civil penalties that may be proposed. According to the testimony of Basil Whiting,[12] in fiscal year 1979, a period during which the Administration of Dr. Bingham attempted to implement a vigorous enforcement policy, the number of federal workplace inspections was 57,937. The average penalty for all serious, willful, and repeat violations was $367.00, and the average for other than serious violations with penalty was $98.00. Further, of the 128,544 violations cited during that fiscal year, more than 83,000 did not involve penalties.[13]

While the Act imposes various constraints on OSHA's flexibility in proposing penalties, the Agency has imposed upon itself additional restrictions limiting the amounts of penalties. Thus, for example, OSHA's instructions to CSHOs say that "all violations of a specific standard found during the inspection of an establishment or worksite shall be combined into one alleged violation."[14] Since only one penalty will be proposed for the "grouped" instances of the violation, the amount of the penalty will necessarily be lower, often substantially lower, than if each violation were penalized separately.[15] Another example is OSHA's policy regarding penalties for failure to abate. The Act provides for penalties of up to $1,000 per day for an employer's failure to abate a hazard in accordance with the requirements of a citation which has become a final order of the Review Commission.[16] Although the statute does not limit the number of days for

[10]29 U.S.C. §666(f).

[11]For a discussion of the importance of surprise in the OSHA inspection program in the context of *ex parte* warrants, see Chapter 12, Section B.

[12]*1980 Senate Oversight Hearings*, *supra* note 8 at 286–87.

[13]The statistics have been provided by the OSHA Office of Management Data Systems, Washington, D.C., with the average calculated by the author. In fiscal year 1982, the average penalty for all serious, willful, and repeat violations was $221.13; and of the total of 97,136 violations cited, more than 72,000 did not involve penalties. 12 OSHR 711 (1983). According to Professor Mark Rothstein the average OSHA penalty in fiscal year 1980 was $192.55. Rothstein, *supra* note 8 at 99 n.164.

[14]Field Operations Manual, ch. V, OSHR [Reference File 77:2906].

[15]The "grouping" policy has been in effect since OSHA's first compliance manual, Compliance Operations Manual, ch. X at 4 (Jan. 1972).

[16]Sec. 17(d), 29 U.S.C. §666(d).

which a daily penalty may be imposed, OSHA, except in "unusual circumstances," will not impose a daily penalty for more than ten days of nonabatement following the day of required abatement. Thus, the nonabatement penalties will normally not exceed a total of $10,000.[17]

Serious questions have been raised as to whether the level of this enforcement effort is sufficient to provide the needed incentive for preinspection abatement. In addition, as Mr. Whiting pointed out, additional problems have been the "delays in OSHA inspection activity occasioned by the warrant requirement of the Supreme Court's *Barlow's* decision"[18] and "the extensive delays in abatement caused by burgeoning litigation of OSHA citations"[19] The Act gives employers the right to contest OSHA citations and penalties and, while the case is pending before the independent Review Commission, the abatement requirements of citations are suspended.[20] This factor was exacerbated as a result of a rapid increase in the contest rate,[21] and delays in the Commission's resolution of cases.[22]

Disappointment with the enforcement of the Act has been voiced by a number of groups. One type of criticism has come from labor unions; an example is the testimony of Nolan Hancock, Citizenship-Legislative Director of the Oil, Chemical and Atomic Workers International Union, one of the most active unions in the field of occupational safety and health, at the 1980 oversight hearings of the Senate Committee on Labor and Human Resources.[23]

---

[17]Field Operations Manual, ch. VI, OSHR [Reference File 77:3106]. This policy was not included in the 1972 compliance manual, but was instituted when the Field Operations Manual was republished in July 1974, ch. XI at 6 (copy on file with OSHA, Washington, D.C.).

[18]*Marshall v. Barlow's, Inc.*, 436 U.S. 307 (1978). This case and the warrant requirement are discussed in Chapter 12.

[19]*1980 Senate Oversight Hearings, supra* note 8 at 296.

[20]Sec. 10(b), 29 U.S.C. §659(b).

[21]Whiting notes that between fiscal years 1977 and 1979, the contest rate "doubled" from 12% to 21%. *1980 Senate Oversight Hearings, supra* note 8 at 297. By 1982, however, the contest rate had fallen by about 50%. According to Assistant Secretary Auchter, this drop was due, firstly, to the philosophy of his administration emphasizing "cooperation rather than confrontation" and, secondly, to the use of informal conferences between employers and OSHA area directors, permitting, whenever possible, for citations to be settled without contest. See *OSHA Oversight—State of the Agency Report By Assistant Secretary of Labor for OSHA: Hearing Before the Subcomm. on Health and Safety of the House Comm. on Education and Labor*, 97th Cong., 2d Sess. 181 (1982). The field memorandum authorizing area director settlements was first issued in 1980 by Dr. Bingham and is currently set forth in the Field Operations Manual, ch. V, OSHR [Reference File 77:2910–:2911]. For further discussion of Mr. Auchter's enforcement policy, see Section E, this chapter.

[22]Professor Rothstein discusses the Review Commission's caseload and "backlog" in detail in *OSHA After Ten Years, supra* note 8 at 115-32. According to a 1978 study of the Comptroller General, administrative law judges of the Review Commission took an average of 169 days to decide cases and the Review Commission itself took an average of 437 days from the date of direction of review to issue a decision. Comptroller General of the United States, Report to the Congress, FPCD-78-25, Administrative Law Process: Better Management is Needed 11 (1978). This total of 606 days was the longest time required by any of the agencies studied. According to Professor Rothstein, the backlog in 1980 remained at between 300 and 400 cases, and as of September 2, 1980, half of them had been pending on review for more than 2 years. Rothstein, *supra* note 8 at 118. In April 1983, however, Chairman Robert Rowland reported that the Review Commission backlog was 160 cases, but he found the age of some of the cases "distressing." 12 OSHR 974 (1983). This drop in Commission caseload and backlog was largely due to the sharp decrease in contest rate, resulting from informal settlements between OSHA and employers.

[23]*1980 Senate Oversight Hearings, supra* note 8 at 873, 875–79.

Statement of Nolan Hancock

***

The basic problem is that the regulatory structure of the Occupational Safety and Health Act is so weak that it makes little difference to employers, especially in the oil and chemical industry, whether the OSHA inspector comes knocking at the door or not.

The reason why employers assume a combative stance with OSHA has nothing to do with a lack of confidence or trust in OSHA. It simply pays to fight OSHA. In the United States, businesses exist in order to make a profit. We have established a regulatory system that makes it profitable for a business to fight OSHA rather than comply. Why, then, should we be surprised at the polarization between industry and OSHA. Again this is not trade union rhetoric. The Interagency Task Force made the same finding in their report:

> employers may profit by contesting (rather than promptly complying) whenever the present use of abatement money is worth more than the proposed penalty plus the legal costs of delay. (p. III-15)*

This point is even more true in an economic climate when money is losing its value at a rate of 18% per year. It pays for a business to utilize its money in order to turn a profit rather than fulfilling a health or safety objective that will show a much lower rate of return.

The profitability in not complying or in delaying compliance with OSHA is manifested in several ways, all of which are rooted in the structure of the Act, not in its administration.

First, the Supreme Court's decision in the *Barlow*['s] case was a significant setback for effectiveness of OSHA. Now, for example, when OSHA responds to a formal employee complaint, OSHA arrives at the workplace and presents the employer with a copy of the complaint. The employer can then refuse to admit the inspector, but the employer now has advance knowledge of the areas in the plant that OSHA wants to inspect. OSHA then returns to the plant with the warrant. Again, the employer can refuse to admit the inspector. Several days to several weeks later, a Federal District Court judge will hold a hearing on the employer's refusal to honor the warrant. Most refusals of entry usually end at this point but the damage to the effectiveness of the inspection has already been inflicted.**

As we indicated earlier, the employer now has advance notice of what is to be inspected and he can manipulate working conditions during the inspection in order to hide potential violations. In the case of an accident investigation, particularly after a fire or explosion, the rubble and other important evidence are already cleaned up by the time the inspector can finally get in which can be several weeks after the incident occurred. This kind of situation recently occurred after two workers were burned to death at the Amoco Oil Company refinery in Texas City, Texas. Amoco successfully held off the OSHA inspectors for three weeks by using these tactics. Such delays diminish the possibility of OSHA inspectors finding serious violations.

Once the inspection begins, management often turns its wrath on employees who cooperate with the OSHA inspectors. Harassment and intimidation, including discriminatory discharge of employees who assist OSHA, are all commonplace. Even though the Act has fairly clear language prohibiting such illegal conduct by employers, it has not deterred them. The rea-

*[Author's note—The Interagency Task Force is discussed in Section C, this chapter.]

**[Author's note—The Supreme Court decision in *Marshall v. Barlow's, Inc*, 436 U.S. 307 (1978) is discussed in Chapter 12.]

son is that the investigatory and enforcement provisions of the Act's anti-discrimination provision have virtually ground to a halt.

In fiscal year 1979, OSHA received 2953 discrimination complaints to be handled by 52 investigators. Going into fiscal year 1980, 1400 cases remain backlogged from 1979. An additional 3300 to 4000 new cases are expected in fiscal year 1980 to be handled by the same inadequate resources. But the most shocking figure is that 429 cases which have been found to be meritorious by the investigators are currently languishing in the Labor Department's Solicitor's Office. No lawsuit has been filed in these cases because of a lack of resources. But equally at fault is the lack of interest that the U.S. District Courts have shown in processing these cases. The process often takes years.

The situation is so bad that the Secretary of Labor recently took the unusual stance of backing an employee who claimed that there was an implied private right of action under the Act's anti-discrimination provision (Section 11(c)). The Sixth Circuit Court of Appeals recently ruled against the employee and the Secretary in that case, but made the following observation:

> The Secretary says he has neither the resources nor the personnel to handle all Section 11(c) complaints adequately. Moreover, he expects the number of such complaints to increase dramatically due to his current campaign to alert employees of their OSHA rights. A private right of action should be implied, the Secretary argues, because individual suits offer the only realistic hope of protecting employees from retaliatory discrimination.
>
> The Secretary should address his arguments to Congress, not the courts. (*Taylor v. Brighton Corporation,* 8 OSHC 1010, at 1016.)

In sum, it can be said without exaggeration that Section 11(c) of the Act no longer works. Employers know that fact. The easiest way to cripple the effectiveness of an OSHA inspection is to quickly retaliate against an employee who cooperates with the inspector. The chilling effect on the willingness of other employees to cooperate is immediate, even in unionized plants. When it becomes apparent that the process to get that person's job back will take years, the chilling effect on that group of workers becomes a deep freeze. OSHA, in turn, has a difficult time in gathering evidence to support a serious violation.

While the difficulties in entry and acts of retaliation work to dilute OSHA effectiveness, efforts to reduce the death and injury rate receive their most crushing blows in the Act's review process. * * *

The heart of the problem with this review process is that Section 10(b) of the Act permits the correction of the cited hazardous condition to be stayed during the pendency of the contest. In other words, a condition that has killed or seriously injured a worker can legally remain unabated while the review process unfolds. * * *

If this Committee is truly interested in reducing the death and injury rate, it is incumbent that it investigate the review process under the Act. In the last three years under Dr. Bingham's leadership, we have personally observed a dramatic improvement in the quality and quantity of citations issued. Today when an inspector goes into a plant, we are confident that serious hazards will be addressed in a serious manner. The significant safety hazards that have plagued our members for years, where management has resisted correction because of the cost, are now being addressed in OSHA citations. Consequently, in the OCAW experience we see a contest rate of 90% of all citations issued against the oil and chemical industry. At this point, the process literally grinds to a halt.

According to the Occupational Safety and Health Review Commission

(OSHRC), it takes the average case nearly three years to go through a hearing before the administrative law judge and then review by the full Commission. This is consistent with our experience. In other words, three years pass[] between issuance of the citation and correction of the hazardous condition.

***

Early in his testimony, Mr. Hancock asks the question, "Why hasn't the [occupational] death and injury rate declined?"[24] In evaluations of the effectiveness of OSHA's enforcement scheme the question inevitably arises: has OSHA succeeded in reducing occupational injuries and illnesses. One of the most common indicators of OSHA's effectiveness has been the annual survey conducted by the Bureau of Labor Statistics (BLS) in the Department of Labor. This survey is based on workplace injury and illness records which most employers are required to keep under the Act and OSHA's implementing regulations.[25] These records relate to occupationally related fatalities, injuries involving lost workdays, injuries not involving lost workdays, and lost workdays. Records are also kept, and a survey conducted, of occupationally related illnesses; however, BLS has often stated that because illnesses often develop years after the employee has left the firm where the illness was contracted, the reporting of illnesses "present[s] some measurement problems."[26] To allow for comparisons between industries and establishments of varying sizes, BLS presents the data in terms of incidence rates, that is, injuries and illnesses in terms of a constant (100 full-time employees working for one year).[27] The results of the BLS survey for calendar year 1982 were summarized in an official news release issued by the Bureau of Labor Statistics.[28]

Occupational Injuries and Illnesses in 1982

Job-related injuries and illnesses declined in 1982, the Bureau of Labor Statistics of the U.S. Department of Labor reported today. The all-industry incidence rate fell to 7.7 injuries and illnesses per 100 full-time workers.

This incidence rate, which is 0.6 below the 8.3 rate reported last year, has declined steadily since 1979. Of this 0.6 decline, about 0.1 may be attributable to a disproportionate drop in 1982 in hours worked in high-risk industries.

---

[24] *Id.*

[25] Sec. 8(c)(1)–(2), 29 U.S.C. §657(c)(1)–(2); 29 C.F.R. pt. 1904 (1983). In 1977, OSHA exempted employers with 10 or fewer employees from most of the requirements under the record-keeping regulations, 42 FED. REG. 38,567 (1977), and in 1982 OSHA further exempted certain employers in classifications with low incidence of worker injury from record-keeping requirements, 47 FED. REG. 57,699 (1982).

[26] BUREAU OF LABOR STATISTICS, U.S. DEP'T OF LABOR, REP. NO. 518, OCCUPATIONAL SAFETY AND HEALTH STATISTICS: CONCEPTS AND METHODS 5 (1978).

[27] *Id.* at 3.

[28] Bureau of Labor Statistics, U.S. Dep't of Labor News Release, Nov. 4, 1983. Available in U.S. Dep't of Labor files.

OTHER KEY RESULTS

• Workplaces employing 11 workers or more in the private sector recorded 4,090 work-related deaths. Nearly 30 percent of these fatalities resulted from car and truck accidents.

• The number of job-related injuries dropped by nearly 530,000 cases, to 4.75 million in 1982. Both lost-time injuries and injuries involving no time loss declined.

• The incidence rate for injuries involving lost worktime dropped from 3.7 per 100 full-time workers in 1981 to 3.4 in 1982.

• About 36.1 million lost workdays resulted from work-related injuries, about 3 million fewer days than in 1981. The number of lost workdays per 100 full-time workers fell from 60.4 in 1981 to 57.5 in 1982.

• The average number of lost workdays per lost workday injury increased to 17, up 1 day from 1981.

• Injury incidence rates increased little or remained unchanged in retail trade, finance, insurance, and real estate, and the services industries and decreased in the other industry divisions.

***

OCCUPATIONAL ILLNESSES

Occupational illness is any abnormal condition or disorder, other than one resulting from an occupational injury, caused by exposure to environmental factors associated with employment. It includes acute and chronic illnesses or diseases which may be caused by inhalation, absorption, ingestion, or direct contact. The incidence of occupational illnesses measured by the annual survey refers to the number of new illness cases occurring during the year and does not measure continuing conditions reported in previous surveys. Thus, illnesses are recorded only for the year in which they are recognized and diagnosed as work-related.

From both statistical and procedural points of view, occupational illness estimates generated from the annual survey provide a valid measure of recognized acute cases. However, the current statistics do not adequately reflect the portion of occupational illnesses which are chronic and long-latent in nature, because of problems of detection and occupational relationship.

About 105,600 occupational illnesses were recorded in 1982, down from 126,100 in 1981. Skin disorders continued to account for the majority of illnesses, about 40 percent. Another category of illness—physical disorders associated with repeated trauma—showed the largest percentage increase in total illnesses, rising from 18 percent in 1981 to 21 percent in 1982.

BACKGROUND OF SURVEY

The Annual Survey of Occupational Injuries and Illnesses is a Federal/State cooperative program in which State agencies participate with the Bureau of Labor Statistics of the U.S. Department of Labor. The 1982 survey, response to which was mandatory, involved a sample of approximately 280,000 units in the private sector.

The occupational injury and illness data reported through the annual survey are based on the records which employers maintain under the Occupational Safety and Health Act of 1970. Excluded from coverage under the Act are working conditions which are covered by other Federal safety and health laws. However, data conforming to OSHA definitions of recordability for coal, metal, and nonmetal mining and railroad activities were provided to the Bureau of Labor Statistics by the Mine Safety and Health Administration of the U.S. Department of Labor and the Federal Railroad Administration of the U.S. Department of Transportation.

The survey covers units in private industries. Excluded are the self-employed; farmers with fewer than 11 employees; private households; and employees in Federal, State, and local government agencies. In a separate reporting system, agencies of the Federal government file reports comparable with those of private industry with the Secretary of Labor.

Estimates based on a sample may differ from figures that would have been obtained had a complete census of establishments been possible using the same schedules and procedures. Relative standard errors are calculated for the estimates generated from the Annual Survey of Occupational Injuries and Illnesses and are made available to the public.

The portion of the decline in the injury and illness incidence rate for the total private sector which may be attributed to the decline in hours worked was calculated by applying the 1982 hours of exposure at the industry division level to the 1981 rates and comparing the result with the total private sector injury and illness rate published for 1981. This procedure does not take into account, however, other factors which may also affect the rate, but which have not been measured, such as the demographic composition of the workforce, worker education, improved safety measures, the role of State and Federal agency compliance programs, technological change, etc.

While injury incidence rates were reduced in 1982, the statistical record since the inception of OSHA has been ambiguous. In 1973, the incidence rate for all injuries and illnesses was 11.0 per 100 full-time workers; for lost-workday cases, 3.4; for non-lost-workday cases, 7.5; and the total lost workdays rate was 53.3. Comparing these statistics with the results of the 1982 survey, the total injury case rate and the non-lost-workday rate went down, while the lost-workday case rate and the lost-workday rate increased. This would suggest that OSHA has been effective in reducing less serious injuries, but that, OSHA notwithstanding, the number of the more serious workplace injuries has increased. Moreover, for a period of several years (1975–1979), the incidence rate for all injury cases increased from 9.1 to 9.5. And, as is apparent, the number of work-related deaths (4,090, including fatalities from traffic accidents), and job-related injuries (4.75 million) continues to be high in absolute terms.[29] Attempts, far from satisfactory, have been made over the years to explain these seeming discrepancies in the impact of OSHA. In any event, it is clear that, so far as appears from these statistics, there has been no dramatic improvement in the occupational safety and health picture during the period that OSHA has been in effect.[30]

[29]These statistics are taken from Table 2 of the Dep't of Labor News Release, Nov. 4, 1983.

[30]There have also been a number of individual studies which have attempted to measure OSHA's impact on workplace injuries and illnesses. Some studies have found a statistically significant improvement in certain areas following OSHA's enactment. See, e.g., J. Mendeloff, REGULATING SAFETY: AN ECONOMIC AND POLITICAL ANALYSIS OF OCCUPATIONAL SAFETY AND HEALTH POLICY 94-120 (1979). Other studies have found no significant difference in injury rates. R. Smith, THE OCCUPATIONAL SAFETY AND HEALTH ACT: ITS GOALS AND ACHIEVEMENTS 70 (1976) (the estimated effects of the OSHA Target Industry Program (TIP) are "virtually nil");

Some commentators have pronounced OSHA to be a failure because of the lack of statistical indicia of reduction in workplace injuries and illnesses. For example, in a well-known study of OSHA, Professors Zeckhauser and Nichols concluded:

> Both politically and practically [OSHA] has been a failure. It has generated fierce antagonism in the business community and is viewed by many as the quintessential government intrusion. And it has had virtually no noticeable impact on work related injuries and illnesses.[31]

Similarly, Senator Schweiker, in introducing a bill which would have made significant amendments in OSHA,[32] stated that the success of OSHA has been "substantially less than overwhelming." He referred particularly to the statistical evidence showing increased serious injuries in the workplace in the period since OSHA was passed.

Statement of Senator Schweiker

125 Cong. Rec. 37,135–36 (1979)

***

Nine years ago, we made a promise in the law "to assure so far as possible every working man and woman in the Nation safe and healthful working conditions."

Today, we can find little evidence that the act has directly improved workplace safety. From 1972 through 1978, the years for which reliable data are available, the overall injury/illness rate has declined from 10.5 to 9.4 per 100 employees, a modest 10-percent decrease. But during the same time period, injuries and illnesses resulting in lost workdays, the more serious cases, increased 28 percent, from 3.2 to 4.1 per 100 workers. These data are not conclusive in assessing the act's effectiveness since they also reflect the impact of recession (which generally lowers injury rates), workers' compensation increases (which may inflate lost workday cases), and other extraneous factors. But what is clear 9 years after enactment is that the success of the act in delivering on its promise has been substantially less than overwhelming.

Not only has the act failed to produce demonstrable benefits in workers'

Smith, *The Impact of OSHA Inspections on Manufacturing Injury Rates,* 14 J. Hum. Resources 145 (1979) (the author compares the injury rates of plants inspected "early" and "late" in the year; in 1973, there was a 16% reduction, in 1974 no statistically significant reduction); Viscusi, *The Impact of Occupational Safety and Health Regulation,* 1978 Bell J. Econ. 117 ("the expected penalty levels imposed by OSHA appear too low to have a perceptible effect on either enterprise decisions on health or safety outcomes"). For a general summary of the studies on OSHA's effectiveness, see Senate Comm. on the Judiciary, The Regulatory Reform Act, S. Rep. No. 284, 97th Cong., 1st Sess. 25–29 (1981). The Senate Committee recognized the conflicting results of these studies and refused to endorse any of them. The Committee, however, stated that the extent of the conflicting information about the achievements of federal regulation "itself provides serious cause for concern," in large measure because it "emasculates the basic mechanism—Congressional oversight—by which the citizenry both authorizes, and checks the exercise of law-making power by unelected officials." *Id.* at 28. The impact of federal regulation on coal mine safety and health is discussed in Lewis-Beck & Alford, *Can Government Regulate Safety? The Coal Mine Example,* 74 Am. Pol. Sci. Rev. 745 (1980). See also generally, Senate Comm. on Governmental Affairs, Benefits of Environmental, Health and Safety Regulation, 96th Cong., 2d Sess. (1980) (prepared for the Committee by the MIT Center for Policy Alternatives); L. Bacow, Bargaining for Job Safety and Health 24–28 (1980).

[31] Nichols & Zeckhauser, *Government Comes to the Workplace: An Assessment of OSHA,* 49 Pub. Interest 39, 42 (1977).

[32] For a discussion of the provisions of the Schweiker bill, see Chapter 13, Section E.

safety, it has created extraordinary public controversy. Nine years ago, the proposed act passed the Senate by a vote of 83 to 3. Not long after that, however, the Occupational Safety and Health Administration, the Agency created to administer the law, had become probably the most despised Federal Agency in existence. In recent years, the erosion of public and political support has been evidenced by several Senate votes in favor of blanket exemptions for broad classes of employers. In a free society, no law can be effective if its administration and enforcement are widely viewed as illegitimate and nonproductive.

Mr. President, the bottom line is this: After 9 years under the act's present safety regulatory scheme, we are left with no demonstrable record that it works and with a bad taste all around from the experience. * * *

***

Supporters of OSHA enforcement have sought to explain the lack of statistical evidence showing improvement in the rate of workplace injuries and illnesses. For example, in 1979, when the BLS statistics showed increases in injury rates in most major areas,[33] Dr. Bingham issued a statement seeking to minimize the importance of these numbers.[34]

Statement of Assistant Secretary Eula Bingham on the Bureau of Labor Statistics' Data for 1979

***

In any analysis of the data released today by BLS, it is important to note that the data largely reflect injuries and deaths resulting from injuries. Data concerning occupationally-related illnesses and diseases are particularly difficult to come by because of the fact that the Nation is only beginning to develop the ability to distinguish between personal and job-related illnesses.

This is important in OSHA's terms because of the great emphasis we have been putting on matters relating to occupational health. Despite this emphasis, it is encouraging to note that there was no material increase in the safety-related statistics.

It thus appears that, by and large, employers across the Nation have been able to devote increasing emphasis to health matters without any sacrifices in the advances they have already made in safety areas.

On the other hand, it is my belief that the absence of any significant statistical improvements in the BLS injury data is directly related to the fact that employers collectively continue to expend too much effort in opposing regulation as a basic strategy instead of getting on with the serious business of improving safety and health conditions.

I can't stress enough that in the final analysis, it is the employer's duty and responsibility to make these improvements and I would encourage all employers to redouble their efforts in the coming years to reduce the terrible toll of death, injury and illness that continues to confront us.

***

[33]The incidence rate for all injury cases rose from 9.4 in 1978 to 9.5 in 1979; the lost-workday injury incidence rate rose from 4.1 to 4.3; and the lost-workday rate rose from 63.5 to 67.7. On the other hand, the non-lost-workday injury rate fell from 5.3 to 5.2.

[34]U.S. Dep't of Labor News Release, Nov. 20, 1980. Available in U.S. Dep't of Labor files.

Finally, I would point out that the BLS data, although a statistically valid indicator of actual industry injury experience provide only a partial picture of the true state of safety and health in the workplace. Even as the measures are made, the target changes. New members join the workforce. Older members retire. Employment increases. Awareness of worker compensation benefits increases. Medical practitioners become more cautious in the face of rising malpractice claims. New industries begin and older ones die out. For all of these reasons it is exceedingly difficult to try to capture any clearcut meaning for one- or two-tenths of a percent shift in data. What is clear from our analysis, however, is that we are correct in our policy of directing our meager inspection and assistance resources towards high hazard industries.

Dr. Bingham noted the absence of reliable data on occupational illnesses and the deficiencies of BLS data, which, she said, provide "only a partial picture of the true state of safety and health in the workplace." She also argued, in a somewhat circular fashion, that the lack of significant improvement was caused by employer intransigence—the expenditure of resources in opposing regulation rather than in promoting safety and health.

More recently, however, Thorne G. Auchter, Assistant Secretary for Occupational Safety and Health has construed the statistical pattern as demonstrating the success of his approach to implementation of the Act. His testimony in November 1983 before a House subcommittee commented at length on the BLS statistics for 1982 which had just been released.[35]

Statement of Thorne G. Auchter

***

The goal of OSHA enforcement is straightforward: to ensure compliance with safety and health regulations and, thus, to reduce the number of workplace deaths, injuries and illnesses.

OSHA enforcement resources are, and always have been, limited. Therefore, OSHA policy decisions are intended to direct those resources to those areas where they can have the greatest positive effect. One measure of the success of these policies is the rates of on-the-job deaths, injuries and illnesses.

By that standard, OSHA has not always been successful in its mission to reduce workplace accidents. In fact, from 1975 through 1979, the rate of lost workday cases, the measure of those injuries serious enough to warrant time off from work, rose every year, according to the Bureau of Labor Statistics. In terms of that statistic, the lost workday case rates for the years 1978, 1979 and 1980 were the three worst in OSHA's history.

I am happy to report today that worker safety has improved substantially since 1980. Just last week, BLS released its 1982 survey of occupational injuries and illnesses, and the findings indicate that the progress made in 1981 continued, or accelerated, through last year. Before I discuss the details of some of our enforcement policies, I would like to briefly men-

[35] *Hearings Before the Subcomm. on Manpower and Housing of House Comm. on Government Operations (1983),* pp. 1–6. (Mimeo) For further discussion of Mr. Auchter's views on OSHA enforcement, see discussion in Section E, this chapter.

tion some of the results of the BLS survey. I will compare the 1982 figures with those of 1980.

> The number of workplace deaths within OSHA's jurisdiction has dropped by 7 percent;
> The number of workers injured has dropped by 750,000, or 14 percent;
> The injury and illness rate has fallen by 11 percent;
> The lost workday case rate has dropped by 13 percent;
> The number of lost workdays has dropped 4.8 million, or 11 percent.
> The lost workday case rate for construction has dropped by 8 percent, and for manufacturing by 17 percent.

It isn't difficult to translate all these numbers into human terms. They mean less suffering, fewer needless tragedies, fewer workers laid up while family income dwindles.

The credit for this good news belongs first and foremost to employers and employees whose daily efforts to recognize and correct workplace hazards are essential to job safety and health. But I also believe that OSHA and its employees—safety engineers, industrial hygienists, scientists, physicians, administrators and other staff—have had the positive impact that President Reagan envisioned when he said, even before taking office, that, "There is a compelling need for an *effective* program to improve safety and health in the workplace for America's working men and women."

OSHA has been a continuing experiment and in times past its enforcement program was inefficient and its antagonistic attitude provoked more resistance than compliance.

The three years immediately preceding this administration graphically illustrate the dismal record of the agency. During those three years—1978–80—record dollar fines were proposed, the rate of contested cases skyrocketed, and the lost workday case rate reached peak levels.

Let me spend a few moments on the contested case rate, which reached 22 percent in both 1979 and 1980. Employers have the right to contest OSHA citations, and they exercised that right with frequency during those years. The result of a contested case, no matter who wins or loses, is wasted time and delay.

When such a case is heard before an administrative law judge, or the Occupational Safety and Health Review Commission, the OSHA compliance officer must take time away from his job to testify. As one of our compliance officers recently told the Buffalo, N.Y., News, "I personally would rather be going into a new establishment than going to court to defend what we've done."

But even more important, when a citation is contested, the hazard cited doesn't have to be corrected until the litigation runs its course, which can take months or years. So a 22 percent of contest really means that OSHA was realizing only a 78 percent rate of prompt hazard abatement. Workplaces were becoming more dangerous, simply, because OSHA was becoming tangled in litigation.

As many others have, OSHA too sought sensible solutions to its growing legal problems. The answer was the area director settlement policy. This concept was developed during the last few months of the previous administration, and has been emphasized during each year of this administration.

The policy gives employers and employee representatives the right to meet informally with the OSHA area director after an inspection to discuss any citations or proposed penalties. The meeting centers on issues of safety and health. Based on information discussed at that time, the area director is authorized to settle cases by combining or downgrading citations and reducing proposed penalties in exchange for one thing: an agreement by the employer to correct the hazards within a certain period of time.

The result of this policy has been extremely positive. The rate of contested cases has fallen to less than 4 percent, which means we are now achieving a better than 96 percent rate of rapid abatement, rather than the 78 percent of only three years ago. Furthermore, we have replaced those old feelings of animosity with a spirit of cooperation, and we are finding that employers are willing to correct hazards when we make the effort to work with them, rather than against them.

Another result of the area director settlement policy has been well publicized. Fines and citations have dropped since 1980. This result has not been a goal of our policy, but I am not displeased with it. Fines and citations are not a measure of success. We prefer to achieve rapid hazard abatement and reductions in the rate of on-the-job injuries and illnesses.

The efficiency of our enforcement effort has increased greatly since 1980. According to our preliminary data for fiscal year 1983, we conducted 68,577 total inspections, over 5100 more than in 1980, and of those, nearly 90 percent were targeted in high hazard industries. In 1980, barely 50 percent of our inspections were targeted in high hazard industries.

Taking federal inspections and consultative visits together, federal OSHA programs visited 98,827 worksites in 1983, a 15 percent increase over 1980.

There are two major reasons for this increased productivity. First: a *ninefold* decrease in time spent in litigation activities by OSHA compliance officers from 1980 to 1983, saving the agency the equivalent services of 30 compliance officers; and second: a *sevenfold* decrease in time spent in follow-up activities by compliance officers, saving the agency the equivalent services of 68 compliance officers. These follow-up activities, we had found, were less useful than new initial visits to different worksites. For example, in 1979 OSHA conducted 11,677 follow-up inspections and found 11 failure-to-abate violations. Today, we are visiting workplaces in high hazard industries that had never before been visited by an OSHA compliance officer.

This increased productivity, Mr. Chairman, has allowed us to target our resources where they can have the greatest positive impact, as measured by a reduction in injuries and illnesses.

***

In testifying before the Senate Labor and Human Resources Committee oversight hearings in 1980,[36] Lloyd McBride, International President of the United Steelworkers of America, looked at the issue of OSHA effectiveness from a broad perspective, minimizing the importance of statistics and emphasizing the "intangible qualities," which, although not easily measurable, represent the real effect of OSHA.[37]

---

[36]*1980 Senate Oversight Hearings, supra* note 8 at 698–99, 745–51.

[37]See also L.I. Boden & C. Levenstein, *Regulating Safety—OSHA's "New Look,"* 3 AM. J. FORENSIC MED. & PATHOLOGY 339, 342 (1982) ("* * * OSHA has substantially increased public awareness of occupational safety and health as important problems, has educated managers and workers about job-related hazards, and OSHA as a symbol has lent support to those people in management and unions who are particularly concerned about health and safety. Company safety professionals are able to use the threat, however remote, of an OSHA inspection and the existence of OSHA standards as tools to improve their ability to use scarce corporate resources in their work. Unions are able to use OSHA standards and the threat of enforcement to gain additional leverage in bargaining for health and safety."). President McBride's favorable comments should be viewed in light of organized labor's often strident criticism of the Department of Labor's implementation of the Act. See e.g., *OSHA: Four Years of Frustration,* AM. FEDERATIONIST, April 1975, at 11–18 (official magazine of AFL-CIO) (criticizing OSHA for "missed deadlines, forgotten timetables and endless needless delay * * * serious questions on inspection capability, the role of state governments and administrative efficiency * * * .").

Statement of Lloyd McBride

***

Much has transpired since the passage of OSHA in 1970. Of course, we are inclined—sometimes compelled—to engage in statistical measurement of the progress. Unfortunately, we find that statistics belie the reality of change, perhaps because the statistical base for evaluation has not yet fully evolved—as you know, there is resistance to recordkeeping or regulatory paperwork—or perhaps because we are dealing with intangible qualities that, once achieved, are no longer measurable; we measure only the occurrence of injury and illness, but removal from exposure to a potential toxic emission or a hazardous situation is not tabulated.

Probably the major impact of the first decade of OSHA has been the development of an occupational safety and health infrastructure which, in turn, generates its own ameliorating influence upon the hard conditions of work and the workplace. Certainly the strength of this infrastructure has been enhanced because it is not limited to the workplace environment. There has been a parallel growth resulting from other environmental protection legislation, like EPA, the Toxic Substance Control Act, hazardous waste control, and so on.

While OSHA as a regulatory agency represents the needed institutionalization of governmental responsibility for the safety and health of workers, there has also evolved an "OSHA movement" which is not confined merely to regulatory aspects of occupational protection. The 1970 act was the necessary ingredient—the leaven, if you please—to bring about a rise in the national consciousness and the development of other private sector institutional responses. We would hope that your committee might be able to reveal this aspect of the safety and health structure which has evolved as a result of the OSHA movement.

The OSHA agency, however, is not the OSHA movement. Therefore, failure for more rapid advances in safety and health cannot be visited upon the agency alone. But more than that, exclusive focusing upon the agency by the Congress will narrow or confine the perspective of what is really occurring in workplace protection. Actually, a distorted or restrictive view might force a diminution of the central influence which the OSHA-NIOSH administrative presence has as a stimulant to private sector voluntary improvement.

The biggest payoff of the 1970 act has been the extraregulatory initiatives, which unfortunately will diminish, not increase, if there is a decrease in OSHA's regulatory presence. We cannot overemphasize the impact of the regulatory presence. Its influence is to be felt not just in the workplaces which have been inspected, but also in the workplaces which might be inspected.

OSHA'S IMPACT ON WORKER ATTITUDES

***

The notion that workers are apathetic about health is a myth. Workers have always cared about health. Unfortunately, there was little they could do about it while employers held absolute power in the workplace. Without the power to clean things up, there was never much reason for workers to study hazard recognition, industrial hygiene, air sampling techniques, and control technology.

All that changed with the advent of OSHA. For the first time, American workers had the power to eliminate occupational health hazards by calling an inspector into the workplace. In the last ten years, union efforts in occupational safety and health have increased dramatically. In 1970, American unions employed two health professionals. Today, they employ more than 60. In 1970, the United Steelworkers of America held 2 conferences on occupational health and safety. In 1979, they held 74. These conferences trained more than 5,000 local union officers, safety committee members, and staff representatives in hazard recognition and control. There is no faster growing segment of union activity.

Much of this activity takes place at the local union level. More than 95% of all United Steelworkers' contracts mandate a joint safety committee. UAW contracts with GM, Ford, and Chrysler provide for full-time union safety representatives in each auto plant. Along with their management counterparts, they receive joint training from the union and the company.

This increased awareness has led to the identification and control of previously unknown or uncontrolled hazards. The fact that the pesticide DBCP could cause male sterility was first discovered by a group of workers in a pesticide plant in Lathrop, California, who contacted an occupational health researcher from the Labor Occupational Health Program at the University of California in Berkeley. A Steelworker staff representative in Missouri was among the first to raise questions about widespread use of dangerous drugs to artificially control blood lead levels in lead workers—a practice that was eventually banned by OSHA's Lead Standard. The new emphasis on occupational health led Steelworkers at the Clairton Coke Plant of U.S. Steel to negotiate a comprehensive program for reducing coke oven emissions in 1974. This program helped provide the basis for the OSHA Coke Oven Standard of 1977.

This worker interest in health hazard control is one of the greatest accomplishments of OSHA. It has occurred only because the Act gives workers the power to improve hazardous working conditions through an OSHA inspection. But workers cannot use that power effectively without adequate training. For the first time, worker education in safety and health can lead to concrete results.

OSHA's Impact on Employer Attitudes

***

Some firms have chosen to resist these costs [of abatement] through Review Commission contests, legal challenges to standards, and Congressional lobbying for weaker health standards. Many employers, however, have made a good-faith effort to comply with the law. Corporate safety and health departments have added staff. Many companies have installed extensive control technology. In some cases, employers have entered into agreements with unions for joint studies of safety and health conditions, most notably the United Rubber Workers and the major tire companies. This new relationship usually comes only after employers realize that their long-term interests are best served by compliance with the law. For example, the United Steelworkers and the Lead Industries Association were bitter opponents during the OSHA lead hearings. The two organizations currently are on opposite sides of the legal challenge to OSHA's new Lead Standard—LIA arguing that it is unnecessarily strict; the Steelworkers that it does not go far enough. Nevertheless, LIA has begun a series of educational sessions for occupational physicians, in which Steelworker representatives were asked to participate. In turn, the Steelworkers invited LIA representatives to a series of regional conferences for lead workers. More extensive cooperation may occur in the future. None of it would have been possible without the OSHA Lead Standard.

OSHA and the Health Professions

The last ten years have seen an explosion in the number of industrial hygienists, occupational physicians, health educators, epidemiologists, and toxicologists. Indeed, OSHA could be renamed "The Full Employment Act for Occupational Health Specialists." Since 1970, membership in the American Industrial Hygiene Association has increased from 1650 to 4700. More and more universities are offering industrial hygiene programs. Medical schools are adding occupational medicine to their curricula. These increases are a direct result of the OSHA Act. Companies are employing health professionals to help them comply with OSHA standards and to flag new hazards before they become entrenched. Union industrial hygienists and physicians act as advisors for local union safety and health committees. Of course, many health professionals also find positions with OSHA and NIOSH. To some extent, these increases have been directly funded under the Act through NIOSH training grants to public health schools.

OSHA AND INDUSTRIAL RELATIONS

***

Before 1970, many brief work stoppages occurred when workers were asked to do dangerous jobs or when management refused to correct unsafe or unhealthful working conditions. These strikes usually lasted for a few hours and affected only one department in a plant. Since the strikes often were illegal for workers and embarrassing to management, many of the details were never recorded, but older workers remember them vividly. At the time, such action was the only way workers could enforce their right to a safe workplace.

OSHA replaced this system with the equivalent of an arbitration procedure. When the union and the company disagree about a health hazard, the union can request an OSHA inspection for an impartial determination. Workers need not walk off the job except in those rare cases where the hazards are so serious and immediate that it is impossible to wait for an inspection.

This system works well. In organized plants, with active safety committees, most hazards are discussed first by the committee. Only where the union and the company disagree is it necessary to call OSHA. In fact, most disputes are settled without an OSHA inspection, since both sides are reluctant to push cases they cannot win. OSHA's power to inspect the workplace has given labor and management a way to resolve health and safety issues without the threat of disruptive strikes or slowdowns.

C. Alternative Strategies: Injury Tax

The dispute over OSHA enforcement structure and the uncertainty of whether OSHA has been successful in achieving its goal has led to numerous suggestions for changes in its approach to the elimination of workplace injuries and illnesses. Some of these recommendations would involve basic revisions of the statutory structure, away from the legal enforcement of standards through penalties, and towards a system of economic incentives for compliance. Other proposals have involved more modest revisions of OSHA strategies, such as changes in targeting workplace inspections and modifications of the penalty structure. One of the most frequently urged recommendations has been for the institution of a program of on-site consultation. This would involve workplace visits resulting in advice to employers on abatement *without* the issuance of citations and penalties. In a significant recent action, OSHA has recently established a program for the encouragement of voluntary private action to abate hazards. A number of these proposals for change, some of which have already been implemented by OSHA, are discussed in the following sections.

Nichols & Zeckhauser, *Government Comes to the Workplace: An Assessment of OSHA*[38]

49 Pub. Interest 39, 64–67 (1977)

***

Incentive mechanisms: Economists frequently argue that where markets fail to achieve desired outcomes, the appropriate solution is not direct control, but rather modifying the incentives faced by market participants. In the 1976 Godkin lectures, Charles Schultze, now Chairman of the Council of Economic Advisors, promoted this approach, using as his title "The Public Use of Private Incentives." In many areas, he argued, incentives would be far preferable to the "command-and-control" systems of direct regulation now used. Incentive schemes secure their efficiency advantage by allowing those who are regulated to select their optimal response. In this way individual differences are respected, freedom of action is enhanced, and the costs of inadequate information on the part of the regulator are substantially diminished.

The principle of levying taxes on offending parties when externalities* are involved is well established in economics, particularly in relation to pollution issues, where the taxes are referred to as "effluent charges." The primary advantage of taxes over a system of uniform standards is that they allow firms to find the most efficient means of reducing the externality, and lead to an efficient outcome, in the sense that for any given level of control, expenditures are minimized. In theory, standards could be set on a firm-by-firm basis to reflect varying costs and benefits; in practice, such a scheme would be unworkable. Taxes also provide firms with an incentive to develop new procedures for reducing hazards still further—unlike standards, which may block the adoption of innovative technologies. Furthermore, a tax system has the advantage that firms that do not or cannot respond to the incentive pay a penalty. Thus even if the injury rate in a hazardous industry is not reduced, the prices of the goods produced will rise, thus shifting demand towards goods produced by less hazardous techniques. (David Lloyd George reportedly noted that workmen's compensation puts this principle to work; "The cost of the product should bear the blood of the working man.")

Incentive mechanisms hold particular promise for occupational safety. As this article was being written, Charles Schultze, Bert Lance, and Stuart Eizenstat gave such an approach a degree of political credibility when in a well-publicized memo they urged President Carter to supplant OSHA's safety standards with financial incentives. In place of virtually all of its several thousand detailed safety standards, OSHA could levy a tax on em-

*[Author's note—The term "externalities" as used by economists, is discussed in Chapter 1.]

[38]Reprinted with permission from THE PUBLIC INTEREST, No. 49 (Fall 1977) pp. 64–67. © 1977 by National Affairs, Inc. A more detailed presentation of the material in this article appears in Zeckhauser & Nichols, *The Occupational Safety and Health Administration: An Overview* in STUDY ON FEDERAL REGULATION, S. DOC. No. 13, 95th Cong., 1st Sess., Vol. VI, 103–248 app (1978).

ployers for each injury sustained by their workers. An injury tax would give firms generalized incentives to improve safety programs, stimulating them to control the whole range of factors that contribute to accidents, not just to control the limited number of physical conditions susceptible to direct regulation. The mix of activities would vary from firm to firm. Some would probably continue to rely solely on the mechanical safeguards required by standards, but others undoubtedly would try innovative approaches, such as safer work practices and new training programs. Moreover, as the cost of accidents rose, safety records would probably become a more important factor in promotion decisions, thus transmitting incentives down to the lowest levels of management. Firms with unusually good safety records, such as Du Pont (whose injury rate is only a small fraction of the chemical industry's average), are generally suffused with tremendous safety consciousness at all levels.

An injury tax would be relatively easy to administer. It no longer would be necessary to inspect individual workplaces on a regular basis. Firms could either report themselves, as they do income taxes. Or the tax system could be tied to workmen's compensation claims, where the administrative mechanism is already in place. Tying the fines to workmen's compensation would give workers an incentive to police compliance by their employers.

Even if it is not coupled with a tax to reflect externalities, workmen's compensation should be modified to enhance the incentives for firms to provide safer working conditions, to "make safety pay." But to promote an appropriate level of safety, a firm's compensation costs must reflect its own accident rate, and not simply the average experience of firms of similar size in the same industry. One way of moving towards this goal would be to require that insurance policies for workmen's compensation include significant deductibles and co-insurance rates, both keyed to the firm's size.[9]

What benefits would come from a well-conceived incentive approach to occupational safety? On the resource side, we could avoid the costs of OSHA's irrelevant impositions. There would be gains for equity from having workmen's compensation pegged at levels that reflect full economic losses. Occupational safety might improve noticeably—but for any reasonable level of incentive, it is unlikely that the accident rate would be cut dramatically.[10] As distasteful as it may be, given the competing claims for resources, we may have to accept a significant level of occupational accidents as a cost of doing business.

Incentive mechanisms, both taxes and workmen's compensation, are a less promising approach to occupational health. Firms should be taxed for the illnesses their workers suffer as a result of employment. But most cases of occupational illness cannot be distinguished from cases resulting from other causes. The asbestos worker who contracts lung cancer, for example, may be the victim of an occupational illness, but he may also be the victim of general air pollution, cigarette smoking, or his genetic inheritance. Quite likely, his illness results from the interaction of several factors. Even if an illness can be identified as occupational in origin, it may be unclear which employer should be taxed. Black lung, for example, is unquestionably an

[9]Laurence Silberman, Undersecretary of Labor at the time of passage of OSH Act, now believes that federalization of workmen's compensation, coupled with a strengthening of its safety incentives, might have curbed political appetites for federal intervention, thus heading off OSHA as we know it. Silberman also reports that he would have pushed hard for such an approach had he then had his present understanding of economics.

[10]Robert Smith [THE OCCUPATIONAL SAFETY AND HEALTH ACT: ITS GOALS AND ACHIEVEMENTS 78–83 (1976)], estimates that a tax of $2,000 per disabling injury would be required to reduce injuries by 8.8 to 12.5 percent. Smith's estimates, however, may be too pessimistic, since they are based on his remarkably high estimates of the wage premiums paid for risk.

occupational disease of coal miners, but if a miner has worked in several different mines, it will probably prove impossible to determine which period of employment "caused" the illness.

Given the difficulties of connecting individual cases of illness with particular firms, a tax, if employed, would probably have to be levied on workers' exposure. This approach would be analogous to the use of effluent charges for environmental pollutants. Such taxes would lack many of the advantages of an injury tax, since it would be necessary to set fee schedules for individual substances, and monitoring of exposure levels would still be required. Like an injury tax, however, an exposure tax would achieve cost-effectiveness in a way that standards cannot, by allowing individual firms to achieve different exposure levels depending on their particular costs. * * *

Taxes and other financial-incentive mechanisms need not totally replace standards, an important feature given the continuing political appeal of direct regulation. For example, an injury tax might be coupled with the most sensible current safety standards. In occupational health, such a combination could be employed for the vast majority of hazardous substances for which there is no positive level of exposure that is absolutely safe. A standard, chosen with careful attention to both benefits and costs, would be set as a rigid upper limit. An exposure tax, or some other type of financial incentive, could then be applied to encourage firms to achieve lower levels of exposure where it was cost-effective to do so. By providing appropriate incentives and then leaving decisions to the firm, efficiency is promoted in a number of ways: Regulatory impositions whose costs are well out of line with the benefits provided are avoided; variations among firms and industries in their costs and capabilities for achieving gains are automatically recognized; all possible methods are increasing workplace safety and health are pursued, including enhanced training, changed work practices, and new technologies; and pressure is maintained to achieve further gains.

While some critics of OSHA have continued to urge that an injury tax or a similar incentive system be adopted, the idea has generally not been considered a viable alternative to OSHA's "command and control" strategy.[39] For example, the Interagency Task Force on Workplace Safety and Health, established by President Carter in August 1977 to consider ways to strengthen the Federal OSHA program, categorically rejected the idea of an injury tax.

Making Prevention Pay[40]

***

Analysis swiftly revealed difficulties with an injury tax. First, pollution charges rest on the assumption that an incremental amount of pollution in a given location is relatively harmless. Thus a firm is allowed flexibility to pollute and pay the charge rather than abate. In the case of injuries, however, the comparable approach of allowing injuries to occur has been called

[39]On "command and control" compliance strategies, see L. Bacow, *supra* note 30 at 12–21.

[40]*Final Report of the Interagency Task Force on Workplace Safety and Health,* Sec. IV-15–17, Dec. 14, 1978. (Draft)

"calculated murder" and "a license to kill." Whether or not one accepts this viewpoint, the injury tax's appearance of allowing injuries to occur when their cost exceeds some specified cost of prevention is politically unacceptable to organized labor, important segments of Congress, and many others.

Second, as directed by the President, any improved economic incentives recommended by this Task Force must supplement direct regulation. In this context it would be difficult, if not impossible, to isolate and evaluate the effect of such an injury tax and employer groups would likely maintain their belief that this approach creates an added cost with little identifiable benefit.

In short, key interest groups show little initial support for the concept of an injury tax, and its impact on injuries would be uncertain and difficult to measure.

Beyond these difficulties, our analysis suggested that the appropriate tax rate would likely be as difficult to establish as the contents of standards. Reported injury rates would be difficult to audit with sufficient accuracy for purposes of a tax, which—unlike inspection targeting based on reported rates—must impose an immediate charge without further investigation. Employers would have a direct financial incentive to err on the side of underreporting. They would also have an increased incentive to contest workers' compensation claims as well as tax assessments, delaying both compensation benefits and the tax payment itself. This incentive to resist rather than comply would likely increase in direct proportion to the tax's bite. Finally, the tax might have the perverse effect of drawing funds away from precisely those firms which most need resources for injury prevention.

An injury tax could increase the cost of injuries to high-rate firms relative to lower-rate firms in the same industry and facilitate least-cost preventive actions. But for the above reasons we have concluded and *recommend* that * * *

an injury tax should not be implemented at the present time.

The Interagency Task Force agreed with Zeckhauser and Nichols that the workers' compensation system should be modified so that the cost of insurance should "more accurately reflect a firm's cost of injuries relative to competitors."[41] In addition, the Interagency Task Force, among numerous other recommendations, suggested that OSHA propose penalties for violations that "would be calculated to offset as closely as possible the net amount saved by cited employers' failure to comply." This would provide, according to the Task Force, a more significant incentive toward compliance by making noncompliance as costly as compliance.[42]

---

[41] *Id.* at IV-6–14.

[42] *Id.* Sec. III at 15–20. Higher OSHA penalties have also been recommended by Professor Mark Rothstein, *supra* note 8 at 109–10 ("OSHA will not be a meaningful deterrent until the cost of noncompliance becomes greater than the cost of compliance."), and by a law student-authored note in the *Yale Law Journal*, which suggested that civil fines should be "dramatically increased." Note, *A Proposal to Restructure Sanctions under the Occupational Safety and Health Act: Limitations of Punishment and Culpability*, 91 YALE L.J. 1446, 1469 (1982) (also proposing that OSHA be given authority to issue broad cease and desist orders against recidivous violations and to impose prospective safety and monitoring programs).

For a discussion of the injury tax proposal as it would be applied to OSHA, see J. Mendeloff, *supra* note 30 at 28–31, 154–57. See also S. Breyer, REGULATION AND ITS REFORM 156–83 (1982); R. Smith, THE OCCUPATIONAL SAFETY AND HEALTH ACT: ITS GOALS AND ACHIEVEMENTS 78–85 (1976). The Council on Wage and Price Stability (COWPS) in commenting on OSHA's

D. On-Site Consultation

The Act is based on the principle that compliance inspections conducted by OSHA are followed, if violations are discovered, by citations and penalties. Under a program of on-site consultation, as it is usually understood, worksite visits take place and information is provided to employers, particularly to small employers, about their abatement responsibilities, without triggering the enforcement mechanisms of the Act. On-site consultation can be undertaken either together with enforcement, or as an alternative to it. Typically, on-site consultation includes a workplace visit by a federal or state consultant, a walk-through of the workplace, and both an oral discussion and a written report by the consultant of the measures the employer must take to come into compliance with the Act. No citations or penalties would normally be issued as a result of the consultation visit. While from the start OSHA has permitted states with approved plans to offer on-site consultation to employers,[43] under an interpretation of the Office of the Solicitor of the Department of Labor, federal personnel were not permitted to conduct "inspections" of workplaces for the purpose of giving advice to employers without issuing appropriate citations and penalties. Bills were introduced in Congress that would have amended the Act to authorize federal on-site consultation; labor unions opposed the bills, however, and they were not passed.[44] In 1974, Congressman Steiger, one of the main sponsors of the Act in 1970, proposed an amendment to the OSHA appropriations bill that would authorize additional funds for the purpose of agreements between OSHA and the states under which OSHA would pay the states, whether or not they had approved plans, to implement on-site consultation programs. Congressman Steiger's floor statement supporting his amendment described the need for on-site consultation, the reasons why it was not being furnished by federal OSHA, and an explanation of the basis for his proposal for state consultation.

Statement of Congressman Steiger

120 Cong. Rec. 21,296 (1974)

***

Three years after the effective date is * * * a good time to take a reflective look at the accomplishments and shortcomings of an act with the size, scope and complexity of the Occupational Safety and Health Act of 1970.

cotton dust proposal suggested as an alternative to the standard a "true" performance standard with fines imposed on employers based on the number of byssinosis cases in the workplace. Comments of the Council on Wage and Price Stability at 35, Proposed Standard on Cotton Dust, Docket No. H-052 (1977) (copy available in OSHA Docket Office.)

[43]Providing for Consultation in a State 18(b) Plan, Program Directive No. 72-27 (Oct. 25, 1972) (copy in author's files).

[44]For a discussion of these bills, see Chapter 21.

The act represents a basic departure from previous State programs which relied on abatement periods after inspection as the primary means of enforcement. Given the limited number of compliance personnel for State programs, the effectiveness was, on a nationwide basis, marginal. The "first instance sanctions" concept of OSHA requires that employers and employees be in compliance before they are inspected, not after. This concept makes sense, but only if information is readily available before inspection on the specific application of the standards in the workplaces. For large employers, particularly those with Federal contracts subject to the Walsh-Healey Act, the safety and health requirements of the insurance companies, and other laws had pretty well prepared them for OSHA.

The record is generally one of progress. A sizable force of generally well-qualified, well-trained compliance personnel has been assembled. Rather comprehensive safety standards have been promulgated, and few can dispute that more has been done about safety and health in the last 3 years than perhaps in the last hundred. A glaring exception to the general progress has been the need for assistance to meet the particular needs of small business.

For the small businessman without attorneys on retainer, or safety and health professionals on their staff, the standards as published in the *Federal Register* might as well be written in a foreign language.

Off-site consultation is available from OSHA and on-site consultation is available in some 18 of the 26 States with federally approved State plans. A wide array of consultation services from insurance companies and private firms in the private sector is available as well. The employer in a State without consultation, however, who requests OSHA to come to his worksite to advise him on the specific application of the standards is correctly told that the law prohibits authorized representatives of the Secretary of Labor from coming onsite without issuing citations and penalties for violations that they may find during the consultation visit.

The legislative history of sections 8 and 9 of the act is clear on this point. This limitation has had a rather chilling effect on voluntary compliance and, in large part, is responsible for the climate of frustration and fear among many small businesses which has generated the large volumes of often outright hostile mail to virtually all Members of Congress. Moreover, it has helped to produce a climate in which rumors, misinformation and outright distortion are often more prevalent than the hard information and sound advice employers and employees need in order to deal with the real hazards in their workplace.

***

The Steiger amendment was passed and on-site consultation programs in states without approved plans were established. These states were authorized to enter into agreements with OSHA under §§7(c)(1) and 21(c) of the Act[45] to conduct sanction-free consultation visits, with 50 percent federal funding. Regulations were issued by OSHA in 1975 to provide a framework for the state on-site consultation program.[46] Twelve states, in addition to those providing on-site consultation under approved state plans, elected to participate in the program, and in order to encourage even broader state participation,

[45] 29 U.S.C. §§656(c)(1), 670(c).

[46] 40 FED. REG. 21,937 (1975) (codified at 29 C.F.R. pt. 1908 (1983)).

the Senate Appropriations Committee directed OSHA to increase the level of federal funding. In 1977, OSHA extensively revised its consultation regulations, among other things, increasing the federal share of the cost of the program to 90 percent.[47]

One of the most controversial issues in the rulemaking was the relationship between consultation activity and enforcement. A basic principle in the on-site consultation program is that the consultation would not lead to citations and penalties. The question arose, however, whether a consultant had an obligation to advise appropriate enforcement officials when he observes an imminent danger situation or a serious hazard during the consultation visit. The competing considerations and OSHA's resolution of the issue were discussed in the preamble to the 1977 regulation.

On-Site Consultation Agreements—Final Rule

42 Fed. Reg. 41,386, 41,387–88 (1977)

***

EMPLOYEE PROTECTION REQUIREMENTS

The discussion under this heading encompasses the provisions in the regulation concerning the requirement that the employer take necessary action to eliminate hazards which present an imminent danger or serious violation. * * *

The majority of the public comments received addressed this issue. The comments ranged from strong objections to any action related to enforcement to acceptance of the concept with questions only on the procedure to be followed. Most comments, however, were opposed to the mandatory referral to enforcement authorities where an employer fails to take action to eliminate a serious violation. * * *

This issue has been the subject of careful consideration. The Agency is cognizant of the need for full employer utilization of the consultation program and is aware of the argument that the requirement for referral might deter some employers from requesting on-site consultation. However, other provisions of the regulation are intended to assure the fundamental separation between the consultation program and enforcement, and would minimize this disincentive. Thus, the regulation requires that the consultation operate independently from OSHA enforcement and that it have its own separate and distinct staff and management. Further, even in the monitoring of a State's performance, the identity of employers receiving on-site consultation is not revealed. In addition, an on-site visit in progress will delay certain types of OSHA inspections, and an employer is not required to make the consultation report available to a compliance officer during a subsequent inspection.

[47]42 FED. REG. 41,386 (1977) (codified at 29 C.F.R. pt. 1908 (1983)).

In October 1983, OSHA proposed various amendments to the consultation regulations. 48 FED. REG. 45,411 (1983). According to OSHA, these amendments would shift the emphasis of State consultation to a "broader concern for the effectiveness of the employer's total management system" in the area of safety and health; would allow for on-site training of employers and employees; and would provide an exemption from general schedule inspections for employers meeting certain consultation conditions. See also OSHA's proposed guidelines for the training of employees in the "recognition, avoidance and prevention" of unsafe and unhealthful working conditions, 48 FED. REG. 39,317 (1983), and OSHA's proposed workplace health programs, 48 FED. REG. 54,546 (1983).

The only situation in which information about a consultation visit is referred to enforcement authorities is if an imminent danger or serious violation is identified and the employer fails to take the necessary action to eliminate the hazard and protect the employees. In the case of a serious violation a reasonable period for the elimination of the hazard is to be provided. Thus, an employer who in good faith seeks consultation advice to identify hazards so that they can be eliminated need have no concern about enforcement action being taken against him or her. It is only in what is likely to be the extremely rare case of an employer who, although aware of the imminent danger or serious violation, fails to act to eliminate them in the workplace that referral will occur. The Agency believes that in these limited circumstances the underlying policies of the Federal OSH Act mandate that the matter be referred for appropriate enforcement action.

Accordingly, the final regulation, although reworded, retains the provision of the present regulation and the proposal requiring referral to enforcement authorities in specified situations.

***

The consultation regulations issued in 1977 require employers to abate imminent dangers found during the consultation visit immediately and to abate serious violations within the time specified in an abatement plan developed by the employer and the consultant; if abatement is not achieved, appropriate OSHA enforcement authorities must be notified by the consultant.[48]

The OSHA on-site consultation program has continued to expand and by 1980, upon request and free of charge, consultative services were available to employers throughout the country as well as in Puerto Rico and the Virgin Islands.[49] In 32 states and the District of Columbia, consultation services were provided under §7(c)(1) agreements with OSHA: under these agreements, OSHA pays 90 percent of the cost of the program. Thirteen states, Puerto Rico, and the Virgin Islands provided consultation under approved state programs, with federal OSHA paying 50 percent of the cost. In the remaining states and jurisdictions, on-site consultation was provided under fully federally funded private contracts.[50]

E. Voluntary Programs

Assistant Secretary Auchter has indicated his continuing support for the on-site consultation program. In addition, he has repeatedly stated that his administration of OSHA will seek to eliminate the "adversary environment [which] has characterized the past OSHA program." His new approach towards the Act was described in

[48] 29 C.F.R. §1908.6(f) (1983). Several changes in these provisions were proposed in 1983 by OSHA "to clarify or revise the relationship between state consultative services and Federal or State OSHA enforcement activities." 48 FED. REG. at 45,413.

[49] *1980 Senate Oversight Hearings, supra* note 8 at 24.

[50] *Id.* at 34. For a discussion of the manner in which OSHA on-site consultation is implemented, see statement of Deputy Assistant Secretary Basil Whiting. *Id.* at 24–29.

a statement before subcommittees of the Senate Labor and Human Resources Committee in September 1981.[51]

Statement of Assistant Secretary Thorne G. Auchter

***

When Congress passed the Occupational Safety and Health Act of 1970 (the Act), it created an agency with the worthiest of purposes and the noblest of goals—to assure safe and healthful working conditions for the workers of America. The results of the program, however, have not matched its ideals. As a recent Library of Congress study points out, the overall workplace injury rate decreased 12 percent from 1972–79, but the rate of lost workday injuries—that is, job-related injuries which result in time away from work—*increased* by 25 percent and the rate of lost workdays per year increased by 34 percent during that period.

Over the years, both labor and management have been critical of OSHA's implementation of the law. In order to move from an adversary situation to one of cooperation and confidence, we are examining that which has been done in the past and determining what ought to be done in the future. We believe that while the American people continue to strongly support the goal of safe and healthy workplaces, they will no longer tolerate regulatory actions that fail to examine, or even ignore, the overall cost to society.

This Administration will provide the leadership and assistance needed to improve workplace conditions, but we believe that this can best be accomplished through cooperation by government with management and labor. Thus, the agency can be most effective through a balanced program of government and private sector activities that afford full and balanced use of the authority granted by Congress. In other words, we see OSHA as more than "policeman"; we see it as a partner lending assistance to those demonstrating a desire to improve workplace conditions and thus enhancing the safety and health of American workers. But, make no mistake—we will effectively enforce the law.

***

In past years, OSHA has placed the predominant emphasis on its inspections and standard-setting powers. The biggest difference between OSHA today and OSHA in the past is that unlike our predecessors we intend to use *all* of the authority granted by the Act. The Occupational Safety and Health Act of 1970 authorizes a wide array of agency activities. These include education, training, consultation, employer-employee voluntary cooperation, self-inspections, standards-setting, and enforcement. We feel that none of these functions should overshadow the others, but rather that all should be utilized and molded together to accomplish our goals.

***

An adversary environment has characterized the past OSHA program. Instead of bringing labor and management together to solve a common problem, the agency's actions often drove them apart. In the future, by employing a variety of assistance programs and encouraging employers and employees to work together, we feel we are more likely to achieve improved workplace conditions.

***

No new policy which places renewed emphasis on cooperation can be truly effective unless those charged with carrying out that policy do so in a

[51]*OSHA Oversight: Joint Hearing Before the Subcomm. on Investigations and General Oversight and the Subcomm. on Labor of the Senate Comm. on Labor and Human Resources,* 97th Cong., 1st Sess. 18–24 (1981).

manner befitting their extraordinary responsibility. We, therefore, recently issued a directive to all OSHA Compliance Safety and Health Officers (CSHO's) to remind them that their demeanor is the basis on which OSHA is judged by the public. Indeed, our ability to build a more cooperative relationship will depend largely on the professional manner in which the agency's compliance officers conduct themselves. They are to give appropriate attention to matters of dress, conduct, and comportment and to use the time of employers and employees wisely. They are to encourage dialogue, offering suggestions for abating workplace hazards.

In the future, CSHO's will explain to both employers and employees that there are, in addition to inspections, additional ways to protect the workforce, such as training, education and consultation. This directive will be reinforced through appropriate management by headquarters and field supervisory personnel.

We trust that the program outlined will both increase the effectiveness of this agency's use of its personnel and at the same time bring further into the national effort to improve safety and health in the workplace the resources and talents of labor and industry.

***

In implementation of this approach,[52] Assistant Secretary Auchter issued, in July 1982, a directive instructing compliance officers to give "general assistance" to employers in identifying methods for correcting safety and health workplace hazards.[53] Another program, undertaken by OSHA on an experimental basis, would exempt employers who had requested and obtained consultation advice from general schedule inspections for a period of one year.[54]

Moreover, another of Assistant Secretary Auchter's major new incentives was the establishment of "voluntary protection" programs

[52]Assistant Secretary Auchter's enforcement policies have not won uniform approval. See P.J. Simon, *Reagan in the Workplace—Unraveling the Health and Safety Net* (ed. by K. Hughes). Published by Center for Responsive Law (1983) ("voluntary," "cooperative," and "nonadversarial" are "clear code words for regulatory abdication." at ii). In November 1983, Margaret Seminario, Associate Director, AFL-CIO Dep't of Occupational Safety, Health and Social Security, testified before a subcommittee of the House Government Operations Committee, charging that the present administration of OSHA had changed a congressionally mandated program of active enforcement into a "voluntary compliance program" which was not working. She presented statistics which, she said, showed that OSHA's inspection force had dropped from 1,289 to 880; citations for serious violations declined 47%; willful violations by 92%, and penalties by nearly 80%. *AFL-CIO News*, Nov. 12, 1983, p. 1. Assistant Secretary Auchter told the OSHA National Advisory Committee in October 1983 that the report of the Center for Responsive Law was "flawed and biased," relying mainly on "inaccurate and misleading statements, undocumented opinions, and unrepresentative anecdotes to support its preconceived conclusions." 13 OSHR 408 (1983).

[53]OSHA Instruction CPL 2.53, OSHR [Reference File 21:8284–:8285]. See also Field Operations Manual, ch. III, OSHR [Reference File 77:2526–:2527] (compliance officers "shall utilize their knowledge and experience in providing the employer with abatement assistance" both during and after the inspection; employers should be informed that OSHA "is willing to work with them" on abatement even after citations are issued and that on-site consultation services are available).

[54]12 OSHR 149, 358 (1982). The program, which covers about 835,000 workplaces employing 12.7 million people in the southern states, was originally scheduled to expire January 1983, but was extended for six months. 13 OSHR 158 (1982). Cf. Lichty, *A Prescription for Improving the Work Environment*, 8 EMPLOYEE REL. L.J. 73, 74 (1982) (failure and deficiencies of OSHA are due to the overuse and institutionalization of litigation to resolve safety and health disputes; positions become polarized; author suggests "cooperative" approach as in Sweden which has "institutionalized cooperation"). On Sweden's methods of regulating safety and health, see generally S. Kelman, REGULATING AMERICA, REGULATING SWEDEN: A COMPARATIVE STUDY OF OCCUPATIONAL SAFETY AND HEALTH POLICY (1981). See discussion of tripartite approach in implementing arsenic standard in Chapter 3, Section B.

as a means of "expanding worker protection" by awarding limited regulatory relief to those employers with good safety and health records. In July 1982, OSHA published a notice of its implementation of three such programs, entitled "Star," "Try," and "Praise."

Notice of Revised Voluntary Protection Programs

47 Fed. Reg. 29,025–26 (1982)

***

Summary: OSHA announces the implementation of three Voluntary Protection Programs. The programs, revised from the January 19, 1982, notice in the *Federal Register* (47 FR 2796), seek out and recognize exemplary safety and health programs as a means of expanding worker protection. Companies, general contractors, and small business organizations which meet specified programmatic safety and health criteria, which go beyond OSHA standards in providing safe and healthful workplaces for their employees, and which want to do more than is required to help the agency accomplish the goals of the Act are the applicants OSHA seeks for these voluntary programs. In return, OSHA will remove participants from general schedule inspection lists and give priority attention to any which request a variance.

The programs are called "Star," "Try," and "Praise." "Star" is aimed at those workplaces having superior safety and health programs that go beyond OSHA standards in providing worker protection, through either employee participation or management initiative efforts. "Star" is designed to demonstrate that good safety and health programs can prevent injury and illness. "Try" is a broader and, in a sense, more flexible program. On one hand, "Try" is designed to evaluate alternative internal safety and/or health systems for the prevention of workplace injuries and illnesses. On the other hand, "Try" allows participation by firms which have good safety records or are anxious to improve them. Finally, "Praise" is a recognition program for employers in low-hazard industries with good safety programs who have been successful in preventing injuries. The unifying purpose of all these programs is injury and illness prevention.

***

Internal complaint mechanisms will be required for "Star" and "Try" programs to give participants an opportunity to resolve complaints without OSHA involvement. Agency and internal complaint records will be reviewed as part of each program's evaluation. Complaints to OSHA from employees whose employer is participating in a voluntary program will be handled in accordance with OSHA procedures. For evaluation purposes the employee will be queried regarding his/her knowledge and use of the internal system.

* * * [A]n OSHA official with technical expertise will be designated as the contact person for each Voluntary Protection Program. Except for construction sites under "Star" and the experimental programs under "Try," the contact person will have no required on-site presence. On-site assistance for the two excepted situations will be arranged before approval.

Pre-approval program reviews will be conducted except where information gathered by an inspection within the last 18 months can be used to verify the information submitted by the applicant. Where reviews are necessary, they will be done by OSHA staff from the national office and field. Information gathered in such reviews will not be made available to enforcement personnel. Each review will be arranged at the applicant's convenience and will take no more than two days. Experience rates are only one factor that OSHA will weigh in considering these programs. These provide an indication, not a conclusive measure, of performance. The other qualifications are spelled out in the program descriptions * * *. Those accepted into "Star" will be evaluated after three years, unless serious problems are identified earlier, and "Try" participants will be evaluated annually.

We have clarified labor-management committee responsibilities for those programs where such committees are used. Assuring abatement is a management prerogative and responsibility, and we have made this clear in the revised programs.

***

OSHA has since approved several plants for participation in these voluntary programs.[55]

[55]E.g., Motorola plant in Boca Raton, Fla. under the "Try" program, 13 OSHR 377–78 (1983); DuPont plant in DeLisle, Miss. under the "Star" program, 13 OSHR 100 (1983); DuPont plant in Circleville, Ohio under the "Star" program, 13 OSHR 313 (1983).

For a discussion of collective bargaining as an alternative to direct government regulation as a means of improving worker safety and health, see L.S. Bacow, BARGAINING FOR JOB SAFETY AND HEALTH, *supra* note 30. See also Kochan, Dyer & Lipsky, THE EFFECTIVENESS OF UNION-MANAGEMENT SAFETY AND HEALTH COMMITTEES (1977). In a study analyzing 400 collective bargaining agreements, representing a cross section of industries, unions, geographical areas, and bargaining unit size, The Bureau of National Affairs, Inc. concluded that joint management-union safety and health committees are called for in 45% of the sample contracts; periodic committee meetings in 65% of the contracts providing for committees; and periodic inspections of the plant, facilities, or equipment in 43%. *Basic Patterns in Safety and Health Provisions in Collective Bargaining Agreements* 12 OSHR 1092–93 (1983). The study also showed that safety and health clauses are included in 87% of the manufacturing agreements and 72% of the nonmanufacturing agreements. *Id.*

12

OSHA and the Fourth Amendment: Impact of the Warrant Requirement on Enforcement

A. The *Barlow's* Decision

Section 8(a)[1] of the Act authorizes representatives of OSHA to enter workplaces "without delay" and to inspect working areas in a reasonable manner and at reasonable times. If an employee complaint is filed meeting the requirements of §8(f)(1)—that is, it is in writing, is signed by an employee or employee's representative, and alleges with "reasonable particularity" that a violation of a standard exists that threatens physical harm or that an imminent danger exists—and if OSHA concludes that there are "reasonable grounds" to believe that such a violation or danger exists, OSHA is *required* to conduct a workplace inspection.[2] In other circumstances, OSHA is *authorized* but *not required* to conduct inspections.[3] Several years after the Act became effective, some employers resisted OSHA inspections on the belief that these inspections violated the Fourth Amendment to the U.S. Constitution which provides in relevant part that "the right of the people to be secure in their persons, homes, papers and effects against unreasonable searches and seizures shall not be violated, and no warrants shall issue but upon probable cause * * * ." The employers claimed that inspections under §8(a) were "unreasonable" and could not be conducted without a warrant. Some district courts rejected the applicability of the Fourth Amendment to OSHA regulatory inspections;[4] others interpreted the Act as requir-

---

[1]29 U.S.C. §657(a) (1976).

[2]29 U.S.C. §657(f)(1).

[3]For a discussion of OSHA policy in selecting workplaces for inspection and in implementing the complaint provisions of §8(f)(1), see Chapter 13.

[4]E.g., *Dunlop v. Able Contractors,* 4 OSHC 1110 (D. Mont. 1975), *aff'd on other grounds, Marshall v. Able Contractors,* 573 F.2d 1055 (9th Cir. 1978), *cert. denied,* 439 U.S. 826 (1978).

ing a warrant if an inspection was resisted.[5] On December 30, 1976, however, a three-judge district court in Idaho held in *Barlow's, Inc. v. Usery,*[6] that the Fourth Amendment applied to OSHA inspections; and refusing to construe §8(a) as incorporating a warrant requirement, the Court held that section unconstitutional and enjoined OSHA from conducting further inspections on the basis of §8(a) authority. The court said:

> [Our] analysis * * * is consistent with that taken by another three-judge panel in *Brennan v. Gibson's Products, Inc. of Plano,* 407 F. Supp. 154 (E.D. Tex. 1976). While we adopt, in general, the similar reasoning employed there, we decline the invitation to judicially redraft an enactment of Congress. Unlike the *Gibson's Products* court, we cannot accept the proposition that the language of the OSHA inspection provisions envision the requirement that a warrant be obtained before any inspection is undertaken. Certainly, Congress was able, had it wished to do so, to employ language declaring that a warrant must first be obtained, the procedures under which it is to be obtained, and other necessary regulations. Congress did not do so and we refuse to accept that duty.[7]

After unsuccessfully requesting the district court to stay, pending appeal, its injunction barring inspections, OSHA appealed the decision to the U.S. Supreme Court and, in the meantime, also asked the Supreme Court to grant a stay. While the request for a stay was pending before the Court, OSHA was confronted with the question of whether the district court injunction applied throughout the country. There was no clear legal precedent on the issue of the geographic scope of the district court injunction and, motivated largely by a strong interest not to terminate all enforcement activity, OSHA discontinued its inspection program only within the state of Idaho. Its refusal, in the face of the district court injunction, to stop inspections throughout the country caused some sharp criticism. Congressman George Hansen of Idaho, a consistent critic of OSHA, urged, as a matter of "equal justice for all citizens" that OSHA inspections be terminated throughout the country.[8] OSHA, however, rejected Congressman Hansen's request.[9] Its response pointed out that other courts had disagreed with the Idaho district court and had not enjoined OSHA inspection activity.[10] OSHA said that "equal justice" includes "equal justice" for employees "whose lives are endangered by unsafe work conditions" and concluded that "to stop OSHA inspections would be to abandon both the Department's duty to protect workers and our mandate under the Act."[11]

---

[5]E.g., *Brennan v. Gibson's Products,* 407 F. Supp. 154 (E.D. Tex. 1976) (three-judge court), *rev'd on other grounds, Marshall v. Gibson's Products,* 584 F.2d 668 (5th Cir. 1978) (reversal was subsequent to the United States Supreme Court's ruling in *Marshall v. Barlow's, Inc.,* 436 U.S. 307 (1978)).

[6]424 F. Supp. 437 (D. Idaho 1976).

[7]*Id.* at 441.

[8]123 CONG. REC. 388 (1977).

[9]Letter from Deputy Assistant Secretary Bert M. Conklin to Congressman George Hansen (Jan. 26, 1977) (copy in author's files).

[10]*Id.*

[11]*Id.*

Congressman Hansen introduced concurrent resolutions which would have barred OSHA inspections, nationwide,[12] but the issue became moot when, on February 3, 1977, Justice Rehnquist granted a stay of the district court judgment except as it applied to Barlow's, Inc., pending a decision by the Supreme Court. Justice Rehnquist said:

> The proposed stay will not affect the respondent [Barlow's, Inc.] in any way, and there are no equities weighing against it which may be asserted by persons actually before the Court. In such a situation where the decision of the District Court has invalidated a part of an Act of Congress, I think that the Act of Congress, presumptively constitutional as are all such Acts, should remain in effect pending a final decision on the merits by this Court.[13]

In its brief to the Supreme Court on the merits, OSHA argued that Congress had made it clear that it intended prompt and unannounced inspections and that a warrant requirement would defeat this legislative purpose and "significantly impede the implementation of OSHA." It also argued that if the Court held that the Fourth Amendment applied to OSHA inspections, the Act should be construed constitutionally to require warrants, rather than held unconstitutional as the district court had decided.

Brief of Department of Labor in Marshall v. Barlow's, Inc.[14]

SUMMARY OF ARGUMENT

I.

***

2. The decision of the three-judge district court that the Secretary cannot conduct an inspection without a search warrant frustrates the clearly articulated intent of Congress and has no valid foundation in the Fourth Amendment decisions of this Court. "The decisions of this Court have time and again underscored the essential purpose of the Fourth Amendment to shield the citizen from unwarranted intrusions into his privacy." *Jones v. United States,* 357 U.S. 493, 498. Thus, whether a search or an inspection without a warrant is constitutionally unreasonable depends upon a determination whether the privacy interest at stake is of such magnitude and the authorized entry so significant an encroachment on that interest that the interposition of a neutral and detached magistrate should be required in the absence of exigent circumstances to approve the search or the inspection.

Here, there is no significant privacy interest at stake that calls for the imposition of the warrant requirement. The areas and equipment within appellee's workplace that the Secretary seeks to inspect are routinely occupied and used by appellee's employees. This critical fact serves to diminish appellee's claims of privacy with respect to the work areas of his business

[12] H.R. Con. Res. 48 & 56, 95th Cong., 1st Sess., 123 CONG. REC. 446, 1049 (1977).

[13] 429 U.S. 1347, 1348 (1977). Previously, on Jan. 25, 1977, Justice Rehnquist had granted a limited stay of the district court order to the extent that it restrained OSHA's conduct outside the district of Idaho.

[14] Brief for the Appellants [Department of Labor] at 12–17, *Marshall v. Barlow's, Inc.,* 436 U.S. 307 (1978).

premises, especially *vis-a-vis* the inspectors who are charged with the responsibility of insuring the health and safety of the employees whom appellee has assigned to such areas. Indeed, the Secretary's specifically focused inspection of an employer's workplace during "regular working hours" when the employees are present (and would be free to report violations of the Act) can hardly be said to intrude upon the employer's right of privacy in the same degree as would a search of his home, office, or person. Thus, in important respects, in the case of an inspection under the Occupational Safety and Health Act, the invasion of privacy "if it can be said to exist, is abstract and theoretical." *Air Pollution Variance Board v. Western Alfalfa Corp.*, 416 U.S. 861, 865.

3. This Court's decisions in *Camara v. Municipal Court,* 387 U.S. 523, and *See v. City of Seattle,* 387 U.S. 541, do not control this case. The privacy interests in *Camara* and *See* that resulted in the imposition of a warrant requirement for local housing and fire code inspections were both of a considerably greater magnitude than appellee's claim of privacy in this case. *Camara* involved a personal residence, necessarily implicating a core privacy interest. And while *See* involved a commercial warehouse, "[t]he warehouse * * * [was] maintained as locked premises and * * * [was] inaccessible to anyone except the defendant" (408 P.2d at 263). Although *See* held Fourth Amendment protections applicable to commercial premises, that decision did not preclude the use of warrantless searches of such premises "[i]n the context of a regulatory inspection system of business premises that is carefully limited in time, place, and scope * * * [pursuant to] the authority of a valid statute." *United States v. Biswell,* 406 U.S. 311, 315.

Three considerations that were significant to the Court's decisions in *Camara* and *See* are absent in this case. First, the Occupational Safety and Health Act does not provide any sanction for simple refusal to consent to an inspection. Second, a magistrate would provide no meaningful safeguard in the present context for an employer's privacy interests because the highly detailed provisions of the Act limit the inspector's discretion to examination of the work areas in order to determine the existence of occupational hazards. There are accordingly no questions of fact or discretion with respect to which the antecedent evaluation of a magistrate would be required or even helpful to safeguard the privacy interests that are the touchstone of the Fourth Amendment.

Finally, unlike the situation in *Camara* and *See,* a warrant requirement would significantly impede the effectuation of the purpose of the Occupational Safety and Health Act. This would be the case whether the warrant need be sought only after access is refused or prior to any attempt to inspect. If an employer could refuse to permit an inspection without a warrant, his refusal would provide him with the functional equivalent of advance notice and he could often temporarily conceal occupational hazards. Moreover, requiring the Secretary to obtain a warrant in advance of each inspection would impede enforcement of the Act by imposing needless additional strain on the Secretary's limited resources to cover nearly five million workplaces with only 1,300 inspectors.

4. This case is governed by this Court's analysis in *United States v. Biswell, supra.* In upholding warrantless inspections as part of a comprehensive federal gun control program, the decision in *Biswell* reflects the Court's recognition that the gun inspection powers at issue were necessary to implement a regulatory system in which important societal interests were at stake. 406 U.S. at 315–316. Congress has similarly determined that the health and safety of the Nation's workers is of great public importance and that unannounced inspections are essential to the enforceability of the statute. Here, as in *Biswell,* "the prerequisite of a warrant could easily frustrate inspection; and if the necessary flexibility as to time, scope, and fre-

quency is to be preserved, the protections afforded by a warrant would be negligible" (406 U.S. at 316). Where, as here, the areas to be inspected are comprehensively regulated, inspectors "kn[o]w with certainty" that those areas contain regulated working conditions and are "within the proper scope of official scrutiny" (*Almeida-Sanchez v. United States,* 413 U.S. 266, 271), and employers are "not left to wonder about the purposes of the inspector or the limits of his task" (*United States v. Biswell, supra,* 406 U.S. at 316), no warrant is required.

II.

Even if the Court should conclude that the Fourth Amendment precludes warrantless safety inspections of comprehensively regulated working areas, the district court erred in declaring 29 U.S.C. 657(a) "unconstitutional and void" (J.S. App. B 11a) and enjoining its enforcement. It should instead have followed this Court's rule that "under familiar principles of constitutional adjudication, our duty is to construe the statute, if possible, in a manner consistent with the Fourth Amendment," *Almeida-Sanchez v. United States, supra,* 413 U.S. at 272, and interpreted the statute to meet Fourth Amendment requirements.

***

On May 23, 1978, the Supreme Court issued its decision in *Marshall v. Barlow's, Inc.* The Court held, in an opinion written by Justice White, that the Constitution requires that OSHA obtain a warrant, based on "probable cause" before conducting a workplace inspection which is not consented to by the employer. Justices Stevens, Blackmun, and Rehnquist dissented, in an opinion written by Justice Stevens.

Marshall v. Barlow's, Inc.

436 U.S. 307 (1978)

MR. JUSTICE WHITE delivered the opinion of the Court.

***

On the morning of September 11, 1975, an OSHA inspector entered the customer service area of Barlow's, Inc., an electrical and plumbing installation business located in Pocatello, Idaho. The president and general manager, Ferrol G. "Bill" Barlow, was on hand; and the OSHA inspector, after showing his credentials, informed Mr. Barlow that he wished to conduct a search of the working areas of the business. Mr. Barlow inquired whether any complaint had been received about his company. The inspector answered no, but that Barlow's, Inc. had simply turned up in the agency's selection process. The inspector again asked to enter the nonpublic area of the business; Mr. Barlow's response was to inquire whether the inspector had a search warrant. The inspector had none. Thereupon, Mr. Barlow refused the inspector admission to the employee area of his business. He said he was relying on his rights as guaranteed by the Fourth Amendment of the United States Constitution.

Three months later, the Secretary petitioned the United States District Court for the District of Idaho to issue an order compelling Mr. Barlow to admit the inspector. The requested order was issued on December 30, 1975, and was presented to Mr. Barlow on January 5, 1976. Mr. Barlow again refused admission, and he sought his own injunctive relief against the warrantless searches assertedly permitted by OSHA. A three-judge court was convened. On December 30, 1976, it ruled in Mr. Barlow's favor. 424 F. Supp. 437. Concluding that *Camara v. Municipal Court,* 387 U.S. 523, 528–529 (1967), and *See v. City of Seattle,* 387 U.S. 541, 543 (1967), controlled this case, the court held that the Fourth Amendment required a warrant for the type of search involved here and that the statutory authorization for warrantless inspections was unconstitutional. An injunction against searches or inspections pursuant to §8(a) was entered. The Secretary appealed, challenging the judgment, and we noted probable jurisdiction. 430 U.S. 964.

I

The Secretary urges that warrantless inspections to enforce OSHA are reasonable within the meaning of the Fourth Amendment. Among other things, he relies on §8(a) of the Act, 29 U.S.C. §657(a), which authorizes inspection of business premises without a warrant and which the Secretary urges represents a congressional construction of the Fourth Amendment that the courts should not reject. Regretfully, we are unable to agree.

***

This Court has already held that warrantless searches are generally unreasonable, and that this rule applies to commercial premises as well as homes. In *Camara v. Municipal Court,* 387 U.S. 523, 528–529 (1967), we held:

> "[E]xcept in certain carefully defined classes of cases, a search of private property without proper consent is 'unreasonable' unless it has been authorized by a valid search warrant."

On the same day, we also ruled:

> "As we explained in *Camara,* a search of private houses is presumptively unreasonable if conducted without a warrant. The businessman, like the occupant of a residence, has a constitutional right to go about his business free from unreasonable official entries upon his private commercial property. The businessman, too, has that right placed in jeopardy if the decision to enter and inspect for violation of regulatory laws can be made and enforced by the inspector in the field without official authority evidenced by a warrant." *See v. City of Seattle,* 387 U.S. 541, 543 (1967).

These same cases also held that the Fourth Amendment prohibition against unreasonable searches protects against warrantless intrusions during civil as well as criminal investigations. *See v. City of Seattle, supra,* at 543. The reason is found in the "basic purpose of this Amendment * * * [which] is to safeguard the privacy and security of individuals against arbitrary invasions by governmental officials." *Camara, supra,* at 528. If the government intrudes on a person's property, the privacy interest suffers whether the government's motivation is to investigate violations of criminal laws or breaches of other statutory or regulatory standards. It therefore appears that unless some recognized exception to the warrant requirement applies, *See v. City of Seattle, supra,* would require a warrant to conduct the inspection sought in this case.

The Secretary urges that an exception from the search warrant requirement has been recognized for "pervasively regulated business[es]," *United*

States v. Biswell, 406 U.S. 311, 316 (1972), and for "closely regulated" industries "long subject to close supervision and inspection." *Colonnade Catering Corp. v. United States,* 397 U.S. 72, 74, 77 (1970). These cases are indeed exceptions, but they represent responses to relatively unique circumstances. Certain industries have such a history of government oversight that no reasonable expectation of privacy, see *Katz v. United States,* 389 U.S. 347, 351–352 (1967), could exist for a proprietor over the stock of such an enterprise. Liquor *(Colonnade)* and firearms *(Biswell)* are industries of this type; when an entrepreneur embarks upon such a business, he has voluntarily chosen to subject himself to a full arsenal of governmental regulation.

Industries such as these fall within the "certain carefully defined classes of cases," referenced in *Camara, supra,* at 528. The element that distinguishes these enterprises from ordinary businesses is a long tradition of close government supervision, of which any person who chooses to enter such a business must already be aware. "A central difference between those cases [*Colonnade* and *Biswell*] and this one is that businessmen engaged in such federally licensed and regulated enterprises accept the burdens as well as the benefits of their trade, whereas the petitioner here was not engaged in any regulated or licensed business. The businessman in a regulated industry in effect consents to the restrictions placed upon him." *Almeida-Sanchez v. United States,* 413 U.S. 266, 271 (1973).

The clear import of our cases is that the closely regulated industry of the type involved in *Colonnade* and *Biswell* is the exception. The Secretary would make it the rule. Invoking the Walsh-Heal[e]y Act of 1936, 41 U.S.C. §35 *et seq.*, the Secretary attempts to support a conclusion that all businesses involved in interstate commerce have long been subjected to close supervision of employee safety and health conditions. But the degree of federal involvement in employee working circumstances has never been of the order of specificity and pervasiveness that OSHA mandates. It is quite unconvincing to argue that the imposition of minimum wages and maximum hours on employers who contracted with the government under the Walsh-Heal[e]y Act prepared the entirety of American interstate commerce for regulation of working conditions to the minutest detail. Nor can any but the most fictional sense of voluntary consent to later searches be found in the single fact that one conducts a business affecting interstate commerce; under current practice and law, few businesses can be conducted without having some effect on interstate commerce.

***

The critical fact in this case is that entry over Mr. Barlow's objection is being sought by a Government agent. Employees are not being prohibited from reporting OSHA violations. What they observe in their daily functions is undoubtedly beyond the employer's reasonable expectation of privacy. The Government inspector, however, is not an employee. Without a warrant he stands in no better position than a member of the public. What is observable by the public is observable, without a warrant, by the Government inspector as well. The owner of a business has not, by the necessary utilization of employees in his operation, thrown open the areas where employees alone are permitted to the warrantless scrutiny of Government agents. That an employee is free to report, and the Government is free to use, any evidence of noncompliance with OSHA that the employee observes furnishes no justification for federal agents to enter a place of business from which the public is restricted and to conduct their own warrantless search.

II

The Secretary nevertheless stoutly argues that the enforcement scheme of the Act requires warrantless searches, and that the restrictions on search discretion contained in the Act and its regulations already protect as much privacy as a warrant would. The Secretary thereby asserts the actual reasonableness of OSHA searches, whatever the general rule against warrantless searches might be. Because "reasonableness is still the ultimate standard," *Camara v. Municipal Court, supra,* at 539, the Secretary suggests that the Court decide whether a warrant is needed by arriving at a sensible balance between the administrative necessities of OSHA inspections and the incremental protection of privacy of business owners a warrant would afford. He suggests that only a decision exempting OSHA inspections from the Warrant Clause would give "full recognition to the competing public and private interests here at stake." *Camara v. Municipal Court, supra,* at 539.

The Secretary submits that warrantless inspections are essential to the proper enforcement of OSHA because they afford the opportunity to inspect without prior notice and hence to preserve the advantages of surprise. While the dangerous conditions outlawed by the Act include structural defects that cannot be quickly hidden or remedied, the Act also regulates a myriad of safety details that may be amenable to speedy alteration or disguise. The risk is that during the interval between an inspector's initial request to search a plant and his procuring a warrant following the owner's refusal of permission, violations of this latter type could be corrected and thus escape the inspector's notice. To the suggestion that warrants may be issued *ex parte* and executed without delay and without prior notice, thereby preserving the element of surprise, the Secretary expresses concern for the administrative strain that would be experienced by the inspection system, and by the courts, should *ex parte* warrants issued in advance become standard practice.

We are unconvinced, however, that requiring warrants to inspect will impose serious burdens on the inspection system or the courts, will prevent inspections necessary to enforce the statute, or will make them less effective. In the first place, the great majority of businessmen can be expected in normal course to consent to inspection without warrant; the Secretary has not brought to this Court's attention any widespread pattern of refusal.[11] In those cases where an owner does insist on a warrant, the Secretary argues that inspection efficiency will be impeded by the advance notice and delay. The Act's penalty provisions for giving advance notice of a search, 29 U.S.C. §666(f), and the Secretary's own regulations, 29 CFR §1903.6 (1977), indicate that surprise searches are indeed contemplated. However, the Secretary has also promulgated a regulation providing that upon refusal to permit an inspector to enter the property or to complete his inspection, the inspector shall attempt to ascertain the reasons for the refusal and report to his superior, who shall "promptly take appropriate action, including compulsory process, if necessary." 29 CFR §1903.4 (1977). The regulation represents a choice to proceed by process where entry is refused; and on the basis of evidence available from present practice, the Act's effectiveness has not been crippled by providing those owners who wish to refuse an initial requested entry with a time lapse while the inspector obtains the necessary process.[13] Indeed, the kind of process sought in this case and apparently

[11]We recognize that today's holding might itself have an impact on whether owners choose to resist requested searches; we can only await the development of evidence not present on this record to determine how serious an impediment to effective enforcement this might be.

[13]A change in the language of the Compliance Operations Manual for OSHA inspectors supports the inference that, whatever the Act's administrators might have thought at the start, it was eventually concluded that enforcement efficiency would not be jeopardized by permitting employers to refuse entry, at least until the inspector obtained compulsory process. The 1972 Manual included a section specifically directed to obtaining "warrants," and one provision of that section dealt with *ex parte* warrants:

"In cases where a refusal of entry is to be expected from the past performance of the employer, or where the employer has given some indication prior to the commencement of the investigation of his intention to bar entry or limit or interfere with the investigation, a

anticipated by the regulation provides notice to the business operator. If this safeguard endangers the efficient administration of OSHA, the Secretary should never have adopted it, particularly when the Act does not require it. Nor is it immediately apparent why the advantages of surprise would be lost if, after being refused entry, procedures were available for the Secretary to seek an *ex parte* warrant and to reappear at the premises without further notice to the establishment being inspected.[15]

Whether the Secretary proceeds to secure a warrant or other process, with or without prior notice, his entitlement to inspect will not depend on his demonstrating probable cause to believe that conditions in violation of OSHA exist on the premises. Probable cause in the criminal law sense is not required. For purposes of an administrative search such as this, probable cause justifying the issuance of a warrant may be based not only on specific evidence of an existing violation* but also on a showing that "reasonable legislative or administrative standards for conducting an * * * inspection are satisfied with respect to a particular [establishment]." *Camara v. Municipal Court, supra,* at 538. A warrant showing that a specific business has been chosen for an OSHA search on the basis of a general administrative plan for the enforcement of the Act derived from neutral sources such as, for example, dispersion of employees in various types of industries across a given area, and the desired frequency of searches in any of the lesser divisions of the area, would protect an employer's Fourth Amendment rights.[17] We doubt that the consumption of enforcement energies in the obtaining of such warrants will exceed manageable proportions.

Finally, the Secretary urges that requiring a warrant for OSHA inspectors will mean that, as a practical matter, warrantless search provisions in other regulatory statutes are also constitutionally infirm. The reasonableness of a warrantless search, however, will depend upon the specific enforcement needs and privacy guarantees of each statute. Some of the statutes cited apply only to a single industry, where regulations might already be so pervasive that a *Colonnade-Biswell* exception to the warrant requirement could apply. Some statutes already envision resort to federal court enforcement when entry is refused, employing specific language in some cases and general language in others. In short, we base today's opinion on the facts and law concerned with OSHA and do not retreat from a holding appropriate to that statute because of its real or imagined effect on other, different administrative schemes.

---

warrant should be obtained before the inspection is attempted. Cases of this nature should also be referred through the Area Director to the appropriate Regional Solicitor and the Regional Administrator alerted." OSHA Compliance Operations Manual (Jan. 1972), at V-7.

The latest available manual, incorporating changes as of November 1977, deletes this provision, leaving only the details for obtaining "compulsory process" *after* an employer has refused entry. OSHA Field Operations Manual, Vol. V at V-1 to V-5. In its present form, the Secretary's regulation appears to permit establishment owners to insist on "process"; and hence their refusal to permit entry would fall short of criminal conduct within the meaning of 18 U.S.C. §§111 and 1114 (1976), which make it a crime forcibly to impede, intimidate or interfere with federal officials, including OSHA inspectors, while engaged in or on account of the performance of their official duties. [Author's note—Recent changes in the Field Operations Manual on this issue are discussed in Section B, this chapter.]

[15]Insofar as the Secretary's statutory authority is concerned, a regulation expressly providing that the Secretary could proceed *ex parte* to seek a warrant or its equivalent would appear to be as much within the Secretary's power as the regulation currently in force and calling for "compulsory process."

[17]The Secretary's Brief, p. 9 n.7, states that the Barlow inspection was not based on an employee complaint but was a "general schedule" investigation. "Such general investigations," he explains, "now called Regional Programmed Inspections, are carried out in accordance with criteria based upon accident experience and the number of employees exposed in particular industries. U.S. Department of Labor, Occupational Safety and Health Administration, FIELD OPERATIONS MANUAL, *supra,* 1 CCH EMPLOYMENT SAFETY AND HEALTH GUIDE ¶43272 (1976)."

*[Author's note—The Supreme Court here quotes §8(f)(1) requiring OSHA to respond to certain employee complaints.]

Nor do we agree that the incremental protections afforded the employer's privacy by a warrant are so marginal that they fail to justify the administrative burdens that may be entailed. The authority to make warrantless searches devolves almost unbridled discretion upon executive and administrative officers, particularly those in the field, as to when to search and whom to search. A warrant, by contrast, would provide assurances from a neutral officer that the inspection is reasonable under the Constitution, is authorized by statute, and is pursuant to an administrative plan containing specific neutral criteria. Also, a warrant would then and there advise the owner of the scope and objects of the search, beyond which limits the inspector is not expected to proceed. These are important functions for a warrant to perform, functions which underlie the Court's prior decisions that the Warrant Clause applies to inspections for compliance with regulatory statutes. *Camara v. Municipal Court, supra; See v. City of Seattle, supra.* We conclude that the concerns expressed by the Secretary do not suffice to justify warrantless inspections under OSHA or vitiate the general constitutional requirement that for a search to be reasonable a warrant must be obtained.

III

We hold that Barlow was entitled to a declaratory judgment that the Act is unconstitutional insofar as it purports to authorize inspections without warrant or its equivalent and to an injunction enjoining the Act's enforcement to that extent.[23] The judgment of the District Court is therefore affirmed.

MR. JUSTICE BRENNAN took no part in the consideration or decision of this case.

MR. JUSTICE STEVENS, with whom MR. JUSTICE BLACKMUN and MR. JUSTICE REHNQUIST join, dissenting:

I

***

Fidelity to the original understanding of the Fourth Amendment, therefore, leads to the conclusion that the Warrant Clause has no application to routine, regulatory inspections of commercial premises. If such inspections are valid, it is because they comport with the ultimate reasonableness standard of the Fourth Amendment. If the Court were correct in its view that such inspections, if undertaken without a warrant, are unreasonable in the constitutional sense, the issuance of a "new-fangled warrant"—to use Mr. Justice Clark's characteristically expressive term—without any true showing of particularized probable cause would not be sufficient to validate them.[5]

[23]The injunction entered by the District Court, however, should not be understood to forbid the Secretary from exercising the inspection authority conferred by §657 pursuant to regulations and judicial process that satisfy the Fourth Amendment. The District Court did not address the issue whether the order for inspection that was issued in this case was the functional equivalent of a warrant, and the Secretary has limited his submission in this case to the constitutionality of a warrantless search of the Barlow establishment authorized by §8(a). He has expressly declined to rely on 29 CFR §1903.4 (1977) and upon the order obtained in this case. Tr. of Oral Arg. 19. Of course, if the process obtained here, or obtained in other cases under revised regulations, would satisfy the Fourth Amendment, there would be no occasion for enjoining the inspections authorized by §8(a).

[5]*See v. Seattle,* 387 U.S. 541, 547 (Clark, J., dissenting).

II

Even if a warrant issued without probable cause were faithful to the Warrant Clause, I could not accept the Court's holding that the Government's inspection program is constitutionally unreasonable because it fails to require such a warrant procedure. In determining whether a warrant is a necessary safeguard in a given class of cases, "the Court has weighed the public interest against the Fourth Amendment interest of the individual * * *." *United States v. Martinez-Fuerte,* 428 U.S., at 555. Several considerations persuade me that this balance should be struck in favor of the routine inspections authorized by Congress.

Congress has determined that regulation and supervision of safety in the workplace furthers an important public interest and that the power to conduct warrantless searches is necessary to accomplish the safety goals of the legislation. In assessing the public interest side of the Fourth Amendment balance, however, the Court today substitutes its judgment for that of Congress on the question of what inspection authority is needed to effectuate the purposes of the Act. The Court states that if surprise is truly an important ingredient of an effective, representative inspection program, it can be retained by obtaining *ex parte* warrants in advance. The Court assures the Secretary that this will not unduly burden enforcement resources because most employers will consent to inspection.

The Court's analysis does not persuade me that Congress' determination that the warrantless-inspection power as a necessary adjunct of the exercise of the regulatory power is unreasonable. It was surely not unreasonable to conclude that the rate at which employers deny entry to inspectors would increase if covered businesses, which may have safety violations on their premises, have a right to deny warrantless entry to a compliance inspector. The Court is correct that this problem could be avoided by requiring inspectors to obtain a warrant prior to every inspection visit. But the adoption of such a practice undercuts the Court's explanation of why a warrant requirement would not create undue enforcement problems. For, even if it were true that many employers would not exercise their right to demand a warrant, it would provide little solace to those charged with administration of OSHA; faced with an increase in the rate of refusals and the added costs generated by futile trips to inspection sites where entry is denied, officials may be compelled to adopt a general practice of obtaining warrants in advance. While the Court's prediction of the effect a warrant requirement would have on the behavior of covered employers may turn out to be accurate, its judgment is essentially empirical. On such an issue, I would defer to Congress' judgment regarding the importance of a warrantless-search power to the OSHA enforcement scheme.

The Court also appears uncomfortable with the notion of second-guessing Congress and the Secretary on the question of how the substantive goals of OSHA can best be achieved. Thus, the Court offers an alternative explanation for its refusal to accept the legislative judgment. We are told that, in any event, the Secretary, who is charged with enforcement of the Act, has indicated that inspections without delay are not essential to the enforcement scheme. The Court bases this conclusion on a regulation prescribing the administrative response when a compliance inspector is denied entry. It provides: "The Area Director shall immediately consult with the Assistant Regional Director and the Regional Solicitor, who shall promptly take appropriate action, including compulsory process, if necessary." 29 CFR §1903.4 (1977). The Court views this regulation as an admission by the Secretary that no enforcement problem is generated by permitting employers to deny entry and delaying the inspection until a warrant has been obtained. I disagree. The regulation was promulgated against the background of a statutory right to immediate entry, of which covered employers are presumably aware and which Congress and the Secretary obviously thought would keep denials of entry to a minimum. In these circumstances, it was surely not unreasonable for the Secretary to adopt an orderly procedure for dealing with what he believed would be the occasional denial of entry. The regulation does not imply a judgment by the Secretary that delay caused by numerous denials of entry would be administratively acceptable.

***

* * * The inspection warrant purports to serve three functions: to inform the employer that the inspection is authorized by the statute, to advise him of the lawful limits of the inspection, and to assure him that the person demanding entry is an authorized inspector. *Camara v. Municipal Court,* 387 U.S. 523, 532. An examination of these functions in the OSHA context reveals that the inspection warrant adds little to the protections already afforded by the statute and pertinent regulations, and the slight additional benefit it might provide is insufficient to identify a constitutional violation or to justify overriding Congress' judgment that the power to conduct warrantless inspections is essential.

The inspection warrant is supposed to assure the employer that the inspection is in fact routine, and that the inspector has not improperly departed from the program of representative inspections established by responsible officials. But to the extent that harassment inspections would be reduced by the necessity of obtaining a warrant, the Secretary's present enforcement scheme would have precisely the same effect. The representative inspections are conducted "'in accordance with criteria based upon accident experience and the number of employees exposed in particular industries.'" *Ante,* at 321 n.17. If, under the present scheme, entry to covered premises is denied, the inspector can gain entry only by informing his administrative superiors of the refusal and seeking a court order requiring the employer to submit to the inspection. The inspector who would like to conduct a nonroutine search is just as likely to be deterred by the prospect of informing his superiors of his intention and of making false representations to the court when he seeks compulsory process as by the prospect of having to make bad-faith representations in an *ex parte* warrant proceeding.

The other two asserted purposes of the administrative warrant are also adequately achieved under the existing scheme. If the employer has doubts about the official status of the inspector, he is given adequate opportunity to reassure himself in this regard before permitting entry. The OSHA inspector's statutory right to enter the premises is conditioned upon the presentation of appropriate credentials. 29 U.S.C. §657(a)(1). These credentials state the inspector's name, identify him as an OSHA compliance officer, and contain his photograph and signature. If the employer still has doubts, he may make a toll-free call to verify the inspector's authority, *Usery v. Godfrey Brake & Supply Service, Inc.,* 545 F.2d 52, 54 (CA8 1976), or simply deny entry and await the presentation of a court order.

The warrant is not needed to inform the employer of the lawful limits of an OSHA inspection. The statute expressly provides that the inspector may enter all areas in a covered business "where work is performed by an employee of an employer," 29 U.S.C. §657(a)(1), "to inspect and investigate during regular working hours and at other reasonable times, and within reasonable limits and in a reasonable manner * * * all pertinent conditions, structures, machines, apparatus, devices, equipment, and materials therein * * *." 29 U.S.C. §657(a)(2). See also 29 CFR §1903 (1977). While it is true that the inspection power granted by Congress is broad, the warrant procedure required by the Court does not purport to restrict this power but simply to ensure that the employer is apprised of its scope. Since both the statute and the pertinent regulations perform this informational function, a warrant is superfluous.

Requiring the inspection warrant, therefore, adds little in the way of protection to that already provided under the existing enforcement scheme. In these circumstances, the warrant is essentially a formality. In view of the obviously enormous cost of enforcing a health and safety scheme of the dimensions of OSHA, this Court should not, in the guise of construing the Fourth Amendment, require formalities which merely place an additional strain on already overtaxed federal resources.

***

Although rejecting OSHA's arguments that the Fourth Amendment did not apply to regulatory inspections under the Act, the Supreme Court agreed that imposing a warrant requirement would tend to undermine the Congressional purpose of unannounced inspections. The majority said, however, that OSHA had already, by regulation, provided that it would "proceed by process where entry is refused."[15] The Court concluded that the Act's effectiveness "has not been crippled by providing those owners who wish to refuse an initial requested entry with a time lapse while the inspector obtains the necessary process."[16] The Supreme Court's reasoning on this point is difficult to follow. As Justice Stevens stated in his dissenting opinion, it was not "unreasonable" for OSHA to establish an "orderly" procedure to deal with what it believed to be occasional denials of entry. This, Justice Stevens reasoned, hardly implies that "delay caused by numerous denials of entry would be administratively acceptable."[17] At bottom, the issue that seemed to concern the majority was that, as a practical matter, an employer could always obtain advance notice of inspection by refusing to allow entry, since OSHA policy was to use legal means rather than force to obtain entry. In its brief to the Supreme Court, OSHA sought to deal with this issue by saying:

> While an employer may, under the Secretary's regulations, claim an alleged right to refuse entry to an inspector and thereby put him to the burden of seeking a court order to enforce his statutory right of entry, a decision by this Court that no warrant is required would presumably reduce an employer's incentive to do so.[18]

The same issue arose during Solicitor General Ward McCree's oral argument before the Court.

> QUESTION: Mr. Solicitor General, it seems to me that the Secretary here by his regulations has indicated that "we will just give him notice," and he has instructed his inspectors, "If refused entry, don't break in and take advantage of an unannounced entry. Go to court."
>
> MR. MCCREE: If the Court please, he first instructs his inspector to seek entry, without notice—
>
> QUESTION: I know, but then he permits him to be turned away and tells him if he is turned away to go to court.

[15]436 U.S. at 317–18. The majority relied on the OSHA regulation at 29 C.F.R. §1903.4 (1977), which provides that if entry is refused, the matter would be referred to the inspector's supervisor, who shall "promptly take appropriate action, including compulsory process, if necessary."

[16]436 U.S. at 318.

[17]*Id.* at 331.

[18]Brief for the Appellants at 39 n.20, *Marshall v. Barlow's, Inc.*, 436 U.S. 307 (1978).

MR. MCCREE: If the Court please, there are regulations which forbid giving advance notice of the intent to inspect, and there are even penalties on the employees for disclosing the intention of inspecting. So it's really not its purpose. The regulation, as the Court appropriately points out, does afford notice once the court is asked to issue such an order. That is the rare case. That is the case where he has been turned down and we hope this Court's ruling today will make it unnecessary to do it.

QUESTION: It is the rare case that he is turned down at all.

MR. MCCREE: It is the rare case that he is denied the right, and we hope if this Court upholds the statutory scheme, it will be even rarer.

QUESTION: What if the Court upholds it and the person, owner of the business establishment, still says don't come in. He is still not subject to any sanction.

MR. MCCREE: The statute has not provided a sanction and then we suspect the Secretary, again, would direct his inspector, or he would, through his inspector, go to court to get an order so that he might use the power of the court to uphold the law. And then, of course, a contempt citation would be sought if he refused to obey it. But that happens in other situations and doesn't necessarily—

QUESTION: If we agree with you, the Secretary is left in precisely the position he is now.

MR. MCCREE: Well, I think he is in the case of recalcitrant persons. I have the belief that we are a nation of law-abiding people and that if this Court issues its pronouncement that employers across the country will abide by it.[19]

OSHA's response—that if the Supreme Court decided that no warrant was required, refusals to permit inspections would become "rarer"—was obviously not satisfactory to the Court majority, which relied heavily on the OSHA regulation requiring process in reaching its decision. On the issue of whether the Supreme Court pronouncement that warrants were required would *increase* the number of refusals, the majority said that it could only await developments to determine "how serious an impediment to enforcement" the *Barlow's* decision might be.[20] OSHA statistics indicate that since *Barlow's*, between two and and three percent of OSHA-attempted inspections result in a refusal of entry.[21] On the basis of approximately 50,000 inspections a year, this would amount to 1,500 warrant proceedings. While obviously this would involve considerable delays in conducting inspections and be a drain on OSHA resources,[22] it is far from clear that the Supreme Court would consider these results a serious enough "impediment to enforcement" to warrant a reexamination of *Barlow's*.

In the *Barlow's* decision, the Supreme Court said that it was not imposing a warrant requirement under any statutory scheme other then OSHA. In June 1981, the Court held that warrantless inspec-

---

[19]Transcript of Proceedings at 41–42, *Marshall v. Barlow's, Inc.*, 436 U.S. 307 (1978).

[20]436 U.S. at 317 n.11.

[21]See *Oversight on the Administration of the Occupational Safety and Health Act, 1980: Hearings Before the Senate Comm. on Labor and Human Resources*, 96th Cong., 2d Sess. pt. 1 at 296 (1980) (statement of Deputy Assistant Secretary Basil Whiting).

[22]See *id.*

tions could constitutionally be conducted under the 1977 Federal Mine Safety and Health Act.[23]

B. *Ex Parte* Warrants

A frequently litigated issue which arose directly out of the Supreme Court opinion in *Barlow's* is whether OSHA is authorized to obtain warrants *ex parte,* that is, without notice to the employer or an opportunity for the employer to be heard on whether the warrant should be issued. Warrants in criminal proceedings, typically, are issued *ex parte;* and in order to minimize delay and to avoid additional advance notice to the employer, OSHA's practice following *Barlow's* was to seek warrants *ex parte*. This practice was not seriously questioned until 1978 when U.S. District Judge Pollak, without issuing a formal decision, ruled that *ex parte* warrants were not authorized by the relevant OSHA regulation, 29 C.F.R. §1903.4.[24] Judge Pollak relied particularly on a footnote in the *Barlow's* majority opinion saying that "a regulation expressly providing that the Secretary could proceed *ex parte* to seek a warrant or its equivalent would appear to be as much within the Secretary's power as the regulation currently in force and calling for 'compulsory process'."[25] Judge Pollak understood that statement to mean that the OSHA regulation, as it was then written, did not authorize *ex parte* warrants.[26] Although OSHA disagreed with the district judge's reading of the Supreme Court decision, in order to eliminate any doubt, in 1978, it amended the regulation to make it clear that *ex parte* warrants were contemplated. This modification was issued, without notice and comment, as an interpretation of the existing regulation.[27] The issue again came before Judge Pollak, who ruled that the amendment to the regulation was not valid on the ground that the amendment was a "legislative rule" requiring notice and comment.[28] OSHA appealed to the Court of Appeals for the Third Circuit which affirmed the district court decision, but on somewhat different grounds.[29] In light of precedent in that Circuit, the Third Circuit accepted OSHA's statement that the rule was "interpretative" rather than "legislative," and therefore was not invalid because of the lack of notice and comment. However, the court refused to accept the substance of OSHA's interpretation that its regulation also contemplated *ex parte* warrants on the grounds that the interpretation was not "consistent

[23]*Donovan v. Dewey,* 452 U.S. 594 (1981). In concurring in that case, Justice Stevens stated that the Court's reasoning in the *Dewey* case was "much closer" to the reasoning of the *Barlow's* dissent than it was to the majority opinion in *Barlow's*. *Id.* at 607.

[24]See *Cerro Metal Prod. v. Marshall,* 467 F. Supp. 869, 871–72 (E.D. Pa. 1979) (discusses Judge Pollak's bench order issued Nov. 10, 1978).

[25]436 U.S. at 320 n.15. The OSHA regulation is relevantly quoted in the Supreme Court decision, this section.

[26]See 467 F. Supp. at 872.

[27]43 Fed. Reg. 59,838 (1978).

[28]*Cerro Metal Prod. v. Marshall,* 467 F. Supp. 869 (E.D. Pa. 1979).

[29]*Cerro Metal Prod. v. Marshall,* 620 F.2d 964 (3d Cir. 1980) (Seitz, C.J., dissenting).

and long-standing." The court reached its conclusion on the basis of its understanding of OSHA's "litigation strategy" before the Supreme Court in *Barlow's* by means of which, the court of appeals said, OSHA "seized on its now-discarded interpretation in an attempt to convince the Court of the unreasonableness of requiring warrants."[30]

While the Court of Appeals for the Fifth Circuit agreed with the result in *Cerro Metal,*[31] other courts of appeals have reached the opposite result.[32]

Although OSHA strongly disagreed with the Third Circuit Court of Appeals decision invalidating *ex parte* warrants, the issue was not appropriate for Supreme Court review since, concededly, OSHA had authority to obtain *ex parte* warrants and the sole issue was whether the regulation properly articulated this authority. This deficiency in the regulation could, of course, be cured by rulemaking and on May 20, 1980, OSHA published a proposed rule to amend §1903.4 expressly to permit *ex parte* warrants.[33] After considering numerous comments received from businesses, unions, trade associations, law firms, and others, OSHA amended its regulations expressly to authorize *ex parte* warrants.[34] The preamble to the amendment to the regulations set forth two basic reasons for *ex parte* warrants: the importance of surprise in conducting inspections[35] and the resource consumption and delay that would result if warrants were obtained only after adversary hearings.

At the same time that it amended its regulations to authorize *ex parte* warrants, OSHA changed the rule to permit it to seek inspection warrants *prior* to attempting entry into the workplace.[36] The purpose of this provision was to conserve OSHA resources in circumstances where it was clear that no inspection would be permitted; for example, the employer had an announced policy of refusing to allow OSHA to enter. In 1982, however, OSHA issued a field instruction which stated that compulsory process "will not be sought before any employer objects to an OSHA inspection or investigation" in order to "use OSHA's time more wisely and to reduce the agency adversarial role."[37] The instruction excepted only a "limited number of in-

[30]620 F.2d at 978. For a discussion, in the context of OSHA walkaround pay, of the issue of "legislative" and "interpretative" rules, see *Chamber of Commerce v. OSHA,* 636 F.2d 464 (D.C. Cir. 1980), discussed in Chapter 16, Section B.

[31]*Donovan v. Huffines Steel Co.,* 645 F.2d 288 (5th Cir. 1981).

[32]E.g., *Rockford Drop Forge Co. v. Donovan,* 672 F.2d 626 (7th Cir. 1982); *Marshall v. Seward Int'l, Inc.,* No. 80-1708 (4th Cir. March 11, 1981) (unpublished); *Stoddard Lumber Co. v. Marshall,* 627 F.2d 984 (9th Cir. 1980); *Marshall v. W & W Steel Co.,* 604 F.2d 1322 (10th Cir. 1979).

[33]45 FED. REG. 33,652 (1980).

[34]45 FED. REG. 65,916 (1980).

[35]See Comment of United Steelworkers of America at 2 and attached affidavits, OSHA Proposal to Obtain Authority to Seek Warrants on an *Ex Parte* Basis, Docket No. W-001 (1980) (describing one case "where the company was notified of the *ex parte* warrant application by the action of the court, [and] supervisory personnel conducted a series of safety inspections to 'correct' long-standing potential hazards in anticipation of an OSHA inspection").

[36]45 FED. REG. at 65,920–21.

[37]Reprinted in *OSHA Oversight—State of the Agency Report by Assistant Secretary of Labor for OSHA: Hearing Before the Subcomm. on Health and Safety of the House Comm. on Education and Labor,* 97th Cong., 2d Sess. 136, 137 (1982).

stances" which receive the authorization of the Regional Administrator.[38]

The validity of the *ex parte* regulations has been upheld by a U.S. district court in *Texas Steel Company v. Donovan.*[39]

C. "Probable Cause"

The Supreme Court in *Barlow's* stated that probable cause "in the criminal law sense" was not required for OSHA inspection warrants. Justice White said that in the OSHA context probable cause could be based "on specific evidence of an existing violation," referring to the Act's complaint provision, or on a showing that the establishment had been selected for inspection "on the basis of a general administrative plan for the enforcement of the Act derived from neutral sources."[40] OSHA has relied on both employee complaints and its targeting plan for selecting workplaces for inspections in obtaining warrants.[41] Two issues in particular have arisen before the Review Commission and the courts in "probable cause" cases involving warrants based on employee complaints: (1) the extent to which OSHA must scrutinize the complaint to determine the credibility of the complainant before conducting an inspection; and (2) the scope of inspections which OSHA may constitutionally conduct where the underlying complaint alleges specific hazards. Conflicting views on each of these issues have been advanced by the Review Commission and the courts of appeals.[42]

In re Establishment Inspection of: Gilbert & Bennett Manufacturing Co.

589 F.2d 1335 (7th Cir. 1979)

SWYGERT, Circuit Judge:

***

The application for the inspection warrant for the Gilbert & Bennett plant alleged the following bases for the issuance of the requested warrant:

***

3. The desired inspection is also in response to an employee complaint that employees are required to climb on palletized stock (wire products), which is handled by forklift trucks. Further, the employee complaint alleges that the employer would "climb" (stack) stock as high

---

[38]*Id.*

[39]11 OSHC 1793 (N.D. Tex. 1984). The current procedures for CSHOs obtaining preinspection compulsory process are in the Field Operations Manual, ch. III, OSHR [Reference File 77:2501]; the procedures in which to gain entry into a workplace once refused are in the FOM, *id.* at 77:2505–:2507.

[40]*Marshall v. Barlow's, Inc.*, 436 U.S. at 320–21 & n.16.

[41]See Chapter 13.

[42]The recent decision of the U.S. Supreme Court in *Illinois v. Gates*, 462 U.S. ______, 51 USLW 4709 (1983), relaxing the "rigid two-pronged test" for determining whether an informer's tip establishes probable cause in the criminal setting, see *Spinelli v. United States*, 393 U.S. 410 (1969); *Aguilar v. Texas*, 378 U.S. 108 (1964), is likely to have some bearing on the basis for determining probable cause in OSHA warrant proceedings involving complaints of hazardous conditions.

as necessary. Both items represent potential violations of section 5(a)(1) of the Act which requires employers to furnish to each of his employees employment and a place of employment which are 'free from recognized hazards that are causing or are likely to cause death or serious physical harm to employees' in that, if conditions are as alleged at said employer's plant, employees may be in danger of injury by falling from palletized stock or hit by wire that has been rendered unsafe by stocking it to great heights, among others.

4. There have been four prior inspections of this employer's workplace, each resulting in OSHA compliance officers response to employee complaints, in each case the responding OSHA compliance officer(s) have found and cited the employer for alleged violations of the Act as a result of their inspections.

Because the criminal law standard of probable cause is not required, Gilbert & Bennett's arguments faulting the contents of paragraph 3 must fail. *Camara* and *Barlow's* do not require that the warrant application set forth the underlying circumstances demonstrating the basis for the conclusion reached by the complainant, or that the underlying circumstances demonstrate a reason to believe that the complainant is a credible person. Nor is there a requirement that the application request be supplemented with a detailed, signed employee complaint. Complainant's names may be deleted from complaints in order to protect them from employer harassment. See, e.g., 29 U.S.C. §657(f)(1); 29 C.F.R. §1903.11(a). The reasonableness of this anonymity is confirmed by the hostility to the compliance inspection expressed by the plant owners in these two appeals.

Here the Secretary's sworn application, detailing the employee's complaint and indicating the bases for concluding that potentially significant hazards to workers were alleged by it, afforded the magistrate sufficient factual data to conclude that a search was reasonable and that a warrant should issue.

Paragraph 4 of the warrant application, which indicates that previous employee complaints had resulted in OSHA investigations and the issuance of citations for violations of the Act, does not provide sufficient evidence for a finding of probable cause when read in isolation. However, when read in conjunction with paragraph 3, it supplied the magistrate with relevant and important background information regarding Gilbert & Bennett's compliance history. *Usery v. Northwest Orient Airlines, Inc.*, 5 OSHC 1617, 1619–1620 (E.D.N.Y. July 10, 1977).

***

Marshall v. Horn Seed Company

647 F.2d 96 (10th Cir. 1981)

SEYMOUR, Circuit Judge:

***

As earlier noted, the Supreme Court has indicated that probable cause in the criminal sense is not to be applied to a warrant application based on specific evidence of an OSHA violation. * * *

But to say that the same degree of probable cause is not required is not to say that no consideration need be given to the concerns focused on in the criminal setting. Relaxation is not the same as abandonment. When the warrant application is grounded not upon conformance with administrative or legislative guidelines but upon "specific evidence" of violations such as an employee complaint, there must be some plausible basis for believing that a violation is *likely* to be found. The facts offered must be sufficient to warrant further investigation or testing.

By necessity, such a determination requires the magistrate to consider the reliability of the information tendered in support of the application. Again, a criminal standard is not imposed. Although a "substantial basis" is not required to credit the information's reliability, there must be some basis for believing that a complaint was actually made, that the complainant was sincere in his assertion that a violation exists, and that he had some plausible basis for entering a complaint. It is not sufficient that the affiant, as was done in this case, simply state that a complaint was received and detail the conditions alleged to be unsafe. The warrant application must, of course, inform the magistrate of the substance of the complaint so that he can determine whether the alleged conditions, if true, constitute a violation. See *Burkart Randall Division of Textron, Inc. v. Marshall,* 625 F.2d 1313, 1319 (7th Cir. 1980) But an application based upon an employee complaint must contain more. In *Marshall v. W & W Steel,* 604 F.2d 1322, 1326 (10th Cir. 1979), we held as sufficient an application containing the following:

> "The magistrate had before him the original signed written complaint of Herman Sedillo which stated that ventilation and respiratory protection measures in the tank welding area within W and W's plant were inadequate. In addition, the magistrate also had the written narrative statement given by Sedillo to OSHA officials, as well as an independent verification of Sedillo's status as an employee of W and W Steel Company. Finally, the compliance officer of OSHA filed with the magistrate his own affidavit setting forth the steps he had taken in an effort to check and verify the complaint. Such is sufficient in our view to support a finding of probable cause based on 'specific evidence of an existing violation,' as mentioned in *Barlow's.*"

Ideally, the affidavit should state whether the complaint was received by the affiant personally or by some other specific OSHA official known to the affiant. While the name of the complainant need not be given, the magistrate should be informed as to the source of the complaint. Is the source an employee, a competitor, a customer, a casual visitor to the plant, or someone else? The magistrate should also be told whatever underlying facts and surrounding circumstances the complainant provided OSHA. If the complaint was received in written form, it should be attached to the application, although the complainant's name may be deleted. The affiant should specify the steps he or other OSHA officials took to verify the information in the complaint. The affiant should relate any personal observations he has made of the premises and the employer's past history of violations. Lastly, as the Supreme Court has said, depending on the circumstances, it may be necessary to provide "the number of prior entries, the scope of the search, the time of day when it is proposed to be made," or other relevant factors. *Michigan v. Tyler,* 436 U.S. 499, 507 (1978). Only with such information can the magistrate actually determine "the need for the intrusion" and execute his duty of "assur[ing] that the proposed search will be reasonable." *Id.*

***

The courts of appeals and the Review Commission have offered a range of views on the issue of the breadth of an inspection that may be conducted constitutionally on the basis of an employee complaint alleging specific hazards. The main opinion of Circuit Judge

Sprecher in the *Burkhart Randall* case represents the most favorable view to OSHA's conducting of broad inspections.

Burkhart Randall Division of Textron, Inc. v. Marshall

625 F.2d 1313 (7th Cir. 1980)

SPRECHER, Circuit Judge:
***

IV

Burkart's final argument is that even if probable cause to issue an inspection warrant was shown in this case, the warrant, which in fact was issued is invalid because it authorizes an overly broad inspection. The warrant authorized an inspection not limited to the areas identified in the employee complaints but rather extending to Burkart's entire Cairo, Illinois facility. Burkart contends that the overly broad warrant permits OSHA to search areas of the facility for which no probable cause has been shown, in clear violation of Fourth Amendment principles. Such a general inspection is, according to Burkart, impermissible where probable cause is established, as in this case, by specific evidence of hazardous or unhealthy conditions within an employer's facility. In addition, a general inspection based upon employee complaints is said to violate section 8(f)(1) of the Act, 29 U.S.C. §657(f)(1).
***

Our examination of these competing positions and of the policies underlying the Act and the warrant requirement convinces us that the better view is that which permits, absent extraordinary circumstances, general inspections in response to employee complaints. We have already held that because administrative plan inspections, by their nature, are not susceptible to scope limitations, general inspections are permissible in that context. *Gilbert & Bennett,* [*In the Matter of: Establishment Inspection of: Gilbert & Bennett Mfg. Co.,* 5 OSHC 1375, 1375–76 (N.D.Ill. 1977), *aff'd,* 589 F.2d 1335, 1343 (7th Cir.), *cert. denied,* 484 U.S. 884 (1979)]. A general inspection is thus permissible in a situation where there is no specific reason to suspect that a particular facility may be housing violations of the Act. It would be anomalous to permit general inspections in this context and yet hold that only a limited inspection may be conducted where there is particularized probable cause to believe that violations will be found in the specific facility to be inspected. The fact that the specific nature of employee complaints will often make scope limitations possible, does not make them necessary or advisable. Indeed, one court has found that because employee complaints are often limited in scope, general inspections are preferred in such cases:

> The fact that the warrant involved in the case at bar is based on specific employee complaints favors a broad, rather than narrow, application of the Secretary's authority, since an employee is normally familiar with limited areas.

In the Matter of: Establishment Inspection of: Wisconsin Steel, No. 79 C4284 (N.D.Ill. Dec. 28, 1979), memo. op. at 2–3.
***

Balanced against the implementation of the statute's purpose is the employer's interest in avoiding invasion of its privacy and disruption of its operations. See *In the Matter of: Worksite Inspection of Quality Products,*

Inc., 6 OSHC 1663, 1666–67 (D.R.I. 1978), *vacated on other grounds,* 592 F.2d 611 (1st Cir. 1979). That interest, however, is adequately protected by the requirement that a nonconsensual inspection be conducted only pursuant to a warrant issued by a neutral Magistrate after a showing of probable cause. The warrant requirement ensures that OSHA inspectors do not conduct mere "fishing expeditions," *In the Matter of: Establishment Inspection of: Chicago Magnet Wire Corp.*, 5 OSHC 2024, 2025 (N.D. Ill. 1977), and apprises the employer of the inspection's scope, a function emphasized by the Supreme Court in *Barlow's, supra,* 436 U.S. at 323, and this court in *Gilbert & Bennett, supra,* 589 F.2d at 1343. The interposition of a neutral Magistrate between inspectors and employers guarantees that inspectors will not exercise unbridled discretion as to when and where to inspect and provides assurances to employers that the inspection is authorized by statute, is permissible under the Constitution, and will be conducted at a reasonable time and in a reasonable manner. Furthermore, to require that OSHA demonstrate probable cause as to every part of the employer's facility

> would render meaningless the Supreme Court pronouncement in *Barlow's* that the Secretary's "entitlement to inspect will not depend on his demonstrating probable cause to believe that conditions in violation of OSHA exist on the premises," 436 U.S. at 320, and would defeat the federal interest in providing workers with safe working places.

Gilbert & Bennett, supra, 589 F.2d at 1344.

We hold, therefore, that where probable cause to conduct an OSHA inspection is established on the basis of employee complaints, the inspection need not be limited in scope to the substance of those complaints. In light of the broad remedial purposes of the Act, the strong federal interest in employee health and safety, and the protections provided by the warrant requirement, it will generally be reasonable in such a case to conduct an OSHA inspection of the entire workplace identified in the complaints. There may be cases in which extraordinary circumstances would render such an inspection unreasonably broad, but this is not such a case.

***

Chief Judge Fairchild concurred in the result that a full scope inspection was authorized in the particular circumstances of this case, but did not agree with Judge Sprecher's broad rationale;[43] Judge Wood dissented.[44] In a 1983 case, *Donovan v. Fall River Foundry Company, Inc.*, the Court of Appeals for the Seventh Circuit explained further its views on the issue of scope of complaint inspections under the Fourth Amendment.

Donovan v. Fall River Foundry Company, Inc.

—F.2d—, 11 OSHC 1570 (7th Cir. 1983)

PELL, Circuit Judge:

***

Two courts of appeals have cited *Burkart Randall* as exemplifying the view that "once an employee complaint establishes probable cause, the in-

[43] 625 F.2d at 1326.
[44] *Id.* at 1326–28.

spection need not be limited in scope to the substance of the complaint." *Donovan v. Burlington Northern, Inc.*, 694 F.2d 1213, 1215 (9th Cir. 1982); accord, *Donovan v. West Point-Pepperell, Inc.*, 689 F.2d 950, 962 [10 OSHC 2057, 2065] (11th Cir. 1982). *Burkart Randall* cannot properly be read, however, as holding that employee complaints *always* justify issuance of a full-scale warrant. Both Judges Wood and Fairchild stated that the possibility violations exist outside the areas identified in employee complaints should be brought to the attention of the magistrate upon application for a warrant. Because the application process is generally an *ex parte* proceeding, it is the responsibility of the Secretary to make this showing.

The Ninth Circuit has twice relied on *Burkart Randall* in concluding that warrants authorizing a full-scale inspection were justified in response to employee complaints. *Donovan v. Burlington Northern, Inc.*, 694 F.2d 1213 (9th Cir. 1982), *cert. denied*, 103 S.Ct. 3538 (1983). *Hern Iron Works, Inc. v. Donovan*, 670 F.2d 838 (9th Cir.), *cert. denied*, 103 S.Ct. 69 (1982). In both cases, the employee complaints included allegations of ventilation defects and the court of appeals held that the magistrate could reasonably have believed that inspection of the entire premises was necessary to detect other ventilation hazards. *Burlington Northern*, 694 F.2d at 1216; *Hern*, 670 F.2d at 841.

Both *Hern* and *Burlington Northern* are consistent with the result in *Burkart Randall*. The alleged ventilation defects in the Ninth Circuit cases, like the complaints of unsanitary conditions in *Burkart Randall*, are conditions that one could reasonably believe indicative of similar hazardous or unhealthy conditions throughout a facility. A warrant authorizing a full plant-wide inspection in such a case is justified.

In the instant case, the areas in which unhealthy conditions existed, according to the employee complaints, are in close proximity to each other and together comprise only 10,522 square feet of a building with more than 60,000 square feet of floor space. These areas, which comprise what we shall term the "complaint area," are separated from the rest of the building by permanent concrete block walls running from floor to ceiling. The complaint area is at the opposite end of the foundry from the administrative offices, which were apparently also subject to search pursuant to the warrant. The complaint area utilizes its own air pressurization system, electrical system, heating system, ventilating system, and water supply system.

For the unhealthy conditions complained of at Fall River to infect the rest of the foundry, they would need either to be spread from the complaint area through the rest of the facility or they would have to reoccur in other portions of the foundry. Given the physical separation of the complaint area, it is unlikely that the unhealthy conditions would have been transmitted to other parts of the facility. Further, the conditions complained of by Fall River employees relate to specific machinery and furnaces located in the complaint area. Unless these machines and furnaces are duplicated in other parts of the foundry, there is little reason to believe that the unhealthy conditions would reoccur outside the complaint area. In this sense, there is a clear contrast between this case and *Burkart Randall* because it is most likely that restrooms, possibly unsanitary, and fire escapes, possibly inadequate, were spread throughout the Burkart Randall facility.

Burkart Randall does not mandate a plantwide inspection of a facility whenever employees complain to OSHA about specific conditions that they believe to be hazardous or unhealthy. If a general warrant is sought, there should be some evidence presented to the magistrate supporting the belief by OSHA that the deleterious conditions may also be present in other portions of the facility. If one reads the employee complaints as enumerated on the standardized OSHA Complaint form, which was attached to the warrant application as Exhibit A, there is little basis on which to conclude that the allegedly unhealthy conditions might exist outside the complaint area.

***

The position of the Review Commission was stated in *Sarasota Concrete Co.*[45] The Commission said:

> * * * We adopt the position that, when probable cause for an inspection is based solely on specific evidence of an existing violation, to accommodate the fourth amendment the inspection generally should be limited to the alleged violative condition. See *In re Central Mine Equipment Co.,* 7 OSHC 1185 (E.D. Mo.), *vacated on other grounds,* 7 OSHC 1907 (8th Cir. 1979); *Marshall v. Pool Offshore Co.,* 467 F. Supp. 978 (W.D. La. 1979). Contra, *Burkart Randall Division of Textron, Inc. v. Marshall,* 625 F.2d 1313 (7th Cir. 1980). A plant-wide inspection is usually permissible when probable cause is established under a general administrative plan. In complaint situations, however, an inspection beyond the scope of the alleged violation is not permissible where the Secretary can determine the precise location of the alleged violation.[18] This position is consistent with "the notion inherent in the fourth amendment that the scope of a warrant shall be tailored to the showing of probable cause." *Burkart Randall Division of Textron, Inc. v. Marshall,* 625 F.2d at 1328 (dissenting opinion).
>
> In this case, the hazard alleged in Storey's complaint—defective concrete trucks—was located in a discrete area of Respondent's facility. Based on the record, we conclude that the Secretary did not possess additional facts to justify the conclusion that an inspection of Respondent's entire facility was reasonable to insure that the safety and health of Respondent's employees would not be jeopardized. The inspection should have been limited to the alleged violations noted in Storey's complaint. In light of the supporting facts, this limited inspection would have been reasonable. Instead the compliance officer inspected Respondent's entire facility and the Secretary issued the citation for twelve other than serious violations, all unrelated to Respondent's trucks. The Secretary presented no evidence to the magistrate other than the employee complaint. Accordingly, we find that the Secretary violated the fourth amendment by exceeding the permissible scope of inspection because only a search related to the trucks was supported by probable cause.[46]

Thus, the Review Commission expressly rejected the broad view, expressed by Judge Sprecher, that, constitutionally, OSHA may conduct a plantwide inspection on the basis of a limited complaint. The Commission did not explicitly deal with the issue of whether, as held by the Court of Appeals for the Seventh Circuit, an inspection of hazards other than those complained of is justified where there is an indication that the hazardous conditions alleged in the complaint "may also be present in other portions of the facility."

[18]We are not confronted in this case with a situation where the Secretary must conduct an investigation to determine the location of alleged violations or where violations other than those alleged in the complaint are in plain view. We therefore need not address the issues arising out of such situations.

[45]9 OSHC 1608 (Rev. Comm'n, 1981), *aff'd,* 693 F.2d 1061 (11th Cir. 1982).

[46]9 OSHC at 1616–17. Commissioner Cottine dissented on other grounds. This case is discussed further in Section D, this chapter.

The constitutional question in the *Burkhart Randall, Fall River,* and *Sarasota Concrete* cases was whether an employee complaint of specific hazards is probable cause for an inspection warrant authorizing an inspection of the entire plant. A separate, and analytically prior question, is whether a broad scope inspection in response to a limited employee complaint is permissible under §8(f) of the Act.[47] On this latter issue, the Court of Appeals for the Third Circuit held that §8(f) requires that an OSHA inspection in response to a complaint "bear some relationship to the alleged violations in the employee complaint."[48] No other court of appeals has subscribed to this view.[49]

The second basis in *Barlow's* for probable cause—a general administrative plan for selection of workplaces—has not been the subject of much litigation. It is generally agreed that an inspection based on an administrative plan may be plantwide in scope; indeed, as the Court of Appeals for the Seventh Circuit noted, no "meaningful limitation" on the scope of an inspection pursuant to a general administrative plan "could be devised."[50] In *Stoddard Lumber Co. v. Marshall,* the Court of Appeals for the Ninth Circuit held that OSHA's general schedule inspection selection plan was not subject to notice and comment procedures before promulgation.[51] Moreover, the Ninth Circuit rejected the employer argument that OSHA must provide to the magistrate injury and illness statistics regarding the specific employer to be inspected to establish probable cause in an inspection scheduled under an administrative plan. On that issue, the court said:

> Stoddard asserts that no probable cause was established since the Secretary made no showing to the magistrate that the selection pro-

[47]29 U.S.C. §657(f).

[48]*Marshall v. North American Car Co.,* 626 F.2d 320, 323 (3d Cir. 1980), *aff'g* 476 F. Supp. 698 (M.D. Pa. 1979).

[49]In his opinion in *Burkhart Randall Div. of Textron, Inc. v. Marshall,* Judge Sprecher expressly rejected the result in *North American Car.* 625 F.2d at 1323.

The application of the Fourth Amendment to OSHA's inspection of documents has been the subject of frequent litigation. In the *Barlow's* decision, the Supreme Court stated that "Delineating the scope of a search with some care is particularly important where documents are involved." The Court expressly rejected OSHA's argument that its compliance officers could inspect records required pursuant to the Act (29 U.S.C. §657(c) and other records which are "directly related to the purpose of the inspection" without a warrant. 436 U.S. 307, 324, n.22. Some courts of appeals have held that documents may be examined pursuant to a warrant only if they are related to employee complaints or kept under regulatory requirements. *Donovan v. Fall River Foundry, supra,* 11 OSHC at 1573–75. See also *Marshall v. W & W Steel Co.,* 604 F.2d 1322, 1326–27 (10th Cir. 1979). At the same time, OSHA continues to have authority to obtain documents pursuant to an administrative subpoena under §8(b), 29 U.S.C. §657(b). *Oklahoma Press Publishing v. Walling,* 327 U.S. 186 (1946); *Donovan v. Lone Steer, Inc.,* 52 USLW 4087 (1984). See also *Union Packing Co. of Omaha v. Donovan,* 11 OSHC 1648 (8th Cir. 1983) (upholds OSHA's use of subpoena to obtain records required to be kept under §8(c) and those necessary to compute lost workday average; distinguishes *Barlow's* decision). Indeed, some courts have held that because of the safeguards provided by the subpoena process, Congress intended that OSHA inspectors only use a subpoena and not rely on a warrant to inspect documents. *In re Kulp Foundry, Inc.,* 691 F.2d 1125, 1130–33 (3d Cir. 1982).

[50]*In re Establishment Inspection of: Gilbert & Bennett Mfg.,* 589 F.2d at 1343.

[51]627 F.2d 984, 986–88 (9th Cir. 1980). Cf. *Establishment Inspection of: Kast Metals Corp.,* 11 OSHC 1266 (W.D. La. 1983) (refusing to follow *Stoddard Lumber* and holding that the OSHA general schedule plan was invalid because it was issued without notice and comment rulemaking).

gram applied to Stoddard's lumber business. Specifically, according to Stoddard, the Secretary was required to provide statistics on the company's accident and illness rate, as well as show the company's position on the worst-first list in order to satisfy the probable cause standard under *Barlow's*.

We find that such a showing is not required. See *Marshall v. Chromalloy American Corp.*, 589 F.2d 1335 (7th Cir.), *cert. denied,* 444 U.S. 884 (1979). In *Chromalloy,* the Seventh Circuit found that a warrant application stating that there was a high incidence of injuries in the foundry industry had stated a basis for the magistrate's finding of probable cause. The court noted that to require individual statistics every time the Secretary wants a warrant would result in a full-blown hearing and "would be imposing on the Secretary an unwarranted 'consumption of enforcement energies' which would 'exceed manageable proportions.' * * * Such a situation was not intended by the Supreme Court in its decision in *Barlow's*." 589 F.2d at 1342. (Citations omitted.)

In his application for a warrant, the Secretary provided the district court with the following information: (1) a detailed explanation of the selection plan; (2) the OSHA area director's sworn affidavit stating that Stoddard was selected pursuant to that plan without influence of any other factors; and (3) that the injury incident rate in the lumber industry was 1.8 times the national injury rate for 1976, the most recent year for which statistics were available. Given this information, it was clear that Stoddard fit within the selection program. Contrary to appellant's contention, the Secretary's failure to provide individual statistics of Stoddard's business does not make the warrant fatally defective. There was sufficient evidence presented to support the district court's finding of probable cause for the issuance of the administrative search warrant.[52]

Applications for OSHA warrants are currently being drafted with great care, as appears below. However, this special effort to avoid litigation is not always successful, and warrant cases frequently are litigated in the Review Commission and the courts, as we have seen.

Brief of Department of Labor in Burlington Northern, Inc. v. Donovan[53]

AFFIDAVIT OF DAVID J. DITOMMASO

I, DAVID J. DiTOMMASO, being first duly sworn, hereby state as follows:

1. I am Acting Area Director for the Occupational Safety and Health Administration, United States Department of Labor. As Acting Area Director I am responsible for the Billings Area Office of said agency and for enforcement of the Occupational Safety and Health Act of 1970 (29 U.S.C. 651, *et seq.*) in the State of Montana.

***

4. On June 24, 1980, I received the Occupational Safety and Health

[52]In *Donovan v. West Point Pepperell,* 689 F.2d 950 (11th Cir. 1982), the Court held that the district judge erred in requiring OSHA to make a prima facie showing of violation in order to establish probable cause and by improperly considering and relying on evidence not presented to the magistrate in deciding the probable cause issue.

[53]Affidavit Supporting OSHA Warrant Application, Brief of Respondent [U.S. Department of Labor] in Opposition, Appendix B, at 6a–11a, *Burlington Northern, Inc. v. Donovan,* No. 82-1780 (filed Apr. 25, 1983), *cert. denied,* 463 U.S. ——, 51 USLW 3919 (1983).

Complaint attached to the affidavit of Daniel W. Wobig, submitted herewith, which alleges the existence of certain specified hazards to an establishment identified as:

Burlington Northern Incorporated
Railroad Maintenance and Repair Facility
Laurel, Montana

and requests inspection thereof by the Occupational Safety and Health Administration pursuant to Section 8(f) of the Act.

5. In reviewing the allegations of the said complaint and additional information furnished by the complainant and evaluating them in accordance with administrative guidelines set forth in OSHA Instruction CPL 2.12A, * * *, I have determined that there are reasonable grounds to believe that violations of applicable occupational safety and health standards exist with respect to the following conditions and areas of the facility:

(a) *Welding, cutting and burning operations conducted in the Big Shop, Coal Shop, One Stop and Rip Track areas of the said facility.* The complainant states that regular employee exposure to toxic lead fume exists when employees weld, cut and burn on railroad cars that are painted with leaded paints. This is corroborated by a Burlington Northern document attached to his affidavit as Exhibit B. The applicable standard, 29 CFR 1910.1025, prohibits lead exposure above specified levels, requires regular exposure monitoring and medical surveillance by the employer, and specifies particular types of mandatory respirator protection, none of which have been observed. It has been my experience that allowable levels for lead fume exposure are quickly exceeded in circumstances similar to those alleged by the complainant. This exposure is occurring regularly to seventy or more employees in the areas mentioned and occurs on a daily basis according to the information furnished by the complainant. Lead exposure can result in both chronic or acute health effects. According to the complainant, he is aware of no employees who suffered acute effects but there is a significant danger that prolonged exposure could cause serious, irreversible damage to certain organ systems including the central nervous system and the kidneys.

In that some welding is performed on galvanized metals, there is a similar hazard presented from zinc fume exposure. Zinc fume exposure is regulated by 29 CFR 1910.1000, which standard sets precise limits for exposure levels, requires specified respirator protection and reduction of exposure through engineering or administrative measures. Zinc fume above the allowable levels is very likely to occur in connection with such welding operations.

(b) *Adequacy of ventilation for welding operations as required by 29 CFR 1910.252 in the Big Shop, Coal Shop and One Spot buildings.* The welding standard, 29 CFR 1910.252, contains specific ventilation requirements for indoor welding operations in terms of flow rate and volume, as well as specific detailed requirements for ventilating welding operations involving exposure to lead fume. Lead fume exposure from welding has been substantiated in the above areas and the complainant states that mechanical ventilation is only provided in one of the three shops. According to the complainant, this ventilation has not been functioning properly. There are therefore reasonable grounds to conclude that violations of the welding standard exist. Inadequate ventilation would substantially increase and prolong welding fume and lead fume exposure to employees in these areas and would likely cause such exposure to employees other than the welders themselves.

(c) *Adequacy of respiratory protection furnished to workers in the Big Shop, Coal Shop, One Spot and Rip Track areas.*

The lead standard, 29 CFR 1910.1025, and the zinc fume standard, 29 CFR 1910.1000, both specify appropriate respirator protection for workers exposed to these substances. In addition, the general respirator standard, 29 CFR 1910.134, requires the use of respirators which are applicable and suitable to the hazard. The AO R90N filter described by the complainant is suitable only for dust exposures and is not suitable for the lead and zinc fume exposure apparent here.

The respirator standard, 29 CFR 1910.134, also requires the selection and use of such respirators pursuant to a detailed written respirator program to insure that they are properly selected, used, maintained and furnished to workers. In that, according to the complainant, respirators are seldom used and only an inappropriate respirator is furnished, there are reasonable grounds to conclude that violations of 29 CFR 1910.134 exist. Such violations would appear to affect the majority of the persons in the three shop areas who do welding, cutting and burning.

6. In obtaining information from the complainant, Daniel W. Wobig, upon which I based my evaluation of the above conditions, I have found him to be knowledgeable and credible as to his description of the working conditions. His allegations concerning lead-based paints are corroborated by the Burlington Northern document referred to above. In addition, in signing the complaint attached to his affidavit, Mr. Wobig recognized that any false statement or representation therein would subject him to criminal penalties.

7. Burlington Northern Incorporated has followed a consistent policy of refusing to permit inspection of its premises by the Occupational Safety and Health Administration and therefore it may be reasonably anticipated that the company will not permit inspection of its Laurel, Montana facility absent an inspection warrant. Most recently, representatives of my office, including myself, have been refused permission to inspect Burlington Northern facilities at Havre, Montana, on January 31, 1980, and at Missoula, Montana, on May 28, 1980.

8. If authorized by the Court, the inspection sought herein will be conducted by Compliance Safety and Health Officers who are authorized to conduct such inspections by the Assistant Secretary of Labor.

9. Said inspection will be conducted in accordance with the limitations as to time, manner and scope set out in paragraph 2. The sole purpose of said inspection will be to determine the status of compliance of Burlington Northern with the Occupational Safety and Health Act and applicable regulations promulgated thereunder with respect to the specific conditions stated above. Civil citations requiring the correction of any observed violations may be issued and civil penalties may be proposed. No evidence of criminal conduct is being sought.

10. Said inspection should take no more than three working days to complete and will be commenced as soon as is practicable after the issuance of an inspection warrant.

David J. DiTommaso
Acting Area Director

D. Exclusionary Rule

A final issue in the warrant area is whether the exclusionary rule is applicable in Review Commission proceedings involving citations based on evidence resulting from an inspection authorized by a warrant which is ultimately held to be invalid. The Review Commission has held that the exclusionary rule is applicable;[54] it rejected OSHA's argument that where the warrant was obtained in good faith, the overriding policy favoring employee protection mandates that the evidence not be excluded. The Courts of Appeals for the Eleventh and Seventh Circuits have differed sharply on the question of the applicability of the exclusionary rule to OSHA proceedings. In *Donovan v. Sarasota Concrete Company,*[55] the Court of Appeals for the Eleventh Circuit held that the application by the Review Commission of the exclusionary rule to OSHA inspections was not improper and that it was also proper for the Commission to refuse to apply the "good faith" exception to the exclusionary rule.[56] The Seventh Circuit Court of Appeals in *Donovan v. Federal Clearing Die Casting Company,* reversed the Review Commission and adopted a good-faith exception to the exclusionary rule.

Donovan v. Federal Clearing Die Casting Company

695 F.2d 1020 (7th Cir. 1982)

CUMMINGS, Chief Judge:

***

On January 7, 1980, Federal [Clearing Die Casting Company] employee Natalio Alamillo severed his hands while operating a hydraulic punch press on Federal's premises. Articles concerning the accident appeared in the *Chicago Sun-Times* on January 9 and 10. Pursuant to the Department of Labor's Occupational Safety and Health Field Operations Manual, IV-B, XVI-C2c(2), the Occupational Safety and Health Administration (OSHA) tried to conduct a safety inspection of Federal's workplace on January 10 but Federal refused to permit the attempted warrantless search. On the same date, U.S. Magistrate John W. Cooley issued a warrant for inspection upon the application of OSHA Compliance Officer John Stoessel.

On the following day, Stoessel and another compliance officer again attempted to conduct an inspection of Federal's premises but they were refused entry this time on the ground that the warrant had been improperly issued. This Court ultimately held the warrant invalid but not until July 29, 1981. 655 F.2d 793 [9 OSHC 2072]. In the interim, OSHA conducted inspec-

[54] 9 OSHC 1608 (Rev. Comm'n 1981) (Commissioner Cottine dissented).

[55] 693 F.2d 1061 (11th Cir. 1982).

[56] The court of appeals also held: (1) that the Review Commission "may determine if sufficient probable cause supported an administrative warrant when the evidence obtained pursuant to the warrant forms the basis of the agency's proceedings," *id.* at 1067; and (2) where an inspection is based solely on an employee complaint relating to a localized condition probable cause exists only to determine whether the complaint was valid, and a search of the entire workplace is unreasonable, *id.* at 1068.

tions of Federal's premises pursuant to orders of the district court, resulting in four May 2, 1980, citations.[2] Instead of holding a hearing on the citations, the ALJ granted Federal's motion to suppress all evidence obtained through the invalid inspection warrant and dismissed the proceedings on January 4, 1982. He relied on a prior ruling of the Commission in *Secretary of Labor v. Sarasota Concrete Co.,* 9 OSHC 1608 (Rev. Comm'n, 1981). * * *

***

III. THE EXCLUSIONARY RULE IS INAPPLICABLE

Federal's argument on the merits is that the evidence discovered by the OSHA compliance officers should be suppressed because the warrant authorizing the inspection was ultimately held to be invalid in July 1981. This argument overlooks the fact that the inspections were made pursuant to the district judge's February 20 and April 3, 1980, orders sustaining the warrant's validity and requiring Federal to permit an OSHA search of its premises (see note[] 2 * * * *supra*). Therefore, the question before us is whether the evidence gathered through OSHA's reasonable and good-faith inspection pursuant to a warrant upheld by the district court and provisionally upheld by this Court must be suppressed under the exclusionary rule because the warrant was invalidated on appeal more than a year thereafter. We hold that a good-faith, reasonable belief exception to the exclusionary rule is appropriate in the circumstances of this case.[6]

Although the Supreme Court has not yet reconsidered the validity of the exclusionary rule in criminal cases, *Taylor v. Alabama,* 50 U.S.L.W. 4783, 4785 (June 23, 1982),[7] it has never applied the rule in a civil proceeding, thus suggesting "that the rule should not be applied to OSHA proceedings." *Todd Shipyards Corp. v. Secretary of Labor,* 586 F.2d 683, 689 (9th Cir. 1978). In a criminal case, *United States v. Williams,* 622 F.2d 830, 841–846 (5th Cir. 1980) (*en banc*), certiorari denied, 449 U.S. 1127, thirteen members[8] of the then Fifth Circuit did narrow the exclusionary rule by recognizing a good-faith exception as follows (622 F.2d at 840):

> Sitting en banc, we now hold that evidence is not to be suppressed under the exclusionary rule where it is discovered by officers in the course of actions that are taken in good faith and in the reasonable,

[2]OSHA alleged the existence of 16 serious, 5 willful, 5 repeated, and 2 other-than-serious violations, and proposed penalties of $35,400 (App. 001-009). Although the citations state that the inspections occurred from January 10, 1980, until May 2, 1980, they did not occur until after this Court denied Federal a stay pending appeal on February 27, 1980, and after the district court issued a further order on April 3, 1980. See Secretary's Br. 3, Supp. App. 5, and 655 F.2d at 795.

[6]Federal and the dissent argue that the Commission may choose to apply the exclusionary rule "as a matter of its own policy pursuant to its supervisory power over the Act's enforcement." But the ALJ applied the exclusionary rule here under the Commission's decision in *Sarasota Concrete Co., supra,* which rested heavily on Fourth Amendment grounds. 9 OSHC at 1612–1615. The Secretary has convincingly responded that the supervisory argument falls in the light of *United States v. Payner,* 447 U.S. 727, 736–737, and *United States v. Williams, infra,* 622 F.2d at 846–847. We hold that the Constitution does not warrant the ALJ's and the Commission's applying the exclusionary rule in the circumstances of this case and that therefore this evidence is not to be suppressed.

[7]The question whether there should be a reasonable and good-faith exception to the exclusionary rule in criminal cases may be resolved by the Supreme Court in *Florida v. Royer,* No. 80-2146, argued October 12, 1982, 51 USLW 3300, or in *Illinois v. Gates,* No. 81-430, 51 USLW 3415.

[8]Judges Hill's and Fay's concurrence joined in the majority rule established in the joint opinion of Judges Gee and Vance in which nine other members of the Fifth Circuit also concurred. 622 F.2d at 840 n. * * *. The Commission was therefore incorrect in stating in *Sarasota Concrete Co., supra,* that only 12 of a 24-judge panel joined in the good-faith exception adopted in *Williams.*

> though mistaken, belief that they are authorized. We do so because the exclusionary rule exists to deter willful or flagrant actions by police, not reasonable, good-faith ones. Where the reason for the rule ceases, its application must cease also. The costs to society of applying the rule beyond the purposes it exists to serve are simply too high—in this instance the release on the public of a recidivist drug smuggler—with few or no offsetting benefits. We are persuaded that both reason and authority support this conclusion.

In the ensuing seven pages, the Court explained why it was adopting a good-faith exception to the exclusionary rule. Perhaps the most significant reason for the exception is that the exclusionary rule can have no deterrent effect when, as here, law enforcement personnel have acted mistakenly, but in good faith and on reasonable grounds. 622 F.2d at 842. Consequently, a majority of the *en banc* Court concluded (622 F.2d at 846–847):

> Henceforth in this circuit, when evidence is sought to be excluded because of police conduct leading to its discovery, it will be open to the proponent of the evidence to urge that the conduct in question, if mistaken or unauthorized, was yet taken in a reasonable, good-faith belief that it was proper. If the court so finds, it shall not apply the exclusionary rule to the evidence.

The Tenth Circuit, while not confronted with the identical issue posed in this case, followed the lead of the Fifth Circuit, ruling that the *Williams* reasoning "is equally applicable to civil OSHA enforcement proceedings." *Robberson Steel Co. v. OSHRC,* 645 F.2d 22, 22 (10th Cir. 1980). In *Robberson* and the earlier case of *Savina Home Indus. v. Secretary of Labor,* 594 F.2d 1358 (10th Cir. 1979), the Court refused to apply the exclusionary rule to OSHA proceedings involving warrantless inspections occurring before *Marshall v. Barlow's, Inc.,* 436 U.S. 307. The Court reasoned that OSHA inspectors could not be charged with knowledge of a warrant requirement before *Barlow's* was decided and that therefore the exclusionary rule's deterrence function could not be served. Accord, *Todd Shipyards Corp. v. Secretary of Labor, supra.* The Court, in effect, recognized a reasonable, good-faith exception to the exclusionary rule.[9]

Just as in criminal cases, there would be "substantial societal harm incurred by suppressing [the] relevant and incriminating evidence" uncovered by the inspections here. See *United States v. Williams, supra,* 622 F.2d at 843. Application of the exclusionary rule in this case would preclude the Commission's issuance of an order requiring abatement of hazardous working conditions at Federal and also preclude the issuance of subsequent citations with enhanced penalties to ensure compliance with the Occupational Safety and Health Act. See *Atlas Roofing Co. v. OSHRC,* 518 F.2d 990, 1002, 1009, 1010 (5th Cir. 1975), affirmed, 430 U.S. 442. The Secretary might be unable to obtain another warrant to reinspect Federal because the original probable cause might be too stale or non-existent now. In addition, the purpose of the Act to regulate "a myriad of safety details that may be amenable to speedy alteration or disguise," *Marshall v. Barlow's, Inc.,* 436 U.S. 307, 316, may have been or could easily be thwarted even though there is no question about the Secretary's good faith and reasonable belief that his action was proper.

[9]The Tenth Circuit did say in dicta that in post-*Barlow's* inspections "the exclusionary rule would be applicable to OSHA proceedings involving inspections violative of the warrant requirements announced in *Barlow's.*" 594 F.2d 1358, 1363. It also noted, citing *Todd Shipyards, supra,* that "the Ninth Circuit has recently suggested the contrary." 594 F.2d at 1363 n.7. However, both cases involved warrantless searches, while the case before us today involved a warrant executed only after the district court had approved the warrant and this Court had denied a stay pending appeal.

We cannot credit Federal's statement that the exclusionary rule should be applied to deter OSHA personnel from agency excesses, because the Secretary has acted reasonably and in good faith. Indeed in all the OSHA warrant cases before this Court after the Supreme Court required a warrant in *Barlow's, Inc., supra,* in 1978, only the warrant to inspect Federal was found by us to be unsupported by probable cause.[10]

Furthermore, the Secretary has reminded us that there are already substantial deterrents to OSHA violations of employers' constitutional rights. Thus a neutral magistrate's approval must be obtained before the Secretary may inspect premises over an employer's objection, and the employer may still move to quash a warrant prior to its execution or refuse entry pursuant to the warrant unless the Secretary prevails in civil contempt proceedings. See *Rockford Drop Forge Co. v. Donovan,* 672 F.2d 626, 631 (7th Cir. 1982).

Finally, as this Court noted in *United States v. Carmichael,* 489 F.2d 983, 988 (1973) (*en banc*), "good faith errors cannot be deterred" by the exclusionary rule. In particular, a good-faith, reasonably based violation of this type cannot be deterred. OSHA would continue to be justified in executing a warrant issued by a magistrate, when as here it was approved by the district court and preliminarily by us. It is noteworthy that the Secretary did not conduct his inspection until the April 3, 1980, district court's second authorizing order, which was more than a month after we had denied Federal's stay request. Such judicial approval certainly demonstrates the Secretary's good faith, the reasonableness of his belief that the warrant was proper, and his caution before executing the warrant.[11]

The Commission's order suppressing the evidence obtained pursuant to the warrant and dismissing the proceedings is reversed, and the cause is remanded for hearing on the merits of the four citations.

PELL, Circuit Judge, dissenting.

The majority opinion holds that a good faith, reasonable belief exception to the exclusionary rule is appropriate in the circumstances of this case. As I view those circumstances, the exclusionary rule should have been applied and I therefore respectfully dissent.

***

I think it is appropriate first to look at the facts leading up to the first effort to inspect in some greater detail than the majority has. A newspaper article appeared in a January 9, 1980, edition of the *Chicago Sun-Times* headed "Hands severed, reattached in 23-hour surgery." The third paragraph stated:

> The injury occurred early Monday at Clearing Die Casting Co., 6220 S. New England, where Alamillo [the surgery patient] was working on a punch press.

The only reference to the cause of the injury was:

> Family members were uncertain how the accident took place, and Clearing Die Casting Officials refused comment.

No reference was made in the article to culpability on anyone's part or to

---

[10]See *Rockford Drop Forge Co. v. Donovan,* 672 F.2d 626, 631 (7th Cir. 1982); *Marshall v. Milwaukee Boiler Mfg. Co.,* 626 F.2d 1339, 1342, 1343 (7th Cir. 1980); *Burkart Randall Division of Textron, Inc. v. Marshall,* 625 F.2d 1313, 1319 (7th Cir. 1980).

[11]The dissent notes that it might be "arguably persuasive to a government agent who has in his grasp a court approved warrant that he was entitled to proceed lawfully and therefore he would be acting in good faith" * * *. The decision in this case rests on even stronger grounds. The warrant was not only approved by the district court but this Court refused to stay the district court's order. See *supra* note [ ] 2 * * *. Reliance on such judicial approval is certainly reasonable.

any OSHA requirements whatsoever. Indeed, it may be reasonably surmised that even hurtful industrial accidents such as the one here involved which do not customarily claim space in a metropolitan newspaper did so on this occasion because of the unusual extended surgery in reattaching the hands which had been severed.

The very next day—January 10, 1980—the same newspaper ran a picture of Alamillo with Dr. James D. Schlenker who headed the surgical team. A five-sentence explanation under the picture included a statement attributed to Thomas Kott, local OSHA area director, who "said the plant where Alamillo was injured had been required by law in 1975 to correct punch press safety violations that led to a citation and a fine" and that "OSHA planned to make an unannounced inspection of the plant Thursday morning." Before that day had ended, OSHA Compliance Officer John Stoessal visited the premises which the newspaper had identified as the scene of the accident and attempted a warrantless search.

Again there was no indication as to the cause of the accident or that Federal Clearing Die Casting Company, identified only as Clearing Die Casting Company in the newspaper article, had provided an unsafe place or condition causing or contributing to the accident. It appears to me that Federal correctly characterized the situation in its brief to this court as that the inspecting agent essentially put all of his training and investigative ability on the shelf, opened the newspaper, rummaged in a file of abated citations, and marched over to the magistrate's office to obtain a search warrant, after being denied his request for a warrantless search.

The majority adverts to this situation only as that OSHA was acting pursuant to the Department of Labor's Occupational Safety and Health Field Operations Manual. This assumption apparently is based upon a statement to that effect in the Secretary's original brief although in the same page in the brief the Secretary in a footnote referred to the manual as providing for investigation of "accidents involving significant publicity." The only significant aspects that I can discern in this situation are with regard to the extraordinary surgical effort and there is not a whisper of a basis for determining that the unfortunate accident resulted from an unsafe working place or condition at the employer's facility.

***

The exclusionary rule has a beneficent effect because it sends a message to the OSHA enforcement officials that the results of their inspections will be nullified if the inspection violates the Fourth Amendment. If the rule were *not* applied in OSHA proceedings, a contrary message would be received with a corresponding detrimental effect upon the employer's "constitutional right to go about his business free from unreasonable official entries upon his private commercial property." *See v. City of Seattle,* 387 U.S. 541, 543 (1967). "The rule is calculated to prevent, not to repair. Its purpose is to deter—to compel respect for the constitutional guaranty in the only effectively available way—by removing the incentive to disregard it." *Elkins v. United States,* 364 U.S. 206, 217 (1960).

It is no well guarded secret that OSHA inspections are anathematic to most operators of industrial facilities. No doubt some of the distaste arises from the fact that certain of these operators realize that safety violations do exist and that inspections will certainly result in their discovery, this in turn to be followed not only by penalties but by costly remedial processes. On the other hand, there seems to be a feeling, apparently not entirely groundless, that the inspection will be followed by a listing of violations even though of questionable merit or justification.

OSHA inspectors cannot be unmindful of this frequently found atmosphere of distrust and dislike as they enter a place of work for inspection. While this is unfortunate, it does not justify a response of bureaucratic arrogance or disregard of reasonable and lawful procedures. It is this failure, when it occurs, which must be deterred in future cases by the application of the exclusionary rule. Further, to the extent that bureaucratic arrogance exists, there is no way of knowing how much it is heightened by a company's resistance to voluntary inspections, a resistance which is lawful and should not be answered by retribution. In the present case, it is noted that the inspection apparently was plantwide and not concerned merely with the operation which resulted in the injury to Alamillo. As a result of the wall-to-wall inspection, there was a listing of sixteen serious, five willful, five repeated, and two other-than-serious violations with proposed penalties of $34,500. One may reasonably surmise that the inspection was painstakingly careful, to say the least.

***

The disagreement between the courts, and between individual judges on the same courts, thus relates not only to the legal question whether a good-faith exception to the exclusionary rule should apply but as well to the largely factual question of what constitutes agency good faith.[57] As his strong dissent makes abundantly clear, Judge Pell entertains serious doubt as to OSHA's good faith, not only respecting this specific proceeding, but in implementing many aspects of the occupational safety and health program. Disagreement has also arisen in the legal literature on the issue of the applicability of the good-faith exception.[58]

[57] In *Illinois v. Gates,* 462 U.S. at _____, 51 USLW 4709, 4710, 4712 (1983), the U.S. Supreme Court, despite expectations of the public, did not reach the issue of whether a good-faith exception should be allowed to the exclusionary rule in the criminal context.

[58] Compare Trant, *OSHA and the Exclusionary Rule: Should the Employer Go Free Because the Compliance Officer Has Blundered?* 1981 DUKE L.J. 667 (favoring the exclusionary rule) with Cochell, *The Exclusionary Rule and Its Applicability to OSHA Civil Enforcement Proceedings,* 12 U. BALT. L. REV. 1 (1982) (favoring a good-faith exception). See also Note, *The Good Faith Exception Rule: Should It Apply to OSHA Enforcement Proceedings,* 9 U. DAYTON L. REV. 95 (1983), and Comment, *OSHA and the Exclusionary Rule: The Cost of Constitutional Protection,* 19 WAKE FOREST L. REV. 819 (1983), both opposing the good-faith exception for OSHA inspections.

There is a split of authority on the issue whether an employer must exhaust its administrative remedies before the Review Commission before challenging the validity of a warrant *after* an inspection has been conducted under the warrant. The majority view is that the employer must exhaust its administrative remedies first and is represented by such cases as *In re Establishment Inspection of: Metal Bank of America,* 700 F.2d 910, (3d Cir. 1983); *Baldwin Metal Co. v. Donovan,* 642 F.2d 768 (5th Cir. 1981), *cert. denied,* 454 U.S. 893 (1981); *In re Worksite Inspection of: Quality Products,* 592 F.2d 611 (1st Cir. 1979). The minority view is represented by *Chromalloy Am. Corp. v. Donovan,* 684 F.2d 504 (7th Cir. 1982). Compare Rader, Lewis & Ehlke, *OSHA Warrants and the Exhaustion Doctrine: May the Occupational Safety and Health Review Commission Rule on the Validity of Federal Court Warrants?* 84 DICK. L. REV. 567 (1980) (supporting view that no exhaustion required) with Note, *Constitutional Challenges to OSHA Inspections: The Exhaustion Problem,* 49 CIN. L. REV. 474 (1980) (supporting exhaustion requirement). If an employer refuses to permit OSHA to conduct an inspection pursuant to a warrant, OSHA's normal procedure would be to seek a civil contempt citation against the employer: the district court will normally allow the employer to purge itself of contempt by permitting the inspection to take place. See, e.g., *Marshall v. Chromalloy Am. Corp.,* 433 F. Supp. 330 (E.D. Wis. 1977), *aff'd sub nom. In re Establishment Inspection of: Gilbert & Bennett Mfg. Co.,* 589 F.2d 1335 (7th Cir. 1979), *cert. denied,* 444 U.S. 884 (1979). In a contempt proceeding, the employer may challenge the validity of the underlying warrant, and because no inspection has been conducted, the exhaustion of remedies issue would not arise. See, e.g., *Blocksom & Co. v. Marshall,* 582 F.2d 1122 (7th Cir. 1978).

For a general discussion of OSHA warrant law, see Rothstein, *OSHA, Inspections After Marshall v. Barlow's, Inc.,* 1979 DUKE L.J. 63.

13

Selecting Workplaces for Inspection

A. Background: Development of OSHA Priorities

The Act covers an estimated five million workplaces and at no time since its effective date has there been more than approximately 2,500 compliance federal and state safety and health officers available for inspection activity. The manner in which OSHA selects workplaces for inspection is, therefore, critical to OSHA's ability to provide the greatest protection for employees. Two statutory provisions are most relevant: §8(a), which gives OSHA general inspection authority, subject, of course, to the provisions of the Fourth Amendment;[1] and §8(f)(1), which *requires* OSHA to conduct an inspection if an employee or a representative of employees files a complaint meeting certain formality requirements and OSHA has reasonable grounds to believe that a violation threatening physical harm or an imminent danger exists.[2] The Senate Labor Committee, in reporting the OSHA bill in 1970, recognized the "possibility of limited inspection manpower in the earlier phases of the program" and stated its expectation that OSHA would "initially place emphasis on inspections in those industries or occupations where the need to assure safe and healthful working conditions is determined to be the most compelling."[3] The Committee's anticipation that in later phases of the program OSHA resources would be adequate proved unduly optimistic, and throughout its history OSHA has faced the difficult task of establishing inspection priorities that would bring OSHA inspectors to the workplaces where the need was most "compelling." In its first compliance manual issued in January 1972, OSHA for the first time gave instructions to its field staff on what it then called "compliance programming," that is, the process of selecting workplaces for inspection.[4]

[1]29 U.S.C. §657(a) (1976). See discussion in Chapter 12.
[2]29 U.S.C. §657(f)(1).
[3]S. Rep. No. 1282, 91st Cong., 2d Sess. 12 (1970).
[4]OSHA 2006, ch. IV at IV-1–IV-4, U.S. Government Printing Office (Jan. 1972).

Compliance Operations Manual

***

2. INSPECTION PRIORITIES

Instructions for each category of inspection activity are contained in the following paragraphs. Priority of accomplishment and assignment of manpower resources shall be as follows:

Priority	Category
First	Catastrophe and/or Fatality
Second	Complaints
Third	Target Industry
Fourth	General inspection and related activities

3. INSPECTION CATEGORY INSTRUCTIONS

a. Catastrophe and/or fatality investigations.

***

b. Complaints.

(1) Complaints must be acted upon as soon as possible * * *.

(2) The Area Director will establish priorities regarding the action to be taken on complaints. Complaints alleging the existence of an imminent danger shall be accorded the highest priority * * *. High priority shall also be given to complaints alleging conditions which appear to be serious.

***

(4) When acting on complaints, the CSHO [Compliance Safety and Health Officer] should inspect the entire facility or workplace if time and resources permit, unless there has been a very recent inspection and the CSHO is assured that a total inspection is not necessary. * * *

c. Target Industry Program (TIP).

(1) To pursue OSHA's goal to deal with the "worst-first" within the resources available to OSHA and because of the need to make an immediate impact on the problems of occupational safety and health, a Target Industries Program (TIP) was developed. In the first year, efforts will be concentrated on the five industries with the highest reported injury-frequency rates. The five industries selected for this year and fiscal year 1972 are:

Industry	SIC	Injury-Frequency Rate	Employees
Longshoring	4463	69.9	118,000
Lumber and Wood Products	24	36.1	520,700
Roofing and Sheet Metal	176	43.0	37,000
Meat and Meat Products	201	38.5	272,500
Mobile Homes and Other Transportation Equipment	379	37.6	74,900

***

d. General inspection and related activities.

(1) Beyond the inspection of catastrophes and/or fatalities, complaints, and the target industries, it is desired to establish OSHA's presence as widely as possible within the framework of available resources.

(2) The general inspection program is designed to provide broad representative inspection coverage of the working conditions of employees in an Area. The selection of establishments to be conducted under this program shall be made in accordance with the following factors:

(a) geographical areas within jurisdiction of Area Office

(b) employee population density; and

(c) available OSHA manpower.

(3) These factors shall be considered in such a manner that employee working conditions in every geographical area within the jurisdiction of the Area Office are inspected in proportion to the density of employees in that area (urban, suburban, rural).

***

O. FOLLOWUP INSPECTIONS

1. POLICY

Strict adherence to the policy set forth in this Section respecting followup inspections is essential in order to protect the safety and health of employees and to make it clear to employers that abatement requirements in citations will be strictly enforced.

2. SCHEDULING

a. Followup inspections are mandatory in the following situations:

(1) Where the court has issued a restraining order in an imminent danger situation; and

(2) Where citations for serious, willful or repeated violations have been issued.

b. In the case of citations for nonserious violations, the Area Director has discretion as to the scheduling of a followup inspection. Where citations for both serious (or willful or repeated) and nonserious violations are issued, the followup inspection should include all violations cited.

c. If a citation contains staggered abatement dates for serious violations, a followup should be scheduled for the earliest abatement date. * * *

***

The basic inspection priorities established in 1972—catastrophes, complaints, target (or high-hazard) industries, general inspections, and follow-up inspections—have remained unchanged. As will be described, however, there have been a number of significant modifications, particularly regarding OSHA's response to complaints and the target-industry program.

B. Catastrophe Inspections

Under OSHA regulations, an employer is required to report to the Agency when a workplace accident results in a fatality to an employee, or in the hospitalization of five employees.[5] This report must be made within 48 hours and may be oral or written.[6] In addition to this source of information, OSHA frequently learns of workplace accidents involving deaths or multiple hospitalizations from employees, the media, or some other source.[7] Under the 1972 compliance manual, field staff were given flexibility in deciding which "catastrophes" to investigate. The manual gave the following as "examples" of events that should be investigated, "if time and resources permit": accidents involving previous complaints alleging imminent dangers or serious conditions; accidents involving "possible" willful or repeated violations, or failure to abate cited violations; occurrences involving fatalities "and/or" the hospitalization of numerous employees; and accidents involving "extensive property damage" that could have involved fatalities or serious injuries if employees had been present.[8]

In the 1980 Field Operations Manual (FOM), OSHA field staff were instructed to investigate accidents that even in the absence of death or injuries, receive "significant publicity," recur on a "frequent basis," are the subject of a "special emphasis program," or involve extensive property damage that "could have" resulted in a large number of deaths or injuries.[9] The 1983 revision of the FOM states that accidents not involving a fatality or catastrophe but involving "significant publicity" shall be given the inspection priority of a complaint or of a referral, depending on the source of the report.[10] These later versions of the field manual emphasized an important policy imperative for catastrophe investigations: where the workplace accident has caused many deaths or injuries and is attended by a great deal of publicity, OSHA, as the federal agency charged with occupational safety and health responsibility, is expected by the public and the Congress to investigate the cause of the accident and to take

---

[5] 29 C.F.R. §1904.8 (1983).

[6] *Id.*

[7] In *Donovan v. Federal Clearing Die Casting Co.*, 655 F.2d 793 (7th Cir. 1981), OSHA sought to conduct an inspection on the basis of two newspaper articles in the *Chicago Sun-Times* describing a workplace accident in which an employee's hands were severed. The inspection was refused, and in the resulting warrant proceeding, the Court of Appeals for the Seventh Circuit held that a newspaper report, standing alone, was not adequate probable cause for a warrant. The proceeding reached the Seventh Circuit a second time on the issue of the application of the exclusionary rule to OSHA proceedings. *Donovan v. Federal Clearing Die Casting Co.*, 695 F.2d 1020 (7th Cir. 1982). See Chapter 12, Section D. Cf. *Chicago Zoological Society v. Donovan*, 558 F. Supp. 1147 (N.D. Ill. 1983) (relying on *Federal Clearing Die Casting* and holding that a warrant application asserting that an accident had occurred failed to satisfy the requirements for probable cause based on the "existing violation" theory because OSHA made insufficient attempts to gain information on a possible violation; but upholding probable cause on the grounds that the inspection was pursuant to the administrative plan for accident investigations set forth in OSHA's Field Operations Manual).

[8] Compliance Operations Manual, ch. VI at VI-5–VI-6 (Jan. 1972).

[9] Field Operations Manual, ch. XVI, OSHR [Reference File 77:4701].

[10] Field Operations Manual, ch. VIII, OSHR [Reference File 77:3501].

appropriate enforcement action. Thus, on several occasions after major workplace catastrophes had occurred, congressional hearings were held to investigate the accident. At these hearings, OSHA was asked why its enforcement activity was not successful in averting the catastrophe. One such hearing followed the death of more than 50 employees of Research Cottrell, Inc. at the site of the collapse of a cooling tower at Willow Island, West Virginia.[11]

Statement of Assistant Secretary Bingham

***

This tragedy has made it clear that we must devote special attention to assuring that the construction of cooling towers be done in a safe manner, even though this program will divert resources from other important activities. We will, therefore, not only inspect this cooling tower when construction resumes, but every other cooling tower being erected throughout the country. The agency's compliance officers have a new detailed instructional guide describing workplace conditions that should be checked during inspections of cooling tower construction.

In addition, OSHA is now conducting a complete inspection of the Willow Island facility to insure that there are no other safety and health hazards at this complex.

Ideally, the agency should inspect this site and every other construction site continuously because of the constantly changing nature of the construction work environment. OSHA had, in fact, conducted 10 inspections of the Pleasants Power Station at Willow Island between September 1974 and April 1977.

As the agency is now staffed, it is impossible for OSHA to maintain a constant presence in any single worksite. Employers on these worksites must exercise daily—indeed, hourly—supervision and vigilance over workplace hazards. Employees must be thoroughly trained, and supervisory personnel have to assure adherence to correct work practices.

The Charleston Area Office, which covers West Virginia, has 17 compliance officers to cover almost 31,000 work establishments in the State. OSHA has about 1,500 Federal compliance officers to cover more than 5 million American workplaces. Our area offices are constantly making very difficult choices in using inspection resources to respond to the most serious workplace problems. More than 90 percent of the inspections conducted by the Charleston office are in response to worker complaints. These inspections are a valuable method of discovering workplace hazards. Resources used for this purpose are not available for other kinds of inspections. OSHA never received a worker complaint regarding Research-Cottrell's operations at Willow Island.

Even if this agency were to double or triple its compliance resources, we could never regularly visit the five million workplaces throughout this Nation.

I certainly do not need to remind you, Mr. Chairman, that the OSH Act clearly intends that the employer have the ultimate responsibility for safeguarding workers. The concern for worker protection must permeate every business from owner to top management to first-line supervisor. Otherwise,

[11] *HA Oversight—Willow Island, West Virginia, Cooling Tower Collapse: Hearing Before the Subcomm. on Compensation, Health and Safety of the House Comm. on Education and Labor,* 95th Cong., 2d Sess. 18–21 (1978).

this Nation will never see a reduction in the tragic toll of injury, illness, and fatalities caused by on-the-job hazards.

Mr. Chairman, I reported to you in January during the hearings on grain elevators that those tragedies were not the last that we would see. And since that time we have seen this tower collapse, and we have seen a refinery explosion, and we have seen a variety of other tragedies throughout this country.

In response to this particular disaster, this agency has acted, we believe, in a professional and responsible manner. We have been subject to criticism, however. It has been alleged that OSHA was forewarned that the scaffold used by Research-Cottrell was unsafe and that if OSHA had inspected the scaffolding or otherwise compelled the company to correct unsafe conditions, the accident might have been averted. We believe this allegation is untrue and actually misleading and has caused unnecessary anguish for members of this community by diverting attention from the true cause of the collapse.

***

Let me conclude by emphasizing that the statutory goal of protecting the Nation's workers cannot be realized unless employers fulfill their responsibility for ensuring hazard-free workplaces regardless of whether or not there has been an OSHA inspection. They must not gamble with the lives and limbs of workers. OSHA can, with present resources, inspect 2 percent of the workplaces annually at best. Workers have the right to expect that their employers—whether inspected or not—will continually carry out their responsibility to maintain a safe work environment. The OSH Act mandates this degree of concern by the Nation's employers.

Mr. Chairman, I pray that I will never have to come before you and the American people to explain the cause of another workplace tragedy. But neither you nor I can ensure this. The Nation's employers must accept their responsibility to prevent further tragedies. I shall do all I can to see that this is done. Thank you once again for providing this opportunity to appear today.

MR. GAYDOS. Thank you, Dr. Bingham.

***

For a moment, forgetting about the inadequacy of the number of inspectors and other personnel that you can call upon, do you at this time have any specific recommendations to the committee regarding some legislative changes that you would advocate or suggest at this stage of your investigation?

DR. BINGHAM. First, let me say that my previous comments to you referred not just to this tragedy but to tragedies like this that occur each day all over America.

To answer your question about legislative changes, I think that the discrimination portion of the act could be strengthened. This would further assure that workers who report unsafe work conditions, can do so without fear of reprisals.

MR. GAYDOS. Would you yield at that point? Do you think if we made that a little more clear and strengthened the language regarding the rights of workers that more workers would sound the alarm if and when it was necessary to do so? Is that what you are talking about?

DR. BINGHAM. Yes, sir.

MR. GAYDOS. Could you give me some specifics in that area, or are you awaiting some recommendations of the staff as to how to do that?

DR. BINGHAM. I cannot give you the precise language for amending the Act, but I can tell you an example that will break your heart. In New Jersey or New York a worker was told to enter and clean a large tank. That worker had been in a similar tank before and had become very dizzy. In this incident the employee told his employer that he felt it was unsafe to go in the tank and that he did not want to work there; he was afraid he would again become dizzy and he would be overcome by the fumes. The employer said, "Well, if you don't want to work there, go home." And that worker did go home that day. He came back on another day and was assigned to the same job. He needed the job. He needed to support his family. He went into the tank, was overcome by fumes and died. That worker feared for his job and because of that fear he lost his life. I believe this example shows why the discrimination section of the act could be reexamined and possibly strengthened.

Another area which may deserve consideration is to provide OSHA with a provision for sealing a worksite until OSHA has completed its investigation. We currently do not have such an explicit provision in our legislation. However, the Mine Safety and Health Act recently considered by your subcommittee does contain such a provision and may serve as a model for OSHA.*

***

In June 1978, OSHA issued citations for 10 willful and 10 serious violations against three employers involved in the Willow Island cooling tower collapse.[12] The proposed penalties totaled $108,000.[13] On November 24, 1978, OSHA announced that it had referred the Willow Island file to the Department of Justice for consideration of criminal prosecution;[14] the civil citations were later settled by OSHA and the Department of Justice declined to initiate criminal proceedings.

A series of workplace catastrophes occurred in December 1977 and January 1978, when six explosions in grain elevators killed 62 persons and injured 53.[15] These tragedies led to a number of actions by OSHA[16] as well as a congressional investigation.[17] In response to the explosions, OSHA in January 1978, issued a hazard alert. This alert was based on existing OSHA standards, industry voluntary consensus standards, and other information from experts; its purpose, according to OSHA, was to provide employers, workers, and

*[Author's note—The provision is contained in §103(j) of the Federal Mine Safety and Health Act of 1977, 30 U.S.C. §813(j) (Supp. V 1981). The provision requires the mine operators to notify the Secretary of Labor of any accident and "take appropriate measures to prevent the destruction of any evidence which would assist in investigating the cause or causes thereof." There have been no amendments to the OSH Act on this or any other issue. The discrimination section of the Act, §11(c), 29 U.S.C. §660(c), is discussed in Chapter 18.]

[12]8 OSHR 60–61 (1978).

[13]*Id.*

[14]*Id.* at 1126.

[15]See COMPTROLLER GENERAL OF THE UNITED STATES, REPORT TO THE CONGRESS, HRD 79-1, GRAIN DUST EXPLOSIONS—AN UNSOLVED PROBLEM 52–63 (Mar. 21, 1979) (containing factual data on the explosions and fires).

[16]OSHA conducted inspections, and citations and civil penalties were issued which were later settled. See, e.g., *Farmer's Export Co.*, 8 OSHC 1655 (Rev. Comm'n 1980) (upholding settlement agreement despite the use of exculpatory language). OSHA also referred to the Justice Department for possible criminal prosecution the results of its investigation of the December 27, 1977, explosion at Farmers Export Co., Galveston, Texas. Eighteen persons died and 22 were injured. See GRAIN DUST EXPLOSIONS—AN UNSOLVED PROBLEM, *supra* note 15 at 54–55. The corporation, Farmers Export, was not charged, and on November 13, 1981, a hung jury resulted in the acquittal of both individual defendants on a 36-count indictment. See Note, *A Proposal to Restructure Sanctions Under the Occupational Safety and Health Act: Limitations of Punishment and Culpability*, 91 YALE L.J. 1446, 1448 n.17 (1982).

[17]*OSHA Oversight—Grain Elevator Explosions: Hearings Before the Subcomm. on Compensation, Health and Safety of the House Comm. on Education and Labor*, 95th Cong., 2d Sess. (1978).

public officials with information concerning grain elevator safety and health.[18] OSHA also greatly increased its inspection rate of grain elevator facilities and gave special training to compliance officers on grain elevator hazards.[19] After an investigation of the explosions and fires, the General Accounting Office issued a report which, among other things, recommended that OSHA evaluate its present standards with a view toward having "enforceable standards to assure implementation of safe practices by employers who would not do so voluntarily."[20]

In February 1980, OSHA published an advance notice requesting public comment and scheduling informal public meetings on a grain elevator standard.[21] After considerable delay and controversy,[22] a proposed standard covering safety hazards associated with grain handling facilities was issued on January 6, 1984.[23]

The usefulness of the files of accident investigations in prioritizing inspection scheduling and standard promulgation was addressed by both the General Accounting Office and the Interagency Task Force on Workplace Safety and Health.[24]

Report of the Comptroller General

***

CONCLUSIONS

Thousands of serious accidents are investigated each year by OSHA* and the States. These investigations are made to determine whether (1) the

*[Author's note—The number of catastrophe inspections has remained fairly constant, ranging between 1,800 and 2,500 inspections a year. See *Oversight on the Administration of the Occupational Safety and Health Act, 1980: Hearings Before the Senate Comm. on Labor and Human Resources,* 96th Cong., 2d Sess. pt. I at 301 (1980) (statement of Deputy Assistant Secretary Basil Whiting) (hereinafter cited as *1980 Senate Oversight Hearings*). In fiscal year 1982, there were 1,879 accident inspections, 12 OSHR 711 (Jan. 20, 1983), and during the first half of fiscal year 1983 there were 694 accident inspections, 13 OSHR 367 (1983).]

[18]See Grain Dust Explosions—An Unsolved Problem, *supra* note 15 at 68–69.

[19]*Id.* at 69. In November 1983, OSHA established a "national emphasis program" for the inspection of grain elevators, 13 OSHR 663–64 (1983). A union official criticized the program as no substitute for an effective grain dust standard, 13 OSHR 695–96 (1983).

[20]See Grain Dust Explosions—An Unsolved Problem, *supra* note 15 at 38.

[21]45 Fed. Reg. 10,732 (1980).

[22]See, e.g., *Review of Grain Elevator Safety: Hearings Before the Subcomm. on Wheat, Soybeans and Feed Grains of the House Comm. on Agriculture,* 97th Cong., 2d Sess. 55, 58 (1982) (statement of Barry White, Director of OSHA Safety Standards); *Mired—Safety Standard Got Lost in the Shuffle,* Washington Post, May 12, 1983 at A11. A substantial controversy had developed between OSHA and OMB over the contents of the proposed grain elevator standard. 13 OSHR 780 (1983).

[23]49 Fed. Reg. 996. Another major workplace catastrophe was the Staten Island, New York gas tank explosion in February 1973, in which 40 workers were killed. *Texas E. Transmission Corp.,* 3 OSHC 1601 (Rev. Comm'n 1975) (held the on-site maintenance and repair contractor liable for violations of the Act, but dismissed the citations against the operators of the natural gas facility, concluding that OSHA authority was preempted under §4(b)(1), 29 U.S.C. §653(b)(1), by the Department of Transportation standards issued under the Natural Gas Pipeline Safety Act of 1968, 49 U.S.C. §§1671–86 (1976 and Supp. V 1981); Chairman Moran dissented as to the contractor, whom he would have also held exempt) 3 OSHC at 1605.

[24]Comptroller General of the United States, Report to the Congress, HRD-79-43, How Can Workplace Injuries Be Prevented? The Answers May Be in OSHA Files 37–39 (May 3, 1979).

accidents could have been avoided had proper safety and health regulations been enforced, (2) standards need to be developed or revised, and (3) violations of standards contributed to the accidents.* The investigations produce the best information available on the specific hazards and related factors causing serious accidents. Despite these investigations, OSHA has not effectively used the data it and States have collected to attain the objectives of the investigations. OSHA does not know (1) to what extent fatal accidents could have been avoided if safety and health regulations had been enforced and followed, (2) what standards need to be developed or revised, and (3) what standards' violations cause most serious accidents and deaths.

The information acquired from accident investigations has not been fully used due to OSHA's practices and methods of recording, classifying, and collecting information. * * *

The identity of serious hazards needing standards['] coverage is contained in accident information obtained during OSHA investigations. However, OSHA has not effectively used this information in selecting standards' projects and establishing priorities. Two-thirds of the standards projects were started to clarify existing standards or eliminate unnecessary detail and do not address specific hazards. Our review of accident cases showed that about 12 percent involved hazards not covered by standards, and 74 percent of these hazards were not being addressed in current standards development projects. The absence of a systematic approach to identifying serious hazards has resulted in standards projects which do not address specific hazards and the creation, suspension, and abandonment of standards development projects without regard to their potential for reducing accidents.

Information from accident investigations has not been used effectively in enforcement activities to optimize the potential for abating hazards which cause serious accidents. The locations of fatal accidents are not considered when targeting inspections. The inspection activity is not monitored by industry and workplace size to assure that inspections are conducted first at locations most frequently having serious accidents. As a result, the types of workplaces where most fatal accidents occur are not receiving a proportionate share of inspections. The industries in which 30 percent of the fatalities occurred received less than 18 percent of OSHA's self-initiated inspections, and small workplaces received a larger share of inspections than their proportion of serious accidents. Also, the probability of inspectors identifying serious hazards at workplaces is reduced because they are not alerted to hazards which have caused serious accidents in each type of workplace.

The greatest potential for accident prevention is in the areas of employee awareness of hazards, training of workers and supervisors, and improvement of employer programs for the continuing identification and abatement of workplace hazards. Although OSHA can contribute in each of these areas, much of this potential has not been realized.

OSHA has not identified the types of workplaces and the occupations where the greatest number of serious accidents could be prevented by each of its programs. A list of hazards has not been used in the training, consultation, and awareness programs to explain and emphasize how to avoid the accidents. Guidance has not been provided to grantees and contractors on where training and education services can produce the greatest benefits in reducing serious accidents. OSHA does not monitor and evaluate grantee and contractor activities to assure that services are provided first to those

*[Author's note—See OSHR [Reference File 77:3501]. The investigation is also to determine if there has been a violation of the general duty clause. Indeed, citations issued in the case of workplace accidents are frequently based on the general duty clause, as in the case of the Staten Island gas tank explosion, *Texas E. Transmission Corp.*, 3 OSHC 1601, 1606–07 (Rev. Comm'n 1975), *supra* note 23.]

with the greatest needs. Information on accident causes has not been provided to employers and labor groups so that they can identify and voluntarily abate serious hazards.

State and corporate accident statistics and specific information on serious accidents have proven useful in directing the safety programs of State and private organizations. Accidents have been significantly reduced by programs in which data from accident investigations were used to identify the location of prevalent accidents and their causes.

***

The Interagency Task Force on Workplace Safety and Health report also concluded that OSHA should make more use of accident investigations in targeting workplace inspections.[25] In its "First Recommendations," the task force recommended that OSHA "increase the numbers and types of accident investigations." According to the report, accident investigations would enable OSHA to "identify the causes of serious injuries and detect accident-causing violations which are transient or otherwise difficult to discover through ordinary inspections." And, the report noted, if OSHA "focused on smaller establishments to make inspections highly probable in the event of serious injury there," this could "sharply enlarge" incentives to correct violations "in smaller, dispersed worksites which current inspections cannot reach."[26]

C. Complaint Inspections

Under the 1972 compliance manual, complaint inspections were to receive the second highest priority among all inspections; and among complaints, those alleging an imminent danger were to receive "highest priority" and those alleging conditions which "appear to be serious" were to be given "high priority." Under §8(f)(1), as interpreted by OSHA's enforcement regulations,[27] OSHA is required to conduct a workplace inspection in response to a complaint if the complaint meets certain requirements. The Act provides that the complaint must relate to a violation that "threatens physical harm or an imminent danger," must be in writing, must be signed by an employee or representative of employees, and must set forth with "reasonable particularity" the grounds for the complaint.[28] The Act also provides that, upon request, the name or names of employees mentioned in the complaint shall not be publicized.

The 1972 compliance manual sets forth the procedures for han-

[25] INTERAGENCY TASK FORCE ON WORKPLACE SAFETY AND HEALTH, MAKING PREVENTION PAY (1978) (Final Report Draft).

[26] *Id.* Ch. III at 4–5.

[27] 29 C.F.R. §§1903.11–.12 (1983).

[28] The language of the enforcement regulations, 29 C.F.R. §1903.11 (1983), is more general, stating only that the complaint must relate to a "violation of the Act [that] exists in any workplace." See discussion this section on OSHA's 1982 field instruction on complaints, which adheres more closely to the statutory language.

dling complaints *not* meeting the formality requirements.[29] These complaints are often referred to as "informal" complaints; complaints meeting the statutory requirements have been referred to as "formal" complaints.

Compliance Operations Manual

***

3. PROCEDURES FOR HANDLING COMPLAINTS NOT MEETING FORMALITY REQUIREMENTS OF 29 CFR SECTION 1903.11

A. ORAL COMMUNICATION IN PERSON OR BY TELEPHONE.

When an OSHA office receives an oral complaint in person or by telephone, the complainant should be advised that if he wants the AD [Area Director] to consider scheduling an inspection of the establishment and if he wants the additional rights noted above, he should submit a formal complaint in accordance with 29 CFR Section 1903.11. * * *

B. WRITTEN COMPLAINT NOT MEETING THE FORMAL REQUIREMENTS OF 29 CFR SECTION 1903.11.

When an OSHA office receives a signed written complaint that does not meet the formal requirements (for example, it does not state with reasonable particularity the nature of the violation), 29 CFR Section 1903.12(b) of the regulations requires that the AD notify the complainant of such fact in writing, also informing the complainant that a new complaint will be considered if the complaint meets these requirements. The complainant should be informed as to what parts of the formality requirements were not met. However, if it appears that the formality requirements can be met by informally advising the complainant of the formal defects of his complaint, that procedure should be followed. The complainant should then resubmit his complaint for consideration. If a complaint contains the complainant's name but not his address and telephone number, an attempt should be made to obtain the information necessary to contact him.

C. COMPLAINTS FROM PERSONS OTHER THAN EMPLOYEES OR THEIR REPRESENTATIVES.

Complaints from persons who are not employees or employee representatives, such as complaints by an employer under inspection against his competitors, do not constitute valid complaints under the criteria set forth in the regulations and in this Chapter. The complainant should be sent an acknowledgment that the complaint cannot be given consideration for this reason.

D. IMMINENT DANGER COMPLAINTS NOT MEETING FORMALITY REQUIREMENTS OF 29 CFR SECTION 1903.11.

If, on the basis of evaluation of a complaint, it appears that an imminent danger situation may be involved, an inspection should

[29]OSHA 2006, ch. VI at VI-2–VI-4 (Jan. 1972).

be conducted immediately without regard to whether the complaint meets the formality requirements for a valid complaint. The OSHA-7 Form contains a space to indicate whether an imminent danger situation is involved; however, the CSHO plainly should not rely on this notation alone as a basis for determining whether there may be an imminent danger situation. * * *

4. PROCEDURES FOR HANDLING COMPLAINTS MEETING FORMALITY REQUIREMENTS OF 29 CFR SECTION 1903.11

A. The complaint should be evaluated to determine whether there are reasonable grounds to believe that the violation or danger complained of exists. In most cases, the submission of a properly completed complaint should be sufficient for evaluation purposes. The CSHO may, if necessary, call or contact the complainant and obtain details of the alleged violation or danger to the fullest extent possible; in some cases, it may be desirable to meet with the complainant in person. If the complainant is contacted, he shall not be told when the inspection will take place, if at all. The employer should not be contacted at this time.

B. If the Area Director determines that an inspection is not warranted because there are no reasonable grounds for believing that the violation or danger exists, he shall notify the complainant in writing of that determination. Such notification shall state briefly the reasons for the Area Director's determination. The complainant should at the same time be notified of his right to informal review of such determination and the procedure for obtaining such review, as contained in 29 CFR Section 1903.12(a) of the regulations.

C. If it is determined that there are reasonable grounds for believing that the violation or danger exists, an inspection shall be scheduled as soon as practicable. * * *

***

E. Normally, if time and resources permit, the inspection should cover not only violations and dangers alleged in the complaint, but the entire establishment of the employer. If time and resources do not permit, the inspection should be limited to the subject matter of the complaint; for example, if a very large establishment is involved and a complete inspection would take several weeks. However, if, in such cases, the CSHO has reason to believe that similar conditions exist in areas of an establishment other than those areas initially brought to his attention in the complaint, he shall broaden the inspection to include such areas. In each case, the CSHO shall advise the employer of any increase in scope of the inspection at the opening conference or at the earliest possible opportunity.

F. At the opening conference with the employer, a copy of the complaint should be given to the employer. Section 8(f)(1) of the Act requires that if the complainant so requests, his name shall be deleted from the copy provided the employer.

***

Several points should be particularly noted in OSHA's first complaint policy. First, informal complaints generally would not result in workplace inspections, except in the case of imminent danger situations. Secondly, the Act requires that OSHA evaluate complaints meeting the formality requirements and decide whether there are "reasonable grounds to believe" that a violation or danger exists. Under the manual, however, this evaluation would be made in most cases on the basis of the complaint itself, without further investigation or communication with the complainant. Finally, if time and resources permit, a complaint inspection would normally cover the entire plant and not only where the violations are alleged to exist.

OSHA complaint policy continued essentially unchanged until the "Kepone incident" in 1975–1976, after which OSHA substantially revised its instructions on handling complaints. The sequence of events comprising the Kepone incident were described in an attachment to a letter, dated March 25, 1976, sent by Assistant Secretary Morton Corn to Chairman Daniels of the House Subcommittee on Manpower, Compensation and Health and Safety.[30]

Letter from Assistant Secretary Morton Corn to Chairman Daniels

THE KEPONE INCIDENT: AN ACCOUNT OF A WORKPLACE TRAGEDY

The revelations in the summer of 1975 of the serious illness suffered by a number of the employees of a small company in Hopewell, Virginia, producing a pesticide under the trade name "Kepone", emphasized once again the grave problem posed by toxic substances in the workplaces of America. The following report is a detailed account of the Occupational Safety and Health Administration's (OSHA) involvement in this incident. The report is intended not as a self-serving defense of OSHA's actions but as a factual description of the Agency's response, including improvements in OSHA policies and procedures undertaken as a result of this incident. The events at Hopewell impinged upon many facets of OSHA's program and pointed up numerous areas for attention. These include, among others: enforcement procedures during health emergencies, employee complaint procedures, information regarding toxic substances, and back-up capability for OSHA field employees.

Kepone is a pesticide patented in the 1950's by Allied Chemical Corporation and produced by that firm from 1965 to 1973. Beginning March 1974, Kepone was produced in the form of powder or concentrated slurry by Life Science Products Company of Hopewell, Virginia under contractual arrangement with Allied. * * * Since Life Science ceased production there are no producers of Kepone in the world.

***

OSHA knew little about the nature and effects of Kepone at the time of the events in Hopewell. Literature references, such as the authoritative Sax's *Dangerous Properties of Industrial Materials,* did not describe the pesticide as particularly dangerous in its application strength. The symptoms observed in the employees of Life Science who were exposed to large amounts of Kepone indicate, however, that the substance produces seriously harmful effects. These included hand tremors, difficulty in walking, weight loss, skin rash, chest and joint pains, abnormal eye movements, and visual

[30] *Oversight Hearings on the Occupational Safety and Health Act: Hearings Before the Subcomm. on Manpower, Compensation and Health and Safety of the House Comm. on Education and Labor,* 94th Cong., 2d Sess. pt. 2 at 667–80 (1976).

difficulties. In addition, a report released by the National Cancer Institute on January 16, 1976, indicates that Kepone is a strong animal carcinogen to rats and mice. Kepone, therefore, may cause cancer in humans.

OSHA became involved with Life Science on July 29, 1975, when the Director of Virginia's Bureau of Industrial Hygiene, Mr. Bryce Schofield, telephoned OSHA's Philadelphia Regional Office to advise that the Commonwealth of Virginia had closed the company's work premises for health reasons. The background of this action included the discovery that month that an employee of Life Science, admitted to the hospital in Petersburg, Virginia, had high levels of Kepone in his blood. After examining ten other employees of the plant and finding seven with symptoms of neurological illness, the Virginia Department of Health was prepared to order the company to cease operations. This became unnecessary since the plant management agreed to shut down full production.

The situation described to OSHA's Regional Administrator, David Rhone, by Mr. Schofield did not seem to warrant immediate attention since Life Science was described as "shut down." It was assumed that employees were no longer being exposed to hazards there. OSHA later learned, however, that under the terms of the "clean-up" agreement negotiated between the Commonwealth of Virginia and Life Science, limited processing continued, with the material previously processed being desiccated prior to packaging in the form of concentrated slurry. This was not a "clean-up" operation as that is known by health professionals.

***

OSHA's inspection of Life Science was conducted from August 11–14, 1975. The complete walk-around inspection revealed visible and considerable Kepone surface contamination throughout the plant. Four citations were prepared for the corporation as well as for Virgil A. Hundofte and W. P. Moore, the co-owners, individually. Two citations were for willful violations of the Occupational Safety and Health Act of 1970 and two citations were for serious violations. Penalties totalling $16,500 were proposed for the alleged violations. Life Science contested the citations and proposed penalties, and at a pre-hearing conference on February 13, 1976, an Administrative Law Judge of the Occupational Safety and Health Review Commission ruled in OSHA's favor on the one disputed point, concerning liability for the violations. It was ruled that Hundofte and Moore are individually liable for the violations found in the plant. This ruling can now be appealed to the full Commission.*

In order to ensure that employees remaining at the plant used proper protective equipment for their own health and that they did not inadvertently threaten the health of others by carrying Kepone outside the facility, OSHA required certain protections. * * *

At the time of these events, OSHA officials believed they represented the first Agency contact with Life Science and that no employee of the company had ever previously alerted OSHA to the danger there. It was subsequently discovered, however, that OSHA's Richmond Area Office had been visited by a former employee of the Life Science, Mr. Orben Ray Dubose, in September 1974. Mr. Dubose had stated that he wished to complain about unhealthful conditions at the plant, and he had filled out an OSHA complaint form. Dubose alleged in his complaint that he was exposed to pesticide fumes and dust while checking the cutting gear of a dryer. Dubose related to the Area Office employee interviewing him that he had allegedly been discharged by his employer for refusing to work on the dryer while it was running. The interview also disclosed that Kepone was being produced

*[Author's note—The citations were affirmed by the Review Commission, *Life Science Product Co.*, 6 OSHC 1053 (1978), and also by the Court of Appeals for the Fourth Circuit, *Moore v. OSHRC*, 591 F.2d 991 (4th Cir. 1979).]

at the plant. Because Dubose was no longer an employee of Life Science, the complaint was treated as a discrimination complaint under section 11(c) of the Act. The failure of the Richmond Office to act on a complaint of a former employee alleging a health hazard in these circumstances was clearly a mistake—a misunderstanding of OSHA policy and procedure. The office determined that there was insufficient evidence to support Mr. Dubose's claim of discriminatory discharge and he was notified of this determination. Recently, the discrimination file has been reopened for further consideration.

Although this matter was not given the priority of an employee complaint under section 8(f)(1) of the Act, it was referred to the Area Office's sole industrial hygienist. The hygienist conducted a literature search of the effects of Kepone in its application strength. On the basis of these data and the facts surrounding the complaint, he determined that the matter did not warrant immediate attention, especially in relation to other situations and workplace chemicals awaiting investigation. It should be noted that the Richmond Office has jurisdiction over approximately 37,000 establishments, of which more than 200 are chemical establishments. The hygienist was later reassigned to another office, and no compliance inspection was conducted at Life Science until August 1975—the inspection previously described.

***

This episode has pointed up a distinct need for improvements in OSHA's response to employee complaints of hazardous workplace conditions. OSHA's primary concern is [to] assure that a prompt inspection is conducted whenever information on hazardous working conditions is brought to its attention. In conformance with this policy, OSHA is issuing a directive to its field staff requiring that all discrimination complaints will now receive a thorough evaluation by the senior safety and health professional in the appropriate OSHA field office. Sufficient information concerning possible safety and health violations will be actively elicited from each complainant to permit identification of possible violations. Any safety or health problems identified during this review process must then be promptly and thoroughly inspected by OSHA's compliance personnel. The safety and health aspects of all section 11(c) complaints will be treated in the same manner as all other employee complaints of hazardous conditions received by OSHA—even when the formality requirements of the Act are not met by the complainant.

***

It is evident that OSHA's health compliance staff faces a wide-ranging problem. Health oriented compliance officers comprise only about twenty percent of the compliance force and more than one-half of these are apprentices or trainees. As a result, less than ten percent of OSHA's inspections last year were health inspections. While OSHA continues to seek personnel from the private sector, there are only about 2000 qualified hygienists in the entire country. OSHA, therefore, has recently turned more attention to training through apprenticeship programs in order to expand the pool of qualified health inspectors.

The paucity of information in OSHA's field office concerning Kepone is another aspect of this problem. There are, at present, minimal tools available to OSHA field staff to deal with the complexities of industrial health. Laboratory back-up facilities and stocks of field instruments are being enhanced to remedy this deficiency, but OSHA's toxicological information is clearly inadequate. Procedures are being established to obtain evaluations from toxicological experts to inform OSHA compliance officers in the field, who are not toxicologists, whether compounds are highly toxic.

***

The tragedy at Hopewell, although involving a relatively small number of workers, is having far-reaching implications. It has compelled OSHA to reexamine various facets of its program and the adequacy of procedures in several areas. Hopewell has also focused public attention upon the magnitude of the potential perils encountered by America's working men and women. Hopefully, this incident will provide the impetus for passage of toxic substances legislation currently under consideration by Congress.* The potential for repetition of incidents of this nature can be reduced only to the extent that some mechanism is adopted for alerting public authorities to the existence of dangerous workplace substances before symptoms of illness are apparent. Lacking such authority, OSHA and other governmental entities charged with protecting the public from environmental dangers are forced, in many instances, to a reactive position, responding only after adverse consequences are observed. But an analysis of the events described above must conclude with the observation that this Nation's workforce will not be complete[ly] protected from dangers such as Kepone without a triad of effective governmental administration of protective legislation, an enlightened employer attitude toward the welfare of the employee, and an awareness of the potential hazards and available recourses by the Nation's employees. These factors constitute the sine qua non for eliminating workplace tragedies.

The Kepone incident reinforced Assistant Secretary Corn's intention to more strongly emphasize OSHA's health enforcement activity. On the issue of the handling of complaints, OSHA quickly issued a field information memorandum making major changes in the existing procedures.[31]

OSHA Field Information Memorandum No. 76-9

March 5, 1976

***

2. PROCESSING 8(f) COMPLAINTS

(a) Complainants will be fully informed of their rights and protections afforded under Section 11(c) of the Act.

(b) Chapter VI, Section C.2 of the Field Operations Manual should be strictly enforced, namely, that a workplace complaint filed by a former employee who alleges he has been discriminatorily discharged should be deemed as a valid complaint and promptly investigated.

(c) All safety and health complaints should be thoroughly evaluated within the receiving OSHA offices for possible 11(c) consequences, and all 11(c) matters should be promptly transmitted to the Operations Review Officers for review or action.

(d) Field offices are to make every effort to respond to 8(f) complaints within the following guidelines:

*[Author's note—The Toxic Substances Control Act was passed by Congress in 1976. Pub. L. No. 94-469, 90 Stat. 2003 (codified at 15 U.S.C. §§2601–29 (1982).]

[31]Reprinted in *Oversight Hearings on the Occupational Safety and Health Act: Hearings Before the Subcomm. on Manpower, Compensation and Health and Safety of the House Comm. on Education and Labor,* 94th Cong., 2d Sess. pt. 2 at 660 (1976).]

Immediately or as soon as administratively feasible, but no later than 24 hours of receipt of the complaint where the Area Director determines that there are reasonable grounds to believe that an imminent danger situation exists.

As soon as administratively feasible, but not later than 3 working days of receipt of the complaint where the Area Director determines that there are reasonable grounds to believe that a serious hazard may exist.

Within 7 working days of receipt of the complaint where the Area Director determines that there are reasonable grounds to believe that a nonserious hazard may exist.

The case file must contain a satisfactory reason explaining why a complaint inspection was not conducted within the above time frames.

(e) Whenever information comes to the attention of the Area Director without regard to its source and without regard to whether it meets the formality requirements of Section 8(f), indicating that safety or health hazards exist at a workplace, an inspection shall be conducted in accordance with the timeframes of 2(d) above.

(f) The Area Director's approval is required in implementing any policy, technical, or procedural determinations in connection with 2(a)–(e) above.

***

The memorandum also contained instructions on procedures for the processing of §11(c) discrimination complaints. The new policy established strict and short time frames for the response to complaints. The memorandum emphasized the policy of treating employees allegedly discharged for discriminatory reasons as current employees for complaint purposes. Also, according to the new policy, if the Area Director receives *any* information indicating hazards at a workplace, an inspection is required in accordance with the strict time frames. The mandatory nature of the inspection was not required by the statute; it was a policy decision directly resulting from the Kepone incident.[32]

The memorandum stated that complaints should be "thoroughly evaluated"; however, the purpose of this evaluation apparently was to determine if any §11(c) violations were arguably present. It was apparently not intended primarily to screen out complaints lacking a "reasonable grounds" basis under §8(f)(1).

The new instructions led to a major focus on OSHA inspection activity, resulting in a far greater percentage of complaint inspections. Thus, in fiscal year 1976, OSHA conducted 9,160 complaint inspections out of a total of 90,321 inspections. In fiscal year 1978, after the change in policy, OSHA conducted 21,485 complaint inspections out of a total of 57,193 inspections; in fiscal year 1978, 20,158

[32]The Review Commission and the courts have consistently held that OSHA has legal authority to conduct an inspection on the basis of a complaint not meeting §8(f) formality requirements. See, e.g., *Aluminum Coil Anodizing Corp.*, 1 OSHC 1508 (Rev. Comm'n 1974); *Burkart Randall Div. of Textron, Inc. v. Marshall*, 625 F.2d 1313 (7th Cir. 1980).

complaint inspections out of a total of 57,910 inspections.[33] This increased complaint load created serious compliance priority problems in the field. As OSHA recognized in a program directive issued in December 1977, "the number of complaints received and investigated in some regions overtaxed the resources available. This burden introduced complaint backlogs, reduced inspection activity at some field offices in several important safety and health programs, and severely decimated planned regional inspection programs."[34] In addition, in April 1979, the Comptroller General issued a report to Congress criticizing OSHA's complaint policy on the grounds that an increasing amount of OSHA resources was being devoted to response to complaints, which did not bring OSHA to high-risk workplaces or result in the detection of serious hazards. Aiming to reduce the complaint workload, the Comptroller General made a number of recommendations to the Secretary of Labor and to the Congress.[35]

Report of the Comptroller General

***

RECOMMENDATIONS TO THE SECRETARY OF LABOR

The Secretary of Labor should direct the Occupational Safety and Health Administration and the States to

—develop criteria for screening safety and health complaints,
—evaluate each complaint and try to resolve nonformal complaints considered less than serious by means other than a workplace inspection,
—identify vague health complaints and use cross-trained safety inspectors to obtain additional information needed,
—develop inspection procedures which require that potentially serious worksite hazards are looked for when an inspector visits a worksite on a complaint inspection,
—make sure that timely complaint inspections are made when the alleged hazards are believed potentially serious, and
—insure that inspectors adequately document the scope and the results of complaint inspections.

***

RECOMMENDATION TO THE CONGRESS

The Congress should amend section 8(f) of the Occupational Safety and Health Act of 1970 to give OSHA authority to resolve formal complaints without inspections when the complaints do not involve potential hazards that can cause death or serious physical harm.

---

[33] *1980 Senate Oversight Hearings,* see Author's note, *supra* p. 406 at 301.
[34] OSHA Program Directive No. 200–69 at 1 (Dec. 1, 1977).
[35] COMPTROLLER GENERAL OF THE UNITED STATES, REPORT TO THE CONGRESS, HRD-79-48, HOW EFFECTIVE ARE OSHA'S COMPLAINT PROCEDURES ii–iii (Apr. 9, 1979).

In recommending that §8(f) of the Act be amended to allow OSHA to resolve certain formal complaints without a workplace inspection, the Comptroller General accepted the OSHA view that it was legally obligated to conduct an inspection if the complaint met the formality requirements of §8(f)(1) and OSHA had reasonable grounds to believe a violation existed. The OSHA view was based on the mandatory language of §8(f)(1): the Secretary "shall" make an inspection, if certain conditions have been met.[36]

One additional factor played an important role in OSHA's further revision of its complaint policy. In 1978, the Supreme Court handed down the *Barlow's* decision requiring warrants for nonconsensual OSHA inspections.[37] The Supreme Court expressly recognized that employee complaints could form a basis for a "probable cause" determination supporting a warrant. However, in determining whether particular complaints constituted probable cause, some courts of appeals looked closely at OSHA's evaluation of complaints to determine if the Agency had sufficient basis to conduct an inspection.[38]

In light of all these circumstances, in August 1979, OSHA made another major revision in its complaint policy.[39] This directive was a comprehensive statement of OSHA policy, making major changes in complaint-handling procedures. According to the new policy, in the case of "nonformal" complaints, OSHA will generally respond by sending a letter to the employer rather than by inspecting the workplace, except if "extremely hazardous conditions" are involved. Also, complaint inspections will be limited in scope, except in high-hazard industries. Careful evaluation of the validity of complaints, including interviews with the complaining party, was also emphasized.

OSHA Instruction CPL 2.12A[40]

***

F. *Background.* * * *

* * * [T]he policy of investigating substantially all complaints has resulted in large backlogs of complaints, many of which do not deal with serious hazards, and the diversion of OSHA resources from inspecting workplaces with serious hazards. It has become clear that the emphasis on investigating all complaints has adversely affected OSHA's ability to concentrate its resources in high hazard areas to conduct programmed inspec-

[36]Drafts of OSHA field directives circulated in 1981 would have provided that an OSHA inspection was required in response to otherwise valid formal complaints only if they involved imminent danger situations, 11 OSHR 187–89 (1981); these directives, however, were not issued. A different directive on complaints was issued in February 1982. See *infra* this section.

[37]*Marshall v. Barlow's, Inc.*, 436 U.S. 307 (1978); see discussion in Chapter 12.

[38]An example of such a strict view on complaint evaluation was articulated by the Court of Appeals for the Tenth Circuit in *Marshall v. Horn Seed Co.*, 647 F.2d 96 (10th Cir. 1981), reprinted in part, Chapter 12, Section C.

[39]An earlier directive, issued in December 1977, had made some changes in procedures, allowing field staff additional flexibility in responding to "nonformal," nonserious complaints. Program Directive No. 200–69, CPL 2.12 (Dec. 1, 1977).

[40]Subject: Safety and Health Complaint Processing Procedures (Sept. 1, 1979), OSHR [Reference File 21:8177–:8183].

tions as outlined in OSHA Instruction PAE 1.1.* This conclusion was reinforced by the General Accounting Office's (GAO) report to the Congress, dated April 9, 1979, criticizing the agency's policy of responding to substantially all complaints. Specifically, the GAO study disclosed numerous complaints which resulted in comparatively few serious violations related to the complained-of hazards. In addition, while recognizing the importance of promptly responding to complaints, the Senate Report on the complaint provision of the OSH Act underscored the necessity of balancing OSHA's multiple priorities: "By requiring that the special (complaint) inspection be made 'as soon as practicable,' the committee contemplates that the Secretary, in scheduling the special inspection, will take into account such factors as the degree of harmful potential involved in the condition described in the request and the urgency of competing demands for inspectors arising from other requests or regularly scheduled inspections."

2. OSHA continues to recognize the importance of responding to complaints from various sources. However, under this instruction the agency's response will take a variety of forms, ranging from an inspection of a complaint to a response by letter, depending upon the formality of the complaint, the nature of the hazard and the abatement response of the employer. The key to the success of this complaint-responsiveness program is the effective, professional screening and evaluation of formal and nonformal complaints to determine the nature and severity of the hazard; the number of employees potentially exposed; and any injuries, illnesses, or symptoms attributable to the hazard. Through this program, OSHA's resources will be focused on a more effective and expeditious investigation of high hazard complaints and, through its general schedule and special emphasis programs, inspection of high hazard industries and workplaces.

***

G. *Evaluating Complaints.* A careful exercise of investigatory techniques is necessary for complete evaluation of all complaints. Only supervisory personnel shall receive and evaluate complaints. * * *

***

H. *Responding to Formal Complaints.* All formal complaints meeting the requirements of Section 8(f)(1) of the Act and 29 CFR 1903.11 shall result in workplace inspections. * * *

1. *Priorities for Responding by Inspections to Formal Complaints.* Inspections resulting from formal complaints shall be conducted according to the following priority:

a. *Imminent Danger.* Any complaint, which in the professional opinion of the Area Director or his designee constitutes an imminent danger as defined in the FOM [Field Operations Manual] Chapter IX, shall be investigated the same day received where possible but not later than 24 hours after receipt of the complaint.

b. All other complaints will be prioritized based upon the gravity of the hazards * * *. Those complaints involving potentially serious hazards should be investigated within 3 working days; those involving other conditions should be investigated within 20 working days.

2. *Circumstances Under Which No Inspection of a Formal Complaint Will be Conducted.* Only under the following exceptional circumstances would a formal complaint not result in an inspection. Instead, the Area Director shall inform the complainant by letter than no inspection is warranted, the reasons therefor, and that he has a right to appeal this decision.

a. A thorough evaluation of a complaint does not establish "reasonable grounds to believe a violation or danger exists"; i.e., if the complaint is so

*[Author's note—This refers to "general schedule" or "high-hazard" inspections, discussed in Section D, this chapter.]

vague and unsubstantiated that the Area Director is unable to make a reasonable judgment as to the nature of the workplace hazard.

b. The complaint concerns a workplace condition which has no direct or immediate relationship to safety or health.

c. As a result of a recent inspection or on the basis of other objective evidence, the Area Director determines that the hazard which is the subject of the complaint is not present or has been corrected.

d. The complaint clearly does not fall within OSHA's jurisdiction, such as a complaint involving conditions inside a mine. Under such circumstances, the complaint should promptly be referred to the appropriate agency.

I. *Responding to Nonformal Complaints.* The procedures described below include a variety of alternative responses to nonformal complaints designed to assure abatement of hazards identified in the complaint.

1. *Procedures for Responding by Letter to Nonformal Complaints.*

a. Upon receipt and evaluation of a nonformal complaint, * * * the Area Director or designee, as soon as possible, shall prepare a letter to the employer advising of the complaint and of the action required. These letters shall be sent by registered mail, return receipt requested.

b. Concurrent with the letter to the employer, a letter to the complainant shall be prepared explaining that the employer has been informed of the complaint and requesting the complainant to notify the Area Director if no corrective action has been taken within 30 calendar days (or less if so indicated in the letter to the employer) or if any adverse or discriminatory action or threats are made against the complainant. A copy of the letter to the employer shall be included with the letter to the complainant.

c. If a response is received from the employer and it appears that appropriate corrective action has been taken, or that no hazard is present, the case file should be closed, subject to random followup inspections as described in paragraph I.2.b. The complainant shall be informed of all responses received from the employer.

2. *Procedures for Responding by Inspection to Nonformal Complaints.*

a. Where the employer fails to respond or submits an inadequate response within the period specified in the letter (See paragraphs I.1.a.–b.), or if the complainant informs OSHA that no corrective action has been taken, the nonformal complaint shall be activated for inspection pursuant to the priorities outlined in paragraph H.1.b. However, where an ambiguity exists, the Area Director may communicate further with the employer or complainant to determine whether the hazard has been fully corrected. Failure of the employer to correct a hazard called to his attention by letter could result in the issuance of a willful citation pursuant to Chapter XI of the FOM.

b. Where employers have sent satisfactory abatement letters, the Area Director shall nevertheless select for a followup inspection every tenth letter of these nonformal complaints. The followup inspection should be performed in accordance with paragraph K. within 30 days of receipt of the employer's letter.

c. The procedures outlined in paragraph I.1.a., involving responses to nonformal complaints by letter, shall not be followed where the nonformal complaint involves *an imminent danger* or an *"extremely serious working condition"* or where other special circumstances arise. These nonformal complaints shall be evaluated and scheduled for an inspection within 24 hours. Examples of "extremely serious working conditions" include: (a) lack of perimeter guarding on a high floor of a building to which a large number of employees are exposed or (b) an exposure to a toxic substance under conditions which strongly suggest a hazard likely to result in serious health effects, such as exposure to high lead concentrations without the use of respirators.

***

The 1979 complaint processing instructions issued by Assistant Secretary Bingham in CPL 2.12A continued in effect until February 1982 when Assistant Secretary Auchter made several changes in the procedures.[41]

OSHA Instruction CPL 2.12B[42]

***

The following significant changes in CPL 2.12A are being made in this instruction.

a. The time frame for response to serious and non-serious complaints is being slightly modified in light of field resources availability. Under CPL 2.12A OSHA attempted to inspect serious complaints within 3 days and other-than-serious complaints within 20 days. Complaints involving serious hazards now will be inspected within 5 working days; those involving other-than-serious conditions will be inspected within 30 working days.

b. Under CPL 2.12A, if a thorough evaluation of a formal complaint did not establish "reasonable grounds to believe a violation or danger exists," no inspection would have been conducted. The new instruction clarifies the original language by saying that, if the complaint does not establish reasonable grounds to believe that a violation threatening physical harm or an imminent danger exists, no inspection shall be conducted.

c. If the complaint alleges a violation of a standard designed to protect against physical harm and the evaluation leads to the conclusion that no physical harm is threatened, no inspection should be conducted.

d. Under CPL 2.12A, in the case of nonformal complaints, a letter would have been sent to the employer advising him or her of the complaint except in the case of imminent danger conditions or "extremely serious" conditions where an inspection would have been conducted. The distinction between "imminent danger" and "extremely serious" conditions has been unclear to field staff and has led to confusion. The agency believes that inspections conducted in response to apparent "imminent dangers" will permit OSHA to deal with those situations where there are immediate threats to employee safety or health requiring inspection. Accordingly, the reference to "extremely serious" conditions has been deleted.

***

f. Pursuant to CPL 2.12A, safety and health complaints filed by former employees who also alleged that they were unlawfully discharged in violation of Section 11(c) of the Act were treated as formal complaints if they otherwise satisfied the requirements of a formal complaint. After evaluating this procedure, it has been determined that, consistent with statutory language, it would be more appropriate to treat such complaints as nonformal unless subsequently referred for inspection by the 11(c) supervisory investigator.

***

[41]The Schweiker bill, introduced in 1979, would have made changes in OSHA's handling of employee complaints. See discussion, Section E, this chapter.

[42]Subject: Safety and Health Complaint Processing Procedures (Feb. 1, 1982), OSHR [Reference File 21:8264–:8270].

The change regarding the handling of complaints from employees who were allegedly discharged discriminatorily completed a cycle that began in 1976. The trigger for the Kepone directive was OSHA's failure to respond to a complaint of an alleged §11(c) discriminatee; the 1977 Kepone directive emphasized that the requirement that inspections be conducted in response to these complaints be "strictly enforced." In 1982, "consistent" with the statutory language, OSHA decided not to respond to these complaints. In addition, the new directive did not provide for response to "formal" complaints by letter to the employer, thus confirming the OSHA view that workplace inspections are mandatory where §8(f) requirements are met.[43]

OSHA policy on the scope of complaint inspections has also undergone significant changes since 1971. The 1972 compliance manual provided that, "if time and resources permit," an inspection based on a complaint should cover the entire establishment of the employer.[44] However, in 1979, the major revision of complaint inspection policy narrowed the scope of complaint inspections, partly because of court warrant cases which, based on constitutional considerations, limited the scope of these inspections.[45] OSHA directive CPL 2.12A provided as follows:

> K. *Scope of Inspection*. The inspection of a complaint in a high-hazard industry should generally extend to the entire facility and should not be limited to the complained-of working condition. Generally, a complaint inspection in a low-hazard industry should be limited to working conditions identified in the complaint. However, if the CSHO believes, on the basis of information indicating the likelihood of hazards in other portions of the plant, that the scope of the inspection should be expanded in a low-hazard industry, he should contact the Area Director or appropriate supervisor. A decision will then be made on the basis of the information whether the inspection should be expanded and the extent to which the inspection should be extended. Whenever such expanded inspections are conducted, the CSHO shall advise the employer and the employee representative of the expanded scope at the opening conference or at the earliest opportunity. The principal reasons for conducting full-scope inspections when responding to complaints in high-hazard industries are: (1) agency experience indicates that complaint inspections result in the discovery of at least as many hazards not related to the complaint as are related to the complained-of working conditions; and (2) efficient utilization of agency resources. The time spent traveling to and from the workplace or in opening and closing conferences must be expended regardless of the scope of the inspection. While these factors would also be applicable in low-hazard industries the agency has determined that in the case of high-hazard industries, which are specially targeted for OSHA enforcement, there is a greater likelihood that a full-scope inspection will disclose serious hazards, thus justifying the additional expenditure of agency resources involved.[46]

[43]These instructions have been incorporated in OSHA's current Field Operations Manual, OSHR [Reference File 77:3701–:3707].

[44]Compliance Operations Manual, ch. VI at VI-3 (Jan. 1972).

[45]See discussion in this section.

[46]CPL 2.12A, *supra* note 40 at 21:8180.

The 1983 revision of the Field Operations Manual further modified the instructions, providing that the scope of a *safety* complaint inspection would be based on the results of the review by the compliance officer of the employer's injury records.[47] *Health* inspections, on the other hand, would normally be a "comprehensive inspection of all areas where a potential serious health problem may exist" if the employer is in an industry listed on the Health Inspection Plan.[48]

D. "Programmed" or "General Schedule" Inspections

OSHA's other major priority for inspections has been the various industry targeting programs. In selecting workplaces for "programmed" or "general schedule" inspections, OSHA has utilized the "worst-first" principle, that is, inspecting the most hazardous workplaces first.[49] The evolution of OSHA inspection targeting, with emphasis on the system adopted during the administration of Dr. Bingham, was described by Basil Whiting, Deputy Assistant Secretary, in 1980 oversight hearings held by the Senate Committee on Labor and Human Resources.[50]

Statement of Deputy Assistant Secretary Basil Whiting

***

In 1971–1972, OSHA established two targeting programs, one for high-hazard industries (Target Industry Program, TIP), the other for selected health hazards (Target Health Hazard Program, THHP), both based on (1) geographic dispersion of firms, (2) numbers of employees, and (3) available injury data.

Two additional targeting programs were initiated in 1973 and 1975 respectively. The first emphasized inspections of trenching operations, work that was known to result in death and serious injuries, but that involved relatively few workers. The second and far more ambitious effort, the National Emphasis Program (NEP) was directed towards foundries.

The NEP was conceived as a testing ground for a variety of experimental approaches to occupational safety and health problems. Representatives of labor and industry were directly involved by OSHA in the initial phases of the program. Actual inspections of foundries did not begin until 1977, but by then it had become apparent that planning for the program had been inadequate. NEP was abandoned later that same year.

[47] See discussion, Section E, this chapter.

[48] Field Operations Manual, ch. IX, OSHR [Reference File 77:3704–:3705]. Industries are listed for health inspections on the basis of their "high potential employee exposures to dangerous substances." *Id.* ch. II at 77:2304. Changes in compliance programming instructions were made on Oct. 1, 1983, CPL 2.45A ch. II at II-1–II-30 (1983) and incorporated in the FOM (establishing, among other things, "Health Establishments Lists" including listing of known establishments by industry with "high potential employee exposures to dangerous substances.").

[49] This approach was in accordance with the principle set down by the Senate Comm. on Labor and Public Welfare in reporting the original OSHA bill, namely, that OSHA will place emphasis on inspections where the need is "most compelling." S. REP. No. 1282, 91st Cong., 2d Sess. 12 (1970).

[50] *Oversight Hearings Before the Senate Comm. on Labor and Human Resources,* 96th Cong., 2d Sess. pt. I at 267–68, 367–68 (1980).

During its earliest years, OSHA was subjected to criticism for (1) haphazard targeting and (2) undue emphasis on numbers of inspections. To the agency's critics, the enforcement emphasis seemed to be on quantity rather than quality. Annual inspections by federal OSHA had in fact increased from approximately 17,000 in 1972, the first full year of the Act's implementation, to a high of 90,000 in 1975, despite the fact that by the latter year, twenty-six States were conducting their own workplace inspections.

***

HIGH HAZARD TARGETING MODEL

Another high priority of this administration [of Dr. Eula Bingham] has been the implementation of a comprehensive and rational targeting policy for effectively utilizing the agency's limited enforcement resources. Because OSHA has only 1575 enforcement personnel to cover approximately 2.4 million firms under federal jurisdiction, we recognize that general schedule inspections must focus on the workplaces that are most likely to have numerous, serious hazards. As part of our "Commonsense" policy, we decided to focus 95 percent of its scheduled inspections on high hazard industries. The remaining five percent are intended to provide the continued presence of OSHA in all covered worksites, thereby underscoring the obligation of all employers to comply with the Act.

Using the most recent BLS [Bureau of Labor Statistics] annual survey, the agency identifies each year the industries in each state that had the highest injury rate. At the beginning of each year, the agency sends to area offices a list of high hazard industries, identified by SIC [Standard Industrial Classification] code and by state. Most area offices are also provided with a list of firms in the high hazard industries in their areas. This forms the basis for our targeting program. In addition, the agency gives special emphasis to construction, logging and maritime sites which are all high hazard industries but are not included in the high hazard list because of their seasonal, regional, or nonstationary nature.

The area office director, however, does not simply go down the high hazard list in scheduling inspections—that is, an area office does not go to all the firms in the most hazardous industry group, then the second most hazardous industry group, and so forth. Rather, the agency has a procedure whereby an area office will go to a small number of firms in each of the high hazard industries, and then go back and do additional inspections in all these industries as resources permit. The idea is to penetrate each of the high hazard industries so that the agency establishes a "presence." We believe this creates a ripple effect. In other words, by going to a few firms in an industry, the word will spread to other firms in the area that OSHA has done an inspection. Other employers in a targeted industry will make a greater effort to improve conditions in expectation of a possible inspection.

Implementation of the high hazard system is a major step toward fulfilling the intent of the OSH Act. The high hazard targeting model focuses on 2 percent of the firms covered under the Act. These firms have 16 percent of the nation's employment and 32 percent of its' recordable injuries. In the federal sector, 51,000 firms are covered by the high hazard model.

More research needs to be done, however, on injury patterns in and among industries to enable us to improve our targeting efforts. We know, for example, that some industries have higher injury rates than others over time. We also know that some firms in the same industry and of the same general size have dramatic changes in injury rates from year to year while other comparably sized firms have relatively constant rates over time. However, we do not know why these trends occur among industries and firms, nor do we know how to predict reliably a firm's injury rate from year to year.

The agency and others are attempting to learn more about firm and industry behavior. Moreover, the agency is exploring ways of refining or supplementing the high hazard industry targeting system to allow a sharper focus on individual firms with poor safety records. We believe very strongly that agency management must maintain the flexibility to change as conditions warrant or new understandings are gained.

HEALTH TARGETING

Targeting health inspections is more difficult than safety inspections because the BLS annual survey cannot accurately assess the extent of occupational illness. The problem is not BLS's statistical methodology, but rather, the long latency period from the time of exposure to the manifestation of many occupational illnesses and the survey's dependence upon employers and employees to identify work related illnesses. Identification of the origin of occupational illness is even more difficult because an employee may leave the establishment or industry where the harmful exposure occurred and work in other establishments or industries under different conditions. In view of this, the agency decided that health targeting must be based on measures of the exposure of employees to known health hazards and that an index for ranking industries by degree of health hazardousness was needed.

A health targeting model has therefore been developed that combines the following elements:

- An inventory of chemicals by industry
- An estimate of the number of persons exposed to each chemical in each industry
- The relative severity of the health effects of each chemical
- The average establishment size in each industry

The NIOSH National Occupational Hazard Survey (NOHS) was used to develop a list of potential chemical exposures by industry and to develop estimates of the numbers of persons potentially exposed to these chemicals. A subset of the chemicals from NOHS, those listed in Chapter II of OSHA's Industrial Hygiene Field Operation Manual, were selected for use in the model. This subset was presented to a panel of experts to classify the severity of the health effects of each chemical. The average size of an establishment in each industry is available from data contained in "County Business Patterns," a Bureau of Census publication.

For each industry, an index is constructed based on the average number of persons exposed to each chemical and the relative hazardousness of that chemical. These indices are summed to give an index for each industry. The industries are then ranked by the average establishment size. These two rankings are summed to give a combined ranking. This combination is used to select industries for inspection.

The new health targeting system is currently being tested in the field and the results are very encouraging. Implementation of this system nationally will begin in the near future. We believe the new system will mean more health protection for workers.*

---

*[Author's note—The targeting procedures utilized during the administration of Dr. Bingham were stated in Scheduling System for Programmed Inspections, OSHA Instruction CPL 2.25 (Dec. 6, 1978).]

E. The Schweiker Bill and Subsequent Developments

OSHA's targeting programs had been based on industry, rather than individual establishment, injury experience rates, in large measure because data on individual establishment injury rates were not readily available.[51] Many groups criticized the Agency for not basing its inspection priorities on the hazardousness of particular establishments. For example, the 1978 Interagency Task Force report recommended this approach.[52] A major effort in that direction was made by the Schweiker bill, introduced in 1979. The purpose of the bill was to significantly reduce OSHA safety inspection activity in "safe" workplaces; whether a workplace was safe would be determined on the basis of the individual employer's lost-workday case rate. In a safe workplace, OSHA would normally not conduct programmed inspections and, except in special circumstances, would respond to employee complaints, including formal complaints, by means of a letter to the employer rather than by an inspection. Senator Schweiker's statement introducing the bill explained its purpose and basic provisions.

Statement of Senator Schweiker

125 Cong. Rec. 37,135–37 (1979)

***

Nine years ago, we made a promise in the law "to assure so far as possible every working man and woman in the Nation safe and healthful working conditions."

***

Mr. President, the bottom line is this: After 9 years under the act's present safety regulatory scheme, we are left with no demonstrable record that it works and with a bad taste all around from the experience. I believe the primary reason for this is the structure Congress established for accomplishment of our ambitious goal. Nine years ago, we cast the Government in a policeman's role in a system of crime and punishment for over 4 million workplaces in the United States. Although there is a valid policeman's role for the Government in very hazardous workplaces, the Government must develop other approaches, particularly for the vast majority of workplaces which are not hazardous. In short, I believe there is a better way for the Federal Government to meaningfully participate in improving the occupational safety and health of America's workers.

The better way is for the Government to redirect its policeman's role to situations where it is really needed to deter and correct grave occupational hazards, and to adopt a new posture of stimulating employer and employee cooperative self-initiative to improve workplace safety and health. The legislation I propose today would adopt this fundamental shift in emphasis in the act's legal framework.

---

[51]For a discussion of the problems involved in the use of state workers' compensation data, see testimony of Dr. Bingham, *infra*. Establishment data collected by the Bureau of Labor Statistics as part of its annual survey of occupational injuries and illnesses was not made available to OSHA for compliance purposes for reasons of confidentiality. See INTERAGENCY TASK FORCE ON WORKPLACE SAFETY AND HEALTH, MAKING PREVENTION PAY ch. III at III-8 (Dec. 1978) (Draft).

[52]*Id.* ch. III at III-6–III-11.

Mr. President, the recent final report of President Carter's Interagency Task Force on Workplace Safety and Health came to many of the same conclusions I have. With respect to the inherent limitations of the act's present enforcement approach, the report stated:

> The task force found that perhaps 25 percent of injuries are preventable by enforcement of present OSHA standards. About three-quarters can probably not be prevented by present enforcement, either because their causes are not covered by standards or because they involve transient conditions that are difficult to detect during inspections.

Moreover, several of the key final recommendations of the task force were:

Making OSHA's direct regulation more cost-effective by prioritizing inspections on detectable serious violations and high injury rate firms.

Focusing on results rather than enforcement and compliance procedures, recognizing low injury rate employers through reduced inspections.

Encouraging the private sector to do the job before Government intervention through cooperation programs.

It is also noteworthy, Mr. President, that Congress has itself already taken significant steps in the direction I propose today, including adoption of an amendment I offered in July to the fiscal year 1980 Labor-HEW Appropriations bill. That amendment dealt with small businesses of 10 or fewer employees and established the firm general rule that OSHA should leave nonhazardous small businesses alone, but preserved essential OSHA authority where really needed. This amendment was agreed to by a vote of 61 to 31, and was subsequently accepted by the Senate-House conference committee and on the House floor. This was the first time such a far-reaching mandate has been written into law.

The bill I offer today seeks to incorporate permanently into the Occupational Safety and Health Act this same positive and responsible approach, but now for businesses of all sizes. Moreover, I propose that Congress write into the law, for the first time, nonpunitive incentives for improving safety and health in the workplace.

More specifically, Mr. President, my bill would offer a partial exemption from OSHA regulation for workplaces with good safety records. The exemption would be from most safety inspections, including all routine inspections and inspections triggered by employee complaints in the majority of cases where the complaint can be resolved without an inspection.

This exemption operates as the targeting mechanism to insure the limited enforcement resources of OSHA are targeted to firms where they are most needed and also provides the incentive for self-initiative activities in the workplace to improve employee safety and health.

Development of a workplace specific targeting mechanism, such as the one I propose, was a major recommendation of President Carter's Interagency Task Force on Occupational Safety and Health to supplement the limited industry-based targeting OSHA now utilizes, and to enable OSHA to redirect inspection resources to high-hazard workplaces.

To qualify for the inspection exemption, a workplace must have had a good safety record during the preceding year, including no employee deaths caused by occupational injury and very few occupational injuries which resulted in one or more lost workdays. The Bureau of Labor Statistics data indicate that approximately 90 percent of all workplaces would qualify under this test.

To facilitate the administrative mechanism for distinguishing those workplaces which have good safety records from those which do not, the bill proposes to utilize data from workers' compensation reports already filed by an employer with the State workers' compensation agency. The State workers' compensation agency will compile a list of employers in the State who reported one or more occupational injuries. Those employers with no injuries reported will automatically qualify for the exemption, without the need for any paperwork from the employer. The Bureau of Labor Statistics estimates that approximately 85 percent of all workplaces would qualify under this method.

Those workplaces not identified as safe through the workers' compensation data, but which meet the safety record test, based on having no deaths and a low-lost-workday-injury rate, will claim their inspection exemption by filing an affidavit with OSHA.

This system of combining limited use of workers' compensation data with the affidavit process effectively identifies safe workplaces with a minimal amount of paperwork for employers, OSHA and State workers' compensation agencies. To assist these State agencies to make any needed modifications in data tabulation procedures, the bill directs the Labor Department to provide financial and technical assistance to the State agencies.

***

Mr. President, having explained what this bill would do, let me now mention what it would not do. Most importantly, my bill does not restrict OSHA authority to deal with occupational health hazards. Such hazards are not revealed by reference to safety performance records (due to long latency periods), and the effectiveness of safety committees and consultation programs in dealing with health hazards is not well established at the present time. However, I am hopeful that experience developed under this legislation would provide a firm basis for expanding the utilization of the self-initiative methods established here.

In addition, this bill specifically preserves essential OSHA authority to deal with other hazards which cannot be identified by reference to safety performance records. Our recent experience with grain elevator explosions proves that some special safety hazards are not forewarned by either industry or individual establishment safety performance records. Therefore, OSHA would retain full authority and responsibility to deal with very special cases of this kind. The bill also preserves full OSHA authority to investigate serious accidents, deal with imminent dangers, and to protect employees against discrimination based on exercising rights under the law.

***

OSHA and labor organizations objected strongly to the Schweiker bill on both philosophical and practical grounds.

Statement of Assistant Secretary Bingham[53]

***

Let us look at how the system proposed by this bill would work in practice. As explained previously, the bill proposes a two-tiered method of determining a firm's record of lost workday injuries and thus its exemption from programed safety inspections: first, an automatic determination through State workers' compensation data; and second, through individual employer affidavits filed with OSHA.

The bill envisions minimizing recordkeeping and paperwork by having the Secretary of Labor enter into agreements with States to furnish data

[53] *Hearings on S. 2153 Before the Senate Comm. on Labor and Human Resources,* 96th Cong., 2d Sess. 33–36 (1980).

from workers' compensation reports already on file with State agencies. The Secretary would be required to provide technical, financial and administrative assistance to States wishing to modify their workers' compensation systems so as to be able to enter into such agreements. Those that agree to cooperate would provide OSHA with lists of employers who had reported 1 or more lost workday injuries in the preceding year. An employer not on such a list would be automatically exempted.

There are many obstacles to the use of State workers' compensation data for exemption purposes, even should the States choose to cooperate: Only two States could come close to furnishing the required complete list of employers, and even they would miss some small groups not covered by the compensation systems in those States.

Federal technical and financial assistance offered to States as an incentive to modify their workers' compensation systems voluntarily in order to meet the data requirements of the bill will doubtless prove inadequate to overcome the present gap, even over the 5-year period allowed in the bill.

Based on past experience, the Department believes that many of the State systems are unlikely to change unless mandated to do so, no matter how much money and effort is expended by Washington. BLS's present difficulties in obtaining full and uniform participation by all States and jurisdictions in its Supplemental Data System, despite years of effort, points out the difficulty of the task.

Even if State participation were mandated, and even were all State systems made uniform, not all lost workday injuries would in fact be reported under State reporting requirements. Consequently, employers whose workers had experienced a significant number of injuries might be exempt from OSHA safety inspections because their firms' names do not appear on lists drawn from these reports. For example, self-insured employers and carriers are not required to report certain uncontested workers' compensation claims in some States.

Also, many lost workday injuries that are not compensable or are not likely to be compensable may not be reported to the agency, and this circumstance will vary greatly from State to State depending on coverage limitations, benefit restrictions, and so forth. In fact, social security data reveal that only 67.3 percent of disabled workers applied for workers' compensation in 1974, because of, among other reasons: ignorance of the law; they may not know that they qualify for workers' compensation; desire to avoid the annoyance of a contested claim; preference for sick leave at full pay; treatment of an injury at the employer's expense, either at the worksite or at a nearby medical facility; or pressure from self-insured employers to forgo application for workers' compensation and seek recompense instead from accident-sickness insurance.

The consensus among Department of Labor experts and others familiar with the intricacies and vagaries of the various State workers' compensations systems is that very few States—possibly none—would be both able and willing to furnish adequate data to comply with the provisions of the exemptions proposed in S. 2153.

It is clear that any attempt to have a legal exemption from OSHA safety inspections based on data from State workers' compensation systems simply would not work.

The irony of this proposal is that while purporting to make more efficient use of agency resources, the bill would require that a significant portion of future agency resources be devoted to administration of the exemption system, and would result in an increase in paperwork both for the agency and small employers.

Verification of affidavits would be difficult, and would again require additional paperwork and investigative work in place of concentration on improving safety and health. Although OSHA could summon an employer's records much as the IRS does in auditing tax returns, on OSHA's case the task would be to check retrospectively the veracity of reports of nonevents—the absence of injuries. Therefore verification of such employer's reports would often inevitably involve visits to the worksite and discussions with employees.

The bill's proposed changes in the procedures for handling complaints would deprive workers of one of their fundamental rights under the act. Indeed, workers themselves have widely expressed the view that these provisions of the bill are the most threatening aspects in the entire proposal.

Taking as its point of departure the agency's new informal complaint policy instituted last September, the bill would, except in extreme circumstances, preclude OSHA from conducting a safety inspection of an exempt employer on the basis of a complaint, even a signed employee complaint involving a serious hazard, unless certain steps were taken first. Thus, OSHA could not respond to a worker complaint until the employer was notified and given a chance to explain or correct the alleged violation. Only if the employer failed to give "satisfactory assurances" of abatement would OSHA be entitled to make an onsite inspection.

The bill would go well beyond present policy in significantly limiting OSHA's right to respond to employee complaints. In the first place, OSHA's current policy of seeking abatement assurances from an employer before an inspection is conducted applies only to informal complaints. An informal complaint is one not in writing or not filed by workers. Under present policy, OSHA in most cases responds by onsite inspection to formal complaints. This difference in treatment accorded to formal complaints is not based only on adherence to the requirements of the statute. Rather, it derives from and embodies a fundamental policy judgment: if an employee actually working in the plant and exposed on a daily basis to hazards cares about his safety and health sufficiently to write the Secretary of Labor about it, and is courageous enough to ask for direct help from OSHA by signing his name to the complaint, that employee deserves and should receive an onsite inspection if the complaint, in OSHA's judgment, appears to have merit.

Under the bill, however, this critical policy judgment would be overturned. Stated simply, in most cases workers would be unable to trigger an immediate OSHA inspection, regardless of the formality and the detail with which they prepare their complaints, and regardless of the fact that OSHA has judged the complaint as having merit. In our view, this change in procedures would foster that very employee criticism regarding occupational safety and health enforcement which this committee sought to avoid in providing explicitly for employee rights under the act. A likely result of this bill would be the drying up of a critical source of information to OSHA on the workplace hazards; workers would consider it not worth the trouble to complain.

It should be emphasized that under current policy OSHA does not inflexibly respond to all formal employee complaints. Under section 8(f)(1), an inspection must be made only if there are reasonable grounds to believe that a violation or danger exists. Field staff have been instructed even in the case of formal complaints not to inspect where the complaint is vague or unsubstantiated; if it deals with workplace conditions which have no "direct" or "immediate" relationship to safety and health; if the complaint is outside of OSHA authority; or if, on the basis of a recent inspection, OSHA decided that a hazard does not exist.

Moreover, priority in scheduling inspections in response to formal complaints is determined by the apparent degree of the hazard involved. Finally, it cannot be argued that responding to formal complaints is an inefficient use of our resources, since an inspector need spend no more time at the workplace than is necessary to determine the extent of the hazard alleged.

In short, the bill cannot be justified as giving OSHA needed flexibility in responding to formal complaints. To the contrary, the bill would deprive the agency of flexibility by prohibiting an inspection response to formal complaints except in egregious circumstances.

This bill goes beyond OSHA's present complaint policy in other significant ways that would pose serious practical difficulties for OSHA enforcement. Under our present policy, an inspection conducted in response to a complaint involving a high hazard industry may involve the entire workplace and not only the area which is the subject of the complaint.

Under the bill, this administrative discretion to inspect beyond the scope of the complaint has been eliminated. Further, the present complaint policy provides for OSHA followup inspection not only where employer assurances of abatement are not satisfactory, but also on a random sample basis to provide a meaningful incentive to employers to respond truthfully to notification letters. This authority for random monitoring of employer assurance letters is not provided by the bill.

***

Three days of hearings were held on the Schweiker bill, and the printed transcript of the hearings encompasses 1,170 pages. Among the major groups testifying were the U.S. Chamber of Commerce, the AFL-CIO, the United Steelworkers of America, and the National Federation of Independent Business. In light of strong Agency and union opposition, the Schweiker bill was never reported out of the Senate Committee on Labor and Human Resources. In 1980, however, Senator Schweiker introduced, and the Congress passed, a "rider" to the OSHA appropriations bill, which barred OSHA from conducting enforcement activity respecting any employer with 10 or fewer employees whose Standard Industrial Classification (SIC) category has an injury and illness incidence rate below a specified number.[54] And, in 1981, Assistant Secretary Auchter modified the OSHA targeting system; under the new plan, upon reaching a workplace, compliance officers would examine the employer's injury and illness log and proceed to inspect only if the employer's incidence rates were at or above the national lost-workday rate for manufacturing, or if the employer did not maintain the required records, or if the accuracy of the records could not be verified.

Individual Establishment Safety Inspection Targeting Plan Announced by OSHA[55]

Beginning Oct. 1, the Department of Labor's Occupational Safety and Health Administration will employ lost workday injury rates to determine which firms in high hazard industries warrant comprehensive programmed safety inspections, the agency announced today.

---

[54]For a discussion of this and other appropriations riders, see Chapter 21, Section C.
[55]U.S. Dep't of Labor News Release, Sept. 23, 1981.

"OSHA has 1200 inspectors to cover approximately 70 million employees at five million workplaces," Assistant Secretary of Labor Thorne G. Auchter said in announcing the new targeting policy. "So part of our strategy for improving OSHA must be to insure the optimum use of our limited inspection resources to provide the greatest protection for workers most likely to be injured on the job," he said.

"Even though an individual firm may be part of an industry which experiences higher than average on-the-job injuries, that firm itself may have an excellent safety record. If so, our inspectors' valuable time is better spent investigating potential hazards at a firm with a poor record," Auchter added.

The new targeting policy includes changes to the procedures of scheduling firms for programmed safety and health inspections in general industry. Firms in the construction and maritime industries will be scheduled for programmed inspections as before. Also, as in the past, OSHA will respond to employee safety and health complaints concerning any firm in any industry.

OSHA area offices will select individual firms for programmed safety inspections in general industry from a high hazard industry list supplied by the national office, with industries ranked according to their lost workday injury rates. Firms with 10 or fewer employees are not required to maintain injury and illness data used in the targeting system; therefore, they will not be included in the list.

No inspections will be planned for firms where a substantially complete inspection conducted during the previous fiscal year turned up no serious violations. Firms will be selected for programmed health inspections in general industry in the same manner, except that the industry lists will be ranked according to each industry's health hazard index, rather than lost workday injury rate.

At the opening conference of the inspection, OSHA will explain to the employer and employee representatives the new targeting system and ask to see copies of injury and illness data which OSHA requires employers with 11 or more employers to maintain. (This procedure applies only to programmed safety inspections in general industry.) Using this information, along with employment figures and/or employee hours worked for the previous three years (two years for firms with more than 20 employees), the inspector will calculate the firm's lost workday injury rate (the number of lost workday injuries per 100 workers). If the rate is below the most recently published Bureau of Labor Statistics national lost workday rate for manufacturing (5.7 lost workday injuries per 100 workers in 1979), the OSHA inspector would not conduct a full scale safety inspection.

The inspector may conduct a limited inspection if the injury data indicate problems with specific processes or areas of a plant, if the inspector observes a serious hazard or imminent danger while at the plant, or if a formal employee safety complaint is filed at the opening conference.

Inspectors will conduct comprehensive safety inspections in firms with lost workday injury rates at or above the national lost workday rate for manufacturing and in companies which have failed to maintain the required injury and illness records or whose records cannot be verified as accurate.

Falsification of required OSHA records can result in a fine of up to $10,000 and/or up to six months in prison.

If a firm's lost workday injury rate is low enough to exclude it from inspection, this will be explained to the employer and employee representatives present at the opening conference. If there is no employee representative present, the OSHA inspector will leave a letter to the employees explaining the exclusion and request the employer to post it. The letter will include the area office address and telephone number for the employees to call if they have any questions about the inspection targeting policy.

The new policy is spelled out in OSHA Program Directive CPL 2.25B, dated Oct. 1, 1981.*

OSHA's new targeting system, which for the first time based a portion of the Agency's inspection activity on individual establishment safety records, was criticized by representatives of the AFL-CIO as "ill-conceived" and "unsound."[56] The union claimed that the plan was "an exemption not a targeting system" that removes over 86 percent of the manufacturing employers from "one of OSHA's most effective compliance tools * * * the threat of a general scheduled safety inspection * * *."[57]

According to OSHA statistics, in fiscal year 1982, out of a total of 61,225 workplace inspections, 8,442 were record inspections; and in the first half of fiscal year 1983, out of a total 32,093 inspections, 5,328 were record inspections. OSHA has stated that a records-check inspection takes an average of 7.5 hours to perform and includes not only a review of injury records and computation of the injury rate, but discussions between the OSHA inspector, employers, and employees regarding occupational safety and health problems at the worksite.[58]

F. Follow-up Inspections

OSHA's first compliance manual provided that field staff were required to conduct follow-up inspections after court restraining orders in imminent danger situations and where citations for serious, willful, or repeated violations have been issued. The emphasis on following up previous inspection activity continued through the administration of Dr. Bingham. Thus, the Field Operations Manual in

---

*[Author's note—Field instructions for the conduct of safety inspections under the new targeting program are included in the Field Operations Manual, ch. III, OSHR [Reference File 77:2510–:2513]. On November 8, 1983, OSHA announced that the new cutoff rate for determining whether a full safety inspection should be conducted at a manufacturing establishment would be a lost workday injury rate at 4.3 per 100 workers. This was the average lost workday injury for manufacturing, based on the 1983 Survey of the Bureau of Labor Statistics (BLS). 13 OSHR 676 (1983). The BLS survey is discussed in Chapter 11, Section B. For statistics on various types of OSHA inspections, see Section F, this chapter.

---

[56] *Oversight on the Administration of the Occupational Safety and Health Act: Joint Hearing Before the Subcomm. on Investigations and General Oversight and Subcomm. on Labor of the Senate Comm. on Labor and Human Resources,* 97th Cong., 1st Sess. 171 (1981) (testimony of Margaret Seminario, Department of Occupational Safety and Health, AFL-CIO).

[57] *Id.*

[58] *OSHA Oversight—State of the Agency Report by Assistant Secretary of Labor for OSHA: Hearing Before the Subcomm. on Health and Safety of the House Comm. on Education and Labor,* 97th Cong., 2d Sess. 181 (1982) [hereafter *1982 State of the Agency Report*].

OSHA INSPECTION PROGRAM CHANGES, FY 1977–1983

	Fiscal Year						
	1977	1978	1979	1980	1981	1982	1983 (first half)
Total Inspections[a]	59,836	57,193	57,910	63,363	56,998	61,225	32,093
Type							
Health[b]	8,943	10,650	11,160	11,871	10,758	9,200	4,953
Safety	50,893	46,543	46,750	51,492	46,234	43,583	21,812
Inspection-Priority Category							
Accident	1,777	2,064	2,301	2,286	2,231	1,879	694
Complaint	19,270	21,485	20,158	16,093	13,448	6,761	2,864
General Schedule	24,847	20,220	23,755	33,320	36,135	42,576	22,486
Follow-up	13,942	13,424	11,696	11,664	5,427	1,567	721
Industry							
Construction	15,563	14,521	17,825	26,223	26,013	29,297	15,021
Manufacturing	31,155	29,953	27,541	27,224	22,685	18,013	9,680
Maritime[c]	1,360	1,333	1,446	1,062	1,099	847	372

[a]The total number of inspections for 1982 includes 8442 record inspections, and for the first half of 1983, 5328 record inspections. See discussion in Section E, this chapter.

[b]In fiscal year 1973, OSHA conducted 3,181 health inspections out of a total of 48,437 inspections; in fiscal year 1974, 3,933 out of a total of 77,089 inspections; in fiscal year 1975, 5,522 health inspections out of a total of 80,945; and in fiscal year 1976, 7,516 health inspections out of a total of 90,231. *1980 Senate Oversight Hearings,* see Author's note, *supra* p. 406, pt. 1, table 2 at 301 (attached to the statement of Deputy Assistant Secretary Basil Whiting). See also Rothstein, *OSHA After Ten Years: A Review and Some Proposed Reforms* 34 VAND. L. REV. 71, 99 (1981) (citing slightly different statistics).

[c]Numbers of inspections in "other" industries have not been included.

1981 provided that follow-up inspections were required for imminent danger, willful, and repeated violations, and that "normally" serious violations would be considered for follow-up, if time and resources permitted.[59]

Assistant Secretary Auchter, however, sharply reduced OSHA's emphasis on follow-up activity. OSHA reported to the Subcommittee on Health and Safety of the House Committee on Education and Labor as follows:

> OSHA has dramatically reduced the number of follow-ups of both health and safety inspections, thereby reducing the total number of health inspections. Whereas in 1980, follow-up inspections made up eighteen percent of OSHA's total inspections, they constituted only about three percent of all inspections during the first half of FY '82. OSHA found that virtually all firms visited on follow-up inspections—that is, inspections to check that hazards previously cited were abated—were in compliance with OSHA standards. Consequently, the agency decided to minimize this type of inspection and use agency resources elsewhere. Since OSHA can conduct approximately three follow-up inspections in the time needed for a single, initial scheduled inspection, a

[59]Field Operations Manual, ch. V (1981).

> reduction in health follow-ups naturally results in a reduction in total health inspections. In our judgment, OSHA compliance officers' time is better spent in conducting initial inspections where hazards are more likely to be found.[60]

The statistics in the chart on p. 433 give a broad view of changes in OSHA's inspection program during the history of the Agency.[61]

[60]*1982 State of the Agency Report, supra* note 58 at 164. In fiscal year 1980, OSHA conducted 11,871 follow-up inspections out of a total of 63,363 inspections; in fiscal year 1982, 1,567 follow-up inspections out of 61,225.

[61]The statistics are based on data furnished by OSHA's Office of Management Data Systems.

14

Violations and Penalties: Legal and Policy Issues

A. Background

The Occupational Safety and Health Act contains detailed provisions setting forth the framework for enforcement activity. In summary, the duties of employers[1] are stated in §5(a).[2] Under §5(a)(1), known as the general duty clause, an employer is required to provide a workplace "free from recognized hazards" that are likely to cause death or serious physical harm to employees.[3] Section 5(a)(2) requires employers to comply with OSHA standards and regulations.[4] Representatives of OSHA—compliance safety and health officers (CSHOs)—conduct physical inspections of covered workplaces, and, if violations are found, citations are issued and penalties may be

[1]The coverage of the Act is pervasive. Under §3(5), 29 U.S.C. §652(5) (1976), employer is defined as a "person engaged in a business affecting commerce who has employees." Government entities are not included as "employers" under Act. *Id.* See also 29 C.F.R. §1975.3(d) (1983) ("Since the legislative history and the words of the statute, itself, indicate that Congress intended the full exercise of its commerce power in order to reduce employment-related hazards * * * it follows that all employments * * * (involving the employment of one or more employees) were intended to be regulated as a class of activities which affects commerce."). See also generally 29 C.F.R. §§1975.1–.6 (1983) ("clarifying which persons are considered to be employers").

The courts have generally agreed with the OSHA view of the coverage of the Act. See, e.g., *United States v. Dye Constr. Co.,* 510 F.2d 78 (10th Cir. 1975). But see *Donovan v. Navajo Forest Products Indus.,* 692 F.2d 709 (10th Cir. 1982), *aff'g* 8 OSHC 2094 (Rev. Comm'n 1980) (holding that the Act does not apply to a business enterprise located on the Navajo reservation and owned and operated by the Navajo Tribe absent express statutory language to that effect; coverage would constitute violation of the Navajo Treaty relating to exclusion of non-Indians from the reservation); *Frank Diehl Farms v. Secretary,* 696 F.2d 1325 (11th Cir. 1983), *rev'g* 9 OSHC 1432 (Rev. Comm'n 1981) (holding that OSHA has authority under the Act to regulate housing only "if company policy or practical necessity force workers to live in employer-provided housing"; court rejects OSHA's coverage test of "directly related to employment" in favor of the "condition of employment test"). See also 29 C.F.R. §1910.142 (1983) (standard covering temporary labor camps). Compare attempts by Congress through appropriations' riders to exempt "small" employers from all or some requirements of the Act, Chapter 21, Section C.

[2]29 U.S.C. §654(a).

[3]29 U.S.C. §654(a)(1).

[4]29 U.S.C. §654(a)(2). Sec. 5(b), 29 U.S.C. §654(b) requires employees to comply with all regulatory requirements applicable to their "own actions and conduct," but no sanctions are provided by the Act against employees. See Chapter 16, Section A.

proposed. OSHA violations are characterized as nonserious, serious, willful, or repeated, with civil penalties in specified increasing amounts for these various types of violations.[5] Since penalties for willful and repeated violations may be as high as $10,000,[6] OSHA's characterization of the various kinds of violations has been an important issue in OSHA enforcement. Under §17(e),[7] an employer who willfully violates a standard, resulting in the death of an employee, is subject to criminal prosecution. This chapter deals with major legal and policy issues relating to the various OSHA violations and the resulting penalties.

B. General Duty Clause

Section 5(a)(1), the general duty clause, provides that each employer "shall furnish to each of his employees employment and a place of employment which are free from recognized hazards that are causing or are likely to cause death or serious physical harm to his employees."[8]

There was considerable controversy in the legislative development of the general duty clause, just as the clause's implementation since 1971 has been controversial. The bill reported by the Senate Labor Committee in 1970 contained a broad general duty clause[9] that would have required employers to provide employees with employment and a place of employment "free from recognized hazards so as to provide safe and healthful working conditions."[10] The committee report explained the need for a general duty clause.

Senate Report No. 1282[11]

91st Cong., 2d Sess. 9–10 (1970)

***

GENERAL DUTY

The committee recognizes that precise standards to cover every conceivable situation will not always exist. This legislation would be seriously deficient if any employee were killed or seriously injured on the job simply because there was no specific standard applicable to a recognized hazard which could result in such a misfortune. Therefore, to cover such circumstances the committee has included a requirement to the effect that employ-

---

[5]See §17, 29 U.S.C. §666.
[6]Sec. 17(a), 29 U.S.C. §666(a).
[7]29 U.S.C. §666(e).
[8]29 U.S.C. §654(a)(1).
[9]S. 2193, 91st Cong., 2d Sess. §5(a)(1) (1970).
[10]S. Rep. No. 1282, 91st Cong., 2d Sess. 27 (1970). The detailed legislative history is discussed in *American Smelting & Ref. Co. v. OSHRC,* 501 F.2d 504 (8th Cir. 1974), reprinted in part this section. For a general discussion of the legislative history of the Act, see Chapter 1.
[11]*Senate Comm. on Labor and Public Welfare,* S. Rep. No. 1282, 91st Cong., 2d Sess. 9–10 (1970).

ers are to furnish employment and places of employment which are free from recognized hazards to the health and safety of their employees.

The committee has concluded that such a provision is based on sound and reasonable policy. Under principles of common law, individuals are obliged to refrain from actions which cause harm to others. Courts often refer to this as a general duty to others. Statutes usually increase but sometimes modify this duty. The committee believes that employers are equally bound by this general and common duty to bring no adverse effects to the life and health of their employees throughout the course of their employment. Employers have primary control of the work environment and should insure that it is safe and healthful. Section 5(a), in providing that employers must furnish employment "which is free from recognized hazards so as to provide safe and healthful working conditions," merely restates that each employer shall furnish this degree of care.

There is a long-established statutory precedent in both Federal and State law to require employers to provide a safe and healthful place of employment. Over 36 states have provisions of this type, and at least three Federal laws contain similar clauses, including the Wa[l]sh-Healey Public Contracts Act, the Service Contract Act, and the Longshoremen's and Harbor Workers' Act.

The general duty clause in this bill would not be a general substitute for reliance on standards, but would simply enable the Secretary to insure the protection of employees who are working under special circumstances for which no standard has yet been adopted. Moreover, the clause merely requires an employer to correct recognized hazards after they have been discovered on inspection and made the subject of an abatement order. There is no penalty for violation of the general duty clause.* It is only if the employer refuses to correct the unsafe condition after it has been called to his attention and made the subject of an abatement order that a penalty can be imposed. Before that is done, the employer would be entitled to a full administrative hearing, followed by judicial review, if he disagrees that the situation in question is unsafe.

***

This broad scope of the general duty clause was challenged by Senator Peter Dominick in the floor debates. Among several other amendments to the Committee bill he submitted, he proposed an alternative general duty clause which would have limited the employer's obligation to "hazards which are readily apparent and are causing or are likely to cause death or serious physical harm to employees." Senator Dominick explained the reasons for this narrower clause in his floor statement.

Statement of Senator Dominick

116 Cong. Rec. 36,531 (1970)

***

We would be remiss in our duty, Mr. President, if any worker were killed or seriously injured on the job because he was unprotected under this

*[Author's note—The Conference Committee, however, adopted the House bill's (H.R. 16785) provision permitting the Secretary of Labor to propose penalties for violations of the general duty clause. H.R. CONF. REP. No. 1765, 91st Cong., 2d Sess. 17 (1970). The Act authorizes penalties for general duty violations under §17(a)-(c), 29 U.S.C. §666(a)–(c).]

bill owing to the fact that no *specific* safety or health standard covered the circumstances which caused his injury. Therefore, in addition to requiring employers to comply with the specific standards promulgated by the Board, the amendment I am offering would require each employer to furnish his employees employment and a place of employment *which are free from any hazards which are readily apparent and are causing or are likely to cause death or serious physical harm to his employees.*

The general duty embodied in this amendment, which applies only to dangerous situations *readily apparent* to an employer, is vastly preferable to the more ambiguous mandate now in the Committee bill.

There are essentially two features which make the substitute's general duty [requirements more acceptable to employers: (1) it is] more clearly spelled out (and would thus be understandable to employers and inspectors) to include specifically those situations which "are causing or likely to cause death or serious physical harm to employees," the Committee bill places no similar limitation on the type of hazard contemplated; and (2) the term "readily apparent" connotes that the hazard must be easily discoverable by a reasonably prudent man under similar circumstances; whereas the Committee bill uses the phrase "recognized hazards" which is rather ambiguous and open-ended, e.g. "recognized" by whom?—a concerned and prudent employer, a safety engineer, a trained doctor, a scientist, etc.?

In summary, Mr. President, the offensive feature of the Committee bill's present general requirement is that it is essentially unfair to employers to require compliance with a vague mandate applied to highly complex industrial circumstances. Under such a requirement, the employer will simply have no way of knowing whether he is complying with the law or not, nor will the inspector have any concrete criteria, either statutory or administrative, to guide him in finding a violation.

The major thrust of the Act contemplates the establishment of specific standards. The existence of a vague general requirement increases the risk that its enforcement will form the basis for the law's enforcement to the detriment of the setting of specific standards. (emphasis in original)

The Dominick amendment was defeated and the Senate adopted the Committee version of the general duty clause. The House of Representatives, on the other hand, passed the substitute submitted by Congressman Steiger, which included a general duty clause identical to the one introduced by Senator Dominick.[12] The compromise adopted by the Conference Committee in December 1970 included the present language of §5(a)(1).[13] In commenting on the conference bill, however, Congressman Steiger construed the Conference Committee language on general duty to apply only to hazards "that can readily be detected on the basis of the basic human senses." He stated that "hazards which require technical or testing devices to detect them" were not intended to be covered.[14]

One of the earliest court of appeals decisions on the general duty clause involved an interpretation of this legislative history. At issue in that case was whether high levels of airborne concentrations of

[12]116 CONG. REC. 38,716, 38,724 (1970).

[13]H.R. CONF. REP. No. 1765, 91st Cong., 2d Sess. 33, 116 CONG. REC. 41, 974 (1970) (statement of the Managers on the Part of the House).

[14]116 CONG. REC. 42,206 (1970).

lead were a "recognized" hazard under §5(a)(1). After a detailed examination of the development of the present language of the general duty clause, the Court of Appeals for the Eighth Circuit upheld the OSHA citations.

American Smelting & Refining Company v. OSHRC

501 F.2d 504 (8th Cir. 1974)

GIBSON, Circuit Judge:

***

The Petitioner, a New Jersey corporation, operates a lead refining plant in Omaha, Nebraska, and employs 390 to 400 workers there. Receiving lead bullion in solid blocks, the plant produces commercial grades of refined lead and lead alloys by separating impurities from the lead. Recovery of the impurities is also accomplished where possible. The Omaha plant, one of the Petitioner's lead refineries, was purchased before 1910 and has since produced refined lead and lead alloys.

Three representatives of the Secretary visited Petitioner's Omaha plant on June 30, 1971, to evaluate the presence of harmful concentrations of airborne lead. The plant's personnel director and plant manager fully cooperated and conducted a tour of the plant. Of the 390 to 400 workers, approximately 95 work in areas that may be affected periodically by high levels of airborne concentrations of lead. Of these 95, 60 work in the lead smelting area, 15 in the retort area (a furnace operation), 10 in the cupel area (another furnace operation), and 10 to 12 in the crane operating area. These are the areas of higher concentrations of airborne lead.

While inspecting the plant, the Secretary's representatives—Charles Adkins, an OSHA Regional Hygienist from Kansas City; Gail Adams, an OSHA Compliance Officer from Omaha; and Howard Ludwig, a United States Public Health Sanitary Engineer from Salt Lake City—observed that all but one of the employees had their company-supplied respirators hanging around their necks, rather than properly wearing them over their noses and mouths. After the tour of the plant, the representatives decided to take air samples of the melting, cupel, retort, and crane areas. Respondent chose seven employees and placed an air sampling pump on each employee. An air sampling pump is a device employed by industrial hygienists to collect contaminants in the air that an individual would normally breathe. Each air sampling pump consisted of a small pump powered by a battery, a rotometer to calibrate the air flow rate through the apparatus, a flow stabilizer device, a transport tube, and a membrane filter or monitor sender from which the sample of air contaminants is taken. The pump and battery were attached to a belt around the worker's waist, the flow stabilizer device was connected to the pump, the tube was connected to the pump, and at the end of the transport tube was connected the membrane filter, which was clipped to the worker's lapel as near the breathing zone of the nose and mouth as possible. The representatives activated the pumps on each of the seven workers, and the pumps were in place for approximately two hours and fifty-seven minutes. This period of time was sufficient to allow each worker to complete at least one complete cycle of his normal work throughout the plant. While the pumps were in place, none of the seven tested employees was wearing his respirator. A pump was not attached to the one worker who was wearing his respirator and working in a large kettle that melts lead. This worker had been chipping solid lead from a dry kettle with an air hammer, and since this job exposes workers to a very high level of lead contaminants, the worker was wisely wearing his respirator. The

Secretary's representatives fairly did not sample the air in this work area.

The pumps were carefully calibrated for air flow at the Public Health Service Laboratory in Salt Lake City and at the plant, and each membrane filter was inserted in the apparatus at the plant. The filters were removed and sent to Salt Lake City to analyze them for the lead contaminant. The following results were obtained:

[Chart omitted. Five of the six samples showed concentrations of lead higher than 0.2 mg/M^3, that is, 0.2 milligrams per cubic meter of air. The court said that mg/M^3 was the "commonly used measure."]

The Respondents throughout these proceedings have maintained that the industry recognizes that any lead concentration above .2 mg/M^3 constitutes a hazardous condition. The Petitioner counters by claiming generally that air sampling is inferior to periodic biological testing of each employee's blood and urine and acting upon those results with medical care and transfers to work areas of lesser lead concentrations. Mainly because of its biological monitoring of each employee, the Petitioner argues that there was not a recognized hazard in the plant causing or likely to cause death or serious physical harm to the employees.

On July 7, 1971, the Kansas City, Missouri, Regional Administrator cited the Petitioner for a serious violation of §5(a)(1) of the Act. 29 U.S.C. §654(a)(1). The citation contained the following description of the alleged violation:

> Airborne concentrations of lead significantly exceeding levels generally accepted to be safe working levels, have been allowed to exist in the breathing zones of employees working in the lead melting area, the retort area, and other working places. Employees have been, and are being exposed to such concentrations. This condition constitutes a recognized hazard that is causing or likely to cause death or serious physical harm to employees.

The citation allowed Petitioner 60 days to complete "the implementation of feasible engineering controls to reduce the concentration of the airborne contaminants" and proposed a civil penalty of $600.

On March 1, 1972, Administrative Law Judge Brennan filed an extensive report of 47 pages. He found that .2 mg/M^3 of airborne lead concentration is generally recognized by the industry as the safe level, that a "recognized hazard" existed in violation of the general duty clause due to concentrations of airborne lead greater than .2 mg/M^3, and that the Petitioner's "preventive program" including biological monitoring, the required use of respirators, and transfers of employees with high levels of lead concentration in blood and urine samples did not negate a finding of a recognized hazard. Having found a violation of the general duty clause, the Administrative Law Judge required the Petitioner to complete feasible engineering controls to reduce the concentration of airborne lead contaminants to .2 mg/M^3, or below within six months from the entry of a final order. He also approved the Secretary's assessment of a $600 fine, specifically finding it was a civil penalty and constitutional.

On August 17, 1973, Commissioner Van Namee, joined by Commissioner Cleary for the majority, upheld the Administrative Law Judge's factual findings and interpretation of the law with one exception. The Commission held that neither it nor the administrative law judges have the authority to rule on the constitutionality of provisions of the Act.

The Petitioner first argues that the work conditions at its Omaha lead refinery did not constitute a "recognized hazard" in violation of the general duty clause. This argument has two facets: (a) the legislative history of the Act purportedly indicates that the general duty clause does not apply to hazards that only can be detected by testing devices; and (b) the Petitioner has a program of protective measures that prevented any likelihood of harm to its employees from airborne concentrations of lead. We reject both arguments and conclude that there is substantial evidence on the record considered as a whole to support the Commission's factual findings and no error of law appears.

Relying on limited though express legislative history, the Petitioner argues that the general duty clause was not intended to cover hazards that can be detected only by testing devices. Since the airborne concentrations of lead in excess of .2 mg/M^3 were discovered by air sampling pumps instead of the human senses, Petitioner argues that no recognized hazard existed. In short, "recognized" only means recognized directly by human senses without the assistance of any technical instruments. The Petitioner relies on a statement made by Representative Steiger on December 17, 1970, just prior to the House roll call enacting OSHA. Representative Steiger said:

> The conference bill takes the approach of this House to the general duty requirement that an employer maintain a safe and healthful working environment. The conference-reported bill recognizes the need for such a provision where there is no existing specific standard applicable to a given situation. However, this requirement is made realistic by its application only to situations where there are "recognized hazards" which are likely to cause or are causing serious injury or death. Such hazards are the type that can readily be detected on the basis of the basic human senses. *Hazards which require technical or testing devices to detect them are not intended to be within the scope of the general duty requirement* * * *. It is expected that the general duty requirement will be relied upon infrequently and that primary reliance will be placed on specific standards which will be promulgated under the Act.

[116 CONG. REC. 42,206 (1970)]

Although Representative Steiger's statement is unequivocally clear concerning his interpretation of "recognized," it does not follow that his interpretation is correct; nor is it indicative of the entire Congress' understanding of "recognized." We must review the legislative history in further detail.

The Administration bill introduced by Senator Dominick required "employers to maintain the workplace free from 'readily apparent' hazards." 3 U.S. CODE CONG. & ADMIN. NEWS 5177, 5222, S. Rep. No. 91-1282 (91st Cong. 2d Sess.1970 (hereinafter 3 U.S. CODE CONG. & ADMIN. NEWS). However, Senator Javits' amendment, adopted by the Senate, substituted the words "recognized hazards" for "readily apparent hazards." 3 U.S. CODE CONG. & ADMIN. NEWS at 5222 (1970). Senator Javits added these remarks in Senate Report No. 91-1282 for the Senate Labor and Public Welfare Committee:

> As a result of this amendment the general duty of employers was clarified to require maintenance of a workplace free from "recognized" hazards. This is a significant improvement over the Administration bill, which requires employers to maintain the workplace free from "readily apparent" hazards. That approach would not cover non-obvious hazards discovered in the course of inspection.

3 U.S. CODE CONG. & ADMIN. NEWS at 5222 (1970).

The House of Representatives also debated the wording of the general duty clause. Representative Daniels proposed that the clause require employers to "furnish to each of his employees employment and a place of employment which is safe and healthful." Committee Print, Legislative History of the Occupational Safety and Health Act of 1970, 92d Cong., 2d Sess. at 726–27, 833 (hereinafter Legislative History). Representative Steiger proposed the "readily apparent" language identical to the Administration bill; however, Representative Daniels, changing his position, proposed the "recognized hazards" language of the Senate bill and said:

> Our modified General Duty * * * differ[s] in important respects from the General Duty provision in the Steiger substitute.
>
> The first difference is that my amendment protects against "recognized" hazards while the Steiger substitute protects only against "readily apparent" ones.
>
> ***
>
> * * * I am afraid that "readily apparent" used in the substitute means apparent without investigation, even though a prudent employer would investigate under the circumstances. A danger, in other words, may be recognized as such in the industry, but may not be apparent to an employer who is ill-informed and does not choose to investigate the danger of the situation. That is not sufficient protection for employees.

Legislative History at 1007, [116 CONG. REC. 38,377].

The House replaced "readily apparent hazards" with "recognized hazards." 3 U.S. CODE CONG. & ADMIN. NEWS at 5229, Conference Report No. 91-1765, Statement of the Managers on the Part of the House (1970). On December 17, 1970, Representative Steiger explained what he thought "recognized hazards" meant, defining it not to include "[h]azards which require technical or testing devices to detect them * * *." 116 CONG. REC. 1189 (daily ed. Dec. 17, 1970) (quoted more fully *supra*). The House went on to pass the bill with the words "recognized hazards," as originally amended in the Senate. 29 U.S.C. §654(a)(1).

***

In this case, the Commission by a divided vote has first expressed its view as follows:

> It is also clear that Congress by rejecting the readily apparent hazards test and accepting the recognized hazards test in its place intended that non-obvious hazards be within the scope of the general duty requirement. There can be no question that non-obvious hazards include those that can only be detected by instrumentation.

Chairman Moran dissented. He first said that Senator Javits' views, including his approach to "non-obvious hazards," were expressed two months before enactment of OSHA and before a conference committee report which "reconciled the differing versions of the general duty requirement." Chairman Moran placed heavy reliance on the fact that "[n]o Member of the House of Representatives or the Senate, at anytime subsequent to the conference committee report took issue with Representative Steiger's view that the general duty clause would not cover hazards requiring technical or testing devices to detect them." (footnote omitted). The fact that no senator or representative expressed a view on Representative Steiger's comments does not necessarily compel a conclusion that Congress agreed with Representative Steiger.

We find Petitioner and Chairman Moran's views unpersuasive. Looking to the words of the Act itself, "recognized hazards" was enacted instead of "readily apparent hazards." From the commonly understood meanings of the terms themselves, "recognized" denotes a broader meaning than "readily apparent." The remarks of Senator Javits demonstrate that non-obvious hazards were intended to fall within the meaning of "recognized hazards." Further, the House earlier had the opportunity to accept Representative Steiger's proposal of "readily apparent," but chose not to. Undoubtedly, it is clear that Representative Steiger desired that the general duty clause be limited to "readily apparent hazards." We think that it is also clear that Representative Steiger according to his comments on December 17, 1970, interpreted "recognized hazards" to mean apparent only to the human senses. In short, we conclude that Representative Steiger interpreted "recognized hazards" in the same manner that he would have interpreted "readily apparent hazards," but that Congress decided otherwise by enacting the words "recognized hazards" with its meaning in the context of Senator Javits' and Representative Daniels' statements.

We further think that the purpose and intent of the Act is to protect the health of the workers and that a narrow construction of the general duty clause would endanger this purpose in many cases. To expose workers to health dangers that may not be emergency situations and to limit the general duty clause to dangers only detectable by the human senses seems to us to be a folly. Our technological age depends on instrumentation to monitor many conditions of industrial operations and the environment. Where hazards are recognized but not detectable by the senses, common sense and prudence demand that instrumentation be utilized. Certain kinds of health hazards, such as carbon monoxide and asbestos poisoning, can only be detected by technical devices. 29 C.F.R. §§1918.83(a) and 1910.93a. The Petitioner's contention, though advanced by arguable but loose legislative interpretation, would have us accept a result that would ignore the advances of industrial scientists, technologists, and hygienists, and also ignore the plain wording, purpose, and intent of this Act. The health of workers should not be subjected to such a narrow construction.

Chairman Moran also based his dissent on the rationale that the use of the general duty clause for the type of alleged violation in this case circumvents the spirit and purposes of the Act, which places much reliance on the promulgation of specific standards. * * *

***

In this case, we cannot agree with either Petitioner or Chairman Moran that the general duty clause should not have been available for citing the alleged violation. On April 28, 1971, the entire Act became effective. On May 29, 1971, the Secretary published in the *Federal Register* the standard of .2 mg/M^3 of maximum allowable concentrations of airborne lead particles, 36 FED. REG. 10466 (1971), which became effective August 27, 1971. 28 C.F.R. §1910.93, Table G-2. The Secretary's representatives inspected the Omaha plant on June 30, 1971; the citation was issued on July 7, 1971; and a pretrial hearing was held on September 28, 1971.

Chairman Moran claims that the Secretary "jumped the gun" in issuing this citation under the general duty clause. That observation may well be apt in certain cases, but not in this one. The evidence in this case strongly demonstrates that .2 mg/M^3 of airborne lead has been a nationally recognized standard of safety for years. Although the Secretary could be said to have "jumped the gun" in relation to enforcing a specific standard under consideration as a violation of the general duty clause, he certainly did not "jump the gun" in recognizing that the national standard of .2 mg/M^3 of airborne lead has been recognized as the safe level. This is not a situation in which a new danger is involved or in which there is any serious debate on a national standard of industrial health.

***

In light of the above evidence, we reject the position that use of the general duty clause in this case destroys the spirit and purpose of the Act. Though a specific standard for concentration of lead was under consideration at the time the Secretary cited the Petitioner for the alleged violation, the standard for airborne lead relied upon by the Secretary had been recognized for years. Dr. [Robert A.] Kehoe's chapter [titled "Industrial Lead

Poisoning" in Industrial Hygiene and Toxicology, Frank A. Patty, editor (1962)] agreeing on that same .2 mg/M[3] of airborne lead was written in 1962. Lead poisoning is an ancient ill, not a new danger for which new standards and technology would have to be developed. The general duty clause should still be available for use by the Secretary, when an employer is violating a standard of health concerning a well-known industrial ill that is recognized as a hazard by the industry through its own nationally accepted criteria, even though a specific standard is under consideration at the time the Secretary cites an employer for violation under the general duty clause.

The Petitioner's second major contention is that it has instituted many protective measures that prevent the likelihood of harm to the employees. We agree with the Administrative Law Judge that Petitioner's program has not reduced the likelihood of serious physical harm.

Petitioner does have a program to attempt to eliminate the dangers of lead poisoning. It has spent over $400,000 since 1960 on various engineering controls to reduce airborne lead contamination. Industrial hygienists and physicians are employed. Respirators are supplied, but admittedly are awkward to wear and seldom used.

Most important in the Petitioner's view is a reliance on a biological monitoring program, which involves the testing of each employee's blood and urine to determine the concentration of lead. * * *

Although a carefully conducted biological monitoring system might prevent the likelihood of lead poisoning harm to employees, we think it was more than reasonable for the Secretary to rely on the effective and efficient air sampling method. In addition, the disadvantages of the biological sampling system are demonstrated in this case. About 10 percent of employees tested from 1970–71 were found to have unsafe levels of lead concentration in their blood and urine, yet generally these employees were not tested frequently enough, according to Petitioner's expert testimony, to ascertain whether they should be changed to another working area in the plant. In fact, the plant manager had no direct involvement with the monitoring plan. The plant's physician did not "follow up" with what happened to individual employees who had high concentrations of lead in their blood and urine. Further, disruption of employees' working habits and the plant operation would result from transferring employees to new positions within the plant where exposure would be lessened. Most significantly, the record does not indicate that any employee was transferred due to high levels of lead concentration discovered by biological monitoring. The biological monitoring did not eliminate or even reduce the hazard; it merely disclosed it. Although testing of the blood and urine is the most important test for each individual, the use of air sampling tests is the most efficient and practical way for the Secretary to check for a hazard likely to cause death or serious physical harm to the workers as a group. We think it also the most efficient manner for the employer to check the existence of a hazard. We do not intend to minimize the importance of the biological monitoring program; obviously, medical examinations of the individual are the most significant manner of assuring safety of the *individual* worker, provided that remedial steps are taken at the first indication of a hazardous concentration of lead accumulating in the blood and urine of a worker. The evidence showed that each human responds differently to exposure to lead. Yet we think that the safety of the workers and the practicality of detecting unsafe levels of airborne lead concentrations are best served by the air sampling method. Workers should not be subjected to hazardous concentrations of airborne lead; biological monitoring should complement an industrial hygiene program for clean or at least safe air, but is not a substitute for a healthful working environment.

In addition, the Petitioner knew or should have known that the respirators would not reduce the likelihood of serious physical harm to the employees. During the unannounced tour of the plant by the Secretary's representatives, only one employee was properly wearing his respirator. The reasonable inference is that employees rarely used the awkward and uncomfortable respirators. It was reasonably foreseeable to the Petitioner that the respirators would not be properly worn.[20] We hold that there was adequate evidence on the record considered as a whole that the biological monitoring program would not prevent a likelihood of harm to employees.

***

As the court opinion notes, the OSHA §6(a) standard regulating lead did not become effective until August 27, 1971. The PEL under the standard was 0.2 milligrams per cubic meter of air ($0.2mg/M^3$) or 200 micrograms per cubic meter. On November 14, 1978, OSHA issued a new standard on lead under §6(b), setting a permissible exposure limit of 0.005 milligrams per cubic meter ($0.005\ mg/M^3$).[15]

The *American Smelting & Refining Company* case illustrates in detail the sequence of events in an OSHA health enforcement case.

1. OSHA representatives conduct a physical inspection of the employer's workplace. In this case, the inspection resulted from a complaint filed by the United Steelworkers of America.[16] Since this was a health inspection, the compliance officers conducted personal air sampling during the tour of the workplace to determine the levels at which employees were exposed to airborne lead.[17]

2. Shortly after the inspection, OSHA issued a citation against American Smelting and Refining Company (ASARCO). The citation issued under §9(a)[18] alleged a serious violation[19] of §5(a)(1); the citation described the violation and allowed ASARCO 60 days to imple-

[20]The Regulations require the use of respirators "[w]hen effective engineering controls are not feasible." 29 C.F.R. §1910.134(a)(1). The control of airborne contaminants must first be "accomplished as far as feasible by accepted engineering control measures." 29 C.F.R. §1910.134(a)(1).

[15]43 FED. REG. 52,952 (1978). The OSHA standard is stated in terms of micrograms (50 ug/m^3).

[16]501 F.2d at 505. OSHA is required to respond to certain complaints under §8(f)(1), 29 U.S.C. §657(f)(1). See discussion in Chapter 12.

[17]In December 1982, OSHA amended its enforcement regulations expressly to authorize compliance officers to use personal sampling devices (i.e., air sampling pumps and noise dosimeters) and to attach these devices to employees during workplace inspections. 47 FED. REG. 55,478 (1982) (codified at 29 C.F.R. §1903.7 (1983)). OSHA stated that there was a "longstanding belief that personal sampling generally provides the most accurate measure of an employee's exposure to noise or air contaminants." *Id.* at 55,478. OSHA's authority to perform personal sampling had previously been drawn into question by court decisions in *Plum Creek Lumber Co. v. Hutton,* 608 F.2d 1283 (9th Cir. 1979) and *In re Establishment Inspection of Metro-East Mfg. Co.,* 655 F.2d 805 (7th Cir. 1981). See also Industrial Hygiene Manual, ch. X, Standard Methods for Sampling Air Contaminants, OSHR [Reference File 77:8301]. OSHA's authority to issue the 1982 regulation was upheld in *Service Foundry Co., Inc. v. Donovan,* 721 F.2d 492 (5th Cir. 1983).

[18]29 U.S.C. §658(a).

[19]Serious violations are defined under §17(k), 29 U.S.C. §666(k). See Section C, this chapter.

ment feasible engineering controls to reduce airborne lead levels. OSHA also proposed a civil penalty of $600.[20]

3. The employer contested the citations and penalty and an evidentiary hearing was held before an administrative law judge of the Occupational Safety and Health Review Commission.[21]

4. After the close of the hearing, the administrative law judge issued a "report" affirming the serious violations and the penalty, and giving ASARCO six months from the entry of the final order to achieve abatement.

5. ASARCO petitioned the Review Commission to review the report of the administrative law judge. The Review Commission agreed to review the report[22] and on August 17, 1973 issued its decision affirming the administrative law judge's decision, except that unlike the judge, it refused to rule on the constitutional issues.[23] Chairman Moran of the Commission dissented.

6. ASARCO then sought a review of the Review Commission decision in the Court of Appeals for the Eighth Circuit.[24] The court, applying the "substantial evidence" test,[25] upheld the Review Commission decision. In the litigation before both the Review Commission and the court of appeals, a major issue was whether the employer was required to abate the hazard by means of engineering controls; ASARCO claimed that its air and biological monitoring program was sufficient with its use of respirators to abate the lead hazard and protect employees. Both the Review Commission and the court rejected ASARCO's contentions.[26]

The *American Smelting & Refining Company* decision effectively settled the issue that §5(a)(1) also covers hazards, if otherwise recognized, whose detection requires monitoring devices, and this policy has consistently been followed by OSHA in its enforcement activity.[27]

The threshold condition for application of the general duty clause is that there is no standard covering the hazard involved. This

[20]The range of OSHA penalties is set forth in §17, 29 U.S.C. §666.

[21]Sec. 10, 29 U.S.C. §659, contains the procedure for contest, and §12, 29 U.S.C. §661, describes the Review Commission and its procedures.

[22]Under §12(j), 29 U.S.C. §661(j), a single Commission member may direct that an administrative law judge report be reviewed.

[23]1 OSHC 1256 (Rev. Comm'n 1973).

[24]See §11(a), 29 U.S.C. §660(a) (judicial review in the courts of appeals is available for "any person adversely affected or aggrieved by an order of the [Review] Commission"). For a discussion of the right of employee representatives to petition for review of Commission decisions, see Chapter 17, Section D.

[25]See 29 U.S.C. §660(a).

[26]OSHA's standards in the past have required engineering controls as the primary means by which employers achieve the permissible exposure limit. But see Advance Notice of Proposed Rulemaking, Methods of Compliance, 48 FED. REG. 7473 (1983) (reexamining the policy favoring the use of engineering controls). On whether engineering controls are required to eliminate general duty violations, see *Kelly Springfield Tire Co.*, 10 OSHC 1970, 1975, n.5 (Rev. Comm'n 1982), *aff'd on other grounds sub nom. Kelly Springfield Tire Co. v. Donovan*, 11 OSHC 1891 (5th Cir. 1984).

[27]Earlier versions of the Field Operations Manual expressly stated that a hazard could be recognized "even if it is not detectable by means of the senses." Field Operations Manual, CPL 2.45 (1980) at VIII-2. This principle is implicit in the 1983 field instructions, Field Operations Manual, ch. IV, OSHR [Reference File 77:2703–:2704].

principle, based on the statement in the report of the Senate Labor and Public Welfare Committee that general duty was designed to protect employees "who are working under special circumstances for which no standard has yet been adopted,"[28] was stated by the Review Commission in an early case[29] and has been uniformly accepted since then. As OSHA stated in its Field Operations Manual,[30] this principle precludes OSHA from using §5(a)(1) to impose stricter requirements than those required by a standard, or an abatement method stricter than that required by a standard where the standard covers the hazard.

The basic requirements for a finding of a violation under the general duty clause were stated as far back as 1973 by the Court of Appeals for the District of Columbia Circuit in *National Realty & Construction Co. v. OSHRC*.[31] The court said that "under the clause, the Secretary must prove (1) that the employer failed to render the workplace 'free' of a hazard which was (2) 'recognized' and (3) 'causing or likely to cause death or serious physical harm.'"[32] This formula has often been restated with approval by the Review Commission[33] and the courts of appeals.[34] The court of appeals in *National Realty* discussed at some length the parameters of a general duty violation, explaining the basic requirements quoted above.

National Realty & Construction Company v. OSHRC
489 F.2d 1257 (D.C. Cir. 1973)

J. SKELLY WRIGHT, Circuit Judge:

***

On September 24, 1971 the Secretary cited National Realty for serious breach of its general duty

> in that an employee was permitted to stand as a passenger on the running board of an Allis Chalmers 645 Front end loader while the loader was in motion.[11]

[11]In addition to referring to the general duty clause by its statute section number, the citation described the violation as follows:

> On or about Sept. 16, 1971, employer did fail to provide his employee employment which was free from recognized hazards which caused and were likely to cause death or serious physical harm to his employees in that an employee was permitted to stand as a passenger on the running board of an Allis Chalmers 645 Front end loader while the loader was in motion. As the loader was proceeding down an unguarded earthen ramp, it went out of control and the employee-passenger jumped from the loader and was fatally injured when the loader rolled over on him.

[28]Reprinted *supra*, this section.
[29]*Brisk Waterproofing Co.*, 1 OSHC 1263 (Rev. Comm'n 1973).
[30]Ch. IV, OSHR [Reference File 77:2705–:2706], reprinted in this section.
[31]489 F.2d 1257 (D.C. Cir. 1973).
[32]*Id.* at 1265.
[33]In *Bomac Drilling Corp.*, 9 OSHC 1681, 1691 (Rev. Comm'n 1981), the Review Commission stated OSHA's burden as showing that "an employer failed to render its workplace free from a hazard that is recognized, the occurrence of an incident was reasonably foreseeable, and the likely consequence in the event of an incident was death or serious physical harm to employees." *Bomac* was reversed in part by *United States Steel Corp.*, 10 OSHC 1752 (Rev. Comm'n 1982) (Cottine, Comm'r, dissenting on this issue), holding that "reasonable foreseeability of an incident" was not a "distinct element" of a general duty violation.
[34]See, e.g., *Titanium Metals Corp. of Am. v. Usery*, 579 F.2d 536 (9th Cir. 1978).

After National Realty filed timely notice of contest, the Secretary entered a formal complaint charging that National Realty had

> permitted the existence of a condition which constituted a recognized hazard that was likely to cause death or serious physical harm to its employees. Said condition, which resulted in the death of foreman O.C. Smith, arose when Smith stood as a passenger on the running board of a piece of construction equipment which was in motion.

At an administrative hearing, held before an examiner appointed by the Commission, William Simms, the Labor Department inspector who cited National Realty, testified in person, and counsel read into the record a summary of stipulated testimony by several employees of National Realty. The evidence is quickly restated.

On September 16, 1971, at a motel construction site operated by National Realty in Arlington, Virginia, O.C. Smith, a foreman with the company, rode the running board of a front-end loader driven by one of his subordinates, Clyde Williams. The loader suffered a stalled engine while going down an earthen ramp into an excavation and began to swerve off the ramp. Smith jumped from the loader, but was killed when it toppled off the ramp and fell on him. John Irwin, Smith's supervisor, testified that he had not seen the accident, that Smith's safety record had been very good, that the company had a "policy" against equipment riding, and that he—Irwin—had stopped the "4 or 5" employees he had seen taking rides in the past two years. The loader's driver testified that he did not order Smith off the vehicle because Smith was his foreman; he further testified that loader riding was extremely rare at National Realty. Another company employee testified that it was contrary to company policy to ride on heavy equipment. A company supervisor said he had reprimanded violators of this policy and would fire second offenders should the occasion arise. Simms, the inspector, testified from personal experience that the Army Corps of Engineers has a policy against equipment riding. He stated he was unaware of other instances of equipment riding at National Realty and that the company had "abated" its violation. Asked to define abatement, Simms said it would consist of orally instructing equipment drivers not to allow riding.

The hearing examiner dismissed the citation, finding that National Realty had not "permitted" O.C. Smith to ride the loader, as charged in the citation and complaint. The examiner reasoned that a company did not "permit" an activity which its safety policies prohibited unless the policies were "not enforced or effective." Such constructive permission could be found only if the hazardous activity were a "practice" among employees, rather than—as here—a rare occurrence. Upon reviewing the hearing record, the Commission reversed its examiner by a 2–1 vote, each commissioner writing separately. Ruling for the Secretary, Commissioners Burch and Van Namee found inadequate implementation of National Realty's safety "policy." Rejecting the hearing examiner's factual findings in part, Commissioner Burch stated that it was "incredible" that an oral safety policy could have reduced equipment riding to a rare occurrence. Commissioner Van Namee reasoned that the Smith incident and the "4 or 5" occurrences shown on the record "put respondent on notice that more was required of it to obtain effective implementation of its safety policy." The majority commissioners briefly suggested several improvements which National Realty might have effected in its safety policy: placing the policy in writing, posting no-riding signs, threatening riders with automatic discharge, and providing alternative means of transport at the construction site. In dissent, Commissioner Moran concluded that the Secretary had not proved his charge that National Realty had "permitted" either equipment riding in general or the particular incident which caused Smith's death.

II. THE ISSUES

Published regulations of the Commission impose on the Secretary the burden of proving a violation of the general duty clause.[23] When the Secretary fails to produce evidence on all necessary elements of a violation, the record will—as a practical consequence—lack substantial evidence to support a Commission finding in the Secretary's favor. That is the story of this case. It may well be that National Realty failed to meet its general duty under the Act, but the Secretary neglected to present evidence demonstrating in what manner the company's conduct fell short of the statutory standard. Thus the burden of proof was not carried, and substantial evidence of a violation is absent.

***

B. *The Statutory Duty to Prevent Hazardous Conduct by Employees*

* * * [T]he Secretary could properly have produced evidence at the hearing on the question whether National Realty's safety policy failed, in design or implementation, to meet the standards of the general duty clause. Under the clause, the Secretary must prove (1) that the employer failed to render its workplace "free" of a hazard which was (2) "recognized" and (3) "causing or likely to cause death or serious physical harm." The hazard here was the dangerous activity of riding heavy equipment. The record clearly contains substantial evidence to support the Commission's finding that this hazard was "recognized"[32] and "likely to cause death or serious physical harm."[33] The question then is whether National Realty rendered its construction site "free" of the hazard. In this case of first impression, the meaning of that statutory term must be settled before the sufficiency of the evidence can be assessed.

Construing the term in the present context presents a dilemma. On the one hand, the adjective is unqualified and absolute: A workplace cannot be just "reasonably free" of a hazard, or merely as free as the average workplace in the industry.[34] On the other hand, Congress quite clearly did not

[23]29 C.F.R. §2200.33 (Jan. 1, 1972).

[32]An activity may be a "recognized hazard" even if the defendant employer is ignorant of the activity's existence or its potential for harm. The term received a concise definition in a floor speech by Representative Daniels when he proposed an amendment which became the present version of the general duty clause:

> A recognized hazard is a condition that is known to be hazardous, and is known not necessarily by each and every individual employer but is known taking into account the standard of knowledge in the industry. In other words, whether or not a hazard is "recognized" is a matter for objective determination; it does not depend on whether the particular employer is aware of it.

116 CONG. REC. (Part 28) 38377 (1970). The standard would be the common knowledge of safety experts who are familiar with the circumstances of the industry or activity in question. The evidence below showed that both National Realty and the Army Corps of Engineers took equipment riding seriously enough to prohibit it as a matter of policy. Absent contrary indications, this is at least substantial evidence that equipment riding is a "recognized hazard."

[33]Presumably, any given instance of equipment riding carries a less than 50% probability of serious mishap, but no such mathematical test would be proper in construing this element of the general duty clause. See Morey, *The General Duty Clause of the Occupational Safety and Health Act of 1970,* 86 HARV. L. REV. 988, 997–998 (1973). If evidence is presented that a practice could eventuate in serious physical harm upon other than a freakish or utterly implausible concurrence of circumstances, the Commission's expert determination of likelihood should be accorded considerable deference by the courts. For equipment riding, the potential for injury is indicated on the record by Smith's death and, of course, by common sense.

[34]Though the Senate and House committee reports on the Act stated that the general duty clause incorporated "common law principles," the standard of care imposed by the clause was not characterized in terms of reasonableness. Rather, the reports speak of a

> general and common duty to bring *no* adverse effects to the life and health of their employees throughout the course of their employment.

H.R. Rep.No.91-1291, [Committee on Education and Labor, 91st Cong., 2d Sess.] 21; S.Rep. No.91-1282, [91st Cong., 2d Sess.] 9, U.S. CODE CONG. & ADMIN. NEWS, 1970, p. 5186. (Emphasis added.) Overtones of the reasonableness standard are to be found only in the Act's definition

intend the general duty clause to impose strict liability: The duty was to be an achievable one.[35] Congress' language is consonant with its intent only where the "recognized" hazard in question *can be* totally eliminated from a workplace. A hazard consisting of conduct by employees, such as equipment riding, cannot, however, be totally eliminated. A demented, suicidal, or willfully reckless employee may on occasion circumvent the best conceived and most vigorously enforced safety regime.[36] This seeming dilemma is, however, soluble within the literal structure of the general duty clause. Congress intended to require elimination only of preventable hazards. It follows, we think, that Congress did not intend unpreventable hazards to be considered "recognized" under the clause. Though a generic form of hazardous conduct, such as equipment riding, may be "recognized," unpreventable instances of it are not, and thus the possibility of their occurrence at a workplace is not inconsistent with the workplace being "free" of recognized hazards.

Though resistant to precise definition, the criterion of preventability draws content from the informed judgment of safety experts. Hazardous conduct is not preventable if it is so idiosyncratic and implausible in motive or means that conscientious experts, familiar with the industry, would not take it into account in prescribing a safety program. Nor is misconduct preventable if its elimination would require methods of hiring, training, monitoring, or sanctioning workers which are either so untested or so expensive that safety experts would substantially concur in thinking the methods infeasible.[37] All preventable forms and instances of hazardous con-

---

of a *serious* violation. That employers must take more than merely "reasonable" precautions for the safety of employees follows from the great control which employers exert over the conduct and working conditions of employees. See H.R. Rep.No.91-1291, *id., and* S.Rep. No.91-1282, *id.*

[35]The Act's purpose is "to assure *so far as possible* every working man and woman in the Nation safe and healthful working conditions * * *." 29 U.S.C. §651. (Emphasis added.) The House committee report, dealing with an earlier but not pertinently distinct version of the general duty clause, stated:

> An employer's duty under Section 5(1) is not an absolute one. It is the Committee's intent that an employer exercise care to furnish a safe and healthful place to work and to provide safe tools and equipment. This is not a vague duty, but is protection of the worker from preventable dangers.

H.R. Rep.No.91-1291, *supra* note [34], at 21. These persuasive indications aside, the very word duty implies an obligation capable of achievement. See Restatement (Second) of Torts §4 (1965).

[36]This is not to say that an employer's statutory responsibility for a hazard vanishes, or is even diminished, because the hazard was directly caused by an employee. The Act provides "that employers and employees have separate but dependent responsibilities and rights with respect to achieving safe and healthful working conditions." 29 U.S.C. §651(b)(2). An employer has a duty to prevent and suppress hazardous conduct by employees, and this duty is not qualified by such common law doctrines as assumption of risk, contributory negligence, or comparative negligence.

> The committee does not intend the employee-duty [to comply with the occupational safety and health standards promulgated under the Act] provided in section 5(b) to diminish in anyway [sic] the employer's compliance responsibilities or his responsibility to assure compliance by his own employees. Final responsibility for compliance with the requirements of this act remains with the employer.

S.Rep. No.91-1282, *supra* note [34], at 10–11, U.S. CODE CONG. & ADMIN. NEWS, 1970, p. 5187. The employer's duty is, however, qualified by the simple requirement that it be achievable and not be a mere vehicle for strict liability.

[37]This is not to say that a safety precaution must find general usage in an industry before its absence gives rise to a general duty violation. The question is whether a precaution is recognized by safety experts as feasible, not whether the precaution's use has become customary. Similarly, a precaution does not become infeasible merely because it is expensive. But if adoption of the precaution would clearly threaten the economic viability of the employer, the Secretary should propose the precaution by way of promulgated regulations, subject to advance industry comment, rather than through adventurous enforcement of the general duty clause. Finally, in an abundance of caution we emphasize that an instance of hazardous employee conduct may be considered preventable even if no employer could have detected the conduct, or its hazardous character, at the moment of its occurrence. Conceivably, such conduct might have been precluded through feasible precautions concerning the hiring, training, and sanctioning of employees.

duct must, however, be entirely excluded from the workplace. To establish a violation of the general duty clause, hazardous conduct need not actually have occurred, for a safety program's feasibly curable inadequacies may sometimes be demonstrated before employees have acted dangerously. At the same time, however, actual occurrence of hazardous conduct is not, by itself, sufficient evidence of a violation, even when the conduct has led to injury. The record must additionally indicate that demonstrably feasible measures would have materially reduced the likelihood that such misconduct would have occurred.

C. *Deficiencies in This Record*

The hearing record shows several incidents of equipment riding, including the Smith episode where a foreman broke a safety policy he was charged with enforcing. It seems quite unlikely that these were unpreventable instances of hazardous conduct. But the hearing record is barren of evidence describing, and demonstrating the feasibility and likely utility of, the particular measures which National Realty should have taken to improve its safety policy. Having the burden of proof, the Secretary must be charged with these evidentiary deficiencies.

The Commission sought to cure these deficiencies *sua sponte* by speculating about what National Realty could have done to upgrade its safety program. These suggestions, while not unattractive, came too late in the proceedings. An employer is unfairly deprived of an opportunity to cross-examine or to present rebuttal evidence and testimony when it learns the exact nature of its alleged violation only after the hearing. As noted above, the Secretary has considerable scope before and during a hearing to alter his pleadings and legal theories. But the Commission cannot make these alterations itself in the face of an empty record. To merit judicial deference, the Commission's expertise must operate upon, not seek to replace, record evidence.

***

Because the Secretary did not shoulder his burden of proof, the record lacks substantial evidence of a violation, and the Commission's decision and order are, therefore, Reversed.

In *National Realty,* the court of appeals addressed a number of general duty issues that continue to be litigated to the present time.

• When is a hazard "recognized"? The court said that "[t]he standard would be the common knowledge of safety experts who are familiar with the circumstances of the industry or activity in question";[35]

• When is a hazard "likely to cause" death or serious physical harm? The court said that the clause encompasses hazards, "which could eventuate in serious physical harm upon other than a freakish or utterly implausible concurrence of circumstances";[36]

• Does the general duty clause cover unpreventable hazards? The court said that the general duty clause did not impose "strict liability" but required only the elimination of "preventable hazards," and that hazardous conduct by employees is not preventable "if it is so idiosyncratic and implausible in motive or means that conscien-

[35]489 F.2d at 1265 n.32.
[36]*Id.* at 1265 n.33.

tious experts familiar with the industry would not take it into account in prescribing a safety program.";[37]

• What steps must an employer take to free his workplace from recognized hazards? The D.C. Circuit Court of Appeals stated that an employer must utilize measures "recognized by safety experts as feasible" to abate general duty hazards and that a safety precaution need not find "general usage" in an industry before its absence gives rise to a general duty violation.[38]

The last-mentioned issue—whether under §5(a)(1) an employer may be required to undertake feasible abatement measures not generally adopted by the relevant industry—has been the subject of sharp disagreement in the courts of appeals. The view of the court of appeals in *National Realty* that an employer's obligations are not limited by industry practice has been accepted by the Review Commission in a number of cases[39] and by several courts of appeals. In *General Dynamics Corp. v. OSHRC*,[40] for example, the court said:

> [W]e cannot agree with [the employer's] position that the measure of the adequacy of its safety program should be that of the industry. Such a standard would allow an entire industry to avoid liability by maintaining inadequate safety training. The purpose of the Act is to require all employers to take all feasible steps to avoid industrial accidents. While the definition of a "recognized hazard" should be made in reference to industry knowledge, by virtue of the definition of the word "recognized" we cannot accept a standard of the precautions which should be taken against such a hazard which is any less than the maximum feasible.[41]

A contrary view has been taken by some courts, notably the Court of Appeals for the Fifth Circuit in a series of cases dealing with the analogous issue of enforcement of OSHA's broadly worded personal protective equipment standards.[42]

S & H Riggers & Erectors v. OSHRC

659 F.2d 1273 (5th Cir. 1981)

GODBOLD, Chief Judge:

In these consolidated cases under the Occupational Safety and Health Act, 29 U.S.C. §651 *et seq.*, we are again presented with the question whether an employer whose conduct is in compliance with the custom and practice of its industry may be found to have violated 29 C.F.R. §1926.-

---

[37]*Id.* at 1265–66. Employee misconduct as a defense to OSHA citations is discussed in Chapter 15, Section E.

[38]*Id.* at 1266 n.37.

[39]E.g., *Bomac Drilling Corp.*, 9 OSHC 1681 (Rev. Comm'n 1981); *Southern Ry. Co.*, 3 OSHC 1657 (Rev. Comm'n 1975).

[40]599 F.2d 453 (1st Cir. 1979).

[41]*Id.* at 464.

[42]29 C.F.R. §1910.132(a) (1983) (for general industry); *id.* at §1926.28(a) (1983) (for the construction industry).

28(a),[1] the primary OSHA regulation pertaining to the provision of personal protective equipment to employees in the construction industry. We have re-examined our earlier cases on this issue, and we reaffirm our conclusion in these cases that, at least in the absence of a clear articulation by the Occupational Safety and Health Review Commission of the circumstances in which industry practice is not controlling, due process requires a showing that the employer either failed to provide personal protective equipment customarily required in its industry or had actual knowledge that personal protective equipment was required under the circumstances of the case.[2]

***

Before turning to the orders of the Commission in these cases it is helpful to examine briefly the history of §1926.28(a) in the Fifth Circuit. This court's struggle to define the requirements of due process in connection with broadly worded OSHA regulations began with *Ryder Truck Lines, Inc. v. Brennan,* 497 F.2d 230 (5th Cir. 1974), a case involving 29 C.F.R. §1910.132(a),[5] the general industry analog to §1926.28(a). In *Ryder* we rejected the employer's argument that §1910.132(a) lacked any ascertainable standard of conduct and was therefore void for vagueness:

> The regulation appears to have been drafted with as much exactitude as possible in light of the myriad conceivable situations which could arise and which would be capable of causing injury. Moreover, we think inherent in that standard is an external and objective test, namely, whether or not a reasonable person would recognize a hazard of foot injuries to dockmen, in a somewhat confined space, from falling freight and the rapid movement of heavy mechanical and motorized equipment, which would warrant protective footwear. *So long as the mandate affords a reasonable warning of the proscribed conduct in light of common understanding and practices, it will pass constitutional muster.*

497 F.2d at 233 (emphasis added).

Our next encounter with this issue was *B&B Insulation, Inc. v. OSHRC,* 583 F.2d 1364 (5th Cir. 1978), a case under §1926.28(a). Adopting the approach we had taken in *Ryder,* we again sustained the regulation against a vagueness attack. In doing so, however, we elaborated on the content of the "reasonable person" standard applicable to §§1910.132(a) and

[1]The regulation provides:

> The employer is responsible for requiring the wearing of appropriate personal protective equipment in all operations where there is an exposure to hazardous conditions or where this part indicates the need for using such equipment to reduce the hazards to the employees.

We have previously recognized that although §1926.28(a) refers only to the "wearing" of personal protective equipment, "merely wearing safety equipment, but not using it, does not comply with [§1926.28(a)]," *Turner Communications Corp. v. OSHRC,* 612 F.2d 941, 944 (5th Cir. 1980), and "that a 'safety belt is only used as a safety device when it is both worn and attached by a lanyard or lifeline to a stationary object.'" *Marshall v. Southwestern Industrial Contractors & Riggers, Inc.,* 576 F.2d 42, 44 (5th Cir. 1978). Accordingly, in this opinion unless the context otherwise requires, references to the "wearing" of safety belts include proper tying off.

[2]The analysis in this opinion applies equally to citations issued under 29 C.F.R. §1910.132(a), the general industry analog to §1926.28(a). See *Owens-Corning Fiberglass Corp. v. Donovan,* No. 79-2516, 659 F.2d 1285 (5th Cir. Oct. 26, 1981), decided this day. See also *Cotter & Co. v. OSHRC,* 598 F.2d 911, 913 (5th Cir. 1979); *B & B Insulation, Inc. v. OSHRC,* 583 F.2d 1364, 1369 (5th Cir. 1978).

[5]29 C.F.R. §1910.132(a) provides:

> Protective equipment, including personal protective equipment for eyes, face, head, and extremities, protective clothing, respiratory devices, and protective shields and barriers, shall be provided, used, and maintained in a sanitary and reliable condition wherever it is necessary by reason of hazards of processes or environment, chemical hazards, radiological hazards, or mechanical irritants encountered in a manner capable of causing injury or impairment in the function of any part of the body through absorption, inhalation, or physical contact.

1926.28(a). We noted that the "reasonable person" standard is borrowed from tort law and that industry custom is not dispositive on the issue of the standard of care in negligence actions. 583 F.2d at 1370. We concluded, however, that rigid application of the tort law concept would be inconsistent with the preventive goals of OSHA and Congress's expressed preference for specific rather than general standards.[6]

> [T]he employer whose activity is not yet addressed by a specific regulation and whose conduct conforms to the common practice of those similarly situated in his industry should generally not bear an extra burden.
>
> Where the Government seeks to encourage a higher standard of safety performance from the industry than customary industry practices exhibit, the proper recourse is to the standard-making machinery provided in the Act, selective enforcement of general standards being inappropriate to achieve such a purpose.

Id. at 1371 (footnotes omitted).

Because we found no evidence that employers in the insulation industry would customarily provide safety belts where B&B Insulation had not, and thus that B&B had no notice that its conduct might be in violation of §1926.28(a), we reversed the Commission's affirmance of the citation.

In *Power Plant Division v. OSHRC,* 590 F.2d 1363 (5th Cir. 1979), we again applied the reasoning of *B&B Insulation* to reverse a Commission decision upholding a citation under §1926.28(a). Unlike the record in *B&B Insulation,* however, in which there was substantial evidence of industry practices and of the employer's compliance with those practices, the record in *Power Plant* contained little or no evidence of industry custom. For that reason we remanded the case to the Commission for further factfinding.

Finally, in *Cotter & Co. v. OSHRC,* 598 F.2d 911 (5th Cir. 1979), we reaffirmed our earlier conclusion in *B&B Insulation* that due process imposed essentially the same requirements on §§1910.132(a) and 1926.28(a) and reversed a citation under the general industry standard. In the process, however, we recognized that where an employer has actual knowledge of the requirements imposed by a regulation the problem of fair notice does not exist, and also recognized that an employer who has actual knowledge that personal protective equipment is necessary to protect its employees from a particular hazard may be found in violation of §1910.132(a) even though its conduct complies fully with the general practice of the industry. We reversed the citation, however, because we found "no evidence in the record of

---

[6]Speaking in support of a substitute bill that influenced the wording of the general duty clause of the Act, 29 U.S.C. §654(a)(1), Congressman Steiger warned of the dangers of rigid application of tort concepts in the OSHA context.

> Another argument offered in support of the committee bill's general requirement is that it is comparable to the general duty of care imposed in the law of torts. This argument is also unpersuasive. Tort law is concerned with providing for after-the-fact payment of damages by one whose negligent act actually caused an injury. To borrow merely the isolated general duty of care from the field of tort law and impose it, along with the sanctions of civil penalties, to provide a before-the-injury method of preventing the occurrence of industrial accidents, strikes me as incongruous. In tort law the general duty of care does not exist in isolation. It is surrounded by other factors which sharply limit it, and thus give it real meaning and practical application in the field of law in which it is used. Centuries of Anglo-American case law have refined the general duty of care through the judicial development of doctrines to serve as guides for the careful application of the general duty. Also, elaborate defenses have been developed to limit its otherwise unjust application. If we are to include any sort of general-care duty in this legislation, Mr. Chairman, we should also limit its terms so that persons upon whom it would impose a duty are not unjustly held accountable for situations of which they are completely unaware.

116 CONG. REC. 38371 (1970).

a specific, confirmed knowledge on Cotter's part regarding a hazard warranting [mandatory personal protective equipment]." *Id.* at 915.

In its opinion in *S&H I* a majority of the Commission for the first time reached an agreement on the standard to be applied in reviewing citations under §1926.28(a). The Commission held that:

> [t]he crucial question in determining whether a hazardous condition exists within the meaning of §1926.28(a) is whether a reasonable person familiar with the factual circumstances surrounding the allegedly hazardous condition, including any facts unique to a particular industry, would recognize a hazard warranting the use of personal protective equipment. * * * Although industry custom and practice are useful points of reference with respect to whether a reasonable person familiar with the circumstances would recognize a hazard requiring the use of personal protective equipment, they are not controlling.

7 OSHC 1260, 1263 (footnote and citations omitted).

***

Due process mandates that an employer receive notice of the requirements of any OSHA regulation before he is cited for an alleged violation. A construction contractor or subcontractor forced to guess at the requirements of §1926.28(a) cannot accurately determine the cost of safety equipment in preparing contract bids. If he believes that certain safety equipment is required when it is not, his bid may be noncompetitive with others that did not include the cost of such equipment, and he may lose the job. *B & B Insulation,* 583 F.2d at 1367 n.4. Conversely, if he determines that certain equipment is not required he may suffer crippling cost overruns if he is later required to provide the equipment. Many, if not most, of the regulations promulgated under OSHA are sufficiently specific concerning the circumstances in which safety precautions must be taken that adequacy of notice is not a significant problem. The generality of §1926.28(a), however, mandates that it be applied only in such a manner that an employer may readily determine its requirements by some objective external referent.

***

* * * [T]he Secretary's argument misses the point of our holding in *B & B Insulation.* We did not hold that industry custom is controlling in all OSHA cases, or that the Secretary cannot impose standards more stringent than those customarily followed in an industry. We merely held that he cannot do so under as general and broadly worded a regulation as §1926.28(a). Moreover, the Secretary overlooks that due process requires not only that employers be aware of a hazard but also that they have notice of what is required of them under the regulation in response to that hazard. The Secretary argues that where an obvious hazard exists, an employer is under an obligation to provide any personal protective equipment that is feasible and that would eliminate or reduce the hazard. While the Secretary is free to pursue this goal through other means, he cannot do so through selective enforcement of §1926.28(a) because that regulation provides insufficient notice of such a requirement.

***

Our adherence to a dispositive industry practice standard in applying §1926.28(a) need not create a significant risk that an entire industry will escape liability under OSHA by failing on an industry-wide basis to provide adequate safety measures, thus endangering the health and safety of their employees and frustrating the will of Congress, because enforcement of an employer's obligation under OSHA is not limited to §1926.28(a). If the Secretary believes that the custom and practices in an industry make inadequate provision for employee safety, he is empowered by the Act to promulgate regulations setting higher safety standards than those currently recognized.[12] All we held in *B&B Insulation*, and all we hold today, is that the Secretary cannot impose standards more stringent than those customarily followed in an industry under as general and broadly worded a regulation as §1926.28(a).

***

The reasoning of the Fifth Circuit Court of Appeals in *S & H Riggers* that due process notice requirements mandate that the broadly worded standard impose only those obligations adopted by the industry would appear to apply to the general duty clause.[43] Substantially the same view was advanced by Robert C. Gombar, formerly general counsel of the Review Commission, who has argued that OSHA must promulgate a standard under §6(b) if it "wants to change an industry's practice."[44] Differing views on OSHA's enforcement of the general duty clause were expressed at a meeting of the National Advisory Committee on Occupational Safety and Health (NACOSH)[45] in January 1980, and in a memorandum submitted to OSHA by an employer association.

[12]We recognize that the decision whether to fill the interstices in a statutory scheme by rulemaking or by ad hoc adjudication "is one that lies primarily in the informed discretion of the administrative agency," *SEC v. Chenery Corp.*, 332 U.S. 194, 203 (1947), although rulemaking is generally to be preferred whenever possible, *id.* at 202. Accordingly, we do not hold that rulemaking is the only available route to imposing greater personal protective equipment requirements than are customary in an industry. It may be that the Commission can, on a case-by-case basis, adequately articulate the circumstances in which §1926.28(a) imposes requirements above and beyond industry practices so as to give employers sufficient advance notice of those requirements. * * *

[43]Other courts have held that the general duty clause was not unconstitutionally vague in light of the requirement that hazards covered must be "recognized." See, e.g., *Bethlehem Steel Corp. v. OSHRC*, 607 F.2d 1069 (3d Cir. 1979). The Court of Appeals for the Third Circuit, disagreeing with the Fifth Circuit view on enforcement of the personal protective equipment standard, has applied the "reasonably prudent person test," which is similar to the test applied by the Review Commission in *S & H Riggers, supra*, and by the court of appeals in *National Realty*, 489 F.2d at 1266 n.37. *Voegele Co. v. OSHRC*, 625 F.2d 1075 (3d Cir. 1980). The same approach was adopted by the First Circuit Court of Appeals in *Cape & Vineyard Div. of New Bedford Gas & Light v. OSHRC*, 512 F.2d 1148 (1st Cir. 1975). Accord *General Dynamics Corp. v. OSHRC*, 599 F.2d 453, 464 (1st Cir. 1979).

A number of Review Commission and courts of appeals cases have construed the term "recognized" as used in §5(a)(1), 29 U.S.C. §654(a)(1). See, e.g., *H-30, Inc. v. Marshall*, 597 F.2d 234 (10th Cir. 1979) (industry recognition); *Titanium Metals Corp. of Am. v. Usery*, 579 F.2d 536, 541 (9th Cir. 1978) (recognition established by voluntary industry standards); *Usery v. Marquette Cement Mfg. Co.*, 568 F.2d 902 (2d Cir. 1977) ("common sense" recognition of hazards); *Brennan v. OSHRC and Vy Lactos Laboratories, Inc.*, 494 F.2d 460 (8th Cir. 1974) (actual employer knowledge of the hazard). See also, 1983 OSHA Field Operations Manual, reprinted in part in this section. In 1980, OSHA issued an instruction to field staff, announcing that it had determined that employee exposure to certain benzidine-based dyes, not covered by standards, were "recognized" hazards under §5(a)(1). The directive stated that this recognition had been established by studies by the National Cancer Institute, and that the Institute's findings had been "widely published and disseminated by employers using the dyes." The directive also described the studies in detail. OSHR [Reference File 21:9163–:9174].

[44]Gombar, *OSHRC Interprets General Duty Clause Too Broadly, Legal Times of Washington*, Nov. 30, 1981 at 13.

[45]NACOSH was established under §7(a), 29 U.S.C. §656(a), to advise and consult with the Secretaries of Labor and Health and Human Services on matters relating to their administration of the Act.

Transcript of Meeting of National Advisory Committee on Occupational Safety and Health[46]

DR. DIXON: I really hope this [is] explored further because I do feel there have been some very bad abuses and companies who have been trying to do a good job have been penalized. I know in our own case of one instance in which one of our plants we had set our own internal standard for exposure to a given chemical that was not on the TLV list or whatnot. We set it extremely low to provide a very wide margin of safety and it was an objective.

And OSHA came in and happened to find us exceeding it by 50 percent at one point. They cited us under the general duty clause as it being a severe hazard likely to cause death or injury. Well, it was no such thing of that, of course. We contested the case and we won our position.

But it pointed out something that was very bad. Some of our plant people now say, gee, we don't want to have standards if OSHA is going to come in and use our own standards which we set in a very [bona fide] effort to be responsible and by some chance we are exceeding it to a minor degree and they cite this under general duty clause as a severe life-threatening hazard, it is nonproductive for us to have that kind of level.

I don't think that should be allowed to exist and I hope we can get this defined very clearly and get it out into the open.

DR. GINNOLD: I think Ernie [Dixon] has a very legitimate concern. My experience with general duty, at least in Region 5 OSHA where I think is probably the most aggressive use of general duty of any region, I haven't seen it used in that way. One of the main uses is to fill the gap in certain major areas which don't have standards. For instance, confined space entry and lockouts. Confined spaces are killing 250 people a year and there is no standard. There hasn't been since 1970.

Lockouts are another major cause of injuries. And application of hazardous pesticides. Those are some of the main examples and I would like in this (sic), I think it is great that we have gotten into this discussion because I think, because of the standard-setting process, we may never be able to write enough standards to cover all the unsafe situations in a workplace, probably not.

I think general duty in a judicious sound way can really provide a lot of safety that standards won't. I would like to get to the concrete on some of our discussion of general duty and maybe if Grover Wrenn* comes up here, I know there are documents floating around of examples of general duty citations. Maybe we could be apprised of some of the major areas where general duty is being used. Because I think it is really interesting.

The important figure on this also I think is not the cases that are contested, but I would say perhaps 80 percent or more of the general duty citations that are issued that are settled voluntarily. That employers accept and they result in correction of a hazard that wouldn't be corrected otherwise. I think that is important.

DR. CHAFFIN: I guess I would like to pick up on the chilling aspect of the overzealous use if I might categorize that potential. It exists, but I see the

*[Author's note—Mr. Wrenn was director of OSHA compliance operations when this meeting took place. He was previously director of health standards.]

[46]At 76–79, Jan. 15, 1980. Transcript available at Department of Labor Occupational Safety and Health Administration, Washington, D.C. The individuals speaking, members of NACOSH, were Dr. Ernest Dixon, Corporate Medical Director, Celanese Corporation; Dr. Richard Ginnold, Assistant Professor, School for Workers, University of Wisconsin; and Dr. Don Chaffin, Professor, Industrial Engineering, University of Michigan. The transcript has been edited in minor aspects to aid in comprehensibility.

real potential here, what Ernie [Dixon] has pointed out as even having an effect on some of the field research that is so necessary here to develop new standards wherein I am finding more and more industries that in the past had a fairly receptive attitude to the point of allowing good epidemiological work to go on and they were even supporting that kind of epidemiological work.

Now saying that one of the companies in their industry group has been cited under general duty for the kind of hazard that the epidemiology is supposed to be working on and all of a sudden data sources are being shut off to a number of epidemiologists. It is very real. I can cite you several cases along those lines.

It is not a hypothetical in that sense. The chilling effect on field research is real.

***

Memorandum from Synthetic Organic Chemical Manufacturers Association to OSHA[47]

***

The general duty clause has been used improperly in at least three ways:

1. to enforce stricter requirements than those in effect in specific standards;
2. to give the force of law to NIOSH recommendations and other unofficial suggested standards;
3. to attack any perceived employer shortcoming, no matter how trivial, that is not covered by a specific standard.

Such misuse causes several major problems:

1. It undercuts OSHA's incentives to promulgate specific standards which address all aspects of particular hazards. Thus, it denies employers and employees the procedural protections afforded them by rulemaking, and it denies OSHA the informed comment of interested persons.
2. It forces employers to second-guess all OSHA standards to determine whether something more is required. Without rulemaking, employers lack notice of the legal requirements they must meet, in violation of their Due Process rights.
3. It arbitrarily places on a single employer, often a small employer, the burden of rebutting OSHA's assertion that a particular requirement is both necessary and feasible.
4. It diverts to wasteful litigation limited resources which would be better spent on protecting employee safety and health.

Despite the prohibitions against use of the general duty clause as a substitute for rulemaking, OSHA continues to issue general duty clause citations for hazards covered by existing standards where OSHA apparently regards those standards as inadequate. OSHA's Industrial Hygiene Field Operations Manual encourages this practice, stating at page II-16, for example:

> In some cases an 8-hour OSHA PEL exists but no OSHA *ceiling* standard exists. If a ceiling standard has been recommended by NIOSH or

[47]The Proper Role of The General Duty Clause in OSHA Enforcement Policy at 7–16 (May 1, 1981).

others, which is equal to or greater than the 8-hour OSHA PEL, a 5(a)(1) citation may be considered for exposures in excess of the recommended ceiling standard (even if the employer is in compliance with the 8-hour OSHA PEL).

In many instances, OSHA field offices go beyond even this Manual in their decisions to issue general duty clause citations for hazards to which standards apply.

The result is that OSHA continues to issue improper general duty clause citations, thereby burdening the cited employers, hurting employees by diverting employer safety and health resources from productive uses, and wasting OSHA's scarce enforcement resources. * * *

Adherence to the statutory limits on use of the general duty clause means, among other things, that OSHA should no longer cite employers under the general duty clause for failure to comply with exposure limits recommended by NIOSH or similar agencies. A NIOSH criteria document, for example, usually meets none of the requirements for a general duty citation.

* * * OSHA should be particularly wary of using the general duty clause to regulate suspect carcinogens. Regulation of these substances involves, among other things, questions of science, technology, and policy that ought to be addressed on a nationwide basis, with the full opportunity for public comment and participation that the rulemaking process provides. In recognition of this, OSHA has several times promulgated standards for substances which it believes to be occupational carcinogens. These rulemakings generated thousands of pages of testimony and comments by hundreds of interested persons. They dramatize the general inappropriateness of OSHA trying to classify substances as carcinogens in individual general duty clause proceedings, each involving only a single employer.

Also, OSHA should no longer cite an employer under the general duty clause for failure to comply with the employer's self-imposed workplace rules. Voluntary adoption of additional safeguards of employee health and safety does not mean that a hazard exists in the employer's workplace which is causing or likely to cause death or serious physical harm. It also does not mean that the employer recognizes the hazard addressed by its own rules to be one which poses such a risk. OSHA has cited employers for failure to enforce completely their internal rules. These general duty clause citations are not only improper; they also take away incentives employers might otherwise have to provide greater protections for employees than OSHA requires. By discouraging voluntary efforts to employers, these general duty clause citations actually tend to diminish health and safety in the workplace.

***

On March 17, 1982, Assistant Secretary Thorne G. Auchter issued a field instruction containing detailed "guidelines-limitations" on use of the general duty clause in OSHA enforcement.[48] The "Background" section of the instruction contains the following statement:

2. In enacting this [general duty] clause, the Congress indicated that: * * * the general duty clause in this bill would not be a general substitute for reliance on standards, but would simply enable the Secretary to insure the protection of employees who are working

[48]OSHA Instruction CPL 2.50, OSHR [Reference File 21:8279–:8284].

under special circumstances for which no standard has yet been adopted. S. Rep. No. 91-1282, 91st Cong., 2d Sess. 10 (1970).
3. Statutory language and legislative history also indicate that the mere existence of a hazard was not intended to establish necessarily an enforceable obligation to abate under the general duty clause. It was Congress' intent to use the general duty clause when the hazard is recognized, serious and capable of being corrected by feasible means.
4. OSHA's use of the general duty clause has grown to about 3,200 citations in Fiscal Year 1980. With this growth, there has been increased criticism of the agency for alleged misuse of Section 5(a)(1) in situations not contemplated by Congress or as a substitute for standard-setting. Although this criticism is not justified in all respects, OSHA has reviewed the 5(a)(1) policy and is now restating this policy to provide more detailed guidance.[49]

The instruction discussed all the major elements of a §5(a)(1) violation, i.e., definition of "hazard," the requirement for "recognition," the likelihood of death or serious physical harm, and the requirement that the hazard be "corrected by a feasible and useful method." It stated the circumstances under which §5(a)(1) may be cited and the type of evidence necessary to support general duty violations before the Commission. The instruction's guidelines were incorporated in OSHA's Field Operations Manual, issued April 18, 1983.[50]

Field Operations Manual

***

a. *Evaluation of Potential 5(a)(1) Situations.* In general, Review Commission and court precedent has established that the following elements are necessary to prove a violation of the general duty clause:

(1) The employer failed to keep the workplace free of a hazard to which employees of that employer were exposed;

(2) The hazard was recognized;

(3) The hazard was causing or was likely to cause death or serious physical harm; and

(4) There was a feasible and useful method to correct the hazard.

b. *Discussion of 5(a)(1) Elements.* * * *

(1) *A Hazard To Which Employees Were Exposed.* * * *

(a) *Hazard.* A hazard is a danger which threatens physical harm to employees.

***

(b) *The Hazard Must Affect the Cited Employer's Employees.* The employees affected by the Section 5(a)(1) hazard must be the employees of the cited employer.

1 An employer who may have created and/or controlled the hazard normally shall not be cited for a Section 5(a)(1) violation if his own employees are not exposed to the hazard.

2 In complex situations, such as multi-employer worksites, where it may be difficult to identify the precise employment relationship between the employer to be cited and the exposed employees, the Area Director shall

[49] *Id* at 21:8280.
[50] Chapter IV—Violations, OSHR [Reference File 77:2702–:2707].

consult with the Regional Administrator and Regional Solicitor to determine the sufficiency of the evidence regarding the employment relationship.

***

(2) *The Hazard Must be Recognized.* Recognition of a hazard can be established on the basis of industry recognition, employer recognition, or "common-sense" recognition. The use of common-sense as the basis for establishing recognition shall be limited to special circumstances. Recognition of the hazard must be supported by satisfactory evidence and adequate documentation in the file as follows:

(a) *Industry Recognition.* A hazard is recognized if the employer's industry recognizes it. Recognition by an industry other than the industry to which the employer belongs is generally insufficient to prove this element of a Section 5(a)(1) violation. Although evidence of recognition by the employer's specific branch within an industry is preferred, evidence that the employer's industry recognizes the hazard may be sufficient. The Area Director shall consult with the Regional Administrator on this issue. Industry recognition of a particular hazard can be established in several ways:

1 Statements by industry safety or health experts which are relevant to the hazard.

2 Evidence of implementation of abatement methods to deal with the particular hazard by other members of the industry.

3 Manufacturer's warnings on equipment which are relevant to the hazard.

4 Statistical or empirical studies conducted by the employer's industry which demonstrate awareness of the hazard. Evidence such as studies conducted by the employee representatives, the union or other employees should also be considered if the employer or the industry has been made aware of them.

5 Government and insurance industry studies, if the employer or the employer's industry is familiar with the studies and recognizes their validity.

6 State and local laws or regulations which apply in the jurisdiction where the violation is alleged to have occurred and which currently are enforced against the industry in question. In such cases, however, corroborating evidence of recognition is recommended.

***

7 Standards issued by the American National Standards Institute (ANSI), the National Fire Protection Agency (NFPA), and other private standard-setting organizations, if the relevant industry participated on the committee drafting the standards. Otherwise, such private standards normally shall be used only as corroborating evidence of recognition. Preambles to these standards which discuss the hazards involved may show hazard recognition as much as, or more than, the actual standards. It must be emphasized, however, that these private standards cannot be enforced like OSHA standards. They are simply evidence of industry recognition, seriousness of the hazard or feasibility of abatement methods.

8 NIOSH criteria documents; the publications of EPA, the National Cancer Institute, and other agencies; OSHA hazard alerts; the IHFOM; and articles in medical or scientific journals by persons other than those in the industry, if used only to supplement other evidence which more clearly establishes recognition. Such publications can be relied upon only if it is established that they have been widely distributed in general, or in the relevant industry.

(b) *Employer Recognition.* A recognized hazard can be established by evidence of actual employer knowledge. Evidence of such recognition may consist of written or oral statements made by the employer or other management or supervisory personnel during or before the OSHA inspection.

1 Company memorandums, safety rules, operating manuals or operating procedures and collective bargaining agreements may reveal the employer's awareness of the hazard. In addition, accident, injury and illness reports prepared for OSHA, workmen's compensation, or other purposes may show this knowledge.

2 Employee complaints or grievances to supervisory personnel may establish recognition of the hazard, but the evidence should show that the complaints were not merely infrequent, off-hand comments.

3 The employer's own corrective action may serve as the basis for establishing employer recognition of the hazard if the employer did not adequately continue or maintain the corrective action or if the corrective action did not afford any significant protection to the employees.

(c) *Common-sense Recognition.* If industry or employer recognition of the hazard cannot be established in accordance with (a) and (b), recognition can still be established if it is concluded that any reasonable person would have recognized the hazard. This theory of recognition shall be used only in flagrant cases.

***

(3) *The Hazard Was Causing or Was Likely to Cause Death or Serious Physical Harm.* * * *

***

(c) In a health context, establishing serious physical harm at the cited levels may be particularly difficult if the illness will require the passage of a substantial period of time to occur. Expert testimony is crucial to establish that serious physical harm will occur for such illnesses. It will generally be easier to establish this element for acute illnesses, since the immediacy of the effects will make the causal relationship clearer. In general, the following must be shown to establish that the hazard causes or is likely to cause death or serious physical harm when such illness or death will occur only after the passage of a substantial period of time:

1 Regular and continuing employee exposure at the workplace to the toxic substance at the measured levels reasonably could occur:

2 Illness reasonably could result from such regular and continuing employee exposure; and

3 If illness does occur, its likely result is death or serious physical harm.

(4) *The Hazard May Be Corrected by a Feasible and Useful Method.* To establish a Section 5(a)(1) violation the agency must identify a method which is feasible, available and likely to correct the hazard. The information shall indicate that the recognized hazard, rather than a particular accident, is preventable.

(a) If the proposed abatement method would eliminate or significantly reduce the hazard beyond whatever measures the employer may be taking, a Section 5(a)(1) citation may be issued. A citation shall not be issued merely because the agency knows of an abatement method different from that of the employer, if the agency's method would not reduce the hazard significantly more than the employer's method. It must also be noted that in some cases only a series of abatement methods will alleviate a hazard. In such a case all the abatement methods shall be mentioned.

***

d. *Limitations of Use of the General Duty Clause.* Section 5(a)(1) is to be used only within the guidelines given in 2.a.

(1) *Section 5(a)(1) Shall Not Be Used When a Standard Applies to a Hazard.* Both 29 CFR 1910.5(f) and legal precedent establish that Section

5(a)(1) may not be used if an OSHA standard applies to the hazardous working condition.

***

(2) *Section 5(a)(1) Shall Not Be Used To Impose a Stricter Requirement Than That Required by the Standard.* For example, if the standard provides for a TLV of 5 ppm, even if data establishes that a 3 ppm level is a recognized hazard, Section 5(a)(1) shall not be cited to require that the 3 ppm level be achieved. If the standard has only a time-weighted average permissible exposure level and the hazard involves exposure above a recognized ceiling level, the Area Director shall consult with the Regional Administrator who shall discuss any proposed citation with the Regional Solicitor and the Director of Field Operations.

(3) *Section 5(a)(1) Shall Normally Not be Used to Require an Abatement Method Not Set Forth in a Specific Standard.* A specific standard is one that refers to a particular toxic substance or deals with a specific operation, such as welding. If a toxic substance standard covers engineering control requirements but not requirements for medical surveillance, Section 5(a)(1) shall not be cited to require medical surveillance.

(4) *Section 5(a)(1) Shall Not Be Used to Enforce "Should" Standards.* If a Section 6(a) standard or its predecessor, such as an ANSI standard, uses the word "should," neither the standard nor Section 5(a)(1) shall ordinarily be cited with respect to the hazard addressed by the "should" portion of the standard.

***

In revising its instructions on §5(a)(1) enforcement, OSHA considered public criticism of its prior policies, including those of the Synthetic Organic Chemical Manufacturers Association (SOCMA). The operations manual gave instructions to avoid use of the general duty clause to impose stricter requirements than those in effect under specific standards and stated that NIOSH criteria documents and similar materials should be used "*only* to supplement other evidence which more clearly establishes recognition" and not independently to establish recognition. Also, the 1982 directive deleted the general duty instructions in the Industrial Hygiene Field Operations Manual.[51] On the other hand, the manual does not directly limit use of the general duty for citing carcinogens and, contrary to the SOCMA suggestion, permits the use of voluntary abatement actions by an employer to establish the "feasibility and usefulness" of an abatement method.[52]

[51]OSHA Instruction CPL 2.50 para. C(1), OSHR [Reference File 21:8279]. SOCMA had objected to the use of §5(a)(1) to cite for excursions above the ceiling where the standard provides an eight-hour time-weighted average.

[52]29 C.F.R. §1910.5(f) (1983) contains another interpretation of the general duty clause, stating that an employer in compliance with a standard shall be deemed to be in compliance with the requirements of §5(a)(1) "to the extent" that the hazard is covered by the standard. A related interpretation in §1910.5(c) (1983) states: "If a particular standard is specifically applicable to a condition, practice, means, methods, operation or process, it shall prevail over any different general standard which might otherwise be applicable to the same condition." The meaning of this provision has been litigated frequently in the context of OSHA's steel erection standards. Employers have argued that under the above interpretation, the requirements of OSHA's general construction industry standards are preempted as to the steel erection industry by OSHA's more specific standard imposing safety requirements for the steel erection industry. The issue has arisen because the steel erection standard only requires fall protection in

An unusual and controversial use of the general duty clause occurred in the *American Cyanamid Company* case, involving an alleged employer policy excluding women between 16 and 50, who had not been surgically sterilized, from employment in areas of the plant involving lead exposure.

American Cyanamid Company

9 OSHC 1596 (Review Comm'n 1981)

BY THE COMMISSION:
***

I

In January 1978, Respondent, American Cyanamid Company, announced to employees of its Willow Island, West Virginia, pigments manufacturing plant that it intended to implement what it called the "fetus protection policy." This policy, as ultimately implemented excluded women aged 16 through 50 from production jobs in the lead pigments department unless they could prove that they had been surgically sterilized. The stated purpose of the policy was to protect the fetuses of women exposed to lead, particularly during early pregnancy when the employee might not know of her pregnancy.[3] Between February and July 1978, five women employed in the lead pigments department submitted to surgical sterilization in a hospital not associated with Respondent. In September 1978, two women who had not been sterilized were transferred out of the lead pigments department. By October 2, 1978, no woman of childbearing capacity was employed in the department.

In response to a complaint filed by the Oil, Chemical and Atomic Workers Union and its Local 3-499 in December 1978, an Occupational Safety and Health Administration ("OSHA") inspection was conducted between January 4 and April 13, 1979. In October 1979, a citation was issued charging Respondent with a willful violation of the general duty clause and proposing a penalty of $10,000. Respondent timely filed a notice of contest. The Secretary's citation and complaint alleged that Respondent failed to

> furnish employment and a place of employment which were free from recognized hazards that were causing or were likely to cause death or serious physical harm to employees, in that: the employer adopted and

[3]The parties stipulated that scientific studies indicate that lead has profoundly adverse effects on the course of reproduction both before and after conception; that fetuses are susceptible at all stages of development, including the first trimester when the mother may not know she is pregnant; and that maternal blood lead levels cannot be reduced instantaneously when pregnancy is discovered.

erections higher than 30 feet, 29 C.F.R. §1926.750(b) (1983), while the general construction standard requires protection for falls of less than 30 feet, 29 C.F.R. §1926.28(a) (1983), and §1926.105(a) requires safety nets for falls of more than 25 feet. The Review Commission, see *Tippens Steel Erection Co.,* 11 OSHC 1428 (Rev. Comm'n 1983), and the courts, see, e.g., *L.R. Willson & Sons, Inc. v. OSHRC,* 698 F.2d 507 (D.C. Cir. 1983), have rejected this argument, holding that the "general" standard "complements" the requirements of the specific standard, quoting *Bristol Steel & Iron Works, Inc. v. OSHRC,* 601 F.2d 717, 721 (4th Cir. 1979). See also OSHA Instruction STD 3-31, OSHR [Reference File 21:8309] which interprets the OSHA standards at 29 C.F.R. §1926.28(a) and §1926.105(a) as they apply to the protection of employees exposed to falling hazards in the construction industry. Under the interpretation, in general, construction falling hazards of 10 to 25 feet not covered by a specific standard are governed by §1926.28(a) while those of more than 25 feet are addressed by §1926.105(a).

implemented a policy which required women employees to be sterilized in order to be eligible to work in those areas of the plant where they would be exposed to certain toxic substances.
***

III

In our view, the critical point is not that the physical impact of the policy on employees, reduced functional capacity due to surgical sterilization, was ultimately achieved outside the workplace. We agree with CRROW [Coalition for the Reproductive Rights of Workers (a non-party intervenor)] that it is employer implementation of the policy that is at issue here and to that extent the policy is a condition of employment within the meaning of the Act. However, we believe that Respondent is correct in arguing that Congress did not intend the Act to apply to every conceivable aspect of employer-employee relations and that due to its unique characteristics this condition of employment is not a hazard with[in] the meaning of the general duty clause.[14]
***

Congressional floor debates, committee reports, and individual and minority views reported in the legislative history are replete with discussions of air pollutants, industrial poisons, combustibles and explosives, noise, unsafe work practices and inadequate safety training, and the like. The effects on employees which Congress hoped to alleviate are described in general terms such as accident, disease, industrial injury, reduced life expectancy, crippling, maiming, disablement and death, and in specific terms such as cancer, allergy, heart disease, respiratory impairment, chemical poisoning, burns, broken bones, and the like. Repeated reference is made to the fact that congressional action with regard to occupational safety and health received its impetus from the vast numbers of on-the-job injuries and deaths reported each year. In the words of Congressman Anderson, "the worker's surroundings and the conditions under which he works are of crucial importance in the whole environmental question for it is in this environment that he spends one third of his day. The air he breathes and the tools and materials he handles can pose a direct threat to his health, safety, and well-being if adequate precautions are not taken. This is really what we are talking about today in considering the need for national industrial health and safety standards." From this it is clear that Congress conceived of occupational hazards in terms of processes and materials which cause injury or disease by operating directly upon employees as they engage in work or work-related activities.[22]

The fetus protection policy is of a different character altogether. It is neither a work process nor a work material, and it manifestly cannot alter the physical integrity of employees while they are engaged in work or work-related activities. An employee's decision to undergo sterilization in order to gain or retain employment grows out of economic and social factors which operate primarily outside the workplace. The employer neither controls nor creates these factors as he creates or controls work processes and materials. For these reasons we conclude that the policy is not a hazard within the meaning of the general duty clause.

[14]In his citation and complaint the Secretary alleges that the fetus protection policy *requires* sterilization as a condition of employment. Inasmuch as it is impossible for an employer literally to compel employees to undergo sterilization, we believe it is more accurate to describe the policy as one excluding women of childbearing capacity from certain employment.

[22]In the ten years of the Act's existence, the Secretary has not, to our knowledge, cited an employer for a condition or practice that did not operate directly upon employees engaged in work or work-related activities.

It is true, as the Secretary points out, that the Act is broad in scope and may be fairly described as intended to protect employees from reduced functional capacity as a result of the work experience. However, it does not follow that the general duty clause applies to an employment policy whose physical impact on employees is indirect and derives not from work processes and materials but from social and economic factors outside the workplace.

Although a statute can operate prospectively so as to include circumstances unknown at the time of enactment, the rule of progressive construction applies only where the language of the statute, as illuminated by legislative history and other extrinsic aids, can be read fairly to include the unforeseen circumstances. Whether an unforeseen circumstance can be read fairly to fall within the ambit of a particular statute depends not only on policy considerations, but on whether that circumstance is "substantially comparable" to those actually contemplated by Congress. This is so because the legislature does not merely enact general policies. "By the terms of a statute, it also indicates its conception of the sphere within which the policy is to have effect." The definition of hazard proposed by the Secretary is not, under any fair reading of the legislative history, substantially comparable to the concept of hazard entertained by Congress in passing the Occupational Safety and Health Act of 1970. Accordingly, we hold that the citation in this case fails to allege a violation cognizable under the general duty clause of the Act and vacate that citation.

***

Cottine, Commissioner, dissenting:

***

The Secretary's allegations in this case are directed toward a policy that the majority acknowledges to be a condition of employment. Nevertheless, my colleagues conclude that because the policy has an indirect physical impact on employees, deriving from social and economic factors outside the workplace that the employer neither creates nor controls, it is not cognizable as a workplace hazard under the Act. Corporate policy that offers employees a choice between jobs and surgical sterilization is comparable to a corporate policy that offers employees a choice between jobs and exposure to sterilizing chemicals. To say one choice is prohibited by the Act while the other is not is to repudiate the letter and spirit of the Act.

The Act's general duty clause, 29 U.S.C. §654(a)(1), was included in the statute to "fill those interstices necessarily remaining after the promulgation of specific safety standards," because "[i]t would be utterly unreasonable to expect the Secretary to promulgate specific safety standards which would protect employees from every conceivable hazardous condition." *Bristol Steel & Iron Works, Inc. v. OSHRC & Marshall,* 601 F.2d 717, 721 & n.11 (4th Cir. 1979). The Senate Committee on Labor and Public Welfare explained the purpose of the clause as follows:

> [P]recise standards to cover every conceivable situation will not always exist. This legislation would be seriously deficient if any employee were killed or seriously injured on the job simply because there are no specific standard[s] applicable to a recognized hazard which results in such a misfortune. Therefore, to cover such circumstances the Committee has included a requirement to the effect that employers are to furnish employment and places of employment which are free from recognized hazards to the health and safety of their employees.

S. Rep. No. 91-1282, 91st Cong., 2d Sess. 9 (1970), *reprinted in Subcomm. on Labor of the Senate Comm. on Labor & Public Welfare,* 92d Cong. 1st Sess. The House Committee on Education and Labor recognized that the absence of a general duty clause "would mean the absence of authority to cope with a hazardous condition * * * for which no standard has been promulgated." H.R. REP. No. 91-1291, 91st Cong., 2d Sess. 21 (1970). The House Committee specifically acknowledged that the general duty clause was "to provide for the protection of employees who are working under such *unique* circumstances that no standard has yet been enacted to cover [the] situation." H.R. REP. No. 91-1291, at 21–22 (emphasis in original).

Incredibly, despite this clear legislative intent, the majority concludes that "due to its *unique* characteristics this condition of employment [the Fetus Protection Policy] is not a hazard within the meaning of the general duty clause." (Emphasis added). This unnecessarily restrictive construction of the general duty clause limits its application and endangers rather than protects the health of American workers. See *American Smelting & Refining Co. v. OSHRC,* 501 F.2d 504, 511 (8th Cir. 1974). My colleagues' narrow view of the reach of the general duty clause is in direct conflict with the remedial purpose of the Act and its legislative history. It is precisely because the legislators deemed it essential that worksites be freed of recognized hazards threatening life or health that the general duty clause was included in the statute. Moreover, in enacting the general duty clause, Congress recognized the impossibility of specifically identifying and regulating all conceivable hazards that existed or would exist. See 29 U.S.C. §655(g).

Though Congress debated occupational hazards primarily in terms of processes and materials that cause injury or disease by operating directly upon employees as they engage in work or work-related activities,[7] it does not follow that Congress intended to exclude from the ambit of the Act an employment policy that offers employees a choice between work and surgical sterilization. The Act was designed to relieve employees of the choice between their jobs and their safety and health. In concluding that "[a] worker should not have to choose between his job and his life," the Sixth Circuit in *Marshall v. Whirlpool Corp.,* 593 F.2d 715, 725 (6th Cir. 1979), *aff'd,* 445 U.S. 1 (1980), noted that "[a] right to a hazard free workplace is implicit throughout the act * * * [and is] specifically contained in the Act's statement of purpose." 593 F.2d at 736.

***

One fact is inescapable in this case. Five American Cyanamid employees have been sterilized. As a matter of law, this irreversible termination of their child-bearing capacity is a material impairment of functional capacity resulting from a condition of employment imposed by their employer. This loss of reproductive capacity, whether actively or passively coerced, runs counter to the stated Congressional policy of assuring "insofar as practicable that no employee will suffer diminished health, functional capacity, or life expectancy as a result of his work experience." 29 U.S.C. §651(b)(7).

The citation and complaint raise triable issues under section 5(a)(1) of the Act and the case should be remanded for further proceedings on the merits.

OSHA did not appeal the adverse Review Commission decision to the court of appeals. The Oil, Chemical and Atomic Workers Inter-

[7]The majority states at n.22 that, to its knowledge, the Secretary has not cited employers previously for conditions or practices that did not operate directly upon employees engaged in work or work related activities. However, the Commission decisions in *C.R. Burnett & Sons, Inc. & Harllee Farms,* 9 BNA OSHC 1009, 1980 CCH OSHD ¶24,964 and *Sugar Cane Growers Cooperative of Florida,* 76 OSAHRC 62/E4, 4 BNA OSHC 1320, 1976–77 CCH OSHD ¶20,795, affirmed violations involving hazardous conditions that did not directly relate to work activities.

[Author's note—But see *Frank Diehl Farms v. Secretary,* 696 F.2d 1325 (11th Cir. 1983), discussed *supra,* note 1.]

national Union, which had become a party to the proceeding, did appeal, however. The threshold issue in the appellate proceeding was whether a union-party could seek review under §11(a) of the Act of an adverse Review Commission decision. The court of appeals held that it could, and the Supreme Court refused certiorari.[53] As of early 1984, the substantive issues in the proceeding are pending before the Court of Appeals for the District of Columbia Circuit.

In issuing a final lead standard in 1978, OSHA found that lead had "profoundly adverse effects" on the course of reproduction in both male and female employees.[54] Although OSHA stated in its preamble that a health standard "should be set which protects all persons affected—male and female workers and the fetus," the lead standard would not do so because of feasibility considerations. OSHA indicated, however, that the lead standard could "minimize" adverse reproductive effects by means of the "action level" for air and biological monitoring, the implementation of medical surveillance protection (MRP), and the education and training of employees.[55]

C. Serious Violations

The Act provides for mandatory penalties of up to $1,000 for serious violations, as defined in §17(k):[56] a violation is "serious" if there is a "substantial probability" that death or serious physical harm "could" result from a workplace condition, unless the employer did not, or could not, with reasonable diligence know of the presence of the violation. Nonserious, or as they are sometimes called, "other than serious" violations, may be penalized up to $1,000.[57] OSHA was

[53]*Oil, Chemical & Atomic Workers v. OSHRC,* 671 F.2d 643 (D.C. Cir. 1982), *cert. denied, sub nom. American Cyanamid Co. v. OCAW,* 51 USLW 3287 (1982).

[54]43 FED. REG. 52,952, 52,965–67 (1978).

[55]*Id.* at 52,966. The issue of the overlapping and potentially conflicting requirements of the OSH Act and Title VII of the Civil Rights Act of 1964, 42 U.S.C. §2000e (1976), respecting reproductive hazards has been extensively discussed. See Nothstein & Ayres, *Sex-Based Considerations of Differentiation in the Workplace: Exploring the Biomedical Interface Between OSHA and Title VII,* 26 VILL. L. REV. 239 (1981); Williams, *Firing the Woman to Protect the Fetus: The Reconciliation of Fetal Protection with Employment Opportunity Goals Under Title VII,* 69 GEO. L.J. 641 (1981); Crowell & Copus, *Safety and Equality at Odds: OSHA and Title VII Clash over Health Hazards in the Workplace,* 2 IND. REL. L.J. 567 (1978); Note, *Exclusionary Employment Practices in Hazardous Industries: Protection or Discrimination,* 5 COLUM. J. ENVTL. L. 97 (1978). See also *Genetic Screening in the Handling of High Risk Groups in the Workplace: Hearings Before Subcommittee on Investigations and Oversight of the House Comm. on Science and Technology,* 97th Cong., 1st Sess., pp. 179–219 (1981) (testimony of Joan E. Bertin, Women's Rights Project, American Civil Liberties Foundation); COUNCIL ON ENVIRONMENTAL QUALITY, CHEMICAL HAZARDS TO HUMAN REPRODUCTION (Jan. 1981). In reviewing OSHA's lead standard under §6(f), 29 U.S.C. §655(f) the District of Columbia Circuit Court of Appeals refused to address what it considered a "hypothetical" Title VII question, noting, however, that fertile women could find protection from discrimination on the basis of §6(b)(5) of the OSH Act, 29 U.S.C. §655(b)(5), requiring that "no" employee suffer material impairment of health. *United Steelworkers v. Marshall,* 647 F.2d 1189, 1238 n.74 (D.C. Cir. 1980). On the issue of the lawfulness under Title VII of the restriction of fertile women from jobs involving exposure to certain chemicals, see *Wright v. Olin Corp.,* 697 F.2d 1172 (4th Cir. 1982) (the court remanded the case to the district court for further proceedings on the applicability of the "business necessity" defense, stated in *Griggs v. Duke Power Co.,* 401 U.S. 424 (1971), to Olin's "fetal vulnerability" program).

[56]29 U.S.C. §666(j).

[57]Sec. 17(c), 29 U.S.C. §666(c).

early confronted with a basic issue in interpreting the language of §17(k): did it mean that there must be a "substantial probability" that an accident could occur which would lead to serious injury; *or* did it mean that if an accident occurred, there must be substantial probability of serious physical injury? The latter construction, which would impose a less stringent evidentiary burden on OSHA in proving serious violations, was adopted in its first compliance manual in 1972. The manual stated:

> In sum, the emphasis in deciding whether a violation is serious should be on the seriousness of the consequences of an accident (or illness) rather than on the statistical probability that the accident (or illness) will occur from the hazard. The question to be determined by the CSHO [Compliance Safety and Health Officer] is whether because of the nature of the hazard there could be an accident (or illness), and whether, as a result of such accident (or illness) it is more probable than not that there will be death or serious physical harm.[58]

This interpretation was uniformly upheld by the Review Commission and the courts. For example, in *California Stevedore & Ballast v. OSHRC,* the court of appeals said that "[w]here violation of a regulation renders an accident resulting in death or serious injury possible, however, even if not probable, Congress could not have intended to encourage employers to guess at the probability of an accident in deciding whether to obey the regulation."[59]

While there was general agreement with OSHA's definition of serious violation, field implementation practices were problematic. In September 1974, the Senate Labor and Human Resources Committee, with the assistance of the General Accounting Office, prepared 17 "issue papers"[60] on areas of enforcement activity in which, in the committee's view, OSHA "seems not to have demonstrated effectiveness."[61] One of these issue papers dealt with "Classification of Violations as Serious and Non-Serious."[62] The committee paper stated:

> In fiscal year 1974 (thru May), a total of 71,774 inspections were made in which 268,385 OSHA violations were found; and some 2,920 violations were cited by inspectors as being of serious nature. This extremely low figure of less than 1% found to be serious (as compared to all violations) is exacerbated when contrasted with results of inspections which have been triggered by reports of accidents involving disability injury or death. In 2,103 accident inspections conducted during FY 1974 (thru May), OSHA issued citations covering 3,737 violations, of which it classified over 17% or 637 as serious.
>
> The disparity between classifying as serious less than 1% of all violations and 17% of those found during investigations of fatal or catastrophic accidents is most striking. Since it must be assumed that not

[58]OSHA Compliance Manual at VIII-5 (Jan. 1972).

[59]517 F.2d 986, 988 (9th Cir. 1975).

[60]The pertinent correspondence and issue papers are printed as appendices to *Occupational Safety and Health Act Review, 1974: Hearings Before the Subcomm. on Labor of the Senate Comm. on Labor and Public Welfare,* 93d Cong., 2d Sess. apps. at 941–1238 (1974).

[61]*Id.* at 945.

[62]*Id.* at 989.

every serious violation has, in fact, caused an accident (even though by definition there is an increased likelihood), the number of total serious violations should be far greater than that which have been reported.

During fiscal year 1974 (thru May), only 0.3% of the more than 100,000 violations cited were classified as serious by the inspectors of OSHA Region #5 (Illinois, Indiana, Michigan, Minnesota, Ohio, and Wisconsin). This is in sharp contrast to over 3.5% serious found in Region #10 (Alaska, Idaho, Oregon, and Washington). The other six regions reporting during this same time period ranged between 0.8% and 2.9%.

This large differential among OSHA regions in the percentage of serious violations to the total number of violations cited suggests a need to further evaluate the Department's concept of decentralization.[63]

The concern of the committee led to a review of OSHA policy in the area which disclosed, according to OSHA, that "a major reason that few violations were being cited as serious was a pattern among field staff of misclassifying serious violations as other-than-serious."[64] As a result, Program Directive No. 200-54, containing "major clarification" of policy on serious violations, was issued in December 1976. As stated by OSHA, the purpose of the directive was "to focus attention on those hazards in the workplace which involve a significant probability of * * * serious physical harm in the event of accident or illness."[65] Two changes were made in the directive to achieve this purpose: (1) "spelling out in detail * * *. [t]he criteria and procedures for classifying a violation as serious or other-than-serious," and (2) lowering the minimum penalty for serious violations.[66] OSHA had concluded that compliance officers were "reluctant" to characterize violations as serious because before 1976 the minimum penalty required for a serious violation was $500.[67] The changes brought about the desired effect. According to statistics submitted by OSHA to a Subcommittee of the House Committee on Education and Labor, in fiscal year 1976, 2.1 percent of all violations were serious, while in fiscal year 1979, 35.5 percent of violations were serious.[68] During fiscal year 1982, 23.2 percent of all violations cited by OSHA were serious. During the first half of fiscal year 1983, 21.8 percent of all violations were serious.

The procedures for determining whether *safety* violations are serious, adopted in 1976, have continued essentially unchanged.[69] Defining serious violations for *health* hazards, on the other hand, has been more difficult. Because of the latency period between the employee's exposure and contracting the illness, a determination of whether there is a "substantial probability" of death or serious physical harm that "could" result from the violation may well involve con-

[63]*Id.* at 992–94.

[64]*Oversight Hearings on OSHA: Hearings Before the Subcomm. on Health and Safety of the House Comm. on Education and Labor,* 96th Cong., 2d Sess. pt. 4 at 679 (1980).

[65]*Id.* at 681.

[66]*Id.*

[67]*Id.*

[68]*Id.* at 679, 700.

[69]See Field Operations Manual, OSHR [Reference File 77:2708–:2709].

sideration of the frequency and the duration of the exposure of the employees involved.[70] The 1974 OSHA Field Operations Manual provided that the compliance officer should conclude that the illness that could result from the exposure to a known carcinogen is cancer;[71] and it indicated that for other air contaminants and physical agents, the inspector "may have to consider the frequency and duration of the exposure in determining the types of illness which it is reasonably predictable could result from the condition."[72] The current Field Operations Manual, issued in 1983, contains somewhat different language, but the import of the instruction appears to be the same.[73]

D. Willful Violations

OSHA has consistently defined a willful violation as one involving voluntary action either done with intentional disregard of, or "plain indifference" to, the statutory requirements; according to OSHA, it is not necessary that a violation be committed with a "bad purpose" or an "evil intent." The presently effective field instructions define willful violations in the following manner:

> **3. Willful Violations.** The following definitions and procedures apply whenever the CSHO suspects that a willful violation may exist:
>
> a. A "willful" violation may exist under the Act where the evidence shows that the employer:
>
> (1) Committed an *intentional and knowing violation of the Act* as contrasted with an inadvertent action, and/or
>
> (2) Was aware that a hazardous condition existed and did not make a reasonable effort to eliminate the condition.
>
> (3) Was aware that the condition violated a standard or other requirement of the Act and was aware of the standard or other requirement violated.
>
> b. It is not necessary that the violation be committed with a bad purpose or an evil intent to be deemed "willful." It is sufficient that the violation was deliberate, voluntary or intentional as distinguished from inadvertent, accidental or ordinarily negligent.
>
> c. The CSHO shall carefully develop and record on the OSHA-1B all evidence available that indicates employer awareness of the disre-

---

[70]See, e.g., *Usery v. Hermitage Concrete Pipe Co.*, 584 F.2d 127 (6th Cir. 1978); *Hydrate Battery Corp.*, 2 OSHC 1719 (Rev. Comm'n 1975).

[71]See *Anaconda Aluminum Co.*, 9 OSHC 1460, 1477 (Rev. Comm'n 1981) ("[T]he standard's purpose in limiting exposure to CTPV [coal tar pitch volatiles] is to protect employees against contracting a life-threatening disease, and Anaconda's failure to provide the [employee] with a respirator suitable to reduce his CTPV exposure to within the limits provided in section 1910.1000 is a serious violation."). See also *id.* at 1488 (Cottine, Comm'r, concurring and dissenting) (the rule should apply not only where OSHA has found a substance to be a carcinogen and promulgated a standard to protect workers from the cancer but also to substances whose carcinogenicity has been identified by presently available evidence even though a standard to abate the newly found hazard has not yet been issued).

[72]Field Operations Manual, ch. VIII, OSHR [Reference File 77:3104].

[73]See ch. IV, OSHR [Reference File 77:2708–:2709]. The sentence in the 1974 manual: "This [considering the frequency and duration of exposure] is not necessary for substances known as cancer-causing" was deleted in the 1983 manual. The CSHO is instead instructed to consider the Substance Toxicity Table found in the Industrial Hygiene Manual, ch. II, OSHR [Reference File 77:8033–:8072], to determine the toxicological properties of substances. The table includes substances that are listed as cancer-causing.

gard for statutory obligations or of the hazardous conditions. Willfulness could exist if an employer is advised by employees or employee representatives regarding an alleged hazardous condition and the employer does not make a reasonable effort to verify and correct the condition. Additional factors which can influence a decision as to whether violations are willful include:

(1) The nature of the employer's business and the knowledge regarding safety and health matters which could reasonably be expected in the industry.

(2) The precautions taken by the employer to limit the hazardous conditions.

(3) The employer's awareness of the Act and of the responsibility to provide safe and healthful working conditions.

(4) Whether similar violations and/or hazardous conditions have been brought to the attention of the employer.

(5) Whether the nature and extent of the violations disclose a *purposeful disregard* of the employer's responsibility under the Act.[74]

The Review Commission and the courts, with the possible exception of the Court of Appeals for the Third Circuit,[75] have agreed with the OSHA definition of willful violations,[76] although there have been a number of cases in which the courts or the Review Commission have disagreed with OSHA on the application of the general rule to the particular facts involved.[77] The discussion by the Court of Appeals for the Ninth Circuit of the willfulness issue in *National Steel & Shipbuilding Company v. OSHRC*[78] summarizes both the view of the majority of the courts and the separate view of the Third Circuit, and explains the reasons why it adopted the majority view.

National Steel & Shipbuilding Company v. OSHRC

607 F.2d 311 (9th Cir. 1979)

CHOY, Circuit Judge:

***

In finding a "willful" violation here, the Commission referred to the First Circuit's definition of "willfully" for §17(a) as enunciated in *F.X. Messina Construction Corp. v. Occupational Safety & Health Review Commission*, 505 F.2d 701 (1st Cir. 1974). The First Circuit defined willfulness as "a conscious, intentional, deliberate, voluntary decision" even if there is no bad motive. *Id.* at 702. The Fourth, Fifth, Sixth, Eighth, and Tenth Circuits have embraced similar definitions which do not require a showing of a bad motive. See, e.g., *Intercounty Construction Co. v. Occupational Safety & Health Review Commission,* 522 F.2d 777, 779–81 (4th Cir. 1975), *cert. denied,* 423 U.S. 1072 (1976); *Georgia Electric Co. v. Marshall,* 595 F.2d 309, 317–19 (5th Cir. 1979); *Empire-Detroit Steel Division v. Occupational Safety & Health Review Commission,* 579 F.2d 378, 385 (6th Cir. 1978); *Western*

[74]Field Operations Manual, ch. IV, OSHR [Reference File 77:2709–:2710].

[75]Compare *Frank Irey, Jr., Inc. v. OSHRC,* 519 F.2d 1200, 1207 (3d Cir. 1974), *aff'd en banc,* 519 F.2d 1215 (3d Cir. 1975), *aff'd on other grounds sub nom. Atlas Roofing Co. v. OSHRC,* 430 U.S. 442 (1977), with *Babcock & Wilcox v. OSHRC,* 622 F.2d 1160 (3d Cir. 1980).

[76]See, e.g., *The Ensign-Bickford Co. v. OSHRC,* 717 F.2d 1419, 11 OSHC 1657 (D.C. Cir. 1983) (Scalia, J., dissenting) (if employer shows "plain indifference" toward safety requirements of general duty clause, no further showing needed to establish willfulness).

[77]See, e.g., *Marshall v. M. W. Watson, Inc.,* 652 F.2d 977 (10th Cir. 1981).

[78]607 F.2d 311 (9th Cir. 1979).

Waterproofing Co. v. Marshall, 576 F.2d 139, 142–43 (8th Cir. 1978), *cert. denied,* 439 U.S. 965 (1979); *United States v. Dye Construction Co.,* 510 F.2d 78, 81–82 (10th Cir. 1975). The majority has also approved the precise verbal formula urged by the Secretary, defining "a willful violation as one involving voluntary action, done either with an intentional disregard of, or plain indifference to, the requirements of the statute." *Georgia Electric Co.,* 595 F.2d at 319; see *Western Waterproofing Co.,* 576 F.2d at 142–43; *Intercounty Construction Co.,* 522 F.2d at 780; *Dye Construction Co.,* 510 F.2d at 81–82. This formula is based upon a Supreme Court decision defining willfulness in the context of a different statute providing for a civil penalty. *United States v. Illinois Central Railroad Co.,* 303 U.S. 239, 243 (1938).

The Third Circuit, the only other Court of Appeals to have addressed this question, has adopted a different definition. It has written:

> It is obvious from the size of the penalty which can be imposed for a "willful" infraction—ten times that of a "serious" one—that Congress meant to deal with a more flagrant type of conduct than that of a "serious" violation. Willfulness connotes defiance or such reckless disregard of consequences as to be equivalent to knowing, conscious, and deliberate flaunting [sic] of the Act.* Willful means more than merely voluntary action or omission—it involves an element of obstinate refusal to comply.

Frank Irey, Jr., Inc. v. Occupational Safety & Health Review Commission, 519 F.2d 1200, 1207 (3d Cir. 1974), *aff'd en banc,* 519 F.2d 1215 (3d Cir. 1975), *aff'd on other grounds sub nom. Atlas Roofing Co. v. Occupational Safety & Health Review Commission,* 430 U.S. 442 (1977). The Third Circuit has subsequently applied this "flaunting" standard in the context of "repeated" violations under §17(a). *Bethlehem Steel Corp. v. Occupational Safety & Health Review Commission,* 540 F.2d 157, 161 (3d Cir. 1976).

We believe that the majority rule—which does not require a bad motive for willfulness—better serves the congressional objectives in enacting OSHA and better reflects the statute. In *Todd Shipyards Corp.* we rejected the "flaunting" test in the context of "repeated" violations, noting that such a requirement would frustrate the purposes of OSHA by placing too great a burden on the Secretary. 586 F.2d at 686. The same criticism may be made of the Third Circuit's "flaunting" standard in the context of "willful" violations. See *Georgia Electric Co.,* 595 F.2d at 319; *Intercounty Construction Co.,* 522 F.2d at 780. Additionally, we find persuasive the other arguments advanced in support of the majority definition of "willfully."

In *Intercounty Construction Co.,* the Fourth Circuit wrote:

> We agree with the position adopted by the Commission in interpreting the statute that "willful" means action taken knowledgeably by one subject to the statutory provisions in disregard of the action's legality. No showing of malicious intent is necessary. A conscious, intentional, deliberate, voluntary decision properly is described as willful, "regardless of venial motive." *F.X. Messina Construction Corp.* * * *.

522 F.2d at 779–80. In thus adopting the First Circuit test upon which the Commission relied in the instant case, the Fourth Circuit explained:

> In reaching this conclusion we are not unmindful of the holding of the Third Circuit which found willfulness in this statute, as connoting the element of "obstinate refusal to comply." *Irey* * * *. However, we

*[Author's note—For a spirited defense of the use of the term "flaunting" by the Court of Appeals for the Third Circuit, see *Babcock & Wilcox v. OSHRC,* 622 F.2d 1160, at 1167 n.6 (3d Cir. 1980). See also Safire, *On Language: In re Flaunt v. Flout, N.Y. Times Magazine,* Sept. 13, 1981 at 16–18.]

> decline to require the Commission to find such a bad purpose before it sustains a citation for willful violation. To require bad intent would place a severe restriction on the statutory authority of OSHA to apply the stronger sanctions in enforcing the law, a result we do not feel was intended by Congress. Rather, we agree with the Commission that willfulness is used in the mere cognitive sense in civil statutes, and connotes bad purpose only when an element of a criminal act.

Id. at 780. The court also indicated that its definition "follows logically from the requirements of the" OSHA. *Id.* It also is consistent with the definition of "willfulness" given in other instances of civil penalties, including that provided by the Supreme Court in *Illinois Central Railroad. Id.* Finally it gives appropriate deference to the Commission's interpretation. *Id.* at 779; see *Udall v. Tallman,* 380 U.S. 1 (1965).

Appellants retort that the Third Circuit's rule is preferable because it better distinguishes between "serious" violations and "willful" violations as those terms are used in §17 of OSHA. However, as the Fifth Circuit recently wrote:

> [T]he "bad purpose" requirement is not necessary to preserve the distinction between serious and willful violations. To provide a willful violation, the Secretary must show that the employer acted voluntarily, with either intentional disregard of or plain indifference to OSHA requirements. To prove a serious violation, a quite different showing need only be made. The gravamen of a serious violation in the presence of a "substantial probability" that a particular violation could result in death or serious physical harm. Whether the employer intended to violate an OSHA standard is irrelevant. The only question relevant to the employer's state of mind is whether he knew or with the exercise of reasonable diligence could have known of the violation. See 29 U.S.C.A. §666(j).

Georgia Electric Co., 595 F.2d at 318–19 [footnote omitted]. We have noted the same distinguishing factor between serious and other violations. *Brennan v. Occupational Safety & Health Review Commission,* 511 F.2d 1139, 1142–44 (9th Cir. 1975). We conclude that the Commission did not err in relying on the majority definition of "willfully."

***

Under §17(e) of the Act,[79] it is a criminal offense for an employer to willfully violate a standard where the violation causes the death of an employee. OSHA has taken the view that the elements of willfulness in civil and criminal contexts are identical.[80] In the only reported case interpreting §17(e), the Court of Appeals for the Tenth Circuit agreed with OSHA in holding that "neither the statute nor the regulation requires that there be moral turpitude. The object of these provisions is prevention of injury or death and its application is not limited to the situation in which the employer entertained a specific intent to harm the employee."[81]

Relatively few criminal matters have been referred by OSHA to

[79]29 U.S.C. §666(e).
[80]Field Operations Manual, OSHR [Reference File 77:2710].
[81]*United States v. Dye Constr. Co.,* 510 F.2d 78, 82 (10th Cir. 1975).

the Department of Justice for prosecution under §17(e).[82] In an article written in 1977 noting the absence of OSHA criminal prosecutions during its early years, Michael H. Levin, formerly head of the OSHA appellate unit in the Solicitor's Office in the Labor Department, recognized that criminal enforcement of agency regulations "has often proven costly and ineffective."[83] He suggested, however, that a "powerful argument can be made justifying Congress' use of OSHA's criminal provision." Levin asserted that "potential jail terms may have a devastating effect on corporate officers of more middle class background" and refers to the successful experience of the Department of Labor in enforcing the debarment sanction of the Walsh-Healey Act[84] against prime coal suppliers.[85] In any event, Levin concluded that OSHA's "prospective and remedial civil sanctions appear to be working relatively well"—a controversial proposition—and that "retrospective and punitive [sanctions] may yet afford a desirable, and perhaps indispensable, complement."[86] In contrast, the author of a Note in the 1982 *Yale Law Journal* concluded that the criminal sanction is the "most spectacular failure" of the Act;[87] the author recommended that the criminal sanction be "abandoned" and its role as OSHA's "teeth" should be filled instead by an "augmented system of civil sanctions and remedies."[88]

E. Repeated Violations

Like willful violations, repeated violations are subject to civil penalties up to $10,000.[89] Unlike willful violations, however, repeated violations have led to great differences of view among OSHA, the Review Commission, and the courts. A narrow reading of the Act on the "repeated" issue was made by the Court of Appeals for the Third Circuit in *Bethlehem Steel Corporation v. OSHRC*[90] in 1976. The court concluded that a violation would be "repeated" only if it occurred more than *twice* and constituted a "flaunting disregard of the requirements of the Act."[91] The view of the Court of Appeals for

[82]As of late 1982, only 12 criminal cases have been prosecuted under the Act since its inception. Note, *A Proposal To Restructure Sanctions Under the Occupational Safety and Health Act: The Limitations of Punishment and Culpability,* 91 YALE L.J. 1446, 1448 n.17 (1982). No incarcerations have resulted from these prosecutions, and in the five cases that have gone to trial, there have been one hung jury, two complete acquittals, and two fines of $5000 and $3500, respectively. *Id.* at 1449–50.

[83]Levin, *Crimes Against Employees: Substantive Criminal Sanctions Under the Occupational Safety and Health Act,* 14 AM. CRIM. L. REV. 717, 738–39 (1977).

[84]41 U.S.C. §37 (1976).

[85]Levin, *supra* note 83 at 739–44. The reference to "costly and ineffective" criminal sanctions is from Goldschmid, *An Evaluation of the Present and Potential Use of Civil Money Penalties as a Sanction by Federal Administrative Agencies.* In 2 *Recommendations and Reports to The Administrative Conference of the United States* 896, 917 (1972).

[86]Levin, *supra* note 83 at 745.

[87]Note, *supra* note 82 at 1448.

[88]*Id.* at 1469. See discussion in Chapter 11.

[89]Sec. 17(a), 29 U.S.C. §666(a). There are no criminal sanctions for repeated violations.

[90]540 F.2d 157 (3d Cir. 1976).

[91]*Id.* at 162. Compare the view of the Third Circuit on the meaning of "willful," stated in *Frank Irey, Jr., Inc. v. OSHRC,* 519 F.2d 1200, 1207 (3d Cir. 1974).

the Third Circuit has not been followed by other courts of appeals[92] or by OSHA.[93] In 1979, the Review Commission articulated a broad interpretation of the meaning of repeated violations in the *Potlatch* case.

Potlatch Corporation

7 OSHC 1061 (Review Comm'n 1979)

Cleary, Chairman:

***

The question of what constitutes a repeated violation has come before the Commission and the courts on a number of occasions. Although several plausible suggestions have been made by individual Commissioners and the courts, no consistent and authoritative answer has emerged. See the various administrative and judicial opinions in *George Hyman Construction Co.*, 77 OSHARC 67/C7, 5 BNA OSHC 1318, 1977–78 CCH OSHD ¶21,774 (No. 13559, 1977), *aff'd,* 582 F.2d 834 (4th Cir. 1978); and *Todd Shipyards Corp. v. Secretary of Labor,* 566 F.2d 1327, 1332 n.1 (9th Cir. 1977) (dissenting opinion) (synopsis of major decisions). Inasmuch as the announcement of authoritative guidelines is an important matter, we have thoroughly reexamined this issue in light of the decisions of the Fourth and Ninth Circuits, and we now announce the following principles.

A violation is repeated under section 17(a) of the Act if, at the time of the alleged repeated violation, there was a Commission final order against the same employer for a substantially similar violation.

The Secretary may establish substantial similarity[8] in several ways. In cases arising under section 5(a)(2) of the Act, which states that each employer shall comply with occupational safety and health standards, the Secretary may establish a prima facie case of similarity by showing that the prior and present violations are for failure to comply with the same standard. It is important to recognize that occupational safety and health standards range from those that designate specific means of preventing a hazard or hazards to those that either do not specify the means of preventing a hazard or apply to a variety of circumstances. Accordingly, in cases where the Secretary shows that the prior and present violations are for an employer's failure to comply with the same specific standard, it may be difficult for an employer to rebut the Secretary's prima facie showing of similarity. This is true simply because in many instances the two violations must be substantially similar in nature in order to be violations of the same standard. However, in cases where both violations are for failure to comply with the same general standard, it may be relatively undemanding for the employer to rebut the Secretary's prima facie showing of similarity. For example, 29 C.F.R. §1926.28(a), one of the most commonly cited construction safety standards, is often alleged to be violated on the grounds that an employer failed to protect its employees from fall hazards by requiring the use of safety belts. On the other hand, the same standard has been cited under dissimilar circumstances, such as an employer's failure to require its employees to use seat belts in an earthmoving vehicle. A prima facie showing of similarity would be rebutted by evidence of the disparate conditions and

[8]Both the Commission and the courts have laid emphasis upon the similarity of the antecedent and present violations. * * *

[92]E.g., *George Hyman Constr. Co. v. OSHRC,* 582 F.2d 834 (4th Cir. 1978); *Todd Shipyards Corp. v. Secretary,* 566 F.2d 1327 (9th Cir. 1977).

[93]The current OSHA instruction on repeated violations is stated in the Field Operations Manual, ch. IV, OSHR [Reference File 77:2711–:2712].

hazards associated with these violations of the same standard. Of course, when the Secretary alleges a repeated violation of a general standard such as §1926.28(a), it is likely that he would introduce evidence of similarity other than that the prior violation and the alleged repeated violation are in contravention of the same standard.

Section 17(a) provides that a repeated violation may be predicated upon violations of "* * * the requirements of section 5 of this Act, any standard, rule, or order promulgated pursuant to section 6 of this Act, or regulations prescribed pursuant to this Act. * * *" A section 5(a)(2) violation may therefore be found to be repeated on the basis of either a prior section 5(a)(1) or section 5(a)(2) violation and a section 5(a)(1) violation may similarly be found to be repeated on the basis of either a prior section 5(a)(1) or section 5(a)(2) violation. There must, of course, be evidence of substantial similarity between the prior and present violations.

We also believe that one prior violation may support a finding of repeated. The common usage of the term "repeatedly" leads us to respectfully reject the "more than twice" concept advanced by the Third Circuit in *Bethlehem Steel v. O.S.H.R.C.*, 540 F.2d 157 (3d Cir. 1976). This view comports with our prior holding in *Hyman,* as well as holdings of the Fourth and Ninth Circuits in *Hyman* and *Todd.*

The Secretary, in order to prove any violation to be repeated, must demonstrate that the earlier citation upon which he relies became a final order of the Commission prior to the date of the alleged repeated violation. As the Fourth Circuit pointed out in its *Hyman* decision, "[B]efore a repeated violation may be found it is essential that the employer receive actual notice of the prior violation." [582 F.2d at 841].

On the other hand we hold that an employer's attitude (such as his flouting of the Act), commonality of supervisory control over the violative condition, the geographical proximity of the violations, the time lapse between the violations, and the number of prior violations do not bear on whether a particular violation is repeated, although these matters will be considered in assessing a penalty.

Section 17(j) of the Act specifically provides that "the good faith of the employer" is to be given due consideration in determining an appropriate penalty. Accordingly, evidence as to aggravated conduct, disregard of the Act, or flouting is relevant only to the assessment of an appropriate penalty.

We do not adopt the view that the same supervisors must control two violative conditions for the subsequent one to be repeated. A corporation as an entity is put on notice of a violation of the Occupational Safety and Health Act by issuance of a citation (and its becoming final), and is obligated to abate cited hazards wherever they may occur in its place or places of business. Corporations commonly administer other aspects of their business and control policy in several or many locations, and we see no reason why compliance with this statute should be fragmented. We recognize that supervisors are key personnel, but they are normally not the policy makers of a corporation, and we do not believe adherence to safety standards by a corporation should depend on localized administration by less than high echelon officials. In short it is not unrealistic to require that an employer observe the law (as with any other statute) in all locations where it transacts business. Finally, as noted by the Fourth Circuit in *Hyman,* basing the classification of repeated upon personal knowledge or ability to control conditions would serve to encourage employers to allocate compliance responsibilities among their supervisors and foremen. [582 F.2d at 837–38]. As with evidence of respondent's attitude, however, we believe that evidence regarding commonality of supervision and an employer's internal distribution of safety responsibility may be indicative of its good faith. Thus, such evidence would be cognizable in assessing an appropriate penalty.

Similarly, we find the fact that the violations occurred at different worksites is not relevant to a determination of a repeated characterization.[12] The geographic proximity of past and present violations may well reflect upon the effectiveness of an employer's internal safety program and its delegation of safety responsibilities as well as its distribution of information. As such, the geographic proximity of past and present violations is indicative of the employer's good faith, and will be considered by the Commission in the assessment of a penalty.

Further, the length of time between the two violations is relevant only to the "good faith" criterion for penalty assessment. While the amount of time having elapsed between prior and present violations does not have a bearing on the similarity of antecedent and present violations, it does reflect upon the degree of an employer's continuing efforts to protect employees against hazards. Accordingly, such evidence will be considered by the Commission in assessing appropriate penalties for repeated violations.

***

[The opinion of Commissioner Barnako, concurring in part and dissenting in part, is omitted]

In reaching its conclusions, the Review Commission relied heavily on opinions of the Courts of Appeals for the Fourth[94] and Ninth Circuits,[95] and explicitly disagreed with the narrow definition of repeated violations proffered by the Third Circuit.[96] In its 1974 manual, as amended in 1980, OSHA stated that for employers with fixed establishments—such as factories, terminals, and stores—no repeated violation would be issued if the second violation occurred at a plant or location other than the one previously cited. In the case of businesses with no fixed establishments, such as construction, repeated citations could be issued based on a prior violation "occurring anywhere within the same state." For maritime employment, the port area would be determinative.[97] This interpretation was reiterated by OSHA in 1982, except that, for nonfixed establishments, the geographic scope was changed from anywhere within the state to the "area office jurisdiction." In addition, under the new instruction, a repeated violation could not be issued more than three years after the original violation became a final order or three years of the final correction date, whichever is later.[98] These policies have been incorporated into the 1983 Field Operations Manual.[99]

[12]Such a consideration, along with the nature of respondent's business, is a factor in the Secretary's decision to allege that a violation is repeated. The Secretary's Field Operations Manual, U.S. Dept. of Labor, Occupational Safety and Health Admin., Chap. VIII B.5e applies different criteria for repeated depending on whether the employer has a fixed establishment. This distinction has withstood a constitutional (equal protection) attack. *Desarrollos Metropolitanos, Inc. v. O.S.H.R.C.* 551 F.2d 874 (1st Cir. 1977). [Author's note—The provisions are presently in the 1983 Field Operations Manual, ch. IV B.5c, OSHR [Reference File 77:2711].]

[94]*George Hyman Constr. Co. v. OSHRC,* 582 F.2d 834 (4th Cir. 1978).
[95]*Todd Shipyards Corp. v. Secretary of Labor,* 566 F.2d 1327 (9th Cir. 1977).
[96]*Bethlehem Steel Corp. v. OSHRC,* 540 F.2d 157 (3d Cir. 1976).
[97]Interim Field Operations Manual, OSHA Instruction CPL 2.45 at VIII-11 (1980).
[98]CPL 2.52, OSHR [Reference File 21:8285].
[99]OSHR [Reference File 77:2711–:2712].

F. Multiemployer Worksites

In 1976, OSHA requested public comment on a recurring issue in enforcement: the policy for the issuance of citations on multiemployer worksites where employees of a number of contractors are present at the same workplace.[100] As the notice stated, the question is identifying the employer to be cited "either because his employees are exposed to a workplace hazard or because he is responsible for that hazard."[101] The OSHA request stated that the Agency was re-evaluating its then current policy of citing only the employer who had exposed his employees "without regard to direct causation or overall responsibility for abatement," particularly in light of several court of appeals cases dealing with the issue.[102] In the *Federal Register* notice, OSHA proposed specific "policy guidelines" for citing employers in multiemployer workplace situations.

Citation Guidelines in Multi-Employer Worksites Request for Public Comment

41 Fed. Reg. 17,639, 17,640 (1976)

***

1. The Secretary may issue citations to an employer on a multi-employer worksite who creates or causes hazardous conditions to which his employees or those of any other contractor or subcontractor engaged in activities on the multi-employer worksite are exposed. By issuing a citation to such employer, the Secretary will be citing the employer directly responsible for the hazardous working condition. For example, a subcontractor who has immediate control over, and responsibility for, all operational phases of a multi-employer worksite, including the responsibility for the erection of guardrails, would be cited for a failure to erect perimeter guarding in violation of 29 CFR 1926.500(d)(1), regardless of whether his employees are actually or potentially exposed to the hazardous working conditions, if any employees of another employer, engaged in worksite activity, are exposed to the hazard.

2. The Secretary may also cite the employer who has the ability to abate the hazardous condition regardless of whether he created the hazard in the first instance. In determining ability to effectuate abatement, the Secretary will consider whether the employer has control over the workers, material and equipment necessary to abate the hazard or whether the employer has the authority to order the abatement of the hazard created by another employer. This latter category includes the employer, such as the general contractor, construction manager, architect, engineering firm, or owner who has safety responsibility for the worksite, as well as subcontractors who have safety responsibility over specific operational aspects of a worksite.

***

---

[100]41 FED. REG. 17,639 (1976).

[101]*Id.*

[102]41 FED. REG. 17,639, citing *Anning-Johnson Co. v. OSHRC,* 516 F.2d 1081 (7th Cir. 1975); *Brennan v. OSHRC (Underhill Constr. Corp.),* 513 F.2d 1032 (2d Cir. 1975); *Brennan v. Gilles & Cotting,* 504 F.2d 1255 (4th Cir. 1974).

3. In those cases involving violative conditions which generally are readily apparent, the Secretary may cite not only the employer who controls or is responsible for the violative conditions, but also the employer who permits his employees to be exposed to the hazardous conditions, regardless of whether the latter employer created the hazard or has the power to abate it. Thus, responsibility based on exposure would be limited to violations which are readily apparent, rather than esoteric and hidden. An employer who exposes his employees to such workplace hazards must, even in the absence of the ability physically to abate, protect his employees from continuing exposure by alerting the employer with the power to abate and by taking affirmative action rendering the hazards inaccessible to his employees.

***

Citations will chiefly be issued under this subparagraph where it is difficult to establish which subcontractor created or controlled the hazardous conditions, where no single subcontractor created the hazard, or where the creating subcontractor does not have the present ability to abate.

***

The view underlying the proposed guideline was that citations should be issued in the first instance against the employer who created or controlled the hazardous condition; and that citations should be issued against the employer whose employees are exposed to the hazard only when, for various reasons, the employer controlling or creating the hazard cannot be cited. The legal predicate for this approach was largely the decision of the Court of Appeals for the Second Circuit in *Brennan v. OSHRC (Underhill Construction Corporation)*.[103] The court there said, "[i]n a situation where * * * an employer is in control of an area, and responsible for its maintenance we hold that to prove a violation of OSHA, the Secretary of Labor need only show that a hazard has been committed and that the area of the hazard was accessible to the employees of the cited employer or those of other employers engaged in a common undertaking."[104] Meanwhile, the Review Commission articulated a comprehensive view on citations in multiemployer worksites in two cases, *Grossman Steel & Aluminum Corporation*[105] and *Anning-Johnson Co.*[106] In these cases, the Review Commission reconsidered its earlier views and ruled, first, that each employer at a construction site is responsible for assuming that its conduct does not create hazards to *any* employees at the site, even if its own employees are not exposed; secondly, that the general contractor is responsible to assure that other contractors fulfill their safety responsibilities affecting the entire site; and that a subcontractor must make a "reasonable effort" to detect hazards to which his or her employees are exposed and "to have them abated" whether or not that employer created or con-

[103] 513 F.2d 1032 (2d Cir. 1975).

[104] *Id.* at 1038. It is well established that to sustain a citation, employees—either of the cited employer or of some other employer—must be actually exposed or "have access" to the hazard. See, e.g., *Daniel Constr. Co.*, 10 OSHC 1549, 1551 (Rev. Comm'n 1982).

[105] 4 OSHC 1185 (Rev. Comm'n 1976) (Moran, Comm'r, dissenting).

[106] 4 OSHC 1193 (Rev. Comm'n 1976) (Moran, Comm'r, concurring and dissenting).

trolled the hazard. A subcontractor whose employees were exposed to a hazard could defend against a citation, according to the Review Commission, if it could show that it did not create or control the hazard and the exposed employees were protected by "realistic" measures alternative to literal compliance or that it did not or could not have known that the condition was hazardous.[107]

OSHA, on the other hand, in its 1980 amendment to the Field Operations Manual, instructed field staff not to cite an employer at a multiemployer worksite whose employees are not exposed or potentially exposed to hazard—"even if that employer created the condition."[108] In 1981, OSHA reiterated the policy that citations would be issued only to employers whose own employees were exposed to hazards, unless that employer could establish "a legitimate defense" to the citation that it had done all it could do but failed to remove the hazard to its employees. If the defense is established, OSHA will cite the employer who is in the best position to control the hazard, whether or not that employer's employees are exposed.[109]

[107] *Grossman Steel,* 4 OSHC at 1188–89; *Anning-Johnson,* 4 OSHC at 1196–99. The Review Commission view has been followed by a number of courts. See, e.g., *Central of Georgia R.R. v. OSHRC,* 576 F.2d 620 (5th Cir. 1978).

[108] OSHA Instruction CPL 2.45, X-11 (1980).

[109] OSHA Instruction CPL 2.49 (1981). See Field Operations Manual, ch. V, OSHR [Reference File 77:2909]. The Review Commission has continued to follow its views in the *Grossman Steel* and *Anning-Johnson* cases. See *Williams Enters.* 10 OSHC 1260, 1262 (Rev. Comm'n 1982). See Note, *The Occupational Safety and Health Act of 1970 as Applied to the Construction Industry: The Multi-Employer Worksite Problem,* 35 WASH. & LEE L. REV. 173 (1978) (employer should not be cited if its employees had access to nonserious dangers that were neither created nor controlled by that employer).

15

Adjudicating OSHA Citations

A. Background

If, as a result of a workplace inspection, OSHA concludes that an employer has violated its obligations under §5, OSHA is required to issue a citation[1] and may issue a proposed penalty.[2] The citation and proposed penalty are served on the employer by certified mail.[3] If the employer files a notice of contest to the citation or penalty, or if the employees contest the time set for abatement, the case is transferred to the Review Commission for adjudication.[4] The case is assigned to a Review Commission administrative law judge authorized to conduct

[1]See Chapter 11, Section A for discussion of basic enforcement procedures under the Act. Sec. 9(a), 29 U.S.C. §658(a)(1976), provides that the Secretary "shall with reasonable promptness" issue a citation. OSHA has construed the "shall" language as mandatory, thus precluding on-site, sanction-free consultation by OSHA representatives. See discussion, Chapter 11, Section D. On the "reasonable promptness" requirement, see *Brennan v. Chicago Bridge & Iron Co.*, 514 F.2d 1082 (7th Cir. 1975) (rejecting Commission's 72-hour rule for issuance of citations); *Coughlan Constr. Co.*, 3 OSHC 1636 (Rev. Comm'n 1975) (Cleary, Comm'r, concurring; Moran, Comm'r, dissenting) (in light of court decision in *Chicago Bridge*, Commission formulated prejudice rule to determine if citation issued with reasonable promptness). Sec. 9(c), 29 U.S.C. §658(c), provides that no citation may be issued "after the expiration of six months following the occurrence of any violation." But see *Yelvington Welding Service*, 6 OSHC 2013 (Rev. Comm'n 1978) (Barnako, Comm'r, concurring and dissenting) (six-month limitation does not apply where OSHA's failure to discover violation in a timely manner was due to the employer's failure to report fatality, as required by OSHA regulations).

[2]Detailed instructions for the calculation of penalties are contained in OSHA's Field Operations Manual, ch. VI, OSHR [Reference File 77:3101–:3108]. Sec. 17(j), 29 U.S.C. §666(i), provides that the Review Commission "shall have authority to assess all civil penalties," considering the size of the business, gravity of the violation, the good faith of the employer, and the history of previous violations. The Commission and the courts, from an early date, have asserted that the Commission is the "final arbiter" of the amounts of penalties. *Brennan v. OSHRC (Interstate Glass Co.)*, 487 F.2d 438 (8th Cir. 1973).

[3]Sec. 10(a), 29 U.S.C. §659(a). There have been inconsistent decisions on the issue of what constitutes adequate service of a citation on an employer. Compare *Buckley & Co. v. Secretary of Labor*, 507 F.2d 78 (3d Cir. 1975) (mailing of citations to superintendent of employer's maintenance shop and garage does not comply with §10(a); notice must be given to official with authority to disburse funds to pay penalty) with *B.J. Hughes, Inc.*, 7 OSHC 1471 (Rev. Comm'n 1979) (Cleary Comm'r, concurring) (Commission refuses to follow *Buckley* and decides that test to be applied is whether service is "reasonably calculated to provide the employer with knowledge" of citation and penalty).

[4]Sec. 10(c), 29 U.S.C. §659(c); see also §12, 29 U.S.C. §661. A considerable body of Review Commission and court law has developed on issues relating to the contest of citations and penalties, such as time for filing, what constitutes a valid notice of contest, and effect of limited notices of contest (penalty only or citation only). One of the early cases involved the question

an evidentiary hearing in the proceeding and to file a report on his determination with the Review Commission.[5] At the hearing, OSHA has the burden of proving by a preponderance of the record that a violation has occurred.[6] The employer has the burden of proving affirmative defenses.[7]

An often-disputed issue in Review Commission proceedings is whether OSHA should be permitted to amend its complaint; this issue frequently arises at the completion of the hearing in the context of a motion to conform the complaint to the evidence adduced at the hearing, and involves an amendment of the allegation of the standard violated.[8] The Review Commission in the past has permitted OSHA amendments more liberally than the courts.[9] If the case is not

whether OSHA's delay in transmitting a notice of contest to the Commission was a basis for dismissing the citation. *Brennan v. OSHRC (Brent Towing Co.)*, 481 F.2d 619 (5th Cir. 1973). The Court of Appeals for the Fifth Circuit reversed the Commission decision dismissing the citation because of the delay, referring to the case as a "procedural dogfight in which the Secretary and the Commission have become ensnarled" and as an "administrative whirlwind of * * * minor proportions." *Id.* at 619–20. On the transmittal of notices of contest, see 29 C.F.R. §2200.32 (1983) (the Secretary "shall" transmit the notice of contest within 7 days of receipt). See also 29 C.F.R. §1903.17 (1983) ("The Area Director shall immediately transmit such notice to the Review Commission in accordance with the rules of procedure prescribed by the Commission."). The Commission has long since changed its position and does not dismiss notice of contests because of delay in transmission by OSHA. See, e.g., *J. Dale Wilson, Bldr.*, 1 OSHC 1146 (Rev. Comm'n 1973) (Moran, Chairman, dissenting).

[5]See Rules of Procedure, 29 C.F.R. §§2200.90–.91.10–.11 (1983). Sec. 2200.2 provides that in the absence of a "specific" provision in the Commission rules, the Federal Rules of Civil Procedure will apply. 29 C.F.R. subpt. D, §§2200.51–.55 deals with prehearing procedures and discovery. Under these regulations, discovery is not permitted in Commission proceedings, except by order of an administrative law judge or the Commission. In proceedings to enforce health standards, OSHA has frequently sought discovery by sending its expert to visit the workplace to inspect operations to determine if engineering controls were feasible. In *Owens-Illinois, Inc.*, 6 OSHC 2162 (Rev. Comm'n 1978) (Barnako, Comm'r, dissenting), the Commission overruled its prior precedent, *Reynolds Metals Co.*, 3 OSHC 1749 (Rev. Comm'n 1975), and *Reynolds Metals Co.*, 6 OSHC 1667 (Rev. Comm'n 1978), and held that OSHA was not barred from conducting discovery inspections of workplaces using nonfederal employee experts; trade secrets could be protected by appropriate protection orders. On the confidentiality of the names of OSHA's informants, see *Stephenson Enter. v. Marshall*, 578 F.2d 1021 (5th Cir. 1978), *aff'g* 2 OSHC 1080 (Rev. Comm'n 1974) (Moran, Chairman, dissenting) (upholding OSHA's refusal to disclose at the administrative hearing the name of nonsupervisory employee interviewed during the course of the inspection; balancing of rights falls in favor of employee's right to be protected from retaliation; right to be protected is "particularly strong in this case").

[6]See 29 C.F.R. §2200.73 (1983). See also *National Realty & Constr. Co. v. OSHRC*, 489 F.2d 1257 (D.C. Cir. 1973), reprinted in Chapter 14, Section B; *Armor Elevator Co.*, 1 OSHC 1409 (Rev. Comm'n 1973).

[7]See, e.g., *Murphy Pacific Marine Salvage Co.*, 2 OSHC 1464 (Rev. Comm'n 1975) (Moran, Chairman, dissenting). The issue of whether an argument is an affirmative defense or part of OSHA's principal case often arises. See, for example, discussion of burden of proof in §4(b)(1) cases, *infra*, note 14.

[8]See, e.g., *McLean-Behm Steel Erectors*, 6 OSHC 2081 (Rev. Comm'n 1978), *rev'd*, 608 F.2d 580 (5th Cir. 1979), discussed *infra* note 9. The need to amend arises in part because evidence not previously available to OSHA is introduced at the hearing and in part because OSHA citations are drafted by nonlawyers and typically are not reviewed by lawyers until close to the hearing. See Field Operations Manual, ch. V, OSHR [Reference File 77:2901] (consultation between area director, regional administrator, and regional solicitor only "where the citation items could involve novel or complex litigation in which the Area Director could expect the investment of major litigation resources").

[9]See, e.g., *McLean-Behm Steel Erectors*, 6 OSHC 2081 (Rev. Comm'n 1978) (Barnako, Comm'r, dissenting) (allowing OSHA to amend complaint to allege violation of 29 C.F.R. §750(b)(1)(ii)—steel erection safety net standard—instead of alleging violation of §1926.28(a)—general fall protection standard—in order to conform evidence to proof, under Rule 15(b) of Federal Rules of Civil Procedure; employer not prejudiced by its lack of opportunity to present a defense to the new charge). The Commission decision was reversed by the Court of Appeals for the Fifth Circuit, *McLean-Behm Steel Erectors v. OSHRC*, 608 F.2d 580 (5th Cir. 1979) (FED. R. CIV. P. 15(b) requires express or implied consent to trial of issues in amended charge; OSHA's "quick reversal of direction would be more appropriate on the football field"; since all OSHA's evidence related to original charge, employer's failure to object does not constitute implied consent). See also *Morgan & Culpepper, Inc. v. OSHRC*, 676 F.2d 1065 (5th Cir. 1982).

settled,[10] the report of the administrative law judge becomes a final order of the Commission, unless within 30 days any member of the Commission directs that the report be reviewed by the entire Commission.[11] Decisions of the Commission and unreviewed decisions of the administrative law judges that have become final orders may be appealed to an appropriate U.S. court of appeals by OSHA or by "any person adversely affected or aggrieved by an order of the Commission."[12] This chapter discusses several major legal issues, particu-

[10]Approximately 90% of Review Commission cases are settled. Rothstein, *OSHA After Ten Years: A Review and Some Proposed Reforms,* 34 Vand. L. Rev. 71, 116 (1980). This is a result of at least two factors: the desire to obtain immediate abatement and to avoid the delays in litigation; and the major shortage of lawyers available to handle OSHA cases and the need to reduce their case load. See *Oversight on the Administration of the Occupational Safety and Health Act, 1980: Hearings Before the Senate Comm. on Labor and Human Resources,* 96th Cong., 2d Sess. pt. 2, at 1239 (1980) (testimony of Lane Kirkland, President of AFL-CIO). One ongoing controversy has revolved around the role of employee representatives in the settlement of cases. See discussion in Chapter 17. Another issue is the propriety of exculpatory language in settlement agreements. In *Farmer's Export Co.,* 8 OSHC 1655 (Rev. Comm'n 1980) (Barnako, Comm'r, concurring; Cottine, Comm'r, dissenting), a case that resulted from citations issued after a grain dust explosion at the employer's facility, the Commission affirmed its supervisory power over settlement agreements, but approved exculpatory language purporting to relieve the employer from the consequence of the violation in any future proceedings, including, apparently, those under the Act; policy of providing for prompt abatement militates for approval of settlements meeting the requirements of Commission Rule 100, 29 C.F.R. §2200.100 (1983). While the settlement in *Farmer's Export* broadly exculpated the employer from enforcement under the OSH Act as well as in other statutory proceedings, the practice of the Office of the Solicitor is not to accept exculpatory language that applies to enforcement under the Act. On the issue of Commission authority over settlements, see discussion in Chapter 17.

[11]Sec. 12(j), 29 U.S.C. §661(i). For a discussion of delays in Review Commission litigation, see Chapter 11, Section B. A major disruption in Review Commission decisional activity occurred in 1975–76 when, because of an internal dispute in the Commission, Commissioner Moran for a period of time directed review of virtually all administrative law judge decisions. See 5 OSHR 1312 (1976), and *Francisco Tower Service,* 4 OSHC 1459 (Rev. Comm'n 1976). The Commission majority in this case affirmed the judge's decision without considering the merits of the case, despite Commissioner Moran's direction for review, because the aggrieved party filed no brief and in the majority view there was an absence of any compelling public interest. Chairman Barnako concurred and Commissioner Moran dissented, saying:

> My colleagues do not know the meaning of the word "review" despite the fact that they are members of a "review" commission. In view of the "disposition" they have ordered in this case and more than one hundred other cases in recent months, it is clear that the time has come for a name change. I think they have earned the right to call themselves the Occupational Safety and Health Dodge, Duck, Skip, Side-Step, and Cut-and-Run Commission. The official seal of OSHDDSSSCRC could be a mammoth loophole and the motto could be "Don't Bother Us If You Can't Hire A Lawyer To File A Brief, Or If Two Of Us Don't Think There Is Any Compelling Public Interest In Your Case, And Don't Bother Asking What We Mean By 'Compelling Public Interest.'" The official bird would, of course, be the ostrich.

Id. at 1461.

[12]Sec. 11(a), 29 U.S.C. §660(a). The Court of Appeals for the District of Columbia Circuit has ruled that an employee representative, in this case a union, which was a party to the Commission proceeding, may seek review of a Commission decision in the courts of appeals. *OCAW v. OSHRC,* 671 F.2d 643 (D.C. Cir. 1982), *cert. denied sub nom. American Cyanamid Co. v. OCAW,* 51 USLW 3287 (1982). See discussion in Chapter 17, Section D.

Section 14 of the Act provides that, except for litigation before the Supreme Court, which is handled by the Office of the Solicitor General, the Solicitor of Labor "may appear for and represent the Secretary in any civil litigation brought under this Act but all such litigation brought under this Act shall be subject to the direction and control of the Attorney General." Shortly after the Act became effective, a sharp dispute developed between the Department of Justice and the Office of the Solicitor in the Labor Department over the issue which group of attorneys would handle OSHA litigation. This dispute, which centered around litigation of cases involving review of standards and review of Review Commission decisions, is described at some length in D.L. Horowitz, The Jurocracy; Government Lawyers, Agency Programs, and Judicial Decisions 109-114 (1977). The Departments of Labor and Justice in 1975 finally reached an agreement which related not only to OSHA litigation but to litigation under the Employee Retirement Income Security Act of 1974 as well, 29 U.S.C. §1132(j), and the Farm Labor Contractor Act Amendments of 1974, 7 U.S.C. §2050a(d). Under the agreement, as im-

larly those with strong policy implications, that have arisen in Review Commission proceedings adjudicating citations.

B. Preemption Under §4(b)(1)—Exercise of Authority by Another Federal Agency

In order to avoid redundancy in enforcement in the field of occupational safety and health, §4(b)(1)[13] provides that OSHA will not apply to "working conditions" with respect to which another federal agency "exercise[s] statutory authority to prescribe or enforce standards or regulations affecting occupational safety or health."[14] Anticipating that implementation of this provision would lead to controversy, Congress required in §4(b)(3)[15] that OSHA, within three years, report to Congress its recommendations "to avoid unnecessary duplication and to achieve coordination" between the Act and other federal laws. The report was filed in December 1980,[16] more than six years late and only after several congressmen had filed suit to compel the Secretary of Labor to comply with his responsibility under §4(b)(3).[17] The report presented an overview of OSHA's experience with §4(b)(1) and concluded that the prevention of duplication of effort and interagency coordination was being "adequately addressed" under existing authority and therefore no legislation was necessary.

Report of the Secretary of Labor[18]

***

II. BACKGROUND

The Occupational Safety and Health Act of 1970 is a comprehensive statute covering all employers who are engaged in a business affecting commerce and who have one or more employees and was designed to "* * * ensure so far as possible every working man and woman in the Nation safe and healthful working conditions * * *." In enacting this comprehensive

plemented, the Department of Labor handled the litigation of all cases under OSHA and the other two programs, except Supreme Court cases and criminal cases. *Id.* at 112; Olson, "Agency Litigating Authority as a Factor in Court Policy Making" 18-23, Paper delivered at 1983 Meeting of American Political Science Association.

[13]29 U.S.C. §653(b)(1).

[14]While OSHA and the Review Commission have viewed §4(b)(1) as an affirmative defense, e.g., *Idaho Travertine Corp.*, 3 OSHC 1535 (Rev. Comm'n 1975), some courts have held that §4(b)(1) relates to subject matter jurisdiction and thus can be raised at any time, e.g., *U.S. Air, Inc. v. OSHR*, 689 F.2d 1191, 1195 (4th Cir. 1982); *Columbia Gas of Pa. v. Marshall*, 636 F.2d 913, 918 (3d Cir. 1980).

[15]29 U.S.C. §653(b)(3).

[16]Report of the Secretary of Labor to Congress Pursuant to Section 4(b)(3) of the Occupational Safety and Health Act of 1970, Coordination in the Administration and Enforcement of This Act and Other Federal Laws Affecting Occupational Safety & Health (1980). See 10 OSHR 797 (1981).

[17]The suit was filed in the U.S. District Court for the District of Columbia by the Washington Legal Foundation on behalf of Rep. Mickey Edwards (R. Okla.) and three other Republican members of the Home of Representatives. 10 OSHR 75 (June 19, 1980). On December 17, 1980, District Judge Charles R. Richey, in an oral ruling, gave the Secretary of Labor until December 24 to file the report, saying: "He is a man of his word. He is just going to do what he is supposed to do." The report was filed on December 23, 1980. See 10 OSHR 797 (1981).

[18]Copy available in files of Department of Labor.

statute, Congress also took into account those provisions of other Federal laws which, to various degrees, deal with worker safety and health issues. Section 4(b)(1) states, in pertinent part, that:

> Nothing in this Act shall apply to working conditions with respect to which other Federal agencies * * * exercise statutory authority to prescribe or enforce standards or regulations affecting occupational safety and health.

Generally speaking, this language means that the OSH Act's protections apply only where there exist no legally enforceable standards or regulations of another Federal agency addressing a particular hazard. Under this framework, gaps in employee protection are avoided while at the same time the regulatory activity of the other agencies is not affected; viewed in this light, section 4(b)(1) places a limit only on the authority of the Department of Labor.

The OSH Act contains a detailed framework of enforcement and educational activities, including numerous employee rights, designed to achieve employee protection. Even where the OSH Act is preempted under section 4(b)(1), there is still an understandable expectation on the part of many employees and their representatives that the quality of protection and the range of rights afforded by the OSH Act would be provided for all workers and in all hazardous working conditions. The Department recognizes this concern and has taken steps to promote a coordinated approach to occupational safety and health programs throughout the Government by means of interagency agreements.

III. COURT AND REVIEW COMMISSION INTERPRETATION OF SECTION 4(b)(1)

A number of decisions of the Occupational Safety and Health Review Commission (OSHRC) and the courts of appeals have interpreted section 4(b)(1). Several courts have ruled that workers may not be denied protection under the OSH Act for *all* hazards they face because their industry happens to be regulated in part by another federal agency which has issued rules for *some* worker hazards.

In other words, these courts hold that industries as such are not preempted from OSHA; rather, the preemption rule of section 4(b)(1) applies only to particular hazards. The courts have also held that in order for preemption to take place, the standard or regulation of the other agency must be in a final rule, not simply a notice of proposed rulemaking or an advance notice of proposed rulemaking.[1] At the same time, the evolving section 4(b)(1) law indicates that this preemption of OSHA does not depend on the efficacy of other agency's standard or its enforcement mechanisms.[2]

These and other cases are shaping current federal policies for worker safety and health. To be sure, there are remaining issues to be decided by the courts which will help further define the roles of the OSH Act and other federal laws in protecting workers. We are satisfied, however, that current case law in general has adequately preserved the primary intent of section 4(b)(1), which is to provide maximum worker protection contemplated by Congress in enacting the comprehensive OSH Act and at the same time to avoid duplication of effort by Federal agencies.

---

[1]See *Southern Pacific Transportation Co. v. Usery,* 539 F.2d 386 (5th Cir. 1976), *cert. denied,* 434 U.S. 874 (1977); *Baltimore and Ohio R.R. Co. v. OSHRC,* 548 F.2d 1052 (D.C. Cir. 1976); *Southern Railway v. OSHRC,* 539 F.2d 335 (4th Cir. 1976), *cert. denied,* 429 U.S. 999 (1976).

[2]See *Mushroom Transportation Co.,* OSHRC Docket No. 1588, 1973–74 CCH OSHD ¶16,881 [1 OSHC 1390] (R.C. 1973), appeal dismissed (3rd Cir. 1974); *Pennsuco Cement and Aggregates, Inc.,* OSHRC Docket No. 15462, 1980 OSHD ¶24,478 [8 OSHC 1378] (R.C. 1980).

IV. QUALITY OF PROTECTION

There are only two federal agencies that deal exclusively with occupational safety and health and both of these are in the Department of Labor: the Occupational Safety and Health Administration and the Mine Safety and Health Administration. No other federal agencies have as their exclusive purpose the protection of workers from on-the-job injuries and illnesses. The statutes administered by other agencies that have an occupational safety and health role are designed primarily or in significant part to protect the public or equipment. Agencies of the Department of Transportation, for example, have as major purposes protecting the public and transportation equipment: i.e., the safe operation of trains (Federal Railroad Administration), aircraft (Federal Aviation Administration), certain vessels (Coast Guard), and certain trucks and buses (Bureau of Motor Carrier Safety). Because those agencies have other statutory functions, they cannot devote exclusive attention and resources to employee protection.

The absence of this primary focus on occupational safety and health has at times resulted in criticisms of other agencies by employees, their representatives, and Congress concerning their programs. These criticisms have related primarily to the efficacy of their enforcement activities.

In a recent example, a General Accounting Office Report, dated July 21, 1980, pointed out that the Department of Energy has not been properly implementing its safety and health programs for employees at its contractor-operated uranium enrichment plants.* In another recent instance, flight attendants contended in Congressional hearings that the Federal Aviation Administration has no effective enforcement program for in-flight occupational hazards even though the FAA has maintained that it has full statutory authority over such hazards.** In another example, the Coast Guard was criticized by Congress for failure to protect merchant seamen from the hazards of asbestos.

In addition to these criticisms that stem from multiple agency functions, many workers have voiced concern about the statutory mechanisms of some of the other agency programs as they relate to occupational safety and health. A number of the other Federal statutes under which workers are covered lack certain important provisions contained in the OSH Act which this Department has repeatedly stated are essential to strong occupational safety and health programs. These include:

- the right of employees to file anonymous complaints about alleged hazards and the responsibility of the agency to respond to those complaints;
- the right of employee representatives to accompany a federal inspector during investigations of working conditions;
- authority of the agency to propose monetary penalties as an incentive for employers to comply voluntarily with safety and health standards;
- authority to seek injunctions to protect employees in imminently dangerous situations; and
- authority to compel the prompt abatement of hazardous working conditions.

In addition, certain industry groups have on occasion complained of duplication and overlap by OSHA and other agencies in regulating and enforcing occupational safety and health concerns.

*[Author's note—See COMPTROLLER GENERAL OF THE UNITED STATES REPORT TO CONGRESS, EMD-80-78 DEPARTMENT OF ENERGY'S SAFETY AND HEALTH PROGRAM FOR ENRICHMENT PLANT WORKERS IS NOT ADEQUATELY IMPLEMENTED (July 11, 1980).]

**[Author's note—OCCUPATIONAL SAFETY AND HEALTH PROTECTION FOR AVIATION INDUSTRY EMPLOYEES—THE CONFLICT BETWEEN FAA AND OSHA, H. REP. No. 393, 97th Cong., 1st Sess. (1981).]

Numerous actions have been taken to deal with these concerns. Agreements between OSHA and other agencies have led to increased understanding, cooperation, reduction of overlap, elimination of recordkeeping duplication and consequent reductions in employers' paperwork burdens, use of OSHA standards by other agencies, and sharing of technical information and enforcement resources. * * * Furthermore, OSHA has cooperated with other agencies in the formulation of legislative proposals dealing with occupational safety and health issues, such as trucking safety, that include compliance provisions parallel to those in the OSH Act.

V. CONCLUSIONS AND RECOMMENDATIONS

Preventing duplication of effort and achieving coordination between the OSH Act and other Federal statutes is, in the Department's opinion, being adequately addressed under existing authority. As the courts continue to refine the interpretation of section 4(b)(1) of the OSH Act, an even clearer picture of each Federal agency's role for worker safety and health will emerge. Federal policies will continue to be adjusted to conform to the courts' interpretations. In addition, cooperative efforts between Federal agencies will continue to expand. The Department sees no need for, and therefore does not recommend, legislation on the subject of occupational safety and health jurisdiction.

This low-key report scarcely reflects the fact that there has been continuing controversy among OSHA, employers, employee representatives, and other federal agencies, notably, the Federal Railroad Administration (FRA), the Federal Aviation Administration (FAA), and the Coast Guard, over the meaning and impact of §4(b)(1). The first critical issue was whether there was an "industry exemption" under §4(b)(1); OSHA argued that there was no industry exemption and that OSHA was preempted only with respect to specific "working conditions" regarding which another federal agency "exercised" statutory authority. The Review Commission and the courts have accepted the basic OSHA view,[19] although there continued to be differences of view in the courts on the precise meaning of the term "working conditions."

Southern Railway v. OSHRC

539 F.2d 335 (4th Cir.), cert. denied, 429 U.S. 999 (1976)

FIELD, Circuit Judge:
***

The facts are undisputed. The petitioner, Southern, a Virginia corporation, operates an interstate common carrier railroad system which includes a facility for maintenance and repair of rolling stock known as the Hayne Shop in Spartanburg, South Carolina. In October of 1973, an OSHA compliance officer made a routine inspection of the facility pursuant to 29 U.S.C.

[19] E.g., *Southern Pac. Transp. Co.*, 2 OSHC 1313 (Rev. Comm'n 1974), *aff'd*, 539 F.2d 386 (5th Cir. 1976), *cert. denied*, 434 U.S. 874 (1977).

§657(a), and on November 2, 1973, the Secretary of Labor (Secretary) cited Southern for ten "non-serious" violations of standards promulgated by the Secretary under the authority of OSHA, proposing certain penalties and requiring abatement of the alleged violations by April of 1974. The citations were affirmed by an administrative law judge and his decision was upheld by a split Commission.

Southern admits that it has not complied with the OSHA standards and regulations alleged to have been violated, but takes the position that it is exempt from compliance by Section 4(b)(1) of the Act * * *.

Southern contends that the exemption is operative because the Secretary of Transportation, acting through the Federal Railroad Administration (FRA), has exercised his authority pursuant to the Federal Railway Safety Act of 1970, 45 U.S.C. §421, *et seq.*, and earlier railway safety acts[4] to promulgate and enforce safety regulations affecting the working conditions of railway employees. Conceding that FRA has not exercised its authority to regulate employee safety in railway shop and repair facilities such as the Hayne Shop, Southern urges that, nonetheless, there has been a sufficient exercise of the regulatory authority to exempt the working conditions of all employees in the railway industry from the OSHA standards. The Secretary admits that under the Federal Safety Act the FRA has authority to regulate all areas[5] of employee safety for the railway industry, but contends that Section 4(b)(1) exempts only those areas of railway employee safety in which FRA has expressly exercised its authority. The Commission, in essence, adopted the Secretary's view.

***

OSHA was enacted in response to an appalling record of death and disability in our industrial environment, and it was the clear intendment of Congress to meet the problem with broad and, hopefully, effective legislation. * * * Accordingly, the exemptive statute should appropriately be construed to achieve the maximum protection for the industrial workers of the Nation.

In our opinion the industry-wide exemption urged upon us by Southern would fly in the face of these principles and objectives. The safety regulations of the Department of Transportation are confined almost exclusively to those areas of the railway industry which affect over-the-road operations such as locomotives, rolling stock, signal installations, road beds and related facilities.[12] While the regulatory program in these areas reflects a concern for the safety of the employees, it is directed primarily toward the general safety of transportation operations. On the other hand, the Department of Transportation and FRA do not purport to regulate the occupational health and safety aspects of railroad offices or shop and repair facilities. To read the exemptive statute in a manner which would leave thousands of workers in these non-operational areas of the railway industry exposed to unregulated industrial hazards would, in our opinion, utterly frustrate the legislative purpose.

Southern suggests that a rejection of its position will, of necessity, constitute an acceptance on our part of the "nook and cranny theory of safety

[4]Some of the Acts of Congress relied upon by Southern are:
The Safety Appliance Acts, 45 U.S.C. §§1–14. The Signal Inspection Act, 45 U.S.C. §26. Train, Brakes Safety Appliance Act, 45 U.S.C. §9. Hours of Service Act, 45 U.S.C. §61, *et seq.* Rail Passenger Safety Act, 45 U.S.C. §502, *et seq.*

[5]45 U.S.C. §431(a) reads in part:
"The Secretary of Transportation * * * shall (1) prescribe, as necessary, appropriate rules, regulations, orders, and standards for all areas of railroad safety supplementing provisions of law and regulations in effect on October 16, 1970, and (2) conduct, as necessary, research, development, testing, evaluation, and training for all areas of railroad safety."

[12]Southern appears to concede that the railway safety statutes referred to in footnote 4, *supra,* are directed solely to the transportation operations of the industry.

regulation"[13] under which the Secretary's jurisdiction would extend into every minute detail of the working environment not covered by a specific standard or regulation of the other Federal agency. While some intra-agency memoranda indicate that the Secretary of Labor has espoused such a theory, the decisions of the Commission have evinced a somewhat ambivalent attitude on the question. In any event, we do not think our choice of alternatives is as limited as Southern would suggest.

The crux of the controversy is the phrase "working conditions" in Section 4(b)(1). The Secretary contends that this phrase means "particular, discrete hazards" encountered by an employee in the course of his job activities. Southern, on the other hand, insists that the term means "the aggregate of circumstances of the employment relationship—that is, the employment itself."

We think both of the parties are a bit wide of the mark. Southern and the Secretary each relies upon *Corning Glass Works v. Brennan,* 417 U.S. 188 (1974), where the Court, in considering the meaning of the term "working conditions" as used in the Equal Pay Act of 1963, stated:

> "* * * the element of working conditions encompasses two subfactors: 'surroundings' and 'hazards'. 'Surroundings' measures the elements, such as toxic chemicals or fumes, regularly encountered by a worker, their intensity, and their frequency. 'Hazards' takes into account the physical hazards regularly encountered, their frequency, and the severity of injury they can cause. This definition of 'working conditions' is * * * well accepted across a wide range of American Industry." (Footnotes omitted) *Id.,* at 202.

We think this aggregate of "surroundings" and "hazards" contemplates an area broader in its contours than the "particular, discrete hazards" advanced by the Secretary, but something less than the employment relationship in its entirety advocated by Southern. The Act was intended both to provide comprehensive coverage to the workers across the country and to avoid duplication of regulatory effort by the various Federal agencies. In the light of these dual objectives, and drawing upon the *Corning* definition, we are of the opinion that the term "working conditions" as used in Section 4(b)(1) means the environmental area in which an employee customarily goes about his daily tasks. We are further of the opinion that when an agency has exercised its statutory authority to prescribe standards affecting occupational safety or health for such an area, the authority of the Secretary of Labor in that area is foreclosed. Such a construction, we think, avoids the confusion and duplication of effort that Section 4(b)(1) of the Act was designed to prevent, and is consonant with the general statutory purpose.
***

The Court of Appeals for the Fifth Circuit in *Southern Pacific Transportation Co. v. Usery*[20] also rejected the industry exemption argument and adopted the view that the other federal agency must exercise its authority in order to preempt OSHA. On the issue of the definition of working conditions, however, the court departed from the view of the Fourth Circuit Court of Appeals. It said:

[13]This phrase was used by Chairman Moran in his dissenting opinion in *Secretary of Labor v. Southern Pacific Transportation Company,* [2 OSHC 1313 (1974)].

[20]539 F.2d 386 (5th Cir. 1976).

Our rejection of the railroads' position does not constitute an acceptance of the theory that every OSHA regulation remains operative until the FRA adopts a regulation of its own on that specific subject. As we have noted, the statutory term "working conditions" embraces both "surroundings," such as the general problem of the use of toxic liquids, and physical "hazards," which can be expressed as a location (maintenance shop), a category (machinery), or a specific item (furnace). Neither OSHA itself nor the existence of OSHA regulations affects the ability of the primary regulatory agency, here the FRA, to articulate its regulations as it chooses. Much of their displacing effect will turn on that articulation. Section 4(b)(1) means that any FRA exercise directed at a working condition—defined either in terms of a "surrounding" or a "hazard"—displaces OSHA coverage of that working condition. Thus, comprehensive FRA treatment of the general problem of railroad fire protection will displace all OSHA regulations on fire protection, even if the FRA activity does not encompass every detail of the OSHA fire protection standards, but FRA regulation of portable fire extinguishers will not displace OSHA standards on fire alarm signaling systems.[10] Furthermore, as the dominant agency in its limited area, the FRA can displace OSHA regulations by articulating a formal position that a given working condition should go unregulated or that certain regulations—and no others—should apply to a defined subject.

We recognize that a regulatory exercise expressed in terms of a category of equipment or a generalized problem may raise questions about whether a given item is covered. Conversely, an exercise expressed in terms of a piece of equipment may create an issue about whether the FRA has regulated the entire category to which that piece belongs. In either situation, the scope of the exemption created by section 4(b)(1) is determined by the FRA's intent, as derived from its articulations.[21]

The Review Commission has also refused to accept the view of the Court of Appeals for the Fourth Circuit that working conditions referred to "environmental area." Thus, in *Allegheny Airlines*,[22] the Review Commission held that OSHA was not preempted from citing Allegheny for locked and blocked exit doors in its departure lounge by FAA's security regulations barring "unauthorized entry" in the same area.[23] The Review Commission said that the "OSHA standard regulates fire protection; the FAA regulatory program regulates criminal violence." And, in disagreeing explicitly with the Fourth Circuit, the Commission insisted that a definition of working conditions must include a "'hazards' component." For, if the inquiry were only on whether the other agency had regulated "the same place" as OSHA and not whether "the same hazard or problem is regulated,"

[10]In *Southern Ry. v. OSHRC*, 539 F.2d 335 (4th Cir. 1976), the court likewise interpreted section 4(b)(1) in a manner which avoided the extreme positions of the Secretary and the railroads. The Fourth Circuit defined "working conditions" as "the environmental area in which an employee customarily goes about his daily tasks." *Id.* at 339. Our conclusion that the operative effect of §4(b)(1) is determined by the manner in which the FRA articulates its exercise of authority seems to us more attuned to the differing possible meanings of the term "working conditions" as it is used elsewhere in the field of industry relations.

[21]*Id.* at 391–92.

[22]9 OSHC 1623 (Rev. Comm'n 1981).

[23]The relevant FAA regulations in force at the time of the alleged violation are quoted in 9 OSHC at 1626.

the Commission said, "the congressional purpose would be frustrated." Allegheny[24] appealed to the Court of Appeals for the Fourth Circuit, which adhered to its earlier view in reversing the Review Commission. The court said:

> It is the Commission's position, as stated in its order sustaining the violation relating to the exit doors in the passenger lounge, that the FAA regulations were directed at protection against terrorists, air pirates and extortioners for both airline customers and aircraft crews and personnel and that such regulations so directed did not preempt the right of the Secretary of Labor to require the doors in the passenger lounge, which the FAA, under its approved program, had required to be blocked or obstructed, to be opened for '[f]ree and unobstructed egress from the departing area" in order to protect employees against fire hazards. In effect, it is declaring that, if the FAA does not require by its regulations precisely the same safety standards against the same discrete hazard as has the OSHA in this departure area of the airport (i.e., safety against the fire hazards), such regulations do not preempt the OSHA's standards. The result of such an argument would be that under FAA procedures mandated by Congress in the interest of passenger and employee safety, Air would be required to keep locked or obstructed the four doors for egress from this one passenger lounge, whereas, under OSHA rules issued validly in the interest of passenger and employee safety, Air would be compelled to keep unlocked or unobstructed the very same four doors in the same passenger lounge. This was the very result—the result created by conflicting rules issued by separate agencies covering safety in a particular working environment—that Congress intended to obviate by including in the Act Section 4(b)(1). Giving to the exemption statute the effect intended by Congress, it is manifest that the exercise of jurisdiction over the doors in the working environment of the passenger lounge involved in this case by the FAA preempted any jurisdiction in the Secretary of Labor, by regulations under the Act, over such area.[25]

Thus, the basic thrust of the Fourth Circuit's approach was to avoid conflict in the regulatory actions by federal agencies. The Review Commission, on the other hand, while not ignoring the importance of avoiding conflict, emphasized the importance of employee protection, that is, the need to avoid gaps in regulatory coverage.[26]

The issue of OSHA authority in relation to the FAA also arose in *Northwest Airlines,*[27] decided by the Review Commission in 1980.

---

[24]At the time of the appellate proceeding, the carrier's name had been changed to U.S. Air, Inc. (Air).

[25]*U.S. Air, Inc. v. OSHRC,* 689 F.2d 1191, 1194 (4th Cir. 1982). The Court of Appeals for the Third Circuit has concurred in this view on the meaning of working conditions. *Columbia Gas of Pa. v. Marshall,* 636 F.2d 913 (3d Cir. 1980).

[26]See COMPTROLLER GENERAL OF THE UNITED STATES, REPORT TO THE CONGRESS, PAD-81-76 GAINS AND SHORTCOMINGS IN RESOLVING REGULATORY CONFLICTS AND OVERLAPS 53–54, 61–62 (1981) (considering some instances of conflict and overlap involving OSHA and concluding that, as a general matter, conflict and overlap between agencies in regulation was not a "major problem" in relation to other regulatory issues, although further improvements are possible"). See also COMPTROLLER GENERAL OF THE UNITED STATES, REPORT TO THE CONGRESS, EMD-80-78 DEP'T OF ENERGY'S SAFETY AND HEALTH PROGRAM FOR ENRICHMENT PLANT WORKERS IS NOT ADEQUATELY IMPLEMENTED 29–32 (1980) (recommending that the Department of Energy give more emphasis to employee complaints and that a system of fines for safety and health violations, similar to that contained in OSHA, was needed).

[27]8 OSHC 1982 (Rev. Comm'n 1980) (Cottine, Comm'r, concurring).

That case involved a safety maintenance manual, prepared by the airline carrier, subject to FAA disapproval under the Federal Aviation Act of 1958.[28] The manual concededly covered the working conditions regulated by OSHA (maintenance employees' ground servicing of landing lights in the wing of a Boeing 747), but OSHA argued the maintenance manual was not a "standard or regulation" within the meaning of §4(b)(1). The Review Commission rejected this argument.

Northwest Airlines

8 OSHC 1982 (Rev. Comm'n 1980)

BY THE COMMISSION:
***

The Secretary's principal argument against the validity of the FAA's manual procedure is that the requirements contained in a manual are not rules of general applicability, approved by the FAA after notice and an opportunity to comment, but are drafted by each individual airline and become effective without any prior FAA approval. According to the Secretary:

> where, as here, each airline controls its own safety matters, subject to disapproval by FAA, there is no guiding force to assure that safety, and not employer interests, are paramount in writing the manual. . . . Clearly a company acting on its own initiative and prerogative should not be able to decide when and to what extent OSHA is preempted.[24]

Under the Secretary's position, in order for a maintenance manual to legally bind airlines, the FAA would have to issue rules for each aircraft that apply uniformly throughout the airline industry. It would have to develop its own manual, presumably using the manufacturer's manual as a starting point, publish or incorporate this manual by reference in the Federal Register, seek comments from interested persons concerning the content of the manual, determine whether the proposed manual should be modified, and ultimately promulgate a final version of the manual before the airplane could be put into service.

There are obvious drawbacks to this approach. Perhaps the most serious is that it would inhibit the very objective the APA seeks to achieve through notice and comment rulemaking proceedings: an opportunity for affected persons to have a meaningful input into rules that affect them. If a manual was promulgated through APA rulemaking proceedings, it could be changed only by similar rulemaking proceedings. Thus, after an aircraft is put into service and airlines and their employees become familiar with the aircraft and the manual on a day-to-day basis, they would be required to

[24]The Secretary states that he does not contend the manual procedure is invalid vis-a-vis the FAA, but only that it is not a sufficient exercise of the FAA's authority to preempt the Act under §4(b)(1). We have concluded, however, that the dispositive question is whether the manual has the force and effect of law, and this in turn depends on whether the FAA has satisfied APA procedures. Thus, if the procedures the FAA follows are inadequate to preempt the Act as to specific working conditions addressed by a manual, they are also inadequate to have any provisions of a manual bind an airline. In this respect, the Secretary's argument for finding the FAA's manual approval procedure inadequate to preempt the Act—that it allows each airline to control its own safety matters—applies with the same strength to matters of flight safety as it does to the safety of workers performing maintenance on an airplane.

[28]49 U.S.C. §§1301–1542 (1976 & Supp. V 1981).

comply with procedures that were developed before they had gained any experience with the aircraft. It is likely that their experience would show that some of these procedures should be modified. If, however, a rulemaking proceeding is necessary before any such change could be instituted, it appears probable that few such changes would ever be implemented; at the very least, an airline would face a substantial delay in implementing a potentially superior procedure. Thus, the industry would tend to be "locked-in" to procedures developed before the persons best able to judge whether better procedures exist have gained the experience necessary to form a judgment. On the other hand, the FAA's manual procedure permits each airline to benefit from its experience with a plane in actual service and to quickly implement the changes to the manufacturer's manual it thinks are necessary, subject only to FAA disapproval. This tends to promote input into an airline's maintenance procedures by those most knowledgeable with the problems and peculiarities of each type of aircraft: the airline's own employees. As one court has stated in rejecting the argument that an airline's manual was not legally binding: "Based on years of experience with the methods of certificated carriers, the regulation is reasonable and practical in substituting the carrier's procedures for the minutiae of individual FAA approval." *United States v. Garrett,* 296 F.Supp. 1302, 1304 (N.D. Ga. 1969), *aff'd,* 418 F.2d 1250 (5th Cir. 1969), *cert. denied,* 300 U.S. 927 (1970).

The effectiveness of the FAA's manual system depends on the airlines acting in good faith and proposing only those changes to the manufacturer's manual that they think will achieve benefits in safety. The Secretary suggests that an airline could follow an entirely different path, scrapping the manufacturer's manual and writing its own so as to promote its own interests over those of its employees and the general public.

The Secretary's argument assumes that an airline would think it beneficial to follow a deliberate course of trying to deceive the government agency responsible for regulating virtually all aspects of the company's operations. It further assumes that the FAA would fail either to recognize if an airline was submitting changes to its manual that were detrimental to safety, or to take appropriate action if it discovered that an airline was doing this. Neither assumption is particularly compelling. But even if there is potential for abuse in the regulatory system that the FAA has chosen, the fact that a procedure may not be ideal is no reason to reject that procedure entirely. As discussed above, there would also be a serious drawback to the FAA's use of APA procedures to promulgate all of the detailed requirements appearing in airline manuals. Moreover, no system that the FAA might follow would guarantee that airlines would always follow safe procedures. Employers have been known to violate even OSHA standards.

***

[The concurring opinion of Commissioner Cottine is omitted.]

The rationale of the Review Commission majority in rejecting OSHA's argument, namely, that an FAA standard is not preemptive unless it goes through a public comment period, is troublesome. In reasoning that notice-and-comment rulemaking would cause "substantial delay" before changes could be made and the air carrier could be "locked-in" to old procedures, the Commission appears to be rejecting the underlying philosophy of public rulemaking which has been almost universally accepted, at least since the Administrative Procedure Act[29] was enacted in 1946.[30]

The Commission decision in *Northwest Airlines* also rejected the OSHA argument that OSHA was not preempted because FAA was not adequately enforcing an occupational safety and health program. Relying on its earlier decision in *Pennsuco Cement & Aggregates, Inc.*,[31] the Commission held that §4(b)(1) does not permit it "to oversee the adequacy of another agency's enforcement efforts."[32]

In *Northwest Airlines,* the Review Commission also stated that in order for the standards or regulations of another federal agency to be preemptive under §4(b)(1), they must have the "force and effect of law."[33] The impact of this requirement was blunted by a 1982 Review Commission decision in *Consolidated Rail Corp.*[34] At issue there was a policy statement issued by FRA in 1978, stating that the agency was exercising its jurisdiction over the safety of railroad employees and that it determined that the hazard of open pits and platforms in railroad repair yards should go unregulated. The Commission majority, overruling its prior *Consolidated Rail* case,[35] held that this policy preempted OSHA standards on open pits and platforms in railroad yards. Commissioner Cottine dissented, saying that the policy statement is "nothing more than an announcement of an 'agency's tentative intentions for the future' that 'does not establish a binding norm' and 'is not finally determinative of the issues or rights to which it is addressed.'"[36]

Issues under §4(b)(1) have also arisen in relation to the authority of the U.S. Coast Guard to regulate most vessels operating on the navigable waters. In 1982 the Review Commission, reversing its earlier decision in *Puget Sound Tug & Barge,*[37] adopted an industry-wide exemption approach on the protection of seamen aboard vessels on the navigable waters.

---

[29] 5 U.S.C. §§551–559 (1982).

[30] Commissioner Cottine concurred on narrower grounds. 8 OSHC at 1993.

[31] 8 OSHC 1378 (Rev. Comm'n 1980).

[32] 8 OSHC at 1990. In September 1982, the FAA, as part of a program of regulatory reform, proposed major changes in its regulation of safety matters. 47 FED. REG. 42,371 (1982). See also HOUSE COMM. ON GOVERNMENT OPERATIONS, OCCUPATIONAL SAFETY AND HEALTH PROTECTION FOR AVIATION INDUSTRY EMPLOYEES: THE CONFLICT BETWEEN FAA AND OSHA, H. REP. NO. 393, 97th Cong., 1st Sess. 7, 9–10 (1981) (discussing the history of the *Northwest Airlines* case and concluding that OSHA, and not the FAA, "has the expertise and willingness needed to protect the health and safety of ground employees"; that §4(b)(1) had created an "ambiguity" as to which agency had authority; that this ambiguity had led to "conflicts, confusion and extensive litigation"; and, therefore, that OSHA and the FAA should enter into a memorandum of understanding that would coordinate and promote the agencies' regulatory activities "with a minimum of duplication").

[33] 8 OSHC at 1989.

[34] 10 OSHC 1577 (Rev. Comm'n 1982) (Cottine, Comm'r, dissenting in part and concurring in part).

[35] 9 OSHC 1258 (Rev. Comm'n 1981).

[36] 10 OSHC at 1582 (quoting *Pacific Gas & Electric Co. v. FPC,* 506 F.2d 33, 38 (D.C. Cir. 1974)).

[37] 9 OSHC 1764 (Rev. Comm'n 1981).

Dillingham Tug & Barge Corporation

10 OSHC 1859 (Rev. Comm'n 1982)

By the Commission:
***

II

In *Puget Sound Tug & Barge,* 9 BNA OSHC 1764 (1981), a divided Commission rejected the argument that section 4(b)(1) precludes the Secretary of Labor from citing an employer for alleged violations of the OSH Act that involve the working conditions of seamen aboard vessels on navigable waters. The majority viewed this argument as seeking an "industry-wide exemption" from the Act and rejected the argument on the basis of precedent holding that "section 4(b)(1) was not designed to create industry-wide exemptions, but rather exemptions only over those working conditions that have been the subject of an exercise of another agency's statutory authority to regulate occupational safety and health." 9 BNA OSHC at 1775 (1981). The majority cited, among other cases, *Southern Pacific Transportation Co. v. Usery,* 539 F.2d 386 (5th Cir. 1976), *cert. denied,* 434 U.S. 874 (1977), and *Southern Railway Co. v. OSHRC,* 539 F.2d 335 (4th Cir.), *cert. denied,* 429 U.S. 999 (1976). Thus, although acknowledging that the Coast Guard possessed broad statutory authority to issue regulations promoting safety on navigable waters and that the Coast Guard has issued regulations pursuant to this authority, the majority concluded that the Coast Guard's exercise of authority did not entirely exempt from the Act the working conditions of seamen aboard vessels on navigable waters. Instead, the majority concluded that section 4(b)(1) only precluded the Secretary from citing those specific working conditions that were the subject of applicable Coast Guard regulations. In dissent, Commissioner Cleary rejected what he referred to as the majority's "nook and cranny" theory. He believed that the Coast Guard's regulations affecting the working conditions of seamen were so comprehensive that it would impair, if not destroy, the fabric of the Coast Guard's regulatory scheme to apply OSHA standards to situations for which the Coast Guard lacked applicable regulations. 9 BNA OSHC at 1782 (1981).

We have reconsidered the Commission's holding in *Puget Sound.* We conclude that section 4(b)(1) in certain circumstances can create industry-wide exemptions from the OSH Act. Having examined the manner in which the Coast Guard has exercised its statutory authority to regulate the health and safety of seamen aboard vessels operating on navigable waters, we conclude that section 4(b)(1) precludes the Secretary from enforcing the OSH Act with respect to such working conditions. We overrule *Puget Sound* to the extent it is inconsistent with this holding.
***

IV

Prior to the OSH Act's passage, Congress had given authority to regulate particular working conditions to federal agencies other than the Secretary of Labor. Congress believed that it would be beneficial to employee safety and health if agencies that had specialized expertise in regulating particular working conditions continued to do so. Thus, Congress intended that other agencies that had been given statutory authority to regulate occupational safety and health would be the dominant agencies in their particular areas, with the OSH Act applying only insofar as the other agencies had not exercised their statutory authority. See *Southern Pacific Transportation Co. v. Usery, supra,* 539 F.2d at 391–92.

As discussed above, the Coast Guard has issued extensive regulations applicable to all vessels operating on navigable waters except certain vessels owned by the United States. The Coast Guard regulations are directed at protecting the health and safety of seamen. *Taylor v. Moore-McCormick Lines, Inc.*, 621 F.2d 88 (4th Cir. 1980); *Clary v. Ocean Drilling and Exploration Co.*, 609 F.2d 1120 (5th Cir. 1980). Thus, employers of seamen, such as Dillingham, must look to the Coast Guard regulations for the requirements they must comply with to protect their employees. Under the approach of the majority in *Puget Sound,* such employers would also be required to compare the Coast Guard regulations with OSHA regulations, determine which conditions of the work environment are not addressed by Coast Guard regulations, and comply with OSHA requirements directed at those working conditions. This approach necessarily assumes that the Coast Guard regulations are insufficient to provide adequate protection to seamen aboard vessels on navigable waters. It is thus inconsistent with the established rule that section 4(b)(1) does not permit an inquiry into the stringency of another agency's exercise of authority. "Once another Federal agency exercises its authority over specific working conditions, OSHA cannot enforce its own regulations covering the same conditions. Section 4(b)(1) does not require that another agency exercise its authority in the same manner or in an equally stringent manner." *Mushroom Transportation Co.*, 1 BNA OSHC 1390, 1392 (1973). See also *Northwest Airlines, Inc.*, 8 BNA OSHC 1982, 1990 (1980); *Pennsuco Cement and Aggregates, Inc.*, 8 BNA OSHC 1378 (1980).

The *Puget Sound* majority's approach also places an unreasonable burden on employers to determine the requirements with which they must comply. As Commissioner Cleary stated in his separate opinion:

> Such an approach also creates a regulatory labyrinth for employers who must review two sets of federal regulations in detail to determine where there are voids in one set of regulations which are susceptible to application of a standard from another set of regulations. In the ensuing uncertainty the goal of providing employees with a safe and healthful workplace can only suffer.

9 BNA OSHC at 1782 (1981).

This case involves an alleged hazard to a seaman arising out of activities aboard a vessel operating on navigable waters. The Coast Guard, an agency with special expertise in maritime hazards, has exercised its statutory authority to prescribe standards and regulations governing the working conditions of seamen. Thus, the Secretary is precluded from enforcing the OSH Act with respect to those working conditions. Moreover, we note that the Coast Guard investigated the incident that led to this case and determined that it did not involve any violations of law or regulation. Nevertheless, the Coast Guard investigation resulted in the recommendation of measures intended to avoid similar incidents in the future. Thus, the dominant agency in maritime safety has taken remedial action. Permitting the Secretary to also seek remedial measures through an enforcement action arising out of the same incident would thus result in the very duplication of federal agency effort that section 4(b)(1) is designed to avoid. * * *

Commissioner Cottine dissented. Relying particularly on the legislative history of the Outer Continental Shelf Lands Act Amendments of 1978,[38] he argued that Congress viewed continued OSHA jurisdiction over "unregulated safety and health matters" as preferred to "the type of industry wide exemption" which the majority was endorsing in its decision in that case.[39]

Attempts have been made by OSHA throughout its history to minimize the disputes under §4(b)(1) by entering into agreements with other federal agencies delineating respective jurisdiction over safety and health. The first jurisdictional agreement was between OSHA and the Federal Railroad Administration, dated May 16, 1972, and it established procedures for the handling of complaints in the railroad industry.[40] In a recent memorandum of understanding intended to "increase consultation and coordination" on occupational safety and health matters regarding personnel working on the Outer Continental Shelf (OCS), OSHA and the Coast Guard recognized Congress' expectation that the Coast Guard "would be the principal Federal agency in matters of occupational safety and health on the OCS."[41] OSHA also entered into a memorandum of understanding with the Mine Health and Safety Administration (MSHA) in the Department of Labor, specifying respective areas of authority, setting forth the factors which would be used by the Secretary of Labor in making determinations "relating to convenience of administration," and providing a procedure for deciding jurisdictional questions and for coordination between MSHA and OSHA.[42]

---

[38]43 U.S.C. §§1331–34, 1337, 1340, 1343–56, 1801–66 (Supp. V 1981).

[39]10 OSHC at 1866–68. See also *Donovan v. Texaco, Inc.*, 720 F.2d 825 (5th Cir. 1983) (OSHA provisions prohibiting retaliatory discharge does not apply to seamen on vessels in navigation under §4(b)(1)).

[40]Copy in Author's files. The agreement was cancelled by Assistant Secretary Stender on December 23, 1974. 4 OSHR 957–58 (1975).

[41]Memorandum of Understanding Concerning Occupational Safety and Health on Artificial Islands, Installations and Other Devices on the Outer Continental Shelf of the United States, Dec. 19, 1979, 45 FED. REG. 9142, 9143 (1980), OSHR [Reference File 21:7101]. On March 8, 1983, OSHA and the Coast Guard entered into a memorandum of understanding under which OSHA concluded, based on its interpretation of §4(b)(1) that "it may not enforce the OSH Act with respect to the working conditions of seamen aboard inspected vessels." 48 FED. REG. 11,365 (1983). The memorandum did not deal with jurisdiction over uninspected vessels, or with jurisdictional issues regarding employer recordkeeping obligations.

[42]Interagency Agreement of April 17, 1979, 44 FED. REG. 22,827, OSHR [Reference File 21:7071]. The Federal Mine Safety and Health Act of 1977, §3(h)(1), 30 U.S.C. §802(h)(1) (Supp. V 1981), authorizes the Secretary of Labor, in deciding what constitutes mineral milling, to give due considerations to the "convenience of administration" resulting from the transfer of mine safety and health authority from the Secretary of the Interior to the Secretary of Labor.

An unusual issue under §4(b)(1) was raised following the action of Congress in December 1981 to bar the use of appropriated funds by the Mine Safety and Health Administration (MSHA) to enforce the Mine Safety and Health Act of 1977, 30 U.S.C. §§801–962 (1976 & Supp. V 1981), with respect to persons engaged in surface mining and milling of stone, clay, collodial phosphate, sand, and gravel; with respect to state and political subdivisions; and against certain independent construction contractors operating in surface mining areas. Appropriations Fiscal Year 1982, Pub. L. No. 97-92, 95 Stat. 1183 (1981), *continued by,* Further Continuing Appropriations, Fiscal Year 1982, Pub. L. No. 97-161, 96 Stat. 22 (1982). OSHA took the position that since MSHA was barred from enforcing the mine safety and health law as to these working conditions, enforcement authority passed to OSHA, and the §4(b)(1) preemption was not applicable. The contrary argument was that since the limitation on MSHA authority was only temporary and the Mine Safety Act was not repealed, OSHA continued to be preempted. The argument favoring preemption, it was asserted, was supported by a letter from the Secretary of Labor to Congressman Natcher, stating: "Since the provisions of the Mine Act and MSHA's standards are neither repealed nor amended by the rider, but only the expenditure of funds to enforce them temporarily prohibited, it is likely that many operators will contest OSHA citations by contending that §4(b)(1) of the OSH Act precludes OSHA authority. Resolu-

C. Challenges to the Validity of Standards in Enforcement Proceedings

Section 6(f)[43] permits preenforcement challenges to OSHA standards within 60 days of their issuance. Unlike the Federal Mine Safety and Health Act of 1977,[44] however, the OSH Act does not provide that §6(f) suits are the "exclusive" means for challenging the validity of standards. The question, therefore, is whether, and to what extent, a cited employer may claim in an enforcement proceeding before the Review Commission (and the courts) that the standard upon which the citation is based is invalid. These challenges may be of two kinds: first, the employer could argue that there were procedural deficiencies in the initial issuance of the standard; and second, that the standard is substantively invalid, typically because compliance with the standard is technologically and economically infeasible. This section deals with the issue whether these challenges may be raised in enforcement proceedings. The question whether the engineering control requirements are feasible in a particular plant or operation is discussed in the following section.

The first case dealing extensively with this issue was *Atlantic & Gulf Stevedores v. OSHRC,* decided in 1976. There, the employer had been cited for a violation of the hard-hat standard and claimed that employees refused to wear hard hats and therefore compliance with the standard was not feasible. The court rejected the employer's defense on the merits and affirmed the citation, but first concluded that the validity of an OSHA standard could be litigated in an enforcement proceeding.

Atlantic & Gulf Stevedores v. OSHRC

534 F.2d 541 (3d Cir. 1976)

GIBBONS, Circuit Judge:

***

* * * [O]ur role [in deciding whether an OSHA standard is valid], whether the standard comes before us in the context of a Sec. 6(f) petition or a Sec. 11(a) enforcement proceeding, would appear to be the same. This conclusion follows from either of two diverging lines of analysis.

I

The two occupational safety and health bills reported out of the respective Labor committees to the floor of the House and Senate originally made

tion of this litigation may take some time. * * * " Reprinted in 127 CONG. REC. S15035 (daily ed. Dec. 11, 1981). An administrative law judge has rejected OSHA's position in a case now pending before the Review Commission, *Daniel Constr. Co.,* OSHRC No. 82-668 (Rev. Comm'n 1983); meanwhile the restriction on MSHA enforcement has been removed, Further Continuing Appropriations, 1983, Pub. L. No. 97-377, 1982 U.S. CODE CONG. & AD. NEWS (96 Stat.) 1883.

[43] 29 U.S.C. §655(f).

[44] 30 U.S.C. §811(d) (Supp. V 1981).

no provision for an Occupational Safety and Health Review Commission. The Senate Bill, S. 2193, and the House Bill, H.R. 16785, each authorized the Secretary of Labor both to issue citations and propose penalties, and to adjudicate the question of liability in the first instance, subject to judicial review in the courts of appeals. Section 11(b) of the House bill described in some detail the procedures to be followed by the Secretary in these administrative adjudications, and concluded by directing that

> "[i]n proceedings under this subsection, the Secretary shall consider, among other things, *the validity of any standard,* rule, order, or regulation alleged to have been violated, and the reasonableness of the period of time permitted for the correction of the violation." (emphasis supplied).

See also H.R. Rep. No. 91-1291, 91st Cong., 2d Sess. 24, 41 (1970). Although the House bill did not provide for pre-enforcement adjudicatory review of standards promulgated by the Secretary, nonstatutory pre-enforcement judicial review could be had under Sec. 10 of the Administrative Procedure Act, 5 U.S.C. Sec. 701–06. See *Abbott Laboratories v. Gardner,* 387 U.S. 136 (1967); *Gardner v. Toilet Goods Association,* 387 U.S. 167 (1967). It is thus clear that the House believed that administrative review of the substantive validity of a safety regulation in an enforcement proceeding was a beneficial and desirable complement to pre-enforcement review in the federal courts. The Senate bill expressly provided for pre-enforcement judicial review of a promulgated standard, and did not in terms authorize substantive administrative review in an enforcement proceeding.

***

* * * We thus believe that as enacted, Sec. 11(a) of OSHA carries forward, if only implicitly, the understanding of the House, reflected in the Committee bill, that the validity of a particular safety standard could preliminarily be determined in a Commission enforcement proceeding.

This conclusion is supported by strong policy considerations as well as the Act's legislative history. As has been seen, Sec. 6(f) provides for pre-enforcement review of a safety standard in the courts of appeals. The provision does not direct the court to expedite the reviewing process, however, so it is likely that there will be substantial lag time between the Secretary's promulgation of a challenged standard and the initial decision on its validity. Because the filing of a pre-enforcement petition to review will not ordinarily operate as a stay of the standard, see Sec. 6(f), 29 U.S.C. Sec. 655(f), enforcement proceedings may commence before even an initial determination of validity has been made. Unless the Commission is authorized to consider defenses of invalidity, it will be reduced to rubber-stamping possibly invalid citations, and employers will for an interim period be deprived of any remedy for sanctions imposed contrary to law.

Moreover, it may become evident that a particular safety and health standard is economically or technologically infeasible, or otherwise unreasonable, only after employers have made good faith efforts to comply. These problems may manifest themselves well after the 60-day period for pre-enforcement review has expired. The Commission would thus be the only available forum for raising the question of invalidity. Because we do not believe that Congress intended to foreclose all possible challenge to the validity of a standard 60 days beyond its effective date, we must conclude that Sec. 11(a) of OSHA empowers the Commission to deny enforcement to a standard determined by it to have been issued in violation of the Act's substantive or procedural requirements.

***

* * * We do not find, from the availability of limited pre-enforcement judicial review permitted under Sec. 6(f) and the silence with respect to

legal issues in Sec. 11(a) an intention to limit the scope of judicial review in the enforcement proceeding. Judicial review at that stage is, after all, the ordinarily preferred method. See *Toilet Goods Association v. Gardner,* 387 U.S. 158, 163–64 (1967). Absent an explicit withdrawal of jurisdiction, we will entertain affirmative defenses attacking the validity of an administrative regulation that is brought to us for enforcement. In *Synthetic Organic Chemical Manufacturers Association v. Brennan,* 503 F.2d 1155 (3d Cir. 1974), *cert. denied,* 420 U.S. 973 (1975) (*Synthetic Organic I*), a Sec. 6(f) case, we set forth in detail the scope of our review of standards promulgated by the Secretary. In the face of Congressional silence the presumption of reviewability at the enforcement stage attaches. We therefore consider that the scope of our review in a Sec. 11(a) case is co-extensive with a Sec. 6(f) case.

***

We hold today that we have jurisdiction to decide the validity of an OSHA regulation in an enforcement proceeding as well as in a direct petition for review. We do not mean to suggest, however, that the posture in which the question is presented to us is irrelevant for all purposes. Indeed, the context in which the challenge to a regulation is made determines the allocation of the burden of proof.

***

Congress barred petitions for Sec. 6(f) review filed more than 60 days after promulgation. We have already held the defense of invalidity is always available in an enforcement proceeding. But we do not believe that the burden of proof in a Sec. 11(a) case is identical with the burden in a Sec. 6(f) case. In an enforcement proceeding invalidity is an affirmative defense to a citation, and the petitioning employer bears the burden of proof on the issue. Thus a petitioner cannot defend solely on the ground that the procedural requirements established in *Synthetic Organic I* have been ignored by the Secretary. To carry its burden the petitioner must produce evidence showing why the standard under review, as applied to it, is arbitrary, capricious, unreasonable or contrary to law. Were we to hold otherwise we would effectively nullify the congressional circumscription of the right to petition for review of an OSHA standard.[13]

***

In *Atlantic & Gulf Stevedores,* the challenge to the validity of the standard was substantive—that its enforcement would be economically infeasible because it would lead to a costly strike. As to the question whether the Third Circuit Court of Appeals would permit procedural challenges to standards, the language of the decision is inconsistent. The court first states that it would not enforce a standard issued in violation of the Act's "substantive or procedural requirements,"[45] thus indicating that procedural challenges would be considered. The court later says, however, that a petitioner cannot

[13]We recognize that in dealing with difficult and complex problems of economic and technological feasibility, the placement of the burden of proof may well be determinative of the question of validity. A standard which, because of the placement of the burden of proof, could not be sustained in a Sec. 6(f) proceeding, may nonetheless be held valid in an enforcement proceeding where the burden shifts to the party asserting invalidity. Whether such an approach is wise is not for us to decide. We believe that Congress, by limiting Sec. 6(f) review to 60 days, has directed employers to be vigilant by raising questions of invalidity prior to actual enforcement.

[45]534 F.2d at 550.

"defend [against enforcement of a standard] solely on the ground that the procedural requirements established in *Synthetic Organic I* [503 F.2d 1155 (3d Cir. 1974)] have been ignored by the Secretary."[46]

A clear distinction between procedural and substantive challenges to standards was made by the Court of Appeals for the Eighth Circuit in *National Industrial Constructors v. OSHRC.*[47] The employer there argued that OSHA's construction standards, adopted as "established Federal standards" under §6(a)[48] in 1971 were procedurally defective. The employer, relying on the requirement in the Administrative Procedure Act[49] that regulations may not become effective for a 30-day period after publication, claimed that OSHA's adoption under §6(a) took place before the end of the 30-day period. The court refused to entertain the procedural challenge. It referred to the two "policy" reasons asserted by the Court of Appeals for the Third Circuit in *Atlantic & Gulf Stevedores* for allowing challenges to standards in enforcement proceedings[50] and said:

> Procedural attacks, on the other hand, implicate different interests. An employer's interest in invalidating an otherwise reasonable regulation solely because the Secretary has failed to comply with the APA's procedural requirements is, in our view, distinguishable from attacks on the reasonableness or feasibility of the regulation. While the unreasonableness of a regulation may only become apparent after a period during which an employer has made a good faith effort to comply, procedural irregularities need not await the test of time and can be raised immediately. The agency's interest in finality, coupled with the burden of continuous procedural challenges raised whenever an agency attempts to enforce a regulation, dictates against providing a perpetual forum in which the Secretary's procedural irregularities may be raised. Were there no limitation upon the time within which procedural attacks could be made, the resulting uncertainty might inhibit employers, otherwise able and willing, from complying with a regulation.
>
> While no court of appeals has specifically decided this question, the Third Circuit has recognized the distinction between substantive and procedural claims of invalidity. See *Atlantic & Gulf Stevedores v. OSHRC,* 534 F.2d at 551–552. There, the Court intimated that, in an enforcement proceeding, an employer may not defend solely on the ground that the Secretary has ignored certain procedural requirements. Rather, it must produce evidence why the regulation under attack is arbitrary, capricious or unreasonable. *Id.* [51]
>
> We think there is merit to the distinction between substantive and procedural claims of invalidity and hold that a challenge to the validity of an OSHA regulation, based solely upon the Secretary's failure to comply with the procedural requirements of the APA, OSHA, or any other applicable statute, may only be raised in a pre-enforcement pro-

[46]*Id.* at 551. These procedural requirements are discussed in Chapter 3, Section B.

[47]583 F.2d 1048 (8th Cir. 1978).

[48]29 U.S.C. §655(a).

[49]5 U.S.C. §553 (1982).

[50]The policy considerations relied on by the Third Circuit Court of Appeals were that enforcement proceedings may commence before the §6(f) challenge is resolved and that the infeasibility may not become apparent until after the 60-day review period under §6(f), 534 F.2d at 549–50, reprinted in this section.

[51]As discussed, the opinion in *Atlantic & Gulf Stevedores* is inconsistent on this issue. See also *Deering Milliken, Inc. v. OSHRC,* 630 F.2d 1094, 1098 n.6 (5th Cir. 1980).

ceeding instituted pursuant to Section 6(f) of OSHA within sixty days from the time the regulation becomes effective. Such attacks may not be raised in an enforcement proceeding. Sixty days, in our view, provides a reasonable period during which procedural challenges can be raised. National is, therefore, barred from raising the procedural invalidity of 29 C.F.R. §1926 in this action.[52]

This distinction between substantive and procedural challenges was rejected outright by the Court of Appeals for the Fifth Circuit in *Deering Milliken, Inc. v. OSHRC.*[53] In that case, OSHA sought to enforce its cotton dust standard, adopted in 1971 under §6(a) as an established federal standard. The employer argued that OSHA's subsequent revision of the standard[54] was invalid because it was completed without regard to the notice and comment requirements of the Administrative Procedure Act[55] and the OSH Act. The court entertained the challenge, disagreeing expressly with *National Industrial Constructors.*

Deering Milliken, Inc. v. OSHRC

630 F.2d 1094 (5th Cir. 1980)

TJOFLAT, Circuit Judge:

***

Deering Milliken readily acknowledges that it did not use [the] pre-enforcement procedure to contest the procedural validity of 29 C.F.R. §1910.1000. The Secretary, relying upon *National Industrial Constructors, Inc. v. OSHRC,* 583 F.2d 1048 (8th Cir. 1978), asserts that this failure on the part of Deering Milliken necessarily bars petitioner from asserting its argument here. See also *A.F.L.-C.I.O. v. Brennan,* 530 F.2d 109, n.5 at 112 (3d Cir. 1975).

In *National Industrial Constructors,* a corporation claimed that certain regulations issued pursuant to section 6(a) of OSHA were invalidly promulgated. The corporation had not raised this issue in a 6(f) proceeding. Under these circumstances, the Eighth Circuit held the procedural attack of the regulation inappropriate. To achieve this result the court distinguished substantive and procedural attacks on the validity of regulations issued under section 6(a). Thus, while "at least for substantive attacks, Congress did not intend pre-enforcement review under Section 6(f) to be the exclusive method for challenging the validity of a regulation," *National Industrial Constructors,* 583 F.2d at 1052, "a challenge to the validity of an OSHA regulation, based solely upon the Secretary's failure to comply with the procedural requirements of the APA, OSHA, or any other applicable statute, may only be raised in a pre-enforcement proceeding instituted pursuant to Section 6(f) of OSHA. * * * Such attacks may not be raised in an enforcement proceeding." *Id.* at 1052–53.

The Eighth Circuit did not indicate any legislative authority for the distinction it drew between procedural and substantive attacks on section

[52]583 F.2d at 1052–53.

[53]630 F.2d 1094 (5th Cir. 1980).

[54]The revision of the standard was published at 36 FED. REG. 15,101 (1971) ("in the interest of greater intelligibility and accuracy").

[55]5 U.S.C. §553 (1982).

6(a) regulations[6], but rather relied upon what it felt to be controlling policy considerations * * *.

Although there is merit in the Eighth Circuit's approach, we find the legislative history of the Act, read in light of the prevailing consensus on the availability of judicial review of agency action, to indicate the contrary result.

This Circuit has consistently held that "[a] long-standing and strong presumption exists that action taken by a federal agency is reviewable in federal court." *Save The Bay, Inc. v. Administrator of EPA,* 556 F.2d 1282, 1293 (5th Cir. 1977). "A statute must demonstrate clear and convincing evidence of an intent to preclude judicial review before courts will cut off an aggrieved party's right to be heard." *Gallo v. Mathews,* 538 F.2d 1148, 1151 (5th Cir. 1976). See *Consolidated-Tomoka Land Company v. Butz,* 498 F.2d 1208, 1209 (5th Cir. 1974). This position is consistent with the Supreme Court's mandate in *Abbott Laboratories v. Gardner,* 387 U.S. 136, 140, 87 S.Ct. 1507, 1511, 18 L.Ed.2d 681 (1967): "[J]udicial review of a final agency action by an aggrieved person will not be cut off unless there is persuasive reason to believe that such was the purpose of Congress."

Section 6(f) is silent concerning its preclusive effect on post pre-enforcement judicial review of section 6(a) regulations. Moreover, the legislative history indicates that "[w]hile [section 6(f)] would be the exclusive method for obtaining pre-enforcement judicial review of a standard, the provision does not foreclose an employer from challenging the validity of a standard during an enforcement proceeding." S.Rep. No. 91-1282, 91st Cong., 2d Sess., 8, *reprinted in* [1970] U.S. Code Cong. & Ad. News, p. 5177, 5184. Mindful that the Supreme Court wishes us to locate a *clear command* before limiting judicial review, see *Barlow v. Collins,* 397 U.S. 159, 167, 90 S.Ct. 832, 838, 25 L.Ed.2d 192 (1970), we simply cannot justify finding that the pre-enforcement review provisions of section 6(f) bar petitioner's procedural attack.

We are aware, while following this path, that the Ninth Circuit has reached the same conclusion. See *Marshall v. Union Oil Co. of California,* 616 F.2d 1113, 1117–1118 (9th Cir. 1980); *Noblecraft Industries, Inc. v. Secretary of Labor,* 614 F.2d 199, 201–202 (9th Cir. 1980). Indeed, the Ninth Circuit has discussed a practical consideration that we wish to reiterate. Many regulations were promulgated summarily through section 6(a). See 29 C.F.R. §1910.1–1910.1500 (1979). Industry would have found it quite burdensome to comb through every 6(a) regulation and object to inappropriate promulgations within sixty days, considering the "multitude of regulations [which] could have been promulgated without notice or hearing within two years of the enactment of OSHA." *Union Oil Co. of California.* 616 F.2d at 1118. Surely it was within Congress's power so to mandate, but the record is devoid of references which would support the finding of such an intent. Arguably, industry was lulled by the legislative requirement that all prop-

[6]The Eighth Circuit, however, did place some reliance on a Third Circuit decision, *Atlantic & Gulf Stevedores v. OSHRC,* 534 F.2d 541 (3d Cir. 1976), which purportedly *intimates,* see *National Industrial Constructors,* 583 F.2d at 1052, that procedural attacks on section 6(a) regulations are improperly raised in an enforcement proceeding. In that same case, however, the Third Circuit stated: "Because we do not believe that Congress intended to foreclose all possible challenge to the validity of a standard 60 days beyond its effective date, we must conclude that §11(a) of OSHA empowers the Commission to deny enforcement to a standard determined by it to have been issued in violation of the Act's [OSHA's] substantive or *procedural* requirements." 534 F.2d at 550 (emphasis added). Although later in the opinion the Third Circuit employs language that casts doubt on the solidity of the above quoted statement, 534 F.2d at 551-52, see also *National Industrial Sand Association v. Marshall,* 601 F.2d 689, 700 n.36. (3d Cir. 1979), we feel the first statement is too emphatic to ignore, and thus read the case as supporting procedural attacks on section 6(a) regulations in enforcement proceedings. See *Noblecraft Industries, Inc. v. Secretary of Labor,* 614 F.2d 199, 202 (9th Cir. 1980) for a similar reading of *Atlantic & Gulf Stevedores.*

erly promulgated 6(a) regulations would be pre-existing and familiar to industry, see S.Rep. No. 91-1282, 91st Cong., 2d Sess., 5–6, *reprinted in* [1970] U.S. Code Cong. & Ad. News, pp. 5177, 5182, and this, combined with the potential number and technical complexity of summarily promulgated regulations, makes it particularly inappropriate to find section 6(f) a bar to procedural attack on 6(a) regulations. In the absence of a statutory scheme clearly indicating a Congressional intent to limit judicial review, we refuse to presume in favor of "administrative absolutism." *Association of Data Processing Service Organizations, Inc. v. Camp,* 397 U.S. 150, 157, 90 S.Ct. 827, 831–832, 25 L.Ed. 2d 184 (1970).

***

In this decision, the court relied particularly on a statement of the Senate Labor Committee in reporting the OSHA bill favorable that the provision for pre-enforcement review of standards "does not foreclose an employer from challenging the validity of a standard during an enforcement proceeding."[56] Despite its thorough analysis of the legislative history on this issue, the Court of Appeals for the Third Circuit in *Atlantic & Gulf Stevedores* failed to mention this critical statement. On the merits in *Deering Milliken,* the court rejected the procedural claim, finding that OSHA had not materially altered the standard in its later revision in 1971 and therefore notice-and-comment rulemaking was not legally required.[57]

Confronted with this conflicting precedent, the Review Commission in *Rockwell International Corporation*[58] made another distinction in an attempt at reconciliation. In *Rockwell,* as in *Deering Milliken,* the employer challenged the validity of a standard,[59] adopted under §6(a) as an established federal standard, on the grounds that OSHA, in adopting the standard, modified the language of the underlying national consensus standard in one respect without notice-and-comment rulemaking.

Rockwell International Corporation

9 OSHC 1092 (Rev. Comm'n 1980)

BARNAKO, Commissioner:

***

Considerations in finality and in avoiding the burden that continuous challenges would impose upon the Secretary's enforcement abilities raise policy issues which should be balanced against an employer's right to challenge the validity of a standard. In *National Industrial Constructors, Inc.,* the employer challenged the procedural validity of the underlying established federal standards on which certain occupational safety and health

[56]S. REP. No. 1282, 91st Cong., 2d Sess. 8, quoted at 630 F.2d at 1099, reprinted in this section.

[57]630 F.2d at 1103.

[58]9 OSHC 1092 (Rev. Comm'n 1980) (Cleary, Chairman, concurring).

[59]The standard involved covered machine guarding and appears at 29 C.F.R. §1910.212(a)(3)(ii) (1983).

standards were based. The right of persons to challenge the procedural validity of underlying established federal standards involves different considerations than those at issue here. Not only was the procedural validity of the underlying established federal standard subject to attack when the standard was promulgated under the prior statute but Congress intended that certain established federal standards be deemed to be occupational safety and health standards under the Act. 29 U.S.C. §653(b)(2). These considerations may weigh heavily against permitting challenges to the procedural validity of an underlying established federal standard in an enforcement proceeding before the Commission.

The challenge here, however, is of a different type. It addresses the very essence of the Secretary's authority to promulgate standards under the Act. Here, the Secretary, without affording notice or an opportunity for comment and without adhering to the other procedural requirements of section 6(b) and the APA, adopted as an "established federal standard" a standard whose wording is dissimilar from the prior standard which the Secretary purported to be reissuing. If the changes were substantive, then the Secretary, as Respondent notes, had no authority to promulgate the standard without following formal rulemaking procedures. See, e.g., *Deering Milliken, Inc. v. OSHRC,* 630 F.2d 1094 (5th Cir. 1980).

The Secretary promulgated the standard in question, section 1910.212(a)(3)(ii), on May 29, 1971, as part of 248 pages of standards which comprised at that time 29 C.F.R. Part 1910, 36 Fed. Reg. 10466–10714 (1971). At the same time, the Secretary adopted as occupational safety and health standards pre-existing maritime and construction safety standards, which occupied 291 pages in the Code of Federal Regulations. 36 Fed. Reg. 10468–10469 (1971), recodified as 29 CFR Parts 1915–1918 and 1926 in 36 Fed. Reg. 25232 (1971). It is unreasonable to require all persons who may have an interest in these standards to examine within sixty days of their issuance this large volume of regulations and ascertain whether the standards had been adopted without any substantial alteration as required under section 6(a) of the Act. As Senator Williams stated, the effect of such a requirement is to "deny small employers and companies their day in court," as they lack the legal resources to thoroughly review the validity of standards within a short time after their adoption. 116 CONG. REC. 37340 (daily ed. Nov. 16, 1979). As a practical matter, large corporations and labor organizations would be hard pressed to review within sixty days standards packages of the magnitude that the Secretary adopted in 1971. Considerations as to finality and the burden of continuous procedural challenges should not preclude employers from raising attacks of the type involved here.

Accordingly, Respondent's challenge may properly be raised in an enforcement proceeding before the Commission.

***

In considering the merits of the challenge, however, the Review Commission held the revised standard valid on the grounds that it "may be read to convey the same meaning" as the originally adopted standard.[60] Thus, the Review Commission accepted the distinction made in *National Industrial Constructors* between substantive and procedural challenges, concluding that substantive challenges may

[60] 9 OSHC at 1097.

be made in enforcement proceedings. It then made a further distinction among procedural challenges: those made to the procedure used by the Department of Labor in issuing the underlying pre-OSHA established federal standard (the situation in *National Industrial Constructors*) would not be considered, and those made to OSHA's procedures in adopting the standard under §6(a) (the *Rockwell* case), or after adoption under §6(b) (*Deering Milliken*) would be considered. OSHA argued this approach before the Court of Appeals for the Fourth Circuit in *Daniel International Corp. v. OSHRC*,[61] involving the same challenge to OSHA's construction standards as had been advanced in *National Industrial Constructors*. The court took a different course, however, adopting a prejudice rule in rejecting the employers' claim that the standard was invalid. The court said:

> Despite the sharp disagreement between *Deering Milliken* and *Union Oil* [616 F.2d 1113 (9th Cir. 1980)] on the one hand, and *NIC* [*National Industrial Constructors*] on the other, the differences are less significant than the opinions themselves suggest. Even if an employer may raise a procedural challenge in an enforcement proceeding, the employer still must demonstrate that it has suffered prejudice because of the administrative action. In both *Deering Milliken* and *Union Oil*, the employer challenged specific standards which the Secretary modified when adopting the standards as OSHA standards. The employers suffered prejudice because the modification of the standards effectively denied them an opportunity to comment on the standards.
>
> No similar prejudice occurred here, or in *NIC*. The purpose of the 30-day notice requirement in §553(d) is to "afford persons affected a reasonable time to prepare for the effective date of a rule or rules or to take any other action which the issuance of rules may prompt." Administrative Procedure Act Legislative History, 79th Cong., 2d Sess. 201 (1946). The Construction Safety Act standards became effective in April 1971, and became effective as OSHA standards in May 1971. The inspections of Daniel's construction site occurred in August and September 1977. Daniel can hardly claim that it was denied a sufficient and reasonable opportunity to prepare for any attack it might, over six years after their promulgation, care to make whether we speak in terms of an April 1971 or a May 1971 effective date. Because Daniel can make no showing that the acceleration of the effective date caused it prejudice, we hold that Daniel's challenge is without merit.[62]

The net effect of these cases is that in enforcement proceedings, challenges to the *procedures* used by OSHA in adopting the standards have generally been rejected either on the merits (*Deering Milliken*) or on jurisdictional grounds (*National Industrial Constructors*).[63] On the other hand, substantive challenges to health standards in enforcement proceedings on such grounds as the infeasibility of the standard have been entertained by the Review Commis-

[61]656 F.2d 925 (4th Cir. 1981).

[62]*Id.* at 930–31.

[63]But see "should-shall" cases, discussed in Chapter 2, which arguably constitute procedural challenges to the standards.

sion and the courts.[64] On the latter issue, Chief Judge Skelly Wright of the Court of the Appeals for the District of Columbia Circuit sought to summarize the state of the law in the *Lead* litigation.

United Steelworkers v. Marshall

647 F.2d 1189 (D.C. Cir. 1980)

J. SKELLY WRIGHT, Chief Judge:

***

* * * Every earlier OSHA standard restricting toxic substances that has created a hierarchy among preferred means of compliance has by its terms built the very concept of feasibility into the standard. The cotton dust standard is typical. It requires employers to meet the PEL solely through engineering and work practice controls "except to the extent that the employer establishes that such controls are not feasible." 29 C.F.R. §1910.1043(e)(1) (1979). Where "feasible engineering and work practice controls" cannot achieve the PEL, the employer may and must use respirators to make up the difference, though he must continue to use engineering and work practice controls to reduce exposure as well as they can. *Id.* §1910.1043(e)(2); see *id.* §1910.1043(f)(1)(iii).

The feasibility test for a standard that only expects feasible improvements from employers may appear circular. But reasonable construction of such a standard avoids circularity. The cases have apparently treated these standards as creating a general *presumption* of feasibility for an industry. A company could not simply refuse to pursue engineering or work practice controls by asserting their infeasibility. Rather, it would have to attempt to install controls to the limits of contemporary technical knowledge and of its own financial resources. Judicial review of feasibility [under §6(f)] would have some meaning, because the court would have to find substantial evidence to justify this presumption—evidence that the technical knowledge to meet the PEL without relying on respirators was available or would likely be available when deadlines arrive, and that enough firms could afford to so meet the PEL that the market structure of the industry would survive. Thus the court would do a preliminary test of feasibility on any pre-enforcement challenge to the rulemaking.

---

[64]The Administrative Conference of the United States recently addressed the issue of "Judicial Review of Rules in Enforcement Proceedings." Recommendation No. 82-7, 1 C.F.R. §305.82-7 (1983). In the preamble to its recommendation, the conference indicated that the purpose of its recommendation was to "distinguish between types of challenges to rules that should ordinarily be covered by any preclusion [of enforcement review] provisions Congress decides to adopt and types of issues that ordinarily should remain available in enforcement proceedings * * *." The conference stated that "sound" principles of administrative law favor "prompt and dispositive resolution of disputed issues" arising out of rulemaking; also, it said, the potential for later conflicting court decisions and the possibility that a rule may be overturned several years after its promulgation "can be extremely disruptive of the regulatory scheme." Finally, it stated, correcting rulemaking defects becomes more difficult as the original record grows "stale" and the situations of the interested parties change. The competing consideration, however, is that those affected by the rule should have a "full and fair opportunity" to challenge it on all available grounds. The conference concluded that Congress should consider the following factors as favoring limiting judicial review to the preenforcement stage: the likelihood that there will be widespread participation in the rulemaking; that the proceeding will involve "intensive exploration of factual issues"; that those affected will incur "substantial and immediate costs" to comply; and the need for prompt compliance with the rule on a national or industrywide basis. The conference recommended that Congress should "ordinarily" limit review to the preenforcement stage of issues "relating to the procedures employed in the rulemaking or the adequacy of factual support in the administrative record." On the other hand, questions of the constitutional basis for the rule and the agency's statutory authority should ordinarily not be precluded at the enforcement stage. See P.R. Verkuil, *Congressional Limitations on Judicial Review of Rules,* 57 TUL. L. REV. 733 (1983). See also Annot., 61 A.L.R. FED. 422 (1983). See discussion in Section D, note 66, on the distinction between challenges based on "general" infeasibility and infeasibility in a particular plant.

But the court always reserves the power to test feasibility again later—in reviewing denial of a temporary variance or, where an employer found such a variance insufficient, in judicial review of an enforcement proceeding under 29 U.S.C. §659 (1976). In the temporary variance proceeding the employer could of course argue that the standard was infeasible for his company in particular—at least at the effective date of the standard. But in an enforcement proceeding the employer could also expect the agency and court to entertain the affirmative defense that the standard had proved *generally* infeasible—even if the court had earlier found otherwise on pre-enforcement review. In any of these proceedings, of course, the employer would bear the burden of proof.[119]

Thus, since the presumption of feasibility remains rebuttable in pre-enforcement review, the court would not expect OSHA to prove the standard *certainly* feasible for *all* firms at *all* times in *all* jobs. But it would have to justify the presumption, and the attendant shift in burden, with reasonable technological and economic evidence and analysis. Assuming this construction of the earlier standards is correct, the question is whether any of it changes under the lead standard.

***

D. Feasibility of Health Standards

OSHA health standards typically provide for the implementation of "feasible" engineering controls. "Feasibility" encompasses both technological and economic feasibility, as has already been discussed.[65] In enforcing OSHA health standards, a major issue has been whether, as to the particular plant or operations involved, engineering controls are feasible.[66]

1. Technological Feasibility

Boise Cascade Corp. v. Secretary

694 F.2d 584 (9th Cir. 1982)

Hug, Circuit Judge:

Boise Cascade Corporation petitions for review of a decision of the Occu-

[119]Because the financial incapacity of a single firm to meet the PEL cannot normally prove a standard infeasible, presumably an employer could only raise a defense of *technological* infeasibility in an enforcement proceeding. Indeed, one OSHA standard, the asbestos standard, generally allows respirators only where engineering and work practice controls prove "technically" infeasible. 29 C.F.R. §1910.1001-(d)(ii) (1979). On the other hand, an employer could attempt to defend against an enforcement citation by showing, through his particular circumstances, that a standard is *generally* economically infeasible. See *Atlantic & Gulf Stevedores v. OSHA*, 534 F.2d at 555.

***

[65]See discussion in Chapter 10.

[66]This issue, whether the engineering control requirement should be applied in a particular plant because of feasibility considerations, must be distinguished from a related but analytically separate issue: whether the standard should be held invalid in an enforcement proceeding because it is "generally" infeasible. This latter issue has been discussed in the prior section; see particularly *United Steelworkers v. Marshall* excerpt reprinted in Section C, this chapter. A significant distinction between the two types of challenges is that if the standard is held to be generally infeasible it would be declared invalid; on the other hand, if it is found to be infeasible in a particular plant, the standard's validity would not be affected but only the citation would be dismissed. This distinction has not generally been clearly recognized by the courts.

pational Safety and Health Review Commission enforcing a citation for violation of occupational noise standards. The Secretary of Labor cited the company for failure to install engineering and administrative controls to abate noise levels and failure to implement a hearing conservation program. We affirm the determination that Boise Cascade violated the standard and the requirement that it institute an effective hearing conservation program. Because the Secretary failed to prove that feasible engineering controls exist, we vacate the portions of the citation requiring implementation of such controls.

Boise Cascade Corporation operates a plant in Turner, Kansas, at which composite cans are formed.

***

The citation ordered implementation of feasible administrative and engineering controls.[9] The ALJ found that administrative controls were not feasible and this finding was implicitly affirmed by the Commission.[10] Therefore, the issue of the feasibility of administrative controls is not before us on Boise's appeal. Feasibility is a question of fact reviewable under the substantial evidence standard. *Donovan v. Castle & Cooke Foods,* No. 77-2565, slip op. at 5393 n.8 [10 OSHC 2169, 2173] (9th Cir. Nov. 19, 1982); see *Diversified Industries Division, Independent Stave Co. v. OSHRC,* 618 F.2d 30, 32 (8th Cir. 1980).

Where the employer is cited for failing to implement feasible engineering controls it is the Secretary's burden to show that engineering controls are both technologically and economically feasible. *Castle & Cooke,* slip op. at 5393, 5395 [10 OSHC at 2173–2175]; *Carnation Co. v. Secretary of Labor,* 641 F.2d 801, 803 (9th Cir. 1981). Neither the ALJ nor the Commission made a specific finding that engineering controls were feasible, but that conclusion was implicit in their decisions. The Commission enforced the regulation without requiring further proof of feasibility, perhaps assuming the Secretary had carried his burden on this issue in promulgating the regulation. It is established, however, that the Secretary must prove the feasibility of controls, both in promulgating regulations and again in enforcing them.

In promulgating a standard, the Secretary is not restricted to the state of the art in the regulated industry. He may impose requirements that force technological development beyond what the industry is presently capable of producing. See *United Steelworkers of America v. Marshall,* 647 F.2d 1189, 1264 (D.C. Cir. 1980), *cert. denied,* 453 U.S. 913 (1981); *Diebold, Inc. v. Marshall,* 585 F.2d 1327, 1333 (6th Cir. 1978); *American Iron & Steel Inst. v. OSHA,* 577 F.2d 825, 838 (3d Cir. 1978), *cert. dismissed,* 448 U.S. 917 (1980); *Society of Plastics Indus., Inc. v. OSHA,* 509 F.2d 1301, 1309 (2d Cir.), *cert. denied,* 421 U.S. 992 (1975). The standard may require imple-

---

[9]Engineering controls are those that reduce the sound intensity at the source of the noise. This is achieved by insulation of the machine, by substituting quieter machines and processes, or by isolating the machine or its operator.

Administrative controls attempt to reduce workers' exposure to excess noise through use of variable work schedules, variable assignments, or limiting machine use.

See United States Department of Labor, *Guidelines to the Department of Labor's Occupational Noise Standards for Federal Supply Contracts* (1970).

[10]The form of administrative controls suggested by the Secretary was the assignment of machine operators to split positions, so that only a portion of each day was spent at [machines exceeding permissible noise levels]. Boise Cascade executives testified that split assignments would violate the seniority provisions of their agreement with the union. Machine operator positions required skill and training, and were desirable because of the pay rate. Qualified workers were allowed to bid for operator positions, which were awarded on the basis of seniority. The company contended that assigning operators to lower-paying, non-operator positions for several hours each day would violate the seniority agreement. In addition, less skilled workers would have to be used as machine operators for part of each day, endangering worker safety. The Secretary offered virtually no evidence to refute these claims, merely noting that the union would have to be "cooperative," and share the burden of worker protection.

mentation of technology which is merely "looming on today's horizon." *American Iron & Steel Inst.*, 577 F.2d at 838. "So long as [the Secretary] presents substantial evidence that companies acting vigorously and in good faith can develop the technology, OSHA can require industry to meet [standards] never attained anywhere." *United Steelworkers*, 647 F.2d at 1264–65 (footnote omitted). See 29 U.S.C. §655(f).

In an enforcement proceeding, the Secretary's burden as to technological feasibility is of a different character. It must be established that specific, technically feasible controls exist to abate the violation. See *Diebold*, 585 F.2d at 1333; *National Realty & Constr. Co. v. OSHRC*, 489 F.2d 1257, 1268 (D.C. Cir. 1973). Nothstein, *The Law of Occupational Safety and Health* 438 (1981) describes the Secretary's burden as follows:

> To prove that an employer violated a standard, the Secretary of Labor must at the very least demonstrate that, at the time of the alleged violation, a technologically feasible method of abatement existed. * * * However * * * the Commission, while holding that employers may not be found in violation of a standard if the technology for compliance does not exist, [has] stated this does not mean that the Secretary of Labor must prove that "off the shelf" controls were available to the employer at the time of the violation. Rather it is sufficient to show the existence of technology that could be adapted to the employer's operations (feasible at the time the employer allegedly violated the standard).

(Citations omitted.) See also *Marshall v. West Point Pepperell, Inc.*, 588 F.2d 979, 981, 985 (5th Cir. 1979) (although it is not necessary that "controls have been fully developed and marketed by disinterested third parties," a control is not feasible if it "exists only in the realm of theory.").

The Secretary can thus satisfy his burden at the time of enforcement by showing that effective controls have been installed on the same or similar equipment by other members of the industry. In *Diversified Industries*, it was held that technological feasibility was established where an expert "testified that she was familiar with woodworking machines generally and with technical literature concerning the abatement of their noise. She had seen engineering controls successfully implemented on a machine similar to Diversified's planer." 618 F.2d at 34. In *RMI Co. v. Secretary of Labor*, 594 F.2d 566 (6th Cir. 1979), the court reluctantly affirmed a finding of technological feasibility on a showing that the Secretary recommended a specific type of acoustical enclosure to be placed around machines. Id. at 569, 571. The Secretary established technological feasibility in *Castle & Cooke* with the testimony of acoustical experts who testified that specific engineering controls were presently available to abate noise levels. Slip op. at 5393 [10 OSHC at 2173].

The Secretary offered no such proof of technical feasibility in this case. His proof of the violation was limited to Stewart's testimony regarding impermissible noise levels. A thorough reading of the record reveals no specific suggestion as to how the winders and seamers could be modified and no indication that other industry members have implemented satisfactory controls. The Secretary contends on appeal that Boise Cascade's own evidence provided proof of feasibility. That evidence, however, principally concerned the now-abandoned punch presses. The only reference to the winders and seamers was made by Carleton Wold, who testified to a "recollection" that "[t]here were some [modifications] put into the other areas of the plant, the winders and some other areas." There is no more detailed description of what modifications were implemented or what their effectiveness was. Clearly the Secretary's reliance on this testimony is misplaced.

We therefore vacate the portion of the citation requiring implementa-

tion of engineering controls, on the ground that the Secretary failed to provide substantial evidence that such controls were technologically feasible.
* * *

As technology develops, engineering controls may well become feasible in the future. Nothing in this opinion should be construed so as to absolve the employer from the continuing obligation to search for feasible engineering controls to protect the hearing of its employees.

The order of the Commission enforcing the citation is affirmed in part, vacated in part.*

2. Economic Feasibility

Unlike the issue of technological feasibility, the question of the economic feasibility of engineering controls has given rise to sharp differences of view among the courts and the Review Commission. The current OSHA view, stated as an instruction to field personnel in the Field Operations Manual, is as follows:

> (3) * * * Economic feasibility means that the employer is financially able to undertake the measures necessary to abate the citations received. The CSHO shall inform the employer that, although the cost of corrective measures to be taken will generally not be considered as a factor in the issuance of a citation, it will be considered during an informal conference or during settlement negotiations.
>
> (a) If the cost of implementing effective engineering, administrative, or work practice controls or some combination of such controls, would seriously jeopardize the employer's financial condition so as to result in the probable shut down of the establishment or a substantial part of it, an extended abatement date shall be set when postponement of the capital expenditures would have a beneficial effect on the financial performance of the employer.
>
> (b) If the employer raises the issue that the company has other establishments or other locations within the same establishment with equipment or processes which, although not cited as a result of the present inspection, nevertheless would require the same abatement measures as those

*[Author's note—The issue of whether the doctrine of "technology-forcing" is also applicable in determining technological feasibility in the enforcement of a standard has divided the Review Commission. See *Samson Paper Bag Co.*, 8 OSHC 1515, 1518–19 (Rev. Comm'n 1980) (Barnako, Comm'r) (OSHA need not show that "off the shelf" controls were available, but must show the existence of technology that could be adapted to the employer's operations); *id.* at 1523 (Cleary, Chairman, concurring) (controls are feasible if they involve not the development of new technology but the adaptation of existing technology to a new use); *id.* at 1526–28 (Cottine, Comm'r, concurring) (technology-forcing applicable in enforcement).

Current OSHA field instructions state that the compliance officer shall determine whether abatement means are technologically feasible; such feasibility "is the existence of technical know-how as to materials and methods available or adaptable to specific circumstances which can be applied to cited violations with a reasonable possibility that employee exposure to occupational health hazards will be reduced." Field Operations Manual, ch. III, OSHR [Reference File 77:2523]. The Commission has held that controls are feasible even though they will not succeed in reducing exposure to the permissible limit; the employer is required to implement the engineering controls to the extent feasible and then use personal protective equipment to reach the PEL. *Continental Can Co.*, 4 OSHC 1541 (Rev. Comm'n 1976). OSHA health standards also require that engineering controls be implemented in these circumstances; see, e.g., 29 C.F.R. §1910.1029(f) (1982) (coke oven emissions).]

under citation, the economic feasibility determination shall not be limited to the cited items alone. In such cases, although the employer will be required to abate the cited items within the time allowed for abatement, the opportunity to include both the cited and the additional items in a long-range abatement plan shall be offered.

(c) When additional time cannot be expected to solve the employer's financial infeasibility problem, the Area Director shall refer the problem to the Regional Administrator who shall contact the Director of Field Operations. A decision will be made at the National Office level and communicated to the Area Director through the Regional Administrator. The citation shall not, however, be delayed beyond 6 months.[67]

Thus, OSHA is clear that it will extend an abatement date where the cost of engineering controls would "jeopardize" an employer's existence.[68] OSHA does not state a clear policy, however, on whether the controls are deemed infeasible where the extended abatement date does not "solve" the problem.[69]

The Review Commission's view, on the other hand, has undergone significant change on this issue. In *Continental Can Company*,[70] decided in 1976, involving enforcement of the OSHA noise standard, the majority of the Commission interpreted the economic feasibility requirement in the noise standard to require that "all the relevant cost and benefit factors * * * be weighed."[71] The issue arose again before the Review Commission in 1980 in *Samson Paper Bag Company*.[72] Each of the Commissioners stated a different view, and, although only Commissioner Barnako voted to reaffirm the *Continental Can* cost-benefit rule, that case remained binding Commission precedent in the absence of a majority view on a different test to be applied in enforcing the noise standard.

In 1981, the legal situation changed significantly when the Supreme Court's *Cotton Dust* decision barred cost-benefit analysis in the promulgation of OSHA health standards.[73] The first major case to deal with the cost-benefit issue in enforcement after the *Cotton Dust* decision was *Donovan v. Castle & Cooke Foods*.[74] In that case, OSHA had cited the employer for violation of the noise standard; the citations, however, were vacated by the Review Commission.[75] The case was pending before the Ninth Circuit Court of Appeals on

[67]Field Operations Manual, ch. III, OSHR [Reference File 77:2523].

[68]OSHA has extensive instructions to field staff on the setting of "long-term" abatement dates for the implementation of engineering controls, and on the setting of multistep abatement requirements. Field Operations Manual, ch. III, OSHR [Reference File 77:2523–:2525].

[69]See *Industrial Union Dep't v. Hodgson*, 499 F.2d 467 (D.C. Cir. 1974), reprinted in Chapter 10. See also *United Steelworkers v. Marshall*, 647 F.2d 1189, 1270 n.119 (D.C. Cir. 1980), reprinted in Section C, this chapter.

[70]4 OSHC 1541 (Rev. Comm'n 1976) (Cleary, Comm'r, dissenting).

[71]*Id.* at 1547.

[72]8 OSHC 1515 (Rev. Comm'n 1980).

[73]*American Textile Mfrs. Inst. v. Donovan*, 452 U.S. 490 (1981).

[74]692 F.2d 641 (9th Cir. 1982).

[75]5 OSHC 1435 (Rev. Comm'n 1977).

OSHA's appeal when the Supreme Court *Cotton Dust* decision was handed down. The court of appeals asked for supplemental briefing on the "impact" of the *Cotton Dust* decision on the case before them. The thrust of the employer's arguments was that the Supreme Court's decision that cost-benefit analysis is prohibited is limited to standards issued under §6(b)(5); it does not apply, the employer claimed, to the noise standard, which was issued under §6(a). OSHA argued that the "comparison of costs and benefits is totally irrelevant to proceedings to enforce the noise standard." It asserted that since the Review Commission's decision in *Continental Can* that "feasible" meant cost-beneficial rested on its view that "feasible" in §6(b)(5) meant cost-beneficial, *Continental Can* should be reversed in light of the *Cotton Dust* decision. The court of appeals handed down its decision in November 1982, agreeing with the employer and holding the Supreme Court's *Cotton Dust* decision inapplicable.

Donovan v. Castle & Cooke Foods

692 F.2d 641 (9th Cir. 1982)

HUG, Circuit Judge:

***

The Secretary argues that because section 6(b)(5), as defined in *American Textile,* and 29 C.F.R. §1910.95(b)(1) both use the term "feasible," there is a "natural presumption" that identical words used in different parts of the same act are intended to have the same meaning. The Secretary may not claim the benefit of this presumption. The regulation is not a legislative enactment. Moreover, critical differences distinguish section 6(b)(5) and section 6(a), from which 29 C.F.R. §1910.95 is derived. The term "feasible," added by Congress to section 6(b)(5), does not appear in the other subsections of section 6. The clear implication is that Congress intended in section 6(b)(5) to give the Secretary the additional grant of authority necessary to protect employees against the most serious health hazards. See *American Textile,* 452 U.S. at 512. As a result, standards promulgated under section 6(b)(5) are a distinct "species of the genus of standards governed by the basic requirement" of the Act. *Industrial Union Department, AFL-CIO v. American Petroleum Institute,* 448 U.S. 607, 642 (1980) (plurality opinion).

The subsections of section 6 are also distinguishable by their relationship to section 3(8), 29 U.S.C. §652(8). The Court concluded in *Industrial Union* that section 3(8)'s definition of a standard was incorporated by reference into section 6(b)(5). 448 U.S. at 642. In *American Textile,* however, it determined that the general requirement of section 3(8) that all standards be "reasonably necessary or appropriate" could not be read to countermand the stricter and more specific requirement imposed by section 6(b)(5). The Court therefore rejected the cotton industry's argument that section 3(8) superimposed a requirement of cost-benefit analysis on the issuance of section 6(b)(5) standards. "Congress did not contemplate any further balancing by the agency *for toxic material and harmful physical agents standards,* and we should not 'impute to Congress a purpose to paralyze with one hand what it sought to promote with the other.'" *American Textile,* 452 U.S. at 513, *quoting Weinberger v. Hynson, Westcott & Dunning, Inc.,* 412 U.S. 609, 631 (1973) (emphasis added).

However, the Court expressly left open the question of how section 3(8)'s "reasonably necessary" requirement might limit the promulgation or enforcement of standards not dealing with toxic substances or harmful physical agents: "[t]his is not to say that §3(8) might not *require* the balancing of costs and benefits for standards promulgated under provisions other than §6(b)(5) of the Act." 452 U.S. at 513 n.32 (emphasis added). Because the Court specifically declined to apply its definition of feasible to regulations promulgated under section 6(a), we conclude that *American Textile* does not control the Commission's interpretation of 29 C.F.R. §1910.95(b)(1) or of the statutes that authorized its promulgation. The Commission was thus free to exercise its authority to interpret the regulations it is charged with enforcing.

We view the Commission's interpretation of "feasible" as a reasonable reconciliation of the regulation's language and the statute's "reasonably necessary" requirement. In requiring employers to implement economically feasible engineering controls, the regulation would require only those controls that are "reasonably necessary and appropriate" to protect employees' health and safety. The Commission has concluded that a control is not reasonably necessary and appropriate if the benefits to be gained by employees do not justify the cost. This interpretation of section 3(8) is neither unreasonable, arbitrary, nor an extension of the authority granted the Commission by the Act.

The Secretary warns that approval of the Commission's test of feasibility will undermine the central goal of the Act, which is "to assure so far as possible every working man and woman in the Nation safe and healthful working conditions * * *." 29 U.S.C. §651(b). We disagree. The test of feasibility advanced by the Secretary, that controls are infeasible only if they threaten the employer's financial viability, offers less protection to the worker than does the Commission's approach. It conditions employee health and safety on the employer's financial viability and thus arguably exempts some employers from providing necessary protection. The Commission's test emphasizes benefits to be gained by employees and does not excuse or exempt marginal employers who cannot adequately protect employees. At the same time, the Commission's test provides a realistic view of the range of hazards from which employees require protection and the alternate means of providing that protection, and recognizes that the Act does not "require employers to provide absolutely risk-free workplaces whenever it is technologically feasible to do so. * * *" *Industrial Union,* 448 U.S. at 641.

The Secretary next contends that, regardless of how "feasible" is to be interpreted, the burden of establishing economic infeasibility should be placed on the employer. The Commission recognized this issue, but did not resolve it.

***

We * * * hold that when the Secretary seeks enforcement of a citation alleging a violation of 29 C.F.R. §1910.95(b)(1), he bears an initial burden of showing that technologically feasible engineering controls are available to the cited employer. Although the Secretary will generally have access to information on the average development and installation cost of the proposed controls, he will not have knowledge of the specific economic impact implementation of the controls will have on the cited employer. Therefore, once the Secretary meets his initial burden, the burden must shift to the employer, who may raise the issue of economic feasibility. The employer may satisfy this burden of production with evidence of the relative cost to him of various methods of noise control. That is, the employer may compare the costs of implementing engineering controls, administrative controls, or personal protective equipment at a specific employment location. If the employer raises the question of economic feasibility in this manner, the burden of proof returns to the Secretary, who must establish that the benefit of the proposed engineering controls justifies their relative cost in comparison to other abatement methods.

***

Shortly thereafter, the Review Commission revisited the economic feasibility issue in *Sun Ship,* reversing *Continental Can* in light of the Supreme Court decision in *Cotton Dust* and disagreeing with the view of the Court of Appeals of the Ninth Circuit in *Castle & Cooke Foods.*

Sun Ship, Inc.

11 OSHC 1028 (Rev. Comm'n 1982)

By the Commission:

***

I.

Sun Ship, Inc. operates a shipyard in Chester, Pennsylvania. In September, 1975, a United States Department of Labor compliance officer inspected the plant and measured the noise to which one blacksmith was exposed during his work shift. Because the compliance officer found excessive noise exposure, Sun Ship received a citation alleging noncompliance with the noise standard, section 1910.95(b)(1).

At the hearing, the Secretary proposed engineering controls to reduce the excessive noise. An expert, a consulting engineer with experience in noise control, testified about the controls and the costs involved in implementing them. He gave a figure representing the cost of fabricating the controls but did not quantify any indirect costs such as any loss of productivity, incidental to using the controls.

***

Sun Ship argues that *ATMI [American Textile Manufacturers Institute, Inc. v. Donovan,* 452 U.S. 490 (1981)] does not affect *Continental Can* because *ATMI* involved a section 6(b)(5) rather than a section 6(a) standard, *ATMI* involved a life-threatening hazard, and under *ATMI* cost-benefit may be used to determine the feasible abatement to reach a given exposure level. The Secretary argues that *ATMI* requires *Continental Can* to be overruled. In the Secretary's view, "feasible" must be given the same meaning in the noise standard as in section 6(5)(b). Also, the Act does not require a cost-benefit analysis for a standard adopted under section 6(a).

II

***

In sum, in *ATMI* the Supreme Court held that, because neither the Act nor its legislative history reveals that Congress intended to require a cost-benefit analysis as a part of rulemaking under section 6(b)(5), the word "feasible" in that section must be given its ordinary meaning—that which is "achievable." Significantly, the Court determined that Congress, in enacting the Act, believed that the costs of employee injuries and illnesses were at least as great as, and likely greater than, the costs of eliminating injuries and illnesses and that, therefore, the benefits of reduced injuries and illnesses generally justified the costs that would be incurred in doing so.

The Court's interpretation of "feasible" is limited, Sun Ship argues in this case, to standards promulgated under section 6(b)(5). Because the noise standard was promulgated under section 6(a) of the Act, * * * and neither Congress nor the Secretary ever considered the balance of costs against benefits for this standard, Sun Ship argues that cost-benefit analysis should be required to enforce the standard. Accordingly, Sun Ship urges that the Commission adhere to *Continental Can*.

We find these arguments unpersuasive in light of the Court's decision. Clearly, the Supreme Court decided only the meaning of section 6(b)(5) and not whether cost-benefit analysis might be required for standards promulgated under other sections. See 452 U.S. at 509 n.29 & 513 n.32. However, the Supreme Court held that, because of the absence of any strong indication that Congress intended "feasible" to have special meaning, "feasible" as used by Congress must be given its ordinary meaning. The Court further held that, under its ordinary meaning of "achievable," "feasible" cannot require cost-benefit analysis. See 452 U.S. at 508–09, 511–12. The Court thus decided that Congress did not intend to require cost-benefit analysis by using the word "feasible." The identical question of legislative intent is presented in interpreting the noise standard.

The pertinent language of the noise standard is: "When employees are subjected to sound exceeding those listed . . ., feasible administrative or engineering controls shall be utilized." The Secretary adopted this language in the noise standard because Congress directed that the Walsh-Healey standards, one of which was the noise standard, be adopted as occupational safety and health standards under section 6(a). * * * Accordingly, the Commission in *Continental Can* examined the Walsh-Healey Act and its history as well as the history of the noise standard to determine whether "feasible" had any special meaning. However, there was no indication that the word was intended to have any meaning other than the common definition. In these circumstances, the Commission concluded, "The standard must be interpreted to effectuate the Congressional purposes underlying the Act." 4 BNA OSHC at 1546.

Because "feasible" had no special meaning under the Walsh-Healey Act, it is appropriate to interpret it in accordance with the Congressional purpose underlying the OSH Act. But, as held by the Supreme Court in *ATMI*, "feasible" under the OSH Act means "achievable" and does not require cost-benefit analysis. Regulations are to be construed consistent with the statutes under which they are promulgated. *United States v. American Trucking Ass'ns,* 310 U.S. 534, 542 (1940). Furthermore, unless a different intent is clearly evident the same statutory terms are to be given the same meaning. *Chugach Natives, Inc. v. Dayon Ltd.*, 588 F.2d 723 (9th Cir. 1979) and cases cited. Accordingly, regulatory language should be given the same meaning as the same language appearing in the statute. See *Baroid Div. of N.L. Industries, Inc. v. OSHRC,* 660 F.2d 439, 447 (10th Cir. 1981), and *RMI Co. v. Secretary of Labor,* 594 F.2d 566 (6th Cir. 1979), where the statutory and regulatory uses of "feasible" are viewed as synonymous. Accordingly, we overrule *Continental Can* to the extent that it holds that "feasible" in the noise standard requires cost-benefit analysis.[9]

[9]The United States Court of Appeals for the Ninth Circuit recently issued its decision in *Donovan v. Castle & Cooke Foods, A Div. of Castle & Cooke, Inc.,* No. 77-2565 [10 OSHC 2169] (9th Cir. Nov. 19, 1982), affirming a 1977 Commission decision that applied the *Continental Can* cost-benefit test. The court concluded that the Commission's interpretation was "neither unreasonable, arbitrary, nor an extension of the authority granted the Commission by the Act," slip op. at 13 [2175], under the applicable standard for judicial review of agency decisions, slip. op. at 6 [2172]. The court deferred to the Commission's expertise in interpreting the standard. This case involves a necessary reevaluation of *Continental Can* and a reinterpretation of the noise standard in light of the Supreme Court's decision in *ATMI*.

The Ninth Circuit considered the Supreme Court's interpretation of the term "feasible" in section 6(b)(5) of the Act to be inapplicable to section 6(a) standards. However, we treat as identical the term "feasible" in the statute and the noise standard. When Congress authorized the Secretary to adopt established federal standards and national consensus standards as occupational safety and health standards, it understood that the Walsh-Healey standards would be the primary source of established federal standards for covered workplace hazards. *General Motors Corp., GM Parts Div.,* 9 BNA OSHC 1331, 1336-37 (1981), *appeal dismissed,* No. 81-

The Supreme Court did not, however, reject the idea that "feasible" includes economic considerations. All parties in *ATMI* agreed that a standard would not be feasible if its costs would be so high as to threaten the economic viability of an industry.[10] The Court did not disagree, expressing the view that inclusion of economic considerations "is certainly consistent with the plain meaning of the word 'feasible.'" 452 U.S. at 530 n.55. In promulgating the standard in *ATMI,* the Secretary had estimated the costs of compliance for the different sectors of the cotton industry affected by the standard and evaluated their ability to absorb the costs. Because he concluded that compliance with the standard would not threaten the "long-term profitability and competitiveness" of the various industrial sectors, the Secretary concluded that the standard was feasible. *Id.*

Thus, "feasibility" under section 6(b)(5) includes consideration of whether the cost of compliance with a standard will be so great as to threaten an industry's long-term profitability and competitiveness. By analogy, considerations of cost must also enter into whether administrative or engineering controls are "feasible" under section 1910.95(b)(1).[11] Generally, administrative or engineering controls would be economically infeasible if their cost would seriously jeopardize the cited employer's long-term financial profitability and competitiveness.[12] There may also, as the Secretary

---

3194 (6th Cir. Sept. 16, 1981). Several of these standards, like the noise standard, regulated exposure to "toxic materials" and "harmful physical agents" and contained feasibility requirements. At the same time that Congress authorized the adoption of section 6(a) standards, it authorized the promulgation of standards dealing with toxic materials or harmful physical agents under section 6(b)(5). Section 6(b)(5) also contains a feasibility requirement. There is no indication that Congress intended the feasibility requirement of existing standards that the Secretary was authorized to implement immediately to be measured by a different criterion than feasibility under section 6(b)(5).

Accordingly, we respectfully decline to acquiesce in the Ninth Circuit's divergent interpretation of the term "feasible." Rather, we view the *ATMI* interpretation to be applicable to the regulation in question. Cf. *Baroid Div. of N.L. Industries, Inc. v. OSHRC,* 660 F.2d 439, 447 (10th Cir. 1981) (definition of "feasible" as economically and technologically capable of being done applied to a section 5(a)(1) violation based on Supreme Court's definition in ATMI).

[10]In an early case involving a challenge by a union to the promulgation of a §6(b)(5) standard, the Court of Appeals for the District of Columbia Circuit held that the Secretary could properly consider the economic impact on an industry in determining whether a standard is feasible. *Industrial Union Department, AFL-CIO v. Hodgson,* 499 F.2d 467 (D.C. Cir. 1974). That holding has been followed in other decisions involving challenges to 6(b)(5) standards. *United Steelworkers of America, AFL-CIO-CLC v. Marshall,* 647 F.2d 1189, 1265 (D.C. Cir. 1980), *cert. denied,* 453 U.S. 913 (1981); *American Iron & Steel Institute v. OSHA,* 577 F.2d 825 (3d Cir. 1978), *cert. dismissed,* 448 U.S. 917 (1980). Thus, by the time *ATMI* was decided by the Supreme Court, there was no longer any real dispute that the Secretary could, and indeed must, consider the economic impact on an industry in determining feasibility. Indeed, in the industry challenge to the cotton dust standard that led to the Supreme Court's decision in *ATMI,* the D.C. Circuit held that the Secretary had not established the economic feasibility of the standard for the cottonseed oil industry. *AFL-CIO v. Marshall,* 617 F.2d 636 (D.C. Cir. 1979). That holding was not in issue before the Supreme Court.

[11]The initial inquiry into feasibility of administrative or engineering controls involves whether such controls are technologically feasible. The Commission has held that, because the standard expressly requires that administrative or engineering controls be used in preference to personal protective equipment, controls which achieve a significant reduction in noise exposure will be deemed technologically feasible even if they do not result in absolute compliance with Table G-16 levels. *Continental Can Co.,* 4 BNA OSHC at 1545–46; see also *Samson Paper Bag Co.,* [8 OSHC 1515 (1980)]; *Turner Co.,* 76 OSHRC 108/A2, 4 BNA OSHC 1554 (1976), *rev'd on other grounds,* 561 F.2d 82 (7th Cir. 1977). This test for technological feasibility is consistent with the ordinary meaning of "feasible" as set forth by the Supreme Court in *ATMI.*

[12]In most cases arising under the noise standard, there will not be a serious question of the employer's ability to afford the cost of technologically feasible controls; economic feasibility will be clear. See, e.g., *Continental Can Co.,* 4 BNA OSHC at 1542 n.4. It has been our experience that the controls suggested by the Secretary in cases arising under section 1910.95(b)(1) are not generally so expensive as to give rise to substantial question of their economic feasibility. Moreover, in most cases it will be possible to alleviate the employer's financial difficulty by extending the time required for the installation of controls. See *Samson Paper Bag Co., supra* (lead and separate opinions). Thus, in cases where financial hardship is asserted, the Commission will consider whether this hardship can be adequately cured through an extended abatement date before concluding that the controls are infeasible.

Commissioner Cleary notes that, when there is a question whether controls will be so

points out, be situations in which a particular employer is lagging so far behind its industry in protecting the health and safety of its employees that it cannot afford to implement controls that are generally feasible throughout the industry. See *Industrial Union Department, AFL-CIO v. Hodgson*, 499 F.2d 467, 478 (D.C. Cir. 1974). In these situations, controls may be feasible even though they are beyond the financial capability of the cited employer. Thus, as part of his proof of feasibility, the Secretary must show either that the cost of engineering and administrative controls will not threaten the cited employer's long-term profitability and competitiveness or that the employer's inability to afford the cost of controls results from the employer lagging behind the industry in providing safety and health protection for employees.[13]

***

Accordingly, we conclude that the Secretary has established a *prima facie* case of violation of the noise standard. The judge's decision granting Sun Ship's motion to dismiss is therefore set aside and the case is remanded for further proceedings.[20]

[Dissenting opinion of Chairman Rowland is omitted.]*

---

costly as to threaten an employer's ability to remain in business, a range of economic factors may merit consideration. These factors could include the competitive structure of the industry involved, the competition that the industry faces from other industries or from other countries, inflation, the ability of the company to raise capital, its ability to alter employees' work schedules and assignments, its other capital investment requirements, energy costs, and the like. See *ATMI; Samson Paper Bag, supra* (Cleary, concurring).

[13]In Commissioner Cleary's view, the Secretary may make out a prima facie showing of economic feasibility by showing that the estimated costs of the controls are small compared with other figures that represent the company's financial ability. For example, if implementation of the controls would require an initial capital investment, the Secretary could show that the company regularly makes capital investments that are substantially larger than implementation of the controls would require. If the controls would involve annual costs, the Secretary could show that the company's annual cash flow or profits, over a representative period of time, are substantially larger than the annual cost the controls would impose. Such information is readily available to the Secretary either through public reports that corporations must file or through discovery.

The employer may then rebut the Secretary's case by showing that the cost figures are incorrect or that additional, indirect, costs will be incurred, or that any of the factors listed above, such as the competitive structure of the industry, make controls infeasible. The employer may also rebut the Secretary's case by showing that while controls for the cited locations are economically feasible, the cost of plant-wide controls would exceed the bounds of economic feasibility. *Carnation Co. v. Secretary of Labor*, 641 F.2d 801 (9th Cir. 1981).

[20]Commissioner Cottine further notes that the dissent's conclusion that the noise standard represents an impermissible delegation depends on the erroneous premise that the noise standard is vague. The dissent characterizes the noise standard as an impermissible delegation to the Commission of the Secretary's authority to promulgate standards and establish enforcement criteria under the Act. This broad characterization ignores the fact that the noise standard establishes a definite criteria for occupational noise exposure—an 8-hour time weighted average of 90 dBA. §1910.95(a). This required performance is to be achieved by "feasible" administrative or engineering controls. §1910.95(b). As previously noted, the term "feasible" permits a fair determination of the required conduct and is capable of application to the factual controversies presented to this adjudicatory forum. Like any factual inquiry "feasibility" is determined by the unique characteristics of an employer's economic conditions and the technological attributes of the abatement methodology. The "feasibility" criterion is ascertainable and the possibility of more than one interpretation does not justify invalidation under the delegation doctrine. *ATMI v. Donovan*, 452 U.S. at 541 n.75. The mere fact that a term requires a factual analysis does not invalidate the standard. Factual determinations are both the heart of adjudication and the focus of delegation. See *Atlas Roofing Co., Inc. v. OSHRC*, 430 U.S. 442, 449–60 (1977). See generally L. Jaffe, *Judicial Control of Administrative Action* ch. 3 (1965). Moreover, a feasibility determination is the type of complex fact-finding regularly committed to an administrative agency for specialized adjudication in the context of the forum's experience and expertise. See, e.g., *Crowell v. Benson*, 285 U.S. 22, 46–47 (1932); *Reconstruction Finance Corp. v. Bankers Trust Co.*, 318 U.S. 163, 170 (1943). Accordingly, Commissioner Cottine rejects the dissent's conclusion that the noise standard is unenforceable because the regulatory use of the term "feasible" results in an impermissible delegation.

***

*[Author's note—See also Martucci, *The Defense of Economic Infeasibility in Enforcement*

E. Unpreventability of the Hazard

Section 17(k)[76] provides that a serious violation exists at a workplace if there is a "substantial probability that death or serious physical harm could result from the condition * * * unless the employer did not, or could not with the exercise of reasonable diligence, know of the presence of the violation." The Act contains no definition of nonserious violations which, accordingly, includes all violations that are not serious, willful, or repeated.[77] On the face of the statute, the requirement that the employer have actual or constructive knowledge "of the presence of the violation" relates only to the question of whether the violation was serious and *not* to the question of whether there was *any* violation. Both the Review Commission and the courts of appeals, however, have applied a "preventability" test in deciding whether there was a violation at all.

The first major decision on this issue was *National Realty & Construction Co. v. OSHRC*,[78] which involved an alleged serious violation of the general duty clause. The Court of Appeals for the District of Columbia Circuit said that Congress "did not intend the general duty clause to impose strict liability" or that unpreventable hazards would be considered "recognized" under §5(a)(1). The court explained its view as follows:

> Though resistant to precise definition, the criterion of preventability draws content from the informed judgment of safety experts. Hazardous conduct is not preventable if it is so idiosyncratic and implausible in motive or means that conscientious experts, familiar with the industry, would not take it into account in prescribing a safety program. Nor is misconduct preventable if its elimination would require methods of hiring, training, monitoring, or sanctioning workers which are either so untested or so expensive that safety experts would substantially concur in thinking the methods infeasible. All preventable forms and instances of hazardous conduct must, however, be entirely excluded from the workplace. To establish a violation of the general duty clause, hazardous conduct need not actually have occurred, for a safety program's feasibly curable inadequacies may sometimes be demonstrated before employees have acted dangerously. At the same time, however, actual occurrence of hazardous conduct is not, by itself, sufficient evidence of a violation, even when the conduct has led to injury. The record must additionally indicate that demonstrably feasi-

Proceedings Under the Occupational Safety and Health Act: An Appraisal of the Decisions of the Occupational Safety and Health Review Commission, 17 New Eng. L. Rev. 1 (1981) (economic considerations are a legitimate concern in OSHA enforcement proceedings and the Review Commission should balance the competing interests).

On November 9, 1983, OSHA issued an enforcement instruction on the noise standard, 29 C.F.R. §1910.95 (1983), addressing the decision of the court of appeals in *Castle & Cooke Foods,* and permitting employer utilization of personal protective equipment and a hearing conservation program rather than engineering controls in certain circumstances to meet the sound permissible exposure level, 13 OSHR 661–62 (1983).]

[76] 29 U.S.C. §666(j).

[77] No citations may be issued for *de minimis* violations, defined as those "which have no direct or immediate relationship to safety or health." Sec. 9(a), 29 U.S.C. §658(a). The Field Operations Manual gives an example of a *de minimis* violation: The standard (29 C.F.R. §1910.27(b)(1)(ii)) requires that rungs of a ladder be no more than 12 inches apart; if the rungs are 13 inches apart, the condition is "de minimis." OSHR [Reference File 77:2712].

[78] 489 F.2d 1257 (D.C. Cir. 1973).

ble measures would have materially reduced the likelihood that such misconduct would have occurred.[79]

The court of appeals in *National Realty* held that to establish the preventability of a violation the record must contain evidence "demonstrating the feasibility and likely utility of the particular measures which National Realty should have taken to improve its safety policy."[80] Since OSHA had not met its burden of proof on the issue, the general duty citation was dismissed. Thus, as early as 1973, the determination of whether a violation was preventable—and therefore should be sustained—hinged in large measure on the factual issue of the adequacy of the employer's safety program.

While *National Realty* analyzed the preventability issue in terms of the "recognition" requirement of the general duty clause, the same principles have been applied to violations of standards as well. In *Brennan v. OSHRC (Alsea Lumber Co.),*[81] the Court of Appeals for the Ninth Circuit expressly applied a preventability test to a nonserious violation. The court said:

> The legislative history of the Act indicates an intent not to relieve the employer of the general responsibility of assuring compliance by his employees. Nothing in the Act, however, makes an employer an insurer or guarantor of employee compliance therewith at all times. The employer's duty, even that under the general duty clause, must be one which is achievable. See *National Realty, supra.* We fail to see wherein charging an employer with a non-serious violation because of an individual, single act of an employee, of which the employer had no knowledge and which was contrary to the employer's instruction, contributes to achievement of the cooperation sought by the Congress. Fundamental fairness would require that one charged with and penalized for violation be shown to have caused, or at least to have knowingly acquiesced in, that violation. Under our legal system, to date at least, no man is held accountable, or subject to fine, for the totally independent act of another. A conspiracy to violate the Act is neither alleged nor reflected in the record before us.[82]

Although OSHA argued and the Commission held that unpreventability—or employee misconduct or isolated event, as the defense was sometimes called—was an affirmative defense, and the employer had the burden of showing the adequacy of its safety program, the *Alsea* court as well as the Court of Appeals for the Fourth Circuit imposed this burden of proof on OSHA.[83] In 1982, however, the Court

[79]489 F.2d at 1266–67. In its decision, the District of Columbia Circuit Court of Appeals distinguished between the preventability criterion, applicable to all violations, and the "knew or could have known requirements" under §17(k), which applied only to serious violations. Under the court's view, a violation may exist because the hazard was "preventable," even though it would not be a serious violation because at the moment of its occurrence the employer could not have known about it. *Id.* at 286 n.41, reprinted in Chapter 14, Section B. Subsequent cases, however, have merged the preventability test with the knowledge requirement of §17(k). See generally Note, *OSHA: Employer Liability for Employee Violations,* 1977 DUKE L.J. 614, and discussion *infra* this section.

[80]489 F.2d at 1267.

[81]511 F.2d 1139 (9th Cir. 1975).

[82]*Id.* at 1144–45. Although the violation in *Alsea* was nonserious, the court spoke in terms of employer "knowledge"; this illustrates the merging by the courts of the preventability test and the §17(k) requirement of employer knowledge. See *supra* note 79.

[83]*Ocean Elec. Corp. v. Secretary of Labor,* 594 F.2d 396 (4th Cir. 1979), *rev'g Ocean Elec.*

of Appeals for the Fifth Circuit decided in *H.B. Zachry Co. v. OSHRC* that the employer had the burden of showing that it "could not reasonably have prevented the isolated, negligent conduct" of an employee. The court's opinion demonstrates as well the manner in which the Review Commission and the courts analyze the issue of the adequacy of an employer's safety program.

H.B. Zachry Co. v. OSHRC

638 F.2d 812 (5th Cir. 1981)

JOHN R. BROWN, Circuit Judge:
***

THE FACTS

The facts which gave rise to the penalty imposed by the Secretary and affirmed by the Commission are uncontroverted. Zachry is a general contractor engaged in worldwide construction with its principal place of business at San Antonio, Texas. Early in 1976, Zachry was engaged in construction at the Sooner Dam and Power Plant project near Pawnee, Oklahoma. At approximately eight o'clock on the morning of February 11, 1976, one of Zachry's crane operators, Raymond Kitchens, was ordered by his immediate supervisors to transport a load of pipe from a storage area to an excavation site—a distance of several hundred feet. Located approximately 28 feet above this pathway were uninsulated energized electrical transmission lines carrying 7,000 to 7,200 watts of electricity.

The load was secured by steel cables attached to the end of the pipe, wrapped around, and then fastened to a hook on the boom of the crane. With this load attached, the boom rested at an angle of about 30° to 35° with a length including the jib of approximately 48 to 50 feet. Two employees, Tobias and Fragu were assigned to assist Kitchens in this moving operation by holding the ends of the pipe to stabilize it. As the load was being transported across the work site, the jib of the mobile crane came into contact with one of the transmission lines fatally electrocuting Tobias and seriously injuring Fragu.

The following morning, Roger Jackson, an OSHA compliance officer commenced an on-site investigation pursuant to 29 U.S.C.A. §657(a). Based on this investigation, the Secretary of Labor issued a citation on February 27, 1976, which alleged that Zachry had committed a serious violation of 29 C.F.R. 1926.550(a)(15)(i) by failing to maintain a minimum clearance of ten feet between energized electrical transmission lines and the crane or its load, thus creating the hazard of electrical shock. The citation was accompanied by the Secretary's recommendation of a $700 penalty and an order of immediate abatement of the hazard.

Zachry timely contested the citation and proposed penalty pursuant to 29 U.S.C.A. §659(c). Both at the administrative hearing on July 20, 1976, and now before this Court, Zachry challenges the citation and penalty * * * [claiming that] the company should not be held liable for an employee's unforeseeable negligence when it has adequately trained and supervised its employees in this particular area of safety.

Corp., 3 OSHC 1705 (Rev. Comm'n 1975) (Moran, dissenting). The Fourth Circuit had previously issued a decision affirming the citation, 5 OSHC 1672 (1977), but a motion for reconsideration was granted, leading to the reversal. Judge Craven, a member of the original panel, died during the pendency of the rehearing. 594 F.2d at 396 n.1.

The Administrative Law Judge (ALJ) by decision and order of October 26, 1976, affirmed the Secretary's citation and $700 penalty. The same results were reached by the Review Commission on January 31, 1980.
***

WAS THE AFFIRMATIVE DEFENSE MADE OUT?

Zachry's * * * principal defense against the Secretary's citation is that it could not reasonably have prevented the isolated, negligent misconduct of crane operator Kitchens because the company had adequately trained and supervised its employees in this area of safety. In order to prevail in this defense, an employer must demonstrate that (i) all feasible steps were taken to avoid the occurrence of the hazard, *General Dynamics v. OSHRC,* 599 F.2d 453, 464 (5th Cir. 1977); *Horne Plumbing and Heating Co. v. OSHRC,* 528 F.2d at 571. This includes the training of employees as to the dangers and the supervision of the work site, *Horne,* 528 F.2d 564 at 569. And (ii) the actions of the employee were a departure from a uniformly and effectively communicated and enforced work rule of which departure the employer had neither actual nor constructive knowledge. See, e.g., *General Dynamics,* 599 F.2d at 465; *Ames Crane and Rental Service, Inc. v. Dunlop,* 532 F.2d 123 (8th Cir. 1976).

When faced with this defense, reviewing courts have consistently looked at the record evidence of the employer's safety program. In the present case, testimony at the hearings by Zachry's safety director described the company's safety program as including (i) the showing of safety films, (ii) regular scheduled safety meetings, and (iii) the distribution of safety bulletins and materials.[14] The manual and materials included work rules, one of which required crane operators to maintain a minimum distance of ten feet from energized overhead wires as well as a rule requiring the assignment of a flag person to assist cranes moving over the job site.

Zachry submitted evidence that Kitchens had signed and received these various company publications and that he had been exposed to numerous safety bulletins and films. On cross-examination, Kitchens testified that he had previously been warned about that particular power line and was familiar with the company's ten foot clearance instruction. Photographs introduced into evidence depict the ten foot clearance warning signs which were painted on the inside and outside of the operator's crane cab and posted around the work site.

Although there is substantial evidence that Zachry had a safety program and work rules requiring cranes to maintain a minimum of ten feet clearance from energized power lines, the ALJ and Commission found deficiencies in the communication and enforcement of the rule. Kitchens testified that he skipped about one-half of the regularly scheduled safety meetings yet would sign an attendance roster for fear that he would lose his job.

An even more glaring deficiency was the inadequacy of supervision given Kitchens and his two helpers the day of the accident. Zachry's "Supervisor's Rules for Safe Operation of Cranes" stated the following principles:

1. Power cranes and excavators are powerful and complex machines. Your complete attention to control is required every minute.
2. You—as the man responsible for all operations under your control—are the crucial key to safe machine performance.

[14]The record reveals that a consistent procedure for distribution of company safety manuals [Zachry's exhibits A, B, C, D(1) and D(3)] was maintained continuously during the various periods the three subject employees worked for Zachry. Each time an employee is hired or rehired, Zachry's testimony stated that each was indoctrinated in safety policies and programs and obtained the employee's receipt of the safety manuals given him [Zachry's exhibit D(4)].

3. Do not operate or allow to be operated any machine under your control until you fully understand the operator's manual.

ZACHRY'S EXHIBIT—B AT 1.

According to the uncontradicted testimony of Kitchens, his supervisor was actually present when he began loading the pipes on the crane. Practically speaking, it is undisputed that the only way the ten foot minimum clearance standard could have been met was by shortening the cables which were wrapped around the pipes. The fact that the supervisor apparently failed to instruct Kitchens to make that adjustment or to specifically designate one of the two helpers to give hand signals to Kitchens, permits the inference that the supervisor lacked familiarity with the manual's rules. This fact becomes particularly significant since this was Kitchens' first attempt to carry a load with this particular crane under these power lines. Apparently, the ALJ and Commission found Kitchens was given little guidance and instruction—evidence which independently demonstrates the ineffective communication and enforcement of the company work rules. Further, this Court has previously stated the "behavior of supervisory personnel sets an example at the work place, an employer has—if anything a *heightened* duty to ensure the proper conduct of such personnel," *Floyd S. Pike Electrical Contractors Inc. v. OSHRC,* 576 F.2d 72, 77 (5th Cir. 1978), *citing National Realty and Construction Company, Inc. v. OSHRC,* 489 F.2d 1257, 1267 n.38 (5th Cir. 1973). Assuming, *arguendo,* that the supervisor was familiar with this particular company rule, his noncompliance with it is additional evidence that the implementation of the rule was lax. *Id.* Thus the company's attempted defense of employee negligent misconduct fails because of the employer's inability to establish to the satisfaction of the factfinder that it effectively communicated and enforced work rules which were necessary to ensure compliance with OSHA standards.[17]

***

The court's view on burden of proof, stated in footnote 17 to the opinion, is that the employer has the burden to prove both that the hazard was unpreventable and that it exercised reasonable diligence under §17(k). And, the court said, if the employer failed to prove unpreventability, it necessarily has failed to prove reasonable diligence. Under this approach, it would seem that the issues whether there is a violation and whether it is serious have become merged, at least in some circumstances. Thus, if the employer shows the ade-

[17]Since Zachry did not qualify for the employee misconduct affirmative defense, its contention that the citation should be vacated is meritless because the record does not demonstrate that the company with the exercise of reasonable diligence could have known of the presence of the violation. An examination of the elements of the employee misconduct affirmative defense reveals that it is designed to demonstrate that an employer exercised reasonable diligence by providing adequate safety training and supervision to its employees. Thus, an employer's inability to establish the adequacy of the safety instructions to his employee shows a failure to exercise reasonable diligence. In a case involving the same standard as the case at bar, the Court rejected the employer's defense of employee misconduct, finding that the company had failed to "take adequate precautionary steps to instruct and train employees to protect against reasonable foreseeable dangers," *Ames Crane and Rental Service,* 532 F.2d at 125. The Eighth Circuit in a case identical to *Ames Crane* stated that an employer "cannot fail to properly train and supervise its employees and then hide behind its lack of knowledge concerning their dangerous work practices." *Danco Construction Co. v. OSHRC and Secretary,* 586 F.2d 1243, 1247 (8th Cir. 1978).

quacy of its safety program, this would lead to the dismissal of the citation altogether; if the employer fails to meet the burden—as the court said here—the serious violation would be affirmed.[84]

F. "Greater Hazard" Defense

One of the more frequently asserted defenses to a citation is the "greater hazard" defense. The essence of the defense is that the hazards of compliance with the standard are greater than the hazards of noncompliance. This defense has been recognized by the Review Commission,[85] the courts,[86] and by OSHA in its Field Operations Manual;[87] however, stringent conditions have been applied before a citation will be dismissed on the basis of the defense.

General Electric Co. v. Secretary of Labor

576 F.2d 558 (3d Cir. 1978)

HIGGINBOTHAM, Circuit Judge:

The General Electric Company has petitioned this Court to review a judgment and order of the Occupational Safety and Health Review Commission (the Commission). That order affirmed a citation charging General Electric with a non-serious violation of §5(a)(2) of the Occupational Safety and Health Act (the Act) for violating the standard of 29 C.F.R. §1910.22(c).[2] We have jurisdiction of this matter pursuant to 29 U.S.C. §660(a). The order of the Commission will be affirmed.

On December 19, an agent of the Secretary of Labor inspected General Electric's plant in Erie, Pennsylvania. As a result of this inspection, a citation was issued alleging a violation of 29 C.F.R. §1910.23(c)(1) pertaining to an unguarded pit in the paintroom of Building 12.[3]

[2]29 C.F.R. §1910.22(c) provides:

This section applies to all permanent places of employment, except where domestic, mining, or agricultural work only is performed. Measures for the control of toxic materials are considered to be outside the scope of this section.

(c) Covers and guardrails. Covers and/or guardrails shall be provided to protect personnel from the hazards of open pits, tanks, vats, ditches, etc.

[3]General Electric had previously been cited with a violation of this same standard with respect to an assembly area in Building 12. This violation had been corrected by the time of the December 19 investigation.

The paintroom which is the subject of the citation in question is used to paint locomotives and transit cars. It is one hundred feet long and twenty feet wide. In the center of the room is a pit ninety-eight feet long, four feet two inches wide and five feet deep which is used to provide access for painting the undersides of the vehicles. There is a permanent steel ladder at one end of the pit for entry and exit.

Normally, only two painters work on one shift. There are three shifts of duty when the plant is at full production. In addition, supervisors may be present. Painting is done mainly from mechanized personnel carriers that move along the top and sides of the vehicles. Only one vehicle is in the paintroom at a time. The carrier comes so close to the vehicle that a permanent railing would not be practical.

When the painting is completed, the vehicle is removed from the room via steel overhead

[84]This is contrary to the views of the *National Realty* court, *supra* note 79. In its 1983 Field Operations Manual, OSHA refers to "unpreventable employee misconduct or 'isolated event'" as an "affirmative defense"; it states that the defense applies if the violative conduct was "unknown to the employer" and "in violation of an adequate workrule which was effectively communicated and uniformly enforced." OSHR [Reference File 77:2908].

[85]E.g., *Industrial Steel Erectors,* 1 OSHC 1497 (Rev. Comm'n 1973).

[86]E.g., *General Electric Co. v. Secretary of Labor,* 576 F.2d 558 (3d Cir. 1978).

[87]Ch. V, OSHR [Reference File 77:2908].

General Electric filed a Notice of Contest pursuant to 29 U.S.C. §659(c). A hearing was held on June 3, 1974 before Administrative Law Judge William E. Brennan. At the hearing, ALJ Brennan granted the Secretary's motion to amend the citation to charge, in the alternative, non-serious violations of 29 C.F.R. §1910.22(c) and 29 C.F.R. §1910.23(a)(5). ALJ Brennan found that none of the standards were applicable and that, if 29 C.F.R. §1919.22(c) were applicable, General Electric had established that compliance with that standard "would diminish rather than enhance the safety of employees." This is referred to as the "greater hazard" defense.

The Secretary of Labor petitioned the Commission to review ALJ Brennan's order insofar as it vacated the citation for violation of §1910.22(c). The Commission held that 29 C.F.R. §1910.22(c) was applicable to the facts. The Commission refused to consider the "greater hazard" defense because General Electric had not sought a variance pursuant to §6(d) of the Act, 29 U.S.C. §655(d), and had not shown that resort to the variance procedure would be inappropriate. General Electric urges us to vacate the Commission's order because of the failure to consider the "greater hazard" defense.

The Commission has recognized that an enforcement action may be successfully defended by proving that compliance with a standard would result in a hazard to employees greater than that resulting from existing procedures. See *Secretary of Labor v. Industrial Steel Erectors, Inc.*, 1 OSHC 1497 (1974). The Commission has stated, however, that the defense is a narrow one. Besides demonstrating that a greater hazard would result from compliance, the employer must also show that "alternative means of protecting employees are unavailable" and that "a variance application under section 6(d) of the Act would be inappropriate." *Secretary of Labor v. Russ Kaller, Inc. t/a Surfa Shield,* 4 OSHC 1758, 1759 (1976); *Secretary of Labor v. Cornell & Company, Inc.*, 5 OSHC 1018 (1977); *Secretary of Labor v. George A. Hormel Co.*, 2 OSHC 1190 (1974).

General Electric has not applied for a variance. The Commission found "no indication that resort to the variance procedure would be inappropriate." General Electric now argues that the Secretary's conduct in pursuing this enforcement action demonstrates that a variance application would be futile and, therefore, inappropriate. We do not accept this argument. The Secretary's vigorous prosecution of this case indicates only that the Secretary is convinced that the standard in question has been violated. Whether the Secretary will grant a variance cannot be known until an application is made. A finding that an exception from the standard should be allowed would not be inconsistent with a finding that the standard has not been complied with. In his thoughtful opinion, ALJ Brennan emphasized that compliance with the standard in question would result in a greater danger to employees than that presented by current practices. General Electric presented both union and management witnesses whose testimony supported this conclusion. Thus, there is no reason to believe that a variance application would be futile or inappropriate.

General Electric argues that exclusion of the "greater hazard" defense under these circumstances is contrary to the purposes of the Act. We conclude that the purposes of the Act are furthered by the Commission's decision.

There is no question that Congress' purpose on enacting the Act was "to assure so far as possible every working man and woman in the Nation safe and healthful working conditions and to preserve our human resources. * * *" 29 U.S.C. §651(b).

doors at either end of the room. The cleanup operation then begins. The two painters stand with their back to the walls and push masking paper, paint cans and other debris left from the painting job into the pit using brooms with five foot handles. The painters then enter the pit, push the debris into a pile and carry it away in boxes. The painting operation takes between twenty-four and sixty-four hours. The cleanup operation takes approximately an hour.

Every employer has the initial obligation to make sure that his working areas comply with all applicable standards. If there is reason to believe that compliance with certain standards may jeopardize his employees, a variance should be sought. If a "greater hazard" defense is allowed at an enforcement proceeding without requiring initial resort to the variance procedures or a showing that such resort would be inappropriate, there would be little incentive for an employer to seek a variance under these circumstances.

General Electric contends that an employer who correctly believes that his working conditions are safer than those prescribed in the standards should not be penalized for bypassing the variance procedures and taking his chances that he will not be cited or that he will prevail in an enforcement proceeding. The flaw in this argument is that some employers will believe *incorrectly* that their working conditions are safer than those prescribed in the standards. By removing this incentive to seek variances, the Commission would be allowing an employer to take chances not only with his money, but with the lives and limbs of his employees. This we cannot do.

The exclusion of the "greater hazard" defense at the enforcement proceeding does not force General Electric to implement procedures which increase the danger to its employees. Counsel for the Secretary made it clear to us at oral argument that General Electric can now apply for a variance.[7] Thus while we agree that no employee should be subjected to greater dangers because of an employer's failure to apply for a variance before a citation was issued, we rely on the Secretary's representation, through his counsel, that this will not be the result of our decision today. General Electric has been assessed a penalty of fifty-five dollars for its non-serious violation of 29 C.F.R. §1910.22(c). Under 29 U.S.C. §666(d), employers who fail to correct violations for which citations have issued within the period prescribed by the Commission may suffer civil penalties of up to one thousand dollars for each day during which the violation continues. The Commission order specifies no time period within which General Electric must comply with the standard here. We assume that General Electric will be given sufficient time to apply for a variance and that no penalty will be imposed while the application, if filed, is being considered. We conclude that, since General Electric may now apply for a variance without incurring additional penalty, the exclusion of the "greater hazard" defense here was not error.

The Commission's order will, therefore, be affirmed. The petition for review will be denied.

In order to successfully assert the greater hazard defense the employer must show, among other things, that a variance application under §6(d)[88] "would be futile" or "inappropriate."[89] Section 6(d) provides for the granting to an employer of a variance, allowing the employer to maintain "conditions" or to implement "practices, means, methods, operations or processes" which are "as safe and healthful as those which would prevail if [the employer] complied

[7]Counsel for the Secretary stated: "I believe that indirectly [the Commission is] telling the employer that, go to the Secretary. You should have made your application before. You can still make it." Later, he reiterated this point: "[O]ur position is that the effect of the Commission's decision is, in effect, it tells the employer we believe that a variance application would still be appropriate." [Author's note—For a recent case in which the greater hazard defense was recognized but rejected, see *True Drilling Co. v. Donovan,* 703 F.2d 1087 (9th Cir. 1983).]

[88]29 U.S.C. §655(d).

[89]*General Elec. Co.,* 576 F.2d at 561, *supra* p. 525.

with the standard." The §6(d) variance may be granted after notice to affected employees and an opportunity to participate in a hearing; at the hearing, the employer must show by the "preponderance of the evidence" that the alternative conditions are equally safe or healthful. OSHA has issued procedural regulations[90] for §6(d) variances as well as "temporary variances" under §6(b)(6)(A)[91] and "national defense" variances under §16.[92]

One of the difficult variance issues has been whether, and to what extent, OSHA personnel are authorized to take enforcement action when they make workplace visits to determine if a variance should be granted.[93] OSHA's field instruction, "Procedures for Entering Workplace for Evaluating Variance Requests," provides as follows:

> 7. If the OVR [OSHA variance representative] sees a hazard, the employer shall immediately be notified. If the employer cannot correct the hazard immediately, he shall be required to provide interim protective measures until the hazard can be corrected. The appropriate Area Director shall be notified of the hazard and of the method of abatement, or of the employer's plan of abatement if the hazard is not corrected prior to the completion of the variance investigation. The Area Director shall take whatever compliance action is deemed necessary to determine that the identified hazard has been corrected.
>
> 8. If the OVR receives a complaint of hazardous conditions from an employee, the OVR shall advise the employee of the special nature of the investigation and provide the complainant with specific instructions on how to contact the Area Director for appropriate action. No inspection of the alleged hazards shall be made by the OVR.[94]

OSHA variance procedures became an issue during the hearing on the Willow Island disaster, when it was suggested that in a variance "inspection" prior to the catastrophe, OSHA had been "forewarned" of hazards at the workplace. Dr. Bingham answered the assertion in her testimony before a congressional oversight committee, by emphasizing that the scope of the variance inspection was unrelated to the integrity of the structure or the reasons for the collapse.[95]

[90]29 C.F.R. §§1905.1–.41 (1983). These regulations provide for the granting of an interim order, without a hearing, granting relief to an employer who has applied or is applying for a variance, to be effective until action is taken on the application. See, e.g., 29 C.F.R. §1905.11(c) (1983).

[91]29 U.S.C. §655(b)(6)(A). For an example of a temporary variance recently issued, see Temporary Variance to 45 Companies from Workplace Lead Standard Provisions on Medical Removal, 48 FED. REG. 4062 (1983). Variances under §6(d) from provisions of the lead and arsenic standards were also granted to major automobile companies. See, e.g., General Motors Corp., Grant of Variance, 45 FED. REG. 46,922 (1980).

[92]29 U.S.C. §665.

[93]Under §6(d), OSHA is authorized to conduct an inspection to determine whether a variance should be granted, "where appropriate."

[94]OSHA Instruction STD 6.2, OSHR [Reference File 21:8211, 21:8212]. For a discussion of the issue of whether consultants should report hazards disclosed during on-site consultation visits, see Chapter 11, Section D.

[95]*OSHA Oversight—Willow Island, West Virginia Cooling Tower Collapse: Hearings Before Subcomm. on Compensation, Health and Safety of the House Comm. on Education and Labor*, 95th Cong., 2d Sess. 9 (1978). The catastrophe is discussed in Chapter 13, Section B. But see also COMPTROLLER GENERAL OF THE UNITED STATES, REPORT TO THE CONGRESS, MWD-76-19 WORKER PROTECTION MUST BE INSURED WHEN EMPLOYERS REQUEST PERMISSION TO DEVIATE FROM SAFETY AND HEALTH STANDARDS (1975) (criticizing OSHA's procedures for granting variances).

Other frequently asserted defenses to OSHA citations are: the standard under which the citation was issued is void for vagueness;[96] compliance with the standard is impossible ("impossibility of performance");[97] and employees were not exposed to the violative condition.[98]

[96]See, e.g., *Kropp Forge Co. v. Secretary,* 657 F.2d 119 (7th Cir. 1981) (holding that the "hearing conservation" requirements to the noise standard (29 C.F.R. §1910.95(b)(3)), before it was amended in 1982, were "unconstitutionally vague" in that they failed to give "reasonable notice of the conduct to be prohibited) (Swygert, J., concurring). But see *Brennan v. OSHRC* (*Santa Fe Trail Trans. Co.*), 505 F.2d 869 (10th Cir. 1974) (first-aid standard upheld against vagueness challenge), and discussion in Chapter 2.

[97]See, e.g., *M.J. Lee Constr. Co.,* 7 OSHC 1140, 1144 (Rev. Comm'n 1979) (to establish impossibility defense, employer must show that compliance with cited standard was "functionally impossible or would preclude performance of required work" and "alternative means of employee protection are unavailable"). This defense is analogous to the technological feasibility defense to health citations, discussed in this chapter, Section D(1).

[98]See, e.g., *Gilles & Cotting, Inc.,* 3 OSHC 2002, 2003 (Rev. Comm'n 1976) (Cleary, Comm'r, concurring; Moran, Comm'r, dissenting) (adopting as test for exposure whether "employees either while in the course of their assigned working duties, their personal comfort activities while on the job, or their normal means for ingress-egress to their assigned workplace, will be, are or have been in a zone of danger").

PART IV

Employee Rights and Responsibilities

16

Employee Participation in the Inspection Process

A. Background

The Act pervasively provides for the participation of employees in the implementation of all aspects of the OSHA program. Among the most important of their rights are: the right to obtain a workplace inspection if a written complaint is filed alleging hazardous conditions and meeting certain formality requirements;[1] the right to accompany the inspector during the physical inspection of the workplace;[2] and the right to participate as a party in proceedings before the Review Commission.[3] In addition, employees are given the right to be informed about safety and health hazards in the workplace.[4] Employee rights are further protected by a broad prohibition against discrimination by any person against any employee for exercising certain specified rights or "any right afforded by this Act."[5]

The importance of employee participation in the OSHA program was clearly articulated by the Senate Committee of Labor and Public Welfare in its favorable report of an OSHA bill in 1970.[6] More recently, in 1980, Secretary of Labor Ray Marshall testified in an oversight hearing before the same, but renamed committee, the Senate Committee on Labor and Human Resources, and a major topic in his testimony was "Workers' Right to Participate." He said:

> If the goals of the OSH Act are to be realized, OSHA must encourage employers and employees to actively participate in efforts to provide safe and healthful working conditions. During its first seven years the Agency dealt almost exclusively with the employer in this regard,

[1]Sec. 8(f), 29 U.S.C. §657(f) (1976).
[2]Sec. 8(e), 29 U.S.C. §657(e).
[3]Sec. 10(c), 29 U.S.C. §659(c).
[4]See discussion in Chapter 18, Section F.
[5]Sec. 11(c), 29 U.S.C. §660(c).
[6]Reprinted in Chapter 1.

with little attention to the role of the worker in recognition and abatement of hazards. The past three years have seen OSHA remedy this and concentrate equally on the contributions all parties can make. The immediate goal of our efforts is to assure that workers have the information they need to aid in protecting themselves.

Workers, by exercising their rights, are in the best position to get hazardous conditions corrected. Leptophos and D.B.C.P. [dibromochloropropane; see discussion in Chapter 4], two pesticides, and E.S.N., a niax catalyst used in the manufacturing of plastics, were identified in three separate instances by sick workers as the toxic agents that caused their illness. These perceptive acts by workers brought the matter to OSHA's attention and by doing so saved thousands of other workers from serious illness.[7]

Secretary Marshall also discussed employee rights relating to inspections, access to injury and illness records, access to medical and exposure records, chemical identification, protection from discrimination, walkaround pay, and the right to refuse hazardous work.[8]

As part of its extensive public education campaign, OSHA has sought to familiarize employees with their responsibilities and rights under the Act. One of OSHA's publications is entitled *All About OSHA,* which contains a section, "Employee Responsibilities and Rights."[9]

All About OSHA

OSHA Leaflet No. 2056 at 42–45 (1980)

***

EMPLOYEE RESPONSIBILITIES AND RIGHTS

Although OSHA does not cite employees for violations of their responsibilities, each employee "shall comply with all occupational safety and health standards and all rules, regulations, and orders issued under the Act" that are applicable.

RESPONSIBILITIES

As an employee, you should:

- Read the OSHA poster at the jobsite.
- Comply with all applicable OSHA standards.
- Follow all employer safety and health rules and regulations, and wear or use prescribed protective equipment while engaged in work.
- Report hazardous conditions to the supervisor.
- Report any job-related injury or illness to the employer, and seek treatment promptly.
- Cooperate with the OSHA compliance officer conducting an inspection if he or she inquires about safety and health conditions in your workplace.
- Exercise your rights under the Act in a responsible manner.

[7] *Oversight on the Administration of the Occupational Safety and Health Act, 1980: Hearings Before the Senate Comm. on Labor and Human Resources,* 96th Cong., 2d Sess. pt. 2 at 1034–44 (1980).

[8] All of these topics are discussed in more detail in this and the following two chapters.

[9] The publication also contains a section, "Employer Responsibilities and Rights" at 39–41.

11(C) RIGHTS: PROTECTION FOR USING RIGHTS

Employees have a right to demand safety and health on the job without fear of punishment. That right is spelled out in Section 11(c) of the Act.

The law says employers shall not punish or discriminate against workers for exercising rights such as:

- Complaining to an employer, union, OSHA or any other government agency about job safety and health hazards;
- Filing safety or health grievances;
- Participating on a workplace safety and health committee or in union activities concerning job safety and health;
- Participating in OSHA inspections, conferences, hearings or other OSHA-related activities.

If an employee is exercising these or other OSHA rights, the employer cannot discriminate against that worker in any way, such as through firing, demotion, taking away seniority or other benefits earned, transferring the worker to an undesirable job or shift or threatening or harassing the worker.

If the employer has knowingly allowed the employee to do something in the past (such as leaving work early), he or she may be violating the law by punishing the worker for doing the same thing following a protest of hazardous conditions. If the employer knows that a number of workers are doing the same thing wrong, he or she cannot legally single out for punishment the worker who has taken part in safety and health activities.

Workers believing they have been punished for exercising safety and health rights should contact the nearest OSHA office within 30 days of the time they learn of the alleged discrimination. A union representative can file the 11(c) complaint for the worker.

The worker does not have to complete any forms. An OSHA staff member will complete the forms, asking what happened and who was involved.

Following a complaint, OSHA investigates. If an employee has been illegally punished for exercising safety and health rights, OSHA asks the employer to restore that worker's job, earnings and benefits. If necessary, OSHA takes the employer to court. In such cases the worker does not pay any legal fees.

If a state agency has responsibility for job safety and health enforcement, employees may file their 11(c) complaint with either federal OSHA or the state agency.

OTHER RIGHTS

As an employee, you have the right to:

- Review copies of appropriate OSHA standards, rules, regulations and requirements that the employer should have available at the workplace.
- Request information from your employer on safety and health hazards in the area, on precautions that may be taken, and on procedures to be followed if an employee is involved in an accident or is exposed to toxic substances.
- Request the OSHA area director to conduct an inspection if you believe hazardous conditions or violations of standards exist in your workplace.
- Have your name withheld from your employer, upon request to OSHA, if you file a written and signed complaint.

- Be advised of OSHA actions regarding your complaint and have an informal review, if requested, of any decision not to inspect or to issue a citation.
- Have your authorized employee representative accompany the OSHA compliance officer during the inspection tour.
- Respond to questions from the OSHA compliance officer, particularly if there is no authorized employee representative accompanying the compliance officer.
- Observe any monitoring or measuring of hazardous materials and have the right to see these records, as specified under the Act.
- Have your authorized representative, or yourself, review the Log and Summary of Occupational Injuries (OSHA No. 200) at a reasonable time and in a reasonable manner.
- Request a closing discussion with the compliance officer following an inspection.
- Submit a written request to NIOSH for information on whether any substance in your workplace has potential toxic effects in the concentrations being used, and have your name withheld from your employer if you so request.
- Object to the abatement period set in the citation issued to your employer by writing to the OSHA area director within 15 working days of the issuance of the citation.
- Be notified by your employer if he or she applies for a variance from an OSHA standard, and to testify at a variance hearing and appeal the final decision.
- Submit information or comment to OSHA on the issuance, modification, or revocation of OSHA standards and request a public hearing.

***

The pamphlet notes that although employees have an obligation under §5(b) to comply with OSHA standards, OSHA does not have authority to cite employees for violation of this obligation. This principle was established in an early case, *Atlantic & Gulf Stevedores v. OSHRC,*[10] and has not been seriously questioned since.[11] The ultimate issue in that case was whether enforcement of OSHA's hard-hat standard was economically infeasible in view of a threatened strike by longshoremen if its requirements were enforced. Review Commissioner Van Namee had expressed in a concurring opinion that since the Commission was authorized to issue a cease and desist order requiring employees to end the strike, there was no merit to the economic infeasibility argument.[12] The court of appeals, with "considerable misgivings," disagreed with Mr. Van Namee.

[10] 534 F.2d 541 (3d Cir. 1976).

[11] But see Comment, *Employee Noncompliance with the Occupational Safety and Health Act: Making the Worker Pay,* 31 AM. U.L. REV. 123 (1981).

[12] *Atlantic & Gulf Stevedores,* 3 OSHC 1003, 1012 (1975) (Cleary, Comm'r, writing the decision to uphold the citations; Van Namee, Comm'r, concurring; Moran, Chairman, concurring and dissenting).

Atlantic & Gulf Stevedores v. OSHRC

534 F.2d 541 (3d Cir. 1976)

GIBBONS, Circuit Judge:

***

Commissioner Van Namee finds the source of such coercive authority [against employees] in a combination of §2(b)(2) of the Act, 29 U.S.C. §651(b)(2), §5(b), 29 U.S.C. §654(b), and §10(c), 29 U.S.C. §659(c). The latter provision authorizes the Commission to issue orders "affirming, modifying, or vacating the Secretary's citation * * * or directing other appropriate relief * * *." Section 2(b)(2), in the section of the Act setting forth congressional findings and declaration of policy, provides:

> (b) The Congress declares it to be its purpose and policy * * * to assure so far as possible every working man and woman in the Nation safe and healthful working conditions and to preserve human resources—
>
> ***
>
> (2) by providing that employers and employees have separate but dependent responsibilities and rights with respect to achieving safe and healthful working conditions * * *.

Section 5(b) provides that

> "[e]ach employee shall comply with occupational safety and health standards and all rules, regulations, and orders issued pursuant to this chapter which are applicable to his own actions and conduct."

According to Commissioner Van Namee the employees' separate responsibilities under §5(b) would be "meaningless and a nullity" if the Commission and this court, in an enforcement proceeding, were powerless to sanction employee disregard of safety standards and commission orders.

In the proceedings before the Commission, the petitioners did not move to join their longshoremen as parties. Nevertheless, at least Commissioner Van Namee and possibly Chairman Moran relied upon the availability of such relief in rejecting the petitioners' challenge to the economic infeasibility of the hardhat standard. At oral argument counsel for the Secretary indicated in response to a question from the court that the Secretary might not oppose granting such relief in appropriate circumstances. After the argument, however, we were advised by letter that this is not the Secretary's position. To the contrary—in two cases now pending before the Commission, the Secretary has taken the position that he has no authority to issue citations, and that the Commission has no authority to issue cease and desist orders, against employees. Our authority under §11(a) of the Act appears to be derivative of that of the Secretary and the Commission.

With considerable misgivings, we conclude that Congress did not intend to confer on the Secretary or the Commission the power to sanction employees. Sections 2(b)(2) and 5(b) cannot be read apart from the detailed scheme of enforcement set out in §§9, 10 and 17 of the Act. It seems clear that this enforcement scheme is directed only against employers. Sections 9(a) and 10(a) provide for the issuance of citations and notifications of proposed penalties only to employers. 29 U.S.C. §§658(a), 659(a). Section 10(a) refers only to an employer's opportunity to contest a citation and notification of proposed penalty. Only after an employer has filed a notice of contest does the Commission obtain general jurisdiction. Employees and their representatives may then elect to intervene under §10(c). The only independent right granted employees by §10(c) is to contest before the Commission the reasonableness of any time period fixed by the Secretary in a citation for the abatement of a violation. Section 17, 29 U.S.C. §666, provides for the assessment of civil monetary penalties only against employers.[17] That the Act's use of the term "employer" is truly generic is made plain in §3, the definitional section, where "employer" and "employee" are separately defined. See 29 U.S.C. §652. We find no room for loose construction of the term of art.

We are likewise unable to find support in §5(b) for the proposition that the Act's sanctions can be directed at employees. Although this provision's injunction to employees is essentially devoid of content if not enforceable, we reluctantly conclude that this result precisely coincides with the congressional intent. The House bill, H.R. 16785, did not even impose this nominal obligation on employees. The Senate version, §5(b) of S. 2193, which was accepted in the Conference Committee, contained what is now §5(b) of the Act. There was virtually no floor debate on the provision in the Senate, and none in the House following the action by the Conference Committee. In such circumstances it cannot be seriously contended that Congress intended to make the amenability of employees to coercive process coextensive with employers. The Senate Report on the employee duty section, quoted in full, says:

> The committee recognizes that accomplishment of the purposes of this bill cannot be totally achieved without the fullest cooperation of affected employees. In this connection, Section 5(b) expressly places upon each employee the obligation to comply with standards and other applicable requirements under the act.
>
> It should be noted, too, that studies of employee motivations are among the research efforts which the committee expects to be undertaken under section 18, and it is hoped that such studies, as well as the programs for employee and employer training authorized by section 18(f), will provide the basis for achieving the fullest possible commitment of individual workers to the health and safety efforts of their employers. It has been made clear to the committee that the most successful plant safety programs are those which emphasize employee participation in their formulation and administration; every effort should therefore be made to maximize such participation throughout industry.
>
> The committee does not intend the employee-duty provided in section 5(b) to diminish in any way the employer's compliance responsibilities or his responsibility to assure compliance by his own employees. Final responsibility for compliance with the requirements of this act remains with the employer.

S. Rep. No. 91-1282 [91st Cong., 2d Sess. 10–11 (1970)].
We simply cannot accept the argument that a remedy for violations of §5(b) can be implied from its terms. All the evidence points in the other direction.[18]

---

[17]Compare in this regard the Federal Coal Mine Health and Safety Act of 1969, 30 U.S.C. §§801–960. Section 109(a)(1)-(2) provides for civil penalties against miners as well as employers. 30 U.S.C. §819(a)(1)-(2). * * *

[18]Our conclusion is fortified by reference to the following post-enactment colloquy between Representative Steiger, a co-sponsor of OSHA, and Representative Hungate:

> Mr. Hungate: Now, employer-employee. We have had a line of testimony about, tell this guy to wear a hardhat, or there are six guys on the job and five of them do and the other guy tosses it out, it is a hot day. And they come through and the employer gets fined and the employee does not. What can we do about that?
>
> Mr. Steiger: Well, Mr. Chairman when this bill was being considered, that was a question on which we spent a considerable amount of time. I would have to say the business community, at the time the bill was under consideration, took a very hard line that they did not want the Federal government to be in the business of disciplining their employees * * *. But on balance, Mr. Chairman, I would not want to see us amend the law to

Nor do we believe that the language in §10(c) authorizing the Commission to issue orders "directing other appropriate relief" can be stretched to the point that it includes relief against employees. Rather, the generality of that language must be deemed limited by its context—relief in connection with the Secretary's citation. The Secretary appears not to have authority to issue a citation against an employee, and the Commission's powers cannot be any broader. "Other appropriate relief" refers to other appropriate relief against an employer.

***

B. Participation in Inspections—Walkaround Rights

The OSHA program grants employees a significant role in inspections of workplaces. As has already been discussed, employees have the right, under §8(f) of the Act, to trigger an OSHA inspection by filing a "formal" complaint with the Agency.[13] The Act also expressly requires that an employee representative be given an opportunity to accompany the compliance officer during his inspection of the workplace.[14] Furthermore, the Agency currently provides for employee participation in the opening, closing, and informal conferences held between OSHA representatives and employers. Although the Act does not specifically require that these conferences be held as part of the inspection and enforcement process, they were instituted by OSHA's enforcement regulations in 1971.[15] These regulations do not mandate employee participation at opening, closing, and informal conferences, but in 1978 OSHA directed its field personnel to provide employees with broad participation.[16]

impose Federal government disciplining on employees. I think that is something left between management and labor.

Mr. Hungate: No, there is not. The law says the employer is responsible * * * but again I want to simply reiterate that I think it would be a mistake to interfere in labor-management relations in terms of who has discipline responsibility * * * I think that is something where there are unions and managements negotiating that can be dealt with in the negotiation process.

Mr. Hungate: Is fining the employer an act of discipline?

Mr. Steiger: Yes, I think it would be.

Mr. Hungate: Then we are interfering in the field of discipline in labor-management relations.

Mr. Steiger: No, we are not.

Mr. Steiger: The employer is fined because, under the law, that is his responsibility. I would hope that the employer would be in the position to deal with his employee who put himself in the position of having his employer fined.

Hearings Before the Subcomm. on Environmental Problems Affecting Small Business of the Select Comm. on Small Business, 92d Cong., 2d Sess. 490–91 (1972).

[13]Chapter 13, Section C.

[14]Sec. 8(e), 29 U.S.C. §657(e).

[15]29 C.F.R. §§1903.7, 1903.19 (1983). The opening conference takes place at the commencement of the inspection; the closing conference at the end of the physical tour of the workplace; and the informal conference after the citation is issued, during the contest period. Area directors are now authorized to enter into settlement agreements at informal conferences. See discussion *supra* p. 340, note 21.

[16]Program Directive No. 200-82, Aug. 15, 1978. (Available in author's files.) These instructions have been incorporated, with changes, into the 1983 Field Operations Manual, ch. III, OSHR [Reference File 77:2508–:2510, 77:2518, 77:2527].

Employee Participation in OSHA Inspections and Enforcement Proceedings

***

3. BACKGROUND

The Field Operations Manual (FOM), Chapter V—Inspection Procedures, is the internal operative guideline which provides for employee participation. The regulations, 29 CFR 1903, outline procedures for conducting inspections and issuing citations and notifications of proposed penalties. Neither the regulations nor the FOM provide sufficient guidance for establishing a uniform and consistent policy for insuring the involvement of affected employees or their representatives in opening, closing and informal conferences.

4. EXPLANATION

a. Employees have a critical interest in OSHA enforcement proceedings, beginning with the opening conference at an employer's worksite and through the closing of the case file. Since workplace hazards directly affect exposed employees who are often in a good position to know the nature and extent of the hazard as well as potential abatement methods, these employees must be encouraged to participate fully in the enforcement process. More specifically, employees should have the opportunity to present their views at conferences and discussions with representatives of the Occupational Safety and Health Administration concerning the enforcement process. In this fashion, OSHA will assure that the views of employee representatives are fully considered before the issuance or modification of citations.

***

5. DEFINITION

For purposes of this directive, except where otherwise stated, an employee representative shall include a collective bargaining representative, any other group or body representing employees, or an individual employee.

6. ACTION

A. PARTICIPATION.

Regional Administrators and Area Directors will assure that employee representatives are afforded the opportunity, and are encouraged, to attend and express their views in discussions relating to workplace inspections and the issuance, amendment or withdrawal of citations. Joint opening and closing conferences should be conducted when practical with all parties represented. Where it is not practical to hold a joint conference, separate conferences for employee representatives and representatives of the employers shall be held. In those instances where separate conferences are held, a written summary of each conference shall be made and the summary made available on request to employee representatives and representatives of the employer. The procedures which follow will assure full participation by employee representatives at every stage of OSHA enforcement proceedings:

(1) OPENING CONFERENCE.

The CSHO should conduct a joint opening conference with the employer and employee representatives. Where it is not practical to hold a joint conference, separate conferences for the employee representatives and the representatives of the employer shall be held. Where separate conferences are necessary, CSHO's shall determine if their conduct will unacceptably delay observation or evaluation of workplace safety or health hazards. In such cases the conferences shall be brief and, if appropriate, reconvened after the CSHO's inspection of the alleged hazards. During the course of the opening conference, pursuant to 29 CFR 1903.8, employer and employee representatives shall be informed of the opportunity to accompany the CSHO during the physical inspection of any workplace. The provisions of the FOM, Chapter V, paragraph D.2 pertaining to the opening conference shall apply to conferences with employee representatives.

(2) WALKAROUND.

The CSHO shall conduct a walkaround inspection in accordance with the provisions of the FOM, Chapter V, paragraphs D.3. and 6. During the walkaround, pursuant to 29 CFR 1910.10, the CSHO may consult with individual employees as well as the employee representative concerning working conditions. If requested, and considered necessary by the CSHO, any additional consultation shall occur during working hours and in private, if so requested.

(3) CLOSING CONFERENCE.

At the conclusion of an inspection, the CSHO shall conduct a joint closing conference with the employer and employee representatives. Where it is not practical to hold a joint closing conference, separate closing conferences shall be held. During the course of the closing conference both the employer and employee representatives shall be advised of their right to participate in any subsequent conferences, meetings, or discussions as described herein. All of the instructions in the FOM, Chapter V, paragraph D.7., which presently are limited to employer participation in closing conferences, shall also be covered for employee representatives. Where closing conferences are delayed pending receipt of sampling data, employee representatives shall be afforded the opportunity to participate in all such delayed conferences.

B. AVAILABILITY OF SAMPLING DATA.

If either the employer or the employee representative requests copies of sampling data, the data shall be provided to both.

C. CITATIONS.

Employee representatives shall receive copies of all Citation and Notification of Penalty issued pursuant to the FOM, Chapter X—Citations. Where there is a collective bargaining agent at the workplace, a copy of each Citation and Notification of Penalty shall be sent to the appropriate collective bargaining representative. (Note: The collective bargaining representative will normally be the local union president for industrial unions and the business manager for building trades and crafts unions). If the workplace inspected does not have a collective bargaining representative, a copy of the Citation and Notification of Penalty shall be forwarded to the employee participating in the walkaround inspection. In those instances where the workplace inspected does not have a collective bargaining representative, and there was no employee participating in the walkaround inspection, the posting of the "Citation and Notification of Penalty" shall be construed as compliance with this paragraph.

D. INFORMAL CONFERENCE.

Pursuant to 29 CFR 1903.19, either the representative of the employer or employee representative may request an informal conference. Whenever an informal conference is requested by the employer, the employee representative will be afforded the opportunity and encouraged to participate fully. To the extent that informal conferences can be held with the employee representative in attendance, they should be arranged. During the conduct of the informal conference, if matters of a delicate nature are brought up by either representative, separate or private discussions shall be permitted.

In any event, the Regional Administrator or Area Director shall not amend or withdraw a citation or penalty without first obtaining the views of the employee representative. When the employee representative disagrees with the proposed amendment or withdrawal of a citation, the proposed disposition may be appealed to the Regional Administrator. However, as the 15-working-day period for filing a notice of contest may affect discussions regarding an amendment or withdrawal of a citation, telephonic communication shall be utilized in order to expedite the resolution of the matter under consideration.

***

Section 8(e) of the Act provides that a representative of the employer and a representative of the employees shall be given an opportunity to participate in the physical inspection of the workplace. The House and the Senate disagreed on the employee walkaround provision; however, the Senate believed the right to be critical, and it was ultimately enacted by the Congress.[17] The importance of the participation of employees in the workplace inspection was articulated by Assistant Secretary Bingham in 1980 in the course of her deposition in a walkaround pay case.

BY MR. STRAUSS [Counsel for General Electric Company, Defendant]:

Q. The statement that says, "During these years, employee participation and cooperation at the inspection stage has proved to be critically important to enforcement effort under the Act."

Again, is that an impression based on talks to inspectors or are there factual data to support it?

A. Well, I think factual data or verbal conversations with compliance officers, compliance officers, inspectors who tell me that it is very important for a worker to tell them about the process, whether or not the process has been slowed down during the inspection or perhaps even speeded up, for whatever reasons, to describe in detail how the job is actually performed.

Sometimes there are pieces of equipment that are over in the corner. Maybe the employer will say they are not used any longer. It is

[17]For a discussion of the legislative history, see Chapter 1. Sec. 8(e) further provides that if there is no employee representative, OSHA "shall consult with a reasonable number of employees concerning matters of health and safety in the workplace."

important to be able to ask the worker are they used, if so, how frequently.

Details about the workplace are extremely important in conducting an inspection so that you can really determine whether or not they are real or imaginary hazards there.

Q. Does the agency have any records or memoranda which would indicate or support the statement of the critical importance of the employee participation in this?

For example, are there any records that the agency has where the inspector has reported that we couldn't make a viable inspection because the employees would not cooperate?

A. I believe that in some inspection files that statement has been written down, that there is not—you know, they couldn't get certain pieces of information, that the employees would not cooperate or talk with them. I am certain that we have discussed cases in this room where this has happened. I am positive that I have had innumerable conversations with compliance officers from all over this United States, where they have told me how important it is for workers to describe what goes on in the workplace.

And let me tell you that the pay issue never arose. But back in the Coke Oven Advisory Committee hearings, which were probably 1974, or '75, somewhere, I would have to get the exact date for the year—I can recall being the Chairman of a session where a question was raised about whether or not coke was handled in a certain way.

We asked the people around the table, who included employer and employee representatives, head of the union, the head of the coke operations for a large company. And they couldn't answer the question. In the back of the room there was a worker who put his hand up, and I called on the worker and he described exactly what happened.

Time and time again, I have been impressed, if you want to know what happens in a workplace, you go to a worker. If I want to know what happens in this workplace, I don't just ask the directors, I go to the people on the third floor, or in the Cincinnati Office, or the Houston Office, and ask them what actually happens and what do you do.

That is how you find out what happens and how the job is done.[18]

OSHA field instructions deal specifically with the manner of selecting walkaround representatives,[19] including instructions to the inspector for handling special situations.

Field Operations Manual[20]

CHAPTER III—GENERAL INSPECTION PROCEDURES

***

5. WALKAROUND REPRESENTATIVES. * * *

***

(b) *Employee Representatives.* One or more employee representatives shall be given an opportunity to accompany the CSHO during the

[18]Deposition of Eula Bingham Mattheis at 62–64, *Marshall v. General Elec. Co.*, No. 79-1821 (E.D. Pa. deposed Apr. 17, 1980).

[19]See also 29 C.F.R. §1903.8 (1983).

[20]OSHR [Reference File 77:2513–14, 77:4302–03].

walkaround phase of the inspection to provide appropriate involvement of employees in the physical inspection of their own places of employment and to give them an opportunity to point out hazardous conditions. 29 CFR 1903.8(b) gives the CSHO authority to resolve disputes as to who represents the employees for walkaround purposes. The following guidelines are suggested in designating employee representatives:

(1) *Employees Represented by a Certified or Recognized Bargaining Agent.* During the opening conference, the highest ranking union official or union employee representative shall designate who will participate in the walkaround.

(2) *Safety Committee.* The employee members of an established plant safety committee may have designated an employee representative for OSHA inspection purposes or agreed to accept as their representative the person designated by the committee to accompany the CSHO during an OSHA inspection. This representative shall have the opportunity to participate in the walkaround.

(3) *No Certified or Recognized Bargaining Agent.* Where employees are not represented by an authorized representative, where [there] is no established plant safety committee, or where employees have not chosen or agreed to an employee representative for OSHA inspection purposes whether or not there is a safety committee, the CSHO shall determine if any other employees would suitably represent the interests of employees on the walkaround. If such selection of employee representatives is impractical, the inspection shall be conducted without an accompanying employee representative; and the CSHO shall consult with a reasonable number of employees during the walkaround in accordance with the provisions of 29 CFR 1903.8 and Section 8(e) of the Act. Selection of random employees to interview shall include individuals knowledgeable about the area or process being inspected.

6. SPECIAL SITUATIONS.

***

d. *Employee Representatives Not Employees of the Employer.* Walkaround representatives authorized by employees will almost always be employees of the employer. If, however, in the judgment of the CSHO, unique circumstances make the presence of a nonemployee third party (industrial hygienist, safety engineer, or other experienced safety or health person) necessary or helpful to the conduct of an effective and thorough physical inspection of the workplace, such a person may be designated by the employees as their representative to accompany the CSHO during the inspection (29 CFR 1903.8(c)). Questionable circumstances, including any unreasonable delays, will be referred to the supervisor. A nonemployee representative shall be cautioned by the CSHO not to discuss matters pertaining to operations of other employers during the inspection.

e. *More Than One Representative.* At establishments where more than one employer is present or in situations where groups of employees have different representatives, it is acceptable to have a different employer/employee representative for different phases of the inspection. More than one employer and/or employee representative may accompany the CSHO throughout or during any phase of an inspection if the CSHO determines that such additional representatives will aid and not interfere with the inspection (29 CFR 1903.8(a)).

***

CHAPTER XII—CONSTRUCTION

***

4. SELECTING EMPLOYER AND EMPLOYEE REPRESENTATIVES. * * *

***

c. *Walkaround Provisions.* The main difficulty in implementing the "walkaround" provisions on construction sites derives from the fact that in the usual situations there will be numerous employers on the job. If all employers and groups of employees selected a different representative to accompany the CSHO on the inspection, the group participating in the inspection could be so large that work on the worksite might be disrupted and the effectiveness of the inspection would be diminished.

(1) An attempt shall be made to encourage employer and employees to select, respectively, a limited number of representatives for accompaniment purposes. It shall be pointed out by the CSHO that this arrangement makes an effective inspection possible without diminishing the accompaniment rights. If any matter comes up during the course of the inspection that requires special knowledge, the representative of the appropriate employer shall be called in to participate in that phase of the inspection.

(2) The CSHO may also divide the inspection into separate phases; e.g., excavation work followed by electrical work, and so forth. If this procedure is followed, the number of employer and employee representatives for each phase of the inspection can be limited to those immediately involved. The CSHO shall avoid, to the extent possible, inspecting the same areas of the worksite more than once.

d. *Too Many Representatives.* The CSHO shall conduct the inspection accompanied by the representatives designated by the employers and employees. However, if during the course of the inspection, the CSHO determines that, because of the large number of persons involved, the inspection is not being conducted in an effective manner or that work is being unduly disrupted, the participants shall be advised that walkaround representation is discontinued and instead a reasonable number of employees will be interviewed. If the participants then agree to a limited number of representatives for accompaniment purposes, the CSHO shall resume the inspection with such representatives.

***

OSHA indicated in a *Federal Register* notice, issued in May 1981, that inspections in establishments with "no authorized representative," that is, nonunion establishments, are "typically" conducted without the walkaround participation of an employee representative since "none exists."[21] OSHA explained that the "selection of employee representatives in a non-union establishment would entail complex and time-consuming procedures that would seriously delay the conduct of inspections and effective and efficient enforcement of the Act." Thus, employee walkaround, as a matter of practice, is limited to union plants. At the same time, OSHA has gone beyond the literal language of §8(e) and "authorized" compliance offi-

[21]Walkaround Compensation, 46 FED. REG. 28,842, 28,844 (1981), discussed in the next section.

cers to interview employees even when accompanied by a walkaround representative.[22]

C. Walkaround Pay

One of the most difficult issues in the area of employee rights is whether employees are entitled to be paid for walkaround time. The Act and the legislative history do not speak directly to this issue, and OSHA policy respecting walkaround pay (as it has been called) has changed several times. The issue was first raised by the Oil and Chemical Workers' Union (OCAW) under §11(c) of the Act. The Solicitor of Labor issued a legal opinion concluding that the refusal by Mobil Oil Corporation to pay for walkaround time was not discriminatory under §11(c).[23] OCAW contended that by refusing to pay for walkaround time, an employer was "effectively interfering with the exercise of a basic statutory right" and that the refusal was therefore *per se* discrimination under §11(c). OCAW also relied on an internal Department of Labor memorandum of July 1, 1971, expressing the view that the time spent by employees in participating in a walkaround inspection during normal working hours was "hours worked" under the Fair Labor Standards Act.

The Solicitor of Labor's opinion rejected both OCAW arguments supporting walkaround pay. It ruled that walkaround time was not "working time" within the meaning of the Fair Labor Standards Act and, further, that the refusal to pay was not discriminatory under §11(c). The two issues are interrelated; in light of the Solicitor's conclusion that walkaround time was not "working time," he determined that the refusal to pay was not "because" of the exercise of a protected right, but, rather, "because" the employee was not "working" at the time. On the other hand, if it were determined that walkaround time was working time, which in fact *was* the Department of Labor's original determination, the discrimination argument under §11(c) would have been considerably more persuasive. Subsequently, on the basis of the same incident, employees of Mobil filed suit in the federal court claiming that they were entitled to walkaround pay under the Fair Labor Standards Act and the applicable collective bargaining agreement. The district court rejected both employee claims, and the Court of Appeals for the District of Columbia Circuit affirmed.[24] Although the suit was based on the provisions of FLSA and not OSHA, the court opinion also dealt with, and rejected, the employee argument that the "policies of OSHA require" that walkaround time be considered "hours worked" under FLSA. The court said:

[22]Field Operations Manual, OSHR [Reference File 77:2517]. The manual contains detailed instructions to compliance officers for the conduct of private employee interviews during an inspection. *Id.* at 77:2517–:2518.

[23]Memorandum from Solicitor of Labor to the Assistant Secretary of Labor on Walkaround Pay, March 1, 1972 (copy in author's file).

[24]*Leone v. Mobil Oil Corp.*, 523 F.2d 1153 (D.C. Cir. 1975), *aff'g* 377 F. Supp. 1302 (D.D.C. 1974).

> * * * Although [the] legislative history reflects Congressional desire to provide employee input into safety inspections, it sheds no light on the issue of compensation for such participation.
> ***
>
> This lack of acknowledgment of the pay issue may indicate that Congress did not consider the problem at the time OSHA was adopted. Whether Congress deliberately or unconsciously omitted provisions for walkaround pay for employees, however, the question remains one properly reserved to the legislative process. We are particularly unwilling to supply this policy decision where Congress has evidenced a continuing interest in the legislation and where the statute provides a method of bringing such problems before Congress through the efforts of the Secretary of Labor.
>
> Although many arguments could be addressed to Congress advocating that compensation be granted, see, e.g., Comment, *OSHA: Unresolved Issues and Potential Problems,* 41 GEO. WASH. L. REV. 309, 309-16 (1972), these policy arguments cannot serve as a basis for judicial supplementation of the expansive statutory scheme. Plaintiffs argue that failure to require employers to compensate their employees for walkaround time will discourage employee participation and thus frustrate expressed Congressional policy. Contrary to this fear, a recent labor union study designed to measure employee participation in OSHA inspections revealed that the Secretary's ruling that walkaround time was not compensable had not significantly affected either compensation or participation among the employees surveyed. The study revealed that 93% were paid by the employers and 4% were paid by the union; the remaining 3% included employees who were paid one-half of their wages by both union and employer or who participated during their own time. Zalusky, *The Worker Views the Enforcement of Safety Laws,* 26 LAB. L.J. 224, 230 (1975). This study is consistent with a statement of a Labor Department spokesman that most employers are paying for walkaround time, *BNA Labor Law Reporter:* LRX 489, as expressly allowed by the regulations: "It should be emphasized that an employer is in no way precluded by the Act from making voluntary agreements to pay employees for time spent by them while participating in walkaround inspections." 29 C.F.R. §1977.21(b) (1975). The survey also points out that the problem is alleviated in large measure because the inspection often follows a grievance and is therefore sometimes covered under the terms of collective bargaining agreements which provide pay for employee time spent investigating grievances. Zalusky, *supra* at 230. The employees' opportunity to protect themselves through providing for compensation via bargaining agreements is another reason for refusing to inject judicial policy-making into the legislative scheme.[25]

The first reversal in policy on walkaround pay took place in 1977. Dr. Eula Bingham, relying on a revised interpretation by the Solicitor of Labor, issued a new policy that an employer's refusal to pay for walkaround time was discriminatory under §11(c). The interpretation also reversed the position of the Wage and Hour Administration that walkaround pay was not "working time" under the Fair Labor Standards Act. The new pay policy was published, without notice or public comment, as a "legal interpretation." Although the *Federal Register* notice was published on September 20, 1977,[26] the change in policy was first discussed on August 10, 1977, in a speech

[25] 523 F.2d at 1160–61.
[26] 42 FED. REG. 47,344 (1977).

made by Dr. Bingham before a convention of the Oil and Chemical Workers International Union.[27]

The 1977 interpretation was soon challenged by the Chamber of Commerce on both substantive and procedural grounds. These challenges were rejected by Judge Green of the U.S. District Court for the District of Columbia.

Chamber of Commerce of the United States v. OSHA

465 F. Supp. 10 (D.D.C. 1978)

June L. Green, District Judge:

***

It is apparent from the words of the statute and from the legislative history that Congress did not address the question of who was to bear the economic burden of employee participation in workplace inspections. Congress only concerned itself with the necessary and desirable extent of employee participation in the enforcement process. See [116 Cong. Rec. 37,340 (1970)]; *Leone v. Mobil Oil Corp.*, 523 F.2d 1153 at 1159–61 (D.C. Cir. 1975). It is also clear that Congress viewed employee participation, particularly in the enforcement area, as an essential aspect of its program for industrial safety and health. 29 U.S.C. §§655(b)(1), (6) and (7), 657(f)(1), 659(c) and 660(a). As the Court of Appeals noted in *Leone,* this "policy decision" regarding employee compensation for participation in walkaround inspections is committed to Congressional discretion and to the expertise and informed judgment of the Secretary of Labor. This Court feels compelled to abide by the Secretary's interpretive ruling. As observed by the Court of Appeals:

> Opinions by the head of an administrative agency, "while not controlling upon the courts by reason of their authority, do constitute a body of experience and informed judgment to which courts and litigants may properly resort for guidance." (citations omitted); this deference is tempered, however, when the ruling is inconsistent with Congressional policy (citations omitted).

Leone, 523 F.2d at 1162–1163. See also *Chisholm v. F.C.C.,* 538 F.2d 349 (1976), *cert. denied,* 429 U.S. 890 (1976); *Marshall v. Daniel Construction Co., Inc.,* 563 F.2d 707 (5th Cir. 1977).

The Secretary's ruling is in no way inconsistent with the Occupational Safety and Health Act or any other expression of Congressional policy.[4] To the contrary, requiring walkaround pay is plainly adapted to the effectuation of employee participatory rights afforded by Section 657(e). One ground for the Secretary's decision is the realization "that management representatives are often paid for the time spent accompanying OSHA representatives during an inspection while the employees participating in the same activity

[4]Plaintiff also claims that the interpretive ruling contravenes the Federal labor policy by intruding on the collective bargaining process. Congress has not, however, indicated that the issue of employee compensation for time spent participating in the physical inspection of a workplace is to be left to collective bargaining. The Court of Appeals in *Leone* was unwilling "to inject judicial policy making into the legislative scheme." *Leone* at 1161. This Court is also of the opinion that the judiciary should leave questions of labor policy in the hands of Congress and the Department of Labor.

[27]Copy of speech in file in *Chamber of Commerce of the United States v. OSHA*, 465 F. Supp. 10 (D.D.C. 1978), *rev'd,* 636 F.2d 464 (D.C. Cir. 1980). The court of appeals referred to the speech in its decision, 636 F.2d at 467, n.4.

are not." 42 FED. REG. 47345 (1977). This unrebutted finding by OSHA, coupled with the Department of Labor's determination that walkaround inspection time constitutes "hours worked" under the FLSA, effectively refutes any argument that non-payment occurs only because the employee is not performing compensable work during a walkaround inspection and leads to the undeniable conclusion that employees who are not paid for walkaround time are being discriminated against within the meaning of Section 660(c).

Moreover, the Secretary has administered the Act for seven years, and based on this experience has determined that "employee participation and cooperation at the inspection stage has proved to be critically important to enforcement efforts under the Act." 42 FED. REG. 47344 (1977). The Secretary has also observed that "loss of pay involved [in a walkaround] would clearly constitute a significant economic disincentive to employees' exercising their express statutory right under section 8(a) to accompany a compliance officer during an inspection." *Id.* at 47344–47345. Accordingly, the Court cannot disagree with the Secretary's conclusion that refusal to compensate employees for time spent in a walkaround "has a twofold and related impact: It is inherently destructive of the employees' right to participate in the walkaround and, consequently, impedes the free flow of information between employees and representatives of the Secretary which is so critical to effective enforcement of the Act." *Id.* at 47344. See *NLRB v. Great Dane Trailers, Inc.*, 388 U.S. 26 (1967); *NLRB v. Erie Resistor Corp.*, 373 U.S. 221 (1963).

***

OSHA's interpretive rule is also challenged for failure to comply with the notice and comment provisions of the Administrative Procedure Act. Interpretive rules and general statements of policy, contrasted with the implementation of independent legal requirements promulgated pursuant to validly delegated legislative authority are, however, specifically exempted from the notice and comment provisions of the Administrative Procedure Act. 5 U.S.C. §553(b)(3)(A) and (d)(2); *Gibson Wine Co. v. Snyder*, 194 F.2d 329 (1952). Accordingly, the Court concludes that the defendants are not barred from reconsidering its own interpretive rules and that a new interpretation regarding discriminatory conduct may be implemented without affording interested parties prior notice and the opportunity to be heard.

***

The argument that an OSHA regulatory action "intrudes" on the collective bargaining process, referred to and answered by the court, has been advanced in several contexts. The district court here rejected the argument,[28] as did the Court of Appeals for the District of Columbia Circuit in ruling that OSHA had authority to impose medical removal protection in the lead standard.[29] A district court in Louisiana reached the same conclusion in upholding the validity of OSHA's records access rule against an objection that the subject matter was reserved to the exclusive jurisdiction of the National Labor Relations Board.[30] However, in announcing its intention to recon-

[28] 465 F. Supp. at 13 n.4.

[29] *United Steelworkers v. Marshall*, 647 F.2d 1189 at 1236 (D.C. Cir. 1980) (stating that "* * * we find nothing in the OSH Act or other labor legislation to suggest that Congress could remove from OSHA the power to create a program instrumental to achieving worker safety simply because such a program could otherwise be created through collective bargaining").

[30] *Louisiana Chemical Ass'n v. Bingham*, 550 F. Supp. 1136 (W.D. La. 1982).

sider its records access rule, OSHA suggested that recent NLRB decisions granting union representatives access to certain employer occupational health records may have made the OSHA requirements for union access unnecessary.[31]

The district court decision upholding OSHA's walkaround pay rule was appealed to the Court of Appeals for the District of Columbia Circuit which reversed the lower court and vacated the walkaround pay rule. The majority opinion was written by Circuit Judge Tamm who earlier had written the *Leone* decision on "hours worked."

Chamber of Commerce of the United States v. OSHA

636 F.2d 464 (D.C. Cir. 1980)

TAMM, Circuit Judge:

***

Despite the Administration's averments, we cannot conclude that it intended its new regulation to be interpretive. In *Leone v. Mobil Oil Corp.*, 523 F.2d 1153 (D.C. Cir. 1975), we held that neither the terms of the Act nor the Act's legislative history nor the policies underlying the employee walkaround right require employers to compensate employees for walkaround time. *Id.* at 1159–61. We would prefer to believe that the Administration acts in good faith; we therefore believe the Administration would not issue an interpretation in flagrant defiance of this court's *Leone* decision. In any event, we are certain the Administration realized that statutory interpretation by an agency is not necessarily controlling, see *Skidmore v. Swift & Co.*, 323 U.S. 134, 139–40 (1944), and thus it knew that although "courts often defer to an agency's interpretive rule they are always free to choose otherwise." *Joseph v. United States Civil Service Commission*, 554 F.2d at 1154 n.26. See *Consumer Product Safety Commission v. GTE Sylvania, Inc.*, 48 U.S.L.W. 4658, 4662 (U.S. June 9, 1980). After this court's ruling in *Leone* that the Act, its legislative history, and its policies do not mandate walkaround pay, an Administration issuance of a differing view solely as a matter of its own interpretation would be inconceivable. Such a rule would be a mere phasm of agency action, "full of sound and fury,/ Signifying nothing," W. Shakespeare, *Macbeth*, act V, sc. v, lines 27–28.

Moreover, the effect of the new regulation exposes the Administration's true intent. Interpretive rules "'are statements as to what the administrative officer thinks the statute or regulation means.'" *Citizens to Save Spencer County v. United States Environmental Protection Agency*, 600 F.2d at 876 (quoting *Gibson Wine Co. v. Snyder*, 194 F.2d 329, 331 (D.C. Cir. 1952)). Such rules only provide a "clarification of statutory language," *Joseph v. United States Civil Service Commission*, 554 F.2d at 1153; the interpreting agency only "'reminds' affected parties of existing duties," *Citizens to Save Spencer County v. United States Environmental Protection Agency*, 600 F.2d at 876 n.153. See *Yale Broadcasting Co. v. FCC*, 478 F.2d 594, 599 (D.C. Cir.), *cert. denied*, 414 U.S. 914 (1973).

The Administration could not be explaining or clarifying the Act's language, for, as we concluded in *Leone*, the Act neither prohibits nor compels pay for walkaround time. There was no "existing duty" to serve as the sub-

[31] 47 FED. REG. 30,420, 30,428 (1982). See *Minnesota Mining & Mfg. Co.*, 261 NLRB 27 (1982). See also *Whirlpool Corp. v. Marshall*, 445 U.S. 1, 17 n.29 (1980).

ject of an Administration reminder. Congress has not "legislated and indicated its will" on the question of walkaround pay, therefore the Administration must have done more than exercise its "'power to fill up the details.'" *United States v. Grimaud,* 220 U.S. 506, 517 (1911) (quoting *Wayman v. Southard,* 23 U.S. (10 Wheat.) 1, 43 (1825)).

Though the walkaround pay regulation does not merely *explain* the statute, the effect of an interpretive rule, the regulation certainly endeavors to *implement* the statute, the effect of a legislative rule. See *Gibson Wine Co. v. Snyder,* 194 F.2d 329, 331 (D.C. Cir. 1952). Accord, *National Association of Insurance Agents v. Board of Governors of the Federal Reserve System,* 489 F.2d 1268, 1270 (D.C. Cir. 1974) (per curiam). The Administration found walkaround pay essential because "the failure [of an employer] to pay for [an employee's] walkaround is inherently destructive * * * of the entire enforcement scheme of the Act." 42 FED. REG. 47344, 47345 (1977). By making this determination, the Administration provided the policy decision Congress omitted—namely, that without walkaround pay there is no walkaround right. It is clear to us that the Administration has attempted through this regulation to supplement the Act, not simply to construe it, and therefore the regulation must be treated as a legislative rule. Cf. *Energy Consumers & Producers Association v. DOE,* 632 F.2d 129 at 141 ((Temp. Emer. Ct. App.) Apr. 4, 1980) (regulation interpretive because meaning of statutory term neither expanded nor contracted in form or substance), *cert. denied,* 449 U.S. 832 (1980).

IV

Because the Administration's rule is an attempted exercise of legislative power, it must be vacated for failure to comply with the procedures specified by the Administrative Procedure Act, 5 U.S.C. §553 (1976). The APA requires the Assistant Secretary to publish a general notice of proposed rulemaking in the Federal Register at least thirty days before the proposed rule is to take effect. *Id.* §553(b), (d). The Assistant Secretary must allow interested parties to submit their comments before a final rule is adopted. *Id.* §553(c).

The Assistant Secretary should not treat the procedural obligations under the APA as meaningless ritual. Parties affected by the proposed legislative rule are the obvious beneficiaries of proper procedures. Prior notice and an opportunity to comment permit them to voice their objections before the agency takes final action. Congress enacted 5 U.S.C. §553 in part to "'afford adequate safeguards to private interests.'" H.R. 1203, 79th Cong., 1st Sess. (Comm. Print June, 1945) (quoting S. Doc. 8, 77th Cong., 1st Sess. 103 (1941) (Final report of Att'y General's Comm. on Ad. Proc.)), reprinted in S. Doc. 248, 79th Cong., 2d Sess. 20 (1946) (official legislative history of the Administrative Procedure Act). Given the lack of supervision over agency decisionmaking that can result from judicial deference and congressional inattention, see Cutler & Johnson, *Regulation and the Political Process,* 84 YALE L.J. 1395 (1975), this protection, as a practical matter, may constitute an affected party's only defense mechanism.

An agency also must not forget, however, that it too has much to gain from the assistance of outside parties. Congress recognized that an agency's "'knowledge is rarely complete, and it must learn the * * * viewpoints of those whom the regulation will affect. * * * [Public] participation * * * in the rule-making process is essential in order to permit administrative agencies to inform themselves * * *.'" H.R. 1203, 79th Cong., 1st Sess. (Comm. Print June, 1945) (quoting S. Doc. 8, 77th Cong., 1st Sess. 103 (1941) (Final report of Att'y General's Comm. on Ad. Proc.)), reprinted in S. Doc. 248, 79th Cong., 2d Sess. 20 (1946). Comments from sources outside of the agency may shed light on specific information, additional policy considerations, weaknesses in the proposed regulation, and alternative means of achieving the same objectives. See *National Petroleum Refiners Association v. FTC,* 482 F.2d 672, 683 (D.C. Cir. 1973), *cert. denied,* 415 U.S. 951 (1974). By the same token, public scrutiny and participation before a legislative rule becomes effective can reduce the risk of factual errors, arbitrary actions, and unforeseen detrimental consequences. See Freedman, *Summary Action by Administrative Agencies,* 40 U. CHI. L. REV. 1, 27–30 (1972).

Finally, and most important of all, highhanded agency rulemaking is more than just offensive to our basic notions of democratic government; a failure to seek at least the acquiescence of the governed eliminates a vital ingredient for effective administrative action. See Hahn, *Procedural Adequacy in Administrative Decisionmaking: A Unified Formulation* (pt. 1), 30 AD. L. REV. 467, 500–04 (1978). Charting changes in policy direction with the aid of those who will be affected by the shift in course helps dispel suspicions of agency predisposition, unfairness, arrogance, improper influence, and ulterior motivation. Public participation in a legislative rule's formulation decreases the likelihood that opponents will attempt to sabotage the rule's implementation and enforcement. See Bonfield, *Public Participation in Federal Rulemaking Relating to Public Property, Loans, Grants, Benefits, or Contracts,* 118 U. PA. L. REV. 540, 541 (1970). See generally *Joint Anti-Fascist Refugee Committee v. McGrath,* 341 U.S. 123, 171–72 & n.19 (1951) (FRANKFURTER, J., concurring).

In holding that this regulation is legislative in nature but improperly promulgated, we intimate no view on whether the Assistant Secretary could reissue the same rule after satisfying the requirements of 5 U.S.C. §553. Only after the full notice-and-comment procedures have run their course will we have a record enabling us to judge whether ordering pay for walkaround time is indeed a statutorily authorized, rational, nonarbitrary, and noncapricious method of supplementing the Act's provisions. See 5 U.S.C. §706(2) (1976). See generally *Natural Resources Defense Council, Inc. v. United States Nuclear Regulatory Commission,* 547 F.2d 633, 658–61 (D.C. Cir. 1976) (Tamm, J., concurring), *rev'd on other grounds sub nom. Vermont Yankee Nuclear Power Corp. v. Natural Resources Defense Council, Inc.,* 435 U.S. 519 (1978); *Greater Boston Television Corp. v. FCC,* 444 F.2d 841, 852 (D.C. Cir. 1970), *cert. denied,* 403 U.S. 923 (1971). We leave that question for another day.

***

BAZELON, Senior Circuit Judge, concurring in the result only:

I write separately because I seriously doubt that application of notice and comment procedures can depend entirely on neat, discrete categories. The labels associated with exceptions to 5 U.S.C. §553 have "fuzzy perimeters"[1] and depend on distinctions "'enshrouded in considerable smog.'"[2] Thus, I find it harder to criticize the agency for not analyzing correctly the import and legal status of the action challenged here.

Nonetheless, the Occupational Safety and Health Administration has announced in no uncertain terms its intention to charge employers with a statutory violation based on its new view of a statutory provision. See majority opinion at [466–67]. This announcement is prospective and definitive, factors which strongly suggest that, at least without more, it is not merely a "general statement of policy" exempt from §553. See, e.g., *American Bus Ass'n v. ICC,* 627 F.2d 525, at 531–533 (D.C. Cir. 1980); *Guardian Federal Savings and Loan Ass'n v. Federal Savings and Loan Insurance Corp.,* 589 F.2d 658, 666 (D.C. Cir. 1978). And, although it serves as an interpretation of existing law, it also effectively enunciates a new requirement heretofore

[1]*Pacific Gas & Electric Co. v. FPC,* 506 F.2d 23, 38 (D.C. Cir. 1974).

[2]*American Bus Ass'n v. ICC,* 627 F.2d 525, at 529 (D.C. Cir. 1980) (quoting *Noel v. Chapman,* 508 F.2d 1023, 1030 (2d Cir. 1975), on distinction between general statement of policy and rule requiring publication).

nonexistent for compliance with the law. In this fashion, OSHA is exercising the authority Congress delegated to it to "'fill up the details' by the establishment of administrative rules and regulations, the violation of which could be punished by fine." *United States v. Grimaud,* 220 U.S. 506, 517 (1911). If left undisturbed by this court, this agency action would wield a significant change in the practices which private employers must follow and in the enforcement steps the agency must take. Under these circumstances, I believe that advance notice and opportunity for public participation are vital if a semblance of democracy is to survive in this regulatory era.*

***

OSHA embarked on notice-and-comment rulemaking and, after receiving public comment, in January 1981 issued a new rule requiring walkaround pay.[32] Unlike the original Bingham interpretation, however, the new rule did not conclude that refusal to pay was discriminatory under §11(c) or base the rule on the "hours worked" principle of FLSA. Rather, the rule was predicated on OSHA's authority under §8(e) to provide for more effective walkaround inspections.

The 1981 regulation was short lived. Its effective date was first delayed by the new Assistant Secretary, Thorne G. Auchter, who then proposed to revoke the regulation.[33] On May 29, 1981, following a period of public comment, the walkaround pay regulation was revoked.

Revocation of Walkaround Compensation Regulation

46 Fed. Reg. 28,842, 28,843–44 (1981)

***

II. BASIS AND PURPOSE OF THE REVOCATION

A. INTRODUCTION

On the basis of a careful re-evaluation of the previous record and consideration of the new comments, the Secretary has determined that the walkaround compensation regulation is not necessary to carry out his re-

*[Author's note—Notwithstanding Judge Tamm's sharp criticism of OSHA's actions on walkaround pay, there has been, as Senior Circuit Judge Bazelon noted, considerable uncertainty in defining the distinction between "legislative" rules—requiring notice and comment—and interpretive rules. Courts have variously applied either the "significant impact" or the "legal effects" test in determining whether a rule was interpretive. The two tests often yielded differing results. See Asimow, *Public Participation in the Adoption of Interpretive Rules and Policy Statements,* 75 MICH. L. REV. 520 (1977). Professor Asimow's report was the basis of the recommendation of the Administrative Conference of the United States, 1 C.F.R. §305.76-5 (1983), that agencies should allow public comment where it issues an interpretation of "general applicability" likely to have "substantial impact" on the public. The issue has also been addressed in regulatory reform legislation passed by the U.S. Senate in 1982. See S. REP. NO. 284, 97th Cong., 1st Sess. 110–14 (1983). See also Recent Development, *Agency Duties to Conduct Rulemaking: Conflicting Approaches After Carter v. Cleveland,* 31 AM. U.L. REV. 177, 183–87 (1981). Cf. *Cerro Metal Prod. v. Marshall,* 620 F.2d 964 (3d Cir. 1980), where the court of appeals vacated the OSHA regulation on *ex parte* warrants, but, unlike the district court, not on the grounds that OSHA had failed to follow notice and comment procedures.]

[32] 46 FED. REG. 3852 (1981).

[33] 46 FED. REG. 18,999 (1981).

sponsibilities under the Act. Therefore, pursuant to sections 8(e) and 8(g)(2) of the Act, the Secretary revokes the regulation.

The rule, when promulgated, was based primarily on the theory that failure to compensate for time spent on the walkaround would chill employee participation in the inspection process. However, on the basis of the record, the agency finds that a walkaround compensation regulation would have only a negligible effect on the number of employees who participate in OSHA inspections and on the Secretary's enforcement efforts. As noted in the January 16, 1981 *Federal Register,* a considerable number of inspections are conducted in establishments in which there are no employee representatives and the OSHA inspector consults with a reasonable number of employees, in accordance with Section 8(e). See 46 FR 3588. In those cases in which an employee representative accompanies the OSHA inspector on the walkaround, the record indicates that the vast majority of representatives are in fact compensated for their time. Furthermore, in those cases in which compensation is not available, the record does not establish that the lack of pay operates as a significant disincentive to employee participation. Thus, it appears that a walkaround compensation regulation will encourage only a few employee representatives to participate in the walkaround. And, even where there is no walkaround representative, in accordance with the requirements of Section 8(e), the compliance officer consults with individual employees at the worksite. In view of its minimal impact on employee participation and on the Secretary's ability to effectively enforce the Act, the agency finds that the regulation is not a necessary exercise of his rulemaking authority.

B. CURRENT WALKAROUND COMPENSATION PRACTICES

The record establishes that the vast majority of employee representatives are in fact compensated for time spent on walkaround. Throughout its ten-year enforcement history, the agency has found that a substantial percentage of employee representatives are paid for time during which they accompany or otherwise assist OSHA compliance officers during inspections. This experience is substantiated by memoranda submitted for the record by various OSHA compliance personnel. Several compliance officers stated that they had never encountered an instance in which an employee was denied walkaround compensation. Other compliance officers noted that in general they had experienced no difficulties with respect to compensation for employee representatives. Many inspectors noted specific instances of compensation by either the employer or a union. On the other hand, although numerous memoranda were submitted for the record by compliance officers, area directors and regional administrators, only approximately 10 instances of non-payment were documented. * * *

***

C. LACK OF DISINCENTIVE

We have discussed the question of whether there is a significant need for a walkaround compensation regulation in view of evidence that a considerable majority of employees are in fact paid for time spent on the walkaround. Quite apart from this question is the issue of whether the regulation would encourage employee participation in the inspection process. As noted in the preamble to the January 16, 1981 rule, the promulgation of the regulation was based to a large extent on the assumption that the loss of pay or other benefits operates as a significant disincentive to employee participation in an inspection. 46 FR 3855–56. After careful reevaluation of the previous record and consideration of the new comments, the agency finds that the record does not adequately support this assumption. Indeed, comments submitted by agency personnel and the general public note fewer than 10 instances in which an employee representative did not participate in the walkaround because of non-payment. * * *

Based on the extremely small number of non-participation cases resulting from non-payment, and on the evidence that a substantial number of employee representatives take part in the inspection process, the record does not adequately establish that lack of compensation is in fact a significant disincentive to employee participation in the walkaround. Moreover, even to the limited extent that the lack of pay might discourage a few employees from participating in the inspection, the agency finds that a walkaround compensation regulation is not necessary since it would not have a significant effect on the Secretary's overall ability to conduct effective and efficient workplace inspections. In those limited cases in which an employee representative does not participate because he is not paid, the OSHA inspectors can rely on consultations with individual employees at the worksite, as provided Section 8(e) of the Act. Indeed, as noted earlier, a considerable number of workplace inspections are conducted in establishments in which there is no employee representative on the walkaround. In such cases, pursuant to the Act and OSHA instructions, compliance officers are required to consult with a reasonable number of employees at the worksite. Accordingly on this record, the agency has decided not to require employers to pay for time spent on the walkaround. Rather, we conclude that the issue of walkaround compensation is best left to voluntary arrangements between employers and employees. In reaching this conclusion, the agency notes that its concern for securing employee participation in the walkaround is satisfied in an overwhelming percentage of inspections in which employee representatives have been designated. Absent evidence that this rate of participation will decline substantially, the agency believes that the policies of the Act would best be effectuated if the parties negotiate payment arrangements between themselves.

Some commentators expressed the view that leaving walkaround compensation to collective bargaining is an inadequate solution in light of the fact that only approximately 20 percent of the U.S. workforce is unionized; thus, the vast majority of workers do not have access to collective bargaining and cannot obtain walkaround benefits. However, inspections of non-union establishments are typically conducted without the participation of employee representatives since none exists. As we have noted in the preamble to the original rule, the selection of employee representatives in non-union establishments would entail complex and time-consuming procedures that would seriously delay the conduct of inspections and effective and efficient enforcement of the Act. See 46 FR 3855. Thus, the issue of walkaround compensation generally does not arise in non-union establishments.

***

OSHA's revocation of the walkaround pay rule was not challenged in court. Thus, a little more than 10 years after the first OSHA expression of views on the issue, the Agency has returned to its original view that walkaround pay is not required. (The only court decision on whether OSHA had legal authority to require walkaround pay was Judge June Green's opinion in the district court case upholding that authority.) While the results of the 1972 and 1981 determinations were the same, their analytic bases were significantly different. In 1972, OSHA apparently assumed that failure to pay discouraged walkaround activity but concluded that this fact was immaterial under §11(c) since the refusal was not "because" of the exercise of statutory rights. This was principally a legal, and not a factual, conclusion. In 1981, OSHA revoked the rule, relying instead on the absence of evidence that, without walkaround pay, employee participation would be significantly diminished or OSHA's ability to effectively enforce the Act undermined. OSHA concluded, therefore, that a pay rule was unnecessary to effectuate walkaround under §8(e).[34] OSHA, during Dr. Bingham's administration, had reached a different conclusion on essentially the same record; OSHA concluded that walkaround was essential to the inspection process and that compensation was "critical to the effectuation of employees' walkaround rights."[35]

[34]The 1981 *Federal Register* revocation does not expressly discuss the issue whether failure to pay for walkaround was discriminatory under §11(c).

[35]The Federal Mine Safety and Health Act of 1977 expressly provides for walkaround pay in §103(f), 30 U.S.C. §813(f) (Supp V. 1981). The report of the Senate Committee on Labor and Human Resources stated:

> To encourage such miner participation [in the walkaround] it is the Committee's intention that the miner who participates in such inspection and conferences be fully compensated by the operator for time thus spent. To provide for other than full compensation would be inconsistent with the purpose of the Act and would unfairly penalize the miner for assisting the inspector in performing his duties.

Federal Mine Safety and Health Act of 1977, S. REP. No. 181, 95th Cong., 1st Sess. 28–29 (1977). Accord *Scott v. Consolidation Coal Co.*, 1 MSHC 2450 (1980).

The Mine Safety and Health Act contains a number of provisions that provide employee rights that were not explicitly authorized under the earlier-enacted OSH Act. In addition to walkaround pay, these rights, notably, are medical removal protection and the right to refuse to work in dangerous conditions. See discussions in Chapter 5, Section D and Chapter 18, Section B. The question whether the OSH Act gave the Agency legal authority to regulate in those areas has been litigated, and one of the arguments made against OSHA authority has been that since Congress in the mine safety legislation expressly provided authority for the agency to take these actions and did not do so under OSHA, this demonstrates Congress' intent to deny OSHA that authority. (This principle is often referred to as *expressio unius est exclusio alterius*.) The argument was rejected by the court of appeals in upholding medical removal protection in the lead standard, *United Steelworkers v. Marshall*, 647 F.2d 1189, 1232 (D.C. Cir. 1980). In *Whirlpool Corp. v. Marshall*, 445 U.S. 1 (1980), the Supreme Court, in upholding the OSHA interpretation on employee refusal to work in imminent danger situations, see 29 C.F.R. §1977.12(a) (1983), reasoned that its decision "conforms to the interpretation that Congress clearly wished the courts to give" to the parallel antidiscrimination provision in the 1977 Federal Mine Safety and Health Act. 445 U.S. at 13 n.18.

On the question whether a citation will be vacated if the OSHA inspector denies or fails to afford an employer representative the right to participate in the walkaround, see, for example, *Marshall v. Western Waterproofing Co.*, 560 F.2d 947 (8th Cir. 1977). Generally, the Review Commission and the courts hold that the citation will not be dismissed for that reason unless the employer demonstrates prejudice in the preparation of its case. The courts and, after a period of uncertainty, the Commission have held that the same demonstration of "prejudice" applies where OSHA allegedly does not issue a citation with "reasonable promptness" under §9(a), 29 U.S.C. §658(a). See *Brennan v. Chicago Bridge & Iron Co.*, 514 F.2d 1082 (7th Cir. 1975). Contra *Chicago Bridge & Iron Co.*, 1 OSHC 1485 (Rev. Comm'n 1974) (Cleary, Comm'r, dissenting).

17

Employee Participation in Commission and Court Litigation

A. The *IMC Chemical* Case

Under §10(c) of the Act,[1] the Review Commission regulations must provide to affected employees or their representatives "an opportunity to participate as parties to hearings" before the Commission.[2] It is well established that the right of employee-parties at Commission proceedings includes, for example, their participation in discovery proceedings, presentation of evidence, and examination of opposing witnesses at hearings.[3] Major disagreements have arisen, however, with respect to the right of employee-parties to object to the withdrawal of a citation by OSHA; their right to object to the settle-

[1]29 U.S.C. §659(c) (1976).

[2]The Review Commission rules, 29 C.F.R. §2200.20 (1983), provide that "affected employees" may elect to participate as parties "at any time" before the commencement of the hearing before the administrative law judge, unless "for good cause" shown, the judge allows an election at a later time. Regulations also provide that if the employee representative, that is, the union, has elected party status, the individual employee represented by that union is not separately entitled to party status. 29 C.F.R. §2200.22 (1983) ("Affected employees who are represented by an authorized employee representative may appear only through such authorized employee representative."). A difficult issue arose, however, when unions did not elect party status, yet individual employees within the unions desired to participate at the hearing. In *United States Steel Corp.*, 11 OSHC 1361 (Rev. Comm'n 1983) (Cottine, Comm'r, dissenting in part on other grounds; Cleary, Chairman, dissenting in part), the Review Commission overruled *Babcock & Wilcox Co.*, 8 OSHC 2102 (Rev. Comm'n 1980), and held that an affected employee could elect party status under 29 C.F.R. §2200.22 when the authorized representative had not elected party status. The majority construed §2200.22's phrase "who are represented by an authorized representative" as meaning "who are represented in the Review Commission proceeding." 11 OSHC at 1363–64.

The Commission has also held that the union representing employees of subcontractors who are exposed to cited hazards, but not employees of the cited employer, is not entitled to party status under 29 C.F.R. §2200.20 (1983). *Brown & Root, Inc.*, 7 OSHC 1526 (Rev. Comm'n 1979). However, the union was permitted to intervene under 29 C.F.R. §2200.21(c) because of the exposure to the represented employees; the union's expertise in the industry involved; and the fact that the intervention would not cause undue delay. *Id.* at 1528 (Barnako, Comm'r, dissenting on this issue).

[3]See *Oil, Chem. & Atomic Workers Int'l Union (OCAW) v. OSHRC*, 671 F.2d 643, 648 & n.4 (D.C. Cir. 1982), *cert. denied sub nom. American Cyanamid Co. v. OCAW*, 51 USLW 3287 (1982), reprinted in Section D, this chapter.

ment of citations and proposed penalties by the attorneys for OSHA; and their right to petition for review of adverse Review Commission decisions under §11 of the Act.

Whether employee-parties have the right to object to OSHA's withdrawal of a citation was considered in the Review Commission decision in *IMC Chemical Group*. The Review Commission ruled, 2–1, that employee-parties have a right to object; OSHA appealed, and the Court of Appeals for the Sixth Circuit reversed the Commission holding.

IMC Chemical Group

6 OSHC 2075 (Rev. Comm'n 1978)

Cleary, Chairman:

This dispute arose on November 2, 1976, when respondent, IMC Chemical Group, Inc., contested two citations that had been issued one week earlier. In lieu of filing a complaint, the Secretary of Labor moved to vacate the first (serious) citation on the ground that respondent had not created the hazard at the time or in the manner alleged in the citation. The Secretary stated in the motion that further prosecution was not intended. Attached to the motion, a copy of which was sent to respondent's counsel, was a notice that objections to the motion should be filed within ten days with the Executive Secretary of the Commission. Ten days after the motion was filed, Local 7-854 of the Oil, Chemical, and Atomic Workers International Union [the union] sent a letter to the Executive Secretary requesting the Commission to affirm the citation. Administrative Law Judge Joseph L. Chalk rejected the union's request and granted the Secretary's motion on the ground that Federal Rule of Civil Procedure 41(a)(1)[2] grants the Secretary unfettered discretion to terminate prosecution before service of an answer or motion for summary judgment, neither of which was filed. The judge's decision was directed for review pursuant to section 12(j) of the Occupational Safety and Health Act of 1970, 29 U.S.C. §651 *et seq* (the Act), to decide whether this case is controlled exclusively by section (a)(1) of Rule 41 and, if the section controls, whether it was applied properly. We do not reach the second issue because we conclude that Rule 41(a)(2), not Rule 41(a)(1), is applicable in this case.

* * * We believe, however, that, for the purposes of Rule 41(a)(1), a notice of contest is analogous to an answer in ordinary civil litigation. Thus, respondent's filing of a notice makes Federal Rule 41(a)(2)[3] applicable to this proceeding.

***

Federal Rule 41(a)(2), as applied to Commission proceedings, permits the Secretary to dismiss an action voluntarily but only with the approval of the Commission. * * *

[2]The Rule provides, in pertinent part:
41. DISMISSAL OF ACTIONS
(a) *Voluntary Dismissal: Effect Thereof*
(1) By plaintiff * * *. Subject to the provisions of * * * any statute of the United States, an action may be dismissed by the plaintiff without order of court (i) by filing a dismissal at any time before service by an adverse party of an answer or of a motion for summary judgment, whichever first occurs. * * *

[3]The Rule provides in pertinent part:
(2) By Order of Court. Except as provided in paragraph (1) of this subdivision of this rule, an action shall not be dismissed at the plaintiff's instance save upon order of the court and upon such terms and conditions as the court deems proper.

The Secretary contends that, regardless of the Federal Rules, the Act grants him complete prosecutorial discretion, including the discretion to terminate prosecution over the objections of affected employees. A divided Commission, relying on an earlier presentation of this argument held that affected employees do not have standing to object to the Secretary's withdrawal of a citation. *Southern Bell Telephone and Telegraph Co.,* 77 OSAHRC 83/D1, 5 OSHC 1405, 1977–78 CCH OSHD para. 21,840 (No. 10340, 1977). Upon reconsideration, we conclude that *Southern Bell* was decided incorrectly and must, therefore, be reversed.

The Secretary's claim to absolute prosecutorial discretion as a ground justifying the Commission's refusal to hear the union is not sound. The decision to conduct an inspection and issue a citation admittedly is within the Secretary's discretion, but even this discretion is not without restraints. Section 657(f)(1) of the Act *requires* the Secretary to consider employee requests for inspections of conditions that threaten physical harm or present an imminent danger, to conduct an inspection if he determines that there are reasonable grounds to believe imminent danger or a violation of the Act exists, and to notify requesting employees in writing if he determines that reasonable grounds do not exist. Section 658(a) of the Act *requires* the Secretary to issue a citation if he or his authorized representative believes that an inspected employer has violated the Act. At the least, the Secretary may not conduct his duties in an arbitrary manner to the detriment of employees he is obligated to protect. Cf. *Dunlop v. Bachowski,* 421 U.S. 560 (1975).

Reliance on the Secretary's prosecutorial authority caused the majority in *Southern Bell Telephone and Telegraph Company, supra,* to conclude that the right of affected employees to full participation as parties vanishes when the Secretary seeks to terminate prosecution because the right, as the majority characterized it, is "vicarious." Implicit in this characterization, and the opinion generally, is the conclusion that affected employees have only a conditional right to intervene in Commission proceedings. This conclusion is not supported by section 659(c) of the Act and, therefore, must be rejected.

In section 659(c) of the Act, Congress directs the Commission to "* * * provide affected employees an opportunity to participate as parties to hearings * * *" conducted by the Commission. This opportunity to participate, which is implemented by Commission Rule 20(a), is an unconditional right on its face, and there is no language in any other provisions of the Act to suggest that Congress intended to create a conditional right. It is also significant that Congress grants the Commission the general authority to adopt rules of procedure (29 U.S.C. §661(f)) but specifically orders the adoption of a rule to permit affected employees an opportunity to protect their interests *as parties.* The right to participate is, therefore, not merely a right to intervene in on-going litigation, it is a right to meaningful participation, which the Secretary has a duty to ensure. *ITT Thompson Industries, Inc.,* OSAHRC Docket Nos. 77-4174 & 77-4175 [6 OSHC 1944] (August 17, 1978). The Secretary cannot be permitted to exercise prosecutorial discretion in a manner that would interfere with the right of affected employees to be heard as parties protecting their interest.[8] This means that the Secretary may be granted permission to withdraw from a case, but the proceedings may continue based on the citation originally issued by the Secretary if affected employees have elected party status. * * *

[8]The "public interest" represented by the Secretary does not always coincide with the interests of affected employees. *American Airlines, Inc.,* 75 OSAHRC 43/F3, 2 OSHC 1391, 1974–75 CCH OSHD para. 19,108 (No. 6087, 1974) (Cleary, Commissioner). Cf. *Trbovich v. United Mine Workers of America,* 404 U.S. 528 (1968).

Marshall v. OSHRC (IMC Chemical)

635 F.2d 544 (6th Cir. 1980)

Phillips, Senior Circuit Judge:
***

We hold that the Act, by its terms, makes the Secretary the exclusive prosecutor of OSHA violations. It follows that, prior to the filing of a complaint and answer, the Secretary has the right to withdraw a contested citation. We conclude that the decision of the Administrative Law Judge, dated January 3, 1977, vacating the citation of the Secretary issued in the present case, was correct.

Therefore, the two-to-one decision of the majority of the Commissioners is erroneous and must be reversed.

V

Since Congress has vested the exclusive prosecutorial responsibility in the Secretary of Labor, the majority decision of the Commission clearly is in error in holding that the Union representing the affected employees can be empowered by the Commission to prosecute the citation when the Secretary has moved to vacate it.

In *Taylor v. Brighton Corporation*, 616 F.2d 256 (6th Cir. 1980), this court held that the Occupational Safety and Health Act does not create a private right of action in favor of employees.

29 U.S.C. §659(c) authorizes any employee of an employer contesting a citation issued by the Secretary to challenge as unreasonable the period of time fixed in the citation for abatement of the violation. This is the *only* provision in the act authorizing an affected employee to participate in OSHA proceedings before the Commission.

In its decision in *Southern Bell Telephone Co.* the Commission placed the following interpretation upon 29 U.S.C. §659(c):

> Only the Secretary of Labor is empowered by the Act to issue a citation, and the issuance thereof is left entirely to his discretion as citations are to be issued when the *Secretary or his authorized representative* "believes" that an employer has violated the Act. 29 U.S.C. §658(a). This and other sections of the Act show that Congress intended to exclusively bestow all enforcement powers under the Act upon the Secretary.
>
> Although the Act authorizes any employee or representative of employees to "request" an inspection of their employer's worksite if it is believed that a violation of a safety or health standard threatens physical harm or presents an imminent danger, the Secretary need not make an inspection unless he determines that such an alleged violation or danger exists. 29 U.S.C. §657(f)(1). This provision does not afford employees the right of self-help or judicial review in the event the Secretary does not make an inspection as the result of an employee complaint. If employees cannot force an inspection, they cannot take over the inspection authority, and it therefore seems clear that they cannot force the Secretary to prosecute or take over the prosecution themselves.
>
> Furthermore, only the Secretary can seek enforcement of the Commission's orders, either through a petition for enforcement in a court of appeals (29 U.S.C. §660(b)), or by issuing a notification of failure to abate. 29 U.S.C. §659(b). The Secretary may also compromise, mitigate, or settle any penalty assessed under the Act. 29 U.S.C. §659(c). See *Dale M. Madden Construction, Inc. v. Hodgson,* 502 F.2d 278 (9th Cir. 1974).

> Congress has provided in 29 U.S.C. §659(c) that employees and their representatives can only contest the reasonableness of the abatement period. The contesting of any other part of a citation is only permitted by the employer. We construe the limitation on employee contests as indicating that Congress intended to preclude employees and their representatives from usurping the Secretary's prosecutorial discretion in all other situations. Therefore, we conclude that they have no standing to object to the Secretary's actions in withdrawing a citation in a case.

In the present case, two members of the Commission voted to reverse the Commission's decision in *Southern Bell*. Commissioner Barnako, in his dissenting opinion, stated that he would follow *Southern Bell* in the instant proceeding. We agree with Commissioner Barnako that the above quotation from *Southern Bell* is a correct statement of the applicable law.

29 U.S.C. §661(f) provides that "unless the Commission has adopted a different rule, its proceedings shall be in accordance with the Federal Rules of Civil Procedure." It is conceded that the Commission had not adopted a specific rule at the time of this action governing the withdrawal of a contested citation. The majority of the Commission applied Fed. R. Civ. P. 41(a)(2). We agree with the dissenting opinion of Commissioner Barnako that Rule 41(a)(1) is controlling.

The majority of the Commission erred in vacating the order of the Administrative Law Judge * * * and in holding that the Union representing affected employees has a right to elect party status to prosecute a citation issued by the Secretary when the Secretary declines to proceed with the citation or to file a complaint.

This does not mean, however, that affected employees are not authorized to perform a substantial role in the enforcement of the Act. Employees are granted a number of substantive and procedural rights under the statute. 29 U.S.C. §659(c) authorizes employees or their representatives to file a notice challenging the reasonableness of the time fixed in a contested citation for abatement of the alleged violation. *International Union v. Occupational Health & Safety Review Commission, supra,* 557 F.2d at 610. 29 U.S.C. §659(c) mandates the Commission to provide in its rules of procedure an opportunity for affected employees or their representatives "to participate as parties" to Commission hearings. The Secretary construes the Act as permitting employees having party status to participate in prehearing discovery and present their own witnesses. 29 U.S.C. §657(e) authorizes an employee representative to accompany an OSHA inspector during workplace inspections. 29 U.S.C. §657(f)(1) authorizes an employee to request inspection of a violation of safety or health standards threatening physical harm or imminent danger. 29 U.S.C. §657(f)(2) gives employees the right to receive a written statement from the Secretary for the final disposition of a case. 29 U.S.C. §669(a)(6) gives employees the right to request hazard evaluations. 29 U.S.C. §660(c) protects employees against discharge of discrimination as a result of exercising rights under the Act. *Whirlpool Corp. v. Marshall,* 445 U.S. 1 (1980), *aff'g* 593 F.2d 715 (6th Cir. 1979). In that case the Supreme Court said:

> To ensure that this process functions effectively, the Act expressly accords to every employee several rights, the exercise of which may not subject him to discharge or discrimination. An employee is given the right to inform OSHA of an imminently dangerous workplace condition or practice and request that OSHA inspect that condition or practice. 29 U.S.C. §657(f)(1). He is given a limited right to assist the OSHA inspector in inspecting the workplace, 29 U.S.C. §§657(a)(2), (e) and (f)(2), and the right to aid a court in determining whether or not a risk of imminent danger in fact exists. See 29 U.S.C. §660(c)(1). Finally, an affected

employee is given the right to bring an action to compel the Secretary to seek injunctive relief if he believes the Secretary has wrongfully declined to do so. 29 U.S.C. §662(d).
(Footnotes omitted) 445 U.S. at 9–10.

There is nothing in the Act, however, authorizing affected employees or their representative to prosecute a contested citation if the Secretary elects not to do so.
***

The position of the Secretary of Labor accepted by the Court of Appeals for the Sixth Circuit was that "sound public policy" militates against the union party sharing in the Secretary's enforcement authority. This was stated in the Secretary's brief to the court, as follows:

> As with the General Counsel under the NLRA [National Labor Relations Act], the determination to proceed on a citation properly belongs with the Secretary and should not hinge on the concurrence of third parties. Only the Secretary can adequately and impartially weigh the factors which constitute the "public interest" considerations while at the same time taking cognizance of the interests of independent private parties. Indeed, the Union's motivation here appears to bear little relationship to the "public interest", since its sole expressed reason for objection to the withdrawal of the citation was to enhance its position in a private arbitration matter with the company.[4]

At the same time, the Secretary's brief went to great lengths to demonstrate that despite the rigors of its legal position denying employees a significant role in deciding on the disposition of citations, the Department of Labor did not believe that views of employees with regard to enforcement are "without worth." The brief stated:

> We do not intimate that the views of employees with regard to OSHA enforcement are without worth. As we have noted, the Secretary should and does take into account the views of employee-parties with regard to the disposition of contested cases. As a matter of *policy,* the Secretary consults with affected employees or their representatives who are parties before withdrawing citations or settling major cases. Regional Solicitors of the Department of Labor have permitted employee-parties to participate in and express their views on the proposed disposition of a contested case involving the withdrawal of a citation (or a change in characterization to *de minimis*), specific abatement requirements or the period of time allowed for abatement. And, while generally, there is no employee-party involvement in dispositions dealing exclusively with characterization of violations (willful, repeated, serious, and other than serious) or penalty amounts, Regional Solicitors are expected to exercise their discretion to decide whether it would be appropriate in any particular case to involve employee representatives. This discretion would be exercised, for example, in cases which have attracted wide public interest, such as catastrophes, those involving novel questions or those where the employee-party has evidenced

[4]Brief for Secretary of Labor at 32, *Marshall v. OSHRC (IMC Chemical)*, 635 F.2d 544 (6th Cir. 1980).

strong interest in the case. This policy, predicated on a recognition of the strong interest of employee-parties in issues relating to safety and health in the workplace, goes well beyond the *legal* requirements for employee involvement in prosecutorial decisions. Our point here is not that the Secretary should not be mindful of employee-party views and interests in exercising his prosecutorial discretion. We do say, however, that once the Secretary has determined the manner in which he wishes to prosecute (or not prosecute a citation) taking into account employee views, employee-parties have no statutory right to block that decision in Commission proceedings except with respect to abatement date.[5]

The insistence by the Secretary, through the Solicitor, on the exclusive prosecutory right of the Secretary, and the consequent restriction on the participatory rights of employee-parties in Review Commission proceedings, was strongly criticized by union representatives. Of particular concern to unions was their perception that attorneys for OSHA were routinely settling cases before the Commission, often reducing penalties, eliminating violations, or reducing the characterization of violations (e.g., from "willful" to "serious") without consultation, or meaningful consultation, with employees. This concern was expressed by Lane Kirkland, President, AFL-CIO, in testimony in 1980 before a Senate oversight committee. He said:

> Responsibility for one of the most serious problems plaguing the enforcement process rests not with OSHA, but with the Solicitor of Labor. Once a case is contested, a Solicitor of Labor essentially takes charge. As Basil Whiting has testified, the contest rate is significant and growing. In 1979, over 20% of all inspections which resulted in violations were contested. A case load of 90 cases for an individual solicitor is not uncommon. Thus, pressure to resolve cases through settlement before a hearing is great. In numerous cases, however, where the union has elected party status in the case, settlements are reached by the solicitor and the company without consulting the union representative or OSHA. Since solicitors have little, if any, real knowledge of workplace hazards, these settlements may not achieve the necessary protection of workers. Similarly, these settlements often involve important cases that could serve as a precedent in future enforcement proceedings, if resolved through the judicial process. This unfortunate practice of two party (solicitor and company) settlements was recently adopted as a formal policy by the Solicitor of Labor under protest of the Assistant Secretary for Occupational Safety and Health and the labor movement. In addition, the Solicitor is presently challenging a decision of the Occupational Safety and Health Review Commission prohibiting two party settlements where workers or their representatives have elected party status. (*Secretary of Labor v. IMC Chemical Group,* OSAHRC Docket No. 76-4761)
>
> The AFL-CIO does not believe that the actions of the Solicitor of Labor are in the best interest of the protection of worker health and safety. We recommend that the Committee call upon the Solicitor to offer a full explanation of these settlement policies, and that the Committee conduct a full investigation of the impact of these practices.[6]

[5]*Id.* at 33–35.

[6]*Oversight on the Administration of the Occupational Safety and Health Act, 1980: Hearings Before the Senate Comm. on Labor and Human Resources,* 96th Cong., 2d Sess. pt. 2 at 1239–40 (1980).

The efforts of the Office of the Solicitor to involve employees in the settlement process were formally recognized in August 1979 when the Solicitor of Labor issued an instruction to field attorneys affording employee-parties a "broad opportunity" to participate in the litigation and disposition of OSHA proceedings.

Memorandum from the Solicitor of Labor to the Regional Solicitors[7]

1. BACKGROUND

The Secretary of Labor, through the Solicitor of Labor, has the exclusive responsibility and discretion for prosecuting civil cases arising from citations and penalties before the Occupational Safety and Health Review Commission. Notwithstanding our view that, as a matter of law, the rights of employee-parties in Commission proceedings are limited, we recognize as a matter of policy the strong interest of employee-parties in issues relating to safety and health in the workplace. Consistent with that view, the Office of the Solicitor in the past has accorded employee-parties a broad opportunity to participate in the litigation or disposition of OSHA proceedings, going beyond the statutory mandate of section 10(c) of the Act under which the Review Commission is authorized to establish procedural rules so that employees may "participate as parties to hearings."

Accordingly, the purpose of this memorandum is to delineate what has generally become the policy regarding the participation of employees in the disposition of contested cases, which includes both the withdrawal of citation items, their reduction to *de minimis,* and the settlement of cases.

Employee-parties have an interest with respect to the disposition of cases raising issues as to whether a violation exists and the time and manner of abatement since these issues directly involve the safety or health of affected employees in the cited workplace. Accordingly, in order to derive the maximum benefit from employee-party participation and to obtain understanding and acceptance of the ultimate disposition of such cases by all parties, full involvement of employee-parties at the earliest practicable stage of the settlement process is warranted. The Regional Solicitor has the authority to enter into a settlement agreement or withdraw a citation with cited employers after thoroughly considering and examining the issues raised by the employee-party during the pendency of the litigation and settlement discussions. Where employee-parties object to a settlement, the matter should be considered personally by the Regional Solicitor with appropriate consultation with the Associate Solicitor regarding cases which involve fatalities, multiple injuries, novel issues or which have evoked widespread public interest.

Issues concerning the characterization of violations (whether willful, repeated, serious or other-than-serious) and the amounts of penalties generally do not involve any question of whether abatement will occur. Thus, whether the cited employers or their employees are deterred from future violative conduct by the assessment of a penalty or whether a violation has been properly characterized are questions of law and policy which do not directly affect the safety and health of affected employees at a cited workplace. It is, therefore, our view that employee-parties need not, as a general matter, be consulted regarding the disposition of contested cases involving only questions of characterization of violations and amounts of penalties.

---

[7]Participation of Employees in Disposition of Commission Cases, Aug. 3, 1979. Copy reprinted in Brief for the Secretary of Labor, addendum B, *Donovan v. International Ass'n of Iron Workers (Brown & Root, Inc.),* 11 OSHC 1840 (5th Cir. 1984).

2. GUIDELINES

(a) Affected employees or their representatives, who have been granted party status pursuant to Commission Rule of Procedure 29 CFR 2200.20(a), (hereinafter referred to as "employee-party") should be given an opportunity by the Regional Solicitor to participate in and express views on the proposed disposition of a contested case involving the withdrawal of a citation (or a reduction to *de minimis*), specific abatement requirements, or the period of time allowed for abatement. With regard to the manner for employee-party involvement, such as, simultaneous three-party negotiation, staggered discussions, telephone communications, exchanges of written documents, or other means, the Regional Solicitor is vested with flexibility, consistent with employee-party participation, in order to insure promptness in settlement of contested cases which fully effectuate the purpose of the Act. However, every effort should be made to involve the employee-party either in negotiations or discussions which parallel in form and substance those dealings undertaken with other parties.

(b) With respect to proposed dispositions dealing exclusively with characterization or penalty amounts (subject to paragraph (c)), Regional Solicitors are not required to follow the procedure outlined above. Commission Rule 29 CFR 2200.100(c) is, of course, applicable.

(c) Regional Solicitors have discretion to determine to what extent employee-parties should be involved in the disposition of cases involving other issues. Thus, for example, in a case with numerous citations or citation-items, some of which involve abatement or withdrawal of violations, and other items involve only penalty issues, it may be desirable for the Regional Solicitor to discuss all items of the proposed settlement with the employee-party, in the manner described in paragraph (a). Similarly, in cases which have attracted wide public interest, such as catastrophes, or those involving novel questions, or where an employee-party has evidenced strong interest in the case, the Regional Solicitor may properly decide to involve employee-parties in all settlement discussions including those dealing with characterization of violations or penalties.

B. The *Sun Petroleum* Case

The issue of the role of employee representative in the disposition of contested cases came up again in the *Sun Petroleum* case,[8] a factually complex proceeding, involving the right of OSHA attorneys to enter into a settlement agreement with a cited employer over the objection of the union party. The Court of Appeals for the Third Circuit, with Judge Pollak dissenting, went further than the *IMC* court in upholding the prosecutorial discretion of the Solicitor of Labor in settling cases. Speaking for the majority, Circuit Judge Aldersirt said:

> We are prepared to accept most of the Secretary's arguments because we agree that in the absence of a contest, neither the Review Commission nor the ALJ has jurisdiction to review a settlement agreement entered into between the Secretary and an employer. Moreover,

[8]*Marshall v. Sun Petroleum Products Co.*, 622 F.2d 1176 (3d Cir.), *cert. denied,* 449 U.S. 1061 (1980).

we agree that even after the employer files a notice of contest, if no employee files a notice or has not acquired party status under 29 C.F.R. §2200.20, then the Commission would lack jurisdiction to review any settlement entered into between the Secretary and the employer. If an employee formally expresses an interest in the proceedings, however, we believe that the Commission would have jurisdiction to review the settlement in order to protect that interest.

The Act provides that the Commission is required to afford an opportunity for a hearing upon the happening of either of two events: (1) if the employer notifies the Secretary that he intends to contest a citation, or (2) if "any employee or representative of employees files a notice with the Secretary alleging that the period of time fixed in the citation for the abatement of the violation is unreasonable." 29 U.S.C. §659(c). If settlement is reached after either of these events occurs, but prior to the scheduled hearing, the ALJ would have jurisdiction to review the settlement, but only for the limited purpose of entertaining objections from the employee or employees' representative that the abatement period proposed by the settlement is unreasonable.

* * *

Under our interpretation we also reject the union's contention that affected employees have a right to be heard on matters other than the reasonableness of the abatement period. The Senate Committee Report makes clear that section 10(c) "gives an employee or representative of employees a right, whenever he believes that the period of time provided in a citation for abatement of a violation is unreasonably long, *to challenge the citation on that ground.*"[17] Moreover, the legislative history discloses no support for the union's position. Indeed, any evidence of congressional intent on this point contradicts the union's assertion that it is entitled to be heard on matters other than the abatement period. We therefore conclude that any challenge advanced by an employee is limited to an attack on the reasonableness of the abatement period.[9]

Judge Pollak dissented from the majority's view that employee parties could not object to settlements on grounds other than the reasonableness of the abatement period. He believed that the unions' objections should be considered and that they should have a right to support these objections through the presentation of evidence, analogizing the status of unions in OSHA to the status of the charging party under the National Labor Relations Act. Judge Pollak said:

This approach comports with procedures applied to the settlement of cases before the National Labor Relations Board. See *Marine Engineers Beneficial Ass'n. v. NLRB,* 202 F.2d 546, 549 (3d Cir. 1953), *cert. denied,* 346 U.S. 819; *Terminal Freight Cooperative Ass'n v. NLRB,* 447 F.2d 1099, 1101 (3d Cir. 1971), *cert. denied,* 409 U.S. 1063 (1972); *Leeds & Northrup Company v. NLRB,* 357 F.2d 527 (3d Cir. 1966). In *Marine Engineers,* this court grappled with the then unfamiliar role of the charging party in Labor Board proceedings:

> The difficulty in this case comes because in changing times and the evolution of administrative procedure our old analogies are not in point. The charging party in a labor case is something like a complaining witness in a criminal case. But he is certainly much more

[17]S. REP. No. 1282, 91st Cong., 2d Sess. 15 (1970) (emphasis added).

[9]*Id.* at 1185–86.

> than that for a complaining witness is certainly not entitled to appeal even when an appeal is allowed for the prosecution in a criminal case. On the other hand, the charging party is not like the ordinary plaintiff in a lawsuit, who does not have to have anybody's permission to go ahead with his action if he can pay the required fees. This is something in between.

202 F.2d at 549. The court went on to hold that the charging party in a Labor Board case had a right to be heard on settlement objections, to present evidence on that issue, and to appeal an adverse ruling. *Id.* Other circuits are generally in accord. See *George Ryan Co. v. NLRB,* 609 F.2d 1249 (7th Cir. 1979); *ILGWU v. NLRB,* 501 F.2d 823 (D.C. Cir. 1974); *NLRB v. OCAW,* 476 F.2d 1031 (1st Cir. 1973); *Concrete Materials of Georgia v. NLRB,* 440 F.2d 61 (5th Cir. 1971); *NLRB v. Electrical Workers Local 357,* 445 F.2d 1015 (4th Cir. 1971).

OSHA procedures, and those under the National Labor Relations Act are, of course, not strictly analogous. The NLRA does not, in terms, provide that the charging party can participate as a party to Board proceedings. Nor does the NLRA provide the charging party with the elaborate protections afforded employees under OSHA. Seemingly of even greater significance, in contrast with the position of the Secretary under the OSHA scheme, the General Counsel of the National Labor Relations Board enjoys the statutory discretion not to issue a complaint even if there exists a violation of the Act. Each of these distinctions would appear to suggest broader rights for employees under OSHA, than under the NLRA. But under Labor Board rules, once the General Counsel issues a complaint, "the charging party is accorded formal recognition: he participates in the hearings as a 'party,' * * * he may call witnesses and cross-examine others, may file exceptions to any order of the trial examiner * * *." *International Union, etc. v. Scofield,* 382 U.S. 205, 219 (1965). It is improbable that Congress, writing against this backdrop of the National Labor Relations Act practice, could have intended any lesser role for the complaining employees under OSHA by explicitly providing them the right "to participate as a party."[10]

C. The *Mobil Oil* Case

Despite the efforts of the Solicitor to deal with the scope of union involvement in settlement negotiations as a policy matter and despite adverse decisions in some courts, unions continued to press their position before the Review Commission and the courts. In August 1982, the Commission, by a 2-1 vote in *Mobil Oil Corporation,*[11] disagreed with the court decisions in *IMC Chemical* and *Sun Petroleum* and held that employee-parties have the right to be heard on their objections to the substance of a settlement agreement; and that §10(c) does not limit their right to be heard on the issue of the period fixed for abatement. According to the majority, when affected employees object to an abatement plan in a settlement agreement,[12] the

[10] *Id.* at 1191 n.5.

[11] 10 OSHC 1905 (Rev. Comm'n 1982) (Cleary, Comm'r, concurring; Rowland, Chairman, dissenting).

[12] Mobil had been issued a serious general duty citation, alleging that it had required bulk plant employees to perform manual gauging sampling of floating roof tanks without proper atmospheric testing, training, or contact with other personnel. A penalty of $540 was proposed. The Secretary and Mobil entered into a settlement agreement to abate the hazards immediately, but the union-party, Petroleum Trades and Employees Union, Local 419 (P.T.E.U.) claimed that the abatement plan was insufficient. See 10 OSHC at 1905–07.

administrative law judge must take these objections into consideration in deciding whether to approve the settlement as consistent with the purposes of the Act.[13] Commissioner Cottine wrote a lengthy lead opinion, considering the language of the Act, its overriding policies, and the legislative history; he particularly emphasized employees' "direct and personal stake in the outcome of litigation before the Commission";[14] the limits of the "prosecutorial discretion" of the Secretary of Labor; and the fact that the statute "read as a whole" establishes a series of "checks and balances" on the Secretary's enforcement discretion.[15] Commissioner Cleary concurred, answering the two major arguments made by the Secretary to limit employee participatory rights.

Mobil Oil Corporation

10 OSHC 1905 (Rev. Comm'n 1982)

COTTINE, Commissioner:
***

D

In spite of the unconditional statutory grant of intervening party status to affected employees and the Commission's implementing rules, the Secretary argues that this statutory language should be restricted to accommodate his "prosecutorial discretion."[20] The Secretary's argument overlooks the fundamental distinction between the initiation of enforcement proceedings and employee participation in adjudicatory proceedings. Certainly, the Secretary is granted discretion to investigate workplace hazards, issue citations, mitigate penalties and enforce the Commission's final orders, 29 U.S.C. §§657(a), 658(a), 655(e), 660(a), respectively. Moreover, the Act re-

[20]"Prosecutorial discretion" can only be defined in reference to the specific governmental authority exercised. It is not a "magical incantation which automatically provides a shield for arbitrariness." *Medical Comm. for Human Rights v. SEC,* 432 F.2d 659, 673 (D.C. Cir. 1970), *dismissed as moot,* 404 U.S. 403 (1972). Enforcement discretion is entirely dependent on the government function assigned, the range of discretion delegated, and the adjudicatory forum involved. * * *

[13]See the Commission rules on settlements, 29 C.F.R. §2200.100(a) (1983).

[14]10 OSHC at 1911.

[15]10 OSHC at 1912–15. The Review Commission has also held that employee-parties have the right to be heard on their objections to the Secretary's withdrawal of a citation. E.g., *Cuyahoga Valley Ry.,* 10 OSHC 2156 (Rev. Comm'n 1982) (Cleary, Comm'r, concurring; Rowland, Chairman, dissenting), *appeal pending,* Nos. 82-3771, 82-3773 (2d Cir. 1983).

Commissioners Cleary and Cottine appear to differ on whether employee representatives may prosecute OSHA citations before the Commission. In *Mobil Oil,* Commissioner Cottine "presumed" that the Secretary would "proceed appropriately" if the Commission decided that the settlement should be set aside. 10 OSHC at 1916, n.30. This apparently means that the Secretary would accept the responsibility and prosecute the citations on behalf of OSHA. Commissioner Cottine therefore found it unnecessary to state any views on the *IMC Chemical* issue of whether employee-parties had an independent right to prosecute citations. Commissioner Cleary's opinion suggested that he believes that employees may "volunteer to take on the financial burden of proving a violation." 10 OSHC at 1927. See also *Cuyahoga Valley Ry.,* 10 OSHC at 2158 (Cleary, Comm'r, concurring). For a discussion of the views of the Court of Appeals for the District of Columbia Circuit, see *Oil, Chemical & Atomic Workers Int'l Union v. OSHRC (American Cyanamid),* 671 F.2d 643 (D.C. Cir. 1982), *cert. denied sub nom. American Cyanamid Co. v. OCAW,* 51 USLW 3287 (1982).

serves to the Secretary the exclusive authority to initiate an enforcement action by issuing a citation under section 9 of the Act. 29 U.S.C. §658. Employees are not authorized to issue a citation. 29 U.S.C. §657(f)(2). Furthermore, the statute does not create any independent cause of action by affected employees against their employers for allegedly violative conditions. 29 U.S.C. §653(b)(4); see, e.g., *United Steelworkers of America v. Marshall,* 647 F.2d 1189 (D.C. Cir. 1980), *cert. denied,* 453 U.S. 913 (1981). Thus, enforcement actions under section 9 may only be initiated by the Secretary.

The citation represents the exercise of the Secretary's discretion under sections 9 and 10 of the Act. However, once a citation is issued and a contest is filed the Commission is the exclusive post-citation forum and the Secretary's exercise of enforcement discretion is the heart of the controversy before the Commission. Parties with interests adverse to the Secretary invoke Commission jurisdiction by their notices of contest. Moreover, their adverse interests confer intervening party status in the event they do not contest the citation but another party does. Commission Rule 20(a) & (b), 29 C.F.R. §2200.20(a) & (b) (1981).

The mere existence of enforcement discretion does not limit the statutory status of the intervening party. This is not to suggest that the Secretary is deprived of enforcement discretion but that this discretion is subject to independent review when it is involved in an enforcement action before the Commission. The interests of all parties, governmental and private, are subject to the scrutiny of adversarial adjudication and "issues cannot be resolved by a doctrine of favoring one class of litigants over another," *Schlagenhauf v. Holder,* 379 U.S. 104, 113 (1964). Unlike the earliest versions of the legislation which vested investigatory, enforcement *and* adjudicatory authority exclusively in the Secretary, Congress ultimately assigned adjudicatory responsibilities to the Commission under section 10 of the Act. H.R. Rep. No. 1765, 91st Cong., 2d Sess. 13 (1970) (conference report). Here the parties' adverse interests are to be adjudicated independent of the investigatory and enforcement discretion assigned to the Secretary. See S. Rep. No. 1282, 91st Cong. 2d Sess. 63 (1970) (minority view of Sens. Dominick & Smith). Thus, the Secretary is deprived of the power to issue an enforceable order once a notice of contest has been filed. Instead, the government must come to an adjudicatory forum to obtain an enforceable final order. See *United States v. Ketchikan Pulp Co.,* 430 F. Supp. at 86 (analogous judicial provision of Federal Water Pollution Control Act).

Furthermore, even in the exercise of the Secretary's investigatory and enforcement discretion, affected parties have distinct rights. Employees are explicitly permitted to file a complaint seeking inspection of a worksite hazard, 29 U.S.C. §657(f)(1), and both employees and employers are authorized to accompany the Secretary's representative during an inspection, 29 U.S.C. §657(e). Moreover, "[i]f the Secretary determines there are no reasonable grounds to believe that a violation or danger exists he shall notify the employees or representatives of the employees in writing of such determination." 29 U.S.C. §657(f)(1). Both the employer and employees are also entitled to be informed that an imminent danger exists and that relief will be sought. 29 U.S.C. §662(c). In addition, affected employees may seek a writ of mandamus if the Secretary "arbitrarily or capriciously fails to seek relief" to correct an imminent danger. 29 U.S.C. §662(d). See generally *Whirlpool Corp. v. Marshall,* 445 U.S. 1 (1980).

The statute read as a whole establishes a series of checks and balances on the Secretary's enforcement discretion, including Commission adjudication of contested enforcement actions. Interpreting the statute to give effect to all its provisions, see *Colautti v. Franklin,* 439 U.S. 379 (1979), including the provision granting party status to affected employees, the Secretary's discretion and the statutory rights of employers and employees co-exist without sacrificing the interests of the private parties to the government's unlimited enforcement discretion.

***

CLEARY, Commissioner, concurring:

***

In the instant case, the argument is made that introduction of yet another stage in the adjudicatory process, i.e., an evidentiary hearing on employee objections to a settlement agreement, hampers employee safety and health by delaying if not actually foreclosing a legal duty to abate; no abatement is required while the parties dispute the abatement plan, and if the hearing ends in disapproval of the settlement agreement, a final order against the Respondent may never issue.[51] The implication is that even if the Secretary's abatement plan is less than perfect it will certainly achieve some improvement in employee safety and health and that a very real gain is being traded away for the chimera of perfect safety. This might be true if the focus of the hearing were whether and to what extent the employees' abatement plan is better than the Secretary's, but that is not the case. The question before the administrative law judge at such a hearing is not whether the employees' plan is better than the Secretary's plan, but whether the Secretary's plan comports with the purposes and objectives of the Act, in other words, whether it is adequate. In the case, as here, of a general duty clause violation, a plan will be deemed adequate if it renders the workplace free of the hazard. The employees' burden will be to prove that the agreement's averment of actual abatement is in fact not true or that the proposed abatement method will not render the workplace free of the hazard. Therefore, the focus of the hearing will be not on the relative merits of the employees' and the Secretary's plan, but on whether the proposed abatement method(s) have rendered or will render the workplace free of the hazard.

A related argument is that a hearing on employee objections does not actually advance employee safety and health interests because it is, in fact, an exercise in futility. The terms of the challenged agreement represent the Secretary's best assessment both of the case and of his own resources, and if his abatement plan is disapproved he will not act to pursue the employees' views. However, as explained above, the settlement agreement will be disapproved only if the employees show that the proposed abatement methods are ineffective. To suggest that even in the face of such evidence the Secretary will refuse to negotiate further or, in the event the employees take over the prosecutorial function and achieve a final order affirming the citation, he will subsequently refuse to issue a failure-to-abate order, is to assume that the Secretary will not act in good faith. There is no foundation for such an assumption.

The Secretary of Labor and Mobil Oil appealed the decision of

[51]My dissenting colleague takes particular note of the fact that the instant case is now four years old. I could not agree more that time is of the essence; no one could seriously argue against speedy resolution of OSHA cases. However, one must question the value of a final, enforceable order which is based on a settlement agreement which will not achieve abatement.

In any case, the question before us is a legal one—what, in light of the Act's legislative history and overall scheme, is the scope of section 10(c) of the Act? The question is not whether cases could be more quickly disposed of if employees were excluded from substantive participation in settlement. Moreover, in the instant case, the four year delay is not the result of anything the employees did or sought to do during settlement, but is the result of the fact that their right to be heard was placed in issue as a legal matter; had the fundamental question of employee rights not been in dispute, a final order in this case would undoubtedly have been issued long ago.

the Review Commission and a unanimous Court of Appeals for the Second Circuit reversed, following the precedent set in the *IMC Chemical* and *Sun Petroleum* cases.

Donovan v. OSHRC (Mobil Oil Corporation)

713 F.2d 918, 11 OSHC 1609 (2d Cir. 1983)

TENNEY, D.J.:

The Secretary of Labor ("the Secretary") has petitioned for review of an order and decision of the Occupational Safety and Health Review Commission ("the Commission") that remanded a proposed settlement between the Secretary and Mobil Oil Corporation ("Mobil") to an Administrative Law Judge ("ALJ") for further review. Mobil had received a citation for creating a hazardous work condition in violation of the Occupational Safety and Health Act of 1970 ("the Act" or "OSHA"), 29 U.S.C. §651 *et seq.* (1976). Both the citation and the proposed settlement called for the immediate abatement of the hazard. The Commission remanded the case in order to give Mobil's employees' representative, the Petroleum Trades Employees Union, Local 419, ("PTEU" or "the union"), an opportunity to present its objections to the methods outlined in the settlement agreement for abating the hazard. Both Mobil and PTEU have been granted intervenor status in this appeal.

The petition for review presents three issues:

1) Whether the Commission's order remanding the settlement to the ALJ is subject to review by this court;

2) Whether the Commission erred in holding that an employee or employee representative has the right to a hearing on his objections to the adequacy of abatement procedures included in a settlement agreement between the Secretary and an employer; and

3) Whether the Commission has the statutory authority to review a settlement between the Secretary and a cited employer where the affected employees have not challenged the reasonableness of the abatement period.

For the reasons stated below we hold that the Commission's remand order is reviewable; that the Commission erred when it remanded the case to the ALJ for a hearing on the union's objections to the proposed methods of abating the hazard; and that the Commission does not have the statutory authority to review a settlement between the Secretary and a cited employer except to hear employee challenges to the reasonableness of the period of time for abating the hazardous or unsafe working condition.

***

III

Having determined that the order is reviewable our task is to decide whether the Commission exceeded its statutory authority by remanding the settlement to the ALJ. This raises issues of a broader dimension concerning the respective roles of the Secretary, employees and the Commission in the enforcement of the Act. In order to decide if the Commission erred, we must determine whether the union has a right under the Act to present its objections to the effectiveness of a proposed abatement plan included in a settlement agreement, and whether the Commission has the independent statutory authority to review abatement plans approved by the Secretary to determine if they are consistent with the purposes and provisions of the Act.

A

The Secretary and Mobil argue that the Commission's interpretation of the term "parties" in §10(c) is contrary to the language of the Act and its legislative history, results in an undue broadening of employees' rights under the Act, and disrupts the enforcement of the Act by infringing upon the Secretary's prosecutorial powers. The Secretary also argues that since §10(c) makes no reference to settlements, the Commission lacks jurisdiction to review any settlement between the Secretary and an employer except to consider employee objections over the abatement period specified in the agreement.[11] * * *

***

PTEU argues that both *Marshall v. OCAW* and *Marshall v. Sun Petroleum Prods. Co.* wrongly concluded an employee or employee representative has no greater rights to contest a settlement as a party in an employer-initiated case than an employee or employee representative in an employee-initiated case. PTEU contends that, notwithstanding the specific limitation on employee contests, the last line of §10(c) establishes an independent basis for employees, as parties, to contest abatement plans. Furthermore, PTEU argues that the legislative scheme of the Act evidences a congressional intent to subordinate the Secretary's prosecutorial discretion to the employees' right to a hearing on their objections to abatement procedures outlined in a settlement.

We disagree, and we find that nothing in the language of the Act or its legislative history indicates that Congress intended to grant employees or their representatives standing to challenge the substantive aspects of abatement plans or methods included in a settlement agreement.

***

Employees have been granted specific rights in the investigatory and rule-making stages of the Act, including the right to participate in rule-making proceedings, *id.* at §655(b)(2), (3), and the right to request that the Secretary conduct a workplace inspection when employees suspect that a violation has occurred or an imminent danger exists, *id.* at §657(f)(1). Nevertheless, employees have a limited role in the enforcement of the Act. Under OSHA, employees do not have a private right of action. *Marshall v. OSHRC, supra,* 635 F.2d at 550–51 [9 OSHC at 1035] (citing *Taylor v. Brighton Corp.,* 616 F.2d 256 [8 OSHC 1010] (6th Cir. 1980)). They may not compel the Secretary to adopt a particular standard, *OCAW v. OSHRC, supra,* 671 F.2d at 649 [10 OSHC at 1349] (citing *National Congress of Hispanic Am. Citizens v. Usery,* 554 F.2d 1196 (D.C. Cir. 1977)). As parties, they may not prosecute a citation once the Secretary decides to withdraw it, *Marshall v. OSHRC, supra;* nor may they continue an appeal of a Commission decision once the Secretary unconditionally asserts that he will not prosecute the citation regardless of the decision by the court of appeals, *OCAW v. OSHRC, supra,* 671 F.2d at 650–51 [10 OSHC at 1350].

Indeed, the Act subordinates the prosecutorial discretion of the Secretary to the rights of employees in only two specific situations: first, employ-

[11]In December 1979, after this case had been presented to the Commission for review, the Secretary changed his policy position regarding the Commission's authority to review settlements. Previously, the Secretary submitted settlements to the Commission for its approval pursuant to Commission Rule 100, 29 C.F.R. §2200.100. The Secretary's current position is that the Commission lacks jurisdiction to review settlements, except to hear employee challenges to the reasonableness of the abatement period. Although this issue was not presented below and PTEU questioned at oral argument whether this was an appropriate issue for this court now to decide, the reasoning we employ in disposing of the question which was contested below necessarily resolves this second, closely related question as well.

ees have the right to challenge the period for abatement noted in a citation, 29 U.S.C. §659(c), and second, employees have the right to bring a mandamus action against the Secretary for his failure to enjoin an imminent danger at their workplace, *id.* at §662(d).

Thus, it is apparent from the detailed statutory scheme that the public rights created by the Act are to be protected by the Secretary, see *Atlas Roofing Co. v. OSHRC, supra,* 430 U.S. at 444–47 [5 OSHC at 1105–06]; *Marshall v. Sun Petroleum Prods. Co., supra,* 622 F.2d at 1187 [8 OSHC at 1427], and that enforcement of the Act is the sole responsibility of the Secretary, *OCAW v. OSHRC, supra,* 671 F.2d at 649 [10 OSHC at 1349]; *Marshall v. OSHRC, supra,* 635 F.2d at 550 [9 OSHC at 1034]; *Marshall v. Sun Petroleum Prods. Co., supra,* 622 F.2d at 1187 [8 OSHC at 1428]; *Dale M. Madden Const., Inc. v. Hodgson, supra,* 502 F.2d at 280 [2 OSHC at 1102–03]. Only he has the authority to determine if a citation should be issued to an employer for hazardous or unsafe working conditions, 29 U.S.C. §658, and only he may prosecute a citation before the Commission, *Marshall v. OSHRC, supra,* 635 F.2d at 550 [9 OSHC at 1034]. A necessary incident to the Secretary's prosecutorial powers is the unfettered discretionary authority to withdraw or settle a citation issued to an employer, *OCAW v. OSHRC, supra,* 671 F.2d at 650 [10 OSHC at 1349]; *Marshall v. Sun Petroleum Prods. Co., supra,* 622 F.2d at 1187 [8 OSHC at 1430], or to settle, mitigate or compromise any assessed penalty, *Dale M. Madden Constr., Inc. v. Hodgson, supra.* A determination that a hazardous or unsafe working condition will be abated by the implementation of specific procedures outlined in a settlement is an enforcement function analogous to the issuance of the original citation, within the Secretary's expertise, see 29 U.S.C. §§659(b), 666(d), and, thus, subject to his exclusive prosecutorial discretion.

"Informal dispositions [are] the 'lifeblood of the administrative process.'" *Local 282, International Brotherhood of Teamsters v. NLRB,* 339 F.2d 795, (2d Cir. 1964) (quoting Attorney General's Committee on Administrative Procedure, Final Report, 35 (1941)). Permitting the Secretary to settle citations issued to employers without hearings before the Commission effectuates the basic remedial purpose of the Act—the rapid abatement of unsafe or unhealthy working conditions. Once an employer files a good faith notice of contest, any requirements to abate the hazardous or unsafe working condition are stayed pending the entry of a final order of the Commission. 29 U.S.C. §659(b). Continuation of the Commission's proceedings after an employer has agreed to withdraw his notice of contest so that an employees' representative or an employee may present objections to a settlement agreement not only puts off the day when abatement should finally occur, but also prevents the Secretary from taking any steps to compel abatement. *Marshall v. OCAW, supra,* 647 F.2d at 387 [9 OSHC at 1586] (citing *American Cyanamid Co.,* 8 OSHC 1346, 1350 (1980) (Comm'r Barnako, dissenting)). Moreover, employers would only be discouraged from entering settlement negotiations with the Secretary if they knew further proceedings before the Commission could be required.

Thus, we find that the legislative scheme of the Act does not evidence a congressional intent to subordinate the Secretary's prosecutorial discretion in reaching settlement agreements to the rights of employees. Indeed, allowing employees to challenge the efficacy of an abatement plan in a settlement would constitute a continued prosecution of the citation by employees and, hence, is proscribed under the Act. See *Marshall v. OSHRC, supra.*

Moreover, we do not agree with the Commission's conclusion that the last line of §10(c) granting employees party status in a Commission hearing establishes a substantive right to challenge abatement plans included in settlements. That section of the Act confers a procedural right on employees;[13] it does not establish an independent entitlement to a hearing on the employees' objections to the substantive elements of an abatement plan. "[T]hat question must turn on what rights, if any, the Act confers upon [employees]." *Local 282, International Brotherhood of Teamsters v. NLRB, supra,* 339 F.2d at 798.

The only substantive right that employees are granted in enforcement proceedings is the limited right to challenge the reasonableness of the "period of time fixed in the citation for the abatement of the violation." 29 U.S.C. §659(c). PTEU, however, contends that Congress did not intend the abatement date language in §10(c) to be a limitation on an employee's rights as a party to challenge abatement plans. The union adopts Commissioner Cottine's theory of why Congress chose to refer only to abatement dates instead of all abatement issues in defining the scope of employee contests. The Commissioner concluded that the language governing employee contests outlined in §10(c) was merely a reflection of the language governing citations issued pursuant to §9(a), 29 U.S.C. §658(a). That section provides in pertinent part that:

> Each citation shall be in writing and shall describe with particularity the nature of the violation, including a reference to the provision of the chapter, standard, rule, regulation, or order alleged to have been violated. In addition, the citation shall fix a reasonable time for the abatement of the violation.

He reasoned that inasmuch as §9(a) makes no reference to abatement requirements or abatement plans, but refers only to the time for abatement, §10(c) provides for employee notices of contest over the only abatement matter an employee would have reason to complain about after reading a citation—the reasonableness of the abatement period.

We do not agree. The Commissioner's construction of the statutory phrase relating to employee-initiated contests ignores the plain meaning of the statute. The time for abatement is not the only abatement factor required to be identified in a citation. Each citation describes "the nature of the violation, including * * * the provision of the chapter, standard, rule, regulation, or order alleged to have been violated." And, if Congress had given the same broad rights to employees as it gave to employers in §10(c), employees, after reading a citation, could contest a number of abatement-related factors. For example, employees could claim that the citation referred to an inapplicable standard and that another standard was more appropriate, cf. *General Elec. Co. v. OSHRC,* 540 F.2d 67, 70 n.2 [4 OSHC 1512, 1514 n.2] (2d Cir. 1976), or that the violations at the work site were incorrectly described in the citation. Hence, the limitation on the scope of employee contests in §10(c) does not parallel the language in §9(a), but rather demonstrates a congressional intent to limit the rights of employees.

[13]As parties to an employer-initiated contest, employees may participate in prehearing discovery and present their own witnesses at Commission hearings. *Marshall v. OSHRC, supra,* 635 F.2d at 552 [9 OSHC at 1036]. The Court of Appeals for the District of Columbia has also noted that "[p]ursuant to the statutory mandate of 29 U.S.C. §659(c), the commission has consistently held that '[an] employee representative electing party status has the right to litigate all the issues raised by the citation and complaint.'" *OCAW v. OSHRC, supra,* 671 F.2d at 648 n.4 [10 OSHC at 1348 n.4] (quoting *Southwestern Bell Tel. Co.,* 5 OSHC 1851, 1852 (1977)).

However, the statute on its face does not provide for employee challenges to the abatement methods an employer would have to use if he were found in violation of the Act. See *Marshall v. B.W. Harrison Lumber Co.,* 569 F.2d 1303, 1308 [6 OSHC 1446 1450] (5th Cir. 1978) ("The statute does not * * * require that the citation specify what corrective measures should be taken, and indeed the employer is often free to correct the violation as he sees fit."); *UAW, supra,* 4 OSHC at 1244 (if the Commission finds an employer in violation of the Act it enters an order affirming the violation without reference to a specific abatement plan that the employer must follow).

We must conclude that "[t]he express allowance of one specific objection suggests a statutory intent to foreclose at least some other objections." *Marshall v. B.W. Harrison Lumber Co., supra,* 569 F.2d at 1307 [6 OSHC at 1448]. If Congress had intended to give employees broader rights in enforcement proceedings before the Commission it would not have limited the grounds upon which employees may challenge a citation. More important, the absence of any reference to abatement plans or methods in §9(a) or §10(c) indicates that Congress did not intend these matters to be contested before the Commission. See *supra* note 13. Indeed, the language of the Act demonstrates that when Congress intended to give employees the right to a hearing to present objections to an employer's abatement plans or methods it so indicated on the face of the statute. See, e.g., §6(b)(6)(A), 29 U.S.C. §655(b)(6)(A).

The legislative history of §10(c) also suggests that Congress intended to limit the rights of employees in the enforcement of the Act. The committee report on the Senate bill, S. 2193, explains the substantive rights granted to employees as follows:

> Section 10(c) also gives an employee or representative of employees a right, whenever he believes that the period of time provided in a citation for abatement of a violation is unreasonably long, to challenge the citation on that ground. Such challenges must be filed within 15 days of the issuance of the citation, and an opportunity for a hearing must be provided in similar fashion to hearings when an employer contests. The employer is to be given an opportunity to participate as a party.

Legis. History, *supra,* at 155. All other evidence of congressional intent on this subject demonstrates that employees are entitled to a hearing before the Commission only on matters relating to the period for abatement.[16]

Furthermore, if we were to accept the union's interpretation of §10(c) an anomalous result would occur. Employees, as parties, would have standing to object to the effectiveness of abatement plans included in settlements between employers and the Secretary after the employer withdraws his notice of contest, but would not have standing in employee-initiated contests to object to abatement plans included in settlements. See *Marshall v. Sun Petroleum Prods. Co., supra,* 622 F.2d at 1185 [8 OSHC at 1429–30]; *UAW v. OSHRC, supra.* There is no justification for such an anomaly under the legislative plan of the Act or its legislative history.[17] Accordingly, we hold that employees or employee representatives may not use their party status under §10(c) as a jurisdictional touchstone to obtain a hearing before the Commission on their objections to the effectiveness of an abatement plan included in a settlement agreement between the Secretary and an employer. Employees do not have a right to this type of hearing under the Act, and the Commission erred in remanding the case to the ALJ for a consideration of the union's objections.

[16]The House Amendment to the Senate bill, S.2193, contained no provision for employee challenges in enforcement proceedings, Legis. History, *supra,* at 1125. Eventually, a compromise was reached; employees were permitted to challenge the reasonableness of the period for abatement of the hazard, but, in exchange, employers were given the right to reopen the proceedings for a rehearing if they were unable to abate the hazard within the period provided for in the citation, *Id.* at 1192, 1202. In light of the nature of this compromise, one can infer that Congress intended to limit the rights of employees.

[17]Finally, we must question whether the interpretation of the Act that the union urges on us would provide employees with an unintended method of obtaining work concessions from employers. To be sure, Congress realized that OSHA regulations or citations issued to employers would be used by unions in bargaining sessions. Legis. History, *supra,* at 1223. But there is no indication that Congress intended to supply employees with a separate forum where they could possibly obtain concessions from employers that they could not get in bargaining sessions. Indeed, all evidence is to the contrary. See *supra* note 16.

B

Nor do we read the Act as giving the Commission the independent statutory authority to review settlement agreements between the Secretary and cited employers who, as a condition to entering settlements with the Secretary, have in essence acknowledged a violation of the Act by withdrawing their notices of contest. Congress gave the Commission the "relatively limited role of administrative adjudication," *General Elec. Co. v. OSHRC,* 583 F.2d 61, 63 n.3 [4 OSHC at 1515 n.3] (2d Cir. 1978), with no direct policy-making functions. See *Marshall v. OSHRC, supra,* 635 F.2d at 547 [9 OSHC at 1034]; *Marshall v. Sun Petroleum Prods. Co., supra,* 622 F.2d at 1183–84 [8 OSHC at 1427–28]; *Dale M. Madden Constr., Inc. v. Hodgson, supra,* 502 F.2d at 280 [2 OSHC at 1102]. But see *Brennan v. Gilles & Cotting, Inc.,* 504 F.2d 1255, 1262 [2 OSHC 1243, 1248] (4th Cir. 1974). The agenda of the Commission is tied strictly to §10 of the Act which, as noted previously, makes no reference to settlements or abatement plans. Its jurisdiction over enforcement proceedings is derivative of the rights granted to employees and employers under the Act, and can be triggered in only one of three ways after an employer has been served with a citation or notification of a penalty. 29 U.S.C. §§658(a), 659(a), (b). An employer may elect to contest the citation or notification of penalty; employees may contest the period for abatement contained in a citation; or an employer may seek a modification of an abatement date established by an earlier final order of the Commission. *Id.* at §659(c). However, if a case is uncontested "the citation * * * shall be deemed a final order of the Commission and not subject to review by any court or agency" including the Commission. *Id.* at §659(a).

When a citation issued pursuant to §9(a) is contested, the Commission has the jurisdiction to determine (1) whether a violation of the Act has occurred; (2) whether a reasonable time for abatement has been fixed in the citation; and (3) whether the penalty, if any, is reasonable. *Id.* at §§658(a), 659(c), 666(i). As part of this process, the Commission is responsible for finding the facts, see *Pratt & Whitney Aircraft v. Secretary of Labor,* 649 F.2d 96, 105 [9 OSHC 1554, 1561] (2d Cir. 1981), and its factual findings must be upheld if supported by substantial evidence. *A. Schonbek & Co. v. Donovan,* 646 F.2d 799, 800 [9 OSHC 1562, 1563] (2d Cir. 1981). Upon completion of the enforcement proceeding, an order "affirming, modifying or vacating the Secretary's citation or proposed penalty" is entered by the Commission. *Id.* at §659(c).

Under §10, the Commission is expressly precluded from reviewing abatement plans in settlements arising out of either uncontested cases, §10(a), *Marshall v. Sun Petroleum Prods. Co., supra,* 622 F.2d at 1185 [8 OSHC at 1428], or employee-initiated contests, §10(c); *UAW v. OSHRC, supra.* Thus, the only inference that we may draw from the statutory scheme is that Congress did not intend to give the Commission the authority to review settlements resulting from the limited number of employer-initiated contests it anticipated.[20] Moreover, the legislative history lends no force to the contention that Congress intended to give the Commission this oversight authority.

A determination whether abatement plans and methods included in a settlement agreement will eliminate the hazardous or unsafe working condition is a policy function beyond the limited jurisdiction of the Commission under §10(c). If Congress had wanted the Commission to undertake this additional task, it would have delegated this authority in the statute. See,

[20]The legislative history indicates that Congress assumed that many citations would not be contested by employers. See Legis. History, *supra,* at 155 ("It is anticipated that in many cases an employer will choose not to file a timely challenge to a citation when it is issued, on the assumption that he can comply with the period allowed in the citation for abatement of the violation.").

e.g., 30 U.S.C. §820(k) (Supp. I 1977) (Federal Mine Safety and Health Review Commission shall approve settlements of proposed penalties); 33 U.S.C. §908(i)(B) (1976) (Secretary may approve settlements relating to claims arising under the Longshoremen's and Harbor Workers' Compensation Act, 33 U.S.C. §901 *et seq.* (1976)); cf. *American Textile Mfrs. Inst., Inc. v. Donovan,* 452 U.S. 490, 510 [9 OSHC 1913, 1921] (1981). And absent a clear indication of congressional intent to the contrary, we will not read into the Act an authorization to review settlements. See, e.g., *Ford Motor Credit Co., v. Cenance,* 452 U.S. 155, 158 n.3 (per curiam) (1981); *National R.R. Passenger Corp. v. National Ass'n of R.R. Passengers,* 414 U.S. 453, 458 (1974).

Thus, we conclude that once an employer withdraws its notice of contest, as part of a settlement agreement with the Secretary, the Commission is ousted of jurisdiction over matters relating to the alleged violation. If the settlement agreement allows for a specific time period within which the hazard must be abated, and employees have filed a separate notice of contest or have become parties, the Commission retains jurisdiction to hear employee complaints over the reasonableness of the abatement period. If, however, the settlement agreement calls for the immediate abatement of the hazardous condition, the Commission has no jurisdiction to review the settlement agreement.

The view of the Court of Appeals for the Second Circuit in *Mobil Oil* was followed by the Court of Appeals for the Fourth Circuit in *Donovan v. United Steelworkers of America.*[16] The Court of Appeals for the Fifth Circuit stated in *Donovan v. Oil, Chemical and Atomic Workers International Union* that if it "were writing on a clean slate" on this issue, it would agree with the Review Commission's decision in *Mobil Oil.* However, the court said, it was "constrained" to adopt the Secretary's position in view of the unanimity of authorities behind it. Judge Alvin B. Rubin said for the court:

> In an administrative territory as vast as is OSHA, the need for uniformity is apparent. Were we to march to the beat of our own drummer, a gross disparity would arise between the Secretary's ability to settle cases in this circuit and his ability to do so elsewhere. Therefore, rather than profess telepathy from the collective mind of the 1970 Congress, we acknowledge that the Act can reasonably be interpreted more than one way, and we adopt a course that at least will not precipitate administrative chaos.[17]

D. The *American Cyanamid* Case

The view of the Secretary of Labor, adopted in the *Sun Petroleum* case by the Court of Appeals for the Third Circuit, that employee participation in Commission proceedings was limited to objections on

[16]722 F.2d 1158 (4th Cir. 1983).

[17]718 F.2d 1341, 1352–53 (5th Cir. 1983). See also *Donovan v. International Ass'n of Bridge, Structural and Ornamental Iron Workers,* 11 OSHC 1840 (5th Cir. 1984) (reversing Commission decision granting intervenor union right to review a settlement agreement between OSHA and the employer). The Court of Appeals for the Eighth Circuit, in *Donovan v. International Union, Allied Indus. Workers of Am.,* 11 OSHC 1737 (8th Cir. 1984), agreed with the decisions of the Court of Appeals for the Third Circuit in *Sun Petroleum* and the Court of Appeals for the Second Circuit in *Mobil Oil.* Circuit Judge Arnold dissented.

the length of the abatement period was taken one step further by the American Cyanamid Company, which argued that the right of a union-party to petition the court of appeals for review of an adverse Commission proceeding was also limited to issues relating to the length of the abatement period. On this issue, however, the Secretary of Labor agreed with the union, and the Court of Appeals for the District of Columbia Circuit rejected the *American Cyanamid* argument.[18]

Oil, Chemical & Atomic Workers International Union v. OSHRC (American Cyanamid)

671 F.2d 643 (D.C. Cir. 1982), cert. denied sub nom. American Cyanamid Co. v. OCAW 51 USLW 3287 (1982)

PER CURIAM:

***

The company contends that section 10(c) limits employee participation in enforcement proceedings to contesting the reasonableness of the abatement period. It reasons that this limit on the initiation of and participation in commission adjudications should be carried over to the instigation of judicial review because Congress intended that the provisions be read together as part of a "detailed statutory scheme." *Whirlpool Corp. v. Marshall,* 445 U.S. 1, 10 (1980). That scheme—says the company—may most consistently be interpreted to limit the rights of employees to initiate and participate in proceedings before the commission and to appeal commission decisions to issues challenging the reasonableness of the abatement period.[1]

***

In sum, the legislative history shows that Congress intended to allow employees to participate as parties in enforcement proceedings in two separate contexts—to initiate contests over the reasonableness of the abatement period and to participate as parties in an employer-initiated contest. We find that Congress did not intend to limit the interest assertable by the union in an employer-initiated proceeding to the length of the abatement period. The scheme of the Act enables employees to translate their concern for workplace safety into a demand for an inspection of the workplace, and, if a violation exists, into a citation. If the employer disputes the inspector's findings and files a notice of contest, the Act entitles employees to "participate as parties" in hearings before the OSHRC. The employees' request for party status confers jurisdiction on the commission to entertain the employees' objections on all matters relating to the citation in question.[4]

[1]The position derives the bulk of its support from a recent decision of the Third Circuit, *Marshall v. Sun Petroleum Products Co.,* 622 F.2d 1176, 1186 (3d Cir.), *cert. denied,* 449 U.S. 1061 (1980). In that decision, the court sharply limited the scope of the employees or their union's party status in a proceeding initiated by the employer to the challenge of the length of the abatement period. It found that "any challenge advanced by an employee is limited to an attack on the reasonableness of the abatement period."

[4]At least one other court has endorsed this interpretation. In *Marshall v. OSHRC and OCAW,* 635 F.2d 544, 552 (6th Cir. 1980), the Sixth Circuit agreed with the position taken by Secretary of Labor and construed the Act "as permitting employees having elected party status to participate in prehearing discovery and present their own witnesses."

Pursuant to the statutory mandate of 29 U.S.C. §659(c), the commission has consistently held that "[an] employee representative electing party status has the right to litigate all the issues raised by the citation and complaint." *Southwestern Bell Telephone Co.,* 5 OSHC 1851, 1852 (1977). See also *Gurney Industries, Inc.,* 1 OSHC 1218 (1973).

[18]The substantive merits of this case—whether the company's "fetus protection policy" was lawful—is discussed in connection with the general duty clause in Chapter 14, Section B.

We therefore hold that where a union or employee has elected party status in proceedings before the OSHRC, the union or employee has a right to appeal the decision of the OSHRC. As a party in an employer-initiated hearing before the OSHRC, the union will be "adversely affected or aggrieved" by an unfavorable OSHRC decision. Accordingly, the union can seek judicial review in a federal court of appeals under section 11(a) of the Act, 29 U.S.C. §660(a).

Our view does not impinge on the Secretary's unique role in administering the Act. As the Supreme Court has noted, the Act creates public rights that are to be vindicated by the Secretary through government management and enforcement of a complex administrative scheme. *Atlas Roofing Co. v. OSHRC,* 430 U.S. 442, 444–47 (1977). We recognize that the Secretary has been vested with considerable discretion in the promulgation of standards. Thus, employees may not compel the Secretary to adopt a standard. *National Congress of Hispanic American Citizens v. Usery,* 554 F.2d 1196 (D.C. Cir. 1977). Similarly, we are persuaded that enforcement of the Act is the sole responsibility of the Secretary. He is the exclusive prosecutor of OSHA violations. *Atlas Roofing Co.,* 430 U.S. at 445–47; *Dale M. Madden Construction, Inc. v. Hodgson,* 502 F.2d 278, 280 (9th Cir. 1974). Necessarily included within the prosecutorial power is the discretion to withdraw or settle a citation issued to an employer, and to compromise, mitigate or settle any penalty assessed under the Act, 29 U.S.C. §655(e).[7] Thus, employees may not contest the representation of a settlement agreement that abatement has occurred. *Marshall v. OCAW and OSHRC,* 647 F.2d 383, 388 (3d Cir. 1981). They may not prosecute a citation before the commission after the Secretary has moved to withdraw it. *Marshall v. OSHRC and OCAW, supra,* 635 F.2d at 552. They may not argue that a particular method of abatement would be inefficacious. *Marshall v. Sun Petroleum Co.,* 622 F.2d 1176, 1184–86 (3d Cir.), *cert. denied,* 449 U.S. 1061 (1980). They may not challenge the substantive provision of an abatement plan. *UAW v. OSHRC, supra,* 557 F.2d at 610–11.

We endorse so broad a reading of prosecutorial discretion under the statute because we believe that such discretion comports with the Congressional intent that the Secretary be charged with the basic responsibilities for administering the Act. We agree with other courts that have considered the issue that the union has no right to challenge the refusal of the Secretary to proceed with a citation or to file a complaint. See, e.g., *Marshall v. OSHRC and OCAW, supra,* 635 F.2d at 550. The company, however, contends that these decisions prohibit employees from challenging the Secretary's prosecutorial decisions before the commission proceedings, except where the length of an abatement period is concerned. It reasons that the decision to seek review of a commission decision is a similar exercise of prosecutorial discretion and concludes that allowing employees to petition for judicial review of a commission decision impermissibly usurps the Secretary's prosecutorial discretion as enforcer of the Act.

We reject this position. We agree with the company that a decision of the Secretary not to appeal the OSHRC ruling is an exercise of prosecutorial discretion. Accordingly, we find that the union has no right to challenge the determination of the Secretary not to appeal. In this instance, however, the OCAW is not seeking review of the Secretary's decision; instead, it is seeking judicial review of an order of the OSHRC, an independent adjudicatory body. Under the Act, we hold that a union has the right to appeal the ruling

[7]The Secretary's prosecutorial discretion to settle cases over employee objections flows from section 10(a) of the Act, 29 U.S.C. §659(a). If an employer does not timely contest a citation, it becomes an enforceable order "not subject to review by any court or agency" including the commission. *Id.* As a part of settlement of a contested citation, an employer withdraws its notice of contest thus ousting the commission of jurisdiction. In either case, employees are only empowered to invoke commission jurisdiction to object to the reasonableness of the abatement period.

of the OSHRC, whether or not the Secretary simultaneously seeks review, as a person "adversely affected or aggrieved" within the meaning of section 11(a) of the Act.

The union's right to appeal OSHRC decisions where it has participated as a party in the commission proceedings is, however, subject to two conditions, derived from the general statutory scheme and purpose of the Act. First, the union must give the Secretary notice of its intention to appeal and must serve him with copies of all of the pleadings. This notice requirement is ordered so that the Secretary is made aware of the litigation and so that he may act to intervene if he deems it appropriate. Second, the case may become moot in those instances when the Secretary, participating in the appeal as an amicus curiae or as an intervenor, provides this court with a clear and unconditional statement that he will not prosecute the claim regardless of the disposition of the appeal by this court. The prosecutorial discretion with which the Secretary is vested empowers him not to renew his prosecution effort even if this court were to find that the citation dismissed by the OSHRC asserted a violation under the Act. *Marshall v. OSHRC and OCAW, supra,* 635 F.2d at 549-52. While we will not require the Secretary to furnish us with a statement of his intent, we may dismiss the appeal as moot in instances where the Secretary has voluntarily proffered such a statement after the union has filed a petition for review.[9]

Several points should be noted about the court's decision. First, the court agreed with the view of other courts of appeals that employee-parties have the right to object solely to the abatement period in settlement agreements.[19] The court also agreed with the *IMC Chemical* court decision that employee-parties have no right independently to prosecute citations before the Review Commission if the Secretary moves to withdraw the citation.[20] Accordingly, while the court recognized the union's right to appeal the citation, it cautioned that the proceeding may become moot "in those instances when the Secretary * * * provides this court with an unconditional statement that he will not prosecute the claim regardless of the disposition of the appeal by this court."[21]

[9]In instances where the Secretary has not furnished the court with a statement of his intent, we will review the decision of the commission. The fact that there is uncertainty as to whether the government will continue to prosecute the case upon remand does not mean that no justiciable controversy exists. See *Dunlop v. Bachowski,* 421 U.S. 560, 575 (1975).

[19]671 F.2d at 650.
[20]*Id.* at 650–51.
[21]*Id.* at 651. This case is discussed in 51 U. CIN. L. REV. 913 (1982). The Commission has applied its view in *Mobil* in a number of subsequent cases, see, e.g., *Whirlpool Corp.,* 10 OSHC 1992 (Rev. Comm'n 1982).

18

Retaliation for Exercise of Protected Rights; Right to Information on Workplace Hazards

A. Background

Section 11(c)[1] prohibits discrimination against an employee because of the exercise of certain specified rights under the Act "or because of the exercise by such employee on behalf of himself or others of any right afforded by this Act." This protection against retaliation for the exercise of rights under OSHA is analogous to similar provisions under the National Labor Relations Act,[2] Title VII of the Civil Rights Act of 1964,[3] and the Fair Labor Standards Act.[4] A number of the issues that have been litigated and decided under these other statutes have also arisen under §11(c), among these are the definition of protected activity, establishment of the causal connection between the discrimination and the protected activity, and various procedural issues. One §11(c) case has reached the U.S. Supreme Court, *Whirlpool Corp. v. Marshall,*[5] involving the question whether employees have the right to refuse to work under conditions that are reasonably believed to be imminently dangerous.

In 1973, OSHA issued a comprehensive regulation establishing procedures for the filing and handling of §11(c) complaints and announcing policies and interpretation of that section.[6]

---

[1]29 U.S.C. §660(c) (1976).
[2]Sec. 8(a)(4), 29 U.S.C. §158(a)(4) (1976).
[3]Sec. 704(a), 42 U.S.C. §2000e–3(a) (1976).
[4]Sec. 15(a)(3), 29 U.S.C. §215(a)(3) (1976).
[5]445 U.S. 1 (1980), discussed in Section B, this chapter.
[6]29 C.F.R. pt. 1977 (1983). No opportunity for public comment was given before the regulation was issued. In *Whirlpool Corp. v. Marshall,* 445 U.S. 1 (1980), the Supreme Court ruled on the scope of this provision, but did not address whether the "interpretation" was invalid because of the absence of notice and comment. 445 U.S. at 11 n.15. For a discussion of the issue when notice and comment is required, see Author's note, p. 553.

B. Definition of Protected Activity: The *Whirlpool* Case

Section 11(c) expressly provides that employees are protected for filing "any complaints," instituting "any proceeding under or related to the Act," or testifying in "such proceedings." In addition, the statute protects employees for "the exercise by such employee on behalf of himself or others of any right afforded by this Act." OSHA has interpreted these provisions expansively, in accordance with the protective purposes of the Act, to include "other rights [which] exist by necessary implication."[7] For example, OSHA has construed this quoted language as including the employee right to request information from OSHA and the right to cooperate through interview with OSHA agents during a workplace inspection. Similarly, OSHA said good-faith complaints about safety and health matters to employers are protected.[8] Also protected are complaints to other federal agencies and state and local agencies relating to safety and health conditions at the workplace "as distinguished from complaints touching only upon general public safety and health."[9] Courts have agreed that complaints filed with employers are included under §11(c).[10]

The major issue in defining protected activity has been whether employees have a right, free from employer sanctions, to refuse to work in conditions perceived to present an imminent danger. The issue was considered by OSHA originally in its 1973 interpretation.

Discrimination Against Employees Exercising Rights Under the Act[11]

38 Fed. Reg. 2681, 2683 (1973)

§1977.12 EXERCISE OF ANY RIGHT AFFORDED BY THE ACT.

(a) In addition to protecting employees who file complaints, institute proceedings, or testify in proceedings under or related to the Act, section 11(c) also protects employees from discrimination occurring because of the exercise "of any right afforded by this Act." Certain rights are explicitly provided in the Act; for example, there is a right to participate as a party in enforcement proceedings (sec. 10). Certain other rights exist by necessary implication. For example, employees may request information from the Occupational Safety and Health Administration; such requests would constitute the exercise of a right afforded by the Act. Likewise, employees interviewed by agents of the Secretary in the course of inspections or investigations could not subsequently be discriminated against because of their cooperation.

---

[7]29 C.F.R. §1977.12(a) (1983).

[8]29 C.F.R. §1977.9(c) (1983).

[9]29 C.F.R. §1977.9(b) (1983).

[10]See, e.g., *Marshall v. Springville Poultry Farm*, 445 F. Supp. 2 (M.D. Pa. 1977) (language referring to complaints filed "under and related to" the Act includes more than just complaints that are filed *under* the Act). See also *Donovan v. R.D. Anderson Constr. Co.*, 552 F. Supp. 249 (D. Kan. 1982) (employee's discussion with newspaper about health hazards at workplace was protected activity).

[11]29 C.F.R. §1977.12 (1983).

(b)(1) On the other hand, review of the Act and examination of the legislative history discloses that, as a general matter, there is no right afforded by the Act which would entitle employees to walk off the job because of potential unsafe conditions at the workplace. Hazardous conditions which may be violative of the Act will ordinarily be corrected by the employer, once brought to his attention. If corrections are not accomplished, or if there is dispute about the existence of a hazard, the employee will normally have opportunity to request inspection of the workplace pursuant to section 8(f) of the Act, or to seek the assistance of other public agencies which have responsibility in the field of safety and health. Under such circumstances, therefore, an employer would not ordinarily be in violation of section 11(c) by taking action to discipline an employee for refusing to perform normal job activities because of alleged safety or health hazards.

(2) However, occasions might arise when an employee is confronted with a choice between not performing assigned tasks or subjecting himself to serious injury or death arising from a hazardous condition at the workplace. If the employee, with no reasonable alternative, refuses in good faith to expose himself to the dangerous condition, he would be protected against subsequent discrimination. The condition causing the employee's apprehension of death or injury must be of such a nature that a reasonable person, under the circumstances then confronting the employee, would conclude that there is a real danger of death or serious injury and that there is insufficient time, due to the urgency of the situation, to eliminate the danger through resort to regular statutory enforcement channels. In addition, in such circumstances, the employee, where possible, must also have sought from his employer, and been unable to obtain, a correction of the dangerous condition.

By this regulation OSHA afforded employees, in certain limited circumstances, the right to refuse to work; while employees were not guaranteed wages for the time they were not working, they were protected against retaliation because of their refusal to work. OSHA's authority to issue the refusal-to-work regulation was quickly challenged, and in November 1977 the Court of Appeals for the Fifth Circuit in *Marshall v. Daniel Construction Company*[12] vacated the regulation as being beyond the Secretary of Labor's statutory authority. OSHA asked the Supreme Court to hear the case, but certiorari was denied.[13] In light of the adverse decision of the Fifth Circuit Court of Appeals, it was for a time uncertain whether OSHA

[12]563 F.2d 707 (5th Cir. 1977). Judge Wisdom dissented vigorously; he concluded that Congress believed that "workers could live with the prescribed processes of this Act * * * [not] that it required workers to die for them." *Id.* at 722. The facts in the *Daniel Construction* case were described by Judge Wisdom in his dissent, as follows:

> For purpose of the motion to dismiss, we assume that the plaintiff would be able to prove the following set of facts which are within the ambit of the facts alleged in the complaint. Daniel Construction Company employed Jimmy Simpson as an ironworker connecting structural steel in the construction of tall buildings. The job required fitting into place heavy steel beams with aid of a crane. One windy day Simpson was working 150 feet above the ground. The wind grew so strong that it imperiled his life. He came down from high on the steel skeleton where he had been working. So did the rest of his crew. A foreman ordered the crew to return to work. Simpson refused. He was fired.

Id. at 717.

[13]439 U.S. 880 (1978) (Brennan & Blackmun, JJ., would have granted certiorari).

could provide legal relief to employees relying on its interpretation in refusing to work in imminent danger situations. This uncertainty is reflected in the guarded advice given by OSHA in 1978 to employees on their "legal rights" in "refusing dangerous work."

> You have the clear legal right to report unsafe or unhealthful working conditions on your job. But it is not yet clear whether the OSHA law protects you if you *refuse* to work under those conditions.
>
> If you walk off the job when faced with a health or safety hazard, you may not be protected under ELEVEN-C from punishment. The right under OSHA's law to refuse to work in imminent danger has not yet been fully recognized by the federal courts.
>
> If you refuse dangerous work, you may be protected under the National Labor Relations Board's law. This is especially true if you refuse in cooperation with or on behalf of other workers. OSHA and the NLRB cooperate in cases involving refusal to work because of workplace dangers.
>
> If you have been punished for refusing dangerous work, you should contact OSHA to discuss your case, and you may want to contact the NLRB as well.[14]

Meanwhile, on February 22, 1979, the Court of Appeals for the Sixth Circuit unanimously disagreed with the Fifth Circuit's *Daniel Construction* decision and upheld the validity of the OSHA regulation.[15] The facts in the case, as found by the district court and adopted by the circuit court, were as follows:

> The Whirlpool Corporation maintains a manufacturing plant at Marion, Ohio where it produces household appliances. The Marion plant has 13 miles of overhead conveyors which transport appliance components throughout the plant. In order to prevent injury should an appliance component fall from one of the overhead conveyors, the Company installed a huge guard screen approximately 20 feet above the plant floor. The guard screen is suspended over one-third of the total plant floor area. As part of their regular duties, maintenance employees must remove fallen parts from the screen and replace paper spread on the screen to catch grease drippings. In addition the overhead conveyors occasionally need maintenance. In order to perform their duties, maintenance workers must step onto the steel mesh screen itself.
>
> The original steel mesh used 16-gauge panels although starting in 1973 the company began to replace these panels with heavier mesh which could better withstand the stresses imposed upon it. The questionable safety of the 16-gauge screens is demonstrated by several incidents where workers fell partly through them and at least one incident where a worker fell to the plant floor below but survived. A number of maintenance employees reacted to these "near-misses" by frequently bringing the unsafe screen conditions to the attention of their foremen. Their complaints were to no avail, and on June 28, 1974, a maintenance employee fell to his death through the guard screen in a section where

---

[14]OSHA No. 3032, OSHA: YOUR WORKPLACE RIGHTS IN ACTION 13 (1978). OSHA had previously entered into a memorandum of understanding with the National Labor Relations Board, 40 FED. REG. 26,083 (1975), under which the two agencies coordinated their overlapping enforcement responsibilities in the area of alleged discrimination against employees for engaging in concerted activity to protect their workplace safety and health. See also *Whirlpool Corp. v. Marshall,* 445 U.S. at 17 n.29.

[15]*Marshall v. Whirlpool Corp.,* 593 F.2d 715 (6th Cir. 1979), *rev'g sub nom. Usery v. Whirlpool Corp.,* 416 F. Supp. 30 (N.D. Ohio 1976).

stronger wire mesh had not yet been installed. Although Whirlpool did respond to the fatality by issuing a general order directing maintenance employees to clean guard screens without walking on them and to effectuate some repairs where the screens were weak, two employees who regularly worked on the screens pointed out that many hazardous areas remained. When the company did not respond to their fears regarding these areas, the two employees asked the company safety director to provide them with the local OSHA area office's phone number. The "safety director" gave them the number but made veiled threats that the men should watch what they were doing and took their names and clock numbers.

The night following this incident the two men punched in on their shift and awaited their assignments. They were ordered onto the screens to do their maintenance work. When the men refused, citing the company safety directive saying such work should be done without stepping on the screens, they were peremptorily ordered to the company personnel office, disciplined and issued written reprimands for insubordination.[16]

The Supreme Court granted Whirlpool Corporation's petition for certiorari. In Whirlpool's brief to the Supreme Court, it pressed its argument that OSHA lacked statutory authority to issue the regulation.

Brief of Whirlpool Corporation in Whirlpool Corp. v. Marshall[17]

***

SUMMARY OF ARGUMENT.

The Act does not expressly afford employees a protected right to refuse work. Congress squarely was presented with a legislative option to extend the protections of the Act to work refusals or stoppages, but rejected all such proposals.

Congress did include within the Act a mechanism to deal with conditions which employees may believe to be imminently dangerous, which mechanism the legislative history clearly shows was adopted in lieu of proposals to protect employees who refuse work for reasons of safety or health.

The Secretary's regulation, purporting to extend the protections of the Act to employee work stoppages, is therefore plainly inconsistent with the will of Congress as expressed in the Act and legislative history. In turn, the Court of Appeals' finding that the Secretary's regulation is consistent with the Act and with Congressional intent as manifested in the legislative history is clearly erroneous as a matter of law. The Court has disregarded the established principle that a patent inconsistency between a regulation and a statute must be resolved in favor of the statute. It has ignored the principle repeatedly recognized by this Court, that a legislative choice to create one form of remedy implies that a decision has been made to exclude other forms of remedy.

Indeed, the Secretary's own interpretation of Section 11(c)(1) begins with the revealing admission that "[r]eview of the Act and examination of the legislative history discloses that, as a general matter, there is no right

[16]593 F.2d at 719. See also *id.* at 719 n.5 (rejecting Whirlpool Corp.'s claim that the findings of fact are clearly erroneous).

[17]Brief of Petitioner at 15–16, 34–35, *Whirlpool Corporation v. Marshall,* 445 U.S. 1 (1980).

afforded by the Act which would entitle employees to walk off the job because of potential unsafe conditions at the workplace." 29 C.F.R. 1977.12 (b)(1), *supra* [593 F.2d at 718].

The Court's decision proceeds from complete disregard of the facts set forth in the record. It fails to recognize that the Secretary's regulation not only is inconsistent with all recorded expressions of Congressional will, but also clearly provides the basis for potential abuse, setting up the potential for endless litigation in Federal Courts over labor matters otherwise relegated to other forums.

The Court of Appeals misstated and/or ignored the undisputed facts of the case to an extreme degree in order to create a factual justification for its decision. The Court created an urgent, no-recourse situation where none existed in fact. Based upon the record facts, the Court should have dismissed the complaint for failing to establish the elements essential to obtaining protection under the regulation.

***

B. RULES OF STATUTORY CONSTRUCTION MILITATE AGAINST IMPLICATION OF THE RIGHT WHICH THE SECRETARY SEEKS TO CREATE.

The well-established principle of statutory construction "expressio unius est exclusio alterius" holds that "a 'legislative affirmative description' [of certain powers] implies denial of the non-described powers," *Continental Casualty Co. v. United States,* 314 U.S. 527, 533 (1942), *citing with approval Durousseau v. United States,* 6 Cranch (U.S.) 307, 314 (1810); that "when a statute limits a thing to be done in a particular mode, it includes the negative of any other mode," *Botany Worsted Mills v. United States,* 278 U.S. 282, 289 (1929); and that "when legislation expressly provides a particular remedy or remedies, courts should not expand the coverage of the statute to subsume other remedies," *National Railroad Passenger Corp. v. National Ass'n of Railroad Passengers,* 414 U.S. 453, 458 (1974).

Applying this rule of statutory construction to the scheme of rights and procedures contained in the Act, it can be seen that the Act specifically describes the rights and powers of the employee, the Secretary, and the federal court in situations allegedly involving danger in the workplace. The employee is empowered to request an inspection, the Secretary is to decide whether the situation should be brought to the attention of the federal district court, and the court is to shut down operations if necessary. The Act limits the shutting down of operations in imminent danger situations to be done only in this particular way. The Act does not create a protected employee right to shut down operations on his own and expressly provides that his particular right is to set in motion the imminent danger procedure which, if necessary, will result in a court-ordered shutdown of business operations.

The absence of a protected right to walk off the job is balanced in the Act by the presence of the provisions which directly involve the employee in almost every aspect of the Act's administration and explicitly protect him for exercising his right to be involved, especially the provision which makes the employee the one who sets in motion the Act's imminent danger procedure. In this regard it is further submitted that the right which the Secretary asks this Court to find implicit in the Act—that an employee has the right to shut down the employer's operations on his own and that the employer is required to continue paying him—would completely disrupt the careful scheme of the Act's request for inspection and imminent danger provisions. The end result of the argument advanced by the Secretary would be that employees, as well as federal judges, would have the power to shut down operations. Such a result cannot be found by implication within the

four corners of the Act without total disruption of its carefully constructed scheme and, as noted above, serious disruption of the NLRA.
***

On February 26, 1980, the Supreme Court unanimously upheld the OSHA regulation.

Whirlpool Corp. v. Marshall

445 U.S. 1 (1980)

MR. JUSTICE STEWART delivered the opinion of the Court:
***

* * * [T]he Secretary is obviously correct when he acknowledges in his regulation that, "as a general matter, there is no right afforded by the Act which would entitle employees to walk off the job because of potential unsafe conditions at the workplace." By providing for prompt notice to the employer of an inspector's intention to seek an injunction against an imminently dangerous condition, the legislation obviously contemplates that the employer will normally respond by voluntarily and speedily eliminating the danger. And in the few instances where this does not occur, the legislative provisions authorizing prompt judicial action are designed to give employees full protection in most situations from the risk of injury or death resulting from an imminently dangerous condition at the worksite.

As this case illustrates, however, circumstances may sometimes exist in which the employee justifiably believes that the express statutory arrangement does not sufficiently protect him from death or serious injury. Such circumstances will probably not often occur, but such a situation may arise when (1) the employee is ordered by his employer to work under conditions that the employee reasonably believes pose an imminent risk of death or serious bodily injury, and (2) the employee has reason to believe that there is not sufficient time or opportunity either to seek effective redress from his employer or to apprise OSHA of the danger.

Nothing in the Act suggests that those few employees who have to face this dilemma must rely exclusively on the remedies expressly set forth in the Act at the risk of their own safety. But nothing in the Act explicitly provides otherwise. Against this background of legislative silence, the Secretary has exercised his rulemaking power under 29 U.S.C. §657(g)(2) and has determined that, when an employee in good faith finds himself in such a predicament, he may refuse to expose himself to the dangerous condition, without being subjected to "subsequent discrimination" by the employer.

The question before us is whether this interpretative regulation[15] constitutes a permissible gloss on the Act by the Secretary, in light of the Act's language, structure, and legislative history. Our inquiry is informed by an awareness that the regulation is entitled to deference unless it can be said not to be a reasoned and supportable interpretation of the Act. *Skidmore v. Swift & Co.,* 323 U.S. 134, 139–140. See *Ford Motor Credit Co. v. Milhollin,* 444 U.S. 555; *Mourning v. Family Publications Service, Inc.,* 411 U.S. 356.

---

[15]The petitioner has raised no issue concerning whether or not this regulation was promulgated in accordance with the procedural requirements of the Administrative Act (APA), 5 U.S.C. §553. Thus, we accept the Secretary's designation of the regulation as "interpretative," and do not consider whether it qualifies as an "interpretative rule" within the meaning of the APA, 5 U.S.C. §553(b)(A).

A

The regulation clearly conforms to the fundamental objective of the Act—to prevent occupational deaths and serious injuries. The Act, in its preamble, declares that its purpose and policy is "to assure so far as possible every working man and woman in the Nation safe and healthful working conditions and to *preserve* our human resources. * * *" 29 U.S.C. §651(b). (Emphasis added.)

To accomplish this basic purpose, the legislation's remedial orientation is prophylactic in nature. See *Atlas Roofing Co. v. Occupational Safety and Health Review Comm'n,* 430 U.S. 442, 444–445. The Act does not wait for an employee to die or become injured. It authorizes the promulgation of health and safety standards and the issuance of citations in the hope that these will act to prevent deaths or injuries from ever occurring. It would seem anomalous to construe an Act so directed and constructed as prohibiting an employee, with no other reasonable alternative, the freedom to withdraw from a workplace environment that he reasonably believes is highly dangerous.

Moreover, the Secretary's regulation can be viewed as an appropriate aid to the full effectuation of the Act's "general duty" clause. That clause provides that "[e]ach employer * * * shall furnish to each of his employees employment and a place of employment which are free from recognized hazards that are causing or are likely to cause death or serious physical harm to his employees." 29 U.S.C. §654(a)(1). As the legislative history of this provision reflects, it was intended itself to deter the occurrence of occupational deaths and serious injuries by placing on employers a mandatory obligation independent of the specific health and safety standards to be promulgated by the Secretary. Since OSHA inspectors cannot be present around the clock in every workplace, the Secretary's regulation ensures that employees will in all circumstances enjoy the rights afforded them by the "general duty" clause.

The regulation thus on its face appears to further the overriding purpose of the Act, and rationally to complement its remedial scheme.[18] In the absence of some contrary indication in the legislative history, the Secretary's regulation must, therefore, be upheld, particularly when it is remembered that safety legislation is to be liberally construed to effectuate the congressional purpose. *United States v. Bacto-Unidisk,* 394 U.S. 784, 798; *Lilly v. Grand Trunk R. Co.,* 317 U.S. 481, 486.*

B

In urging reversal of the judgment before us, the petitioner relies primarily on two aspects of the Act's legislative history.

***

[18]It is also worth noting that the Secretary's interpretation of 29 U.S.C. §660(c)(1) conforms to the interpretation that Congress clearly wished the courts to give to the parallel antidiscrimination provision of the Federal Mine Safety and Health Act of 1977, 30 U.S.C. §801 *et seq.* (1976 ed. and Supp. II). The legislative history of that provision, 30 U.S.C. §815(c)(1) (1976 ed., Supp. II), establishes that Congress intended it to protect "the refusal to work in conditions which are believed to be unsafe or unhealthful." S. Rep. No. 95-181, p. 35 (1977). See *id.,* at 36; 123 CONG. REC. 20043–20044 (1977) (remarks of Sen. Church, Sen. Williams, Sen. Javits).

*[Author's note—Whirlpool Corporation noted the inclusion of the right to refuse work in the Mine Safety Act and argued that if Congress had intended the same right of refusal under the OSH Act, it would have included that right under the original Act or amended the Act accordingly. Brief of Petitioner at 20 n.15. See also Chapter 5, Section D; *United Steelworkers v. Marshall,* 647 F.2d 1189, 1232 (D.C. Cir. 1980) (court of appeals rejects the application of the principle of *expressio unius est exclusio alterius* in deciding that OSHA has authority to require medical removal protection).]

1

Representative Daniels of New Jersey sponsored one of several House bills that led ultimately to the passage of the Act. As reported to the House by the Committee on Education and Labor, the Daniels bill contained a section that was soon dubbed the "strike with pay" provision. This section provided that employees could request an HEW examination of the toxicity of any materials in their workplace. If that examination revealed a workplace substance that had "potentially toxic or harmful effects in such concentration as used or found," the employer was given 60 days to correct the potentially dangerous condition. Following the expiration of that period, the employer could not require that an employee be exposed to toxic concentrations of the substance unless the employee was informed of the hazards and symptoms associated with the substance, the employee was instructed in the proper precautions for dealing with the substance, and the employee was furnished with personal protective equipment. If these conditions were not met, an employee could "absent himself from such risk of harm for the period necessary to avoid such danger without loss of regular compensation for such period."

This provision encountered stiff opposition in the House. Representative Steiger of Wisconsin introduced a substitute bill containing no "strike with pay" provision. In response, Representative Daniels offered a floor amendment that, among other things, deleted his bill's "strike with pay" provision. He suggested that employees instead be afforded the right to request an immediate OSHA inspection of the premises, a right which the Steiger bill did not provide. The House ultimately adopted the Steiger Bill.

The bill that was reported to and, with a few amendments, passed by the Senate never contained a "strike with pay" provision. It did, however, give employees the means by which they could request immediate Labor Department inspections. These two characteristics of the bill were underscored on the floor of the Senate by Senator Williams, the bill's sponsor.[26]

After passage of the Williams bill by the Senate, it and the Steiger bill were submitted to a conference committee. There, the House acceded to the Senate bill's inspection request provisions.

The petitioner reads into this legislative history a congressional intent incompatible with an administrative interpretation of the Act such as is embodied in the regulation at issue in this case. The petitioner argues that Congress' overriding concern in rejecting the "strike with pay" provision was to avoid giving employees a unilateral authority to walk off the job which they might abuse in order to intimidate or harm their employer. Congress deliberately chose instead, the petitioner maintains, to grant employees the power to require immediate administrative inspections of the workplace which could in appropriate cases lead to coercive judicial remedies. As the petitioner views the regulation, therefore, it gives to workers precisely what Congress determined to withhold from them.

We read the legislative history differently. Congress rejected a provision that did not concern itself at all with conditions posing real and immediate threats of death or severe injury. The remedy which the rejected provision furnished employees could have been invoked only after 60 days had passed following HEW's inspection and notification that improperly high levels of toxic substances were present in the workplace. Had that inspection revealed employment conditions posing a threat of imminent and grave harm, the Secretary of Labor would presumably have requested, long before

[26]"[D]espite some wide-spread contentions to the contrary, . . . the committee bill does not contain a so-called strike-with-pay provision. Rather than raising a possibility for endless disputes over whether employees were entitled to walk off the job with full pay, it was decided in committee to enhance the prospects of compliance by the employer through such means as giving the employees the right to request a special Labor Department investigation or inspection." 116 CONG. REC. 37326 (1970).

expiration of the 60-day period, a court injunction pursuant to other provisions of the Daniels bill. Consequently, in rejecting the Daniels bill's "strike with pay" provision, Congress was not rejecting a legislative provision dealing with the highly perilous and fast-moving situations covered by the regulation now before us.

It is also important to emphasize that what primarily troubled Congress about the Daniels bill's "strike with pay" provision was its requirement that employees be paid their regular salary after having properly invoked their right to refuse to work under the section.[29] It is instructive that virtually every time the issue of an employee's right to absent himself from hazardous work was discussed in the legislative debates, it was in the context of the employee's right to continue to receive his usual compensation.

When it rejected the "strike with pay" concept, therefore, Congress very clearly meant to reject a law unconditionally imposing upon employers an obligation to continue to pay their employees their regular pay checks when they absented themselves from work for reasons of safety. But the regulation at issue here does not require employers to pay workers who refuse to perform their assigned tasks in the face of imminent danger. It simply provides that in such cases the employer may not "discriminate" against the employees involved. An employer "discriminates" against an employee only when he treats that employee less favorably than he treats other similarly situated.[31]

2

The second aspect of the Act's legislative history upon which the petitioner relies is the rejection by Congress of provisions contained in both the Daniels and the Williams bills that would have given Labor Department officials, in imminent danger situations, the power temporarily to shut down all or part of an employer's plant. These provisions aroused considerable opposition in both Houses of Congress. The hostility engendered in the House of Representatives led Representative Daniels to delete his version of the provision in proposing amendments to his original bill.[33] The Steiger

[29]Congress' concern necessarily was with the provision's compensation requirement. The law then, as it does today, already afforded workers a right, under certain circumstances, to walk off their jobs when faced with hazardous conditions. See 116 CONG. REC. 42208 (1970) (Rep. Scherle) (reference to Taft-Hartley Act). Under Section 7 of the National Labor Relations Act (NLRA), 29 U.S.C. §157, employees have a protected right to strike over safety issues. See *NLRB v. Washington Aluminum Co.*, 370 U.S. 9. Similarly, Section 502 of the Labor Management Relations Act, 29 U.S.C. §143, provides that "the quitting of labor by an employee or employees in good faith because of abnormally dangerous conditions for work at the place of employment of such employee or employees [shall not] be deemed a strike." The effect of this section is to create an exception to a no-strike obligation in a collective-bargaining agreement. *Gateway Coal Co. v. United Mine Workers*, 414 U.S. 368, 385.

The existence of these statutory rights also make clear that the Secretary's regulation does not conflict with the general pattern of federal labor legislation in the area of occupational safety and health. See also 29 CFR §1977.18.

[31]Deemer and Cornwell were clearly subjected to "discrimination" when the petitioner placed reprimands in their respective employment files. Whether the two employees were also discriminated against when they were denied pay for the approximately six hours they did not work on July 10, 1974, is a question not now before us. The District Court dismissed the complaint without indicating what relief it thought would have been appropriate had it upheld the Secretary's regulation. The Court of Appeals expressed no view concerning the limits of the relief to which the Secretary might ultimately be entitled. On remand, the District Court will reach this issue.

[33]* * * As Representative Daniels explained:

> [B]usiness groups have expressed great fears about the potential for abuse. They believe that the power to shut down a plant should not be vested in an inspector. While there is no documentation for this fear, we recognize that it is very prevalent. The Courts have shown their capacity to respond quickly in emergency situations, and we believe that the availability of temporary restraining orders will be sufficient to deal with emergency situations. Under the Federal rules of civil procedure, these orders can be used *ex parte*. If the Secretary uses the authority that he is given efficiently and expeditiously, he should be able to get a court order within a matter of minutes rather than hours.

116 CONG. REC. 38378.

bill that ultimately passed the House gave the Labor Department no such authority. The Williams bill, as approved by the Senate, did contain an administrative shutdown provision, but the conference committee rejected this aspect of the Senate bill.

The petitioner infers from these events a congressional will hostile to the regulation in question here. The regulation, the petitioner argues, provides employees with the very authority to shut down an employer's plant that was expressly denied a more expert and objective United States Department of Labor.

As we read the pertinent legislative history, however, the petitioner misconceives the thrust of Congress' concern. Those in Congress who prevented passage of the administrative shutdown provisions in the Daniels and Williams bills were opposed to the unilateral authority those provisions gave to federal officials, without any judicial safeguards, drastically to impair the operation of an employer's business. Congressional opponents also feared that the provisions might jeopardize the Government's otherwise neutral role in labor-management relations.

Neither of these congressional concerns is implicated by the regulation before us. The regulation accords no authority to government officials. It simply permits private employees of a private employer to avoid workplace conditions that they believe pose grave dangers to their own safety. The employees have no power under the regulation to order their employer to correct the hazardous condition or to clear the dangerous workplace of others. Moreover, any employee who acts in reliance on the regulation runs the risk of discharge or reprimand in the event a court subsequently finds that he acted unreasonably or in bad faith. The regulation, therefore, does not remotely resemble the legislation that Congress rejected.

C

For these reasons we conclude that 29 CFR §1977.12(b)(2) was promulgated by the Secretary in the valid exercise of his authority under the Act. Accordingly, the judgment of the Court of Appeals is affirmed.

Whirlpool's arguments were based largely on two critical items of legislative history: Congress' refusal to enact the "strike with pay" provision and its rejection of OSHA's administrative shutdown authority. As the unanimous opinion makes clear, the court found these arguments unpersuasive. Later, while arguing in the court of appeals' *Lead* proceeding[18] against OSHA authority to provide for medical removal protection (MRP), the Lead Industries Association (LIA) relied on the Supreme Court's rationale in *Whirlpool* rejecting the relevance of the "strike with pay" legislative history as it applied to the right to refuse work. According to LIA, the significant element of the Supreme Court's rationale in finding the legislative history irrelevant was that strike with pay included compensation while OSHA's refusal-to-work interpretation did not; therefore, it argued, MRP, which did involve continuation of earnings, was barred because of the legislative history. The court of appeals in the *Lead* case rejected the argument, saying that the critical objection in the strike-with-

[18] *United Steelworkers v. Marshall*, 647 F.2d 1189 (D.C. Cir. 1980).

pay debate was the fact that employees would leave their job after a "self-initiated" decision that the workplace was dangerous. This was not true in the case of MRP; the removal was based on the employees' blood-lead level or a medical determination, and, therefore, *Whirlpool* "does not bear on MRP."[19]

In *Whirlpool,* the Supreme Court also noted that under §7 of the National Labor Relations Act,[20] employees had a protected right to strike over safety issues, and that under §502 of the Labor Management Relations Act,[21] good-faith quitting of labor would not be deemed a strike if it took place because of "abnormally dangerous conditions of work." The Supreme Court concluded that the existence of these statutory rights makes it clear that the OSHA regulation "does not conflict with the general pattern of labor legislation in the area of occupational safety and health."[22] Whirlpool had argued, to the contrary, that OSHA's regulation would "remove employee-employer disputes from the collective bargaining arena envisioned by the NLRA."[23] The Court of Appeals for the District of Columbia Circuit rejected a similar argument by LIA in upholding medical removal protection.[24] The issue of alleged redundancy between union rights under the NLRA and rights granted under OSHA also arose in the context of the OSHA rule on access to medical and exposure records.[25]

Both the *Whirlpool* and *Daniel Construction* cases dealt with safety hazards: potential falls from dangerous heights. The question has been raised on the applicability of the OSHA refusal-to-work regulation to health violations. The regulation states that the "condition causing the employee's apprehension of death or injury must be of such a nature that a reasonable person under the circumstances then confronting the employee would conclude that there is a real danger of death or serious injury and there is insufficient time due to the urgency of the situation to eliminate the danger through resort to regular statutory enforcement channels."[26] Since the commonly ac-

[19] *Id.* at 1233 n.69.
[20] 29 U.S.C. §157 (1976).
[21] 29 U.S.C. §143 (1976).
[22] 445 U.S. at 17 n.29.
[23] Brief of Petitioner at 20, *Whirlpool v. Marshall,* 445 U.S. 1 (1978).
[24] 647 F.2d at 1236, appears in Chapter 5, Section D.
[25] See discussion in Chapter 5, Section F. The right of employees under NLRA to quit work for safety and health reasons has been the subject of numerous decisions of the NLRB and the courts of appeals. E.g., *NLRB v. Washington Aluminum Co.,* 370 U.S. 9 (1962); *NLRB v. Northern Metal Co.,* 440 F.2d 881 (3d Cir. 1971); *Daniel Constr. Co.,* 264 NLRB No. 104 (1982); *Comet Fast Freight,* 262 NLRB 430 (1982); *Alleluia Cushion Co.,* 221 NLRB 999 (1975). The Supreme Court recently ruled that an employee's reasonable refusal to drive an unsafe truck was concerted activity under §7 of the NLRA, where it was grounded on a collective bargaining agreement. *NLRB v. City Disposal Systems, Inc.,* 52 USLW 4360 (1984). The leading case interpreting §502 of the LMRA is *Gateway Coal Co. v. United Mine Workers,* 414 U.S. 368 (1974).

Clauses in some collective bargaining agreements have provided procedures for handling disputes over work situations involving hazards beyond those inherent in the operation in question. See, e.g., Agreement between United States Steel Corp. and United Steelworkers of America, §14(c) (March 1, 1983). For a full discussion of the subject see Ashford & Katz, *Unsafe Working Conditions: Employee Rights Under the Labor Management Relations Act and the Occupational Safety and Health Act,* 52 NOTRE DAME LAW. 802 (1977).
[26] 29 C.F.R. §1977.12(b)(2) (1983).

cepted view is that health hazards cause illness only after an extended period of time, it may be questioned whether exposure to a toxic substance is the type of "urgent" situation referred to in the regulation justifying a refusal to work. At the same time, OSHA has said that the cancer process can begin with a single exposure.[27] If so, the right to refuse work might be equally available when employees are confronted with some health hazards. This issue is analogous to the question whether the imminent danger provision[28] may be applicable to health hazards; that is, do any health hazards constitute a danger that "could reasonably be expected to cause death or serious physical harm immediately or before the imminence of such danger can be eliminated" through the Act's enforcement procedures. The OSHA Field Operations Manual states:

> For a health hazard there must be a reasonable expectation that toxic substances or other health hazards are present and exposure to them will cause harm to such a degree as to shorten life or cause substantial reduction in physical or mental efficiency even though the resulting harm may not manifest itself immediately.[29]

The Industrial Hygiene Field Operations Manual (IHFOM) provides as follows:

> An imminent danger condition is one in which "a danger exists which could reasonably be expected to cause death or serious physical harm immediately or before the imminence of such danger can be eliminated through the enforcement procedures otherwise provided by this Act" (FOM Chapter [VII]). Examples would include acute exposures to life-threatening concentrations of gases (e.g., hydrogen sulfide and hydrogen cyanide), and levels of radiation capable of causing irreversible damage. Certain exposure to suspect human carcinogen may constitute imminent danger; these cases shall be referred to the Regional office as soon as possible.[30]

These definitions, though differing in emphasis, leave open the possibility that OSHA would proceed on imminent dangers involving health hazards; but OSHA has proceeded in court in a very limited number of imminent danger situations,[31] and probably in none involving health.

In the *Whirlpool* decision, the Supreme Court left open the question when employees who refuse to work in circumstances covered by

[27]See, e.g., Carcinogen Policy, 45 FED. REG. 5002, 5023 (1980).
[28]Sec. 13, 29 U.S.C. §662.
[29]Ch. VII, OSHR [Reference File 77:3301].
[30]Ch. II, OSHR [Reference File 77:8023].
[31]Professor Mark Rothstein noted that there had been nine reported cases in which OSHA had sought imminent danger injunctions as of 1977; relief was granted in four cases, denied in one, and consent agreements entered in four cases. Rothstein, OCCUPATIONAL SAFETY AND HEALTH LAW 319–20 (1978). See Note, *A Proposal to Restructure Sanctions Under the Occupational Safety and Health Act: The Limitations of Punishment and Culpability,* 91 YALE L.J. 1446, 1462, n.79 (1982). The second edition (1983) of Professor Rothstein's book did not include a listing of imminent danger cases; he stated that "Most employers have abated [imminent dangers] voluntarily, thereby eliminating the need for other action." Rothstein, OCCUPATIONAL SAFETY AND HEALTH LAW 322 (1983). OSHA instructions to field staff on handling imminent danger situations are contained in the Field Operations Manual, ch. VII OSHR [Reference File 77:3301–4].

the OSHA regulation are entitled to back pay.[32] The issue was later considered by the federal district court, on remand of the principal case. It said:

> This Court agrees with the defendant that back pay should not be awarded in the typical "strike with pay" situation. It makes sense that an employee who refuses to perform hazardous work should be required to perform safe alternative tasks if they are available. However, recovery of lost wages is appropriate where the evidence shows that the denial of pay was a result of discrimination prohibited by 29 C.F.R. 1977.12. See *Whirlpool Corp. v. Marshall,* 445 U.S. 1, 19 n.31; *Marshall v. N.L. Industries,* 618 F.2d 1220 (7th Cir. 1980).
>
> In the present case, the facts show that the complainants Deemer and Cornwell were not given an opportunity to perform safe alternate work. The testimony adduced at the evidentiary hearing showed that, upon refusing to climb on the screen, complainants were given written reprimands and ordered to punch out without working or being paid for the remaining six hours of the shift. *Usery v. Whirlpool Corp.,* 416 F. Supp. 30, 32; *Whirlpool Corp. v. Marshall,* 445 U.S. 1, 7. The evidence also showed that the complainants were sent home because of their refusal to perform the work they perceived to be unreasonably dangerous. Thus, the facts are clear that it was defendant's own retaliatory and discriminatory conduct, unlawful under §1977.12, which deprived the employees of their opportunity to render any further services. Under these circumstances, this Court finds that an award of lost wages is entirely justified.[33]

C. Causal Connection

As has been the case under antidiscrimination provisions in other statutes, a critical issue under OSHA's §11(c) is whether, as required by that subsection, the discrimination was "because" of the protected activity. OSHA first faced the issue of defining the necessary causal connection to establish a §11(c) violation in its interpretive regulation. It said that, to constitute a violation under §11(c), the protected activity must have been a "substantial reason" for the employer's action, or "the discharge or other adverse action would not have taken place 'but for' engagement in protected activity."[34] The ultimate determination in these cases, as the regulation recognizes, "will have to be * * * on the basis of the facts in the particular case." Illustrative is the decision of the U.S. District Court for the District of Massachusetts involving an allegedly discriminatory discharge under OSHA.

---

[32]445 U.S. 1 at 19 n.31, reprinted in this section.

[33]*Marshall v. Whirlpool Corp.,* 9 OSHC 1038, 1039–40 (N.D. Ohio 1980). With respect to the underlying hazard involved in the *Whirlpool* case, OSHA issued a general duty citation, which was affirmed by the Review Commission, 8 OSHC 2248 (1980), but vacated and remanded by the court of appeals, *Whirlpool Corp. v. OSHRC,* 645 F.2d 1096 (D.C. Cir. 1981).

[34]29 C.F.R. §1977.6 (1983). In *NLRB v. Transportation Mgmt. Corp.,* 51 USLW 4761 (1983), the Supreme Court upheld the NLRB rule that a discharge or other adverse action is discriminatory if the employee's protected conduct "was a substantial or motivating factor" in the adverse action, unless the employer establishes that the action would have taken place "regardless of * * * forbidden motivation." This has been known as the *Wright Line* policy. See also *Mount Healthy City Bd. of Ed. v. Doyle,* 429 U.S. 274 (1977). For a discussion of the possible distinction between the "substantial reason" and "but for" tests in the context of the Age Discrimination in Employment Act, 29 U.S.C. §§621–34 (1976 & Supp. V 1981), see *Loeb v. Textron, Inc.,* 600 F.2d 1003, 1019 (1st Cir. 1979).

Marshall v. Commonwealth Aquarium

469 F. Supp. 690 (D. Mass.), aff'd, 611 F.2d 1 (1st Cir. 1979)

CAFFREY, Chief Judge:

This case arises from the discharge on May 15, 1976 of Jeffrey Boxer from his position as manager of Commonwealth Aquarium, a pet store located in Brookline, Massachusetts. On May 26, 1976 Boxer filed a complaint with the Secretary of Labor alleging that he was terminated by the defendant in retaliation for the exercise of his rights under the Occupational Safety and Health Act, 29 U.S.C.A. §651 *et seq.* (OSHA). Boxer's complaint to the Secretary claimed that his employment was terminated because he had reported to the National Institute for Occupational Safety and Health his fear that a potential health hazard existed at Commonwealth Aquarium. The Secretary determined that Boxer had indeed been discharged as a result of exercising his rights under OSHA and brought the instant action against the defendant in 1977. The jurisdiction of this Court is invoked under 29 U.S.C.A. §660(c)(2).

In August, 1975 Boxer was hired to manage defendant's Brookline store and was trained for that position by Richard Lerner, his immediate supervisor. The parties agree that from August 1975 to December 1975 Lerner and Boxer enjoyed an excellent working relationship. Lerner often expressed his satisfaction with Boxer's performance and Boxer received a substantial pay raise a few months after he had begun to work for the defendant. In fact, Lerner testified that he gave Boxer a silk tie as a Christmas gift.

It is in regard to the period following Christmas 1975 that the testimony becomes widely divergent. Lerner testified that there was a sudden and drastic change in Boxer's performance and that as a result of Boxer's tardiness, absenteeism, and mismanagement, their relationship deteriorated rapidly. However, although Lerner testified that he was seriously dissatisfied with Boxer's performance, he also testified that he kept hoping that Boxer would "return to the Jeffrey that [he] had once hired." There was no evidence tending to prove that he had ever considered terminating Boxer prior to May 1976.

Boxer testified and I find that the silk tie had been presented to him by Lerner not as a Christmas gift in December of 1975 but at a celebration of Boxer's birthday at Lerner's home on April 18, 1976 at which time, according to Boxer, Lerner told Boxer that he had a good future working for the defendant. Springer, a former employee of defendant, testified and I find that Lerner had continued to express his satisfaction with Boxer's work up to two weeks before the incidents culminating in Boxer's discharge.

I find that as late as mid-April 1976 the working relationship between Lerner and Boxer was good and that Lerner had no grounds for and had not considered discharging Boxer prior to May 1976.

In early May, 1976 Dr. Marjorie McMillan, a veterinarian at Angell Memorial Animal Hospital, performed an autopsy on a bird from Commonwealth Aquarium. The autopsy disclosed lesions which were highly suggestive of psittacosis, although no definite diagnosis could be made for three weeks. Psittacosis is a contagious respiratory ailment which affects birds and which when contracted by humans is potentially fatal.

A second bird acquired from the store about this time died with similar symptoms and four persons who had come in contact with the two birds thereafter contracted respiratory ailments.

Dr. McMillan called Boxer at Commonwealth Aquarium to warn him of the possible presence of psittacosis in his bird population and the doctor also informed him that, although her medical data was highly suggestive of psittacosis, the diagnosis was only tentative and would not be confirmed for several weeks. Boxer immediately checked all the birds in the store and saw no signs of sickness. He testified that he called Lerner to express his concern over the possible health hazard and was told to carry on with business as usual. Springer testified that he overheard a telephone conversation between Boxer and Lerner in which Boxer informed Lerner of a potential health hazard. Lerner refused to admit that such a problem existed.

The next morning, Boxer found the mate of one of the dead birds lying in its cage breathing very rapidly. He killed the bird and delivered it to an avian bacteriologist to be autopsied. Boxer later informed Lerner that there was a good chance that the third bird also had psittacosis. Boxer testified and I find that Lerner became angry and told Boxer that he should either continue to sell birds or he would be fired.

Boxer then began to add tetracycline to the birds' drinking water and he contacted Dr. John Lewis, an employee of the National Institute of Occupational Safety and Health, to whom he explained the situation. As a result of that conversation, Dr. Lewis came to inspect the store.

Dr. Lewis recommended to both Lerner and Boxer that tetracycline-impregnated bird food should be purchased and that the birds should be quarantined in a separate area. He suggested that if possible no birds be sold to the public. Lerner admitted that he spoke with Dr. Lewis and that although he had acknowledged Dr. Lewis' recommendations he did nothing to effectuate them. He testified that he examined the birds and believed them to be healthy. Boxer testified and I find that despite Dr. Lewis' advice he was told by Lerner to conduct business as usual. Boxer also testified and I find that no tetracycline-impregnated food was purchased.

Shortly thereafter Mr. Daley of the State Department of Labor told Lerner to buy masks for the employees to wear and the health inspector for the Town of Brookline instructed Lerner to quarantine the birds, stop selling them to the public for a few days, and to put tetracycline in their water rather than in their food. It is obvious therefore that although the authorities conflicted as to the specific remedial measures called for, they all believed that a potential health hazard existed at Commonwealth Aquarium. Lerner, however remained firm in insisting that the birds were healthy.

The birds were quarantined and Mr. Boxer left for his weekend off. Upon returning to work on Monday morning, he testified that he was told to conduct business as usual. When he discovered later that day that a bird had been sold in his absence over the weekend, he notified the Brookline health inspector who ordered the store closed.

Although Lerner testified that he never threatened to discharge anyone for refusing to carry on business as usual, Springer testified, and I find that he was approached by Lerner at some point during early May 1976 as to the possibility of his (Springer's) assuming Boxer's managerial duties at Commonwealth Aquarium. At that time Lerner told Springer that Boxer was "causing problems" and that things were getting "blown out of proportion." Boxer was terminated within a week after the store was closed by health officials.

***

It is undisputed that Boxer complained to an OSHA official about the potential health hazard at Commonwealth Aquarium, that Lerner knew of Boxer's complaint, and that Boxer was discharged shortly thereafter. It is necessary to determine therefore whether Boxer has met his burden and established that he was discharged because of that complaint.

In *Mount Healthy City School District Board of Education v. Doyle,* 429 U.S. 274 at 285–86 (1977) the Supreme Court rejected a rule of causation which relied solely upon a determination of whether the protected activity had played a substantial part in the employer's decision, the rationale being that to apply such a rule would place the employee in a *better* position than if he had not engaged in the protected activity. The Court ruled that the right to engage in protected activity is sufficiently vindicated if the employee is placed in *no worse* a position than if he had not engaged in the protected activity.

In so ruling the Court stated that once the plaintiff establishes that his activity was protected and that the protected activity was a substantial factor in the employer's decision, the burden then shifts to the employer to establish by a preponderance of the evidence that it would have reached the same decision even in the absence of the protected conduct. See also *Givhan v. Western Line Consolidated School District,* 47 USLW 4102 (January 9, 1979).

Applying that distribution of the burden to the facts of this case I rule that the plaintiff must prevail. Having in mind Lerner's claimed disbelief that a health hazard existed at Commonwealth Aquarium on May, 1976, his reluctance to cooperate with health officials, his resentment at their interference with his business activities and his belief that Boxer was overreacting, the fact that Boxer was discharged on May 15, 1976 and that Lerner had not considered discharging Boxer before May, 1976, I find that Boxer's complaint to OSHA in May, 1976 was a significant and substantial proximate causative factor for Lerner's decision to discharge him.

Apart from Lerner's testimony as to the deterioration of his relationship with Boxer prior to May 1976 which has been discussed and disbelieved above there is nothing in the record upon which the Court could base a finding that Lerner would have reached the same decision even in the absence of Boxer's complaints regarding the safety of his working conditions.

Lerner testified that the Friday after the store was closed down Boxer asked him if he could be laid off for the Summer so that he could do some fishing. Boxer's version of that conversation was that in light of Lerner's repeated threats to terminate his employment he was trying to ascertain whether he was to be fired or laid off so that he could determine whether he would be eligible for unemployment benefits. Furthermore, Lerner's contention that he decided to discharge Boxer because Boxer refused to work on a scheduled work day is unpersuasive. Boxer testified and I find that he never refused to come into the store when requested to do so.

I rule that after plaintiff established that improper motivation played a substantial part in Lerner's decision to discharge Boxer defendant failed to prove by a preponderance of the evidence that he would have discharged Boxer even in the absence of the protected activity.

***

The required relationship between the employer's adverse action and the protected activity in §11(c) cases was particularly troublesome in the walkaround pay issue.[35] The question there was: if the employer refuses to pay the employee for walkaround time, is the refusal "because" of the exercise of the walkaround right, concededly protected under §8(e)?[36] This issue is closely related to the question whether walkaround time is "working time" for purposes of the Fair Labor Standards Act. OSHA first determined that no working time was involved and that the refusal to pay was not because of the exercise of walkaround rights. OSHA later changed its position, but the Court of Appeals for the District of Columbia Circuit reversed the

[35]This issue is discussed in Chapter 16, Section C.
[36]29 U.S.C. §657(e).

Agency on procedural grounds, and may well have disagreed on the merits as well. OSHA's later resolution of the issue was not based on §11(c) discrimination grounds but on its authority under §8(e), and the regulation requiring walkaround pay was eventually withdrawn.

D. Procedural Issues Under §11(c)

Under §11(c), upon receiving an employee complaint of discrimination, which must be filed within 30 days of the alleged violation, OSHA is required to conduct an appropriate investigation.[37] If, upon investigation, OSHA determines that §11(c) has been violated, the Secretary "shall" bring an action in a U.S. district court.[38] Section 11(c)(3) further provides that within 90 days of receipt of the complaint, the Secretary "shall" notify the complaining party of his determination.[39] District courts have jurisdiction to restrain violations and to order "all appropriate relief," including reinstatement and back pay.[40]

In 1980, the Court of Appeals for the Sixth Circuit rejected an employee argument, supported by OSHA as *amicus curiae,* that a private right of action exists under §11(c). The statute provides only for suits to be brought by OSHA; the question is whether an aggrieved employee may also sue to redress discrimination.

Taylor v. Brighton Corporation

616 F.2d 256 (6th Cir. 1980)

PHILLIPS, Senior Circuit Judge:

The principal question raised on this appeal is whether the Occupational Safety and Health Act, 29 U.S.C. §651–678, creates an implied private right of action whereby an employee discharged in retaliation for reporting safety violations to OSHA may maintain a suit against his former employer. We affirm the district court's decision that it does not.

I.

The plaintiffs-appellants are former employees of Brighton Corporation who allege they were discharged in retaliation for reporting safety violations to the Occupational Safety and Health Administration (OSHA) or for opposing the company's retaliatory and discriminatory treatment of such employees.

***

On June 27, 1975, the appellants complained to the Secretary of Labor

[37]Sec. 11(c)(2), 29 U.S.C. §660(c)(2).
[38]*Id.*
[39]*Id.* at §660(c)(3).
[40]Sec. 11(c)(2), 29 U.S.C. §660(c)(2). OSHA's regulations deal expressly with procedures for §11(c) cases, 29 C.F.R. §§1977.15–18 (1983), as does Ch. X of the Field Operations Manual, OSHR [Reference File 77:3901–:3903].

that Brighton Corporation had discharged them in violation of §11(c)(1).[2] In October 1976, the Secretary notified appellants Herring and Taylor that he intended not to file suit on their complaints. While appellant Hinners' case was still open at that time, the Secretary had taken no action on his complaint.[3]

On February 7, 1977, the appellants filed this suit in the district court for the Southern District of Ohio. Claims 1, 2 and 4 of their complaint allege that Brighton discharged them in violation of §11(c). * * * The appellees moved to dismiss.

* * * With respect to claims 1, 2, and 4 of the complaint, Judge Hogan held that OSHA does not grant employees a private right of action to redress retaliatory dismissals that violate §11(c). * * *

This appeal followed. We affirm the district court's order of dismissal. Part II of this opinion holds that there is no implied private right of action under OSHA §11(c). * * *

In addressing the difficult question whether §11(c) implies a private right of action, we are guided by the discussion of the Supreme Court in *Cort v. Ash,* 422 U.S. 66, 78 (1975):

> In determining whether a private remedy is implicit in a statute not expressly providing one, several factors are relevant. First, is the plaintiff "one of the class for whose *especial* benefit the statute was enacted," *Texas & Pacific R. Co. v. Rigsby,* 241 U.S. 33, 39 (1916) (emphasis supplied)—that is, does the statute create a federal right in favor of the plaintiff? Second, is there any indication of legislative intent, explicit or implicit, either to create such a remedy or to deny one? See, e.g., *National Railroad Passenger Corp. v. National Assn. of Railroad Passengers,* 414 U.S. 453, 458, 460 (1974) *(Amtrak).* Third, is it consistent with the underlying purposes of the legislative scheme to imply such a remedy for the plaintiff? See, e.g., *Amtrak, supra; Securities Investor Protection Corp. v. Barbour,* 421 U.S. 412, 423 (1975); *Calhoon v. Harvey,* 379 U.S. 134 (1964). And finally, is the cause of action one traditionally relegated to state law, in an area basically the concern of the States, so that it would be inappropriate to infer a cause of action based solely on federal law? See *Wheeldin v. Wheeler,* 373 U.S. 647, 652 (1963); cf. *J. I. Case Co. v. Borak,* 377 U.S. 426, 434 (1964); *Bivens v. Six Unknown Federal Narcotics Agents,* 403 U.S. 388, 394–395 (1971); *id.,* at 400 (Harlan, J., concurring in judgment).

The parties are in agreement that the appellants are members of the class of persons, employees who report OSHA violations, whom §11(c) was intended to benefit. Nor is there any argument that retaliatory-discharge actions have traditionally been relegated to state law; the cause of action, if one exists, is solely federal. The questions we must address under *Cort v. Ash* are whether there is any indication in the legislative history of OSHA that Congress intended either to create or to deny a private remedy and whether implying one is consistent with the legislative plan.

All the *Cort v. Ash* factors, and particularly those concerning the legislative history of the statute and a private remedy's consistency with the legislative scheme, are signposts that guide the court's larger inquiry: Did Congress intend to create a private right of action in this situation? See *Transamerica Mortgage Advisors, Inc. v. Lewis,* 444 U.S. 11, 13 (1979) (here-

[2]Since all the appellants were discharged more than 30 days before they complained to the Secretary, Brighton argues that the appellants' suit is barred regardless of whether §11(c) creates an implied private right of action. Because we find no private right of action, we need not decide this issue.

[3]The Secretary has since filed suit on behalf of appellant Hinners.

inafter cited as *TAMA v. Lewis*). Congress created the statutory right not to be discharged in retaliation for filing OSHA complaints, and Congress could limit that right's corresponding remedy as it saw fit. Moreover, as the Supreme Court's recent cases make clear, "the fact that a federal statute has been violated and some person harmed does not automatically give rise to a private cause of action in favor of that person." *Cannon v. University of Chicago,* 441 U.S. 677, 688 (1979). "The central inquiry remains whether Congress intended to create, either expressly or by implication, a private cause of action." *Touche Ross & Co. v. Redington,* 442 U.S. 560, 575 (1979). With that caveat in mind, we turn to the issue presented in the case at bar.

***

Two points are evident from the statutory language. First, Congress nowhere mentioned private suits to enforce §11(c). Second, Congress explicitly provided an alternative means of redressing §11(c) violations. "In view of these express provisions for enforcing the [statutory prohibition], it is highly improbable that 'Congress absentmindedly forgot to mention an intended private action.'" *TAMA v. Lewis, supra,* 444 U.S. at 20 (*quoting Cannon v. University of Chicago, supra,* 441 U.S. at 742 (Powell, J., dissenting)).

The legislative history of §11(c) suggests that Congress intended suits by the Secretary of Labor to be the exclusive means of redressing violations. The language of §11(c) as finally enacted is the result of a compromise between the Senate version, which contained an administrative enforcement procedure, and the House version, which provided only civil and criminal penalties. The evolution of the Senate's administrative enforcement plan, which formed the basis for the section as enacted, reveals much about the intent of Congress.

As introduced in both the Senate and the House of Representatives, the Occupational Safety and Health bill included no express retaliatory-discharge provision. Rather, the bill contained only a general prohibition against interference with OSHA-related activity[.]

***

There was concern, however, that the possibility of retaliatory discharge might inhibit employees from reporting OSHA violations. See, e.g., H.R. Rep. No. 91-1291, 91st Cong., 2d Sess. 27 (1970). As a result, both the House and the Senate inserted provisions prohibiting discrimination against employees who report OSHA violations. Their enforcement provisions, however, were quite different.

The Senate's predecessor version of §11(c) established an administrative procedure whereby victims of retaliatory discrimination could obtain relief, including reinstatement and back pay. As reported by the Committee on Labor and Public Welfare, the Senate bill not only gave the Secretary of Labor the power to investigate complaints, but required him to provide the opportunity for a public hearing on the record and in accordance with 5 U.S.C. §554, the Administrative Procedure Act. If the Secretary found a violation, he was directed to issue a decision and order requiring the person committing the violation to take such affirmative action as might be appropriate to abate the violation, including but not limited to rehiring or reinstatement with back pay. See S. Rep. No. 91-1282, 91st Cong. 2d Session., *reprinted in* [1970] U.S. CODE CONG. & AD. NEWS 5177, 5211. Any person, including the complaining employee, adversely affected by the Secretary's decision could obtain review in the court of appeals for the circuit where the violation was alleged to have occurred.

When the Senate considered the bill, however, it made some important changes. Most significantly, it adopted an amendment creating the Occupational Safety and Health Review Commission as a means of separating the prosecutorial and adjudicative functions which the Committee's bill had combined in the Secretary. * * *

* * * [T]he Senate also gave the Review Commission the task of conducting hearings on retaliatory-discharge complaints. As under the Committee's version, the Secretary had the duty to investigate complaints, but for the first time his investigation was to precede rather than include a hearing. "If upon such investigation, the Secretary determines that the provisions of this subsection have been violated, he shall so notify the Commission and the Commission shall afford an opportunity for a hearing provided in subsection (c)." S. 2193, 91st Cong., 2d Sess. §10(f) (1970). Subsection (c) directed the Commission to "afford an opportunity for a hearing (in accordance with section 554 of title 5, United States Code, but without regard to subsection (a)(3) of that section)," and stated that the Commission's rules of procedure "shall provide affected employees * * * an opportunity to participate as parties to hearings under this subsection." S. 2193, 91st Cong., 2d Sess. §10(c) (1970). As in the version reported by the Committee on Labor and Public Welfare, parties aggrieved by the hearing's result could obtain review in the court of appeals.

The Senate's decision to have retaliatory-discrimination hearings conducted by the Review Commission rather than by the Secretary narrowed employees' rights considerably. Under the version of the bill reported by the Committee on Labor and Public Welfare, every complaining employee had been entitled to request and receive both a record hearing and judicial review of the Secretary's decision whether to order relief. Under the version passed by the Senate, however, an employee's only right was to complain to the Secretary. Only those employees whose complaints the Secretary deemed meritorious were entitled to request a hearing. Moreover, judicial review was available only from the Review Commission's decision, not the decision of the Secretary; employees whose complaints the Secretary deemed frivolous had no right to appeal.

The change in the retaliatory-discharge hearing procedure is not explainable as a manifestation of the Senate's desire to avoid a potential conflict between the Secretary's prosecutorial and adjudicative roles. No such potential for conflict existed under the Committee's predecessor version of §11(c). Unlike its safety-violation review procedures, the Committee bill's retaliatory-discharge provision directed the Secretary to conduct a hearing as part of his investigation, not after his personnel had made a preliminary determination that a violation had occurred. Since the hearing was to be a de novo inquiry rather than a review of an inspector's prior decision, there was no need to commit the decision-making function to an independent panel. Furthermore, even under the version passed by the Senate, the Secretary was directed only to investigate and notify the Review Commission of meritorious complaints; there was no requirement that he argue the employee's case before the Commission. Thus, the concerns that prompted the Senate to create the Review Commission do not explain its decision to have retaliatory-discharge hearings conducted by the Commission in meritorious cases only, rather than by the Secretary in all cases.

A more plausible explanation for the change, we think, is that the Senate wanted the Secretary to screen out frivolous complaints so as not to overburden the hearing body. Under the Committee's predecessor version, a record hearing was a potential part of the Secretary's investigation in every case, whether meritorious or not. By removing the adjudicative function to the newly created Review Commission, the Senate was able to confine record hearings to cases the Secretary found meritorious. In further recognition of the potential for frivolous retaliatory-discharge complaints, the Senate bill provided that the Secretary need only "cause such investigation to be made as he deems appropriate." Finally, the Senate made no provision for appeal from the Secretary's determination that a complaint was frivolous. These changes suggest that the Senate removed the hearing function to the Review Commission in retaliatory-discharge cases as a means of eliminating the necessity for hearing in cases not found by the Secretary to be meritorious.

By contrast to the Senate's remedial plan, the House's predecessor version of §11(c) specified that violators would be "assessed a civil penalty by the Commission of up to $10,000" and might also "be subject to a fine of not more than $10,000 or imprisonment of a period not to exceed ten years or both." H.R. 16785, 91st Cong., 2d Sess. §17(g) (1970). The House bill contained no provision whereby employees injured by retaliatory discrimination could obtain relief. See Conf. Rep. No. 91-1765, 91st Cong., 2d Sess., *reprinted in* [1970] U.S. CODE CONG. & AD. NEWS 5228, 5235.

The Senate and House conferees agreed to adopt the Senate's remedial approach to retaliatory-discharge violations rather than the House's civil and criminal penalty approach. The House conferees did insist, however, that the Senate's version be amended to specify that the Secretary would prosecute §11(c) actions and that he would do so in the district courts rather than before the Review Commission. *Id.*

The development of §11(c), particularly in the Senate, evidences a progressive narrowing of both the employee's right to a hearing on his claim that he was the victim of retaliatory discrimination and his role in securing relief from the alleged violation. Originally conceived as an investigative forum available to every complaining employee, the hearing evolved first into an administrative procedure applicable only to pre-selected cases and finally into a formal lawsuit in which the Secretary, not the employee, is the plaintiff. Such a legislative narrowing of the individual employee's rights and role under §11(c) indicates a Congressional intent to deny alternative remedies. In short, §11(c)'s legislative history cuts against a private right of action for retaliatory discharge.

Supporting our conclusion that Congress did not intend to create a private right of action under §11(c) is the inconsistency of such a remedy with the section's enforcement provisions. Congress directed the Secretary of Labor to investigate and file suits to secure relief for employees injured by retaliatory discrimination. In order to find that discharged employees have the right to bring private actions under §11(c), we would have to conclude that the enforcement mechanism Congress specified was not intended to be exclusive.

> Yet it is an elemental canon of statutory construction that where a statute expressly provides a particular remedy or remedies, a court must be chary of reading others into it. "When a statute limits a thing to be done in a particular mode, it includes the negative of any other mode." *TAMA v. Lewis, supra,* 444 U.S. at 18 (*quoting Botany Mills v. United States,* 278 U.S. 282, 289 (1929)).

See *Securities Investor Protection Corp. v. Barbour,* 421 U.S. 412, 420–23 (1975); *National Railroad Passenger Corp. v. National Association of Railroad Passengers,* 414 U.S. 453, 458 (1974) (hereinafter cited as *Amtrak*). We conclude it to be unlikely that Congress, having deliberately interposed the Secretary's investigation as a screening mechanism between complaining employees and the district courts, intended to permit those employees whose claims are screened out to file individual actions in those same courts. A private cause of action is simply inconsistent with the enforcement plan provided by Congress.

***

* * * While the absence of any indication in the legislative history that Congress intended to create a private cause of action under §11(c) would not

necessarily preclude private suits,[7] the existence of a Congressional purpose to deny such a cause of action is controlling. See *Cort v. Ash, supra,* 422 U.S. at 82. Compare *Chumney v. Nixon,* 615 F.2d 389, No. 77-1370 (Jan. 24, 1980, 6th Cir.). (*Chumney* found an implied private right of action under 18 U.S.C. §113, which provides criminal penalties for personal assaults on any aircraft within the special aircraft jurisdiction of the United States. The statute contains no alternative to private damage actions as a means of relieving victims of illegal assaults, and this court found no indication of Congressional intent either to create or to deny a private right of action.)

The Secretary of Labor filed an amicus brief urging this court to find an implied private right of action under §11(c). The Secretary says he has neither the resources nor the personnel to handle all §11(c) complaints adequately. Moreover, he expects the number of such complaints to increase dramatically due to his current campaign to alert employees of their OSHA rights. A private right of action should be implied, the Secretary argues, because individual suits offer the only realistic hope of protecting employees from retaliatory discrimination.

The Secretary should address his arguments to Congress, not the courts. In *Touche Ross & Co. v. Redington, supra,* the Supreme Court rejected similar policy arguments for the implication of a private damage remedy under §17(a) of the Securities Exchange Act of 1934:

> SIPC and the Trustee, contend that the result we reach sanctions injustice. But even if that were the case, the argument is made in the wrong forum, for we are not at liberty to legislate. If there is to be a federal damage remedy under these circumstances, Congress must provide it. "[I]t is not for us to fill any *hiatus* Congress has left in this area." *Wheeldin v. Wheeler,* 373 U.S. 647, 652, 83 S.Ct. 1441, 1446, 10 L.Ed.2d 605 (1963). Obviously, nothing we have said prevents Congress from creating a private right of action on behalf of brokerage firm customers for losses arising from misstatements contained in §17(a) reports. But if Congress intends those customers to have such a federal right of action, it is well aware of how it may effectuate that intent. 442 U.S. at 579, 99 S. Ct. at 2490-91.

"The dispositive question" is not whether a private right of action under §11(c) is desirable, but "whether Congress intended to create any such remedy. Having answered that question in the negative, our inquiry is at an end." *TAMA v. Lewis, supra,* 444 U.S. at 24.

Accordingly, we hold that there is no private right of action under OSHA §11(c), 29 U.S.C. §660(c). Claims 1, 2, and 4 of the complaint were properly dismissed.

***

The court of appeals' decision, while based largely on the particular legislative history of §11(c), also reflects the fact that the Supreme Court had increasingly construed statutes strictly and was

[7]But see *Cannon v. University of Chicago, supra,* 441 U.S. at 717, 99 S. Ct. at 1968 (Rehnquist, J., (author of the opinion in *Touche Ross v. Reddington, supra*) concurring):

> Not only is it "far better" for Congress to so specify when it intends private litigants to have a cause of action, but for this very reason this Court in the future should be extremely reluctant to imply a cause of action absent such specificity on the part of the Legislative Branch.

Mr. Justice Stewart, author of the opinion in *TAMA v. Lewis, supra,* joined in the concurring opinion of Mr. Justice Rehnquist.

refusing to imply private rights of action.[41] OSHA's *amicus* brief urged the court to find a private right of action on the grounds that the Agency had "neither the resources nor personnel to handle all section 11(c) complaints adequately."[42] The court rejected the argument, saying that it should more properly be addressed to Congress; however, OSHA's statement in its brief was relied on later by a union in congressional testimony to establish that the §11(c) program "no longer works."[43]

Section 11(c) requires that the employee complaint be filed "within thirty days after such violation occurs." In construing this limitations period, OSHA said that "recognized equitable principles or * * * strongly extenuating circumstances" could justify tolling the 30-day period.[44] This doctrine of equitable estoppel, as it has been called, was applied by a district court to a complaint filed more than 30 days after the employee's termination.[45] The court said:

> Defendant argues that because Kidd failed to file a complaint with OSHA within thirty days of his termination he is barred from pursuing this action. The thirty-day filing period is prescribed by 29 U.S.C. §660(c)(2). The Court concludes, however, that this period was tolled and began to run only on March 15, 1980, when Kidd first learned of his termination.
>
> An analogous case was recently decided by the Supreme Court. In *Zipes v. Trans World Airlines, Inc.*, 455 U.S. 385 (1982), the Court examined the statutory time limit for filing claims under Title VII of the Civil Rights Act of 1964. The Court unanimously held that the time limit was not a jurisdictional requirement, but acted as a statute of limitations which was subject to waiver, estoppel and equitable tolling.
>
> In reaching this holding, the Court relied on the statute and the congressional policy underlying it. While the structure of the Occupational Safety and Health Act differs from that of Title VII in that the jurisdictional provision in the Act if contained in the same section as the timely filing provision, the Court believes that the underlying congressional purpose and the absence of evidence that Congress intended to make timely filing a jurisdictional requirement lead to a conclusion that the requirement is like a statute of limitations and is subject to waiver, estoppel and equitable tolling.
>
> An examination of the legislative history of the Occupational Safety and Health Act shed no light directly on this issue. The overall policy of the Act, however, is clearly to protect the American work force. This policy would be thwarted if the Court were to consider the limit jurisdictional so as to prevent the application of the doctrines of waiver,

[41]See *Transamerica Mortgage Advisors, Inc. v. Lewis*, 444 U.S. 11 (1979) (decided shortly before *Taylor v. Brighton Corp.* was argued).

[42]616 F.2d at 263.

[43]*Oversight on the Administration of the OSH Act, 1980: Hearings Before the Senate Comm. on Labor and Human Resources*, 96th Cong., 2d Sess. 878 (1980) (statement of Nolan W. Hancock, Citizenship-Legislative Director, OCAW) [hereinafter cited as *1980 Senate Oversight Hearings*]. On the right of employees to prosecute citations before the Review Commission independently, see discussion in Chapter 17, Section A. On implication of private rights of action generally, see Siegel, *The Implication Doctrine and the Foreign Corrupt Practices Act*, 79 COLUM. L. REV. 1085 (1979).

[44]29 C.F.R. §1977.15(d)(3) (1983).

[45]*Donovan v. Hahner, Foreman & Harness, Inc.*, 11 OSHC 1081 (D. Kan. 1982).

estoppel and equitable tolling. This would encourage employers to engage in tactics such as the defendant's in this case in order to evade the provisions of the law. In holding that the filing requirement is subject to waiver, estoppel and tolling, the Court honors the purpose of the Act without undue damage to the purpose of the timely filing requirement, that of prompt notice to the employer.

***

In applying the finding of facts to this legal conclusion, the Court finds that the running of the time period was equitably tolled by defendant's concealment from Kidd of the fact that Kidd had been fired and not just laid off. Kidd first learned of his termination on March 15, 1980, and the filing period began to run on that date. As Kidd filed within thirty days of that date, he clearly fulfilled the timely filing provision of 29 U.S.C. §660(b)(2).[46]

Another court, however, refused to apply equitable estoppel principles where the employee's timely complaint was filed with a state agency on a related but different issue.[47] The Court of Appeals for the Seventh Circuit has held that the Secretary of Labor's failure to notify the complaining party of his determination within 90 days, as required by §11(c)(3), does not bar a later filing of a suit;[48] and a district court held that a delay by OSHA of 31 months after discovery of the violation before bringing suit does not constitute laches barring the suit.[49]

Two other major §11(c) procedural issues were decided by courts of appeals relying on Supreme Court precedent under Title VII of the Civil Rights Act of 1964.[50] In *Marshall v. N.L. Industries,*[51] the Court of Appeals for the Seventh Circuit held that submission by an employee to a contractual grievance-arbitration procedure of a claim

[46]11 OSHC at 1084. See also *Marshall v. Lummus Co.*, 663 F.2d 1072 (6th Cir. 1981), *vacating* 8 OSHC 1358 (N.D. Ohio 1980) (applying equitable estoppel where employee made timely oral §11(c) complaint to OSHA office, but written complaint was beyond 30-day period).

[47]*Marshall v. Certified Welding Corp.*, 7 OSHC 1069 (10th Cir. 1978).

[48]E.g., *Marshall v. N.L. Industries*, 618 F.2d 1220 (7th Cir. 1980).

[49]*Dunlop v. Bechtel Power Corp.*, 6 OSHC 1605 (M.D. La. 1977). But see *Chadsey v. United States*, 11 OSHC 1198 (D. Or. 1983) (holding United States liable under Federal Tort Claims Act to injured employee where Department of Labor attorney negligently failed to produce records in response to a court order and to oppose a motion to dismiss in a §11(c) case that was dismissed). Suits based on the Federal Tort Claims Act have been brought against OSHA compliance officers, alleging that their negligent inspections caused injuries to the plaintiff-employee. A much-discussed case on this issue is *Blessing v. United States*, 447 F. Supp. 1160 (E.D. Pa. 1978) ("discretionary function" exception to the Tort Claims Act for claims arising out of regulatory activities does not warrant dismissal of suit for want of jurisdiction).

Although §4(b)(4), 29 U.S.C. §653(b)(4), provides that the Act shall not be "construed to supersede or in any manner affect any workmen's compensation law or to enlarge or diminish or affect in any manner the common law or statutory rights, duties or liabilities of employers and employees," OSHA has had some impact on private personal injury litigation. See, e.g., *Mandolidis v. Elkins Industries, Inc.*, 246 S.E. 2d 907 (W. Va. 1978) (Court's conclusion that employer was responsible for employee injury under intentional injury exception to workers' compensation exclusivity rule based in part on prior OSHA citations). This subject is discussed in Rothstein, OCCUPATIONAL SAFETY AND HEALTH LAW, 439–58 (1983). See also, Amchan, *"Callous Disregard" for Employee Safety: The Exclusivity of the Workers Compensation Remedy Against Employers,* 34 LAB. L.J. 683 (1983) (proposes amendment to §4(b)(4) to allow private suits against employers where "willful, intentional or grossly negligent" conduct causes death or permanent disability). Sec. 4(b)(4) is also discussed as it relates to medical removal protection in Chapter 5, Section D.

[50]42 U.S.C. §2000e–3(a) (1976).

[51]618 F.2d 1220 (7th Cir. 1980).

dealing with the same subject matter as a later §11(c) complaint does not bar relief under §11(c).[52] And in *Marshall v. Intermountain Electric Co.,*[53] the Court of Appeals for the Tenth Circuit held that a state statute of limitations could not be relied on as a bar to the filing of a §11(c) suit by OSHA.[54]

E. Evaluation of §11(c) Program

In 1980, testifying before a Senate oversight hearing, Secretary of Labor Marshall gave a frank appraisal of the strengths and weaknesses of OSHA's program on §11(c).[55]

Statement of Secretary of Labor Marshall

***

In administering Section 11(c) of the Act, the Agency encountered a number of difficult problems in the past, including:

- a basic resource problem in terms of availability of qualified personnel to conduct field investigations, and in terms of having sufficient staff to service the increasing discrimination caseload;
- lack of established case law; and
- the complexity of issues raised.

The problem concerning the availability of qualified personnel to conduct investigations has been ameliorated. During the formative years of the Agency's existence, compliance officers conducted 11(c) investigations. While well qualified to conduct safety and health inspections, their expertise was less well suited to discrimination investigations. The time spent by them also diminished the time available for enforcement activities. Faced with such difficulties, OSHA recruited professional investigators for 11(c) functions. As a consequence, the quality of 11(c) investigations has significantly improved.

We currently have a backlog of uninvestigated cases. To reduce this backlog and to improve our response time to complainants, the Agency initiated an experimental program designed to streamline the investigative process. The experiment is an accelerated settlement program designed to help workers and their employers resolve questions of discrimination without long and costly adversary proceedings.

---

[52]The court of appeals relied principally on the leading case under Title VII of the Civil Rights Act, *Alexander v. Gardner-Denver Co.*, 415 U.S. 36 (1974), which held, in the language of the court of appeals in *N.L. Industries,* that "an arbitrator's decision under a collective bargaining agreement to deny relief does not bar a later suit in federal court under Title VII, even if the discrimination question was presented in the arbitration proceedings." 618 F.2d at 1222. In its regulations on §11(c), 29 C.F.R. §1977.18 (1983), OSHA recognizes the national policy favoring voluntary resolution of disputes under collective bargaining agreements, and specifies the limited circumstances under which it would defer to the outcome of other proceedings or postpone its §11(c) determination pending the outcome of other proceedings. See Comment, *The Assertion of Statutory Rights Under FLSA and OSHA: Expand or Limit the* Gardner-Denver *Rationale,* 1981 B.Y.U. L. REV. 361.

[53]614 F.2d 260 (10th Cir. 1980). Accord: *Donovan v. Square D Company,* 709 F.2d 335 (5th Cir. 1983).

[54]The leading Title VII case on this issue is *Occidental Life Ins. Co. v. EEOC*, 432 U.S. 355 (1977). For a general discussion of retaliation under the Civil Rights Act, see B. Schlei & P. Grossman, EMPLOYMENT DISCRIMINATION LAW, ch. XV at 533–63 (1983).

[55]*1980 Senate Oversight Hearings, supra* note 43 at 1040–42.

If this experiment is successful, OSHA will attempt through this system to resolve a complaint before a full-scale discrimination investigation is initiated. Workers are spared a lengthy period of uncertainty about their job status; employers will learn of discrimination complaints sooner and will have an opportunity to avoid costly litigation and large back-pay awards.*

The statute's requirement that violations found be resolved through judicial action in U.S. District Court creates additional problems. Most court dockets are crowded; and since the Act does not provide for expedited review in such cases, complainants whose cases cannot be settled are faced with a lengthy judicial process before their complaint is resolved. The Agency faces the inherent problem of the lack of established case law in the 11(c) area. This necessitates the implementation of a careful legal strategy in order to establish a sound legal foundation for future 11(c) actions.

Finally, the type of issues raised by 11(c) complaints continue to increase in complexity, including such issues as the employee's refusal to perform an imminently dangerous task, "protective discrimination" affecting reproductive rights of workers, employees' rights to accompany compliance officers during the OSHA inspection, and others.

Future actions in the program will be taken with a view toward resolving the problem issues outlined, particularly improving the timeliness of investigations. The Agency will continue to develop innovative programs such as the accelerated settlement program to handle the growing workload efficiently, protect worker rights effectively, and give employers an opportunity to resolve disputes quickly.**

The Oil, Chemical, and Atomic Workers Union concurred in expressing disappointment in OSHA's implementation of the §11(c) program[56] and suggested, among other things, that the enforcement of employees' §11(c) rights should be handled through the Review Commission rather than in the federal district courts.[57] The Federal Mine Safety and Health Act of 1977 adopted this method of enforcement[58] and broadened the scope of protection by expansively defining protected activities for purposes of that statute.[59] The Mine Safety Act also contains several new procedural features in the antidiscrimination clause, which are designed, in the words of the Senate Committee on Labor and Human Resources, to make the "national mine safety and health program * * * truly effective."

*[Author's note—See Field Operations Manual, ch. X, OSHR [Reference File 77:3902], on preinvestigation settlements of §11(c) complaints.]

**[Author's note—In 1982, Assistant Secretary Auchter reported that the backlog of §11(c) complaints had dropped from about 1,000 at the beginning of the year to 650 on September 30, 1982. *Oversight Hearings on OSHA—Occupational Safety and Health for Federal Employees: Hearings Before the Subcomm. on Health and Safety of the House Comm. on Education and Labor,* 97th Cong., 2d Sess. 456 (1982). He noted that during fiscal year 1982, OSHA completed investigation of 2,721 §11(c) complaints, half of which were dropped after preliminary screening because of procedural or legal defects.]

[56]*1980 Senate Oversight Hearings, supra* note 43 at 876–78.

[57]*Id.* at 886–87.

[58]Sec. 105(c), 30 U.S.C. §815(c) (Supp. V 1981).

[59]*Id.* at §815(c)(1). The legislative history of the Mine Act explicitly refers to protection for refusals to work in unsafe and unhealthful working conditions and protection against layoff or termination in the context of medical removal protection.

Federal Mine Safety and Health Act of 1977[60]

***

The bill provides that a miner may, within 60 days after a violation occurs, file a complaint with the Secretary. While this time-limit is necessary to avoid stale claims being brought, it should not be construed strictly where the filing of a complaint is delayed under justifiable circumstances. Circumstances which could warrant the extension of the time-limit would include a case where the miner within the 60 day period brings the complaint to the attention of another agency or to his employer, or the miner fails to meet the time limit because he is misled as to or misunderstands his rights under the Act.

The Secretary must initiate his investigation within 15 days of receipt of the complaint, and immediately file a complaint with the Commission, if he determines that a violation has occurred. The Secretary is also required under section 106(c)(3) [§105(c)(3) as enacted, 30 U.S.C. §815(c)(3)] to notify the complainant within 90 days whether a violation has occurred. It should be emphasized, however, that these time-frames are not intended to be jurisdictional. The failure to meet any of them should not result in the dismissal of the discrimination proceedings; the complainant should not be prejudiced because of the failure of the Government to meet its time obligations.

The Secretary's investigation of matters alleged in the complaint must commence within fifteen days of receipt of the complaint. Upon determining that the complaint appears to have merit, the Secretary shall seek an order of the Commission temporarily reinstating the complaining miner pending final outcome of the investigation and complaint. The Committee feels that this temporary reinstatement is an essential protection for complaining miners who may not be in the financial position to suffer even a short period of unemployment or reduced income pending the resolution of the discrimination complaint. To further expedite the handling of these cases, the section requires that upon completion of the investigation and determination that the provisions of this section have been violated, the Secretary must immediately petition the Commission for appropriate relief.

In proceedings before the Commission brought by the Secretary, miner, applicants, or their representatives, may present additional evidence in their own behalf. In addition, under section 106(c)(3), if the Secretary determines that no violation has occurred, the complainant has the right within 30 days of receipt of the Secretary's determination, to file an action on his own behalf before the Commission. If the Secretary has provided a procedure for the miner to appeal a negative determination within the Department of Labor, the thirty-day period will not commence until such appeal procedures have been exhausted. Further, as mentioned above in connection with the time for filing complaints, this thirty-day limitation may be waived by the court in appropriate circumstances for excusable failure to meet the requirement. The Committee also intends to afford a complainant the right to institute an action on his own behalf before the Commission if the Secretary, in the exercise of his discretion, settles a case brought under section 106(c)(2) on terms unsatisfactory to the complainant.

It is the Committee's intention that the Secretary propose, and that the Commission require, all relief that is necessary to make the complaining party whole and to remove the deleterious effects of the discriminatory conduct including, but not limited to reinstatement with full seniority rights, back-pay with interest, and recompense for any special damages sustained as a result of the discrimination. The specified relief is only illustrative.

[60]SENATE COMM. ON HUMAN RESOURCES, S. REP. NO. 181, 95th Cong., 1st Sess. 36–37 (1977).

Thus, for example, where appropriate, the Commission should issue broad cease and desist orders and include requirements for the posting of notices by the operator.
***

The antiretaliation provision recommended by the committee and passed by Congress went further than the OSHA antidiscrimination protection in these major respects: (1) the definition of protected activity was explicitly broadened in scope; the Mine Act includes, for example, protection against termination or layoff in the medical removal protection context; (2) the time limitation for the filing of employee complaints is 60 days rather than 30 days as under OSHA; (3) a miner is entitled to temporary reinstatement if his or her complaint appears to have merit ("not frivolously brought"); and (4) if the Secretary of Labor determines that the employee complaint has no merit, the employee has the right within 30 days to file an action in his or her own behalf before the administrative agency.[61]

F. Employees' Right to Information on Safety and Health

In his testimony before an oversight hearing in 1980, Secretary of Labor Marshall emphasized that if employees and their representatives were to play a meaningful role in occupational safety and health programs, it was "vitally important to ensure that workers have the information they need to protect themselves."[62] Several provisions in the Act are designed to achieve this goal. Section 9(b)[63] requires that a citation issued to an employer be "prominently posted * * * at or near" the place where the violation occurred. (The citation must include a description of the violation and the abatement date.) In addition, §6(b)(7)[64] requires that OSHA standards prescribe the use of "labels or other appropriate forms of warning" to apprise employees of the hazards to which they are exposed; and further, that employees' physicians have access to their medical records. Moreover, §8(c)(3)[65] provides that employees must have an opportunity to observe workplace monitoring and to have access to monitoring records.[66]

A number of Agency actions have amplified the rights of employees to information on safety and health hazards. In July 1978, OSHA amended the regulations dealing with the recording and reporting of

[61]SENATE COMM. ON HUMAN RESOURCES, FEDERAL MINE SAFETY AND HEALTH ACT OF 1977, S. REP. NO. 181, 95th Cong., 1st Sess. 35–37 (1977).
[62]29 U.S.C. §658(b). *1980 Senate Oversight Hearings, supra* note 43 at 969.
[63]29 U.S.C. §658(b). See also 29 C.F.R. §1903.16 (1983).
[64]29 U.S.C. §655(b)(7).
[65]29 U.S.C. §657(c)(3).
[66]Other employee rights in the enforcement context, such as the walkaround right, also serve in part to inform employees of workplace hazards.

occupational injuries and illnesses to give employees, former employees, and their representatives access to the log and summary of all recorded occupational injuries and illnesses.[67] Health standards regulating specific substances have routinely provided for employee observation of workplace monitoring and access to monitoring records as well as access to medical records.[68] In 1980, OSHA issued a "generic" regulation providing for employee access to existing employer-maintained exposure and medical records relating to employees exposed to toxic substances and harmful physical agents.[69] The access regulation was upheld by a U.S. district court;[70] however, OSHA has reopened the regulation's record for reconsideration.[71]

The issue of whether employers should be required to identify chemical hazards in the workplace and communicate the information to employees has been before OSHA virtually throughout its existence. The history of OSHA's efforts to regulate in this area was set forth in the preamble to OSHA's 1982 proposed standard on hazard communication.

Hazard Communication

47 Fed. Reg. 12,092–93 (1982)

***

I. BACKGROUND

A. HISTORY OF OSHA'S PROPOSED HAZARD COMMUNICATION STANDARD

OSHA's involvement in the identification and communication of hazards in the workplace began some years ago. In 1974, the Standards Advisory Committee on Hazardous Materials Labeling was established under section 7(b) of the OSH Act to develop guidelines for the implementation of section 6(b)(7) of the Act with respect to hazardous materials. On June 6, 1975, the Committee submitted its final report which identified issues and recommended guidelines for categorizing and ranking chemical hazards. Labels, material safety data sheets, and training programs were also prescribed.

The National Institute for Occupational Safety and Health (NIOSH) published a criteria document in 1974 which recommended a standard to OSHA. The document, entitled "A Recommended Standard * * * An Identification System for Occupationally Hazardous Materials," included provisions for labels and material safety data sheets.

In 1976, Congressman Andrew Maguire (New Jersey) and the Health Research Group petitioned OSHA to issue a standard to require the labeling of all workplace chemicals. The House of Representatives' Committee on

[67] 43 FED. REG. 31,329 (1978); 29 C.F.R. §1904.7(b) (1983).

[68] See, e.g., standard on vinyl chloride, Chapter 3, Section B. But see hearing conservation amendment, 46 FED. REG. 42,622, 42,624–25 (1981).

[69] 45 FED. REG. 35,212 (1980). See Chapter 5, Section F for discussion of employee access to medical records. One of the major issues in the employee access rule is whether the access, particularly to exposure records, will impair the competitive value of employers' trade secrets. See 45 FED. REG. at 35,237–40 (1980).

[70] *Louisiana Chem. Ass'n v. Bingham,* 550 F. Supp. 1136 (W.D. La. 1982).

[71] 47 FED. REG. 30,420 (1982).

Government Operations in 1976 and 1977 recommended that OSHA should enforce the health provisions of the OSH Act by requiring manufacturers to disclose any toxic ingredients in their products, and by requiring employers to disclose this information to workers.

On January 28, 1977, OSHA published an advance notice of proposed rulemaking on chemical labeling in the *Federal Register* (42 FR 5372). The notice requested comments from the public regarding the need for a standard that would require employers to label hazardous materials. Information was also requested regarding the provisions to be included in such a standard to assure that employees are apprised of the hazards to which they are exposed.

A total of eighty-one comments were received from a variety of federal, state, and local government agencies, trade associations, businesses, and labor organizations. In general, there was support for the concept of a hazard communication standard. A number of commenters said that such a standard should be comprehensive in scope, but not too complex in design. Many expressed the opinion that OSHA's standard should be compatible with the standards of other regulatory agencies with labeling authority, such as the Department of Transportation (DOT), and with existing voluntary labeling standards, such as that of the American National Standards Institute (ANSI). A few commenters expressed concerns about protection of trade secret information and about labeling chemical intermediates.

Various suggestions were put forth for determining which materials should be considered hazardous and thus covered by such a standard. Some commenters thought that chemicals that meet specified definitions or other classifications should be regulated. Others preferred that a list of substances to be regulated be provided, for example, those substances in 29 CFR 1910.1000 (OSHA's list of air contaminants), in the NIOSH Registry of Toxic Effects of Chemical Substances (RTECS), or in the DOT hazardous materials list.

Virtually all commenters recognized the need for labels in the workplace, and for inclusion of warnings and descriptive information. However, opinions varied as to what form these labels and information should take, or whether an existing system should be adopted. Similarly, there was general recognition of the need to inform employees of the hazards to which they are exposed by means of data sheets and training programs, although suggestions as to content and format varied.

On January 16, 1981, OSHA published a notice of proposed rulemaking (NPRM) entitled "Hazards Identification" (46 FR 4412). The NPRM would have required employers to assess the hazards in their workplaces using specified determination procedures. Labels including extensive information about these hazards would have been required on all containers within the workplace (including pipes), as well as on containers leaving the workplace.

OSHA withdrew the NPRM on February 12, 1981 for further consideration of regulatory alternatives (46 FR 12214). An Agency task force was formed in April to review the withdrawn proposal, and to develop alternatives as necessary. This task force has since met with a number of interested parties to solicit their views on issues related to hazard communication, reviewed the proposal and other information obtained by the Agency, and developed the standard proposed herein.

***

As noted in the preamble, the House Government Operations Committee has long been involved in pressing for OSHA regulation of toxic substances, including the labeling of chemicals; it has conducted several hearings[72] and issued reports on the subject. In a 1976 report, the committee stated in its "findings and conclusions" the reasons why a hazard communication standard was needed:

FINDINGS AND CONCLUSIONS

1. Identifying and controlling toxic substances in the workplace is becoming progressively more difficult as more chemicals, chemical processes, and chemical products are used in industry. Tens of thousands of trade-name products, whose chemical contents are not disclosed, are used daily. Lack of knowledge about exposure hampers the identification of occupationally caused diseases, illnesses, and deaths and is a major impediment to preventing them.

2. Both employers and employees are often un[a]ware of the toxic chemicals in the trade-name products that they buy and use. An extensive NIOSH survey shows that toxic chemicals are found in almost half of the trade-name products and that 90 percent of the time the chemical composition of trade-name products is not known to the buyer or user.

3. Manufacturers and formulators of trade-name products often do not disclose the contents of their products on the grounds that such information is a trade secret or proprietary information. These claims are made for products that contain toxic chemicals, including known carcinogens.

4. Attempts at self-regulation by the chemical industry have not generated adequate information for buyers and users about toxic chemicals in industrial products. Voluntary labeling guidelines developed by the chemical manufacturing industry are directed primarily to the avoidance of injury from single, accidental exposures and do not address the health hazards caused by chronic low-level exposure to toxic chemical substances.[73]

The committee recommended that OSHA should develop a "mandatory system of identifying toxic substances in the workplace." The committee explained that "[k]nowledge of workplace dangers should not wait for the tortuous process of issuing standards on an agent-by-agent basis."[74]

On November 25, 1983, OSHA issued a final standard on hazard communication.[75] OSHA summarized the contents of the standard as follows:

* * * The standard requires chemical manufacturers and importers to assess the hazards of chemicals which they produce or import, and all employers having workplaces in the manufacturing division, Standard Industrial Classification (SIC) codes 20 through 39, to provide information to their employees concerning hazardous chemicals by means of hazard communication programs including labels, material safety data

[72]E.g., *Toxic Substances in the Workplace: Hearings Before a Subcomm. of the House Comm. on Government Operations*, 94th Cong., 2d Sess. (1976). The preamble does not mention the suit brought in 1976 by Congressman McGuire of the House Committee on Government Operations and the Public Citizen Health Research Group to force OSHA to issue a standard on hazard communication. The suit was dismissed on jurisdictional grounds. *Public Citizen Health Research Group v. Marshall*, 485 F. Supp 845 (D.D.C. 1980), discussed in Chapter 6, Section C.

[73]HOUSE COMM. ON GOVERNMENT OPERATIONS, CHEMICAL DANGERS IN THE WORKPLACE, H. REP. No. 1688, 94th Cong., 2d Sess. 4 (1976).

[74]*Id.* at 5.

[75]48 FED. REG. 53,280 (1983). OSHA concluded that the rule was a "standard" and not a "regulation." 48 FED. REG. at 53,320–21.

> sheets, training and access to written records. In addition, distributors of hazardous chemicals are required to ensure that containers they distribute are properly labeled, and that a material safety data sheet is provided to their customers in the manufacturing division SIC Codes.
>
> Implementation of this final standard will reduce the incidence of chemically related occupational illnesses and injuries in employees of the manufacturing division. Increased availability of hazard information will assist employers in these industries to devise appropriate protective measures, and will give employees the information they need to take steps to protect themselves.

Among the major issues dealt with in the final standard is "the treatment of hazardous chemicals that are considered trade secrets by the chemical manufacturer or employer."[76] OSHA stated that its responsibility was to "strike a particularly fine and creative balance" in order to "accommodate" the "health interest in limited trade secret disclosures and the economic interest in trade secret protection." This balance has been struck, said OSHA, by "narrowly [defining] the circumstances under which specific chemical identity must be disclosed and to authorize the use of confidentiality restrictions that are necessary to protect the value of the trade secret to its holder.[77]

OSHA also considered the legal question whether it had statutory authority to require not only that employers disclose hazard information to their employees but also that manufacturers disclose the information to employers to whom the chemicals are shipped, so that this information will be communicated to employees of these downstream employers.[78] In concluding that statutory authority existed, OSHA relied primarily on *American Petroleum Inst. v. OSHA,*[79] the decision vacating the benzene standard, in which the Court of Appeals for the Fifth Circuit said:

> Placing the responsibility to warn downstream employees of concealed hazards on those upstream employers who create the hazards and know of the hazards is consistent with the remedial purposes of the Act and is within OSHA's broad authority to prescribe warning labels.[80]

The AFL-CIO was not satisfied with the "fine and creative" balance that OSHA struck in the hazard communication standard, however. It criticized the standard as being "loophole-ridden"[81] and petitioned for review of the standard in the Court of Appeals for the Third Circuit.[82]

[76] 48 FED. REG. at 53,312.

[77] *Id.* See §15 of the Act, 29 U.S.C. §664, "Confidentiality of Trade Secrets." Procedures for the handling of trade secrets during inspections appear in 29 C.F.R. §1903.9 (1983). Trade secret issues arise frequently in litigation over OSHA citations. See, e.g., the discussion of *Reynolds Metals Co.,* 6 OSHC 1667 (Rev. Comm'n 1979) and related cases in Chapter 15, Section A.

[78] 48 FED. REG. at 53,322.

[79] 581 F.2d 493 (5th Cir. 1978), *aff'd on other grounds sub. nom. Industrial Union Dep't, AFL-CIO v. American Petroleum Inst.,* 448 U.S. 607 (1980).

[80] 581 F.2d at 510. The issue of the preemptive effect of the hazard communication "standard" is discussed in Chapter 20, Section C.

[81] AFL-CIO *News,* Nov. 26, 1983 at 1.

[82] *United Steelworkers of America v. Auchter,* Nos. 83-3554, 3561, etc. (3d Cir. Nov. 22, 1983). Three states have intervened in the proceeding. See discussion in Chapter 20, Section C.

Part V

OSHA and the States

19

Another Uneasy Partnership: OSHA, the States, and the Benchmarks Litigation

A. Background

Section 18[1] of the Act sets forth the legal framework for state participation in the OSHA program. The basic elements in this framework are as follows: state jurisdiction over occupational safety and health is generally preempted by the federal program;[2] if a state wishes to exercise occupational safety and health jurisdiction, it must submit to the federal OSHA a plan meeting the requirements of §18(c);[3] the most important of these requirements is that the state standards and their enforcement be "at least as effective" in providing safe and healthful working conditions as the federal program;[4] in particular, the states must have "qualified personnel necessary for the enforcement of such standards."[5] If OSHA approves the state plan, state preemption is avoided, and the state may enforce its program; for a period of at least three years, the federal OSHA has "concurrent" enforcement authority in that state.[6] OSHA is authorized to pay up to 50 percent of the cost of an approved state plan[7] and is required to continue to monitor the effectiveness of the state plan.[8] At any time after three years from approval, OSHA may grant "final approval" to the state plan, at which time most federal enforcement authority ends, unless it is restored after a formal proceeding withdrawing state plan approval.[9] Finally, in order to avoid preemption

[1] 29 U.S.C. §667 (1976).
[2] Sec. 18(a), 29 U.S.C. §667(a). See discussion pp. 666–67.
[3] 29 U.S.C. §667(c).
[4] Sec. 18(c)(2), 29 U.S.C. §667(c)(2).
[5] Sec. 18(c)(4), 29 U.S.C. §667(c)(4).
[6] Sec. 18(e), 29 U.S.C. §667(e).
[7] Sec. 23(g); 29 U.S.C. §672(g).
[8] Sec. 18(f); 29 U.S.C. §667(f).
[9] Secs. 18(e)–(g); 29 U.S.C. §667(e)–(g).

of state authority while the states are preparing to submit their plans and federal OSHA is developing a national program, §18(h)[10] authorizes OSHA to enter into agreements allowing states to continue their enforcement activity for a period of no more than two years after the date of enactment of OSHA. As of January 1984, there were 24 states and jurisdictions with approved plans.[11]

Labor unions have traditionally been skeptical of the ability of states to mount effective programs, and have generally opposed transfer of federal enforcement authority to the states.[12] Their strongest opposition, of course, has been directed to those state programs they view as being significantly less effective than federal OSHA; one such state is Indiana.[13]

Business groups have viewed state programs with various levels of confidence. Richard F. Boggs, Director of Occupational Safety and Health of the Organization Resources Counselors, presented a generally favorable view to state programs at a congressional hearing in 1981.

> To summarize, the success of the State programs requires the meeting of certain basic criteria. There must be consistency of standards nationwide. The level of funding for the programs, both from Federal and State sources, must be adequate to attract and hold skilled administrative and technical personnel. There must be active Federal monitoring to insure a consistent framework. It is our view that this continuing Federal monitoring is necessary to prevent many of the problems described above. In particular, noncomparable regulations and erratic enforcement must be minimized.
>
> I should also emphasize that State occupational safety and health enforcement programs can have a significant advantage over Federal programs when properly coordinated. The States should be more famil-

[10]Sec. 18(h); 29 U.S.C. §667(h).

[11]One of the states, Connecticut, has an approved plan covering only its public employees. 29 C.F.R. §§1956.40–.43 (1983). OSHA regulations on state plans applicable to public employees only appear at 29 C.F.R. pt. 1956 (1983).

[12]The proper role of the states was a significant issue during the early development of federal occupational safety and health legislation. Testifying in 1968, Secretary of Labor Willard Wirtz said that the record of state programs was one "of a significant variability in results"; he urged that a major role be assigned to the federal government in the area. *Hearings on S. 2864 Before Subcomm. on Labor of the Senate Comm. on Labor and Public Welfare,* 90th Cong., 2d Sess. 72 (1968), reprinted in Chapter 1. Employer representatives, on the other hand, arguing in 1968 against federal intervention, stated that the "perfunctory dismissal" of state occupational safety and health programs was a "superficial and unfair judgment." These representatives endorsed the provisions of the bills that authorized federal assistance to the states to encourage them "to measure up" to their responsibilities. See *id.* at 351 (Statement of Leo Teplow, Vice President, American Iron & Steel Inst.), reprinted in Chapter 1. The representative of the AFL-CIO was highly critical of the adequacy of the state programs, pointing out, for example, that in Arizona there were 55 game wardens and only 2 safety inspectors. *Id.* at 485 (Statement of Clinton M. Fair, Legislative Rep., on behalf of Andrew J. Biemiller, Director, Dep't of Legislation, AFL-CIO). The administration bills in 1968 authorized the Secretary of Labor to decline to exercise federal occupational safety and health authority over an issue when he determines that state enforcement would "reasonably carry out the objectives of the Act." S. 2864, reprinted, *id.* at 14. The AFL-CIO also criticized this formulation and urged, as an alternative, a provision that would have required the states to submit plans, to be approved by the Secretary, which "are or will be substantially as effective" as the federal program. *Id.* at 487. The proposal of the AFL-CIO was the basis for the state plan provision reported favorably by the Senate Committee of Labor and Public Welfare in 1970, S. REP. No. 1282, 91st Cong., 2d Sess. 18, 36 (1970), which was ultimately enacted as §18, 29 U.S.C. §667.

[13]OSHA began proceedings to withdraw its approval of the Indiana State Plan but later discontinued these proceedings. See discussion in Chapter 20, Section B.

iar with the industrial environment and the local needs of their own particular areas. Being closer to the work settings, they should be in a better position to provide consultation and assistance to address local needs.[14]

The states, on the other hand, have been critical of federal OSHA for not treating them as equal partners in the program and for imposing unnecessary and burdensome requirements on state programs. During its early years, OSHA actively encouraged the states to submit plans under §18. During the administration of Dr. Bingham, which began in 1977, state programs were scrutinized more carefully, and some withdrawal proceedings were initiated. Assistant Secretary Auchter stated at the beginning of his administration that he intends to "fully integrate the states into the overall OSHA program" and that "in the last analysis, local problems are best addressed by those closest to them."[15]

B. Developmental Plans and Preemption

In October 1971, OSHA issued comprehensive regulations on state plans.[16] These regulations established detailed procedures for the submission, approval, and rejection of state plans; they also included interpretations of the "at least as effective" criteria in the statute. One of the major innovations of these regulations was the policy on "developmental plans." Under this policy, OSHA would approve a state plan even though, when submitted, it did not "fully meet" the statutory criteria, so long as the state committed itself to meeting those requirements within three years. At the end of that period, if the developmental steps were met, the state program would be "certified," and the three-year intensive monitoring period leading to final approval would begin.[17]

The developmental plan policy was strongly objected to by unions, who believed that the approach was not authorized by the Act and was a step by federal OSHA towards abdicating its enforcement responsibility to the states. The issue of developmental plans was vigorously debated at the meeting of the National Advisory Committee on Occupational Safety and Health (NACOSH) at its meeting of September 24, 1971. Participants were Assistant Secretary George G. Guenther; John Sheehan, legislative representative of the United Steelworkers of America; and Associate Solicitor Benjamin W. Mintz.

[14] *State Implementation of Federal Standards: Hearings Before the Subcomm. on Intergovernmental Relations of the Senate Comm. on Governmental Affairs,* 97th Cong., 1st Sess. pt. 2 at 41 (1981). For a discussion of employer concern about the issuance under state plans of standards which are stricter than the federal standards, see Chapter 20, Section C.

[15] *Id.* at 25. For a critical discussion of the role of the states in the OSHA program, see Note, *Cooperative Federalism and Worker Protection: The Failure of the Regulatory Model,* 60 TEX. L. REV. 935 (1982) (arguing that states have "substantial incentives" to evade their responsibilities for worker protection: (1) it costs less not to enforce than to enforce; (2) vigorous enforcement can damage the local economy; and (3) enforcement officials are vulnerable to pressure from local business and industrial interests).

[16] 36 FED. REG. 20,751 (1971) (codified at 29 C.F.R. pt. 1902).

[17] 29 C.F.R. §1902.2 (1983).

Proceedings of National Advisory Committee on Occupational Safety and Health[18]

***

MR. GUENTHER: * * * [T]he concept of a developmental plan was created for the precise purpose of taking into consideration the kind of indefinite factors that * * * have [been] mentioned, the difficulty in moving legislation through legislatures, the difficulty in ensuring appropriations, and so forth. Given the fact that state legislatures do not meet regularly each year in many cases * * * we felt that it was required that we come up with this kind of approach to provide an opportunity for those states which would otherwise be denied such an opportunity by virtue of their legislative calendars to consider participation in the program.

A most important element in the developmental plan concept is the requirement that the Secretary shall, upon approving such a plan, at the same time indicate his intention with regard to ongoing, continuing, concurrent jurisdiction in that particular state during the development of that particular state plan.

If it is not clear, it should be made clear that it is our very definite intention in cases where states have developmental plans to continue our jurisdiction to ensure that in the interim, while that developmental action is taking place and is being watched, that protection will be provided at the same time.

* * * [A]t the conclusion of the three-year maximum developmental period when, assuming that the state has in fact met its stated timetables for the developmental plan, an additional period of one year of actual specific operations will be required and will be monitored by the Department to ensure that the plan not only has the statutory standards and other regulatory bases but that it is in fact an operational plan.

***

MR. SHEEHAN: Yes, Governor. [Chairman Pyle] On the development plan, it's a non-plan as far as I can understand it. And to the use the term "developmental plan" has been expressed on the basis of some assurance that the governor in presenting legislation to the legislature that that in itself will assure its passage, we all know that that doesn't always follow through. And I think the fact that the state may not be able to come in with a plan because the state legislature did not do so hits up against the deadline in the act. That may be unfortunate, but everybody was talking about what was in the act and what was not in the act. And if we allow the developmental plan to take the place of an actual plan, that certainly seems to be far beyond what the act is saying.

In addition, there is a section further on in the guidelines that, in order for the Secretary to take back jurisdiction which he may have given as a result of the developmental plan, he has to go through all the procedures that he would have had to go through if the state would have come in with a plan and failed to live up to it. In other words, they would have to have the whole question of a public hearing.

***

Recently the EPA, which is a sister agency of the government anyhow, in talking in terms of the recent clean air act of 1970, went to great extent in talking to its state administrators, the people that are going to relate to the federal act in terms of the state involvement with the clean air act. And they were very specific in saying you are faced with deadlines, statutory deadlines in your legislature that may or may not be acting. But there's Ruck-

[18]Washington, D.C., Sept. 24, 1971 at 119–26. On file with OSHA, Washington, D.C.

elshaus* saying I'm telling you now you get busy; if you need some help from us, we will provide help to the state legislature, technical assistance to tell the state legislature what is necessary for the state to come into compliance so that we can, in effect, cooperate with the state. And they don't talk in terms of developmental plans here. They talk in terms of plans that have to be submitted prior to the statutory deadline and that you had better get your state attorney general and you get your legislature in order and we'll help you out.

***

Certainly the comment that the Secretary shall maintain enforcement authority—he has it anyhow. So, there's nothing new here. He has it even when a plan is accepted. He can in a three-year period retain enforcement authority. I sort of suspect, therefore, that if it's stated here that the Secretary will not relinquish his enforcement authority [when he approves a] developmental plan, [he's] going to exercise [his] discretion and relinquish all enforcement authority where there are non-developmental plans and that also would be a tragedy.

***

CHAIRMAN PYLE: We have had a rather interesting exchange here. Mr. Mintz would like to make a contribution.

MR. MINTZ: What I think is a contribution.

On the subject of developmental state plans, the discussion seems to be proceeding almost as though this is an idea, a strategy, which we thought ourselves. In fact, it has a rather clear base in the statutory language. I refer you specifically to Section 18(c)(2). The plan provides for the development and enforcement of safety and health standards and so forth [by the states], which standards and the enforcement of which standards, * * * *are or will be* at least as effective in providing safe and healthful employment as those under the federal act.

* * * Also in this connection, 18(c)(4) contains satisfactory assurances that such agency or agencies have or will have the legal authority and qualified personnel necessary for the enforcement of such standards. To give some meaning to the words of the law, the language "or will," looking to the future more than suggests, almost dictates we think, a construction that the state plan may be acceptable if it does not now—are at least as effective—but will be as effective, if the state is able to produce the assurances that will convince us that at some point the state procedures and the standards will be as effective as those under the act.

***

OSHA's purpose in adopting the idea of developmental plans was to encourage the submission of state plans, to facilitate their approval, and thus to allow the states to begin implementing their own programs, with 50 percent federal support, at the earliest time. In response to objections by unions, OSHA representatives argued that no prejudice to employees would result from the approval of state plans at a developmental stage since federal enforcement would continue undiminished during the developmental period. This commitment, however, was not honored.[19] OSHA's goal in encouraging the

*[Author's note—William D. Ruckelshaus at that time was Administrator of the Environmental Protection Agency. He resumed that post in 1983.]

[19]See Section C, this chapter. OSHA's policy of approval of developmental plans was upheld by a U.S. district court in *Robinson Pipe Cleaning Co. v. Department of Labor & Indus.*, 2 OSHC 1114 (D.N.J. 1974).

submission of state plans was largely achieved, and, as 1972 drew to a close, most states, the District of Columbia, and four territories had submitted plans to OSHA for approval. At about that time, however, OSHA developed an even more controversial strategy. The policy was to extend the effectiveness of §18(h) agreements by means of temporary order.

After the Act was passed, 47 states and various jurisdictions had signed 18(h) agreements, allowing them to continue enforcement pending approval of their plans; however, even with the policy of developmental plans, most states did not have approved plans by December 28, 1972 (the cutoff date of 18(h) agreements). Thus, the states were threatened with preemption of their enforcement activity. To avoid this preemption, OSHA proposed regulations that would preserve state enforcement by issuing temporary orders extending the effectiveness of the 18(h) agreements. These regulations were strongly opposed by union representatives, and again OSHA's National Advisory Committee provided a forum for one of the sharpest legal and policy debates in the history of OSHA.

Proceedings of National Advisory Committee on Occupational Safety and Health[20]

***

MR. GUENTHER: I would just like to say, by way of general comment on the proposed rule on Temporary Orders, that this matter has received a great deal of attention from us, from all angles.

***

* * * [I] would say that the two-year period, the basic two-year period, that Congress contemplated, as it turns out, was not a bad estimate. But, as in all estimates, it was off by a matter of months, so far as we are concerned, in terms of the time necessary to conceive, put together, and submit State plans on the State's part, and for us to conduct the exhaustive review process necessary to make the kind of determinations which stem from the general, at least as effective, requirements.

What became clear to us was the fact that unless some action were taken, as of the end of this calendar year, the 28th of December, we would find ourselves in a position where quite a substantial number of States' estimated plans, some of which may have passed through the review process and been approved or disapproved, but the great majority of which would not, by that point in time, have been able to be given the kind of thorough review that we consider necessary before final action on a plan is taken.

My judgment is that the additional amount of time for review of the substantial number of plans coming in is a matter of a few months.

***

We are concerned, however, with the fact that if we follow an absolutely strict interpretation of the Act, as it stands now, that between the 28th of December and the time of approval or disapproval, as the case may be—but particularly assuming approval, and there will undoubtedly be approvals—that there would be a protection gap in terms of the activities of States who, for all practical purposes, are prepared to launch upon activities under a plan but who, within absolutely strict interpretation, could be put out of

[20]Washington, D.C., Nov. 16, 1972 at 44–46, 81–85. Copy on file with OSHA, Washington, D.C.

business as far as State programs are concerned between the 28th of December and the commencement of activities under approved State plans for a very brief interval of time.

We, therefore, conceived of the proposed rule involving temporary orders whereby States, who have submitted a plan to us, can overcome this apparent protection gap subsequent to the 28th of December and prior to the approval of a plan by way of authority under a Temporary Order.

And our primary justification for this approach is what we consider to be the basic intent of the Act, and that is to provide continuance and growing protection for American workers.

In our judgment, we would be guilty of subverting a most basic and fundamental intention of the Act if we permitted the situation to exist whereby States who were operating in good faith and we, at the Federal level, were operating in good faith, to move in the direction of State participation as provided for by the Act under legitimately submitted State plans, were to permit a protection gap period of a few months to exist, during which, no State activity would be ongoing to continue worker protection.

***

MR. SHEEHAN:

***

Now, it seems to me that the temporary orders are being used as—the word "guise" [is] too hard—justification, certainly, for the continued inability of the administration, OSHA administration, to meet its responsibility in the field.

***

Another comment that I want to make has reference to the gap. You will pardon my abruptness—I'll say, "what gap?" We don't maintain there will be a gap, if the States are preempted on December the 28th.

As a matter of fact, the original Act of the Congress would have preempted that State, on the date it was signed or the date it became effective, I would assume. 18(h) had the other purpose we were talking about.

We do not maintain that workers' interests will be affected by the States being unable to enforce these functions and rushing them through. There is on-site consultant service here, but that would be extraneous.

In the State of Mississippi, there are two inspectors in the whole State. We do not feel there will be a loss of protection. As a matter of fact, and I think it is to the credit of the OSHA administration, the massive attack on the Act was in the area of the effectiveness under which it was enforced, and I think that is to the credit of the administration.

We do not feel that there would be a gap, and many of the functions that the States are forming—Iowa was mentioned—certainly can continue to be performed and the State can, at any time, come in and assume the inspection functions.

***

I must say that our main preoccupation here is not just for some minor proposal of the administrators but, rather, extends to the fact that it both indicates a philosophy of long-term attitude about the Act and the fact that it will now be used again as a justification for the lack of requesting funds from the Congress to provide the work force of the administrators—of the administration—to get into the work places.

We feel it very seriously, Mr. Chairman, and I should like to put a motion on the table that the Advisory Committee expresses its disagreement with this particular proposal.

Unfortunately, I think we are in that area of it, because it is very evident the Committee was going to be asked its advice and so we might as well, in a very formal way, through a motion, express it.

***

Mr. Sheehan's motion that NACOSH disapprove temporary orders was not passed;[21] the committee, however, adopted a recommendation that the temporary orders be valid for only six months, rather than a year as proposed. Despite the strong union opposition, OSHA decided to issue the rule authorizing temporary orders for six months.[22] Immediately, the AFL-CIO and the United Steelworkers of America sued in federal district court to have the new rule set aside as beyond OSHA authority. The unions' complaint stated that the nation's working men and women face "irreparable injury" because of the temporary orders. The states, the complaint asserted, would be "permitted to continue to enforce their own laws even though they have not yet been approved by the Secretary and, therefore, there is no assurance that the standards and enforcement procedures are at least as effective as those required by OSHA."[23]

On January 2, 1973, Judge Barrington Parker decided that the injunction, as requested by the unions, should be granted.

AFL-CIO v. Hodgson

1971–1973 OSHD (CCH) ¶15,353 (D.D.C. 1973)

PARKER, District Judge:

***

4. OSHA was enacted on December 29, 1970 (29 U.S.C. Section 651; Sec. 34, Pub. L. 91-596). It represents the most comprehensive piece of federal legislation enacted to date covering the subject area of occupational safety and health. It is designed specifically to achieve on a uniform, nationwide basis the far reaching goal of "assur[ing] so far as possible . . . safe and healthful working conditions" for all employees working in establishments engaged in interstate commerce (29 U.S.C. §651(b)).

5. Among the important factors which led to the enactment of OSHA was the recognition by Congress that federal action was needed with respect to this subject area because State regulation historically had proven to be totally inadequate.

6. The Act authorizes and directs the Secretary of Labor to perform a broad spectrum of preventive and enforcement functions which, in turn, provide workers with substantive protections with respect both to the promulgation and enforcement of standards which significantly exceed those contained under state laws. 29 U.S.C. §§651(b)(3), 655, 657, 658, and 659.

7. Section 18(a), OSHA, provides that once the Secretary promulgates a federal standard, federal jurisdiction over the implementation and enforcement of that standard is exclusive; State agencies and courts are pre-empted from asserting jurisdiction under State law over any occupational safety or health issue with respect to the subject matter encompassed by that federal

[21]The procedure adopted by OSHA at the NACOSH meeting was strongly criticized by the Health Research Group. Letter from Bertram R. Cottine, Staff Associate of Public Citizen Health Research Group to Assistant Secretary of Labor George Guenther (Oct. 23, 1972) (criticizing OSHA for "effectively" foreclosing "essential rebuttal" at the meeting and making it a forum for the Office of the Solicitor "to persuasively *argue* the merits of the case").

[22]37 FED. REG. 25,711 (1972).

[23]Complaint for Injunctive Relief and Declaratory Judgment at ¶22, *AFL-CIO v. Hodgson*, 1971–1973 OSHD (CCH) ¶15,353 (D.D.C. 1973).

standard. On April 12, 1971, the Secretary of Labor published in the Federal Register certain regulations covering "Procedures for State Agreements" wherein he stated: "Section 18(a) of the Act is read as preventing any State agency or court from asserting jurisdiction under State law over any occupational safety or health issue with respect to which a Federal Standard has been issued under * * * the Act" (29 C.F.R. 1901.2).

8. The Act provides two exceptions to exclusive federal jurisdiction after federal standards are promulgated—one is a temporary alternative (Section 18(h), OSHA) which expired by its terms on December 28, 1972, and the other is a permanent alternative (Section 18(b), OSHA) which requires approval of a State plan which satisfies certain criteria set forth in 18(c) to assure that the plan is "at least as effective as" the protections accorded under the Act.

(a) The Secretary of Labor stated in the regulations covering "Procedures for State Agreements" published in the Federal Register on April 12, 1971: "Section 18(h) permits the Secretary to provide an alternative to the exclusive federal jurisdiction on such occupational safety and health issue. This alternative is temporary . . . in that [Agreements under 18(h)] cannot continue beyond December 28, 1972, or the date of approval of a State plan under Section 18(c), whichever occurs first (29 C.F.R. §§1901.2 and 1901.5(b)).

(b) On August 16, 1972, the defendants reiterated this same analysis of the Act in a "Program Directive" sent to the Regional Administrators and Assistant Regional Administrators for State Programs.

9. On October 3, 1972, defendant Guenther published in the Federal Register (37 F.R. 20728) a notice of a proposed rule which would authorize the Secretary to issue "temporary orders which would preserve State authority to enforce standards covering occupational safety and health issues contained in a proposed plan submitted to the Assistant Secretary of Labor for Occupational Safety and Health for approval under Part 1902 of Title 29, Code of Federal Regulations." Under the proposal any temporary order would remain in effect until the State plan was approved or one year from date of issuance, whichever period is shorter.

In the explanatory comments which accompanied the Notice, defendant Guenther stated (37 F.R. 20728):

> The temporary order would have the effect of preserving State authority to enforce, in accordance with State Law, existing standards covering issues contained in the proposed State plan. Further, the order would allow the States to enforce any additional such standards promulgated in accordance with the provisions of the proposed plan and to utilize any additional enforcement authority which becomes effective within the scope of the plan.

10. Thereafter, on December 2, 1972, defendant Guenther, acting on behalf of the Secretary of Labor, published a formal regulation in the Federal Register (37 F.R. 25711–25712) which added a new Section 1902.16 ("Temporary orders") to Part 1902—State Plans for the Development and Enforcement of State Standards. In publishing this regulation defendant Guenther purported to act essentially in pursuance of "the general regulatory authority of section 8(g)(2) of the Act [29 U.S.C. §657(g)(2)]."

The regulation contained essentially the same provision as that quoted above from the notice of a proposed rule except that the maximum period of time that any such temporary order could remain in effect was shortened to six (6) months. Pursuant to §1902.16(b):

> To be the subject of a temporary order under this section, a proposed State plan must have been approved by the Governor of the State

> * * *. The proposed plan must also *address itself* to each of the criteria for State plans and indices of effectiveness prescribed in Sub Part B of this Part 1902. The temporary order may be issued ex parte (emphasis added).

A temporary order does not constitute approval of a State plan.

11. Objections to the proposed rule were voiced by Senator Harrison A. Williams (D.-N.J.), one of the authors of OSHA, the Industrial Union Department, AFL-CIO, and numerous affiliated International Unions, including plaintiff Steelworkers. The objectors urged that the proposed rule exceeded the Secretary's authority because it was directly contrary to the express language of the Act which mandated exclusive Federal jurisdiction after December 28, 1972, unless a State plan has been approved by that date.

12. As a consequence of the promulgation of §1902.16 the States with unapproved plans will continue to enforce their own occupational safety and health standards resulting in delay of the application of the more stringent and effective approach to occupational safety and health provided under the Act with respect to both standards and enforcement. (Affidavit of John J. Sheehan, Legislative Director of the United Steelworkers of America, AFL-CIO).

13. At present approximately 36 States are awaiting decisions on submitted plans and there are indications that such decisions might not be forthcoming within the next six months. (Affidavit of Chain Robbins, Administrator of Occupational Safety and Health Administration, United States Department of Labor). To date the plans of only two States, South Carolina and Montana, have been finally approved.

14. The issuance of an injunction would not seriously disrupt State enforcement of occupational safety and health programs over which they retain statutory jurisdiction. The "Program Directive" issued by the Administrator, Occupational Safety and Health Administration sets forth several important activities which would not be affected by the Secretary's implementation of his responsibilities under the statute. Moreover, under §7(c)(1) of the act, there exists an alternative method whereby the Secretary can utilize the services of State inspectors to supplement the federal compliance effort under federal standards. Pursuant to those provisions the Secretary has already negotiated agreements with 14 States. By virtue of such agreements, 98 selected and trained State safety and health inspectors are already empowered to conduct inspections for and under the direction of the federal government consistent with the provisions of OSHA.

***

A preliminary injunction should be granted.

OSHA did not pursue an appeal from Judge Parker's decision, and temporary orders were abandoned.[24]

[24]OSHA's argument in justifying temporary orders was that absent an approved state plan, state enforcement would be preempted. There have been court decisions on this issue of preemption of state enforcement under §18(a), 29 U.S.C. §667(a), where the state does not have an approved plan. See, e.g., *Columbus Coated Fabrics v. Industrial Comm'n of Ohio,* 1 OSHC 1361 (S.D. Ohio 1973), *appeal dismissed,* 498 F.2d 408 (6th Cir. 1974) (holding that federal OSH Act preempts Ohio enforcement, and preliminary injunction should issue); *Five Migrant Farmworkers v. Hoffman,* 1975–1976 OSHD (CCH) ¶20,057 (Sup. Ct. N.J. 1975) (holding that New Jersey has "neither the authority nor the obligation" to make preoccupancy inspection of farm labor camps in light of fact that N.J. state plan had been withdrawn).

C. Approvals and Concurrent Enforcement

One of OSHA's first approvals of a state plan was that of South Carolina's on December 6, 1972.[25] The preamble to the approval decision discussed various issues relating to whether the plan was "at least as effective" as the federal program. One of the more difficult issues was whether the South Carolina counterpart to the 11(c) program was equally effective, even though employees had to press their own claims of discrimination; under federal OSHA, the government initiates §11(c) suits. The preamble concluded, with some effort, that the South Carolina remedies were "as effective as Federal remedies." The preamble also listed the developmental steps that South Carolina must take to achieve certification, including various legislative amendments that had to be adopted. And, finally, the decision on approval stated that federal concurrent enforcement activity would continue undiminished in South Carolina.

Notice of OSHA Approval of South Carolina Plan

37 Fed. Reg. 25,932–33 (1972)

***

1. *Background.* The South Carolina State plan was submitted to the Assistant Secretary on May 8, 1972. Notice of receipt of the plan was published in the *Federal Register* on May 24, 1972 (37 F.R. 10535). A public hearing on the plan was held on July 10 and 11, 1972, in Columbia, S.C.

South Carolina submitted amendments to the plan on September 13, 1972. Notice of receipt of the amendments and an invitation for public comments thereon was published in the *Federal Register* on September 28, 1972 (37 F.R. 20289). Comments on the amended plan were received from the American Federation of Labor-Congress of Industrial Organizations (AFL-CIO).

***

2. *Issues.* The public comments and the national office review of the plan raise * * * significant issues. The first is whether South Carolina has an adequate sanction against unauthorized advance notice of inspections. Section 1902.3(f) of Chapter XVII of the Code of Federal Regulations states that it shall be a criterion for State plan approval that the plan contain a prohibition against unauthorized advance notice of inspections. The South Carolina plan has such a prohibition.

The requirement of a prohibition has been interpreted to mean a prohibition backed up by a reasonable sanction. The South Carolina plan does have sanctions. The Commissioner of Labor may discipline any State employee who gives unauthorized advance notice. A $1,000 civil penalty may also be assessed against any employer who violates the prohibition.

These sanctions differ from the Federal criminal sanction of $1,000 or 6 months imprisonment or both for any person who violates the advance notice prohibition. However, South Carolina does have a prohibition against

[25] 37 FED. REG. 25,932 (1972).

advance notice which is backed up by a reasonable sanction applicable to the classes of persons who would most likely be in a position to give unauthorized advance notice, State employees and employers. Hence, the plan meets the requirement of §1902.3(f).

***

Another significant issue is whether South Carolina provides adequate remedies to employees who suffer discrimination because they exercise rights afforded them under the State occupational safety and health program. Under the Federal program, an employee who believes he has suffered such discrimination may file a complaint with the Secretary, who shall make an appropriate investigation. If the investigation reveals a violation, the Secretary shall bring an appropriate action in a U.S. district court. The district courts shall have jurisdiction to restrain violations and order all appropriate relief including reinstatement with back pay.

In South Carolina, employees who suffer discrimination will have the same remedies available. They will also be entitled to the additional remedy of punitive damages where appropriate. However, employees will have to press their own claims either in a Court of Common Pleas or in an administrative hearing before the Commissioner. This may require employees to make an expenditure for attorneys fees. The South Carolina amendment will authorize recovery of attorneys fees by a successful litigant. However, the public comments express concern that the initial expenditure of pressing a claim and the uncertainty of recovery will discourage employees who suffer discrimination from seeking redress.

Upon full consideration, we do not agree with this argument. Arguably, a private right of redress is less effective than a public right. However, authorizing recovery of attorneys fees and punitive damages, and the presumption of unlawful discrimination of an employee who is adversely treated within 90 days of the exercise of any rights afforded to him under the plan are all factors which make the South Carolina discrimination remedies as effective as Federal remedies. Hence, the South Carolina system for sanctioning employees and providing relief to employees who suffer discrimination is effective and meets the requirement of §1902.4(c)(2)(v) of this chapter. Furthermore, the authority of the Secretary to bring action, under section 11 of the Act, to remedy discrimination against employees for exercising rights under State law included within the scope of the plan would not terminate, even upon determination by the Secretary under section 18(e) that a State plan, upon the basis of actual operations, is applying the criteria set forth in section 18(c). Such rights are related to the Occupational Safety and Health Act within the meaning of section 11. Appropriate rules will be issued in the near future dealing with the relationship between the Federal and State remedies.

Another significant issue is whether the South Carolina procedure for judicial review of administratively reviewed contested citations will comply with due process and afford affected employees an adequate right of participation. Under the South Carolina system, only parties to the administrative review will be afforded the right to initiate or participate in a subsequent judicial review. However, the plan amendments provide that "Any employer, employee or their representative shall have the right to appear as a party in any review proceedings * * *."

Hence employees and their representatives will always have the right to participate in judicial review if they have exercised their right to participate in the administrative review. This procedure neither violates due process nor constitutes an unreasonable requirement on employees.

3. *Decision.* After careful consideration of the South Carolina plan and comments submitted regarding the plan, the plan is hereby approved under section 18 of the Act and Part 1902.

This decision incorporates requirements of the Act and implementing regulations applicable to State plans generally. It also incorporates our intentions as to continued Federal enforcement of Federal standards in areas covered by the plan and the State's Developmental Schedule as set out below.

Pursuant to §1902.20(b)(iii) of Title 29, Code of Federal Regulations, the present level of Federal enforcement in South Carolina will not be diminished. Among other things, the U.S. Department of Labor will continue to inspect catastrophes and fatalities, investigate valid complaints under section 8(f), continue its target safety and target health programs, and inspect a cross-section of all industries on a random basis.

Within 9 months following this approval, an evaluation of the State plan, as implemented, will be made to assess the appropriate level of Federal enforcement activity. Federal enforcement authority will continue to be exercised to the degree necessary to assure occupational safety and health protection to employees in the State of South Carolina.*

***

OSHA policy at this time was "to seek the active participation of the states in bringing about safe and healthful working conditions."[26] The approvals of state plans, all of which were "developmental," continued. By the end of 1973, the plans of 22 states and jurisdictions had been approved; in 1974, four more approvals took place, making a total of 26.[27] At this time, there emerged yet another controversial state plans issue: the extent to which federal OSHA would continue to assert its concurrent enforcement authority in states with approved developmental plans. Although under the Act OSHA legally could discontinue its enforcement any time after a state plan was initially approved, OSHA had committed itself not to withdraw its enforcement while the approved state plans were still in the developmental stage, since the states concededly had not yet achieved equal effectiveness during this period.[28] In 1973, however, Assistant Secretary Stender expressed concern about redundant enforcement. His policy was to enter into "operational" agreements with the states. This new enforcement policy was articulated in the President's 1974 OSHA report in the following language:

> Generally, OSHA follows a policy of withdrawing Federal enforcement activity in States with approved plans (except with regard to com-

*[Author's note—Decisions by OSHA to approve state plans are published in the *Federal Register*. See, e.g., Approval of Virginia Plan, 41 FED. REG. 42,655 (1976). The approvals have been codified at 29 C.F.R. §§1952.100–.384 (1983). Each subpart in this codification describes the plan, states the level of federal enforcement, and lists the "developmental steps" in the state plan and when these steps are scheduled to be completed. The South Carolina plan appears at 29 C.F.R. §§1952.100–.104 (1983); the Virginia plan at 29 C.F.R. §§1952.370–.374 (1983).]

[26]The President's Report on Occupational Safety and Health at 25 (1973).

[27]The President's Report on Occupational Safety and Health at 39–40 (1974). Later, several states whose plans had been approved withdrew their plans, including New Jersey, New York, Illinois, and Wisconsin. This was due in part to budgetary considerations and partly to the campaign undertaken by the unions to fight state plans at the state level. In addition, large multistate employers that originally had supported state enforcement wavered in their support because of the burden of having to comply with varying state standards. See Note, *supra* note 15 at 945 n.66. On the issue of varying state standards, see discussion in Chapter 20, Section C.

[28]*Supra* pp. 620–21.

plaints concerning employee discrimination and recently promulgated standards prior to State adoption) even though the Federal Government has discretion in this regard pending successful completion by the State of the 3-year period of initial approval. Withdrawal is done on an issue-by-issue basis, upon OSHA determination that a State has achieved "operational status" with respect to a particular issue. An "issue" is an industrial, occupational, or hazard grouping for which Federal standards have been promulgated under the Act. Key elements in determining whether or not operational status has been achieved include the existence of required legislative action, standards at least as effective as those of the Federal program, a sufficient number of qualified enforcement personnel, and a method for review of enforcement actions. OSHA and the States formalize favorable determinations by concluding "Operational Status Agreements."

OSHA assistant regional directors are responsible for recommending when a State should be deemed operational in any or all issues under its plan. Such a recommendation is made after careful consideration of the degree to which the State satisfies the key requirements described and consideration of factors brought out by evaluation or which otherwise have come to the attention of the assistant regional director.

When, by evaluation, it is determined that the State is carrying out its program in the manner anticipated in the State plan and when Operational Status Agreement and final approval is granted, Federal enforcement staff not engaged in requirement monitoring activities may be redeployed to other States within the region or transferred to other regions. Sufficient Federal staff will remain in the State to discharge residual Federal enforcement and evaluation responsibilities.[29]

OSHA's new policy drew strong objection not only from labor organizations and public interest groups,[30] but also from a NACOSH subcommittee on state plans. The subcommittee, under the chairmanship of Frank R. Barnako,[31] issued a report criticizing Assistant Secretary Stender's policy of withdrawing federal enforcement.[32]

Report of the Subcommittee on State Programs

***

On October 29, 1971, the Secretary issued rules for state plans (29 C.F.R. 1902). These established the procedures by which the Secretary would be guided in approving or rejecting state plans and sets forth criteria for concurrent enforcement during the 3-year period as they apply to developmental and complete plans. An early Program Directive, #72-26, of September 26, 1972, established guidelines for the exercise of concurrent authority between the federal government and the state during the 3-year period described above. The Department also, in 1972, developed methods and procedures for evaluating state plan implementation and for monitoring. In substance, 29 C.F.R. 1902 and Directive #72-26, together with the system for monitoring state plans, established a method for determining the

[29]President's Report on Occupational Safety and Health at 42 (1974). See 29 C.F.R. 1954.3 (1983), for OSHA's statement of present policy on its exercise of discretionary authority.

[30]See, e.g., 3 OSHR 932 (1973) (criticism by Bertram R. Cottine on behalf of the Public Citizen Health Research Group).

[31]Mr. Barnako later became a Commissioner and Chairman of the Occupational Safety and Health Review Commission.

[32]Report on file with OSHA, Washington, D.C.

level of federal enforcement tailored to complement the state program. No date or series of events was established which would predetermine when federal enforcement would diminish or terminate during the interval between first approval and final determination.

At a meeting of NACOSH on December 6, 1973, the Assistant Secretary advised that a new method for determining the level of federal enforcement and date of withdrawal from the state was under consideration. This method contemplated that an approved state plan would include: 1. enabling legislation, including operational review procedures, 2. promulgated standards whether or not they have been approved as "at least as effective", and 3. competent enforcement personnel; OSHA enforcement would then cease and OSHA field personnel would shift to monitoring and training activity.

This was a significant departure from the guidelines of 29 C.F.R. 1902 and Directive #72-26, and some original approval notices published in the Federal Register, such as that of South Carolina. Rather than a gradual withdrawal based upon periodic evaluation, these three criteria would automatically trigger withdrawal of OSHA enforcement. The Committee questioned the wisdom of this policy and whether the Act permits implementation of this policy. On motion then this Subcommittee was requested to discuss the issues involved and to report to NACOSH.

***

To assist the Subcommittee, the Department prepared the * * * "Options on Exercise of Concurrent Federal Enforcement Authority". Four options are outlined. In brief these are:

A. A full federal enforcement for three years concurrent with state enforcement until an 18(e) determination has been made.

B. Upon commencement of enforcement by a state with enabling legislation, promulgated standards, and competent enforcement personnel, the federal authority would shift to monitoring and training without any federal enforcement.

C. Following an initial evaluation of actual operations under an approved state plan, the level of federal enforcement would depend on (1) the results of the evaluation and (2) the extent to which the state has completed its developmental steps. The level of federal enforcement would be "high", "moderate", or "low" dependent on what phases of the developmental plan are operational, and federal enforcement would cease at such time as the Secretary determines the state operations are "at least as effective". There would be a detailed examination of the state activity during this period of concurrent enforcement, and the level of federal enforcement would be modified on the basis of further evaluation of state activity, and completion of developmental steps, or correction of deficiencies in the operational as well as developmental aspects of the program.

D. On approval of a plan, federal enforcement would continue only in areas or issues not covered by the state plan. Federal enforcement on any issue covered by the approved plan would cease entirely. The federal role would be confined to monitoring.

The Subcommittee considered all four of these options. Plan A, it was found, would create a dual enforcement and inspection authority which would amount to harassment of employers. It would also result in a waste of manpower resources since even in a state with an effective inspection and enforcement program the federal level would continue, thus inhibiting release of federal compliance officers to states without a state plan or in which it was not effective.

Plan B, it was decided, would establish an instant and unrealistic removal of federal authority, without an adequate evaluation of the enabling legislation, the standards promulgated by the state, and an evaluation of the competency of state enforcement personnel.

Plan D, it was noted, removed federal presence, other than monitoring, without any evaluation of a plan.

Plan C would continue federal enforcement until an initial evaluation of actual operations under the state plan, and thereafter federal enforcement would diminish or increase consistent with that evaluation and the extent to which the state had completed its developmental steps. The Subcommittee recommends that the level of federal enforcement should be determined by (a) what is in fact operational in the state plan, (b) an evaluation of the effectiveness of the state plan, (c) an analysis of the number of, and competency of, state staff, including industrial hygienists, (d) a specific determination that the standards promulgated by the state are, in fact, "as effective as" the federal standards.

Despite NACOSH objection, Assistant Secretary Stender continued his policy,[33] and by the end of 1975 there were 13 operational agreements in effect in the 23 states with approved developmental plans. In 1982 the process was completed by Assistant Secretary Auchter, and operational agreements were in effect in *all* states with approved plans.[34]

Most of the major components of the federal-state OSHA relationship were in place by 1975. This rather complex structure, embodying both statutory provisions and a regulatory superstructure, was explained in some detail in a brief filed by OSHA in the Court of Appeals for the Tenth Circuit in 1981.

Brief of the Secretary of Labor[35]

***

2. STATUTORY AND REGULATORY BACKGROUND

***

B. RELATIONSHIP OF STATE AND FEDERAL ENFORCEMENT UNDER THE ACT; INITIAL APPROVAL OF STATE PLANS

The Federal Act generally preempts State enforcement of all occupational safety and health laws which deal with issues regulated by Federal OSHA. However, Section 18(b) of the Act provides an exception to this general rule of preemption, in that States desiring to assume responsibility for the development and enforcement of safety and health standards may submit a "State plan" for such enforcement. If the Secretary, applying criteria set forth in Section 18(c), and 29 C.F.R. Part 1902, finds that the state plan

---

[33]The criticism by NACOSH of the policies of Assistant Secretary Stender elicited a sharp response. In March 1974, Mr. Stender urged that NACOSH abolish its subcommittees and concentrate its attention on training and education issues. 3 OSHR 1324 (Mar. 28, 1974). The late Congressman Steiger responded, saying that a weakened NACOSH would be a "tragic mistake." 3 OSHR 1667 (1974). On the conflict between Assistant Secretary Stender and Congressman Steiger, see J. Mendeloff, REGULATING SAFETY 51 (1979).

[34]See Chapter 20, Section B.

[35]At 3–8, *Environmental Improvement Div. of the New Mexico Health and Environment Dep't v. Marshall,* 661 F.2d 860 (10th Cir. 1981).

provides or will provide for protection of employees at least as effective as Federal protection, initial Federal approval may be granted and the State may then conduct its own enforcement program, with concurrent Federal enforcement by OSHA. The statute specifically provides that, after initial approval, but before final approval is granted, the Secretary "may, but shall not be required to" exercise concurrent enforcement authority. After initial approval, up to 50 percent of the operating expenses of the State program are funded by Federal OSHA.

Prior to initial approval, a State plan must be submitted for an extensive review by OSHA to determine whether the plan meets the statutory criteria of Section 18(c). The plan must, among other things, designate a State agency responsible for administering the plan; assure that the State agency will have adequate funding, personnel, enforcement mechanisms, and legal authority to accomplish the Act's purposes; and assure that the safety and health standards enforced by the State are at least as effective as those enforced by Federal OSHA. If the Assistant Secretary determines that the plan meets such criteria, he may grant initial approval, and publish the decision in the Federal Register, setting forth OSHA's rationale and a description of the initially-approved State plan. If OSHA determines that a plan submitted by a State does not satisfy the statutory and regulatory requirements, a notice of proposed rejection is published, leading to a formal hearing pursuant to the Administrative Procedures Act. A State may seek judicial review of a rejection decision in the United States Court of Appeals for the circuit in which the State is located.

C. EFFECT OF INITIAL PLAN APPROVAL

The issuance of a decision granting initial approval initiates a second phase in the administrative process. Initial approval relates only to basic plan components, and after initial approval OSHA must monitor and evaluate the State's program to determine whether, on the basis of actual operations, the State's plan operates at the level of effectiveness required for final approval under Section 18(c) of the Act. Therefore, before final approval can be granted, the State must develop its plan beyond the level required for initial approval and must put into effect a fully developed, complete and effective safety and health program, see 29 C.F.R. §1953.10. OSHA must continually monitor State operations to determine the State's progress toward the effectiveness level required for final approval, and must provide evaluation reports on a regular basis, see generally 29 C.F.R. pt. 1954. The fact-finding and evaluation process required prior to a final approval determination must last a statutory minimum of three years, and in the case of a developmental state program requiring extensive modification and supplementation, the process may be considerably longer.[1] When monitoring discloses that some element of a program adversely affects State effectiveness, the State will be required to amend or supplement its plan, 29 C.F.R. §1953.30. In addition, States are required to keep pace with changes in the Federal enforcement program, see 29 C.F.R. §1953.20, and any substantive changes in a State's program must be approved by OSHA, see 29 C.F.R. §1953.40. Whenever the Assistant Secretary determines that the State, in administering its program, has failed to comply substantially with the State

[1]The term "developmental" refers to the fact that all essential components of a plan need not be in place before initial approval can be granted. States are permitted up to three years to develop elements such as legislative amendments, regulations, operations manuals, staffing levels, etc. A state's developmental schedule is set out in the initial approval notice. A State which has met all of its developmental commitments and assurances is "certified" and an appropriate notice published in the Federal Register and codified in the Code of Federal Regulations. See 29 C.F.R. §1902.33. After such certification the State program must be monitored for a minimum of one more year before final approval may be granted under [Section 18(e)] 29 C.F.R. §1902.34(a).

plan, a proceeding shall be instituted for withdrawal of initial plan approval. Grounds for such action would include, *inter alia,* failure to provide effective enforcement of State standards; failure of the State program to keep pace with changes and improvements in the Federal program; or failure to provide for the exercise by employees of employee rights provided under the Federal Act. Procedures for withdrawal are similar to those required for plan rejection, and are also reviewable by the Courts of Appeals. See [Section 18(f),] 29 C.F.R. pt. 1955.

D. CONCURRENT ENFORCEMENT

After initial approval of a State plan, and while OSHA monitors the developing plan for final approval, the Act provides for concurrent Federal/State jurisdiction over occupational safety and health issues covered under the State plan. This concurrent jurisdiction continues to exist until formal determination of final approval is made. Since the Act provides that "the Secretary may, but shall not be required to" exercise this authority, the nature and extent of Federal enforcement is a matter of discretion; however, because the Act requires the Secretary to "assure so far as possible every working man and woman in the Nation safe and healthful working conditions," this concurrent jurisdiction must be exercised to the degree necessary to provide maximum protection to employees in States whose programs have not yet reached the level of effectiveness required for final approval. In determining the appropriate level of Federal enforcement, OSHA must consider the effectiveness of State enforcement and the current worker protection needs in the State; and, as a practical matter, the availability of Federal personnel to conduct supplementary inspections.

If monitoring shows that a State program has advanced to a degree sufficient to justify a low level of Federal enforcement activity, regulations at 29 C.F.R. §1954.3 provide that OSHA may enter into an "operational status agreement" with the State, setting forth areas of Federal and State enforcement responsibility, 29 C.F.R. §1954.3(f). Any finding of operational status, and any procedural agreement based upon such a finding, must be approved by the Assistant Secretary. Notice of the agreement is published in the Federal Register, and a description of the level of Federal enforcement is published in the Code of Federal Regulations. Because of OSHA's continuing responsibility to protect workers while State plans are being monitored for final approval, operational status agreements may be terminated by the Assistant Secretary, and full Federal enforcement activity may be reinstated upon a determination that such action is necessary to assure occupational safety and health protection to employees, 29 C.F.R. §1954.3(c)(3).

E. FINAL APPROVAL

A formal determination of final approval pursuant to Section 18(e) of the Act may only be made in accordance with procedures found at 29 C.F.R. pt. 1902D, which provides for Federal Register notice of [the] proposed approval, public comment, and the opportunity for legislative type hearings before an administrative judge. A notice of the final determination is then published in the Federal Register and codified in the Code of Federal Regulations. Final approval may be withdrawn by OSHA upon a showing of substantive failure to comply with any provision of the plan. At the present time, no State plan has received final approval.

Environmental Improvement Division of the New Mexico Health & Environment Department v. Marshall[36] involved a challenge by New Mexico to OSHA's exercise of concurrent enforcement authority in that state. The New Mexico plan was approved as a developmental plan in 1975.[37] In May 1978, an acting OSHA regional director entered into an agreement with New Mexico which purported to withdraw federal enforcement authority in that state. In December 1978, the new regional director indicated that federal concurrent enforcement would continue, particularly since the state's developmental plan had not been completed. The state brought an action to compel OSHA to honor the initial agreement and to withdraw concurrent enforcement authority. The Court of Appeals for the Tenth Circuit upheld the district court's granting of summary judgment in favor of OSHA, concluding that under §18(e)[38] OSHA continued to have concurrent enforcement authority, and that the original agreement did not abrogate the Secretary's statutory authority to resume enforcement.[39]

D. "Benchmarks" Litigation

The so-called benchmarks litigation, which began in 1974 and continued until 1983, when the case was dismissed, involved the federal numerical requirements for state compliance personnel under approved plans. Section 18(c)(4)[40] requires that a state plan contain "satisfactory assurances" that the state will have "qualified personnel necessary for the enforcement" of the state's safety and health standards, described in §18(c)(2) as being "at least as effective" as federal standards. In construing §18(c)(4), OSHA had said that the Act required the states only to provide staffing and funding levels "at least as effective as" those that would be provided by the federal government in the absence of a state program. Since OSHA, because of budget restrictions, had only limited resources available for its program, this meant less stringent staffing requirements for the states than would optimally be required. The AFL-CIO challenged the OSHA interpretation, noting that the phrase "at least as effective" does not appear in §18(c)(4), and arguing that the state must

[36]661 F.2d 860 (10th Cir. 1981).

[37]29 C.F.R. §§1952.360–.364 (1983).

[38]29 U.S.C. §667(e).

[39]The legal effect of OSHA's having entered into an "operational agreement" with a state has arisen in the adjudication of citations and penalties before the Review Commission. In *General Motors Corp., Chevrolet Motor Div.*, 10 OSHC 1293 (Rev. Comm'n 1982), the employer contended that while the case was pending before the administrative law judge, OSHA determined that the developmental plan was "operational" and, therefore, the "jurisdictional" limitations of §18(e) were "triggered." The Review Commission rejected the contention, concluding that this determination of "operational status did not limit OSHA authority as a matter of law under section 18(e)." *Id.* at 1296. See also *General Motors Corp., Central Foundry Div.*, 8 OSHC 1298 (Rev. Comm'n 1980).

[40]29 U.S.C. §667(c)(4).

provide adequate funds and staff "to ensure that normative standards are in fact enforced."[41] The Court of Appeals for the District of Columbia Circuit, in an opinion by Judge Leventhal, agreed with the union and, in one of the major OSHA decisions during the 1970s, held that, while the existing approvals were valid, the personnel and funding benchmarks must be "part of a coherent program to realize a fully effective enforcement effort at some point in the foreseeable future." The court remanded the case so that the Secretary could establish criteria for the states in accordance with the decision.

AFL-CIO v. Marshall

570 F.2d 1030 (D.C. Cir. 1978)

LEVENTHAL, Circuit Judge:

***

A. BACKGROUND

There are two stages to the state plan approval process, initial approval and final approval, and the implementation of an approved plan is monitored at all times. Section 18(c) of OSHA sets out criteria for determining plan acceptability. "The Secretary shall approve the plan submitted by a State * * * if such plan in his judgment" satisfies the stated criteria. These initial approvals do not necessarily cede federal jurisdiction to the state nor affect the federal enforcement program's operation within the state. Instead, there is a concurrent jurisdiction period of at least three years during which the Secretary monitors the state program for compliance with §18(c) and may enforce the federal program. See 29 U.S.C. §667(e), (f) (1970).

This case requires us to focus on the standards for evaluating state plans set forth in §18 of OSHA, and particularly the requirement of (c)(2), that the state standards be "at least as effective" as the federal standards, and of (c)(4) and (5), that there be adequate assurances that the state agency administering the plan will have sufficient "qualified personnel" and "adequate funds" to enforce the state standards.

The AFL-CIO challenged the regulations of the Secretary that interpreted "adequate funds" and "qualified personnel necessary for enforcement of [the state] standards." It claimed the regulations only parroted the language of the statute, established no rational criteria and guidelines for evaluating the sufficiency of a state's plan in terms of effective inspection and enforcement, and have resulted in state plans with wide disparities in manpower commitments and fund allotments.

The District Court upheld the Secretary of Labor's regulations on motions for summary judgment. It held that "the Secretary has promulgated rational, ascertainable standards for personnel and funding." The District Court interpreted those standards to require that the state effort be at least comparable to what the federal effort would have been in the absence of an approved state plan. The District Court also held that the Secretary of Labor had consistently and properly applied those standards. The AFL-CIO has appealed that decision to this court.

***

The AFL-CIO contends that the Secretary exceeded the scope of his authority by interpreting the requirements of §18 to mandate only the provision of staff and funding levels "at least as effective as" the Federal en-

[41]The AFL-CIO argument is stated in *AFL-CIO v. Marshall*, 570 F.2d 1030, 1034 (D.C. Cir. 1978).

forcement program. It emphasizes that (4) and (5) do not contain the "at least as effective as" criterion explicitly used in (2) and (3) of §18(c) and it argues that "adequate funds" and "qualified personnel necessary for enforcement of such standards" mean force and funding levels sufficient to ensure that the normative standards are in fact enforced.

To demonstrate the invalidity of use of an "at least as effective as" standard for subsections (c)(4) and (5), the AFL-CIO notes that the federal benchmarks that have been employed by the Secretary are predicated on federal enforcement levels that are artificially low because the Secretary deliberately withheld commitment of adequate resources until he knew the full extent of likely state participation. It cites testimony given by Assistant Secretary of Labor George Guenther before the Senate Subcommittee on Appropriations on April 16, 1972.

> [W]e have consciously attempted to control the level of our staff resources until we get a better reading of State participation.
>
> We have had some pressure on us, as you are aware, to substantially increase our forces immediately. We have refrained from requesting those kinds of increases for the reason that we don't believe that it's responsible to beef up the Federal program to such a level, and then find the States coming onboard, and have to withdraw from that staff level.

The Secretary maintains that this attitude was prompted by a lack of funding and "by Congress' repeated expressions of concern regarding possible State dislocations produced by precipitous Federal movement in this delicate area."

The Secretary argues that subsections (4) and (5) must be read in light of subsection (2), which he characterizes as the "overriding criterion." Subsection (2) states that the Secretary shall approve the state plan if it, *inter alia,* "provides for the development and enforcement of safety and health standards relating to one or more safety or health issues, which standards (and the enforcement of which standards) are or will be at least as effective in providing safe and healthful employment and places of employment as the standards promulgated under section 6." The Secretary maintains that with the inclusion of the parenthetical phrase, Congress indicated that state plan approvals were to be governed by direct comparison of state funding and personnel with that of the federal program which would otherwise be operative.

C. COURT'S VIEW OF OSHA §18(C)

After consideration of the text, legislative history and purpose of OSHA, we reach a conclusion as to the meaning of the statute that is different from that of either party—but that comports, we think, both with an effective enforcement program and a dynamic federal-state partnership in occupational health and safety matters. We hold that in referring to personnel and funding levels in terms of adequacy or sufficiency, Congress clearly intended that the states assure effective enforcement programs. In granting initial approval to state plans, the Secretary can consider—as interim federal benchmarks—whether the state is willing to provide personnel and funding "at least as effective as" that provided by the federal government. But such interim federal benchmarks must be part of an articulated, coherent program calculated to achieve a fully effective program at some point in the foreseeable future. As we discuss in part D of this opinion, the interim federal benchmarks that have been used for initial approval of state plans are not part of such a program. To comply with the mandate of the statute, the regulations and program directives establishing the benchmarks must be supplemented.

We now set forth the several pertinent considerations that have led us to our interpretation of §18(c).

1. STATUTORY LANGUAGE

We think, along with plaintiffs, that there is significance in the fact that the "at least as effective as" criterion was explicitly set forth in §18(c)(2) as to *standards* in state plans, but was omitted when Congress moved from standards to enforcement in subsections (4) and (5). The omission of the "at least as effective as" language from (4) and (5), dealing with personnel and funds, is the more conspicuous in view of insertion of these words in subsection (3). The Secretary's argument that the use of the "at least as effective as" language in subsection (2) is an overriding criterion, with no need for explicit repetition in (4) and (5), is difficult to reconcile with the existence of such repetition in (3) of §18(c).

In interpretation, we must serve the legislative purpose. The purpose of OSHA, as explicitly set forth in §2(b)(10), is "to assure so far as possible every working man and woman in the Nation safe and healthful working conditions and to preserve our human resources * * * by providing an effective enforcement program." This clear statement of purpose in §2 gives context for the use of such terms as "adequate" and "necessary" in §18(c), the context of providing "an effective enforcement program." "It is our duty 'to give effect, if possible, to every clause and word of a statute' * * * rather than to emasculate an entire section."

2. LEGISLATIVE HISTORY

It is a well-known canon of construction that the language of a statute is the best indication of legislative intent. *Browder v. United States,* 312 U.S. 335, 338 (1941). Here we find that the legislative history also comports with our interpretation of the statutory language.

Subsections (2), (4) and (5) of 18(c) moved from introduction to passage virtually without change, and were not the subject of extensive discussion in Congressional committees or in floor debates. The Senate Report states:

> Moreover, whenever a State wishes to assume responsibility for developing or enforcing standards in an area where standards have been promulgated under this act, the State may do so under a state plan approved by the Secretary of Labor. The plan must contain assurances that the State will develop and enforce standards at least as effective as those developed by the Secretary, that the State will have the legal authority, personnel and funds necessary to do the job * * *.

The House Report contains a virtually identical passage.

The language of both committee reports carefully distinguished between standards and the means to enforce them. As to the latter, once the standards were set there were to be assurances that the state would have the resources "necessary to do the job." The language straightforwardly expresses a Congressional intention that approved state programs assure sufficient resources to effectively enforce the standards set forth in those programs.

The foregoing recognizes a requirement for state plans. It does not, however, mandate a rigid approach toward state efforts that would make the perfect the enemy of the good. The legislative history develops another consideration of material importance—and that is time. Congress was aware that both the states and the federal government would need time to bring a program of this type on-line. Thus §18(c)(4) states that a state plan must give "satisfactory assurances that such agency or agencies *have or will have* the legal authority and qualified personnel necessary for the enforcement of such standards." Time is needed to staff up enforcement efforts. Moreover, switching jurisdiction from the state to federal authority entails significant transition costs. Concern was expressed on the floor of the Senate lest precipitate federal preemption mean the loss of the services of the state occupational safety and health machinery to the detriment of the overall effort.

3. OVERALL ASCERTAINMENT OF LEGISLATIVE PURPOSE

If we were controlled by the text alone, it might be argued that the "at least as effective as" concept is completely irrelevant for subsections (c)(4) and (5), where those words are omitted. An overall ascertainment of legislative purpose, however, bids us take into account practical necessities, including the concern for precipitate preemption brought out in the legislative history. In the application of a new federal statute, there is a likely Congressional contemplation of maximizing use of existing state machinery. "Federalism" cannot be invoked categorically where it would defeat the national objective. A sensitivity to state participation has value, however, where the state administration can be integrated with federal application—either as a supplement, or as a stopgap measure, or as a delegation under federal control. We have been instructed that a federal statute should not be given a rigid interpretation that would "prevent States from undertaking supplementary efforts toward [the] very same end." *New York State Department of Social Services v. Dublino,* 413 U.S. 405, 419 (1973).

Taking all these considerations into account, we hold that it is reasonable for the Secretary to employ federal "at least as effective" benchmarks under (4) and (5) of §18(c), if those objectives are integrated with the objectives of "necessary" personnel and "adequate" funds. The resultant of these vectors of analysis is the approval of current federal levels of personnel and funds as benchmarks for state plans, provided they are part of a coherent program to realize a fully effective enforcement effort at some point in the foreseeable future. These benchmarks must be pragmatic in terms of time frame but not lax in goal, and sustain state efforts that are reasonably adapted to realize the Act's objectives within the same time frame as the federal effort.

D. ANALYSIS OF REGULATIONS

***

In the last analysis, the "OSHA System for Enforcement Planning and Resource Allocation" does nothing to justify the standards employed in granting initial plan approval. It remains the case that however poor the federal effort, a state plan is approved if it is marginally better. For the years 1972 to 1974 this was clearly the case. For 1975, again, the state effort is compared with the federal effort "as is." The federal effort is now described as a percentage of what is really needed, but all that is required of an approved state plan is that its percentage be at least as high as that of the federal effort. Although we think Congress conceded the Secretary some practical flexibility in administering OSHA, this rudderless scheme of plan approval cannot be reconciled with the mandate that state plans assure "necessary" personnel and "adequate" funds. The Secretary must have some plan to reach that objective and the states must subscribe to that timetable as a precondition to implementation of the state plan.

We disagree with the District Court that the Secretary's regulations on number of personnel and funding provide the kind of specificity required for administering a state plan approval program in a consistent and rational manner. * * *

***

As to the requirement of "adequate funds" we agree with the Secretary that the "overall sufficiency of a State's funding is inevitably tested by the personnel sufficiency criteria." "[E]nforcement staff and necessary support salaries and fringe benefits constitute the principal cost element in any enforcement system." Once the Secretary articulates adequate personnel criteria, acceptable "adequate funding" requirements will follow.

We affirm the District Court's holding that the regulations concerning qualifications of state inspectors are valid. We find detail in these regulations sufficient to permit the Secretary to approve plans in a manner consistent with the mandate of the Act.

E. RELIEF

We declare that the Secretary has a duty to establish criteria that are part of an articulated plan to achieve a fully effective enforcement effort at some point in the foreseeable future. The "OSHA System for Enforcement Planning and Resource Allocation" has been drawn up with ideal benchmarks. Accordingly, with a planning horizon and an appreciation for administrative lags, the Secretary should be able to articulate a series of benchmarks clearly related to eventual achievement of the goals of the Act.

This is an action for equitable relief wherein considerations of the public interest weigh heavily. We decline to enjoin implementation of any of the state plans that have been approved to date by the Secretary. The problems we have identified do not necessarily mean that the Secretary erred in giving an initial approval based on present "as effective as" criteria. What those problems require is that in addition there be an articulated and coherent program moving from an initial "as effective as" approach toward the personnel and funding levels needed for satisfactorily effective enforcement. These criteria can be provided by supplements to the approval process. We contemplate that once the Secretary has articulated his program goals and objectives in terms of personnel and funding in accordance with this opinion, the necessary resources will be sought from Congress, and the underlying criteria will be applied in the state plan approval process.[40]

The case will be remanded for entry of a decree not inconsistent with this opinion.

[Judge MacKinnon's concurring opinion is omitted.]

[40]We realize that the Secretary may draw up his plan and find that at some point in the future Congress is unwilling to appropriate the funds requested. The significance of such actions will depend on the nature of the Congressional action and its history. It may reflect a Congressional estimate that the same goals may be achieved with less expenditure, through greater efficiency. However, it may evince a legislative judgment that the administrative goal was excessive. If Congress approves only X% of the Secretary's benchmark, a legitimate corollary would be the Secretary's approval of state plans that provide detailed assurances to meet at least X% of the Secretary's benchmark during the appropriate time frame. As this court noted in *NRDC v. Train,* 166 U.S.App.D.C. 312, 333, 510 F.2d 692, 713 (1975) (footnotes omitted):

> We think the court can forebear the issuance of an order in those cases where it is convinced by the official involved that he has in good faith employed the utmost diligence in discharging his statutory responsibilities. The sound discretion of an equity court does not embrace enforcement through contempt of a party's duty to comply with an order that calls him "to do an impossibility."

After further litigation before the U.S. District Court for the District of Columbia and negotiations between the parties, OSHA, with the substantial concurrence of the AFL-CIO, on April 25, 1980, filed benchmarks for state staffing of safety and health compliance officers in order to achieve a "fully effective" program, as required in the court of appeals decision. For state safety inspectors, OSHA would have required an increase for all states with approved plans from 849 to 1,154 over a five-year period. For health inspectors, OSHA would have required an increase from 332 to 1,683; however, because this constitutes such a great increase and there was a shortage in health personnel, OSHA asked the court to extend the time frame for completion of the health-inspector increase from 5 to 10 years.[42] In filing these benchmarks, OSHA also submitted a lengthy report with the district court describing the procedure it utilized to determine the new benchmarks as well as the available mechanisms for adjustment of the benchmarks. The OSHA submission was summarized in a Department of Labor news release on April 25, 1980.

OSHA Submits Guidelines to Court for "Fully Effective" Compliance Staffing for State Job Safety and Health Programs[43]

The Labor Department's Occupational Safety and Health Administration today submitted to the U.S. District Court for the District of Columbia, the agency's response to the Court's order to establish benchmarks (number of personnel required) for a "fully effective" enforcement program in states with OSHA-approved programs for employee safety and health protection.

In its submission, OSHA presents: the background of the court case; the design of the benchmark model and limitations associated with its development and use; staffing requirements generated by the model; and a process for revising the established benchmarks when required.

The need for development of benchmarks arose as a result of a suit filed in 1974 by the AFL-CIO. The suit challenged OSHA's criteria for approval of state plans for administering their own job safety and health programs which required state enforcement staffing and funding levels to be "at least as effective" as those of the federal program. In practice, this meant that state funding and personnel must be roughly parallel to those in the federal sector for like workloads.

The U.S. District Court upheld the OSHA criteria, but on appeal by the AFL-CIO to the U.S. Court of Appeals, the lower court decision was overturned and the case remanded to the U.S. District Court for issuing an order directing development of "fully effective" enforcement benchmarks. That order was issued December 5, 1978. OSHA has complied with the terms of that order by its submission, but the agency emphasizes its recognition that inspections alone are not the sole instrument to provide worker protection.

[42]On December 17, 1980, the District Court denied OSHA's request, and, instead, ordered that any reduction in the benchmarks for industrial hygienists would have to be negotiated by the parties, and in the event of disagreement, submitted to the court. 9 OSHC 1235 (D.D.C. 1980).

[43]U.S. Dep't of Labor, Office of Information, Occupational Safety and Health Administration, News Release, Apr. 25, 1980.

Ultimate success in reducing or eliminating on-the-job threats to life and health also depends heavily on the best efforts of employers as well as the availability and effectiveness of other efforts such as on-site consultation, training and education programs, strengthening of workers' rights, broader public understanding and appreciation of the problems of job safety and health, and the use of other innovative approaches such as labor-management safety and health committees.

In designing the benchmark model, OSHA remained consistent with its philosophy of concentrating its resources on those industrial sectors with the gravest safety and health hazards while not formally exempting any establishment or worker from the duties or protections afforded by the Occupational Safety and Health Act of 1970. The task was to establish for each state with its own job safety and health program the "ideal" number of safety and health inspectors necessary to provide a "fully effective" enforcement level. This number would function as a planning guide to be achieved over time and to the extent that Congressional appropriations would allow. While such an estimate involved a number of assumptions and measures, the central question was how frequently should an establishment receive a general schedule, routine safety or health inspection; that is, those inspections not conducted in response to a worker complaint, to investigate an accident, or to check on compliance with previous inspection results.

To aid in making this judgment, OSHA consulted the National Advisory Committee on Occupational Safety and Health, which advised the agency to convene [a] panel of experts to recommend frequency-of-inspection rates in accordance with industry classifications and the size of establishments. Accordingly, two panels of 14 experienced members each were constituted. Each panel had eight federal and six state experts. One panel was for safety and one for health. Each panel examined the various industries in the Nation and decided, based on available data and their own experience, how frequently such establishments should receive a general schedule inspection. In addition, the panels estimated how long general schedule inspections should last to be fully effective in each industry in each of three size break-downs—establishments with fewer than 50 employees, those between 50 and 250, and those larger than 250 employees.

The results of the panel's deliberations reflected the wide variance in the hazardous nature of American workplaces. For instance, the safety panel recommended that firms in Standard Industrial Classification Code (SIC) 32 (Primary Metal Products) should be inspected on the average of once every 1.7 years, but that those in most of the various service industries only once every half century or less. The health panel felt that firms in SIC's 28 and 29, (Chemical and Allied Products and Petroleum and Coal Products) should be inspected once every 2.1 and 1.9 years, respectively, while those in other industries only once every 116 years. In addition, OSHA would inspect those relatively non-hazardous, non-manufacturing establishments with 10 or fewer employees on only a random, spot-check basis consisting of one-tenth of one percent of them each year.

Multiplying the results of the above calculations by the number of establishments in each industry in each state, incorporating estimates of time required to complete other-than-scheduled inspections and applying a standard for available person hours per year, the final number of safety and health inspectors needed for a "fully effective" program for each state was derived.

As a result, many states may have to increase staffing totals appreciably. The total change would require the addition of 305 safety personnel and 1,351 health personnel to the state program staff. The court order specified that the plan submitted should provide for these increases over a five year period. However, in a separate submission, OSHA is petitioning for a change in the order to require 10 percent increases over a 10-year period for industrial hygienists. The report also reflects the court's recognition that the attainment of these additions is contingent on the willingness and ability of Congress to provide the needed funds. To the extent that the funds are not made available in any fiscal year, the annual incremental benchmark would be appropriately reduced and the plan would be extended an additional period of time.

The OSHA submission, in compliance with the court order, also provides a plan for adjustment of the benchmarks. The plan recognizes, first, that adjustments may be made unilaterally by the Labor Department based on such actions as court decisions, Congressional action, or reevaluation of the effect and impact of such activities as the employee complaint response system, inspection targeting systems, and the like. Second, since the benchmarks were based upon uniform criteria for all states, the OSHA submission recognizes that individual states may wish to petition to adjust their benchmarks to take into account local considerations such as geography, industry mix, inspection histories of various companies and other factors.

Although the Court required the development of benchmarks only for the state plan states, OSHA is committed to making every effort to reach a parallel level of personnel necessary for a "fully effective" program in states with federally operated programs. This should ensure that both federal and state programs provide comparable levels of employee protection. The Department will continue to fund state plans at the current 50 percent level.

While the AFL-CIO was satisfied generally with the benchmarks submitted to the district court, representatives of the states were not, particularly with the stringent staffing requirements for health inspectors. The states, in a brief as "friend of the court," argued that the benchmarks were improperly adopted without notice and comment; were not based on the "best available information"; and the "best available techniques" were not utilized.[44] Testifying in 1980, Assistant Secretary Eula Bingham defended the benchmarks, which she had submitted, saying:

> Admittedly, achievement of the benchmarks will be a costly undertaking and one that must be made within the context of national budgetary policy as determined by the President and the Congress through your final action on funding requests. But I believe the workplace penetration rates on which the benchmarks are based do not reflect an excessive or unreasonable amount of coverage—in fact we have heard objections from certain workers' groups that they provide inadequate coverage. For example, the benchmarks contemplate general schedule health inspection of textile mills only once every 3.9 years and safety inspection of printing firms only once every 5.5 years. The least hazardous industries would require inspection once every 35–50 years. That is hardly excessive coverage.
>
> In Europe, the Federal Republic of Germany has 3,000 workplace inspectors for a worker population of 30 million—a ratio of 1 inspector

[44]Brief of State Amici Curiae Concerning Report to the Court of April 25, 1980, *AFL-CIO v. Marshall,* 570 F.2d 1030 (D.C. Cir. 1978).

per 10,000 employers. There are enough inspectors in the United Kingdom to allow for inspection of 20 percent of all workplaces in any given year, and in Sweden 50 percent of all establishments are inspected annually. Even under the full benchmark, OSHA would only be able to inspect less than ten percent of the nation's establishments each year.[45]

The OSHA view on benchmarks changed markedly in 1981, when Assistant Secretary Auchter, testifying early in his administration on state plans, saw the high levels of the benchmarks as an impediment to the final approval of state plans and promised that they would be recalculated "more closely [to] reflect current agency policies."[46] Mr. Auchter later indicated that OSHA would seek a rider to its appropriations bill in order to offset the decision of the court of appeals in *AFL-CIO v. Marshall* requiring OSHA to develop staffing benchmarks looking toward a "fully effective [state] enforcement effort." This action, he said, was based on the statement in the court opinion that, "at some point in the future Congress [may be] unwilling to appropriate the funds requested" for the benchmarks, which "may evince a legislative judgment that the administrative goal was excessive."[47] OSHA's effort was initially successful, and, in enacting a supplemental appropriations bill, Congress provided that none of the funds may be expended to require "as a condition for initial, continuing or final approval of state plans" any "staffing levels which are greater than" levels determined by the Secretary to be equivalent to federal staffing levels.[48] The conference report stated that this language would permit OSHA to grant final approval to state plans that "have staffing levels that are at least equivalent to Federal staffing levels."[49]

This rider effectively nullified the effect of the District of Columbia Circuit Court of Appeals decision and on November 5, 1982, OSHA published a notice in the *Federal Register* proposing revised, and significantly lower, staffing benchmarks.[50] Under this proposal, OSHA first determined the "ratio of Federal staff to the workload for which it is responsible." In order to satisfy the revised benchmark, the ratio of the state's "currently allocated compliance safety and health officers to the benchmark for that state" set in the 1980 report to the court must be at least the same as the federal "ratio" referred to above. Using this approach, OSHA proposed a revised benchmark for all states with approved plans of 463 safety inspectors and 289

[45] *Oversight Hearings on OSHA—Occupational Safety and Health for Federal Employees, Part 4: State Plans: Hearings Before the Subcomm. on Health and Safety of the House Comm. on Education and Labor,* 96th Cong., 2d Sess. 492–93 (1980).

[46] *State Implementation of Federal Standards: Hearing Before the Subcomm. on Intergovernmental Relations of the Senate Comm. on Governmental Affairs,* 97th Cong., 1st Sess. 19 (1981), reprinted in Chapter 20, Section B.

[47] 570 F.2d at 1042 n.40. Mr. Auchter's statement appears in an interview with a representative of the Bureau of National Affairs, 12 OSHR 859 (1982).

[48] Supplemental Appropriations Act of 1982, Pub. L. No. 97-257, tit. I, ch. IX, 96 Stat. 818, 844 (1982).

[49] H. REP. No. 747, 97th Cong., 2d Sess. 36 (1982).

[50] 47 FED. REG. 50,307 (1982).

health inspectors.[51] On December 21, 1982, Congress passed a continuing appropriations bill that deleted the state-plans rider.[52]

In May 1983, Assistant Secretary Auchter testified before the Subcommittee on Health and Safety of the House Committee on Education and Labor, comprehensively setting forth his position on the benchmarks issue.[53] He said that in the last analysis, the benchmarks issue was a "funding" issue; that the impasse over benchmarks might cause some state legislatures to discontinue funding their programs; and, if that occurred, it would cost the federal government $110 million to maintain the present complement of state-plan compliance officers now being funded on a 50 percent basis by the federal government.[54] Mr. Auchter stated four possible solutions to the benchmarks "impasse":

1. An appropriations rider, such as was rejected by Congress in 1982.
2. Petitions by individual states for adjustments in the 1980 benchmarks. Mr. Auchter said that the benchmarks litigation would be a "lengthy" process and states should not have to wait "that long" in order to be eligible for final approval.
3. Unilateral adjustment by OSHA of the benchmarks. He said that the Agency has "resumed consideration" of a "comprehensive revision" of the benchmark numbers, a revision that "will better reflect current Agency policies, as well as differing conditions in the various states (that is, State-specific data)." He noted that any proposal to revise the benchmarks would be published for public comment.
4. A "slight" modification in the language of §18, by way of amendment, to explicitly provide that state compliance staffing levels be "at least as effective" as federal levels. It was this "ambiguity" in the language of §18(c)(4) that was largely relied on by the court of appeals in ruling in favor of the AFL-CIO in the benchmarks case.[55]

Finally, Assistant Secretary Auchter pointed out that in the meantime, OSHA was proceeding with the "performance evaluations" of

[51] *Id.* at 50,309. In its original benchmarks, OSHA would have required for all approved states 1,154 safety inspectors and, over a 10-year period, 1,683 health inspectors. As discussed in Chapter 21, Section C, OSHA has traditionally opposed appropriations riders, both because of their substance and on the grounds that legislative changes should not be enacted as part of the appropriations process. It would appear that footnote 40 in the court of appeals decision, this section, addressed itself to the possibility that Congress would appropriate insufficient funds for OSHA to provide its 50% share of the cost, in which event the benchmark would have to be adjusted. The idea of congressional action rejecting the principle of "fully effective" state plans apparently did not occur to the court at the time of its decision.

[52] Further Continuing Appropriations Act, 1983, Pub. L. No. 97-377, 1982 U.S. CODE CONG. & AD. NEWS (96 Stat.) 1881. See also H. REP. No. 980, 97th Cong., 2d Sess. 184 (1982) (stating that because of the "complexity" of the issue it would be reexamined in the next session of Congress).

[53] *Hearings Before the Subcomm. on Health and Safety of the House Comm. on Education and Labor*, 96th Cong., 1st Sess. (1983) (Mimeo).

[54] *Id.* at 5–6.

[55] *Id.* at 11–17.

the state plans of Alaska, Hawaii, and the Virgin Islands to consider whether they may be granted final approval. These plans, unlike most other state plans, appear to meet the 1980 benchmarks, according to Mr. Auchter.[56]

Testifying in response to Mr. Auchter before the same subcommittee, George H. Cohen, counsel to the AFL-CIO, disagreed with Mr. Auchter, saying that the OSHA view "[g]oes so far as to berate the Court of Appeals on the grounds that Congress could *not* have intended to precondition final approval of state plans upon states having in place a 'fully effective' enforcement capability * * *."[57] Mr. Cohen also argued that even a "minor" change in the language of §18 would "drastically alter the substantive obligations upon the states."[58]

The benchmarks case was dismissed by District Judge John Lewis Smith on May 26, 1983. The parties, however, made it clear to the court that the court order and the benchmarks submitted under that order remained in effect.

[56]*Id.* at 17. See, e.g., 48 FED. REG. 20,434 (1983) (inviting public comment on whether final approval should be given to the Virgin Islands state plan).

[57]*Hearing Before the Subcomm. on Health and Safety of the House Comm. on Education and Labor,* 98th Cong., 1st Sess. 9 (1983) (mimeo) (emphasis in original).

[58]*Id.* at 10–11.

20

Evaluation of State Operations and Evolving Federal Policies

A. Monitoring of State Operations

Section 18(f)[1] of the Act requires OSHA to make a "continuing evaluation" of how the state is "carrying out" its approved plan. Federal OSHA's monitoring policy has been another area of serious friction between federal and state OSHA agencies. The main components of the existing monitoring system, described in Chapter XVI of OSHA's Field Operations Manual,[2] are: (1) case-file reviews, (2) "spot-check" visits by federal monitors of establishments previously inspected by state enforcement personnel,[3] and (3) accompanied visits by federal monitors of actual state inspections or consultation visits. In addition, OSHA regulations authorize interested persons to file complaints, known as CASPAs (complaints against state plan administration), concerning any aspect of the operations of a state plan.[4] The monitoring results, based on all sources of information, are transmitted to the state in an annual evaluation report. This evaluation includes federal OSHA's recommended changes in state operations necessary for the state to be "at least as effective" as the federal program. A summary by federal OSHA of the 1981 annual evaluation of the State of Washington's approved plan provides an example of such a report.[5]

[1]29 U.S.C. §667(f) (1976).

[2]OSHR [Reference File 77:5001–:5051]. See also p. 659, note 17.

[3]An appropriations rider attached to the fiscal year 1980 appropriation limited OSHA's spot-check monitoring activity. See Chapter 21, Section C.

[4]29 C.F.R. §1954.20 (1983). OSHA regulations, 29 C.F.R. §§1954.1–.22 (1983), deal generally with the procedures for its evaluation and monitoring of approved state plans. The procedures for the withdrawal of approval of state plans appear in 29 C.F,R. §§1955.1–.47 (1983).

[5]Covering the period May 1, 1980–April 30, 1981, "Evaluation Summary" at ii–iii. Copy on file with OSHA, Washington, D.C.

State of Washington Annual Evaluation Report

***

The evaluation report addresses several areas with a considerable amount of discussion in regard to the compliance side of the States' program. One of the major components to an effective enforcement program is the proper scheduling of inspections into workplaces that have been identified as having the most hazardous conditions. WISHA [Washington Industrial Safety & Health Administration] has implemented, and is currently holding public hearings on their safety scheduling system known as the Washington Inspection Targeting System (WITS). This system was submitted to OSHA Region X during this reporting period for an advisory review. The WITS safety scheduling program establishes an objective scheduling system for establishments already identified from among selected high hazard industries. This program differs from OSHA's system of scheduling establishments for inspection of high hazard industries. WISHA uses a variety of information available through workmen's compensation statistics, combined with the industries injury-illness incidence rates, to direct their compliance effort at specific establishments actually experiencing occupational injuries that may have been prevented by adherence to State standards. Through the process of inspecting those establishments experiencing injuries, WISHA is covering a broad spectrum of business activities in the State rather than concentrating on relatively few industries. Preliminary monitoring observations have identified that the WITS safety scheduling program meets the intent of OSHA instructions relating to scheduling and conducting inspection in high hazard industries/establishments. WISHA has been in the process of developing a comprehensive programmed safety and health inspection scheduling system over the past several years; however, the total system has not been finalized and submitted to OSHA as required by Federal program changes. The State has yet to submit their programmed scheduling system for health and the various specialty sections. Such a system would help to ensure that more productive and meaningful inspections would be conducted in high risk workplaces. Thus, the report contains several recommendations urging WISHA's management to formally adopt and implement a total programmed scheduling system.

Another area that affects the compliance side of the program and, more directly, health scheduling is the subject of complaints and referrals. WISHA's policy for handling complaints has been to respond to almost all complaints, whether formal or nonformal, with an inspection. Although WISHA's complaint policy is very comparable to OSHA's, the State's procedure does not provide for a letter to be sent to an employer when a nonformal complaint is received alleging general (nonserious) violations. As a consequence the health division conducted 57 percent of their inspections in response to complaints with a resultant compliance rate of 53 percent. OSHA monitoring, through a special study directed primarily at one WISHA regional office, identified that approximately fifty percent of the health complaints receiving an inspection could have been addressed through a letter to the employer. Additionally, under the health division, referrals from the safety section accounted for approximately one-third of the scheduled or programmed health inspection workload. Referrals from the safety section had a compliance rate of 57 percent, and the referred hazards were determined not to be a violation of a State standard in 78 percent of the cases. Due to these findings, and in reference to our previous recommendation to adopt a programmed health scheduling system, a recommendation has also been offered for the State to adopt a complaint procedure that would allow a letter to be sent to an employer when the conditions warrant; plus, the need for WISHA supervisors to ensure that referrals are adequately documented in support of a potential health hazard.

The Washington plan for general industry inspections—based on establishment data—anticipated OSHA's adoption of a similar approach in 1981.[6] Because a state has better access to establishment data through the workers' compensation system, it may more flexibly adopt new targeting approaches than federal OSHA. As to the evaluation report's comments on Washington's complaint system, OSHA was pressing the state to respond by letter to nonformal, nonserious complaints, a procedure adopted by federal OSHA shortly before the evaluation on August 1, 1979.[7]

B. Congressional Hearings on State Programs

OSHA monitoring efforts have resulted in considerable state resentment. State hostility culminated in 1980 in oversight hearings held by the Subcommittee on Health and Safety of the House Committee on Education and Labor. The subcommittee, chaired by Congressman Gaydos of Pennsylvania, heard testimony from representatives of 11 states with approved plans, criticizing OSHA's monitoring activity and its "incomplete communications" with the states. Typical of this testimony was that of Michael Ragland, Executive Director, Occupational Safety and Health Program, Kentucky Department of Labor.[8]

Statement of Michael Ragland

***

MONITORING

As has been indicated in previous testimony from state representatives, nearly all states with state programs are closer to achieving some ideal level of compliance field personnel than is presently the case in areas without state programs which utilize federal personnel. Nevertheless, federal OSHA

[6]Under the OSHA procedure, the inspector would examine the employer's injury and illness records at the workplace before beginning the inspection, and, if that employer's record was below a certain cut-off, the inspection would be discontinued. See generally Field Operations Manual, ch. III, OSHR [Reference File 77:2510–:2513]. See also discussion in Chapter 13, Section E.

[7]For a discussion of the Washington state plan, see Dick, *Washington Job Safety Legislation*, 9 GONZ. L. REV. 457 (1974); Note, *An Alternative to Federal Preemption: The Washington Plan*, 9 GONZ. L. REV. 615 (1974); Note, *A Survey of WISHA's First Months of Operation*, 9 GONZ. L. REV. 639 (1974).

[8]*Oversight Hearings on OSHA—Occupational Safety and Health for Federal Employees, Part 4: State Plans: Hearings Before the Subcomm. on Health and Safety of the House Comm. on Education and Labor*, 96th Cong., 2d Sess. 417–20 (1980). Also testifying at the hearing, among others, were the heads, or assistant heads, of the state OSHA programs in Michigan, North Carolina, South Carolina, Wyoming, Oklahoma, and Minnesota, and Governor Dalton of Virginia. No union representative testified at the state plan hearings.

continues to tie up vast resources of personnel merely to monitor operational state programs, most of which have been in operation over six years. While one can readily agree that some form of monitoring is essential, it is hard to imagine that these federal compliance officers are best serving the declared purpose of the Act—reducing injuries and illnesses in the workplace—by continually monitoring the performance of state personnel, rather than actually conducting inspections in hazardous workplaces in areas of federal jurisdiction where unnecessary complaint backlog exist.

In Kentucky, for instance, there is an Area Office staffed with a full-time Area Director, two full-time Safety Compliance Officers and one full-time Industrial Hygienist, as well as office personnel, who do absolutely nothing but monitor the activities of the Kentucky Program. Obviously, in states larger than Kentucky the number of federal personnel is much greater. It would certainly appear reasonable to assume that this number of people working full-time could have made some type of determination after six or seven years in intense monitoring as to whether or not the state program is performing its functions adequately.

As it presently exists, however, federal monitoring for the most part is no different for a state program with seven years of operation than one just commencing. Furthermore, many of the federal monitors possess less experience than the state personnel they are assigned to monitor. In fact, often times these are merely state personnel lured away from state programs by more attractive, higher federal salaries. There appears to be no training for a federal monitor other than that required for a federal compliance officer. Surely one could easily recognize that even though a monitor should possess those skills, there are other qualifications and skills necessary to effectively function as a monitor of an approved state program.

Perhaps, yet, the most unfortunate aspect of the monitoring of state programs by federal OSHA is that most federal monitors ultimately are the final judge and jury of a particular state's performance. Almost without exception recommendations which are conceived by individual monitors go unchallenged by Regional and National OSHA personnel even though the state may continually rebut, with documentation, a particular allegation. Perhaps further complicating this issue is the fact that many federal monitors view their ability to remain assigned to a particular state contingent upon their ability to convince their superiors that continued monitoring is necessary for the well-being of the workers in that particular state.

We have previously made recommendations to federal OSHA concerning what would seem to be a more appropriate utilization of personnel in the area of monitoring. All monitoring of state programs could emanate either from the Regional or National level and thus eliminate the cost of maintaining 23 area offices and reduce significantly the number of personnel presently assigned responsibilities as full-time monitors of state programs. One of the first steps to accomplish this objective would be to devise a functional quarterly report to replace the presently existing one which is, for the most part, totally nonusable for evaluating the statistical performance of a state program or a federal area office. Certainly final monitoring determinations could not be based on mere compilation of statistics; however, if meaningful statistics were compiled and presented on a quarterly, or even monthly, basis, these could be utilized as indicators of potential deficiencies in specific areas. Once these apparent deficiencies were detected, monitoring personnel could be sent on-site, if necessary, to fully evaluate the state's performance in a particular area of concern.

COMMUNICATIONS

The key to improving the federal/state OSHA relationship lies in the area of communications. It appears that this one area has been the basis for most of the problems previously discussed. The failure to communicate directly with the state is a problem that prevails at the Area, Regional and National levels. For instance, the Area Office has never discussed our scheduling system with us, and yet they have proposed a recommendation concerning this system * * *. In a past evaluation report, the Regional Office submitted a recommendation concerning our Management Information System based on "unofficial information" which later was discovered to be no more than an unfounded rumor—the contents of which had never been discussed with us prior to receiving the evaluation report. It is not uncommon for us to learn via the news media that the National Office has adopted a National Emphasis Program for grain elevators, or developed new policies concerning carcinogens. The National Office continually holds press conferences, confers with management and labor leaders, and issues policy statements and directives affecting all employers and employees, as well as state programs, without ever consulting with any state program officials.

These are mere examples of the ever-continuing problems of incomplete communications. This problem obviously extends to the benchmark issue * * *, and is the overriding cause for the unfortunate relationship that exists in the area of monitoring. On the few occasions that the Assistant Secretary has met with State Designees, the issue has been raised with grave concern with repeated assurances that improvements will be forthcoming. Unfortunately, the communications problem has continually worsened in the last few years. Certainly there appears to be no conception of a "partnership" between federal OSHA and the state programs, even though the state programs are responsible for the administration of the occupational safety and health regulations in 21 states and 2 territories, and annually conduct more inspections and consultations than all of federal OSHA combined.

***

Much of the state criticism at the 1980 oversight hearings was directed at the then Assistant Secretary, Dr. Eula Bingham. At the hearings, Dr. Bingham presented a detailed justification of her state plan policy. She stated that, as Assistant Secretary, she had a responsibility to "take some positive action to deal with those states that were clearly operating deficient programs" and that federal OSHA has a "leadership role which does not lend itself to a traditional partnership of equality as I believe a number of States desire."[9]

Statement of Dr. Bingham

***

When this Administration took office in 1977, a number of State plans had been in operation for four years and had been certified. They felt they were ready for "final approval" with the political and public image of maturity and independence that goes with such approval. At the same time a series of petitions from outside parties had been filed requesting the agency to withdraw the approval of three State plans. OSHA's monitoring of State performance had become more definitive after four years of experience, and we were able at that time to conclude that some States were demonstrating a capacity to run a credible and effective occupational safety and health

[9] *Id.* at 476–97.

program while some others were not fulfilling their promise, much less the expectations of the Act.

***

The Federal program, itself, had matured, was moving rapidly into an expanded occupational health role, and was becoming more sophisticated in its technology and methods. This necessitated a parallel movement away from the "old ways" of doing things on the States' part.

It was in this environment of changing requirements and expectations that I took office. Much of the dissatisfaction that has been voiced before you in the recent testimony of State designees can be attributed, I believe, to the thwarted expectations deriving from this confluence of events.

DEFICIENT STATE PROGRAMS

While it was the desire of the agency to continue to encourage those States willing and able to provide effective programs, it was also absolutely essential to take some positive action to deal with those States that were clearly operating deficient State programs. Petitions from outside sources requesting withdrawal of State-plan approval in three States have further heightened the urgency of this problem. As a result action on plan withdrawal has been or is being considered in several States. Petitions requesting the initiation of withdrawal of plan approval have been submitted by both the State and National AFL-CIO in Indiana; the Oil, Chemical and Atomic Workers Union (OCAW), the Fairfax County Board of Supervisors, the State AFL-CIO, and the Building and Construction Trades Department of the AFL-CIO in Virginia. The Carolina Brown Lung Association and the AFL-CIO earlier petitioned for the withdrawal of the South Carolina plan. In addition, OSHA determined that the Wyoming program was operating under legislative authority that could no longer be considered acceptable.

On April 3, 1980, I initiated formal plan withdrawal action in Indiana because of the inadequacy of its health enforcement program, the inability of its staff to recognize safety hazards, the proposal of less than meaningful penalties, and denial of worker rights.* Indiana responded by denying all charges and requesting a formal hearing on the issues. I intend to proceed with this withdrawal action. * * *

In Virginia, the adequacy of the State's system for review of contested cases through its General District Courts has been in question. No decision on plan withdrawal action has yet been made, pending full review of the implementation of amendatory legislation enacted in 1979. Governor Dalton testified before this subcommittee in April that the revised Virginia legislation cured many of the problems OSHA found in Virginia's substitution of judicial review of contested cases for the Federal administrative system. Subsequent to that testimony an internal State review, directed by the Governor, identified continuing State deficiencies. The State instituted procedures to correct them, and we currently are considering their effectiveness. Nonetheless, Virginia proceeded to bring action in the U.S. District Court in Roanoke to preclude any potential plan withdrawal action by OSHA. Briefs have been filed in this case, but it has not as yet been decided.

If Virginia's alternative occupational safety and health adjudicatory system is found deficient, it will be my responsibility to the workers in the State to take appropriate action. We have given the State repeated notice of our concerns with the judicial system and were most disappointed when

*[Author's note—Proposed withdrawal of the Indiana plan, 46 FED. REG. 3919 (1981). Upon taking office, Assistant Secretary Auchter first deferred the effective date of the proposal, 46 FED. REG. 11,253 (1981), and later withdrew the complaint that had initiated the withdrawal proceeding. 46 FED. REG. 19,000 (1981).]

proposed amendatory legislation that would have completely cured the deficiencies failed to gain enactment.*

Coincident with its petition for withdrawal of the Virginia State plan's approval, Fairfax County offered the services of its County building inspectors as an adjunct to Federal and State construction enforcement activity in Northern Virginia. Because of the questionable efficacy of the State's enforcement mechanism and a concern for the increasing number of construction accidents and fatalities in the area, we accepted this offer. We have entered into an arrangement whereby Fairfax County building inspectors have been trained in OSHA's standards and refer to us for follow-up enforcement action unabated violations they find in the course of their routine activity. As you have no doubt seen in local press coverage, the arrangement is working more than satisfactorily with most employers voluntarily abating hazards brought to their attention by County inspectors. I believe this demonstrates the flexibility of the Federal program and our willingness to work with the States and local jurisdictions to experiment with new, creative approaches for dealing with occupational hazards.

***

In all such actions my concern is with assuring effective State performance. When that is demonstrated, OSHA will support the State plan. A worker under State program jurisdiction should have no less protection than he or she would under the Federal program, and an employer should have no fewer obligations.

MONITORING AND EVALUATION

In the States where the continuation of plan approval is at issue, the evidence of deficient commitment and performance has been clear and supported by sources both within and beyond our internal monitoring system. At the same time, we have concluded that the monitoring system we inherited can be improved to provide better information for evaluating State programs.

OSHA maintains an Area Office for each State plan which, together with the Regional Office, is responsible for on-site monitoring of the State plan. Annual evaluation reports with recommendations for corrective action are prepared and transmitted to the State designees. The agency's monitoring and evaluation procedures are contained in Chapter XVII of the Field Operations Manual. These procedures were developed early in the program without the benefit of much actual experience. In reviewing these procedures from the perspective of 8 years of State plan operations, we have learned that some data collected do not provide meaningful insight into program operations, while data on other important indicators are not available. A project is now underway within the Agency to revise this monitoring system. The first phase of this project included an assessment of alternative monitoring techniques. The results of this study will be used by a workgroup of OSHA field and National Office staff to develop a more efficient and effective monitoring and evaluation system. The States will, of course, be asked to comment and advise us on any resulting proposals.

***

COMMUNICATIONS

A second concern raised by the States to this Committee involved communications between OSHA and the States.

---

*[Author's note—On August 19, 1981, OSHA denied the petition for withdrawal of the Virginia state plan. 46 FED. REG. 36,141 (1981).]

Under the Federal regional system, we rely primarily on our Regional Administrators and Area Directors to keep State officials informed of OSHA program developments and to offer assistance in addressing their needs. The States are included in the same system for distributing policy and informational materials as are our Regional and Area staff.

Our Regional and Area staff generally have daily contact with State officials. Regional Administrators and Area Directors are always available to discuss national policy issues or specific problems with State designees. When issues requiring National Office guidance are raised, regional staff contact the Director of Federal Compliance and State Programs, whose Office of State Programs staff coordinates the review and approval of State plans and plan revisions as well as the monitoring and evaluation of State programs. In some cases, the State officials contact National Office staff. Issues requiring higher-level executive decision are brought to the attention of Basil Whiting or myself.

The volume of OSHA policy development makes it infeasible to involve the States extensively in many of the decisions we must make. However, on major issues affecting the States, we try to give them an opportunity for input. For example, as I have already indicated, there has been a great deal of discussion with the States on the benchmarks. Although they could not be involved in the actual staff work and deliberation at the Departmental level, we kept them informed of all aspects of the benchmarks process and sought their input where appropriate.

In the development of standards, some States would like to be involved at every stage of our research and deliberations. In some cases, we have included State experts on standards advisory committees and have sought State input early in the development of standards in which the State has special expertise. However, the complexity and sensitivity of the standard-development process usually precludes the release of draft materials prior to their publication for general public comment. We therefore encourage the States to participate in the hearing process and to offer written comments on proposed standards. A review of the record indicated that a few States have commented on each of the major standards we have proposed but that most have not done so. We hope that more States will make fuller use of this opportunity.

In addition to the contact between Regional and National Office staff with State officials, I have met personally with the State designees as a group on several occasions, as has Basil Whiting, to hear their general concerns. We have met with individual State designees more frequently to discuss their specific concerns. We have addressed groups at the request of the designees. We have not always been able to meet with them as a group when they have requested; but we have tried to take account of the concerns they have expressed, within the constraints of our responsibilities under the Act.

PARTNERSHIP

Finally, many of the States who have testified before you have complained that the Act envisioned a partnership which has not been honored by this Administration. The Act does instruct me, as Assistant Secretary, to encourage State plans for the enforcement of standards but clearly that encouragement does not mean acceptance of State performance which results in inferior worker protection. The Act also charges the agency with devising the national program for occupational safety and health as well as monitoring the State's performance and deciding on its acceptability. The Federal agency is given a leadership role which does not lend itself to a traditional partnership of equality as I believe a number of the States desire. It has been my continued intent and desire to maintain a cooperative and amicable relationship with the States, seeking their input and advice on issues and promoting the continuance of effective State Occupational Safety and Health plans.

***

During the questioning of Dr. Bingham at the oversight hearings, Chairman Gaydos summarized the views of the states towards federal OSHA in the following language:

> [Y]ou get accused time and time again, the administration does generally, that you are insensitive to the States needs, your requirements are prohibitive, it is self-serving on your part, that what you want to do is preserve complete Federal jurisdiction and cut every State out and that the concept that there is a partnership between State and Federal Government has become abhorrent for want of a better name to the administration.[10]

Soon after Thorne G. Auchter assumed the position of Assistant Secretary for OSHA, he testified in Congress on his views towards state plans. He addressed the major issues of "benchmarks," monitoring, and the overall relationship between federal OSHA and the states, and made it clear that major changes were contemplated in order to "fully integrate the states into the overall OSHA program."[11]

Statement of Assistant Secretary Auchter

***

TWENTY-THREE STATE PLANS TO DATE:

Early in its existence, the Occupational Safety and Health Administration (OSHA) made a concerted effort to encourage the development of State programs. Of the fifty-six States and jurisdictions included under the Act, fifty have at some time submitted a plan for review. Thirty-one of these took the steps needed to meet OSHA's requirements for initial approval of a plan. At present, however, only twenty-three States and jurisdictions conduct comprehensive job safety and health programs, while a twenty-fourth—Connecticut—has retained a partial plan covering only State and local government employees, and a twenty-fifth—New York—is developing a partial plan similar to Connecticut's.

There are a variety of reasons why thirty-three States and jurisdictions have chosen not to administer their own job safety and health programs. For example, the Act permits the Federal government to *fund no more than* 50 percent of State's costs of operating a program. Many States today face serious budget constraints which make it even more difficult for them to consider initiation of a new program, for which they must provide at least 50 percent of the funding—especially a program which will otherwise be carried out by the Federal government. For others, the fear exists that the Federal share may be lowered once a State program is fully established.

[10] *Id.* at 515–16.

[11] *State Implementation of Federal Standards: Hearings Before the Subcomm. on Intergovernmental Relations of the Senate Comm. on Governmental Affairs,* 97th Cong., 1st Sess. 14–25 (1981).

Moreover, at the time of the Act's passage, some States had only minimal programs on which to build. And finally, political considerations may have figured in the inclination of some States to disassociate themselves from the sometimes unpopular enforcement procedures mandated by the Occupational Safety and Health Act of 1970. In other States, the opposition of organized labor to the concept of State programs may have influenced the decision to leave responsibility for occupational safety and health with the Federal government.

Nonetheless, twenty-three States have found compelling reasons to operate their own programs.

PROGRESS OF STATE PLANS:

To assume this responsibility, they first had to submit Developmental Plans for OSHA's initial approval. Thereafter, to gain final approval of their programs, they had to continue to develop and implement their Plans—a process which, theoretically, could have been completed within three years of initial approval. None of the Plans now in existence, however, has yet received final approval, though most of them won initial approval over eight years ago.

***

Since we have been in office we have taken steps to encourage the orderly progress of State Plans toward final OSHA approval. For example, we have eliminated redundant Federal enforcement authority in several States with satisfactory occupational safety and health Plans and have shifted Federal inspectors to Federally covered States. Operational status agreements confirming these arrangements will be signed shortly with all States still without such agreements.

We have also expedited the review process leading to certification. Of the nine State Plans which had not received certification by January 20, 1981, four have now been certified, and the remainder should receive certification in this fiscal year, assuming they meet the necessary criteria.

"BENCHMARKS," AN IMPEDIMENT TO FINAL APPROVAL:

A significant obstacle, however, stands in the way of final approval of State Plans. Although eleven Plans have been certified for over a year, OSHA has been unable to grant final approval because of stringent requirements for high levels of enforcement staff—levels known as "benchmarks," which have been imposed upon the States as the result of litigation. These benchmarks were developed by the previous administration in response to a decision of a Federal appeals court.

I have directed that the "benchmark" formula* be recalculated to establish staffing requirements which more closely reflect current agency policies. * * *

MONITORING:

As noted, the Occupational Safety and Health Act of 1970 requires that Federal monitoring continue even after a State Plan receives final approval. It is the goal of this Administration to keep such monitoring to an essential and effective minimum. An improved system of monitoring now under development would allow States as full autonomy as possible so long as the requirements of the Act are met.

---

*[Author's note—See discussion in Chapter 19, Section D on Assistant Secretary Auchter's actions on the benchmarks issue.]

Heretofore, Federal monitoring has relied heavily on a Federal presence in the States. In the early developmental stages of a State Plan, it was helpful, no doubt, to have OSHA monitors accompany State inspectors on visits to workplaces; these visits served a training, as well as a monitoring, function. As State programs gained experience, however, there was less need for hands-on monitoring. By retaining Federal monitors in States with mature Plans, the agency failed to make the most efficient use possible of its limited complement of safety and health professionals. Moreover, the old system did not focus adequately on the critical aspects of State performance.

Consequently, in the past few months we have reduced the number of Federal monitors by one-half. At the same time we are revising our entire system of monitoring State Plans. Our purpose is to develop an efficient and objective system which relies more on State-submitted information than on a Federal presence.

To achieve this goal, we are developing objective indicators of program effectiveness—in part to let the States know the performance goals that are expected of them in part to provide Federal OSHA with a running guide to State performance. We are also taking steps to obtain uniform State data for monitoring purposes. Plan States will be required to provide Federal OSHA with statistical information about their programs, preferably by participating in OSHA's computerized Management Information System. Our overall monitoring system should then be able to rely in large part on analysis of State-submitted information and on our newly developed indicators of effectiveness. Such a system should reduce the need for on-site Federal monitoring.

FEDERAL PREEMPTION—UNITY IN DIVERSITY:

One of the unique features of the Federal-State relationship under the OSH Act is the preemption by Federal OSHA of State enforcement until such time as a State obtains approval of a job safety and health Plan in accordance with the requirements of the Act. This feature of the OSH Act serves to eliminate duplication of effort by Federal and State governments and also to promote a general level of uniformity among the Federal and State programs.

At the same time, the States are allowed some latitude under the Act in setting standards and enforcement procedures. For example, the State of California has pioneered in a voluntary compliance agreement at a large nuclear plant construction site in San Onofre where a labor-management safety and health committee is being used, apparently successfully, as a self-inspection mechanism.

At times, the discretion allowed the States can create difficulties. A few States, notably California and Michigan, do not routinely adopt Federal standards. Although State standards and enforcement procedures must be at least as effective as the Federal ones, variations in standards and procedures in different States can cause difficulties for employers with establishments in more than one State.

At present we are considering various solutions to this problem. Our goal is to encourage a reasonable degree of uniformity in standards nationwide, while still allowing States flexibility to meet local needs. For the future, at least as far as new standards are concerned, early State participation in the standards-setting process should help alleviate the problem.

INCREASED COOPERATION:

Whenever possible, we have acted to increase Federal-State cooperation and reduce friction between State officials and Federal OSHA. Since assuming office, I have met twice with all the administrators of State occupational safety and health programs as a group, as well as meeting with many of them individually. For the first time in OSHA's history, a special assistant has been appointed to advise the Assistant Secretary on State-plan matters and to maintain liaison with the States. We have also taken steps to include State Plan representatives in the revision of the monitoring system now under way. Finally, I have directed my staff to develop mechanisms to assure State participation in all stages of the agency's decisionmaking process on all issues with a direct impact on State programs.

***

It is our firm intention to resolve differences that have existed in the past between the States and Federal OSHA and to develop a management and policy framework that in the future will fully integrate the States into the overall OSHA program. It is my belief—and the belief of this administration—that in the last analysis, local problems are best addressed by those closest to them.

***

Pursuing his "partnership" approach to the states, Assistant Secretary Auchter withdrew OSHA's complaint against Indiana's state plan,[12] denied the petition for withdrawal of the Virginia plan,[13] proposed revisions of the state benchmarks,[14] and entered into operational agreements with the remaining states with approved plans.[15] On the issue of state participation in the standards-making process, OSHA told a House subcommittee in 1982 the following:

> OSHA is taking several steps to include States more fully in the Federal standards-setting process. The Director of Federal Compliance and State Programs will serve on the Executive Committee which coordinates all OSHA standards-setting activities. He will be the contact for the States designees on all issues related to OSHA standards. He will receive and pass on any proposals for standards from the state designees, obtain information which they request, and ensure that they are informed of pending and planned actions. The Director will also invite the States to comment on all new standards-actions at the "concept" stage and on any draft standards which are being submitted to an OSHA standards advisory committee for review. The States will continue to be encouraged, as in past Notices and Rulemaking, to submit formal comments in response to OSHA's Advance Notices of Proposed Rulemaking.[16]

And, on the new monitoring system, OSHA explained as follows:

> OSHA is now in the process of developing a revised monitoring system which will rely primarily on State-submitted statistical data rather than onsite monitoring. Quantitative measures (Effectiveness Measures) have been developed to compare State to Federal performance.

[12] 46 FED. REG. 19,000 (1981).
[13] 46 FED. REG. 36,141 (1981).
[14] 47 FED. REG. 50,307 (1982); see discussion in Chapter 19, Section D.
[15] 47 FED. REG. 25,323 (1982) (codified in 29 C.F.R. pt. 1952 (1983)).
[16] *OSHA Oversight—State of the Agency Report by Assistant Secretary of Labor for OSHA: Hearing Before the Subcomm. on Health and Safety of the House Comm. on Education and Labor*, 97th Cong., 2d Sess. 218 (1982).

Comparable data will be generated by State participation in a Unified State/Federal Management Information System. Onsite monitoring will occur only when statistical analysis cannot provide an explanation of State performance that falls outside established parameters. Special onsite studies using case file reviews, accompanied visits (worksite visits on which a Federal monitor accompanies the State inspector) or spot check monitoring visits will be used only as a last resort and will never be used to pass judgment on individuals or a particular aspect of a State plan but rather as an indicator of overall State program performance.[17]

C. The Ethylene Dibromide Controversy

Mr. Auchter's testimony before the Senate Committee on Intergovernmental Relations that "variations in standards and procedures in different states can cause difficulties for employers in more than one state"[18] raised an issue that is emerging as critical in the evolving relationship between federal OSHA and the states. The relevant statutory provision is §18(c)(2),[19] which requires that state standards must be at least as effective as federal standards, contains a proviso stating that "when [the state standards are] applicable to products which are distributed or used in interstate commerce, [the standards must be] required by compelling local conditions and * * * not unduly burden interstate commerce." The legislative history of this "products clause," as it has been called, indicates that one purpose of the clause was to make certain that states with approved plans would not unduly burden interstate commerce by setting standards differing from the federal standards.[20] While the application of the products clause to safety standards regulating products is relatively clear, considerable debate has surrounded the application of the clause to health standards.

The issue was presented most sharply in relation to California's standard regulating ethylene dibromide (EDB). One of the main uses of EDB is as a fumigant applied to harvested fruits and vegetables to prevent the spread of insects such as the Mediterranean fruit fly.[21] California first issued an emergency temporary standard in 1981 and later a permanent standard in 1982 which differed markedly from the parallel federal standard. The federal standard set a 20-parts-per-million (ppm) limit for an average 8-hour-a-day exposure and a

[17] *Id.* at 213. In August 1983, OSHA issued a lengthy directive, completely revising its system of state monitoring, largely replacing on-site monitoring with statistical comparison of state performance. OSHA Instruction STP 2.22 (1983).

[18] Quoted *supra* Section B.

[19] 29 U.S.C. §667(c)(2).

[20] 116 CONG. REC. 37,622 (1970) (statement of Sen. Saxbe) ("If * * * states were permitted to set differing safety standards for dirt movers, it would place a tremendous burden on interstate commerce. This amendment provides that they may do so because there may be circumstances that would so require but the words are put in, 'compelling local conditions' and, second that they 'do not burden interstate commerce.'").

[21] See OSHA Recommended Procedures to Minimize Worker Exposure to Ethylene Dibromide, 11 OSHR 535 (1981). California has an approved plan.

ceiling of 30 ppm for any 15-minute period.[22] The California standard set a 130-parts-per-billion (ppb) average exposure limit with a 130 ppb ceiling for any 15-minute period.[23] The California standard also contained specific provisions on monitoring, methods of compliance, protective clothing, and others similar to the provisions in OSHA health standards issued after §6(b) rulemaking[24] California's action was based largely on new experimental data from the National Cancer Institute on the carcinogenic effects of EDB at 10 ppm.[25] In November 1981, OSHA asked for public comment on three issues: (1) whether the California EDB standard is applicable to products within the meaning of §18(c)(2); (2) whether it is compelled by local conditions; and (3) whether it unduly burdens interstate commerce.[26] The AFL-CIO urged that federal OSHA approve the California standard; various business groups, on the other hand, argued for disapproval.

Comments of AFL-CIO on the Proposed Supplement to California State Plan[27]

***

It is the opinion of the AFL-CIO that the California ETS on ethylene dibromide is a necessary and appropriate response to protect against the serious health risks posed by EDB; that the Cal-OSHA standard is not a product standard as defined by section 18(c) of the OSHA Act; and that no restraint of trade has resulted from the imposition of the emergency temporary standard.

Available scientific evidence on ethylene dibromide shows unequivocally that the chemical is a potent carcinogen in animals at levels one half the federal OSHA standard of 20 ppm, and thus poses a serious potential carcinogenic risk to humans. Moreover, additional studies have documented that EDB is a mutagen, a teratogen as well as a potent testicular toxin and that reduction in the current PEL is necessary to protect against these toxic effects.

With an increase in the use of EDB resulting from fruit fumigation

[22] 29 C.F.R. §1910.1000 at table Z-2 (1983). On December 18, 1981, OSHA denied a petition for a federal emergency standard on EDB filed by the International Brotherhood of Teamsters and other unions. OSHA concluded that the "present situation does not fulfill the statutory criteria necessary for the imposition of an emergency temporary standard." 46 FED. REG. 61,671, 61,672 (1981). In the same document, OSHA published an advance notice of proposed rulemaking on worker exposure to EDB. In October 1983, OSHA issued a proposed ethylene dibromide standard. 48 FED. REG. 45,956. The proposed standard would reduce the permissible exposure level from 20 ppm to 0.1 ppm.

[23] See 48 FED REG. 8610 (1983).

[24] See *id.* at 8611. See also discussion Chapter 3, Section C.

[25] See 46 FED. REG. at 61,673 (1981).

[26] While the OSHA review process was taking place, the California EDB standard was subject to litigation. In 1981, a federal district court refused to issue a preliminary injunction barring California from enforcing the EDB emergency standard until approved by OSHA. *Florida Citrus Packers v. California,* 10 OSHC 1137 (N.D. Cal. 1981). When in February 1982 California commenced enforcing the same EDB standard as a permanent standard, the district court again decided that California was permitted to enforce the standard pending OSHA review. *Florida Citrus Packers v. California,* 545 F. Supp. 216 (N.D. Cal. 1982). See also *Florida Citrus Packers v. California,* 549 F. Supp. 213 (N.D. Cal. 1982) (OSHA, not the court, must decide if the regulation is a product standard.).

[27] OSHA Docket No. H-111, Dec. 21, 1981. Comments on file in OSHA Docket Office, Washington, D.C.

against the mediterranean fruit fly, the potential for worker exposure to this toxic chemical increased as well. According to State of California estimates, over 12,000 California workers may be exposed to EDB in fumigation operations or from transport and handling of off-gasing fruit. Faced with increased worker exposure to this highly toxic chemical, the State of California prudently initiated an emergency temporary standards action to require a reduction in the permissible exposure limit to 130 ppb and to impose monitoring, medical surveillance, posting, training and personal protective equipment requirements as well.

* * * [T]he Florida Growers Association continues to seek relief from the federal Occupational Safety and Health Administration on grounds that the Cal-OSHA EDB standard is a product standard under section 18(c)(2) of the OSHA Act and that the promulgation of the ETS has unduly burdened interstate commerce.

A reading of the legislative history of section 18(c)(2) of the OSHA Act clearly indicates that California's EDB standard in no way constitutes a product standard under this section of the Act. The floor debate on the OSHA Act shows that the term product standard as used in the section was intended to apply to such items as machinery, tractors, farm vehicles etc. and was introduced to prevent undue restriction on the sale and use of such equipment across state lines. In introducing the amendment in the Senate Mr. Saxbe said:

> this is an amendment which applies to those who manufacture machinery products that got into interstate commerce, specifically applying to those who manufacture movable equipment such as dirt movers, tractors, and other heavy equipment. If, after the expiration of the initial stage of this bill, each of the States were permitted to set differing safety standards for dirt movers, it would place a tremendous burden on interstate commerce. This amendment provides that they may do so because there may be circumstances that would so require, but the words are put in, 'compelling local conditions,' and, second, that they 'do not unduly burden interstate commerce.'
>
> This amendment is offered so that we do not have differing safety regulations on equipment moving from State to State in interstate commerce and to try to prevent States from making unreasonable limitations on certain type of equipment. [116 Cong. Rec. 37,622 (1970).]

Similarly when the amendment was introduced in the House, it was equally clear that term product standard applied to design specifications for such equipment as forklifts, tractors or lawn mowers. [116 Cong. Rec. 38,703 (1970).]

The California ethylene dibromide standard in no way prescribes the design or specification of any product—the ethylene dibromide itself or the fruit which has been fumigated with EDB. Rather, the standard is directed at worker exposure and prescribes certain requirements for the reduction of exposure levels, monitoring of exposure and health conditions and training and notification of exposed workers.

Even if the Cal-OSHA standard did constitute a product standard under section 18(c)(2), promulgation of the standard is still permissible if it does not pose an undue burden on interstate commerce and even then it is still valid if required compelling local conditions. The growers have argued that the California EDB standard is responsible for the refusal of certain groups of transport workers, and longshore workers to handle EDB fumigated fruit. This work refusal along with a decision by certain grocery retailers not to buy fumigated fruit has resulted in restraint of trade according to the growers.

While members of the International Brotherhood of Teamsters and the

International Longshoremen's and Warehousemen's Union have refused to handle EDB fumigated fruit, the work refusal is in response to the *hazards* posed by EDB not the Cal-OSHA standard. Workers have refused to handle fruit because they fear that protection is inadequate. Similarly retail grocers have declined to buy the fruit because they fear a risk is posed to workers and consumers from the off-gasing EDB. Staying or overturning the California EDB standard will not reduce or eliminate the hazards posed by EDB and the resultant refusals to work with or buy the fruit. Indeed, if anything, a stay of the standard will result in more widespread refusals to handle EDB fruit, since protection in the absence of a standard will clearly be inadequate.

***

In its comment on the OSHA proposal, the Hawaii Papaya Industries Association argued, as did other business groups, that:

> * * * California's emergency standard has had a direct and devastating impact on interstate commerce in papayas.
>
> California's erroneous interpretation of section 18(c)(2) is demonstrated by the fact that Congress considered and flatly rejected a proposal to limit OSHA review power to state standards for the design of machines. Robert L. Barrett of John Deere & Co., in a letter of September 26, 1969, to Senator Harrison A. Williams, proposed language specifically exempting "the design or construction of machines, apparatus, devices and equipment" from state product safety regulation. *Bills on Occupational Safety and Health, Hearings on S. 2191 and S. 2788 Before the Subcommittee on Labor of the Senate Committee on Labor and Public Welfare,* 91st Cong., 1st & 2d Sess. 1163.
>
> Congress rejected Barrett's proposal, however, and adopted a much broader provision. The amendment that it adopted, the current section 18(c)(2), does not, as Barrett proposed, exempt only one particular kind of product; and it is not, as Barrett suggested, limited to safety standards.
>
> It is clear that Congress did not intend section 18(c)(2) to be limited to product design specifications. For instance, in announcing that either he or a colleague would introduce an amendment to the House bill parallel to Senator Saxbe's amendment in the Senate, Representative Erlenborn stated:
>
>> * * * [A] similar or identical amendment, I believe, will be offered in the House that would limit the variety of requirements in the *State health and safety laws,* and the State requirements in their diversity only if the standards that they applied were necessary to provide a greater degree of safety, and not unduly burden interstate commerce." *Subcommittee on Labor of Senate Committee on Labor and Public Welfare, Legislative History of the Occupational Safety and Health Act of 1970,* at 1018 (June 1971). (Emphasis supplied).
>
> Although Representative Erlenborn also cited the problems posed by a variety of design specifications for machinery, he clearly did not believe that the Saxbe-Railsback amendment would deal only with state product design standards.

To exclude the California emergency standard from scrutiny under section 18(c)(2) would contravene the express language and legislative history of that section. The fact that the California EDB standard on its face does not prohibit the importation, handling or sale of fruit emitting EDB is irrelevant. Health standards seldom directly regulate products that contain toxic substances, though they are applicable to those products. In view of the fact that California distributors will not handle EDB-treated papaya because the State regulation imposes intolerable reporting, monitoring and posting requirements, it violates any sensible meaning of the statute to say that the EDB standard is not applicable to papaya products.[28]

On March 1, 1983, OSHA published a notice approving the permanent California EDB standard. The Agency concluded that the standard "was applicable to a product which is used and distributed in interstate commerce"; however, OSHA further determined that compelling local conditions required the issuance of the standard and there is insufficient evidence that the standard results in a burden to interstate commerce.

Notice of Approval of California Standard on Ethylene Dibromide

48 Fed. Reg. 8610, 8611–12 (1983)

***

2. APPLICABILITY OF THE PRODUCT CLAUSE

Under section 18(c)(2), an examination of whether a state standard poses an undue burden on interstate commerce and is required by compelling local conditions is required if the standard is "applicable to products which are distributed or used in interstate commerce." Written comments submitted by the State of California and by several labor organizations assert, in summary, that the California EDB standard is not "applicable to products" because the standard does not directly regulate the construction or design of a product. The evidence in the record, however, clearly demonstrates that the standard establishes conditions and procedures which restrict the "manufacture, reaction, packaging, repackaging, storage, transportation, sale, handling and use" of the chemical product, ethylene dibromide (EDB), as well as the handling and exposures which may result after EDB has been applied as a fumigant to fruit products. The standard affects producers as well as users in the manufacture, application, and distribution of these products. In addition, both EDB and the fumigated fruit containing EDB unquestionably move in interstate commerce as demonstrated in comments from various produce growers' shippers', and labor organizations and the State of California. OSHA, therefore, must conclude that the California standard, GISO 5219, is applicable to a product which is used and distributed in interstate commerce. The approvability of the standard is determined, however, by virtue of the findings discussed below that

[28]Comments of Hawaii Papaya Indus. Ass'n on the proposed supplement to California plan, OSHA Docket No. H-111 (Dec. 21, 1981).

the standard is required by compelling local conditions and does not unduly burden interstate commerce.*

3. COMPELLING LOCAL CONDITIONS

The Assistant Secretary has determined that compelling local conditions exist which permit the issuance of GISO 5219 by the State of California.

EDB is a chemical that has a number of beneficial uses. For example, EDB is widely used as a component in antiknock compounds in leaded gasoline. Additionally, EDB is used as a fumigant to destroy the adult, egg, and larval stages of fruit flies in fresh fruits and vegetables to meet national and international quarantine requirements. These quarantine requirements generally specify that produce being shipped from areas in which fruit flies or other pests have been discovered be subjected to some approved deinfestation treatment. Written comments of the United States Department of Agriculture suggest that there is no effective or practical alternative to EDB for all quarantine purposes.

The current OSHA standard for EDB was adopted in 1971 as a national consensus standard, under Section 6(a) of the Act. The source of the standard was the American National Standards Institute (ANSI) 1970 recommendation for acceptable concentrations of EDB (ANSI Z37.31-1970). The ANSI exposure limits were intended to protect workers from injury to the lungs, liver, and kidneys which had been observed from excessive, acute, or chronic exposures to EDB in humans and experimental animals. The potential for EDB to cause cancer or reproductive damage was not a basis for the establishment of the current exposure limits for EDB.

On December 18, 1981, OSHA published an Advance Notice of Proposed Rulemaking in the Federal Register (46 FR 61671) concerning proposed changes to the existing EDB standard. In this notice, OSHA summarized much of the current evidence concerning the health hazards associated with EDB. While noting that such evidence does not legally support the adoption of an emergency temporary standard, OSHA acknowledged that toxicological studies indicate that the present permissible exposure limit (PEL) for EDB may not be sufficiently protective of those working with and exposed to EDB.

California is uniquely situated with respect to both the interstate and international distribution of fresh fruit. For example, California is the largest market for Hawaiian papayas. An estimated seventy-eight percent of papayas shipped to the continental United States from Hawaii are normally routed through California. Comments of the Hawaiian Papaya Industry Association and the State of Hawaii Department of Agriculture. Historically, California has been a primary market for Texas and Florida citrus as well. Texas and Florida citrus growers ship as much as twenty percent of their crop for consumption in California. Comments of the American Farm Bureau Federation and Texasweet Citrus Advertising, Inc. Additionally, significant quantities of citrus are routed through California for export to Japan. Comments of the Indian River Citrus League and Texasweet Citrus Advertising, Inc. Thus, it is clear that by virtue of California's size and geographical location, substantial quantities of fresh citrus pass through the State's borders for domestic consumption, distribution to other states or export to foreign countries.

*[Author's note—California state plan officials have said that they intend to comment indicating their concern over OSHA's shift of position in determining for the first time that a toxic-substance standard may be a product standard under §18(c)(2) and, therefore, that states are restricted in their right to issue toxic standards differing from the federal standards. 12 OSHR 832 (Mar. 10, 1983).]

As a result of quarantine requirements of the United States Department of Agriculture and the State of California Department of Agriculture, virtually all fresh citrus (plus some other produce, such as papayas) entering California from areas in which Mediterranean fruit flies (medfly) or similar pests have been discovered, including Hawaii, Texas, and Florida, must be subjected to an approved fruit fly deinfestation treatment. Post-harvest fumigation of citrus and other produce with EDB is generally considered the only effective and feasible method for meeting quarantine requirements. Comments of the Hawaii Papaya Industry Association, Papaya Administrative Committee, State of Hawaii Department of Agriculture, Indian River Citrus League, American Farm Bureau Federation, State of Florida Department of Citrus, Texas Department of Agriculture and the United States Department of Agriculture. In addition, until last May the Japanese government had quarantined the entire State of California and required that nearly all agricultural commodities exported to Japan from California be treated against medfly. For many commodities, this requirement necessitated fumigation with EDB prior to export. Comments of the California-Arizona Citrus League. As a result of this widespread fumigation of citrus and other produce entering the State, commentors alleged that greater than 12,000 California workers will be exposed to EDB. Comments of the Hawaii Papaya Industry Association and the AFL-CIO.

On the basis of this evidence concerning the hazards associated with EDB and evidence concerning the degree of worker exposure to EDB within California, OSHA has determined that adoption of GISO 5219 is sufficiently justified by compelling local conditions.

4. UNDUE BURDEN

The final inquiry in the analysis under section 18(c)(2) is whether the State standard results in an undue burden on interstate commerce. In this regard, OSHA has determined that there is insufficient evidence to support the conclusion that the California standard results in such a burden.

As discussed above, GISO 5219 differs from the current federal EDB standard in several respects. In addition to establishing a more stringent permissible exposure limit than the federal standard, the California standard contains requirements relating to: reporting of use; exposure monitoring; protective clothing; training; posting and notification of shipments; recordkeeping; and medical surveillance which are not included in the federal standard. On the basis of the comments, it does not appear that compliance with these requirements is either infeasible or unreasonably costly.

A number of the comments indicate that levels of EDB off-gassing from fumigated produce can feasibly be maintained below the permissible exposure level of 130 ppb. Comments of Texasweet Citrus Advertising, Inc.; Hawaii Papaya Industry Association; Papaya Administrative Committee; and the State of Hawaii Department of Agriculture. Accordingly, it appears that imposition of the more stringent exposure limitation is not by itself unduly burdensome.

The most prevalent objection to GISO 5219 is that the ancillary requirements of the standard, such as the posting requirement, have resulted in a significant disruption of the interstate movement of citrus into California. It is alleged that adoption of the standard has caused California and foreign purchasers to cancel citrus orders and longshoremen in California to refuse to handle shipments of treated citrus. Comments of the State of Florida Department of Citrus; Indian River Citrus League; Hawaii Papaya Industry Association; Seald-Sweet Growers, Inc.; and DNE Sales International. Citrus industry comments suggest that resultant economic losses range between six and thirty-two million dollars; however, there is no concrete or consistent data from which to assess the actual extent of any loss.

Although OSHA does not dispute the various allegations that losses have been suffered by the citrus industry, it is unable to conclude on the basis of the record that these losses were sustained as a result of any specific provision of GISO 5219. The standard's requirements appear reasonably related to the hazard involved. Additionally, compliance with the requirements of GISO 5219 does not appear to be overly burdensome. In this regard it should be noted that OSHA received no comments suggesting that compliance with GISO 5219 is either infeasible or impractical from California employers who are actually subject to its requirements. Additionally, comments of the AFL-CIO suggest that any refusals by workers to handle EDB-fumigated fruit are in response to the hazards posed by EDB, not the Cal-OSHA standard. On the basis of the record it appears equally likely that any losses sustained by the citrus industry were the result of general fear of the hazards associated with EDB rather than the burden of compliance with the California standard. Accordingly, based on a consideration of all the evidence in the record, OSHA concludes that GISO 5219 does not unduly burden interstate commerce.

***

The issue of the validity of state standards that differ from federal standards has arisen also in the context of OSHA's hazard communication standard.[29] In its preamble to that standard, OSHA noted that approximately 12 states and 6 local governments have some kind of regulation related to the identification of hazardous substances and that about 16 additional jurisdictions have proposed such legislation at one time or another. The preamble further stated that industry representatives had "acknowledged" that the "potential for conflicting or cumulatively burdensome state and local laws" was "immense."[30] In order to "reduce the regulatory burden posed by multiple state laws," OSHA stated that its standard preempts laws in states without approved plans that "deal with hazard communication requirements for employees in the manufacturing sector."[31] States with approved plans would have to submit their proposed hazard communication requirements to OSHA for approval.[32] OSHA made it clear that it considered the hazard communication standard to be applicable to "products" within the meaning of §18(c)(2) "in the sense that it [the standard] permits the distribution and use of hazardous chemicals in commerce only if they are in labeled containers accompanied by material safety data sheets." For this reason, and

[29]The standard is discussed more fully in Chapter 18, Section F. On the issue whether the hazard communication rule is a "standard" or a "regulation," see discussion in preamble, 48 FED. REG. 53,280, 53,320–21 (1983), and discussion of *Louisiana Chemical Ass'n v. Bingham*, 657 F.2d 777 (5th Cir. 1981) (holding OSHA's access rule to be a regulation), Chapter 5, Section F.

[30]See also proposed hazard communication standard, 47 FED. REG. 12,092, 12,095 (1982).

[31]48 FED. REG. at 53,284. The federal standard applies only to employees in manufacturing. 48 FED. REG. at 53,284–87. Under §18(a), states *without* approved plans may assert jurisdiction over any "occupational safety and health issue" on which no federal standard has been issued. Cf. statement of Mr. Auchter, p. 667.

[32]See 29 C.F.R. §§1953.40–41 ("State Initiated Change Supplements;" procedures) (1983).

because of the "strong policy justification for uniform application throughout the distribution system of a national hazard communication standard," OSHA said it would "scrutinize carefully" any state requirements submitted and approve only those meeting the provisions of the "product proviso" of §18(c)(2).[33]

The alleged burdens of differing state hazard communication standards on employers doing business in more than one state was explained by OSHA in 1982 in the following response to questions posed by the Subcommittee on Health and Safety of the House Education and Labor Committee.

> Five [sic] plan States (California, Connecticut, Michigan, and Washington), as well as several other Federal OSHA States, have adopted their own chemical labeling standards or "worker-right-to-know" laws. Dupont contends that the diverse State laws "have a disruptive effect on interstate commerce." The Atlantic Richfield Company has written to OSHA that "this multiplicity of differing and conflicting State and local hazard communication laws imposes an undue burden on products moving in interstate commerce and on multi-state employers." Atlantic Richfield also points out that some States, such as California, use different hazardous substance lists, thereby creating a recordkeeping burden for multi-state employers. Moreover, State laws differ one from the other, compelling the multi-state employer to keep constant track of the different, and sometimes changing, requirements in each State. Atlantic Richfield also contends that different State specifications for contents of Material Safety Data Sheets create an additional cost for manufacturers. The National Paint and Coatings Association, Inc. submitted similar comments to OSHA.[34]

The scope and timing of the preemptive effect of the federal hazard communication standard may be a difficult issue. The federal standard does not become effective until November 1985 and applies only in the manufacturing area. Assistant Secretary Auchter stated that OSHA believes that in states without approved plans, the state hazard communication regulations would be preempted "in all occupational settings" and "at least" when the federal standard becomes effective and "arguably it could happen even sooner." Auchter said that while OSHA would not take preemptive action, it would support the credible arguments of "some other party" in a court of law.[35]

---

[33]This preamble discussion appears at 48 Fed. Reg. at 53,284, 53,322–23.

[34]*State of the Agency Report, supra* note 16 at 223.

[35]Address before the CMA Occupational Safety and Health Committee, Dec. 15, 1983 (mimeo), 11–12; see also 13 OSHR 795 (1983). Three states, New York, New Jersey, and Connecticut, which do not have approved state plans (Connecticut has an approved public employee plan), have joined in the challenge to OSHA's hazard communication standard. The states, among other things, stated that they will argue that the federal standard does not preempt their hazard communication regulations. 13 OSHR 811–12 (1984). See Chapter 18, Section F for a discussion of the substantive provisions of the hazard communication standard.

Part VI

OSHA and Congress

21

Congressional Oversight, Amendments, and Appropriations Riders

It has been frequently observed that in the past forty years Congress has delegated immense regulatory authority to administrative agencies often with only the broadest kind of standards limiting the exercise of that authority.[1] Increasingly, however, there are indications that Congress is insisting on more direct and continuing control over the exercise of powers it delegated to administrators. One of the most controversial and debated of these methods of control is the legislative veto, which involves review by the Congress of regulatory rulemaking actions by an agency.[2] More broadly, the term "congressional oversight" has been used to describe "all activities undertaken by Congress to enhance its understanding of and influence over the implementation of legislation it has enacted."[3] The purpose of this chapter is to discuss the relationship between OSHA—the Agency—and the Congress subsequent to the enactment of the Occupational Safety and Health Act in 1970.

[1]See, e.g., Javits & Klein, *Congressional Oversight and the Legislative Veto: A Constitutional Analysis,* 52 N.Y.U. L. REV. 455 (1977); McGowan, *Congress, Court and Control of Delegated Power,* 77 COLUM. L. REV. 1119, 1133–62 (1977). See also Stewart, *The Reformation of American Administrative Law,* 88 HARV. L. REV. 1669, 1693–97 (1975) (discussing constraints on Congress' ability to specify regulatory policy in meaningful detail).

The judicial doctrine banning unduly broad delegations from Congress to agencies has not been applied by the Supreme Court since the 1930s. See *A.L.A. Schechter Poultry Corp. v. United States,* 295 U.S. 495 (1935). Justice Rehnquist, however, concurring in the Supreme Court decision in the *Benzene* case, would have revived the doctrine and held the first sentence of §6(b)(5) of the Act, 29 U.S.C. §655(b)(5), unconstitutional. *Industrial Union Dep't v. American Petroleum Inst.,* 448 U.S. 607, 672 (1980). See also *American Textile Mfrs. Inst. v. Donovan,* 452 U.S. 490, 543 (1981) (Rehnquist, J., dissenting).

[2]See, e.g., Javits & Klein, *supra* note 1 (favoring veto); McGowan, *supra* note 1 (opposing veto); Bruff & Gellhorn, *Congressional Control of Administrative Regulation: A Study of Legislative Vetoes,* 90 HARV. L. REV. 1369 (1977) (opposing veto). In June 1983, the Supreme Court ruled that legislative veto provisions were unconstitutional. *Immigration & Naturalization Service v. Chadha,* 51 USLW 4907 (1983) (Powell, J., concurring; Rehnquist & White, JJ., dissenting). See also Kaiser, *Congressional Action to Overturn Agency Rules: Alternatives to the Legislative Veto,* 32 AD. L. REV. 667 (1980).

[3]Javits & Klein, *supra* note 1 at 460.

A. Oversight

Both the Senate and House of Representatives labor committees have regularly conducted general oversight hearings on OSHA. Other committees, such as the House Government Operations Committee and the Small Business Committee, from time to time have also held hearings on particular aspects of the OSHA program touching upon their interests. In some cases, hearings were held to investigate how OSHA handled a particular event, for example, the Willow Island cooling tower collapse that occurred in 1978 and the Kepone incident in 1976.[4] In other cases, hearings have dealt with broad occupational safety and health issues. The House Government Operations Committee has evinced a particular interest in the OSHA program for federal employees and the protection of employees against chemical hazards in the workplace,[5] and in 1980 the House Labor Committee conducted a hearing to consider OSHA state-plans policy.[6] Finally, the general oversight hearings of the labor committees frequently cover a broad range of issues relating to Agency activity.[7]

Oversight hearings give members of the public an opportunity to state their criticism, and less frequently their praise, of OSHA. On behalf of the Agency, the Assistant Secretary of OSHA usually testifies, and sometimes the Secretary is also asked to testify.[8] On occasion, an oversight hearing results in a committee report with specific recommendations for action by the Agency.[9] Frequently no report is issued, as was the case following the Kepone hearings. In either event, the committee's point of view on the issue at hand is usually quite clear to OSHA, and a response, frequently favorable, from OSHA often takes place. A striking example is OSHA's change in its policy in response to informal complaints that followed the Kepone hearings of the House labor committee.[10] In sum, Congress' oversight

---

[4]These hearings are discussed in Chapter 13, Section B.

[5]*Safety in the Federal Workplace: Hearings Before the Manpower and Housing Subcomm. of the House Comm. on Government Operations,* 94th Cong., 1st Sess. (1975); *Control of Toxic Substances in the Workplace: Hearings Before the Manpower and Housing Subcomm. of the House Comm. on Government Operations,* 94th Cong., 2d Sess. (1976).

[6]*Oversight Hearings on OSHA—Occupational Safety and Health for Federal Employees, Part 4: State Plans: Hearings Before the Subcomm. on Health and Safety of the House Comm. on Education and Labor,* 96th Cong., 2d Sess. (1980). These hearings are discussed in Chapter 20, Section B.

[7]See, e.g., *Occupational Safety and Health Act Review: Hearings Before the Subcomm. on Labor of the Senate Comm. on Labor and Public Welfare,* 93d Cong., 2d Sess. 47–48 (1974). See discussion *infra* this section.

[8]See, e.g., *Oversight on the Administration of the Occupational Safety and Health Act, 1980: Hearings Before the Senate Comm. on Labor and Human Resources,* 96th Cong., 2d Sess. 952–1082 (1980) (testimony of both Secretary of Labor Marshall and Assistant Secretary of Labor Bingham).

[9]See, e.g., SUBCOMM. ON ENERGY, ENVIRONMENT, SAFETY AND RESEARCH OF THE HOUSE COMM. ON SMALL BUSINESS, IMPACT OF THE ADMINISTRATION OF THE OCCUPATIONAL SAFETY AND HEALTH ACT ON SMALL BUSINESS, H.R. REP. NO. 757, 95th Cong., 1st Sess. ch. 5 at 49–51 (1977) (containing recommendations to OSHA, NIOSH, EPA, the Review Commission, the General Accounting Office, the Small Business Administration, appropriate committees of Congress, and Congress on OSHA's program for small business).

[10]The Kepone hearings and the OSHA response are discussed in Chapter 13, Section C. See also the discussion of OSHA's response to the investigation by the Senate Labor Committee and the Comptroller General of its policy on issuing serious citations, in Chapter 14, Section C.

activity has functioned, undoubtedly as intended, as a continuing and mostly effective monitoring device over OSHA's policies implementing the program.

Each assistant secretary of OSHA, beginning with George C. Guenther in 1971 to the present, has been called on by Congress to present detailed explanations of what OSHA was doing and why. One of the earliest OSHA oversight hearings was held in 1972 before the Subcommittee on Agricultural Labor of the House Education and Labor Committee. Assistant Secretary Guenther, accompanied by Deputy Assistant Secretary Chain Robbins and Associate Solicitor Mintz, were questioned closely by Chairman O'Hara of Michigan on the Agency's policies for protecting farm workers.[11]

Testimony of Assistant Secretary Guenther

***

MR. O'HARA. Mr. Guenther, in your statement you make reference to your authority to issue temporary emergency standards and you quote provisions of 6(c)(1) of the act which provide "that temporary standards shall be issued," and that is mandatory, not permissive, "if the Secretary determines, that employees are exposed to grave danger from exposure to substances or agents determined to be toxic or physically harmful or from new hazards and that, (b) such emergency standard is necessary to protect employees from such danger."

Have you promulgated any emergency standards at all, agriculture or otherwise, under this authority?

MR. GUENTHER. Yes, Mr. Chairman, we have. One temporary emergency standard on asbestos. That was last December, which became a permanent standard in June of this year.

MR. O'HARA. That was based on indisputable evidence that asbestos workers had a very high incidence of cancer of the lung and that it was pretty well established that the asbestos fibers lodged in the lung and their presence there combined with other carcinogenic factors, led to the development of lung cancer at a higher rate than among the general population, right?

MR. GUENTHER. That is a fair statement, Mr. Chairman.

MR. O'HARA. With respect to pesticides, are there not some pretty well established hazards associated with contact with certain pesticides?

MR. GUENTHER. Mr. Chairman, I am not a pesticide expert so that it is impossible for me to give you the knowledgeable answer. My limited experience would lead me to believe that there are, in fact, some pesticides, exposure to which is highly hazardous.

MR. O'HARA. In any event, you indicate that you plan to propose permanent pesticide standards this winter; is that right?

MR. GUENTHER. Yes, sir.

MR. O'HARA. And you feel that farm equipment standards are of equal significance to the pesticide standards, so let me ask you when do you hope to have farm equipment standards?

MR. GUENTHER. I would say, Mr. Chairman, that within the next 6 to 9 months, somewhere in that neighborhood, we should have a new proposed standard for farm equipment.

[11]*Farm Workers Occupational Safety and Health: Oversight Hearings Before the Subcomm. on Agricultural Labor of the House Comm. on Education and Labor,* 92d Cong., 2d Sess. 15–20 (1972).

MR. O'HARA. Let me ask you about this permanent standard procedure. You are hoping to have permanent standards proposed for pesticides this winter?

MR. GUENTHER. Yes, sir.

MR. O'HARA. And proposed for farm equipment or at least some farm equipment in 6 to 9 months? What is the procedure after you propose the standard?

MR. GUENTHER. A standard is proposed by publication. I think Mr. Mintz, my legal adviser here, who is most expert in this area, might want to answer that question.

MR. MINTZ. A standard is proposed by publication in the Federal Register. Thirty days is given for public written comment. In addition, a member of the public has an opportunity to request a public hearing. If such hearing is requested, we would hold a public hearing. Then the Assistant Secretary would evaluate the written testimony, the written comments, the transcript of the public hearing and any recommendations of an advisory committee; and at that point he would publish the final standard in the Federal Register.

MR. O'HARA. The hearing procedure, is that provided under the Administration Procedures Act?

MR. MINTZ. The hearing procedure is provided in some detail in section 6 of the Occupational Safety and Health Act.

MR. O'HARA. Assuming that the public comments and public hearing is requested, about how long would it be from the closing of standards to their final promulgation?

MR. MINTZ. There are some rather specific time limits provided for in the act and we calculated it would be several months from the initial proposal to the final promulgation. We could add up those numbers and come up with a more specific figure.

***

MR. O'HARA. Mr. Guenther, I am somewhat concerned. If maximum times are to be taken to develop standards, as you know, the committee can take 270 days to develop the standards, 60 days for publication of the proposed rule, 30 days for notice of hearing, 60 days after the completion of the hearing to issue a rule, and then up to 90 days' delay to acquaint employers with the standard. That's 17 months.

So, it could run on into a very considerable period of time and I would like to have from you a statement that you are going—the Department is going to accelerate those periods of time in connection with this pesticide standard.

Is that your intention?

MR. GUENTHER. That is our intention as stated in our testimony.

MR. O'HARA. There are a couple of other points that I would like to make.

In the first place, I really think that an emergency standard might be appropriate and I would like to suggest for your consideration, Mr. Guenther, once you have a good proposed standard which you hope to have within the next couple of months on pesticide, that you promulgate it as an emergency standard, because I am sure you will find the health hazards, while programs not as severe as asbestos, in connection with some of the pesticides that have been used as substitutes for DDT, which are highly toxic, that you consider the promulgation of an emergency standard with respect to them while the hearing process and so forth continues. Does that make any sense to you?

MR. GUENTHER. Yes, it does, Mr. Chairman, and we have at no instance indicated that we have out of hand dismissed the possibility of a temporary emergency standard. I don't believe our testimony states that. That is still an open issue within the Department.

MR. O'HARA. One of the things about the way in which agriculture has been handled that concerns me is that when we had the act before us, there was a good deal of talk about the "worst-first" idea, that first we would go into the most hazardous areas and indeed, in part, you have done that. Your five target industries include some quite hazardous ones, right? Such as construction and longshoring.

MR. GUENTHER. Obviously, we could not make construction a target industry with the limited number of personnel in the early going. [T]hat would have been impossible. The purpose of the target industry program is to have an impact and we hope to be able to show that we have had an impact in relatively precise industry areas to serve as an example, a proving ground, and a pilot program.

MR. O'HARA. So, taking the three most hazardous industries, which are mining, construction and agriculture, mining is not in your bill, right? Mining is in the Coal Mine Safety Act and in the Metal and Non-Metallic Mine Safety Act, neither of which is administered by the Department.

MR. GUENTHER. That is correct.

MR. O'HARA. So that leaves you with the two most dangerous industries, which are the construction industry and the agriculture industry. The construction industry, you have made some approaches to that and you have put them on your target list.

MR. GUENTHER. Not as a complete industry. We have roofing and sheet metal which is one of the most hazardous elements of the construction industry, but we have been inspecting across the board.

MR. O'HARA. But there is no aspect of agriculture, including pesticide application or indeed working in fields in which pesticides have recently been applied, no area of agriculture is on that target list, right?

MR. GUENTHER. That is correct for inspection purposes.

MR. O'HARA. For inspection purposes.

MR. GUENTHER. That is correct. We have had agriculture as far as our program objectives and priorities in the standards setting area. It has been clearly one of the areas of need from the very beginning and we have addressed ourselves among many concerns handed to us by this very complicated piece of legislation to begin the work in agriculture which we recognize must be done.

MR. O'HARA. That is my concern, that you perhaps have not given agriculture the priority that the occupational hazard associated with it would suggest it deserves and I am going to urge sure that you do.

***

Further, there isn't any question about the toxicity of some of these pesticides, is there? The questions have to do with the proper reentry times. There is no question about the hazards associated with contact and breathing of the vapor or contact with the skin of these agents, is there?

MR. GUENTHER. It would depend, Mr. Chairman, in much the same sense as exposure to carbon monoxide in a tightly confined room can be quickly fatal, but exposure to it walking down a street is not. It depends on the type of application, the type of exposure and the concentration of the substance as used in a particular application.

MR. O'HARA. I think that there is some misunderstanding. I mean with respect to the particular substance. There are questions about the reentry time, about the conditions under which one works, whether he uses gloves or doesn't, but there is not any question but what some of these pesticides are highly toxic if the vapor is breathed or inhaled or if there is direct contact with the skin, especially with broken skin. There isn't any question about that, they are highly toxic, right?

MR. ROBBINS. Yes. It depends on the chemical and there are so many of them.

MR. O'HARA. Parathion, for example.
MR. GUENTHER. The answer is yes. You are correct.
MR. O'HARA. Mr. Guenther, I want to thank you very much. I want you to know that it is not my intention that you leave this hearing with the impression that we want you to slow down on your enforcement of OSHA in agriculture.
MR. GUENTHER. Mr. Chairman, your impression is a clear one.
***

This hearing involved a scenario that has often been repeated in OSHA oversight hearings. A committee in Congress is interested in the occupational hazards of a particular industry—for example, agriculture—and presses OSHA for prompt standards and enforcement action. Because of delays in the standards' promulgation process and in light of the asserted seriousness of the hazards involved, the committee calls on OSHA to promulgate an emergency temporary standard. The committee also urges that OSHA assign more compliance resources to conduct inspections in the industry. Partially in response to the subcommittee's interest in a pesticide standard, OSHA eventually did promulgate an emergency temporary standard covering pesticide reentry hazards. The standard, however, elicited much public and congressional criticism and was stayed, and later vacated, by the Court of Appeals for the Fifth Circuit.[12]

Congress itself has not always been so inclined to encourage OSHA enforcement in agriculture. In 1976, Congress enacted a rider to the OSHA appropriations' act prohibiting OSHA enforcement in small farms.[13]

Assistant Secretary Stender, serving from April 1973 to mid-1975, also appeared on several occasions at oversight hearings. On July 30, 1974, for example, he appeared before the Subcommittee on Labor of the Senate Committee on Labor and Public Welfare and was questioned on a broad range of topics, including the protection of employees of U.S. Steel's Clarendon Works against coke oven emission hazards, and OSHA's distribution of (or failure to distribute) publications to familiarize employers with the requirements of standards.[14] Of particular concern to Senator Schweiker of Pennsylvania was OSHA's delay in setting up an advisory committee to make recommendations on a proposed coke oven standard. The following discussion took place:

SENATOR SCHWEIKER. Why has it taken so long to set up the advisory committee for these particular standards? Earlier this spring I contacted your Department and I see in your appendix [listing the status of

[12] *Florida Peach Growers Ass'n v. Department of Labor,* 489 F.2d 120 (5th Cir. 1974). See Chapter 4 for a discussion of OSHA experience in regulating pesticides.

[13] See Section C, this chapter. See also *OSHA Regulations Affecting Agriculture: Hearing Before Subcomm. on Department Operations, Investigations and Oversight of the House Comm. on Agriculture,* 94th Cong., 2d Sess. 15–47 (1976).

[14] *Occupational Safety and Health Act Review, 1974: Hearing Before the Subcomm. on Labor of the Senate Comm. on Labor and Public Welfare,* 93d Cong., 2d Sess. 221–50 (1974).

standards] that you still have not appointed a committee up until this coming month. I have a hard time understanding why just the advisory committee proceeding takes so long.

MR. STENDER. Well, of course, first we have to find out from industry and labor who they would nominate from their particular interests. Both industry and labor are involved, as you know, in the advisory committees under the statute. We have to determine who they nominate and get an input from them as to who they have confidence in and can well represent their interest in the advisory committee.

Then we have the public members and we have to make some determination as to who will serve, who is competent to serve from the public sector. Then we have to determine who are the professionals in the health and safety field that are competent to serve and will in fact accept a nomination if it is offered to them. So taking that all together takes a considerable amount of time. We wish that it could be done much more quickly but, unfortunately, that's the fact here in this particular situation as well as in every one of them.

I'm an impatient man. I'm not a bureaucrat. I came here from the private sector. I'm used to having things done more quickly and I can't get used to why it takes so long and you have asked and that's the reason it does.

SENATOR SCHWEIKER. I'm glad it bothers you anyway. It bothers me too.[15]

Dr. Morton Corn, who became Assistant Secretary for OSHA late in 1975, was the first health professional to head the Agency. He was quickly confronted with House Labor Committee hearings relating to the Kepone incident held in Hopewell, Virginia on January 30, 1976.[16] Several months later, the same committee held hearings in Washington, D.C. on OSHA enforcement of the lead standard issued under §6(a) and on OSHA's proposed new lead standard. The colloquy between Chairman Daniels and Dr. Corn also dealt with the issue of exposure of women of child-bearing age to toxic substances in the workplace.[17] Congressman Sarasin of Connecticut also participated.

Testimony of Assistant Secretary Corn

***

MR. DANIELS. I realize that in the past our efforts have been directed to the male in the workplace but you have just mentioned that 36 million women are now engaged in our workplace also. You must bear in mind that there has been a substantial increase in the last decade. I think the number of working women has jumped from 24 million to 36 million. Now, in view of the fact that women are playing such an important part in our business and commercial activity today, I think that OSHA should look into this question to see what effect this type of occupationally-related disease will have upon them.

---

[15]*Id.* at 244–45. OSHA issued a coke oven emissions standard in 1976, 41 FED. REG. 46,742 (1976), which was upheld by the Court of Appeals for the Third Circuit, *American Iron & Steel Inst. v. OSHA*, 577 F.2d 825 (3d Cir. 1978). After the Coke Oven Standard Advisory Committee, which was chaired by Dr. Bingham, OSHA used standards advisory committees much less frequently. See Chapter 10, Section B.

[16]*Oversight Hearings on the Occupational Safety and Health Act, Occupational Health Hazards: Hearings Before the Subcomm. on Manpower, Compensation, and Health and Safety of the House Comm. on Education and Labor*, 94th Cong., 2d Sess., pt. 2 at 144–63 (1976).

[17]*Id.* at 410–12.

DR. CORN. Mr. Chairman, I could not agree with you more, but at the same time I am cognizant of the enormous ramifications of the possible regulatory action in this field. For example, if we deny the right of women to work in certain environments, we may be in conflict with concurrent efforts to assure them of equal employment opportunity. Will we request that they demonstrate they are no longer planning or are capable of childbearing? Will we insure retention of wage rates if we remove them from a hazardous job to a lower paying job? Will OSHA regulate these things?

I think these questions are in our minds as we gather the evidence on this issue. While I do not wish to delay action on this matter, I am very appreciative of the fact that we are into an area of social regulation, if you will, that has enormous ramifications.

MR. DANIELS. I agree with you that we cannot nor should not prohibit women from entering the workplace, but I do believe that regulations may be established and standards promulgated for their protection.

DR. CORN. Yes; but, Mr. Chairman, if this is done and we find that exposure of women of child-bearing age to harmful agents must be kept at a lower level, the very first response to this regulation on the part of employers may be not to hire women. That is predictable. In fact, I have also heard by word of mouth that this is occurring in certain isolated places today with respect to lead. So the promulgation of the standard is only the first step. What I am saying is that an enlightened regulator should try to anticipate and deal with the moves that will follow the issuance of a standard, and we are trying to do that.

MR. DANIELS. Do you think that your proposed lead standard will be sufficient to protect individuals with sickle cell trait?

DR. CORN. That is another matter I cannot answer positively. We are seeking further information on that point. The limit of exposure we are recommending in the proposed standard is one-half of the present one. However, we do flag certain issues where more information would be very helpful and this is such an issue.

MR. DANIELS. Dr. Corn, the promulgated regulation establishes 200 micrograms per cubic meter of air as determined on an 8-hour time-weighted average. Now you propose to change that regulation by reducing the 200 micrograms to 100 micrograms, is that true?

DR. CORN. Yes.

MR. DANIELS. Do you think that is a safe level?

DR. CORN. With the exception of the uncertainties of what I call special risk groups, women and those predisposed to sickle-cell anemia, I would answer yes. With respect to the healthy adult male at the workplace, I think 100 micrograms is a well-chosen level.

If you recall, our research agency, the National Institute of Occupational Safety and Health, advised us that they did not have sufficient hard data at their disposal to focus in any closer than somewhere between 50 and 150 micrograms per cubic meter. They informed us that this is the range which should be considered on the basis of the shakiness of the evidence. We have proposed and documented our reasons for selecting 100 micrograms, which happens to be the mid-point of the range recommended by NIOSH.

MR. DANIELS. Thank you very much.

Mr. Sarasin, do you have any questions?

MR. SARASIN. Yes, Mr. Chairman. Thank you very much.

I would like to join Chairman Daniels in complimenting Dr. Corn for his evidence in this field. I think we are very fortunate, Mr. Chairman, to have a man of Dr. Corn's qualifications and also his determination to now head the Occupational Safety and Health Administration.

***

In issuing a final lead standard in November 1978 imposing a 50-microgram permissible exposure level (PEL), OSHA expressly directed its attention to the concern raised by Chairman Daniels on the need to protect women employees from fetal hazards resulting from lead.[18] The Agency concluded that lead affected both men and women and, in order to protect against the hazard, the PEL should be no greater than 30 micrograms. Based on feasibility considerations, however, OSHA set the 50-microgram level and stated that the adverse reproductive effects of lead would be minimized by means of: (1) an action level of 30 micrograms; (2) the medical surveillance program, which included medical removal protection; and (3) education and training of employees.[19]

Dr. Eula Bingham appeared frequently at oversight hearings during her term of office between March 1977 and January 1981. On several occasions, she was compelled to justify her policies before hostile committees.[20] Early in her administration, Dr. Bingham appeared before a generally sympathetic subcommittee of the House Government Operations Committee and was questioned, among other things, on her views on health enforcement and on the use of OSHA emergency authority under §6(c) by Congressman Andrew McGuire of New Jersey, and by Chairperson Cardiss Collins of Illinois, as well as by Joseph C. Luman, Staff Director of the subcommittee.[21]

Testimony of Assistant Secretary Bingham

***

MR. MAGUIRE. You are at the beginning of a rather important and I suspect difficult enterprise.

As you know, criticisms have been made of OSHA for having concentrated almost entirely on safety issues as opposed to health issues. As you

[18] 43 FED. REG. 52,952, 52,959–60 (1978). The Court of Appeals for the District of Columbia Circuit upheld the standard for the most part against an industry challenge. *United Steelworkers v. Marshall,* 647 F.2d 1189 (D.C. Cir. 1980), *cert. denied sub nom. Lead Industries Ass'n v. Donovan,* 453 U.S. 913 (1981). One of the major issues in the litigation was whether OSHA had given adequate notice of the 50-microgram level. See Chapter 8, Section C. Dr. Corn's statements at this 1976 hearing suggest that, at least in the mind of the then Assistant Secretary, a 50-microgram lead PEL was an issue under consideration.

[19] 43 FED. REG. 52,952, 52,965–66 (1978). See Chapter 5 on the implementation of the medical surveillance provisions. OSHA also attempted to use the general duty clause to cite an employer for allegedly requiring women employees of childbearing age to undergo sterilization to avoid workplace hazards, but this citation was vacated by the Review Commission in the *American Cyanamid* case. 9 OSHC 1596 (Rev. Comm'n 1981), *appeal pending, OCAW v. American Cyanamid Co.,* No. 81–1687 (D.C. Cir. 1984). The subject of reproductive hazards is discussed in Chapter 14, Section B. Dr. Corn's concern with the "ramifications" of regulatory action in this field, particularly the potential "conflict" with principles of equal employment opportunity, anticipated considerable debate on the issue among commentators on both OSHA and Title VII of the Civil Rights Act of 1964, 42 U.S.C. §2000e to e-17 (1976 & Supp. V 1981). See Chapter 14, note 55, and works cited there.

[20] One example was her testimony before a subcommittee of the House labor committee explaining her policies on state plans. *Oversight Hearings on OSHA—Occupational Safety and Health for Federal Employees, Part 4: State Plans: Hearings Before the Subcomm. on Health and Safety of the House Comm. on Education and Labor,* 96th Cong., 2d Sess. 500–43 (1980).

[21] *Performance of the Occupational Safety and Health Administration: Hearings Before the Subcomm. on Manpower and Housing of the House Comm. on Government Operations,* 95th Cong., 1st Sess. 77–78, 92 (1977).

also know, some one in four Americans alive today is going to get cancer in some form. As I am sure you also know, the best scientists also now tell us that 70 to 90 percent of cancers are caused by environmental pollutants of one sort or another. Of course, the workplace is one of the places where people may be exposed to and possibly receive concentrated doses of pollutants.

In view of this, I wonder if you could elaborate on your own opening statement with respect to the point about environmental health hazards versus safety in terms of allocation of resources, time, energy, and personnel that you anticipate for OSHA in the future.

DR. BINGHAM. First, let me emphasize that we do not intend to stop enforcing safety standards. There are employees in the construction industry and in other industries who are still meeting with accidental death. It would be a mistake, therefore, to stop enforcing safety standards.

MR. MAGUIRE. Let me hasten to indicate that I am not advocating that enforcement of safety standards be abandoned.

DR. BINGHAM. When you discuss health standards, people often assume that OSHA intends to stop conducting safety inspections. That is not my intention at all.

As you know, my own background has been in the health area. I was concerned with occupational safety and health problems before there was an OSHA Act. My primary interest is the elimination of occupational health problems, and my second concern is the elimination of cancer-causing chemicals in the environment. Similarly, I feel it is imperative that we eliminate exposure to materials in the workplace that induce irreversible disease, such as dusts that irreversibly damage the lungs and pesticides which cause nonregenerative nerve damage.

We are now going out and actively recruiting scientific and medical personnel. I plan to place great emphasis on health in the future. We are also exploring new ways of promulgating generic standards which cover more than one chemical substance at a time. In the meantime we will continue to promulgate standards on single substances. Quite honestly, I plan to stretch the resources of the agency in putting out health standards, and I intend to use the emergency temporary standard whenever employees are exposed to grave danger.

MR. MAGUIRE. I am very glad to hear that.

DR. BINGHAM. All I can say is watch the *Federal Register*.

***

MRS. COLLINS. We note also that you intend to use the emergency temporary standards for not only unregulated substances, but also where present standards are inadequate. GAO noted OSHA's refusal to issue such standards where it would take more than 6 months to develop permanent standards.

What is your position on this?

DR. BINGHAM. I do know that OSHA, in the past, has issued very few emergency temporary standards.

MRS. COLLINS. What are you planning to do in the future?

DR. BINGHAM. We are going to issue emergency temporary standards whenever necessary. We are going to press to meet the 6-month time limit. I think it is a debatable legal point as to whether an emergency temporary standard runs out in 6 months or whether it actually stays in effect for longer periods of time. I am willing to test it to see whether or not we can make it stay in effect for longer periods of time. I am willing to test a number of things in court. Perhaps some of the attorneys will not be too thrilled about that.

MR. LUMAN. The GAO noted that there was a legal difference of opinion in the way they read it and the way that OSHA read it. I noticed that you have the top legal man with you. I wondered if he would care to comment.

MR. MINTZ. As a general matter, let me say that the OSHA lawyers will support the Assistant Secretary. [Laughter.]

We have read the legislative history behind section 6(c) and our conclusion was that the standard does expire at the end of the 6-month period. However, GAO has presented another point of view. The issue is an important one. We would like to sit down with the GAO people and anyone else to see exactly what other possible constructions of that language and legislative history are present.

Surely from a policy and a protection point of view there is much advantage to making sure that where grave danger exists, exposed employees will receive adequate protection for as long of a period of time as is necessary.

MR. LUMAN. It would appear from the GAO reading that there are two legal interpretations. So if the Assistant Secretary wanted to choose one instead of the other, she could do so.

MR. MINTZ. Mr. Luman, the GAO report as I read it, and I think I am correct, had a conclusion but there was no specific analysis. I believe that it would be very useful to have a dialog on GAO's analysis of the issue.

***

Dr. Bingham continued, and strengthened, Dr. Corn's policy in placing increased emphasis in the area of health. Thus, for example, in fiscal year 1975, there were 5,522 OSHA health inspections out of a total of 80,945 inspections; in fiscal year 1977, there were 8,943 health inspections out of a total of 59,836; and in 1979, 11,160 health inspections out of 57,910.[22] And if the committee members had "watched" the *Federal Register,* as urged by Dr. Bingham, they would have seen notice of the issuance of a series of health standards covering benzene, arsenic, cotton dust, lead, DBPC, and acrylonitrile, as well as the Carcinogens Policy and the rule on access to exposure and medical records.

The House committee's interest in the use of ETSs was shared by Dr. Bingham who issued three ETSs during her administration: in May 1977 for benzene; in September 1977 for DBCP; and in January 1978 for acrylonitrile. The benzene ETS was stayed by the Court of Appeals for the Fifth Circuit and never went into effect. The acrylonitrile and DBCP emergency standards, however, remained in effect for the entire six-month period and were replaced by permanent standards. Whether an ETS may remain in effect beyond the six-month statutory period was never litigated, and OSHA has accepted the view that these standards expire at the end of six months.[23]

On occasion, the questioning of an Assistant Secretary at an oversight hearing becomes quite hostile. An example is the exchange between Senator Edward Kennedy of Massachusetts and Assistant Secretary Auchter at a joint oversight hearing before two subcommittees of the Senate Labor Committee on September 23, 1981.[24]

[22]Statistics from OSHA's Office of Management Data Systems (Mar. 1980).

[23]See discussion on emergency standards in Chapter 5.

[24]*Oversight on the Administration of the Occupational Safety and Health Act, Joint Hearing Before the Subcomm. on Investigations and General Oversight and the Subcomm. on Labor of the Senate Comm. on Labor and Human Resources,* 97th Cong., 1st Sess. 79–82 (1981).

Testimony of Assistant Secretary Auchter

***

SENATOR KENNEDY. Thank you, Madam Chairman.

At the conclusion of the other period of questioning, I was talking about the nature of some of these substances and the number of workers who were affected. I will just mention the numbers for the record at this time, I think we were down to cadmium, which is 100,000; chromium, 1.8 million; nickel, 2.7 million; anesthetic gas, 214,000.

I understand that NIOSH has recommended that OSHA promulgate health standards for over 100 toxic substances. Is that correct?

MR. AUCHTER. I do not know, Senator. I do not know the answer to that.

SENATOR KENNEDY. Well, don't you think you should know?

MR. AUCHTER. I expect I will know, about 30 minutes after this hearing is over, but I really do not know right now. I know we have a priority list. The first thing that I did with regard to standards was to see what was in our promulgation process when I first got here. Those were the first things that I had to bring to fruition.

[The following information was supplied for the record:]

NIOSH RECOMMENDATIONS FOR HEALTH STANDARDS

In response to Senator Kennedy's question on the number of NIOSH criteria documents, I would like the record to show the following:

NIOSH has submitted 104 criteria documents to OSHA covering approximately 150 toxic substances.

SENATOR KENNEDY. Well, I would be less than candid if I did not say that I am troubled if you do not know the scope of the problem or the number of workers who are affected, or the recommendations that are even being made within the department on this subject matter.

I do not know whether the millions of workers in this country would feel that their interest or their health is being adequately protected.

How many standards for the substances that I have mentioned here were issued under the Carter administration; do you know?

MR. AUCHTER. No, I do not, Senator.

SENATOR KENNEDY. Do you know which of these standards you have delayed since assuming your responsibilities?

MR. AUCHTER. Yes, sir.

SENATOR KENNEDY. And how many have you suspended?

MR. AUCHTER. Well, the standards that were promulgated, I have suspended one, walkaround pay—that is, of standards that were finalized. Of the major health standards that were promulgated—

SENATOR KENNEDY. Health standards that have been delayed by you that were promulgated by the Carter administration in the area of these toxic substances; how many have you delayed in whole or in part?

[The following information was supplied for the record:]

PREVIOUS HEALTH STANDARDS FOR PARTICULAR SUBSTANCES

I am providing the following information in response to Senator Kennedy's question regarding standards covering particular substances promulgated by the previous administration:

OSHA has had standards for chromium, cadmium and nickel since 1971. During the Carter Administration, the agency had tentative plans for revisions to these standards and for a proposal governing waste anesthetic gases, but none of these actions was ever formally proposed in the Federal Register.

OSHA has promulgated 11 health standards for which NIOSH had previously submitted criteria documents. The agency has proposed, but not finalized, standards for an additional six substances or classes of substances.

MR. AUCHTER. It would depend on how you define it, I guess. Hazardous labeling, I have delayed that; the hearing conservation amendment, I delayed, reviewed, and then reissued most sections of the amendment. I think that is about it.

SENATOR KENNEDY. Well, I think in the presentation that you made, the cancer policy, you have delayed; you have delayed lead; cotton dust, you have delayed; noise, you have delayed; access to medical records, you have delayed—all of which have been listed as high priority, all of which have been subject to extensive scientific examination and legal review before this President became President.

MR. AUCHTER. Senator, I delayed nothing in cotton dust. That is an error. On lead, in fact, I recently extended medical removal protection for lead to certain industries. The access to medical records standard in construction, I stayed on April 28, because it had not been through the Construction Advisory Committee. They reviewed it, made about 13 recommendations; I accepted 10 or 11. That stay was lifted some week or 10 days ago, after the conclusion of a public comment period.

The promulgation process, certainly, is an ongoing process, and there are always reasons for reopening the record, reassessing what is going on. We are not dealing with a static world or a static society. New information is always coming up, and new problems, especially when you are talking about the area of health standards, especially when you are dealing in subjects such as cotton dust and lead, that are as difficult and as complicated as anything we have, and they embody all of the different parts of OSHA's area of responsibility, the forefront of health regulations, the question of engineering controls, personal protective equipment, extending time periods for various and sundry industries, numbers of employees affected, and so forth.

So I think it is absolutely appropriate that we do continue to reopen the record when evidence warrants that.

SENATOR KENNEDY. Well isn't it true, in effect, that you have effectively suspended those standards for arsenic and asbestos, access to employee exposure?

MR. AUCHTER. No sir.

SENATOR KENNEDY. Well, you have, as I understand, assigned two-thirds of your staff—these are 13 project officers and 7 supporting professionals—to review all of those standards, for over a 2-year period—according to this document. This is a document you have signed. This is part of a brief, or the material that was submitted with the brief, which you evidently signed. You set out the priorities, and then you continue—and I will just read—

> Several of these standards address serious health hazards affecting an extremely large number of employees: access to employee exposure, medical records, noise, cancer problems, labeling, respirators, relationship of personal protective equipment. The remaining four address life-threatening hazards and a large number of employees, arsenic, asbestos, lead, and cotton dust. These high-priority standards impose large costs on employers.

I suppose this is where some people have thought that you were bending to the employers' interests.

> OSHA will review them with the view of reducing the unnecessary burdens in accordance with the President's directive. These high-priority standards also may impose substantial economic impact on small business, and OSHA has decided to review them in order to minimize the impact. OSHA has considered that these high-priority standards are, in most cases, major rule-making projects to which OSHA has been committed for several years.

So you cannot say that these are fully in effect—at least, according to this document.

MR. AUCHTER. Senator, of course we can, because the standards that are in effect remain in effect. We continue to enforce them. We continue to gather data all the time. There are new studies that are being done all the time. The Regulatory Flexibility Act requires us, by the way, to prioritize our existing standards for review, and certainly, that is part of our process. We have to continue to reprioritize on a regular basis, based on the information that comes in.

SENATOR KENNEDY. Well, let us understand, then. You have not delayed those particular standards which were promulgated under the Carter administration that deal with those particular substances; they are in effect, then, today, are they?

MR. AUCHTER. Those that were in effect remain in effect. I did withdraw some that were in the promulgation process for reconsideration. But if I am going to abolish a standard, if I am going to eliminate a standard, I have to go to "Notice and Comment." There is a procedure that I have to go through to reopen the record, to build a data base upon which to base a decision. I cannot walk in arbitrarily with the stroke of a pen and change regulations. That would be totally inappropriate. Data has to tell us which way to move.

SENATOR KENNEDY. And I guess two-thirds of your staff are assigned to that kind of a review? That is what it says in this document. If you have got something else—but the time is moving on.

MR. AUCHTER. Well, I guess that is the case.

SENATOR KENNEDY. You say,

> The remainder of the OSHA health standards staff, three project officers and four supporting personnel, will assist in the development of highest priority standards detailed above and will also examine other standards which OSHA believes merit consideration. The second group includes cadmium, anesthetic gases, chromium, nickel, radio frequency radiation. Five of these hazards pose potential life-threatening risks of cancer.

Then it goes on,

> It is estimated that the dedication of OSHA's health standards staff to the aforementioned activities will occupy the full time and energies of the staff for at least two years.

Now, what are you going to tell the workers in those areas that will be affected—the areas I have just mentioned—relative to the protection of their health over the period of the next 2 years?

MR. AUCHTER. I would tell them the same thing that I have told other people and other organizations, that we are a regulatory agency, that we expend resources given to us by Congress, paid for by the American people. If we are going to take action, we need to have results to show for that action. The courts have told us that is what we must do.

***

On the other hand, Assistant Secretary Auchter has received enthusiastic reviews from some members of Congress.[25] For example, in 1982 Congressman Eugene Johnston of North Carolina told Mr. Auchter:

> I would like to say to the Secretary that I serve on several committees and subcommittees, and I wish I could think that every committee with which I am involved, which has an oversight function, could make me as happy with the performance of an agency as I am with OSHA.
>
> This is probably the last meeting of the 97th Congress with you, Mr. Secretary. I would like to express my opinion that you have done an outstanding job in applying to this agency a quantified approach, a rationale and what I call a hard-scale of achievement or non-achievement, to the performance of the agency, which a lot of other agencies within the Government would do well to emulate.
>
> I wish you continued success in your efforts over the next year and hope that managers such as you, with the techniques which you brought from the private sector, can continue to be enticed into government service and will bring the same hardnosed, rational approaches to performance which obtain in the private sector. I commend you for a job well done.[26]

As we have seen, Congress may make its views known to OSHA on a particular issue in a number of ways. The message may be implicit, or even explicit, in a floor debate[27] or at a hearing.[28] From time to time, Congress will make its views known explicitly through statements in committee reports[29] or in letters to the administration from committee chairpersons.[30] OSHA, of course, must respond to the "messages" from Congress, and this response, not unexpectedly, is often favorable. Thus, for example, OSHA made changes in its policy on informal complaints, and regarding citations for serious violators, in large part as a result of congressional involvement. And while the connection between congressional opposition to the field sanitation proposal and OSHA's discontinuing the rulemaking has not been established, the time sequence suggests that the two events were not unrelated.[31]

---

[25]Assistant Secretary Auchter's various actions relating to health standards are discussed in other parts of this work. See particularly discussions of asbestos emergency standard, Chapter 4; hazard communications standard, Chapter 18, Section F; and proposals on ethylene oxide, Chapter 6, Section C, and ethylene dibromide, Chapter 20, Section C.

[26]*Occupational Safety and Health for Federal Employees: Hearings Before the Subcomm. on Health and Safety of the House Comm. on Education and Labor*, 97th Cong., 2d Sess. 445 (1982).

[27]Congress' hostility to the OSHA field sanitation proposal became apparent from the floor debate on an appropriations bill; see Section C, this chapter.

[28]The questioning of Assistant Secretary Guenther on the agriculture standard clearly evidenced the committee's desire for more protection in that area, *supra* this section.

[29]See, e.g., the recommendations of a subcommittee of the House Committee on Small Business in its 1977 report, *supra* note 9; the appropriations committee statement urging OSHA to pay a greater share of state on-site consultation costs, Chapter 11, Section D. In 1979, the OSHA appropriations act directed the Agency to prepare a report on cotton dust, exploring "viable alternatives which are less costly and more technologically feasible." The report was submitted May 14, 1979. See discussion in Chapter 10, Section B.

[30]See, e.g., Letters from Chairman Harrison Williams of the Senate Comm. on Labor and Human Resources to Secretary of Labor Peter Brennan, reprinted in, *Occupational Safety and Health Act Review, 1974: Hearings Before the Subcomm. on Labor of the Senate Comm. on Labor and Public Welfare,* 93d Cong., 2d Sess. 943–61 (1974) (raising numerous questions, based on investigations of the General Accounting Office, on OSHA's enforcement policies).

[31]See Chapter 6, Section C.

Another important point of contact between Congress and the executive departments is the reports of the General Accounting Office (GAO). GAO, headed by the Comptroller General, was set up by Congress in order, among other things, to investigate the executive departments, as may be requested by the Congress or its committees.[32] GAO has issued approximately 20 reports dealing with OSHA; these reports have dealt with such issues as delays in the setting of health standards[33] and OSHA's complaint policy.[34] The typical procedure of GAO is to conduct its investigation, submit a draft report to the agency, which offers comments, and then, after appropriate changes, publish the report, including the agency comments, and submit it to Congress. The GAO recommendations, while not binding on the agency, often are influential in directing agency policy. For example, OSHA's redirection of its informal complaint policy in 1979, requiring field staff to conduct inspections in most cases in response to informal complaints, was the result, in part, of a GAO report.[35]

Another GAO report issued in 1978 and entitled "Workplace Inspection Program Weak in Detecting and Correcting Serious Hazards"[36] influenced OSHA policy. The report dealt with the inadequacies of OSHA inspections in identifying serious hazards. It stated the following general conclusions:

> Although the Occupational Safety and Health Administration and the States have made over 1 million inspections, workers have not been able to rely on these inspections to identify hazards which could lead to serious physical harm or death. Also, inspections do not assure workers that all hazards observed are cited as violations. Hazards not detected or cited can remain in the workplace and cause injury or death.
>
> Our review at five OSHA offices and three State offices showed that OSHA and State compliance officers did not cite many serious hazards. Inspections would be improved if OSHA and the States (1) provided compliance officers information on the serious hazards to look for in particular industries, (2) reviewed inspection files to insure that compliance officers inspect plant operations or conditions which could cause serious harm, and (3) periodically monitored the work of their compliance officers. Better supervisory review could also result in more violations being cited when serious hazards are found even though compliance officers are unaware that a standard exists or believe a standard is unenforceable.
>
> Unless OSHA and the States improve their inspection programs, they will be unable to determine the effectiveness of their inspections, and compliance officers will continue to miss serious hazards.[37]

The draft GAO report was sent to OSHA, and, by the time the final GAO report was issued, OSHA had already begun to make changes in its inspection program. As stated in the final report:

---

[32]31 U.S.C. §53 (1976 & Supp. V 1981).
[33]See Chapter 3, Section B.
[34]See Chapter 13, Section C.
[35]See Chapter 13, Section C.
[36]COMPTROLLER GENERAL OF THE UNITED STATES, REPORT TO CONGRESS, HRD-78-34 (May 19, 1978).
[37]*Id.* at 4.

The Department of Labor, in an April 3, 1978, letter commenting on a draft of this report agreed that improvements were needed in the workplace inspection program, especially in the areas of hazard identification and program evaluation. Labor said that many of our recommendations have been addressed by new programs and policies, and policies relating to other major concerns of the report are being considered.

***

Labor said that OSHA believes the best means of ensuring that serious hazards are not overlooked during inspections is to better educate compliance officers in hazard recognition, classification, and documentation.

To help compliance officers better identify serious hazards, OSHA said it had begun a study of the incidence and causes of fatalities it investigates to determine the major causes.

Labor said OSHA is considering developing hazard identification and abatement guidelines for inspecting the 10 industries having the potential to produce the greatest number and the most severe injuries and illnesses. This would be a good first step, but identification of potentially serious worksite hazards is needed for all industries where workers are exposed to serious hazards.

***

OSHA said that a regional audit program is being developed which would center audit and evaluation responsibility in each region to ensure that a thorough annual program evaluation can be made of every regional and area office. On-the-job evaluations and case file reviews will be important parts of the audit program. OSHA is also considering making spot-checks of its inspections.

***

Since our review, OSHA said it had established supervisory positions in most area offices to review inspections. We believe supervisory review of inspections is necessary in all area offices. Also, OSHA should develop procedures for effective supervisory review of inspection case files, including determining what the compliance officer looked for, what he found, and whether the inspections were adequate.[38]

B. Amendments to OSHA

Numerous bills to amend OSHA have been introduced in Congress since 1970. The large majority of these would have limited OSHA authority, either by restricting enforcement activity or by exempting classes of employers from the Act's requirements. OSHA has opposed these weakening amendments; indeed, because of the potentially strong opposition to OSHA in Congress and the uncertainty as to what would occur if the Act were opened for amendment on the floor of Congress, OSHA, with rare exception, has opposed all amendments of the Act, even those that otherwise would have been deemed desirable. This position has been supported by the labor unions; the House and Senate Labor Committees, concurring in this point of view, have consistently refused to report out bills to amend OSHA. As a result, the Act has never been amended.

[38] *Id.* at 14–16.

Only three bills to amend OSHA have made any serious progress in Congress. In 1975, the House of Representatives passed a bill, introduced by Chairman Daniels of the House Labor Committee, dealing with on-site consultation. The bill would have authorized federal OSHA to perform on-site consultation activity without sanctions attached in order to assist employers, particularly small employers, in complying with the multitude of OSHA standards.[39] Although Chairman Daniels had generally opposed OSHA amendments, his concern was that, absent an on-site consultation bill, even more damaging amendments would be enacted. As indicated in Chairman Daniels' floor statement, the Department of Labor supported the bill.

Statement of Chairman Daniels

121 Cong. Rec. 36,905–06 (1975)

***

MR. DOMINICK V. DANIELS. Mr. Speaker, I yield myself such time as I may consume.

Mr. Speaker, I rise in support of H.R. 8618, a bill to amend the Occupational Safety and Health Act to provide consultation and education services to employers, upon request, at their workplaces.

As sponsor of the original bill and as chairman of the subcommittee which has jurisdiction over OSHA, I want to reiterate my support for this landmark piece of legislation enacted in 1970 to protect the lives, health, and safety of 65 million working men and women.

I realize that no law is perfect, that there comes an appropriate time to amend a law in order to improve it. Today, the time has come for us to amend and improve the original Occupational Safety and Health Act. Accordingly, I urge my colleagues to support H.R. 8618, which authorizes the Department of Labor to visit the workplace of employers for the purpose of affording consultation and advice. This onsite consultation service will be voluntary, based solely upon an employer's request for consultation to assist him in interpreting standards as they apply to his workplace. The bill directs the Labor Department to give priority to filling requests to small businesses and hazardous workplaces.

A consultative visit authorized by this amendment is not an inspection. No citation shall be issued nor any civil penalties imposed except in the case of an uncorrected imminent danger or an unabated serious hazard.

The bill clearly separates consultation from enforcement proceedings, and provides for separation of consultation and compliance staffs.

The Secretary is directed to establish programs for education and training of employers and employees, which shall be conducted in the local communities.

This amendment does not weaken nor diminish the vital enforcement provisions of OSHA, including first instance sanctions. I believe that this bill will, in fact, improve working conditions because employers who in good faith request consultations will surely act upon the consultant's advice and improve safety and health practices.

[39]Earlier in 1972 and 1974, Congressman Steiger had introduced similar legislation. Those bills, however, were strongly opposed by labor unions and were not enacted. See, e.g., 2 OSHR 475 (1972). For a discussion of the Steiger amendment to the OSHA appropriations bill earmarking federal funds for state on-site consultation, see Chapter 11, Section D.

The chances of each of the 5 million workplaces covered by OSHA being inspected are statistically remote. Therefore, programs of onsite consultation in every jurisdiction will have two major benefits. First, it will assist employers in complying with the act, and second, it will upgrade working conditions for employees.

The most persuasive case for Federal onsite consultation comes from employers doing business in 23 jurisdictions where onsite consultation is not available. These businessmen have every reason to be concerned because Congress has not responded to the need for consultation in every State.

Specifically, under the two existing consultation programs employers in 34 jurisdictions can receive consultation within their workplaces. Employers in the other 22 jurisdictions covered by OSHA cannot. The purpose of H.R. 8618 is simple—it will rectify this glaring inequity.

***

Critics of OSHA have complained that standards are complex and confusing. Clearly onsite consultative services which offer an employer an opportunity to apply OSHA standards to his own workplace will remedy this situation.

Consultative services will also assist employers in understanding sophisticated health hazards, which, without such services, would have gone undetected.

Further, onsite consultation to employers will facilitate the dissemination of information under the act by placing in the law a mechanism for imparting information directly to employers, pointing out specific hazards in the workplace, and providing advice on elimination of hazards.

Onsite consultation to employers has been considered at great length by the Manpower Compensation, and Health and Safety Subcommittee. Thirty-one days of oversight hearings have been held on OSHA since its enactment. In July 1975, 4 days were devoted exclusively to consultation.

* * *

I hold no pride in this legislation simply because it bears my name. But I do hold a very deep and abiding concern for the American worker—a concern for the life and health of millions of working men and women.

In summary, may I state that it is my considered judgment that enactment of H.R. 8618 is a responsible and necessary step in improving the 1970 OSHA legislation:

It will improve the mechanism to secure voluntary compliance.

It does not weaken or diminish the enforcement provisions of OSHA.

It does not eliminate first instance sanctions which is the heart of OSHA.

It provides onsite consultation to employers in every State.

It will mitigate the criticism leveled at OSHA.

OSHA will gain greater acceptance in all 56 jurisdictions covered by the act and thereby increase worker safety.

***

I would be happy to yield to the gentleman from Louisiana.

Mr. Waggonner. Mr. Speaker, I thank the gentleman for yielding to me.

I think it is a matter of understatement that in administering the Occupational Safety and Health Act that there has been an undue burden, in a relative way, placed upon what we commonly refer to as the small businessman or the small contractor as opposed to the large one, or the small manufacturing facility as opposed to the large manufacturing facility. As I view the gentleman's proposal it is only intended that these people who do not have the financial or other resources to comply with OSHA to the extent that a larger firm does be provided a mechanism by which one who is really interested in complying with occupational standards can find out for themselves what their problems are and then be able to do something about it.

It would seem to me that any small businessman who asked this help, who makes a valid request, is acting in good faith and wants to provide for safety.

Mr. Dominick V. Daniels. I agree with the gentleman.

Mr. Waggonner. So I see absolutely nothing wrong in this amendment to try to relieve the burden on the small businessman because, in the final analysis, he wants to provide for safety, he wants to know what he has got to do to comply. He simply needs some help and some capability that he does not have himself. It would just seem the better part of wisdom and in the interest of safety in small business establishments for us to allow this new proposal to become effective and I support the distinguished gentleman's proposal.

Mr. Dominick V. Daniels. I want to thank the gentleman from Louisiana for his contribution and may I further add that this amendment specifies that priorities shall be given small businessmen on their requests for consultation and also to those industries which are deemed hazardous.

While OSHA at that time supported a federal bill on on-site consultation, representatives of labor unions did not. Chief among these was Mr. John Sheehen, legislative director of the United Steelworkers of America. He said:

> I think it is evident that much of the momentum behind the effort to place consultative services in the Federal OSHA comes from those who are not seeking a legitimate educational goal, but rather are seeking an avenue through which to weaken OSHA's crucial procedure of enforcement through first instance citations. That is the root of our steadfast opposition to consultation by OSHA personnel. The mere fact that the act needs to be amended in order to allow this type of service is indicative of the fact that it countervenes the enforcement principle.[40]

The Daniels bill passed the House, but the Senate Labor Committee took no action on the bill, which never became law.[41]

In 1978, Senator Dewey Bartlett of Oklahoma submitted an OSH Act amendment to a pending bill dealing with small business. The amendment sought to limit, or eliminate, OSHA inspections of employers, particularly small employers, with low injury/illness records[42]—an issue of continuing concern to Congress. Under the Bartlett bill, employers with fewer than 10 employees would have been exempted from OSHA coverage if their injury/illness rate was below a certain level.

[40] *On-Site Consultation: Hearings Before the Subcomm. on Manpower Compensation and Health and Safety of the House Comm. on Education and Labor,* 94th Cong., 1st Sess. 158 (1975).

[41] On-site consultation services were ultimately provided by the states, either under approved plans or under agreements with federal OSHA, with federal OSHA paying 90% of the cost. See Chapter 11, Section D.

[42] See discussion of amendment submitted by Senator Schweiker in 1979, Chapter 13, Section E, and the Schweiker appropriations rider, enacted in 1980, Section C, this chapter.

Statement of Senator Bartlett

124 Cong. Rec. 23,841–42 (1978)

***

Mr. Bartlett. Mr. President, the amendment before us is simple but significant. It provides that small businesses which employ 10 or less people shall be exempt from coverage under the Occupational Safety and Health Act. At the same time, it recognizes that certain small businesses, primarily in the mining, manufacturing, construction and petrochemical industries engage in activities which have a history of high occupational injury/illness incidence rate and are considered hazardous by many. Therefore the amendment specifies that those in the category of 10 or fewer employees which have an occupational injury/illness incidence rate of 7 or more per 100 full-time workers based upon the annual Bureau of Labor Statistics survey for two-digit Standard Industrial Classification (SIC) code industries shall remain subject to the provisions of the act.

I would point out that nine rather than seven is considered by OSHA as the breakover point for high risk business categories. So this is written in a very conservative way that does recognize that there are a number of business classifications which are hazardous, and that the provisions of OSHA and the responsibility that OSHA has would not be interfered with in any way by the hazardous nature of the business.

***

When Dr. Eula Bingham assumed control over the Occupational Safety and Health Administration she announced her intention to redirect enforcement priorities. She said that 95 percent of OSHA's inspection resources would be devoted to those industries with the most serious safety and health problems, such as construction, manufacturing, transportation, and petrochemicals. No more than 5 percent would be devoted to such areas as wholesale and retail trade, finance, or the service industries.

This policy has lessened the likelihood of a small business inspection by OSHA. It also has defused the criticism by small businessmen that they were being unfairly singled out for inspection. My amendment simply insures the continuation of this constructive policy and provides that Congress take a position on this, to extend the policies of Dr. Eula Bingham in the future, after administrations and OSHA leadership have changed.

Mr. President, the current pace of inflation makes it increasingly difficult for small businesses to compete effectively in the marketplace. Excessive Government regulation exacerbates their problems.

***

My intention here is only to provide relief to those small businesses which are burdened most by the plethora of OSHA regulations, but which have compiled the best records so far as occupational health and safety are concerned. The proviso to my amendment assures employees whose work has proven to be hazardous in the past that they will continue to receive the necessary support in their effort to keep the workplace reasonably safe.

The approach I have suggested is a positive one. It provides incentive for those small business establishments that border on the line of exemption to improve their record of safety. But it penalizes those employers who are careless with the health and safety of their workers.

The result will be better and safer working conditions for employees and less Government interference for those employers who are most deserving. I welcome and encourage the support of my colleagues.

The Bartlett bill was opposed by OSHA, which has consistently argued against bills that would exempt categories of employers from OSHA coverage. OSHA has agreed that inspections should be targeted to high-hazard workplaces,[43] but has asserted that this should be achieved by administrative priorities, which could be modified as needed, rather than by rigid statutory exemptions. The policy, and morality, argument against exemptions from OSHA was stated by Senator Jacob Javits of New York in the same debate as follows:

> But, Mr. President, I cannot in conscience and morality agree with a proposition which will simply on a wholesale basis, eliminate these workers just because they happen to work for small business. That does not in any way change the danger and damage to them. Yet this amendment is pressed now with what I consider to be very clever, but very easily ascertainable exemptions, which exempt them all, whether they have a good record or not, because there are no figures to substantiate the specific exemption the Senator wants to make. Therefore he exempts them all * * *.[44]

The Bartlett amendment was passed by the Senate, but not by the House of Representatives, and was ultimately deleted from the bill by the conference committee in the closing days of the 95th Congress.[45]

In 1979 Senator Schweiker attempted to exempt from most OSHA safety inspections those employers with "good" injury records.[46] Extensive hearings on the bill were held by the Senate Labor Committee in 1980, but, in the face of strong opposition from OSHA and the labor unions, the bill was not reported out of committee. No serious attempts to amend OSHA have been made since.

A bill affecting OSHA, but not technically amending the OSH Act, was passed in 1975. The bill made the U.S. Postal Service subject to §19 of the OSH Act, which establishes a special program for the safety and health of employees of the federal government.[47]

C. Riders to OSHA Appropriation Bills

Because the possibility of amending the Act has effectively been blocked by the Senate and House Labor Committees, proponents of amendments to the Act have resorted to other means to bring about change. Notable among these has been the process of attaching riders to appropriation bills. Since riders need not be reported favorably by the congressional committee with jurisdiction over the subject mat-

[43]See, e.g., *Oversight on the Administration of the Occupational Safety and Health Act, 1980: Hearings Before the Senate Comm. on Labor and Human Resources*, 96th Cong., 2d Sess. 273–76 (1980) (statement of Deputy Assistant Secretary Basil Whiting).

[44]124 CONG. REC. 23,844 (1978).

[45]124 CONG. REC. 32,725 (1978).

[46]Senator Schweiker's proposed amendments are discussed in Chapter 13.

[47]Postal Service Executive Pay, Pub. L. No. 94-82, 89 Stat. 419 (1975) (amending 39 U.S.C. §410(b)(7)). The status of Postal Service employees under OSHA had been unclear. It had been sometimes argued that these employees were federal employees, sometimes, that they were private sector employees, and sometimes, that they were neither and entitled to no safety and health protection.

ter involved (indeed, they can be added from the floor during debate), they have afforded a practical means for OSHA opponents in Congress to circumvent the labor committees and thus to limit OSHA authority. An appropriations rider technically limits the way OSHA can spend money; for example, a rider may provide that no funds may be expended by OSHA to assess penalties for nonserious violations. A rider must be renewed each time an appropriations bill is passed. As a practical matter, however, these riders, once enacted, have in virtually all cases[48] become a permanent part of OSHA appropriations. At present, there are seven riders that are part of the OSHA appropriations law, the first two going back to 1976.

In 1976 Senator Robert Taft of Ohio explained the reasons why many members of Congress had used the rider method for making changes in the OSHA program. The on-site consultation amendment had passed the House of Representatives in 1975,[49] but no action had yet been taken by the Senate Labor Committee on the House-passed legislation and, indeed, never was taken. Senator Taft, expressing his frustration, stated that the committee inaction was the reason that Congress was pressing forward on appropriations riders to provide relief for small business from OSHA.

Statement of Senator Taft

122 Cong. Rec. 21,207–08 (1976)

***

On Thursday, June 24, [1976], the House passed two devastating amendments dealing with OSHA exemptions in its consideration of the Labor/HEW appropriations bill.* As I understand the situation, the sponsors of these amendments intend that their amendment will exempt agricultural and other employers with 10 or fewer employees from the coverage of the act. I want to make my position clear that I have never supported, nor will I support, any OSHA exemptions based upon the numbers of workers an employer employs. I do not think this is a reasoned approach to OSHA reform. And yet, I can see that the House is extremely frustrated by the fact that the onsite consultation bill which was overwhelmingly passed in November 1975, H.R. 8618, has not been acted upon by the Senate.

It is currently pending before the Labor Subcommittee and the hearings which we have been having do relate to it. However, no action appears to be imminent on it. As most of my colleagues are aware, I have done what I can to move S. 3182, an identical OSHA reform bill, through the Senate without much success thus far other than persuading almost one-third of this body to cosponsor the legislation. We now have 29 cosponsors.

We also have broad backing from business in this connection. In this respect, for instance, the position of the U.S. Chamber has been against any

*[Author's note—These riders—dealing with small farms and nonserious penalties—are discussed in this section. They were enacted and are still in effect.]

[48]The rider passed in 1974 limiting the recordkeeping obligations of small employers was deleted from the appropriations acts when the principle was adopted by OSHA in its regulations. (29 C.F.R. §1904.15 (1983)). The rider relating to state plan benchmarks was deleted after being in effect a relatively short time. See discussion in Chapter 19, Section D.

[49]See Section B, this chapter.

of the amendments that have been offered over the years to exempt many employers of small numbers of employees from OSHA, and correctly so, because they want to protect safety. But the problem has become so serious that they are now shifting their position to say they will support exemptions such as the House has enacted because there has been no action on the onsite consultation bill. This point was made very clearly by Congressman Findley in his debate on the amendments that were put in for the 10 employees and the agricultural employees, stating that the amendment was only necessary because the Senate has not passed an onsite consultation bill.

Mr. President, I do believe that the Congressman's statement is part of the consideration relied upon by Members of the House when they overwhelmingly approved both the Skubitz and Findley amendments.

At that time, Mr. Findley stated:

> There is no prospect that that bill is going to move unless the Congress, by this means of a limitation amendment on an appropriation bill, brings pressure to get the attention of Senators to make them recognize that consultation is an absolute essential for a small business firm which cannot employ a high-cost consultant and expert on compliance with Federal health and safety regulations.

Mr. President, it seems to me that the House is thus experiencing a sense of disappointment and irritation that the onsite consultation proposal has not become law and, therefore, cannot be used by businessmen, including small businessmen and farmers, to assist them in their efforts to comply with OSHA's voluminous regulations.

***

Senator Harrison Williams, Chairman of the Senate Labor Committee, sought, unsuccessfully from the point of view of Senator Taft, to explain the delay in the on-site consultation bill.

> MR. WILLIAMS. All I can say to the Senator from Ohio is that within the schedules that are available to us, we certainly have his consultation bill on our schedule for consideration, and hope again that our committee will have increasing opportunities to meet. Right now we are foreclosed insofar as meeting at any reasonable hours is concerned. We set a meeting for 8 o'clock this morning, and thought we would not be meeting during a Senate session, but the Senate session was moved up to 8 o'clock, and, we of course could not have the meeting.
>
> MR. TAFT. Mr. President, I thank the Senator for his comment. I do not feel very encouraged: therefore, I think we ought to go ahead with the Durkin amendment.[50]

When an appropriations rider is introduced, the threshold question that often arises is whether the amendment is in order. An amendment to an appropriations bill is out of order if its effect is to amend existing law by imposing additional duties not previously required on the executive department. In some cases, OSHA appropriations riders have been ruled out of order by the chair; for example, in 1977, a rider introduced by Congressman George Hansen, before the

[50] 122 CONG. REC. 21,209 (1976). The Durkin amendment would have restricted OSHA from proposing a penalty for any first-instance nonserious violation, regardless of the size of the business. 122 CONG. REC. 21,203 (1976). The rider was not enacted in that form.

Supreme Court decision in *Barlow's*,[51] that would have required OSHA to obtain warrants before conducting inspections was ruled out of order by the presiding officer of the House of Representatives on the grounds that it would require "affirmative action" by an executive officer.

House of Representatives Floor Debate

123 Cong. Rec. 19,373 (1977)

POINT OF ORDER

MR. FLOOD. Mr. Chairman, I make a point of order against the amendment.

THE CHAIRMAN. The gentleman will state it.

MR. FLOOD. Mr. Chairman, this, in our opinion, is clearly legislation again upon an appropriation bill. It imposes additional duties upon the executive branch, because it requires OSHA to obtain a search warrant before an inspection. The basic law does not now require this. This amendment goes beyond the basic law.

Also, Mr. Chairman, this amendment is what we refer to as a double negative, and therefore gives affirmative direction.

THE CHAIRMAN. Does the gentleman from Idaho desire to be heard?

MR. HANSEN. Yes, I do desire to be heard.

Mr. Chairman, regarding the double negative, I have checked this thoroughly with parliamentary procedure under Legislative Precedents, chapter 25, section 13, in which "Deschler's Procedure" states:

> An amendment to a general appropriation bill, providing that no funds therein may be used for the seizure of mail (in connection with income tax investigations) without a search warrant, was held to be a limitation and in order.

We checked this language out, and we find the language used in my amendment is directly parallel to the language that was used in the precedent cited in "Deschler's Procedure." So, I think that this answers the problem there.

As far as legislation, Mr. Chairman, this is not legislation. I can tell the Chair that the courts across the land—and there have been 15 court cases in the past few months in a row—have decided against warrantless searches, and they have decided this in a situation where they could not interpret the will of the Congress. I think it is time that the will of the Congress is expressed, so what we are doing here is stipulating something that ought to be apparent in the first place. That is, these searches ought to be conducted with a search warrant, properly procured, with due reason, before a magistrate.

So, I think it is entirely in order as far as procedure is concerned, and as far as philosophy and content is concerned.

THE CHAIRMAN. The Chair is prepared to rule. The Chair would first like to acknowledge that this is a difficult question. The Chair has not only examined the citations to previous rulings, but he had thought about the Constitution and examined a series of other rulings, one of which he will cite at the end of his statement.

There is a difference in the kind of warrant required in the pending

[51] *Marshall v. Barlow's, Inc.*, 436 U.S. 307 (1978), discussed in Chapter 12.

amendment and the one in the ruling cited which dealt with a Treasury and Post Office appropriation bill, and provided no funds were to be used for seizure of mail without a search warrant as authorized by law. That search warrant apparently was an administrative warrant. The warrant required in the case is a judicial warrant based on probable cause. It applies also to any inspection in the United States, whether it is voluntarily agreed to by the proprietor or not.

On that basis, it appears to the Chair that the precedent which appears on page 678, volume 7, Cannon's Precedents, section 1686, "A limitation upon an appropriation must not be accompanied by provisions requiring affirmative action by an executive in order to render the appropriation available" should prevail.

The Chair, therefore, sustains the point of order, for the reasons cited.

Chairman Daniel Flood, head of a subcommittee of the House Appropriations Committee who consistently opposed riders, argued that the appropriations process should not be distorted as a means of making substantive changes in the legislation. Although he succeeded in blocking the Hansen rider, his procedural arguments were often rejected by the House of Representatives.

Frequently, procedural challenges were not made to OSHA riders and, when made, were overruled. An example is the ruling of the presiding officer in 1980 that a rider introduced by Congresswoman Beverly Byron of Maryland to limit OSHA inspections of "safe" employers was in order.

Statement of the Chairman of the House

126 Cong. Rec. 7993–94 (1980)

***

The gentleman from Pennsylvania [Congressman Gaydos] makes a point of order against the amendment offered by the gentlewoman from Maryland on the ground that it constitutes legislation on an appropriation bill. In reviewing the amendment, it would prohibit the use of funds in the bill to enforce standards or rules under the Occupational Safety and Health Act with respect to certain employers, except for enumerated functions and activities authorized under such Act. The amendment applies to employers with 10 or fewer employers whose business falls within a category having an injury work loss day rate less than the national average as indicated by statistics published by the Bureau of Labor pursuant to law. The amendment does not require individual findings of injury rates in each separate business, but only a determination as to the category into which the business falls.

The Chair has reviewed the set of statistics that is required by section 673 of the OSHA law, and finds that the determination as to what category that the business relates to and the relationship between the average rate for that category and the average rate for all business is very easily ascertainable and is now being undertaken under OSHA regulations.

Furthermore, the amendment does not require the compilation and publication of such statistics but refers to the most recent publication of the statistics already compiled pursuant to statute. The exemptions from the limitation refers only to those functions and authorities which are provided by law in title 29 of the United States Code.

No new duties or determination[s] are hereby required, and the final proviso, while requiring findings as to the temporary status of a farm labor camp, is already in the bill and the amendment does not add legislation to that permitted to remain in the bill.

The gentleman from Pennsylvania (Mr. Gaydos) has argued in essence that the amendment seeks to amend existing law by restricting executive discretion to perform certain activities under the Occupational Safety and Health Act. The Chair cannot agree with the assertion of the gentleman that the amendment requires OSHA to perform its authorized function in a certain manner.

The amendment restricts the use of funds to carry out part of the authorized activity while allowing but not requiring the agency to use funds in the bill to carry out other authorized activities. While an amendment to an appropriation bill may not directly curtail executive discretion delegated by law, it is in order to limit the use of funds for an activity or a portion thereof authorized by law if the limitation does not require new duties or impose new determinations.*

The chair overrules the point of order.* * *

In ruling on these procedural objections, the issue is usually, as was stated by the Chairman of the whole House, whether the rider requires "new duties or determinations." If it does, the rider would be subject to a point of order. The chair in ruling on the Byron rider concluded that the determinations which OSHA must make in order to decide whether to inspect were "very easily ascertainable"; in other words, while the Agency would have a "new" duty, it would be insubstantial and therefore an insufficient basis to render the rider procedurally objectionable. On the other hand, in ruling on the Hansen rider on warrants, the chair distinguished between administrative and judicial warrants, concluding that the latter requires "affirmative action" by the Agency. The basis for this distinction is not entirely clear, and, in sum, these rulings suggest that the chairmen have considerable discretion in deciding whether a rider is out of order. As a general matter, however, a limitation on OSHA activity, e.g., a prohibition on penalties, will be in order, while a requirement for new activity, e.g., issuing a standard, will not.

The first OSHA rider was considered by the Congress in 1972 for the 1973 fiscal year appropriations bill. As originally introduced, the rider would have exempted employers with 25 or fewer employees from OSHA. The rider was opposed by OSHA, but was ultimately passed by Congress, exempting employers with three or fewer employees. The appropriations bill, however, was vetoed, and the rider never became law. In 1974, a rider was enacted into law exempting employers with 10 or fewer employees from OSHA recordkeeping

*[Author's note—The Byron rider, which was identical to the rider introduced by Senator Schweiker, was enacted into law. See *infra* this section.]

requirements. The rider was incorporated into OSHA regulations in 1977[52] and subsequently dropped from the appropriations bills.

In 1976, a major debate on OSHA was held during the appropriations process, and two riders were added to the OSHA appropriations. The first, introduced by Congressman Skubitz, exempted small farms from OSHA enforcement. The sharp debate on the rider demonstrates the hostility to OSHA of some members of Congress who indicated particular opposition to the proposed field sanitation standard,[53] and irritation at OSHA's publication on beef cattle. The second rider, introduced by Congressman Findley, eliminated OSHA authority to impose penalties against establishments which are cited for fewer than 10 nonserious violations during an inspection. Congressman Sarasin responded to Congressman Skubitz and sought, without notable success, to defend OSHA's actions.

Statements of Congressmen Skubitz and Sarasin

122 Cong. Rec. 20,366–67, 20,372 (1976)

MR. SKUBITZ. Mr. Chairman, the amendment which I present is a simple one, all it does is to exempt farm operators with 10 or fewer employees from the requirements of OSHA.

As you will recall the Occupational Safety and Health Administration was created December 29, 1970.

It was the intent of Congress to create with the Department of Labor a cadre of experts effective in the improvement of the safety and health of our country's workplaces.

But we did not create experts—we did not create improvements—we created a monster, a monster which does not have the guts to question big business but centers upon small business that can not afford to—or are afraid to—strike back.

What started out to be a laudable program has turned into a nightmare, in part because of arrogant inspectors who feel they have not done a job unless they find something wrong in every little plant.

Now, OSHA has begun to expand its horizons. It wants to grow, be powerful, because with size comes higher grades in Government and more prestige.

Several weeks ago I introduced a bill exempting all farms that employ less than 25 persons.

The Fort Scott Tribune of Fort Scott, Kans., called OSHA for its comments on my proposal.

Let me read what a safety engineer with the National Standards Office of the Department is reported in the Fort Scott paper as saying:

> Robert Bailey, the engineer, said last week in a telephone interview that the Skubitz bill was feasible only if you want to castrate OSHA.

Believe me, my colleagues, I do not want to castrate OSHA because if I do it might grow more rapidly.

---

[52] 29 C.F.R. §1904.15 (1982).

[53] The history of OSHA's field sanitation proposal is discussed in Chapter 6, Section C.

And yet, if we do not do something it will produce more rapidly and destroy our small farmers.

But if castration is the only solution I would sooner castrate the zealots who are drawing up regulations at OSHA than let them destroy the smaller farmers of America.

After consultation with the various farm organizations and cattle organizations I have reduced the number of farm employees from farms of 25 persons or less to 10 persons or less.

Now what are the reasons for the steps I have taken?

I am sure most of you are familiar with the Earth-shaking story carried by the Washington Post June 18, 1976, announcing that OSHA had made the amazing discovery that manure is slippery.

I am sure that pearl of wisdom caught every farmer by surprise.

The article was entitled "Manure Slippery, U.S. Warns."

The Washington Post story stated:

> The half million dollars worth of pamphlets prepared by OSHA are designed to help farmers and farm hands understand new safety rules.

Let me read a few more gems of wisdom from the OSHA pamphlet:

> The best way to stop an accident is to prevent it.

That must have taken days, weeks, months to figure out. Here is another:

> When floors are wet and slippery with manure you could have a bad fall.

Now, this is not a "shoot-from-the-hip" type of conclusion from OSHA—it is a carefully researched conclusion costing around $119,000.

Perhaps you also read the editorial in last night's Washington Star entitled "Answering OSHA's Call." I call your attention to the opening lines of that editorial.

> The slippery manure caper is an absurdity of howling dimensions: it contributes to the notion that the Federal bureaucracy has difficulty pouring milk out of a boot.

I suspect they have already hired the mayor of the small town in Florida who decreed that all horses using the streets of his little community must be properly diapered.

No doubt powder to ease diaper rash will follow.

Here is another proposal by OSHA—

OSHA proposes that any farm having five or more employees must have a toilet within 5 minutes walking distance. I ask, How far is 5 minutes walking distance?

My guess is the distance one could cover in 5 minutes would depend upon the age of the person and the urge to go.

Let us just assume a man could cover one-half mile in 5 minutes.

Now a mile section of cultivated land is not unusual in Kansas.

That means that on every square mile a farmer would be required to construct a minimum of nine privies.

For years the great wheat plains of Kansas have been dotted with those great towers of productivity, the grain elevators:

Kansans point to them with pride and refer to them as the "great cathedrals of the plains."

Under the proposed regulations OSHA has decreed that we should have more "temples" on the plains, not only in Kansas but in all agricultural areas.

OSHA the mandator of the "privy on the plains."

Now OSHA tells us that this little farmer that employ[]s over five men can arrange for a caterer or a concessioner to take care of the job by placing a portable privy and a washstand on a small moving vehicle that would follow the men about the field.

Or maybe a taxi service could be provided.

It is not clear whether the toilet would be a pay-as-you-enter toilet or if taxi transportation is provided who picks up the chip.

Let OSHA have its way—and they are going to make port-a-johns this country's biggest business—all at the expense of the small farmer.

Now do not misunderstand me, safety and health are important, and it is something about which we should all be concerned. But when an elite corps of Government experts decree that safety and health is better served by putting up a privy in any wheat field I say, enough. It is time to draw the line.

Here is another example of OSHA's meddling:

The requirement that the employer shall provide each tractor with seat belts.

But that is not the worst of it. It goes on to say the employer shall insure that each employee uses the seat belt while the tractor is moving.

Did you ever ride a tractor for half a day? If so you know that the operator moves into a dozen different positions to relieve his "tired bottom." He sits down, stands up, leans over, you name it.

What happens if our friendly OSHA expert[] visits the farm, finds the seat belt removed and the driver standing up? Who is liable? Not the driver, the farmer.

Oh yes, and now OSHA is going to require rollover bars—so if the tractor operator enters a drag race they will be protected if a tractor overturns.

Now permit me to let Congressman Esch tell you about the regulation dealing with ladders:

> In his newsletter dated June 9, 1976, Congressman Esch says:
>
> What is a ladder? Webster defines a ladder as a structure for climbing trees—up or down—consisting of two long side pieces joined at intervals by cross-pieces on which we can step.

Just 23 words.

Then Congressman Esch's letter goes on to say:

> It takes the occupational safety and health administration 64 pages in the Federal Register to define and outline regulations pertaining to construction use and safety of the simple ladder.

Then these gems appear in Congressman Esch's letter:

> Its a good thing Jacob had his ladder when he did because it probably wouldn't pass OSHA's standards.
>
> Sixty-four pages in the Register—impossible.
>
> I looked it up—here it is.

I could go on for hours telling you about OSHA and their plans to destroy small farmers; feed bin construction requirements, requirements to conceal belts and chains, and so forth.

Let me close by pointing out that when the Fort Scott Tribune asked Mr. Morton Corn, the head of OSHA, and an Assistant Secretary of Labor—to comment upon my statement:

That "there has been no appreciable decline in injury since OSHA's inception."

The Fort Scott Tribune reports that Mr. Corn admits what he calls "Skubitz' charge" was partially true, saying the decline in injury occurred only among those establishments that OSHA's inspectors have visited.

Now that is a "pot of crock" and Mr. Corn knows it.

From time to time the CONGRESSIONAL RECORD is filled with glorious speeches in support of the family farm, commending the contribution of American agriculture to our balance of trade, and expressing our concern for the plight of the hungry thousands who would so greatly benefit from increased agriculture production.

I support those sentiments and that is why I introduced this amendment.

***

MR. SARASIN. Mr. Chairman, I wish to join my distinguished colleagues in opposing the amendment offered by my colleague, the gentleman from Kansas (MR. SKUBITZ), and also the amendment offered by my colleague, the gentlewoman from New Jersey (MRS. FENWICK).

With regard to the problems of our farm labor sectors throughout the country, I am in a rather unique position in that I serve on both the House Subcommittee on Manpower, Compensation and Health and Safety, which has jurisdiction over OSHA, and on the Agriculture-Labor Subcommittee, which has been examining the difficulties facing farm workers. The evidence I have heard through 17 separate days of overnight [sic] hearings on OSHA and the innumerable meetings and sessions on the issue of farm labor, does not substantiate the contentions made by my distinguished colleague.

First, we must consider the procedural issues involved here. Time and time again we appear to be legislating through the appropriations process. This only serves to deny interested parties the opportunity to present their views to the Congress, and prevents all of us from making rational decisions on the basis of fact. Such a situation can be termed no less than an abrogation of due process and a violation of the purpose of our branch of the government.

Second, I would like to stress the point raised by the gentleman from New Jersey (MR. DOMINICK V. DANIELS) the chairman. Any exemptions, regardless of the number of employees, regardless of the type of industry, creates a second class of American workers. Farmworkers are entitled to the same protections of the law as are workers in factories, on construction sites, and in retail establishments.

The potential impact of this legislation is not, under any terms, small. We are not talking just about a couple of farmers, we are talking about 87 percent of all of the farms in this country. We are actually discussing the elimination of the protection of the law for thousands of workers. We are not talking about a hazard-free industry.

I do not think we should move in this direction, or in this manner that we are going forward without the benefit of hearings, and that we are revoking the process that we have labored so hard and so long to provide. I do not think we can tell the farmworkers that their health and safety is less important than that of any other American workers. If we have difficulties with the administration of the law, and I think we have, then Congress has adequate procedures to address the specific issues involved.

Much has been made of the little pamphlet that was put out with what appears to be very dubious notions on how to maintain farm safety. I read it, and I agree it is rather clumsily worded, and it seems to be a little strange, but I think the thing we have to remember is that those of us who read it and laughed about it in the cloakroom the other day should realize that we are Members of the Congress who, hopefully, can read well. We must remember that this particular pamphlet was prepared by Purdue University just to take care of those 23 million Americans who have only a low literacy capability, meaning that they cannot read a want ad, or to handle simple transactions, so that that was the purpose of that pamphlet. I repeat that I agree that as we look at it, it looked rather strange, but it was not written for us.

Mr. Skubitz. Mr. Chairman, if the gentleman will yield. I want to say that I will hand him a few more pamphlets which make excellent reading.

Mr. Sarasin. I have seen the pamphlets. I would suggest to my colleague, and I will agree that they all seem to be rather ridiculous also but we are talking about the 23 million Americans who are functionally illiterate and those pamphlets are deliberately designed for those individuals. Similar pamphlets are written in ordinary English and Spanish as well.

***

Congress, having passed two substantive OSHA appropriations riders in 1976, followed the same legislative route in subsequent years. The OSHA appropriations bill for 1979 continued the earlier riders with limited changes and added two new riders. The first eliminated OSHA jurisdiction over recreational fishing and hunting, and the second restricted OSHA imposition of penalties for small employers if they had made good faith efforts to implement prior consultant advice to abate hazards. The fiscal year 1980 appropriations bill included three new riders. The first, introduced by Senator Schweiker, served a purpose similar to his bill introduced late in 1979 but not passed; the rider limited OSHA safety inspections of small employers to those in high-hazard industries.[54] The second rider restricted OSHA spot-check monitoring of state plans,[55] and the third limited OSHA jurisdiction over employers operating on the Outer Continental Shelf.[56]

As of early 1984, there are seven appropriations riders relating to OSHA in effect. These are:

1. Restriction on the assessment of first-instance civil penalties when an establishment has fewer than 10 nonserious violations during an inspection.
2. Restriction on the inspection and the enforcement of stand-

[54]The Schweiker bill is discussed in Chapter 13, Section E. In introducing his bill, Senator Schweiker urged that OSHA targeting be based on injury/illness data of individual establishments. His bill sought to achieve that purpose. In introducing the rider, however, Senator Schweiker apparently gave up on the possibility of establishment-specific data and based his exemptions on industry averages.

[55]For OSHA monitoring of state plans, see Chapter 20, Section A.

[56]For a discussion of OSHA disputes over jurisdiction with the U.S. Coast Guard, see Chapter 15, Section B.

ards on farms that do not have temporary labor camps and that employ 10 or fewer workers.[57]

3. Exemption from general schedule safety inspections of firms in industrial classifications with low injury/illness rates which employ 10 or fewer workers.
4. Restriction on OSHA enforcement in connection with recreational hunting, shooting, and fishing.
5. Restriction on other-than-serious penalties for employers with 10 or fewer employees who follow consultants' recommendations for dealing with particular hazards.
6. Prohibition on OSHA spot-check monitoring visits within six months of a state-plan inspection except in certain circumstances.
7. Definition of jurisdiction of OSHA and the U.S. Coast Guard in enforcing safety and health requirements on the Outer Continental Shelf.

[57]Difficult interpretation issues arise on occasion in implementing the appropriations riders. Indeed, it is understandable that the meaning of the riders would be uncertain since they are often introduced on the floor of Congress and passed without committee consideration. For example, an issue was presented on the manner of calculating whether a farm employer "employs 10 or fewer employees." OSHA's field instruction in implementing the rider said:

> Farms may be inspected provided one of the following conditions is present:
>
> a. 11 or more workers are employed for all farm operations at the time of the inspection or a comparable number have been employed by the employer *at any point* during the 12-month period prior to the inspection
>
> b. 10 or fewer workers are employed for all operations and the farm maintains an occupied temporary labor camp, or one that can reasonably be expected to be occupied during the subsequent 12-month period.

OSHA Instruction CPL 2.38, OSHR [Reference File 21:8176] (emphasis added). OSHA special instructions on the inspection of temporary labor camps are included in the Field Operations Manual, ch. XI OSHR [Reference File 71:4101].

Appendix 1

Table of Health Standards Proceedings

Note: The following abbreviations are used in this table:

PEL—permissible exposure limit
TWA—time weighted average
TLV—threshold limit value
ETS—emergency temporary standard
ppm—parts per million
ppb—parts per billion
mg/m^3—milligrams per cubic meter
ug/m^3—micrograms per cubic meter
f/cc—fibers per cubic centimeter

Some of these terms are further discussed, *supra*, p. 40, note 24, and pp. 90–91.

Access to Medical and Exposure Records

29 C.F.R. 1910.20 (1983)

1980

May 23. OSHA publishes a final rule requiring access by employees, their representatives, and OSHA to certain employer-maintained medical and exposure records of employees. The rule also requires long-term preservation of records and contains provisions for the protection of trade secrets. 45 FED. REG. 35,212. The rule is challenged in a district court and in courts of appeals.

1981

September 30. The Court of Appeals for the Fifth Circuit decides that access rule is a "regulation" and its validity should be determined by a district court. *Louisiana Chem. Ass'n v. Bingham,* 657 F.2d 777.

1982

July 13. OSHA proposes modifications to the 1980 standard. The most important of these relate to the definition of "toxic substances", the right of unions representing employees to exposure data, and the protection of trade secrets. 47 Fed. Reg. 30,420.

November 5. The U.S. District Court for the Western District of Louisiana upholds the 1980 access rule in all respects. *Louisiana Chem. Ass'n v. Bingham,* 550 F. Supp. 1136. The decision is appealed to the Court of Appeals for the Fifth Circuit.

1984

May 16. Court of Appeals for the Fifth Circuit affirms the district court decision in a one-sentence *per curiam* opinion, 11 OSHC 1922.

Asbestos

29 C.F.R. §1910.1001 (1983)

1971

May 29. OSHA adopts as established federal standard Walsh-Healey asbestos standard with PEL of 12 fibers of asbestos per cubic centimeter of air (f/cc). 36 Fed. Reg. 10,466.

December 7. OSHA issues an emergency temporary standard (ETS) mandating an 8-hour time weighted average (TWA) of 5 f/cc, with 10 f/cc ceiling for any 15-minute period. 36 Fed. Reg. 23,207. The standard is not challenged in court.

1972

June 7. After rulemaking, OSHA issues a "permanent" standard, with 5 f/cc TWA, effective immediately, and a 2 f/cc TWA, effective in 4 years; the standard also contains a ceiling limit of 10 f/cc. 37 Fed. Reg. 11,316.

1974

April 15. The Court of Appeals for the District of Columbia Circuit, in a landmark opinion, upholds the standard in large part, but orders OSHA to justify (1) the delayed effective date for the 2-fiber level as it applies to all industries; and (2) the requirement that employers retain employee exposure records for only 3 years. *Industrial Union Dep't v. Hodgson,* 499 F.2d 467.

1975

October 9. OSHA proposes to lower the PEL to 0.5 f/cc and the 15-minute ceiling to 5 f/cc. 40 Fed. Reg. 47,652. No further action is taken on this proposal.

1976

March 19. Pursuant to the D.C. Circuit Court order, OSHA extends retention period for exposure records to 20 years. 41 Fed. Reg. 11,504. (OSHA does not respond to the court's order regarding the 2-fiber effective date prior to July 1, when the lower PEL takes effect under the standard.)

1983

November 4. OSHA issues an ETS lowering the PEL to 0.5 f/cc. 48 Fed. Reg. 51,086.

1984

March 7. The Court of Appeals for the Fifth Circuit permanently stays the ETS. *Asbestos Information Ass'n v. OSHA,* 727 F.2d 415.

April 10. OSHA proposes two alternative 8-hour TWAs: 0.2 f/cc and 0.5 f/cc. The proposal would apply to all industries covered by the Act. 49 Fed. Reg. 14,110. Public hearings were held on the proposal in June.

Acrylonitrile

29 C.F.R. §1910.1045 (1983)

1971

May 29. OSHA adopts national consensus standard for acrylonitrile with 8-hour time weighted average (TWA) of 20 parts per million (ppm). 36 FED. REG. 10,466.

1978

January 17. Based on new data on carcinogenicity of acrylonitrile, OSHA issues ETS, with 2 ppm TWA and ceiling of 10 ppm for 15-minute period. 43 FED. REG. 2586. ETS is challenged in the Court of Appeals for the Sixth Circuit by employers.

March 28. The Court of Appeals for the Sixth Circuit refuses to stay ETS, *Vistron v. OSHA*, 6 OSHC 1483, and challenge is discontinued.

October 3. OSHA issues "permanent" standard after rulemaking containing same limits as ETS. 43 FED. REG. 45,762. Standard is not challenged.

Benzene

29 C.F.R. §1910.100, Table Z-2

1971

May 29. OSHA adopts national consensus standard for benzene with following PELs: 8-hour TWA—10 ppm; 15-minute ceiling—25 ppm; maximum peak level—50 ppm for 10-minute interval. 36 FED. REG. 10,466.

1977

May 3. OSHA issues ETS lowering 8-hour TWA to 1 ppm and 15-minute ceiling to 5 ppm. 42 FED. REG. 22,516.

May 20. The Court of Appeals for the Fifth Circuit stays the effective date of ETS and OSHA later withdraws the ETS when litigation over venue continues for five months. *Industrial Union Dep't v. Bingham,* 570 F.2d 965 (D.C. Cir.).

1978

February 10. After rulemaking, OSHA issues "permanent" benzene standard, containing same PELs as ETS. The standard also prohibits skin contact with liquid benzene. 43 FED. REG. 5918. Standard is stayed by the Court of Appeals for the Fifth Circuit.

October 5. The Fifth Circuit vacates the standard, holding, *inter alia,* that OSHA failed to demonstrate measurable benefits that would result from lowering the PEL and to balance these benefits against the costs of the standard. The court also sets aside skin contact provisions. *American Petroleum Inst. v. OSHA*, 581 F.2d 493.

1980

July 2. The Supreme Court affirms the Fifth Circuit decision vacating the benzene standard. The Court concludes that the Act requires OSHA to find regulations address a "significant risk"; Court does not reach cost/benefit issue. *Industrial Union Dep't v. American Petroleum Inst.,* 448 U.S. 607.

1983

July 8. OSHA announces that it is considering lowering the PEL for benzene to either 5, 1, or 0.5 ppm, and requests data from the public. 48 FED. REG. 31,412.

14 Carcinogens

29 C.F.R. §§1910.1003–.1016 (1983)

1973

May 3. OSHA issues an ETS regulating 14 carcinogens. The ETS sets no exposure limits and emphasizes work practices to protect employees. 38 Fed. Reg. 10,929. The 14 chemicals regulated are:

(1) 2-Acetylaminofluorene
(2) 4-Aminodiphenyl
(3) Benzidine
(4) 3,3′-Dichlorobenzidine
(5) 4-Dimethylaminoazo-benzene
(6) alpha-Naphthylamine
(7) beta-Naphthylamine
(8) 4-Nitrobiphenyl
(9) N-Nitrosodimethylamine
(10) beta-Propiolactone
(11) bis-Chloromethyl ether
(12) Ethyleneimine
(13) Methyl Chloromethyl ether
(14) 4,4′-Methylene(bis)-2-Chloroaniline

October 4. The Court of Appeals for the Third Circuit vacates the ETS relating to 3,3′-Dichlorobenzedine (DCB) and Ethyleneimine (EI), finding that OSHA failed adequately to state the reasons for its conclusion that an ETS was necessary, contrary to §6(e). *Dry Color Mfrs. Ass'n v. Department of Labor,* 486 F.2d 98.

1974

January 29. OSHA issues identical "permanent" standards on 14 carcinogens, including DCB and EI. The provisions are essentially the same as those of the ETS. 39 Fed. Reg. 3756.

August 26. The Third Circuit upholds the "permanent" carcinogens rule as applied to EI, except the provisions regarding the use of EI in laboratories which it vacates because of lack of proper notice. *Synthetic Organic Chem. Mfrs. Ass'n v. Brennan,* 503 F.2d 1155.

December 17. The Third Circuit vacates the "permanent" standard as it applies to 4,4-Methylene(bis)-2-Chloroaniline (MOCA) because OSHA published the proposal *before* it received the recommendations of the advisory committee. The court upholds the standards for the remaining 12 carcinogens, except for the laboratory provisions. In addition, the court remands the medical surveillance provisions in standards for further consideration of specific protocols and diagnostic tests. *Synthetic Organic Chem. Mfrs. Ass'n v. Brennan,* 506 F.2d 385, *cert. denied,* 423 U.S. 830.

Carcinogens Policy

29 C.F.R. Part 1990 (1983)

1980

January 22. OSHA issues Carcinogens Policy designed to facilitate process of issuance of substance-specific carcinogens standards. 45 Fed. Reg. 5002. Policy is challenged in court and, after litigation (*American Petroleum Inst. v. OSHA,* 8 OSHC 2025 (5th Cir.)), venue is determined to lie in the Fifth Circuit Court of Appeals (No. 80-3018), where case is pending.

August 12. Pursuant to the new Carcinogens Policy, OSHA publishes a "candidates list" of 204 suspected carcinogens which OSHA plans to study in greater depth. 45 FED. REG. 53,672. (Carcinogens Policy calls for the development of "priority lists" of selected substances drawn from the candidates list that pose a "relatively high degree of hazard" to employees. Substances will be chosen in turn from the priority lists for full scientific review "with a view to subsequent rulemakings". See 29 C.F.R. §§1990.121–.133.)

1981

January 19. In light of Supreme Court *Benzene* decision, 448 U.S. 607 (1980), OSHA deletes provisions of the Carcinogens Policy that automatically (1) set the PEL for carcinogens at the lowest feasible level; and (2) characterize suspect carcinogens as presenting a "grave danger" to employee health. 46 FED. REG. 4889.

1982

January 5. Advance notice published on reconsideration of Carcinogens Policy. 47 FED. REG. 187.

1983

January 4. OSHA issues administrative stay on several provisions of the Carcinogens Policy including those on the issuance of candidate and priority lists, pending overall reconsideration of the rule. 48 FED. REG. 241. OSHA states that it will propose revisions in the Carcinogens Policy in 1984.

Chemicals, Hazard Communication

29 C.F.R. §1910.1200

1981

January 16. OSHA issues proposed rule requiring employers and chemical manufacturers to identify chemical hazards in the workplace and to communicate this information to employees. 46 FED. REG. 4412.

February 12. Newly appointed Assistant Secretary withdraws the proposal to consider "regulatory alternatives". 40 FED. REG. 12,214.

1983

November 25. After rulemaking, OSHA issues final hazard communication rule differing from the 1981 proposal as to coverage, protection of trade secrets, and in other respects. 48 FED. REG. 53,280. Labor organizations challenge the rule in the Court of Appeals for the Third Circuit, and several states intervene in the proceeding. See *United Steelworkers v. Auchter,* Nos. 83-3554, 3561, etc. (3d Cir. Nov. 22, 1983). A major issue in the litigation is the extent of OSHA's preemption of state and local "right to know" laws.

1984

Litigation pending in the Court of Appeals for the Third Circuit.

Coke Oven Emissions

29 C.F.R. §1910.1029 (1983)

1971

May 29. OSHA adopts as established federal standard Walsh-Healey TLV for coal tar pitch volatiles of 0.2 milligrams per cubic meter (mg/m^3). 36 FED. REG. 10,466.

1976

October 22. After rulemaking, OSHA issues new standard regulating the benzene-soluble fraction of total particulate matter present during the coking process. Standard lowers the 8-hour TWA to 0.15 mg/m^3 (or 150 ug/m^3). 41 Fed. Reg. 46,742. Standard is challenged by employer association in the Court of Appeals for the Third Circuit.

1978

March 28. The Third Circuit upholds the standard, finding the TWA necessary for the protection of employees and technologically and economically feasible. However, court vacates the research and development provisions and those requiring quantitative fit tests for respirators. Court also remands for further proceedings the issue of application of the standard to noncoke employers. *American Iron and Steel Inst. v. OSHA,* 577 F.2d 825. Employer association files petition for certiorari in Supreme Court, certiorari is granted, but petition is later withdrawn, 448 U.S. 917 (1980).

Cotton Dust

29 C.F.R. §1910.1043

1971

May 29. OSHA adopts national consensus standard for cotton dust with 8-hour TWA of 1,000 micrograms per cubic meter (ug/m^3). 36 Fed. Reg. 10,466.

1978

June 23. After rulemaking, OSHA issues revised standard, with three different TWAs, to be measured by vertical elutriator:
—yarn manufacturing operations—200 ug/m^3
—slashing and weaving operations—750 ug/m^3
—all other covered operations—500 ug/m^3
43 Fed. Reg. 27,350. The standard is challenged in court, and after venue dispute, transferred to Court of Appeals for the District of Columbia.

1979

October 24. The D.C. Circuit upholds most provisions of standard, but vacates provisions covering cottonseed oil industry on feasibility grounds. *AFL-CIO v. Marshall,* 617 F.2d 636.

1981

June 17. The Supreme Court affirms the D.C. Circuit decision, but remands the wage-guarantee provision to OSHA for further consideration. Court holds that the Act does not require OSHA to balance costs and benefits in promulgating standards issued under §6(b)(5). *American Textile Mfrs. Inst. v. Donovan,* 452 U.S. 490.

1983

June 10. OSHA proposes modifications of cotton dust standard, including removal of certain industries from coverage and delay of compliance dates for some other industries. Agency points out that because of judicial and administrative stays, 1978 cotton dust standard is in effect only for the textile industry (except knitting). 48 Fed. Reg. 26,962.

1,2-Dibromo-3-Chloropropane (DBCP)

29 C.F.R. §1910.144

1977

September 9. OSHA issues ETS for DBCP, establishing 8-hour TWA of 10 parts per billion (ppb) and 15-minute ceiling level of 50 ppb. 42 FED. REG. 45,536. (No DBCP standard had been adopted by OSHA under §6(a) in 1971.) The ETS is not challenged.

1978

March 17. After rulemaking, OSHA issues "permanent" DBCP standard, lowering 8-hour TWA to 1 ppb, with no ceiling limit. 43 FED. REG. 11,514. The standard is not challenged in court.

Diving Operations

29 C.F.R. §§1910.401–.441 (Subpart T)

1976

June 15. In response to petition filed by United Brotherhood of Carpenters and Joiners of America, AFL-CIO, OSHA issues ETS covering diving operations. 41 FED. REG. 24,272.

August 11. The Court of Appeals for the Fifth Circuit issues an indefinite stay of the ETS in a suit brought by several diving contractors. *Taylor Diving and Salvage Co. v. Department of Labor,* 537 F.2d 819.

November 5. OSHA withdraws ETS. 41 FED. REG. 48,742.

1977

July 22. OSHA issues "permanent" standard which, among other things, contains novel provisions on medical surveillance of diving employees. 42 FED. REG. 37,650.

1979

July 16. The Court of Appeals for the Fifth Circuit vacates medical surveillance provisions, holding that three-physician review procedure beyond OSHA authority. Court, however, upholds provisions on access to employer records against challenge based on inadequate notice. *Taylor Diving and Salvage Co. v. Department of Labor,* 599 F.2d 622.

1982

February 26. OSHA publishes advance notice of proposed rulemaking on reconsideration of diving standard which was targeted for review by President's Task Force on Regulatory Relief. 47 FED. REG. 8379.

November 26. OSHA amends diving standard by exempting operations "performed solely for marine scientific research and development purposes by educational institutions." 47 FED. REG. 53,357. Amendment is challenged in the Court of Appeals for the District of Columbia by Carpenters Union. *United Bhd. of Carpenters v. Department of Labor* (No. 82-2509).

1984

April 4. D.C. Circuit remands proceeding to OSHA for supplemental proceedings on issue of whether Carpenters Union has standing to challenge validity of scientific diving standard (11 OSHC 1905).

Ethylene Dibromide

29 C.F.R. §1910.1000, Table Z-2

1971

May 29. OSHA adopts national consensus standard on ethylene dibromide containing 8-hour TWA of 20 parts per million (ppm); ceiling limit of 30 ppm for any period of time; and 5-minute "maximum peak" of 50 ppm. 36 FED. REG. 10,466.

1981

September 2. International Brotherhood of Teamsters petitions OSHA for ETS for ethylene dibromide based on new carcinogenicity data. OSHA denies petition and, on *December 18,* publishes advance notice of proposed rulemaking. 46 FED. REG. 61,671.

1983

March 1. OSHA approves standard on ethylene dibromide issued by California under its approved state plan, finding that the standard meets the requirements of the proviso to §18(c)(2) of the Act. 48 FED. REG. 8610.

October 7. OSHA proposes revised federal standard on ethylene dibromide containing an 8-hour TWA of 0.1 ppm and a short term exposure limit (STEL) of 0.5 ppm. 48 FED. REG. 45,956.

1984

March Public hearing held on proposal.

Ethylene Oxide (EtO)

29 C.F.R. §1910.1000, Table Z-1

1971

May 29. OSHA adopts as established federal standard Walsh-Healey EtO establishing an 8-hour TWA of 50 parts per million (ppm). 36 FED. REG. 10,466.

1981

August. Public Citizen Health Research Group petitions for ETS on ethylene oxide relying in part on new carcinogenicity data. OSHA denies petition.

1982

January 26. OSHA publishes advance notice, announcing its intention to reevaluate the existing EtO standard. 47 FED. REG. 3566.

1983

January 5. The District Court for the District of Columbia, in suit brought by Health Research Group, orders OSHA to issue an ETS within 20 days of its decision. *Public Citizen Health Research Group v. Auchter,* 554 F. Supp. 242.

March 15. The Court of Appeals for the District of Columbia issues *per curiam* decision, reversing District Court on ETS but ordering OSHA to issue a *proposed* standard within 30 days and stating that it "expects" a final rule within a year. *Public Citizen Health Research Group v. Auchter,* 702 F.2d 1150.

April 21. OSHA issues proposed EtO standard containing 8-hour TWA of 1 ppm and an action level of 0.5 ppm as TWA. 48 FED. REG. 17,284.
July. Public hearings held on the proposal.

1984
June 22. OSHA issues final EtO standard. 49 FED. REG. 25,734. OSHA states that issue of whether standard should contain STEL will remain open for expert review and public comment.

Inorganic Arsenic

29 C.F.R. §1910.1018 (1983)

1971
May 29. OSHA adopts as established federal standard Walsh-Healey TLVs as follows: arsenic and compounds-500 micrograms per cubic meter (ug/m^3); lead arsenate-150 ug/m^3; calcium arsenate-1000 ug/m^3. 36 FED. REG. 10,466.

1978
May 5. OSHA issues new arsenic standard, lowering the 8-hour TWA for all inorganic arsenic compounds to 10 ug/m^3. 43 FED. REG. 19,584.

1981
April 7. At the request of OSHA, the Court of Appeals for the Ninth Circuit remands the record to OSHA on the 1978 standard to perform a risk assessment consistent with the Supreme Court's *Benzene* decision. 448 U.S. 607 (1980). The Ninth Circuit retains jurisdiction over the case and the standard remains in effect (except as previously stayed respecting engineering controls), pending the outcome of the risk assessment proceeding. *ASARCO, Inc. v. OSHA*, 647 F.2d 1.

1983
January 14. OSHA issues a supplemental statement of reasons, concluding that the 10 ug/m^3 TWA is "needed to substantially reduce a significant risk of lung cancer". 48 FED. REG. 1864.

1984
Litigation still pending in court of appeals on validity of standard (Nos. 78-1959, 2764, 3038).

Lead

29 C.F.R. §1910.1025

1971
May 29. OSHA adopts national consensus standard with 8-hour TWA of 0.2 milligrams per cubic meter (mg/m^3). 36 FED. REG. 10,466.

1978
November 14. After rulemaking, OSHA issues new lead standard, reducing PEL to 50 micrograms per cubic meter (ug/m^3). Standard requires use of engineering controls, but provides varying compliance deadlines (up to 10 years) for affected industries. Standard also contains provision for

medical removal protection (MRP). 43 Fed. Reg. 52,952. Industry and union groups challenge the standard in courts of appeals.

1979

Union and industry petitioners and OSHA dispute proper venue of proceeding. Case is transferred to the Court of Appeals for the District of Columbia. *United Steelworkers v. Marshall,* 592 F.2d 693 (3d Cir.).

1980

August 15. D.C. Circuit, in a lengthy opinion, upholds new PEL and MRP provisions, but directs the agency to determine the feasibility of engineering controls for 38 industries and occupational categories. Standard goes into effect for all industries covered, but 38 remand industries given flexibility to use respirators to implement the PEL. *United Steelworkers v. Marshall,* 647 F.2d 1189, *cert. denied,* 453 U.S. 913 (1981).

1981

January 21. Pursuant to court order and after rulemaking, OSHA reexamines the compliance capabilities of the 38 industries listed by the court (and 8 additional industries). Agency issues a supplemental statement of reasons, concluding that most industries can comply with standard, a few will require extra time to implement engineering controls, and others may not be able to implement engineering controls at all. 46 Fed. Reg. 6134.

December 11. After additional rulemaking, OSHA issues revised supplemental statement of reasons, announcing amendments to the lead rule that allow greater flexibility in using respirators rather than engineering controls and include delays in the effective date for certain affected industries. 46 Fed. Reg. 60,758.

1982

December 3. OSHA stays the deadline for the primary and secondary lead smelting and battery manufacturing industries for the submission of compliance plans to the Agency. OSHA indicates that, in light of its reconsideration of the entire lead standard, development of the plans would be a "costly exercise that could later prove unnecessary." 47 Fed. Reg. 54,433. Steelworkers Union challenges stay because interim stay not preceded by notice and comment.

1984

April 17. The Court of Appeals for the District of Columbia vacates stay as of June 1, 1984, without prejudice to OSHA's conducting rulemaking on issue of the effective date for filing compliance plans by industries subject to stay. *United Steelworkers v. Auchter,* 11 OSHC 1920.

April 24. OSHA lifts stay and asks for comment on new effective date for filing compliance plans. 49 Fed. Reg. 17,545.

Noise

29 C.F.R. §1910.95

1971

May 29. OSHA adopts as established federal standard Walsh-Healey standard on noise, containing a scale of permissible noise exposure limits, including a 90-decibel limit for an 8-hour day. 36 Fed. Reg. 10,466.

1974

October 24. OSHA proposes major revision on noise standard, raising issue as to appropriate permissible limits and methods of compliance. 39 FED. REG. 37,773.

1981

January 16. OSHA issues "Hearing Conservation Amendment" to noise standard, imposing mandatory program of audiometric testing, exposure monitoring, and training for employees exposed in excess of 85 decibels for an 8-hour day. 46 FED. REG. 4078.

Newly appointed Assistant Secretary stays the effective date of the "Hearing Conservation Amendment" and in August publishes an interim revised "Hearing Conservation Amendment." 46 FED. REG. 42,622.

1983

March 8. Final "Hearing Conservation Amendment" issued, revising Amendment issued in 1981 in a number of respects. 48 FED. REG. 9738.

1984

Amendment challenged in the Court of Appeals for the Fourth Circuit by employer association. *Forging Indus. Ass'n v. Secretary of Labor* (No. 83-1420). Suit is pending.

Pesticides

1973

May 1. OSHA issues ETS specifying permissible field reentry times for 21 organophosphate pesticides. 38 FED. REG. 10,715.

1974

January 9. The Court of Appeals for the Fifth Circuit vacates ETS, finding that OSHA finding of "grave danger" not supported by substantial evidence. *Florida Peach Growers Ass'n v. Department of Labor,* 489 F.2d 120. OSHA concedes primary EPA authority on field reentry times (39 FED. REG. 16,888), and is sued by farmworkers for not issuing standard.

1975

October 9. The Court of Appeals for the District of Columbia decides that OSHA is preempted under §4(b)(1) by EPA on regulating pesticide reentry times. *Organized Migrants in Community Action v. Brennan,* 520 F.2d 1161.

Vinyl Chloride

29 C.F.R. §1910.101 (1983)

1971

May 29. OSHA adopts national consensus standard for vinyl chloride containing ceiling limit of 500 parts per million (ppm). 36 FED. REG. 10,466.

1974

April 15. OSHA issues ETS for vinyl chloride based on new data on its carcinogenicity. ETS lowers ceiling level to 50 ppm. 39 FED. REG. 12,341. ETS is not challenged in court.

October 4. After rulemaking, OSHA issues "permanent" vinyl chloride standard containing a 1 ppm 8-hour TWA and a 5 ppm 15-minute ceiling limit. 39 Fed. Reg. 35,890. Employer association challenges standard in the Court of Appeals for the Second Circuit.

1975

January 31. The Second Circuit upholds vinyl chloride standard in all respects, relying in part on principle of "technology forcing" in determining feasibility of standard. *Society of the Plastics Indus. v. OSHA,* 509 F.2d 1301, *cert. denied,* 421 U.S. 992.

Appendix 2

Occupational Safety and Health Act of 1970

Public Law 91-596
91st Congress, S. 2193
December 29, 1970

An Act

84 STAT. 1590

To assure safe and healthful working conditions for working men and women; by authorizing enforcement of the standards developed under the Act; by assisting and encouraging the States in their efforts to assure safe and healthful working conditions; by providing for research, information, education, and training in the field of occupational safety and health; and for other purposes.

Be it enacted by the Senate and House of Representatives of the United States of America in Congress assembled, That this Act may be cited as the "Occupational Safety and Health Act of 1970".

Occupational Safety and Health Act of 1970.

CONGRESSIONAL FINDINGS AND PURPOSE

SEC. (2) The Congress finds that personal injuries and illnesses arising out of work situations impose a substantial burden upon, and are a hindrance to, interstate commerce in terms of lost production, wage loss, medical expenses, and disability compensation payments.

(b) The Congress declares it to be its purpose and policy, through the exercise of its powers to regulate commerce among the several States and with foreign nations and to provide for the general welfare, to assure so far as possible every working man and woman in the Nation safe and healthful working conditions and to preserve our human resources—

(1) by encouraging employers and employees in their efforts to reduce the number of occupational safety and health hazards at their places of employment, and to stimulate employers and employees to institute new and to perfect existing programs for providing safe and healthful working conditions;

(2) by providing that employers and employees have separate but dependent responsibilities and rights with respect to achieving safe and healthful working conditions;

(3) by authorizing the Secretary of Labor to set mandatory occupational safety and health standards applicable to businesses affecting interstate commerce, and by creating an Occupational Safety and Health Review Commission for carrying out adjudicatory functions under the Act;

(4) by building upon advances already made through employer and employee initiative for providing safe and healthful working conditions;

(5) by providing for research in the field of occupational safety and health, including the psychological factors involved, and by developing innovative methods, techniques, and approaches for dealing with occupational safety and health problems;

(6) by exploring ways to discover latent diseases, establishing causal connections between diseases and work in environmental conditions, and conducting other research relating to health problems, in recognition of the fact that occupational health standards present problems often different from those involved in occupational safety;

(7) by providing medical criteria which will assure insofar as practicable that no employee will suffer diminished health, functional capacity, or life expectancy as a result of his work experience;

(8) by providing for training programs to increase the number and competence of personnel engaged in the field of occupational safety and health;

84 STAT. 1591

(9) by providing for the development and promulgation of occupational safety and health standards;

(10) by providing an effective enforcement program which shall include a prohibition against giving advance notice of any inspection and sanctions for any individual violating this prohibition;

(11) by encouraging the States to assume the fullest responsibility for the administration and enforcement of their occupational safety and health laws by providing grants to the States to assist in identifying their needs and responsibilities in the area of occupational safety and health, to develop plans in accordance with the provisions of this Act, to improve the administration and enforcement of State occupational safety and health laws, and to conduct experimental and demonstration projects in connection therewith;

(12) by providing for appropriate reporting procedures with respect to occupational safety and health which procedures will help achieve the objectives of this Act and accurately describe the nature of the occupational safety and health problem;

(13) by encouraging joint labor-management efforts to reduce injuries and disease arising out of employment.

DEFINITIONS

Sec. 3. For the purposes of this Act—

(1) The term "Secretary" mean the Secretary of Labor.

(2) The term "Commission" means the Occupational Safety and Health Review Commission established under this Act.

(3) The term "commerce" means trade, traffic, commerce, transportation, or communication among the several States, or between a State and any place outside thereof, or within the District of Columbia, or a possession of the United States (other than the Trust Territory of the Pacific Islands), or between points in the same State but through a point outside thereof.

(4) The term "person" means one or more individuals, partnerships, associations, corporations, business trusts, legal representatives, or any organized group of persons.

(5) The term "employer" means a person engaged in a business affecting commerce who has employees, but does not include the United States or any State or political subdivision of a State.

(6) The term "employee" means an employee of an employer who is employed in a business of his employer which affects commerce.

(7) The term "State" includes a State of the United States, the District of Columbia, Puerto Rico, the Virgin Islands, American Samoa, Guam, and the Trust Territory of the Pacific Islands.

(8) The term "occupational safety and health standard" means a standard which requires conditions, or the adoption or use of one or more practices, means, methods, operations, or processes, reasonably necessary or appropriate to provide safe or healthful employment and places of employment.

(9) The term "national consensus standard" means any occupational safety and health standard or modification thereof which (1), has been adopted and promulgated by a nationally recognized standards-producing organization under procedures whereby it can be determined by the Secretary that persons interested

84 STAT. 1592

and affected by the scope or provisions of the standard have reached substantial agreement on its adoption, (2) was formulated in a manner which afforded an opportunity for diverse views to be considered and (3) has been designated as such a standard by the Secretary, after consultation with other appropriate Federal agencies.

(10) The term "established Federal standard" means any operative occupational safety and health standard established by any agency of the United States and presently in effect, or contained in any Act of Congress in force on the date of enactment of this Act.

(11) The term "Committee" means the National Advisory Committee on Occupational Safety and Health established under this Act.

(12) The term "Director" means the Director of the National Institute for Occupational Safety and Health.

(13) The term "Institute" means the National Institute for Occupational Safety and Health established under this Act.

(14) The term "Workmen's Compensation Commission" means the National Commission on State Workmen's Compensation Laws established under this Act.

APPLICABILITY OF THIS ACT

SEC. 4. (a) This Act shall apply with respect to employment performed in a workplace in a State, the District of Columbia, the Commonwealth of Puerto Rico, the Virgin Islands, American Samoa, Guam, the Trust Territory of the Pacific Islands, Wake Island, Outer Continental Shelf lands defined in the Outer Continental Shelf Lands Act, Johnston Island, and the Canal Zone. The Secretary of the Interior shall, by regulation, provide for judicial enforcement of this Act by the courts established for areas in which there are no United States district courts having jurisdiction.

67 Stat. 462.
43 USC 1331 note.

(b)(1) Nothing in this Act shall apply to working conditions of employees with respect to which other Federal agencies, and State agencies acting under section 274 of the Atomic Energy Act of 1954, as amended (42 U.S.C. 2021), exercise statutory authority to prescribe or enforce standards or regulations affecting occupational safety or health.

73 Stat. 688.

(2) The safety and health standards promulgated under the Act of June 30, 1936, commonly known as the Walsh-Healey Act (41 U.S.C. 35 et seq.), the Service Contract Act of 1965 (41 U.S.C. 351 et seq.), Public Law 91–54, Act of August 9, 1969 (40 U.S.C. 333), Public Law 85–742, Act of August 23, 1958 (33 U.S.C. 941), and the National Foundation on Arts and Humanities Act (20 U.S.C. 951 et seq.) are superseded on the effective date of corresponding standards, promulgated under this Act, which are determined by the Secretary to be more effective. Standards issued under the laws listed in this

49 Stat. 2036.
79 Stat. 1034.
83 Stat. 96.
72 Stat. 835.
79 Stat. 845;
Ante, p. 443.

paragraph and in effect on or after the effective date of this Act shall be deemed to be occupational safety and health standards issued under this Act, as well as under such other Acts.

Report to Congress.

(3) The Secretary shall, within three years after the effective date of this Act, report to the Congress his recommendations for legislation to avoid unnecessary duplication and to achieve coordination between this Act and other Federal laws.

84 STAT. 1593

(4) Nothing in this Act shall be construed to supersede or in any manner affect any workmen's compensation law or to enlarge or diminish or affect in any other manner the common law or statutory rights, duties, or liabilities of employers and employees under any law with respect to injuries, diseases, or death of employees arising out of, or in the course of, employment.

DUTIES

Sec. 5. (a) Each employer—

(1) shall furnish to each of his employees employment and a place of employment which are free from recognized hazards that are causing or are likely to cause death or serious physical harm to his employees;

(2) shall comply with occupational safety and health standards promulgated under this Act.

(b) Each employee shall comply with occupational safety and health standards and all rules, regulations, and orders issued pursuant to this Act which are applicable to his own actions and conduct.

OCCUPATIONAL SAFETY AND HEALTH STANDARDS

80 Stat. 381; 81 Stat. 195. 5 USC 500.

Sec. 6. (a) Without regard to chapter 5 of title 5, United States Code, or to the other subsections of this section, the Secretary shall, as soon as practicable during the period beginning with the effective date of this Act and ending two years after such date, by rule promulgate as an occupational safety or health standard any national consensus standard, and any established Federal standard, unless he determines that the promulgation of such a standard would not result in improved safety or health for specifically designated employees. In the event of conflict among any such standards, the Secretary shall promulgate the standard which assures the greatest protection of the safety or health of the affected employees.

(b) The Secretary may by rule promulgate, modify, or revoke any occupational safety or health standard in the following manner:

Advisory committee, recommendations.

(1) Whenever the Secretary, upon the basis of information submitted to him in writing by an interested person, a representative of any organization of employers or employees, a nationally recognized standards-producing organization, the Secretary of Health, Education, and Welfare, the National Institute for Occupational Safety and Health, or a State or political subdivision, or on the basis of information developed by the Secretary or otherwise available to him, determines that a rule should be promulgated in order to serve the objectives of this Act, the Secretary may request the recommendations of an advisory committee appointed under section 7 of this Act. The Secretary shall provide such an advisory committee with any proposals of his own or of the Secretary of Health, Education, and Welfare, together with all pertinent factual information developed by the Secretary or the Secretary of Health, Education, and Welfare, or otherwise available, including the results of research, demonstrations, and experiments. An advisory committee shall submit to the Secretary its recommendations regarding the rule to be promulgated within ninety days from the date of its appointment or within such longer or shorter period as may be prescribed by the Secretary, but in no event for a period which is longer than two hundred and seventy days.

(2) The Secretary shall publish a proposed rule promulgating, modifying, or revoking an occupational safety or health standard in the Federal Register and shall afford interested persons a period of thirty days after publication to submit written data or comments. Where an advisory committee is appointed and the Secretary determines that a rule should be issued, he shall publish the proposed rule within sixty days after the submission of the advisory committee's recommendations or the expiration of the period prescribed by the Secretary for such submission.

Publication in Federal Register.

(3) On or before the last day of the period provided for the submission of written data or comments under paragraph (2), any interested person may file with the Secretary written objections to the proposed rule, stating the grounds therefor and requesting a public hearing on such objections. Within thirty days after the last day for filing such objections, the Secretary shall publish in the Federal Register a notice specifying the occupational safety or health standard to which objections have been filed and a hearing requested, and specifying a time and place for such hearing.

Hearing, notice.

Publication in Federal Register.

(4) Within sixty days after the expiration of the period provided for the submission of written data or comments under paragraph (2), or within sixty days after the completion of any hearing held under paragraph (3), the Secretary shall issue a rule promulgating, modifying, or revoking an occupational safety or health standard or make a determination that a rule should not be issued. Such a rule may contain a provision delaying its effective date for such period (not in excess of ninety days) as the Secretary determines may be necessary to insure that affected employers and employees will be informed of the existence of the standard and of its terms and that employers affected are given an opportunity to familiarize themselves and their employees with the existence of the requirements of the standard.

(5) The Secretary, in promulgating standards dealing with toxic materials or harmful physical agents under this subsection, shall set the standard which most adequately assures, to the extent feasible, on the basis of the best available evidence, that no employee will suffer material impairment of health or functional capacity even if such employee has regular exposure to the hazard dealt with by such standard for the period of his working life. Development of standards under this subsection shall be based upon research, demonstrations, experiments, and such other information as may be appropriate. In addition to the attainment of the highest degree of health and safety protection for the employee, other considerations shall be the latest available scientific data in the field, the feasibility of the standards, and experience gained under this and other health and safety laws. Whenever practicable, the standard promulgated shall be expressed in terms of objective criteria and of the performance desired.

Toxic materials.

(6) (A) Any employer may apply to the Secretary for a temporary order granting a variance from a standard or any provision thereof promulgated under this section. Such temporary order shall be granted only if the employer files an application which meets the requirements of clause (B) and establishes that (i) he is unable to comply with a standard by its effective date because of unavailability of professional or technical personnel or of materials and equipment needed to come into compliance with the standard or because necessary construction or alteration of facilities cannot be completed by the effective date, (ii) he is taking all available steps to safeguard his employees against the hazards covered by the standard, and (iii) he has an effective program for coming into compliance with the standard as quickly as

Temporary variance order.

practicable. Any temporary order issued under this paragraph shall prescribe the practices, means, methods, operations, and processes which the employer must adopt and use while the order is in effect and state in detail his program for coming into compliance with the standard. Such a temporary order may be granted only after notice to employees and an opportunity for a hearing: *Provided*, That the Secretary may issue one interim order to be effective until a decision is made on the basis of the hearing. No temporary order may be in effect for longer than the period needed by the employer to achieve compliance with the standard or one year, whichever is shorter, except that such an order may be renewed not more than twice (I) so long as the requirements of this paragraph are met and (II) if an application for renewal is filed at least 90 days prior to the expiration date of the order. No interim renewal of an order may remain in effect for longer than 180 days.

Notice, hearing.

Renewal.

Time limitation.

(B) An application for a temporary order under this paragraph (6) shall contain:

(i) a specification of the standard or portion thereof from which the employer seeks a variance,

(ii) a representation by the employer, supported by representations from qualified persons having firsthand knowledge of the facts represented, that he is unable to comply with the standard or portion thereof and a detailed statement of the reasons therefor,

(iii) a statement of the steps he has taken and will take (with specific dates) to protect employees against the hazard covered by the standard,

(iv) a statement of when he expects to be able to comply with the standard and what steps he has taken and what steps he will take (with dates specified) to come into compliance with the standard, and

(v) a certification that he has informed his employees of the application by giving a copy thereof to their authorized representative, posting a statement giving a summary of the application and specifying where a copy may be examined at the place or places where notices to employees are normally posted, and by other appropriate means.

A description of how employees have been informed shall be contained in the certification. The information to employees shall also inform them of their right to petition the Secretary for a hearing.

(C) The Secretary is authorized to grant a variance from any standard or portion thereof whenever he determines, or the Secretary of Health, Education, and Welfare certifies, that such variance is necessary to permit an employer to participate in an experiment approved by him or the Secretary of Health, Education, and Welfare designed to demonstrate or validate new and improved techniques to safeguard the health or safety of workers.

Labels, etc.

(7) Any standard promulgated under this subsection shall prescribe the use of labels or other appropriate forms of warning as are necessary to insure that employees are apprised of all hazards to which they are exposed, relevant symptoms and appropriate emergency treatment, and proper conditions and precautions of safe use or exposure. Where appropriate, such standard shall also prescribe suitable protective equipment and control or technological procedures to be used in connection with such hazards and shall provide for monitoring or measuring employee exposure at such locations and intervals, and in such manner as may be necessary for the protection of employees. In

Protective equipment, etc.

addition, where appropriate, any such standard shall prescribe the type and frequency of medical examinations or other tests which shall be made available, by the employer or at his cost, to employees exposed to such hazards in order to most effectively determine whether the health of such employees is adversely affected by such exposure. In the event such medical examinations are in the nature of research, as determined by the Secretary of Health, Education, and Welfare, such examinations may be furnished at the expense of the Secretary of Health, Education, and Welfare. The results of such examinations or tests shall be furnished only to the Secretary or the Secretary of Health, Education, and Welfare, and, at the request of the employee, to his physician. The Secretary, in consultation with the Secretary of Health, Education, and Welfare, may by rule promulgated pursuant to section 553 of title 5, United States Code, make appropriate modifications in the foregoing requirements relating to the use of labels or other forms of warning, monitoring or measuring, and medical examinations, as may be warranted by experience, information, or medical or technological developments acquired subsequent to the promulgation of the relevant standard.

Medical examinations.

80 Stat. 383.

(8) Whenever a rule promulgated by the Secretary differs substantially from an existing national consensus standard, the Secretary shall, at the same time, publish in the Federal Register a statement of the reasons why the rule as adopted will better effectuate the purposes of this Act than the national consensus standard.

Publication in Federal Register.

(c)(1) The Secretary shall provide, without regard to the requirements of chapter 5, title 5, United States Code, for an emergency temporary standard to take immediate effect upon publication in the Federal Register if he determines (A) that employees are exposed to grave danger from exposure to substances or agents determined to be toxic or physically harmful or from new hazards, and (B) that such emergency standard is necessary to protect employees from such danger.

Temporary standard. Publication in Federal Register. 80 Stat. 381; 81 Stat. 195. 5 USC 500.

(2) Such standard shall be effective until superseded by a standard promulgated in accordance with the procedures prescribed in paragraph (3) of this subsection.

Time limitation.

(3) Upon publication of such standard in the Federal Register the Secretary shall commence a proceeding in accordance with section 6(b) of this Act, and the standard as published shall also serve as a proposed rule for the proceeding. The Secretary shall promulgate a standard under this paragraph no later than six months after publication of the emergency standard as provided in paragraph (2) of this subsection.

(d) Any affected employer may apply to the Secretary for a rule or order for a variance from a standard promulgated under this section. Affected employees shall be given notice of each such application and an opportunity to participate in a hearing. The Secretary shall issue such rule or order if he determines on the record, after opportunity for an inspection where appropriate and a hearing, that the proponent of the variance has demonstrated by a preponderance of the evidence that the conditions, practices, means, methods, operations, or processes used or proposed to be used by an employer will provide employment and places of employment to his employees which are as safe and healthful as those which would prevail if he complied with the standard. The rule or order so issued shall prescribe the conditions the employer must maintain, and the practices, means, methods, operations, and processes which he must adopt and utilize to the extent they

Variance rule.

differ from the standard in question. Such a rule or order may be modified or revoked upon application by an employer, employees, or by the Secretary on his own motion, in the manner prescribed for its issuance under this subsection at any time after six months from its issuance.

Publication in Federal Register.

(e) Whenever the Secretary promulgates any standard, makes any rule, order, or decision, grants any exemption or extension of time, or compromises, mitigates, or settles any penalty assessed under this Act, he shall include a statement of the reasons for such action, which shall be published in the Federal Register.

Petition for judicial review.

(f) Any person who may be adversely affected by a standard issued under this section may at any time prior to the sixtieth day after such standard is promulgated file a petition challenging the validity of such standard with the United States court of appeals for the circuit wherein such person resides or has his principal place of business, for a judicial review of such standard. A copy of the petition shall be forthwith transmitted by the clerk of the court to the Secretary. The filing of such petition shall not, unless otherwise ordered by the court, operate as a stay of the standard. The determinations of the Secretary shall be conclusive if supported by substantial evidence in the record considered as a whole.

(g) In determining the priority for establishing standards under this section, the Secretary shall give due regard to the urgency of the need for mandatory safety and health standards for particular industries, trades, crafts, occupations, businesses, workplaces or work environments. The Secretary shall also give due regard to the recommendations of the Secretary of Health, Education, and Welfare regarding the need for mandatory standards in determining the priority for establishing such standards.

ADVISORY COMMITTEES; ADMINISTRATION

Establishment; membership.

Sec. 7. (a)(1) There is hereby established a National Advisory Committee on Occupational Safety and Health consisting of twelve members appointed by the Secretary, four of whom are to be designated by the Secretary of Health, Education, and Welfare, without regard to the provisions of title 5, United States Code, governing appointments in the competitive service, and composed of representatives of manangement, labor, occupational safety and occupational health professions, and of the public. The Secretary shall designate one of the public members as Chairman. The members shall be selected upon the basis of their experience and competence in the field of occupational safety and health.

80 Stat. 378.
5 USC 101.

(2) The Committee shall advise, consult with, and make recommendations to the Secretary and the Secretary of Health, Education, and Welfare on matters relating to the administration of the Act. The Committee shall hold no fewer than two meetings during each calendar year. All meetings of the Committee shall be open to the public and a transcript shall be kept and made available for public inspection.

Public transcript.

(3) The members of the Committee shall be compensated in accordance with the provisions of section 3109 of title 5, United States Code.

80 Stat. 416.

(4) The Secretary shall furnish to the Committee an executive secretary and such secretarial, clerical, and other services as are deemed necessary to the conduct of its business.

(b) An advisory committee may be appointed by the Secretary to assist him in his standard-setting functions under section 6 of this Act. Each such committee shall consist of not more than fifteen members and shall include as a member one or more designees of the Secretary of Health, Education, and Welfare, and shall include among its members an equal number of persons qualified by experience and affiliation to present the viewpoint of the employers involved, and of persons similarly qualified to present the viewpoint of the workers involved, as well as one or more representatives of health and safety agencies of the States. An advisory committee may also include such other persons as the Secretary may appoint who are qualified by knowledge and experience to make a useful contribution to the work of such committee, including one or more representatives of professional organizations of technicians or professionals specializing in occupational safety or health, and one or more representatives of nationally recognized standards-producing organizations, but the number of persons so appointed to any such advisory committee shall not exceed the number appointed to such committee as representatives of Federal and State agencies. Persons appointed to advisory committees from private life shall be compensated in the same manner as consultants or experts under section 3109 of title 5, United States Code. The Secretary shall pay to any State which is the employer of a member of such a committee who is a representative of the health or safety agency of that State, reimbursement sufficient to cover the actual cost to the State resulting from such representative's membership on such committee. Any meeting of such committee shall be open to the public and an accurate record shall be kept and made available to the public. No member of such committee (other than representatives of employers and employees) shall have an economic interest in any proposed rule.

80 Stat. 416.

Recordkeeping.

(c) In carrying out his responsibilities under this Act, the Secretary is authorized to—

(1) use, with the consent of any Federal agency, the services, facilities, and personnel of such agency, with or without reimbursement, and with the consent of any State or political subdivision thereof, accept and use the services, facilities, and personnel of any agency of such State or subdivision with reimbursement; and

(2) employ experts and consultants or organizations thereof as authorized by section 3109 of title 5, United States Code, except that contracts for such employment may be renewed annually; compensate individuals so employed at rates not in excess of the rate specified at the time of service for grade GS-18 under section 5332 of title 5, United States Code, including traveltime, and allow them while away from their homes or regular places of business, travel expenses (including per diem in lieu of subsistence) as authorized by section 5703 of title 5, United States Code, for persons in the Government service employed intermittently, while so employed.

Ante, p. 198-1.

80 Stat. 499;
83 Stat. 190.

INSPECTIONS, INVESTIGATIONS, AND RECORDKEEPING

SEC. 8. (a) In order to carry out the purposes of this Act, the Secretary, upon presenting appropriate credentials to the owner, operator, or agent in charge, is authorized—

(1) to enter without delay and at reasonable times any factory, plant, establishment, construction site, or other area, workplace or environment where work is performed by an employee of an employer; and

(2) to inspect and investigate during regular working hours and at other reasonable times, and within reasonable limits and in a reasonable manner, any such place of employment and all pertinent conditions, structures, machines, apparatus, devices, equipment, and materials therein, and to question privately any such employer, owner, operator, agent or employee.

Subpoena power.

(b) In making his inspections and investigations under this Act the Secretary may require the attendance and testimony of witnesses and the production of evidence under oath. Witnesses shall be paid the same fees and mileage that are paid witnesses in the courts of the United States. In case of a contumacy, failure, or refusal of any person to obey such an order, any district court of the United States or the United States courts of any territory or possession, within the jurisdiction of which such person is found, or resides or transacts business, upon the application by the Secretary, shall have jurisdiction to issue to such person an order requiring such person to appear to produce evidence if, as, and when so ordered, and to give testimony relating to the matter under investigation or in question, and any failure to obey such order of the court may be punished by said court as a contempt thereof.

Recordkeeping.

(c) (1) Each employer shall make, keep and preserve, and make available to the Secretary or the Secretary of Health, Education, and Welfare, such records regarding his activities relating to this Act as the Secretary, in cooperation with the Secretary of Health, Education, and Welfare, may prescribe by regulation as necessary or appropriate for the enforcement of this Act or for developing information regarding the causes and prevention of occupational accidents and illnesses. In order to carry out the provisions of this paragraph such regulations may include provisions requiring employers to conduct periodic inspections. The Secretary shall also issue regulations requiring that employers, through posting of notices or other appropriate means, keep their employees informed of their protections and obligations under this Act, including the provisions of applicable standards.

Work-related deaths, etc.; reports.

(2) The Secretary, in cooperation with the Secretary of Health, Education, and Welfare, shall prescribe regulations requiring employers to maintain accurate records of, and to make periodic reports on, work-related deaths, injuries and illnesses other than minor injuries requiring only first aid treatment and which do not involve medical treatment, loss of consciousness, restriction of work or motion, or transfer to another job.

(3) The Secretary, in cooperation with the Secretary of Health, Education, and Welfare, shall issue regulations requiring employers to maintain accurate records of employee exposures to potentially toxic materials or harmful physical agents which are required to be monitored or measured under section 6. Such regulations shall provide employees or their representatives with an opportunity to observe such monitoring or measuring, and to have access to the records thereof. Such regulations shall also make appropriate provision for each employee or former employee to have access to such records as will indicate his own exposure to toxic materials or harmful physical agents. Each employer shall promptly notify any employee who has been or is being exposed to toxic materials or harmful physical agents in concentrations or at levels which exceed those prescribed by an applicable occupational safety and health standard promulgated under section 6, and shall inform any employee who is being thus exposed of the corrective action being taken.

(d) Any information obtained by the Secretary, the Secretary of Health, Education, and Welfare, or a State agency under this Act shall be obtained with a minimum burden upon employers, especially those operating small businesses. Unnecessary duplication of efforts in obtaining information shall be reduced to the maximum extent feasible.

(e) Subject to regulations issued by the Secretary, a representative of the employer and a representative authorized by his employees shall be given an opportunity to accompany the Secretary or his authorized representative during ~~the physic~~al inspection of any workplace under subsection (a) for the purpose of aiding such inspection. Where there is no authorized employee representative, the Secretary or his authorized representative shall consult with a reasonable number of employees concerning matters of health and safety in the workplace.

(f)(1) Any employees or representative of employees who believe that a violation of a safety or health standard exists that threatens physical harm, or that an imminent danger exists, may request an inspection by giving notice to the Secretary or his authorized representative of such violation or danger. Any such notice shall be reduced to writing, shall set forth with reasonable particularity the grounds for the notice, and shall be signed by the employees or representative of employees, and a copy shall be provided the employer or his agent no later than at the time of inspection, except that, upon the request of the person giving such notice, his name and the names of individual employees referred to therein shall not appear in such copy or on any record published, released, or made available pursuant to subsection (g) of this section. If upon receipt of such notification the Secretary determines there are reasonable grounds to believe that such violation or danger exists, he shall make a special inspection in accordance with the provisions of this section as soon as practicable, to determine if such violation or danger exists. If the Secretary determines there are no reasonable grounds to believe that a violation or danger exists he shall notify the employees or representative of the employees in writing of such determination.

(2) Prior to or during any inspection of a workplace, any employees or representative of employees employed in such workplace may notify the Secretary or any representative of the Secretary responsible for conducting the inspection, in writing, of any violation of this Act which they have reason to believe exists in such workplace. The Secretary shall, by regulation, establish procedures for informal review of any refusal by a representative of the Secretary to issue a citation with respect to any such alleged violation and shall furnish the employees or representative of employees requesting such review a written statement of the reasons for the Secretary's final disposition of the case.

Reports, publication.

(g)(1) The Secretary and Secretary of Health, Education, and Welfare are authorized to compile, analyze, and publish, either in summary or detailed form, all reports or information obtained under this section.

Rules and regulations.

(2) The Secretary and the Secretary of Health, Education, and Welfare shall each prescribe such rules and regulations as he may deem necessary to carry out their responsibilities under this Act, including rules and regulations dealing with the inspection of an employer's establishment.

CITATIONS

Sec. 9. (a) If, upon inspection or investigation, the Secretary or his authorized representative believes that an employer has violated a requirement of section 5 of this Act, of any standard, rule or order promulgated pursuant to section 6 of this Act, or of any regulations prescribed pursuant to this Act, he shall with reasonable promptness issue a citation to the employer. Each citation shall be in writing and shall describe with particularity the nature of the violation, including a reference to the provision of the Act, standard, rule, regulation, or order alleged to have been violated. In addition, the citation shall fix a reasonable time for the abatement of the violation. The Secretary may prescribe procedures for the issuance of a notice in lieu of a citation with respect to de minimis violations which have no direct or immediate relationship to safety or health.

(b) Each citation issued under this section, or a copy or copies thereof, shall be prominently posted, as prescribed in regulations issued by the Secretary, at or near each place a violation referred to in the citation occurred.

Limitation.

(c) No citation may be issued under this section after the expiration of six months following the occurrence of any violation.

PROCEDURE FOR ENFORCEMENT

Sec. 10. (a) If, after an inspection or investigation, the Secretary issues a citation under section 9(a), he shall, within a reasonable time after the termination of such inspection or investigation, notify the employer by certified mail of the penalty, if any, proposed to be assessed under section 17 and that the employer has fifteen working days within which to notify the Secretary that he wishes to contest the citation or proposed assessment of penalty. If, within fifteen working days from the receipt of the notice issued by the Secretary the employer fails to notify the Secretary that he intends to contest the citation or proposed assessment of penalty, and no notice is filed by any employee or representative of employees under subsection (c) within such time, the citation and the assessment, as proposed, shall be deemed a final order of the Commission and not subject to review by any court or agency.

(b) If the Secretary has reason to believe that an employer has failed to correct a violation for which a citation has been issued within the period permitted for its correction (which period shall not begin to run until the entry of a final order by the Commission in the case of any review proceedings under this section initiated by the employer in good faith and not solely for delay or avoidance of penalties), the Secretary shall notify the employer by certified mail of such failure and of the penalty proposed to be assessed under section 17 by reason of such failure, and that the employer has fifteen working days within which to notify the Secretary that he wishes to contest the Secretary's notification or the proposed assessment of penalty. If, within fifteen working days from the receipt of notification issued by the Secretary, the employer fails to notify the Secretary that he intends to contest the notification or proposed assessment of penalty, the notification and assessment, as proposed, shall be deemed a final order of the Commission and not subject to review by any court or agency.

(c) If an employer notifies the Secretary that he intends to contest a citation issued under section 9(a) or notification issued under subsection (a) or (b) of this section, or if, within fifteen working days of the issuance of a citation under section 9(a), any employee or representative of employees files a notice with the Secretary alleging that the period of time fixed in the citation for the abatement of the violation is unreasonable, the Secretary shall immediately advise the Commission of such notification, and the Commission shall afford an opportunity for a hearing (in accordance with section 554 of title 5, United States Code, but without regard to subsection (a)(3) of such section). The Commission shall thereafter issue an order, based on findings of fact, affirming, modifying, or vacating the Secretary's citation or proposed penalty, or directing other appropriate relief, and such order shall become final thirty days after its issuance. Upon a showing by an employer of a good faith effort to comply with the abatement requirements of a citation, and that abatement has not been completed because of factors beyond his reasonable control, the Secretary, after an opportunity for a hearing as provided in this subsection, shall issue an order affirming or modifying the abatement requirements in such citation. The rules of procedure prescribed by the Commission shall provide affected employees or representatives of affected employees an opportunity to participate as parties to hearings under this subsection.

80 Stat. 384.

JUDICIAL REVIEW

SEC. 11. (a) Any person adversely affected or aggrieved by an order of the Commission issued under subsection (c) of section 10 may obtain a review of such order in any United States court of appeals for the circuit in which the violation is alleged to have occurred or where the employer has its principal office, or in the Court of Appeals for the District of Columbia Circuit, by filing in such court within sixty days following the issuance of such order a written petition praying that the order be modified or set aside. A copy of such petition shall be forthwith transmitted by the clerk of the court to the Commission and to the other parties, and thereupon the Commission shall file in the court the record in the proceeding as provided in section 2112 of title 28, United States Code. Upon such filing, the court shall have jurisdiction of the proceeding and of the question determined therein, and shall have power to grant such temporary relief or restraining order as it deems just and proper, and to make and enter upon the pleadings, testimony, and proceedings set forth in such record a decree affirming, modifying, or setting aside in whole or in part, the order of the Commission and enforcing the same to the extent that such order is affirmed or modified. The commencement of proceedings under this subsection shall not, unless ordered by the court, operate as a stay of the order of the Commission. No objection that has not been urged before the Commission shall be considered by the court, unless the failure or neglect to urge such objection shall be excused because of extraordinary circumstances. The findings of the Commission with respect to questions of fact, if supported by substantial evidence on the record considered as a whole, shall be conclusive. If any party shall apply to the court for leave to adduce additional evidence and shall show to the satisfaction of the court that such additional evidence is material and that there were reasonable grounds for the failure to adduce such evidence in the hearing before the Commission, the court may order such additional evidence to be taken before the Commission and to be made a part of the record. The Commission may modify its findings as to the facts, or make new findings, by reason of additional evidence so taken and filed, and it shall file such modified or new findings, which findings with respect to questions of fact, if supported by substantial evi-

72 Stat. 941;
80 Stat. 1323.

dence on the record considered as a whole, shall be conclusive, and its recommendations, if any, for the modification or setting aside of its original order. Upon the filing of the record with it, the jurisdiction of the court shall be exclusive and its judgment and decree shall be final, except that the same shall be subject to review by the Supreme Court of the United States, as provided in section 1254 of title 28, United States Code. Petitions filed under this subsection shall be heard expeditiously.

62 Stat. 928.

(b) The Secretary may also obtain review or enforcement of any final order of the Commission by filing a petition for such relief in the United States court of appeals for the circuit in which the alleged violation occurred or in which the employer has its principal office, and the provisions of subsection (a) shall govern such proceedings to the extent applicable. If no petition for review, as provided in subsection (a), is filed within sixty days after service of the Commission's order, the Commission's findings of fact and order shall be conclusive in connection with any petition for enforcement which is filed by the Secretary after the expiration of such sixty-day period. In any such case, as well as in the case of a noncontested citation or notification by the Secretary which has become a final order of the Commission under subsection (a) or (b) of section 10, the clerk of the court, unless otherwise ordered by the court, shall forthwith enter a decree enforcing the order and shall transmit a copy of such decree to the Secretary and the employer named in the petition. In any contempt proceeding brought to enforce a decree of a court of appeals entered pursuant to this subsection or subsection (a), the court of appeals may assess the penalties provided in section 17, in addition to invoking any other available remedies.

(c) (1) No person shall discharge or in any manner discriminate against any employee because such employee has filed any complaint or instituted or caused to be instituted any proceeding under or related to this Act or has testified or is about to testify in any such proceeding or because of the exercise by such employee on behalf of himself or others of any right afforded by this Act.

(2) Any employee who believes that he has been discharged or otherwise discriminated against by any person in violation of this subsection may, within thirty days after such violation occurs, file a complaint with the Secretary alleging such discrimination. Upon receipt of such complaint, the Secretary shall cause such investigation to be made as he deems appropriate. If upon such investigation, the Secretary determines that the provisions of this subsection have been violated, he shall bring an action in any appropriate United States district court against such person. In any such action the United States district courts shall have jurisdiction, for cause shown to restrain violations of paragraph (1) of this subsection and order all appropriate relief including rehiring or reinstatement of the employee to his former position with back pay.

(3) Within 90 days of the receipt of a complaint filed under the subsection the Secretary shall notify the complainant of his determination under paragraph 2 of this subsection.

THE OCCUPATIONAL SAFETY AND HEALTH REVIEW COMMISSION

Establishment; membership.

Sec. 12. (a) The Occupational Safety and Health Review Commission is hereby established. The Commission shall be composed of three members who shall be appointed by the President, by and with the advice and consent of the Senate, from among persons who by reason

84 STAT. 1604

of training, education, or experience are qualified to carry out the functions of the Commission under this Act. The President shall designate one of the members of the Commission to serve as Chairman.

(b) The terms of members of the Commission shall be six years except that (1) the members of the Commission first taking office shall serve, as designated by the President at the time of appointment, one for a term of two years, one for a term of four years, and one for a term of six years, and (2) a vacancy caused by the death, resignation, or removal of a member prior to the expiration of the term for which he was appointed shall be filled only for the remainder of such unexpired term. A member of the Commission may be removed by the President for inefficiency, neglect of duty, or malfeasance in office.

Terms.

(c)(1) Section 5314 of title 5, United States Code, is amended by adding at the end thereof the following new paragraph:

80 Stat. 460.

"(57) Chairman, Occupational Safety and Health Review Commission."

(2) Section 5315 of title 5, United States Code, is amended by adding at the end thereof the following new paragraph:

Ante, p. 776.

"(94) Members, Occupational Safety and Health Review Commission."

(d) The principal office of the Commission shall be in the District of Columbia. Whenever the Commission deems that the convenience of the public or of the parties may be promoted, or delay or expense may be minimized, it may hold hearings or conduct other proceedings at any other place.

Location.

(e) The Chairman shall be responsible on behalf of the Commission for the administrative operations of the Commission and shall appoint such hearing examiners and other employees as he deems necessary to assist in the performance of the Commission's functions and to fix their compensation in accordance with the provisions of chapter 51 and subchapter III of chapter 53 of title 5, United States Code, relating to classification and General Schedule pay rates: *Provided*, That assignment, removal and compensation of hearing examiners shall be in accordance with sections 3105, 3344, 5362, and 7521 of title 5, United States Code.

5 USC 5101, 5331.
Ante, p. 198-1.

(f) For the purpose of carrying out its functions under this Act, two members of the Commission shall constitute a quorum and official action can be taken only on the affirmative vote of at least two members.

Quorum.

(g) Every official act of the Commission shall be entered of record, and its hearings and records shall be open to the public. The Commission is authorized to make such rules as are necessary for the orderly transaction of its proceedings. Unless the Commission has adopted a different rule, its proceedings shall be in accordance with the Federal Rules of Civil Procedure.

Public records.
28 USC app.

(h) The Commission may order testimony to be taken by deposition in any proceedings pending before it at any state of such proceeding. Any person may be compelled to appear and depose, and to produce books, papers, or documents, in the same manner as witnesses may be compelled to appear and testify and produce like documentary evidence before the Commission. Witnesses whose depositions are taken under this subsection, and the persons taking such depositions, shall be entitled to the same fees as are paid for like services in the courts of the United States.

(i) For the purpose of any proceeding before the Commission, the provisions of section 11 of the National Labor Relations Act (29 U.S.C. 161) are hereby made applicable to the jurisdiction and powers of the Commission.

61 Stat. 150;
Ante, p. 930.

Report.

(j) A hearing examiner appointed by the Commission shall hear, and make a determination upon, any proceeding instituted before the Commission and any motion in connection therewith, assigned to such hearing examiner by the Chairman of the Commission, and shall make a report of any such determination which constitutes his final disposition of the proceedings. The report of the hearing examiner shall become the final order of the Commission within thirty days after such report by the hearing examiner, unless within such period any Commission member has directed that such report shall be reviewed by the Commission.

80 Stat. 453.

Ante, p. 198-1.

(k) Except as otherwise provided in this Act, the hearing examiners shall be subject to the laws governing employees in the classified civil service, except that appointments shall be made without regard to section 5108 of title 5, United States Code. Each hearing examiner shall receive compensation at a rate not less than that prescribed for GS-16 under section 5332 of title 5, United States Code.

PROCEDURES TO COUNTERACT IMMINENT DANGERS

Sec. 13. (a) The United States district courts shall have jurisdiction, upon petition of the Secretary, to restrain any conditions or practices in any place of employment which are such that a danger exists which could reasonably be expected to cause death or serious physical harm immediately or before the imminence of such danger can be eliminated through the enforcement procedures otherwise provided by this Act. Any order issued under this section may require such steps to be taken as may be necessary to avoid, correct, or remove such imminent danger and prohibit the employment or presence of any individual in locations or under conditions where such imminent danger exists, except individuals whose presence is necessary to avoid, correct, or remove such imminent danger or to maintain the capacity of a continuous process operation to resume normal operations without a complete cessation of operations, or where a cessation of operations is necessary, to permit such to be accomplished in a safe and orderly manner.

28 USC app.

(b) Upon the filing of any such petition the district court shall have jurisdiction to grant such injunctive relief or temporary restraining order pending the outcome of an enforcement proceeding pursuant to this Act. The proceeding shall be as provided by Rule 65 of the Federal Rules, Civil Procedure, except that no temporary restraining order issued without notice shall be effective for a period longer than five days.

(c) Whenever and as soon as an inspector concludes that conditions or practices described in subsection (a) exist in any place of employment, he shall inform the affected employees and employers of the danger and that he is recommending to the Secretary that relief be sought.

(d) If the Secretary arbitrarily or capriciously fails to seek relief under this section, any employee who may be injured by reason of such failure, or the representative of such employees, might bring an action against the Secretary in the United States district court for the district in which the imminent danger is alleged to exist or the employer has its principal office, or for the District of Columbia, for a writ of mandamus to compel the Secretary to seek such an order and for such further relief as may be appropriate.

84 STAT. 1606

REPRESENTATION IN CIVIL LITIGATION

SEC. 14. Except as provided in section 518(a) of title 28, United States Code, relating to litigation before the Supreme Court, the Solicitor of Labor may appear for and represent the Secretary in any civil litigation brought under this Act but all such litigation shall be subject to the direction and control of the Attorney General.

80 Stat. 613.

CONFIDENTIALITY OF TRADE SECRETS

SEC. 15. All information reported to or otherwise obtained by the Secretary or his representative in connection with any inspection or proceeding under this Act which contains or which might reveal a trade secret referred to in section 1905 of title 18 of the United States Code shall be considered confidential for the purpose of that section, except that such information may be disclosed to other officers or employees concerned with carrying out this Act or when relevant in any proceeding under this Act. In any such proceeding the Secretary, the Commission, or the court shall issue such orders as may be appropriate to protect the confidentiality of trade secrets.

62 Stat. 791.

VARIATIONS, TOLERANCES, AND EXEMPTIONS

SEC. 16. The Secretary, on the record, after notice and opportunity for a hearing may provide such reasonable limitations and may make such rules and regulations allowing reasonable variations, tolerances, and exemptions to and from any or all provisions of this Act as he may find necessary and proper to avoid serious impairment of the national defense. Such action shall not be in effect for more than six months without notification to affected employees and an opportunity being afforded for a hearing.

PENALTIES

SEC. 17. (a) Any employer who willfully or repeatedly violates the requirements of section 5 of this Act, any standard, rule, or order promulgated pursuant to section 6 of this Act, or regulations prescribed pursuant to this Act, may be assessed a civil penalty of not more than $10,000 for each violation.

(b) Any employer who has received a citation for a serious violation of the requirements of section 5 of this Act, of any standard, rule, or order promulgated pursuant to section 6 of this Act, or of any regulations prescribed pursuant to this Act, shall be assessed a civil penalty of up to $1,000 for each such violation.

(c) Any employer who has received a citation for a violation of the requirements of section 5 of this Act, of any standard, rule, or order promulgated pursuant to section 6 of this Act, or of regulations prescribed pursuant to this Act, and such violation is specifically determined not to be of a serious nature, may be assessed a civil penalty of up to $1,000 for each such violation.

(d) Any employer who fails to correct a violation for which a citation has been issued under section 9(a) within the period permitted for its correction (which period shall not begin to run until the date of the final order of the Commission in the case of any review proceeding under section 10 initiated by the employer in good faith and not solely for delay or avoidance of penalties), may be assessed a civil penalty of not more than $1,000 for each day during which such failure or violation continues.

84 STAT. 1607

(e) Any employer who willfully violates any standard, rule, or order promulgated pursuant to section 6 of this Act, or of any regulations prescribed pursuant to this Act, and that violation caused death to any employee, shall, upon conviction, be punished by a fine of not more than $10,000 or by imprisonment for not more than six months, or by both; except that if the conviction is for a violation committed after a first conviction of such person, punishment shall be by a fine of not more than $20,000 or by imprisonment for not more than one year, or by both.

(f) Any person who gives advance notice of any inspection to be conducted under this Act, without authority from the Secretary or his designees, shall, upon conviction, be punished by a fine of not more than $1,000 or by imprisonment for not more than six months, or by both.

(g) Whoever knowingly makes any false statement, representation, or certification in any application, record, report, plan, or other document filed or required to be maintained pursuant to this Act shall, upon conviction, be punished by a fine of not more than $10,000, or by imprisonment for not more than six months, or by both.

65 Stat. 721; 79 Stat. 234.

(h)(1) Section 1114 of title 18, United States Code, is hereby amended by striking out "designated by the Secretary of Health, Education, and Welfare to conduct investigations, or inspections under the Federal Food, Drug, and Cosmetic Act" and inserting in lieu thereof "or of the Department of Labor assigned to perform investigative, inspection, or law enforcement functions".

62 Stat. 756.

(2) Notwithstanding the provisions of sections 1111 and 1114 of title 18, United States Code, whoever, in violation of the provisions of section 1114 of such title, kills a person while engaged in or on account of the performance of investigative, inspection, or law enforcement functions added to such section 1114 by paragraph (1) of this subsection, and who would otherwise be subject to the penalty provisions of such section 1111, shall be punished by imprisonment for any term of years or for life.

(i) Any employer who violates any of the posting requirements, as prescribed under the provisions of this Act, shall be assessed a civil penalty of up to $1,000 for each violation.

(j) The Commission shall have authority to assess all civil penalties provided in this section, giving due consideration to the appropriateness of the penalty with respect to the size of the business of the employer being charged, the gravity of the violation, the good faith of the employer, and the history of previous violations.

(k) For purposes of this section, a serious violation shall be deemed to exist in a place of employment if there is a substantial probability that death or serious physical harm could result from a condition which exists, or from one or more practices, means, methods, operations, or processes which have been adopted or are in use, in such place of employment unless the employer did not, and could not with the exercise of reasonable diligence, know of the presence of the violation.

(l) Civil penalties owed under this Act shall be paid to the Secretary for deposit into the Treasury of the United States and shall accrue to the United States and may be recovered in a civil action in the name of the United States brought in the United States district court for the district where the violation is alleged to have occurred or where the employer has its principal office.

STATE JURISDICTION AND STATE PLANS

SEC. 18. (a) Nothing in this Act shall prevent any State agency or court from asserting jurisdiction under State law over any occupational safety or health issue with respect to which no standard is in effect under section 6.

(b) Any State which, at any time, desires to assume responsibility for development and enforcement therein of occupational safety and health standards relating to any occupational safety or health issue with respect to which a Federal standard has been promulgated under section 6 shall submit a State plan for the development of such standards and their enforcement.

(c) The Secretary shall approve the plan submitted by a State under subsection (b), or any modification thereof, if such plan in his judgment—

(1) designates a State agency or agencies as the agency or agencies responsible for administering the plan throughout the State,

(2) provides for the development and enforcement of safety and health standards relating to one or more safety or health issues, which standards (and the enforcement of which standards) are or will be at least as effective in providing safe and healthful employment and places of employment as the standards promulgated under section 6 which relate to the same issues, and which standards, when applicable to products which are distributed or used in interstate commerce, are required by compelling local conditions and do not unduly burden interstate commerce,

(3) provides for a right of entry and inspection of all workplaces subject to the Act which is at least as effective as that provided in section 8, and includes a prohibition on advance notice of inspections,

(4) contains satisfactory assurances that such agency or agencies have or will have the legal authority and qualified personnel necessary for the enforcement of such standards,

(5) gives satisfactory assurances that such State will devote adequate funds to the administration and enforcement of such standards,

(6) contains satisfactory assurances that such State will, to the extent permitted by its law, establish and maintain an effective and comprehensive occupational safety and health program applicable to all employees of public agencies of the State and its political subdivisions, which program is as effective as the standards contained in an approved plan,

(7) requires employers in the State to make reports to the Secretary in the same manner and to the same extent as if the plan were not in effect, and

(8) provides that the State agency will make such reports to the Secretary in such form and containing such information, as the Secretary shall from time to time require.

(d) If the Secretary rejects a plan submitted under subsection (b), he shall afford the State submitting the plan due notice and opportunity for a hearing before so doing.

Notice of hearing.

(e) After the Secretary approves a State plan submitted under subsection (b), he may, but shall not be required to, exercise his authority under sections 8, 9, 10, 13, and 17 with respect to comparable standards promulgated under section 6, for the period specified in the next sentence. The Secretary may exercise the authority referred to above until he determines, on the basis of actual operations under the

State plan, that the criteria set forth in subsection (c) are being applied, but he shall not make such determination for at least three years after the plan's approval under subsection (c). Upon making the determination referred to in the preceding sentence, the provisions of sections 5(a)(2), 8 (except for the purpose of carrying out subsection (f) of this section), 9, 10, 13, and 17, and standards promulgated under section 6 of this Act, shall not apply with respect to any occupational safety or health issues covered under the plan, but the Secretary may retain jurisdiction under the above provisions in any proceeding commenced under section 9 or 10 before the date of determination.

Continuing evaluation.

(f) The Secretary shall, on the basis of reports submitted by the State agency and his own inspections make a continuing evaluation of the manner in which each State having a plan approved under this section is carrying out such plan. Whenever the Secretary finds, after affording due notice and opportunity for a hearing, that in the administration of the State plan there is a failure to comply substantially with any provision of the State plan (or any assurance contained therein), he shall notify the State agency of his withdrawal of approval of such plan and upon receipt of such notice such plan shall cease to be in effect, but the State may retain jurisdiction in any case commenced before the withdrawal of the plan in order to enforce standards under the plan whenever the issues involved do not relate to the reasons for the withdrawal of the plan.

Plan rejection, review.

(g) The State may obtain a review of a decision of the Secretary withdrawing approval of or rejecting its plan by the United States court of appeals for the circuit in which the State is located by filing in such court within thirty days following receipt of notice of such decision a petition to modify or set aside in whole or in part the action of the Secretary. A copy of such petition shall forthwith be served upon the Secretary, and thereupon the Secretary shall certify and file in the court the record upon which the decision complained of was issued as provided in section 2112 of title 28, United States Code. Unless the court finds that the Secretary's decision in rejecting a proposed State plan or withdrawing his approval of such a plan is not supported by substantial evidence the court shall affirm the Secretary's decision. The judgment of the court shall be subject to review by the Supreme Court of the United States upon certiorari or certification as provided in section 1254 of title 28, United States Code.

72 Stat. 941; 80 Stat. 1323.

62 Stat. 928.

(h) The Secretary may enter into an agreement with a State under which the State will be permitted to continue to enforce one or more occupational health and safety standards in effect in such State until final action is taken by the Secretary with respect to a plan submitted by a State under subsection (b) of this section, or two years from the date of enactment of this Act, whichever is earlier.

FEDERAL AGENCY SAFETY PROGRAMS AND RESPONSIBILITIES

SEC. 19. (a) It shall be the responsibility of the head of each Federal agency to establish and maintain an effective and comprehensive occupational safety and health program which is consistent with the standards promulgated under section 6. The head of each agency shall (after consultation with representatives of the employees thereof)—

(1) provide safe and healthful places and conditions of employment, consistent with the standards set under section 6;

(2) acquire, maintain, and require the use of safety equipment, personal protective equipment, and devices reasonably necessary to protect employees;

(3) keep adequate records of all occupational accidents and illnesses for proper evaluation and necessary corrective action;

Recordkeeping.

(4) consult with the Secretary with regard to the adequacy as to form and content of records kept pursuant to subsection (a)(3) of this section; and

(5) make an annual report to the Secretary with respect to occupational accidents and injuries and the agency's program under this section. Such report shall include any report submitted under section 7902(e)(2) of title 5, United States Code.

Annual report.

80 Stat. 530.

(b) The Secretary shall report to the President a summary or digest of reports submitted to him under subsection (a)(5) of this section, together with his evaluations of and recommendations derived from such reports. The President shall transmit annually to the Senate and the House of Representatives a report of the activities of Federal agencies under this section.

Report to President.

Report to Congress.

(c) Section 7902(c)(1) of title 5, United States Code, is amended by inserting after "agencies" the following: "and of labor organizations representing employees".

(d) The Secretary shall have access to records and reports kept and filed by Federal agencies pursuant to subsections (a)(3) and (5) of this section unless those records and reports are specifically required by Executive order to be kept secret in the interest of the national defense or foreign policy, in which case the Secretary shall have access to such information as will not jeopardize national defense or foreign policy.

Records, etc.; availability.

RESEARCH AND RELATED ACTIVITIES

SEC. 20. (a)(1) The Secretary of Health, Education, and Welfare, after consultation with the Secretary and with other appropriate Federal departments or agencies, shall conduct (directly or by grants or contracts) research, experiments, and demonstrations relating to occupational safety and health, including studies of psychological factors involved, and relating to innovative methods, techniques, and approaches for dealing with occupational safety and health problems.

(2) The Secretary of Health, Education, and Welfare shall from time to time consult with the Secretary in order to develop specific plans for such research, demonstrations, and experiments as are necessary to produce criteria, including criteria identifying toxic substances, enabling the Secretary to meet his responsibility for the formulation of safety and health standards under this Act; and the Secretary of Health, Education, and Welfare, on the basis of such research, demonstrations, and experiments and any other information available to him, shall develop and publish at least annually such criteria as will effectuate the purposes of this Act.

(3) The Secretary of Health, Education, and Welfare, on the basis of such research, demonstrations, and experiments, and any other information available to him, shall develop criteria dealing with toxic materials and harmful physical agents and substances which will describe exposure levels that are safe for various periods of employment, including but not limited to the exposure levels at which no employee will suffer impaired health or functional capacities or diminished life expectancy as a result of his work experience.

(4) The Secretary of Health, Education, and Welfare shall also conduct special research, experiments, and demonstrations relating to occupational safety and health as are necessary to explore new problems, including those created by new technology in occupational safety and health, which may require ameliorative action beyond that

which is otherwise provided for in the operating provisions of this Act. The Secretary of Health, Education, and Welfare shall also conduct research into the motivational and behavioral factors relating to the field of occupational safety and health.

Toxic substances, records.

(5) The Secretary of Health, Education, and Welfare, in order to comply with his responsibilities under paragraph (2), and in order to develop needed information regarding potentially toxic substances or harmful physical agents, may prescribe regulations requiring employers to measure, record, and make reports on the exposure of employees to substances or physical agents which the Secretary of Health, Education, and Welfare reasonably believes may endanger the health or safety of employees. The Secretary of Health, Education, and Welfare also is authorized to establish such programs of medical examinations and tests as may be necessary for determining the incidence of occupational illnesses and the susceptibility of employees to such illnesses. Nothing in this or any other provision of this Act shall be deemed to authorize or require medical examination, immunization, or treatment for those who object thereto on religious grounds, except where such is necessary for the protection of the health or safety of others. Upon the request of any employer who is required to measure and record exposure of employees to substances or physical agents as provided under this subsection, the Secretary of Health, Education, and Welfare shall furnish full financial or other assistance to such employer for the purpose of defraying any additional expense incurred by him in carrying out the measuring and recording as provided in this subsection.

Medical examinations.

Toxic substances, publication.

(6) The Secretary of Health, Education, and Welfare shall publish within six months of enactment of this Act and thereafter as needed but at least annually a list of all known toxic substances by generic family or other useful grouping, and the concentrations at which such toxicity is known to occur. He shall determine following a written request by any employer or authorized representative of employees, specifying with reasonable particularity the grounds on which the request is made, whether any substance normally found in the place of employment has potentially toxic effects in such concentrations as used or found; and shall submit such determination both to employers and affected employees as soon as possible. If the Secretary of Health, Education, and Welfare determines that any substance is potentially toxic at the concentrations in which it is used or found in a place of employment, and such substance is not covered by an occupational safety or health standard promulgated under section 6, the Secretary of Health, Education, and Welfare shall immediately submit such determination to the Secretary, together with all pertinent criteria.

Annual studies.

(7) Within two years of enactment of this Act, and annually thereafter the Secretary of Health, Education, and Welfare shall conduct and publish industrywide studies of the effect of chronic or low-level exposure to industrial materials, processes, and stresses on the potential for illness, disease, or loss of functional capacity in aging adults.

Inspections.

(b) The Secretary of Health, Education, and Welfare is authorized to make inspections and question employers and employees as provided in section 8 of this Act in order to carry out his functions and responsibilities under this section.

Contract authority.

(c) The Secretary is authorized to enter into contracts, agreements, or other arrangements with appropriate public agencies or private organizations for the purpose of conducting studies relating to his responsibilities under this Act. In carrying out his responsibilities under this subsection, the Secretary shall cooperate with the Secretary of Health, Education, and Welfare in order to avoid any duplication of efforts under this section.

(d) Information obtained by the Secretary and the Secretary of Health, Education, and Welfare under this section shall be disseminated by the Secretary to employers and employees and organizations thereof.

Delegation of functions.

(e) The functions of the Secretary of Health, Education, and Welfare under this Act shall, to the extent feasible, be delegated to the Director of the National Institute for Occupational Safety and Health established by section 22 of this Act.

TRAINING AND EMPLOYEE EDUCATION

SEC. 21. (a) The Secretary of Health, Education, and Welfare, after consultation with the Secretary and with other appropriate Federal departments and agencies, shall conduct, directly or by grants or contracts (1) education programs to provide an adequate supply of qualified personnel to carry out the purposes of this Act, and (2) informational programs on the importance of and proper use of adequate safety and health equipment.

(b) The Secretary is also authorized to conduct, directly or by grants or contracts, short-term training of personnel engaged in work related to his responsibilities under this Act.

(c) The Secretary, in consultation with the Secretary of Health, Education, and Welfare, shall (1) provide for the establishment and supervision of programs for the education and training of employers and employees in the recognition, avoidance, and prevention of unsafe or unhealthful working conditions in employments covered by this Act, and (2) consult with and advise employers and employees, and organizations representing employers and employees as to effective means of preventing occupational injuries and illnesses.

NATIONAL INSTITUTE FOR OCCUPATIONAL SAFETY AND HEALTH

Establishment.

SEC. 22. (a) It is the purpose of this section to establish a National Institute for Occupational Safety and Health in the Department of Health, Education, and Welfare in order to carry out the policy set forth in section 2 of this Act and to perform the functions of the Secretary of Health, Education, and Welfare under sections 20 and 21 of this Act.

(b) There is hereby established in the Department of Health, Education, and Welfare a National Institute for Occupational Safety and Health. The Institute shall be headed by a Director who shall be appointed by the Secretary of Health, Education, and Welfare, and who shall serve for a term of six years unless previously removed by the Secretary of Health, Education, and Welfare.

Director, appointment, term.

(c) The Institute is authorized to—

(1) develop and establish recommended occupational safety and health standards; and

(2) perform all functions of the Secretary of Health, Education, and Welfare under sections 20 and 21 of this Act.

(d) Upon his own initiative, or upon the request of the Secretary or the Secretary of Health, Education, and Welfare, the Director is authorized (1) to conduct such research and experimental programs as he determines are necessary for the development of criteria for new and improved occupational safety and health standards, and (2) after consideration of the results of such research and experimental programs make recommendations concerning new or improved occupational safety and health standards. Any occupational safety and health standard recommended pursuant to this section shall immediately be forwarded to the Secretary of Labor, and to the Secretary of Health, Education, and Welfare.

(e) In addition to any authority vested in the Institute by other provisions of this section, the Director, in carrying out the functions of the Institute, is authorized to—

(1) prescribe such regulations as he deems necessary governing the manner in which its functions shall be carried out;

(2) receive money and other property donated, bequeathed, or devised, without condition or restriction other than that it be used for the purposes of the Institute and to use, sell, or otherwise dispose of such property for the purpose of carrying out its functions;

(3) receive (and use, sell, or otherwise dispose of, in accordance with paragraph (2)), money and other property donated, bequeathed, or devised to the Institute with a condition or restriction, including a condition that the Institute use other funds of the Institute for the purposes of the gift;

(4) in accordance with the civil service laws, appoint and fix the compensation of such personnel as may be necessary to carry out the provisions of this section;

(5) obtain the services of experts and consultants in accordance with the provisions of section 3109 of title 5, United States Code;

80 Stat. 416.

(6) accept and utilize the services of voluntary and noncompensated personnel and reimburse them for travel expenses, including per diem, as authorized by section 5703 of title 5, United States Code;

83 Stat. 190.

(7) enter into contracts, grants or other arrangements, or modifications thereof to carry out the provisions of this section, and such contracts or modifications thereof may be entered into without performance or other bonds, and without regard to section 3709 of the Revised Statutes, as amended (41 U.S.C. 5), or any other provision of law relating to competitive bidding;

(8) make advance, progress, and other payments which the Director deems necessary under this title without regard to the provisions of section 3648 of the Revised Statutes, as amended (31 U.S.C. 529); and

(9) make other necessary expenditures.

Annual report to HEW, President, and Congress.

(f) The Director shall submit to the Secretary of Health, Education, and Welfare, to the President, and to the Congress an annual report of the operations of the Institute under this Act, which shall include a detailed statement of all private and public funds received and expended by it, and such recommendations as he deems appropriate.

GRANTS TO THE STATES

Sec. 23. (a) The Secretary is authorized, during the fiscal year ending June 30, 1971, and the two succeeding fiscal years, to make grants to the States which have designated a State agency under section 18 to assist them—

(1) in identifying their needs and responsibilities in the area of occupational safety and health,

(2) in developing State plans under section 18, or

(3) in developing plans for—

(A) establishing systems for the collection of information concerning the nature and frequency of occupational injuries and diseases;

(B) increasing the expertise and enforcement capabilities of their personnel engaged in occupational safety and health programs; or

(C) otherwise improving the administration and enforcement of State occupational safety and health laws, including standards thereunder, consistent with the objectives of this Act.

(b) The Secretary is authorized, during the fiscal year ending June 30, 1971, and the two succeeding fiscal years, to make grants to the States for experimental and demonstration projects consistent with the objectives set forth in subsection (a) of this section.

(c) The Governor of the State shall designate the appropriate State agency for receipt of any grant made by the Secretary under this section.

(d) Any State agency designated by the Governor of the State desiring a grant under this section shall submit an application therefor to the Secretary.

(e) The Secretary shall review the application, and shall, after consultation with the Secretary of Health, Education, and Welfare, approve or reject such application.

(f) The Federal share for each State grant under subsection (a) or (b) of this section may not exceed 90 per centum of the total cost of the application. In the event the Federal share for all States under either such subsection is not the same, the differences among the States shall be established on the basis of objective criteria.

(g) The Secretary is authorized to make grants to the States to assist them in administering and enforcing programs for occupational safety and health contained in State plans approved by the Secretary pursuant to section 18 of this Act. The Federal share for each State grant under this subsection may not exceed 50 per centum of the total cost to the State of such a program. The last sentence of subsection (f) shall be applicable in determining the Federal share under this subsection.

(h) Prior to June 30, 1973, the Secretary shall, after consultation with the Secretary of Health, Education, and Welfare, transmit a report to the President and to the Congress, describing the experience under the grant programs authorized by this section and making any recommendations he may deem appropriate.

Report to President and Congress.

STATISTICS

SEC. 24. (a) In order to further the purposes of this Act, the Secretary, in consultation with the Secretary of Health, Education, and Welfare, shall develop and maintain an effective program of collection, compilation, and analysis of occupational safety and health statistics. Such program may cover all employments whether or not subject to any other provisions of this Act but shall not cover employments excluded by section 4 of the Act. The Secretary shall compile accurate statistics on work injuries and illnesses which shall include all disabling, serious, or significant injuries and illnesses, whether or not involving loss of time from work, other than minor injuries requiring only first aid treatment and which do not involve medical treatment, loss of consciousness, restriction of work or motion, or transfer to another job.

(b) To carry out his duties under subsection (a) of this section, the Secretary may—

(1) promote, encourage, or directly engage in programs of studies, information and communication concerning occupational safety and health statistics;

(2) make grants to States or political subdivisions thereof in order to assist them in developing and administering programs dealing with occupational safety and health statistics; and

(3) arrange, through grants or contracts, for the conduct of such research and investigations as give promise of furthering the objectives of this section.

(c) The Federal share for each grant under subsection (b) of this section may be up to 50 per centum of the State's total cost.

(d) The Secretary may, with the consent of any State or political subdivision thereof, accept and use the services, facilities, and employees of the agencies of such State or political subdivision, with or without reimbursement, in order to assist him in carrying out his functions under this section.

Reports.

(e) On the basis of the records made and kept pursuant to section 8(c) of this Act, employers shall file such reports with the Secretary as he shall prescribe by regulation, as necessary to carry out his functions under this Act.

(f) Agreements between the Department of Labor and States pertaining to the collection of occupational safety and health statistics already in effect on the effective date of this Act shall remain in effect until superseded by grants or contracts made under this Act.

AUDITS

Sec. 25. (a) Each recipient of a grant under this Act shall keep such records as the Secretary or the Secretary of Health, Education, and Welfare shall prescribe, including records which fully disclose the amount and disposition by such recipient of the proceeds of such grant, the total cost of the project or undertaking in connection with which such grant is made or used, and the amount of that portion of the cost of the project or undertaking supplied by other sources, and such other records as will facilitate an effective audit.

(b) The Secretary or the Secretary of Health, Education, and Welfare, and the Comptroller General of the United States, or any of their duly authorized representatives, shall have access for the purpose of audit and examination to any books, documents, papers, and records of the recipients of any grant under this Act that are pertinent to any such grant.

ANNUAL REPORT

Sec. 26. Within one hundred and twenty days following the convening of each regular session of each Congress, the Secretary and the Secretary of Health, Education, and Welfare shall each prepare and submit to the President for transmittal to the Congress a report upon the subject matter of this Act, the progress toward achievement of the purpose of this Act, the needs and requirements in the field of occupational safety and health, and any other relevant information. Such reports shall include information regarding occupational safety and health standards, and criteria for such standards, developed during the preceding year; evaluation of standards and criteria previously developed under this Act, defining areas of emphasis for new criteria and standards; an evaluation of the degree of observance of applicable occupational safety and health standards, and a summary of inspection and enforcement activity undertaken; analysis and evaluation of research activities for which results have been obtained under governmental and nongovernmental sponsorship; an analysis of major occupational diseases; evaluation of available control and measurement technology for hazards for which standards or criteria have been developed during the preceding year; description of cooperative efforts undertaken between Government agencies and other interested parties in the implementation of this Act during the preceding year; a progress report on the development of an adequate supply of trained manpower in the field of occupational safety and health, including estimates of future needs and the efforts being made by Government and others to meet those needs; listing of all toxic substances in industrial usage for which labeling requirements, criteria, or standards have not yet been established; and such recommendations for additional legislation as are deemed necessary to protect the safety and health of the worker and improve the administration of this Act.

NATIONAL COMMISSION ON STATE WORKMEN'S COMPENSATION LAWS

SEC. 27. (a) (1) The Congress hereby finds and declares that—

(A) the vast majority of American workers, and their families, are dependent on workmen's compensation for their basic economic security in the event such workers suffer disabling injury or death in the course of their employment; and that the full protection of American workers from job-related injury or death requires an adequate, prompt, and equitable system of workmen's compensation as well as an effective program of occupational health and safety regulation; and

(B) in recent years serious questions have been raised concerning the fairness and adequacy of present workmen's compensation laws in the light of the growth of the economy, the changing nature of the labor force, increases in medical knowledge, changes in the hazards associated with various types of employment, new technology creating new risks to health and safety, and increases in the general level of wages and the cost of living.

(2) The purpose of this section is to authorize an effective study and objective evaluation of State workmen's compensation laws in order to determine if such laws provide an adequate, prompt, and equitable system of compensation for injury or death arising out of or in the course of employment.

Establishment.

(b) There is hereby established a National Commission on State Workmen's Compensation Laws.

Membership.

(c) (1) The Workmen's Compensation Commission shall be composed of fifteen members to be appointed by the President from among members of State workmen's compensation boards, representatives of insurance carriers, business, labor, members of the medical profession having experience in industrial medicine or in workmen's compensation cases, educators having special expertise in the field of workmen's compensation, and representatives of the general public. The Secretary, the Secretary of Commerce, and the Secretary of Health, Education, and Welfare shall be ex officio members of the Workmen's Compensation Commission:

(2) Any vacancy in the Workmen's Compensation Commission shall not affect its powers.

(3) The President shall designate one of the members to serve as Chairman and one to serve as Vice Chairman of the Workmen's Compensation Commission.

Quorum.

(4) Eight members of the Workmen's Compensation Commission shall constitute a quorum.

Study.

(d)(1) The Workmen's Compensation Commission shall undertake a comprehensive study and evaluation of State workmen's compensation laws in order to determine if such laws provide an adequate, prompt, and equitable system of compensation. Such study and evaluation shall include, without being limited to, the following subjects: (A) the amount and duration of permanent and temporary disability benefits and the criteria for determining the maximum limitations thereon, (B) the amount and duration of medical benefits and provisions insuring adequate medical care and free choice of physician, (C) the extent of coverage of workers, including exemptions based on numbers or type of employment, (D) standards for determining which injuries or diseases should be deemed compensable, (E) rehabilitation, (F) coverage under second or subsequent injury funds, (G) time limits on filing claims, (H) waiting periods, (I) compulsory or elective coverage, (J) administration, (K) legal expenses, (L) the feasibility and desirability of a uniform system of reporting information concerning job-related injuries and diseases and the operation of workmen's compensation laws, (M) the resolution of conflict of laws, extraterritoriality and similar problems arising from claims with multistate aspects, (N) the extent to which private insurance carriers are excluded from supplying workmen's compensation coverage and the desirability of such exclusionary practices, to the extent they are found to exist, (O) the relationship between workmen's compensation on the one hand, and old-age, disability, and survivors insurance and other types of insurance, public or private, on the other hand, (P) methods of implementing the recommendations of the Commission.

Report to President and Congress.

(2) The Workmen's Compensation Commission shall transmit to the President and to the Congress not later than July 31, 1972, a final report containing a detailed statement of the findings and conclusions of the Commission, together with such recommendations as it deems advisable.

Hearings.

(e)(1) The Workmen's Compensation Commission or, on the authorization of the Workmen's Compensation Commission, any subcommittee or members thereof, may, for the purpose of carrying out the provisions of this title, hold such hearings, take such testimony, and sit and act at such times and places as the Workmen's Compensation Commission deems advisable. Any member authorized by the Workmen's Compensation Commission may administer oaths or affirmations to witnesses appearing before the Workmen's Compensation Commission or any subcommittee or members thereof.

(2) Each department, agency, and instrumentality of the executive branch of the Government, including independent agencies, is authorized and directed to furnish to the Workmen's Compensation Commission, upon request made by the Chairman or Vice Chairman, such information as the Workmen's Compensation Commission deems necessary to carry out its functions under this section.

(f) Subject to such rules and regulations as may be adopted by the Workmen's Compensation Commission, the Chairman shall have the power to—

(1) appoint and fix the compensation of an executive director, and such additional staff personnel as he deems necessary, without regard to the provisions of title 5, United States Code, governing appointments in the competitive service, and without regard to the provisions of chapter 51 and subchapter III of chapter 53 of such title relating to classification and General Schedule

80 Stat. 378.
5 USC 101.

5 USC 5101, 5331.

84 STAT. 1618

pay rates, but at rates not in excess of the maximum rate for GS–18 of the General Schedule under section 5332 of such title, and

Ante, p. 198-1.

(2) procure temporary and intermittent services to the same extent as is authorized by section 3109 of title 5, United States Code.

80 Stat. 416.

Contract authorization.

(g) The Workmen's Compensation Commission is authorized to enter into contracts with Federal or State agencies, private firms, institutions, and individuals for the conduct of research or surveys, the preparation of reports, and other activities necessary to the discharge of its duties.

Compensation; travel expenses.

(h) Members of the Workmen's Compensation Commission shall receive compensation for each day they are engaged in the performance of their duties as members of the Workmen's Compensation Commission at the daily rate prescribed for GS–18 under section 5332 of title 5, United States Code, and shall be entitled to reimbursement for travel, subsistence, and other necessary expenses incurred by them in the performance of their duties as members of the Workmen's Compensation Commission.

Appropriation.

(i) There are hereby authorized to be appropriated such sums as may be necessary to carry out the provisions of this section.

Termination.

(j) On the ninetieth day after the date of submission of its final report to the President, the Workmen's Compensation Commission shall cease to exist.

ECONOMIC ASSISTANCE TO SMALL BUSINESSES

SEC. 28. (a) Section 7(b) of the Small Business Act, as amended, is amended—

72 Stat. 387; 83 Stat. 802. 15 USC 636.

(1) by striking out the period at the end of "paragraph (5)" and inserting in lieu thereof "; and"; and

(2) by adding after paragraph (5) a new paragraph as follows:

"(6) to make such loans (either directly or in cooperation with banks or other lending institutions through agreements to participate on an immediate or deferred basis) as the Administration may determine to be necessary or appropriate to assist any small business concern in effecting additions to or alterations in the equipment, facilities, or methods of operation of such business in order to comply with the applicable standards promulgated pursuant to section 6 of the Occupational Safety and Health Act of 1970 or standards adopted by a State pursuant to a plan approved under section 18 of the Occupational Safety and Health Act of 1970, if the Administration determines that such concern is likely to suffer substantial economic injury without assistance under this paragraph."

(b) The third sentence of section 7(b) of he Small Business Act, as amended, is amended by striking out "or (5)" after "paragraph (3)" and inserting a comma followed by "(5) or (6)".

(c) Section 4(c)(1) of the Small Business Act, as amended, is amended by inserting "7(b)(6)," after "7(b)(5),".

80 Stat. 132. 15 USC 633.

(d) Loans may also be made or guaranteed for the purposes set forth in section 7(b)(6) of the Small Business Act, as amended, pursuant to the provisions of section 202 of the Public Works and Economic Development Act of 1965, as amended.

79 Stat. 556. 42 USC 3142.

ADDITIONAL ASSISTANT SECRETARY OF LABOR

SEC. 29. (a) Section 2 of the Act of April 17, 1946 (60 Stat. 91) as amended (29 U.S.C. 553) is amended by—

75 Stat. 338.

84 STAT. 1619

(1) striking out "four" in the first sentence of such section and inserting in lieu thereof "five"; and

(2) adding at the end thereof the following new sentence, "One of such Assistant Secretaries shall be an Assistant Secretary of Labor for Occupational Safety and Health.".

80 Stat. 462.

(b) Paragraph (20) of section 5315 of title 5, United States Code, is amended by striking out "(4)" and inserting in lieu thereof "(5)".

ADDITIONAL POSITIONS

Sec. 30. Section 5108(c) of title 5, United States Code, is amended by—

(1) striking out the word "and" at the end of paragraph (8);

(2) striking out the period at the end of paragraph (9) and inserting in lieu thereof a semicolon and the word "and"; and

(3) by adding immediately after paragraph (9) the following new paragraph:

"(10) (A) the Secretary of Labor, subject to the standards and procedures prescribed by this chapter, may place an additional twenty-five positions in the Department of Labor in GS–16, 17, and 18 for the purposes of carrying out his responsibilities under the Occupational Safety and Health Act of 1970;

"(B) the Occupational Safety and Health Review Commission, subject to the standards and procedures prescribed by this chapter, may place ten positions in GS–16, 17, and 18 in carrying out its functions under the Occupational Safety and Health Act of 1970."

EMERGENCY LOCATOR BEACONS

72 Stat. 775.
49 USC 1421.

Sec. 31. Section 601 of the Federal Aviation Act of 1958 is amended by inserting at the end thereof a new subsection as follows:

"EMERGENCY LOCATOR BEACONS

"(d) (1) Except with respect to aircraft described in paragraph (2) of this subsection, minimum standards pursuant to this section shall include a requirement that emergency locator beacons shall be installed—

"(A) on any fixed-wing, powered aircraft for use in air commerce the manufacture of which is completed, or which is imported into the United States, after one year following the date of enactment of this subsection; and

"(B) on any fixed-wing, powered aircraft used in air commerce after three years following such date.

"(2) The provisions of this subsection shall not apply to jet-powered aircraft; aircraft used in air transportation (other than air taxis and charter aircraft); military aircraft; aircraft used solely for training purposes not involving flights more than twenty miles from its base; and aircraft used for the aerial application of chemicals."

SEPARABILITY

Sec. 32. If any provision of this Act, or the application of such provision to any person or circumstance, shall be held invalid, the remainder of this Act, or the application of such provision to persons or circumstances other than those as to which it is held invalid, shall not be affected thereby.

84 STAT. 1620

APPROPRIATIONS

Sec. 33. There are authorized to be appropriated to carry out this Act for each fiscal year such sums as the Congress shall deem necessary.

EFFECTIVE DATE

Sec. 34. This Act shall take effect one hundred and twenty days after the date of its enactment.

Approved December 29, 1970.

Table of Cases

Cases presented in text or partial text are in italic type. Other cases—those discussed or merely cited—are in roman type.

D

E

F

G

H

I

K

L

Q

R

S

T

U

V

W

Index

F

G

M

P

Q

R

S